# CONTACTS TO SEMICONDUCTORS

# CONTACTS TO SEMICONDUCTORS

## Fundamentals and Technology

Edited by

**Leonard J. Brillson**

Xerox Corporation
Joseph C. Wilson Center for Technology
Webster, New York



np

**NOYES PUBLICATIONS**
Park Ridge, New Jersey, U.S.A.

Library of Congress Catalog Card Number: 93-26692
ISBN: 0-8155-1336-4
Printed in the United States

Published in the United States of America by
Noyes Publications
Mill Road, Park Ridge, New Jersey 07656

10 9 8 7 6 5 4 3 2 1

Library of Congress Cataloging-in-Publication Data

Contacts to semiconductors : fundamentals and technology / edited by
Leonard J. Brillson.
p. cm.
Includes bibliographical references and index.
ISBN 0-8155-1336-4
1. Electrical Contacts. 2. Semiconductors--Junctions.
I. Brillson, L. J.
TK7872.C68C68 1993
621.3815'2--dc20 93-26692
CIP

## MATERIALS SCIENCE AND PROCESS TECHNOLOGY SERIES



### Electronic Materials and Process Technology

HANDBOOK OF DEPOSITION TECHNOLOGIES FOR FILMS AND COATINGS, 2nd Edition: edited by Rointan F. Bunshah

CHEMICAL VAPOR DEPOSITION FOR MICROELECTRONICS: by Arthur Sherman

SEMICONDUCTOR MATERIALS AND PROCESS TECHNOLOGY HANDBOOK: edited by Gary E. McGuire

HYBRID MICROCIRCUIT TECHNOLOGY HANDBOOK: by James J. Licari and Leonard R. Enlow

HANDBOOK OF THIN FILM DEPOSITION PROCESSES AND TECHNIQUES: edited by Klaus K. Schuegraf

IONIZED-CLUSTER BEAM DEPOSITION AND EPITAXY: by Toshinori Takagi

DIFFUSION PHENOMENA IN THIN FILMS AND MICROELECTRONIC MATERIALS: edited by Devendra Gupta and Paul S. Ho

HANDBOOK OF CONTAMINATION CONTROL IN MICROELECTRONICS: edited by Donald L. Tolliver

HANDBOOK OF ION BEAM PROCESSING TECHNOLOGY: edited by Jerome J. Cuomo, Stephen M. Rossnagel, and Harold R. Kaufman

CHARACTERIZATION OF SEMICONDUCTOR MATERIALS, Volume 1: edited by Gary E. McGuire

HANDBOOK OF PLASMA PROCESSING TECHNOLOGY: edited by Stephen M. Rossnagel, Jerome J. Cuomo, and William D. Westwood

HANDBOOK OF SEMICONDUCTOR SILICON TECHNOLOGY: edited by William C. O'Mara, Robert B. Herring, and Lee P. Hunt

HANDBOOK OF POLYMER COATINGS FOR ELECTRONICS, Second Edition: by James Licari and Laura A. Hughes

HANDBOOK OF SPUTTER DEPOSITION TECHNOLOGY: by Kiyotaka Wasa and Shigeru Hayakawa

HANDBOOK OF VLSI MICROLITHOGRAPHY: edited by William B. Glendinning and John N. Helbert

CHEMISTRY OF SUPERCONDUCTOR MATERIALS: edited by Terrell A. Vanderah

CHEMICAL VAPOR DEPOSITION OF TUNGSTEN AND TUNGSTEN SILICIDES: by John E. J. Schmitz

ELECTROCHEMISTRY OF SEMICONDUCTORS AND ELECTRONICS: edited by John McHardy and Frank Ludwig

HANDBOOK OF CHEMICAL VAPOR DEPOSITION: by Hugh O. Pierson

DIAMOND FILMS AND COATINGS: edited by Robert F. Davis

ELECTRODEPOSITION: by Jack W. Dini

HANDBOOK OF SEMICONDUCTOR WAFER CLEANING TECHNOLOGY: edited by Werner Kern

CONTACTS TO SEMICONDUCTORS: edited by Leonard J. Brillson

HANDBOOK OF MULTILEVEL METALLIZATION FOR INTEGRATED CIRCUITS: edited by Syd R. Wilson, Clarence J. Tracy, and John L. Freeman, Jr.

HANDBOOK OF CARBON, GRAPHITE, DIAMONDS AND FULLERENES: by Hugh O. Pierson

## Ceramic and Other Materials—Processing and Technology

SOL-GEL TECHNOLOGY FOR THIN FILMS, FIBERS, PREFORMS, ELECTRONICS AND SPECIALTY SHAPES: edited by Lisa C. Klein

FIBER REINFORCED CERAMIC COMPOSITES: by K. S. Mazdiyasni

ADVANCED CERAMIC PROCESSING AND TECHNOLOGY, Volume 1: edited by Jon G. P. Binner

FRICTION AND WEAR TRANSITIONS OF MATERIALS: by Peter J. Blau

SHOCK WAVES FOR INDUSTRIAL APPLICATIONS: edited by Lawrence E. Murr

SPECIAL MELTING AND PROCESSING TECHNOLOGIES: edited by G. K. Bhat

CORROSION OF GLASS, CERAMICS AND CERAMIC SUPERCONDUCTORS: edited by David E. Clark and Bruce K. Zoitos

HANDBOOK OF INDUSTRIAL REFRACTORIES TECHNOLOGY: by Stephen C. Carniglia and Gordon L. Barna

CERAMIC FILMS AND COATINGS: edited by John B. Wachtman and Richard A. Haber

## *Related Titles*

ADHESIVES TECHNOLOGY HANDBOOK: by Arthur H. Landrock

HANDBOOK OF THERMOSET PLASTICS: edited by Sidney H. Goodman

SURFACE PREPARATION TECHNIQUES FOR ADHESIVE BONDING: by Raymond F. Wegman

FORMULATING PLASTICS AND ELASTOMERS BY COMPUTER: by Ralph D. Hermansen

HANDBOOK OF ADHESIVE BONDED STRUCTURAL REPAIR: by Raymond F. Wegman and Thomas R. Tullos

CARBON–CARBON MATERIALS AND COMPOSITES: edited by John D. Buckley and Dan D. Edie

# Preface

Contacts to semiconductors have been a topic of active interest for researchers and technologists for over five decades. A number of books written over this period have focused either on fundamental physical properties or technological device features associated with the metal-semiconductor junction. Fundamental research efforts have centered on elucidation of charge transfer mechanisms by which the contacting media reach electronic equilibrium, using sophisticated preparation techniques to analyze junction properties on an atomic scale. From a device point of view, the control of barrier heights for a constrained set of semiconductor and metal components is critical to the design of high performance micro- and opto-electronics. Technological progress in growing and processing semiconductor device structures has provided several useful methods for lowering or raising effective barrier heights. As device dimensions shrink below fractions of a micron, features of atom and charge redistribution on a nanometer scale can render many of these metallurgical approaches ineffective. Instead, a deeper understanding of the interface physics and chemistry—coupled with advanced growth and processing techniques—will be required to design and prepare metal-semiconductor junctions. Thus, issues of device performance, reliability, and reproducibility are becoming more closely linked than ever before with the basic chemical, geometrical, and electronic properties of the interface on an atomic scale.

The composition of this book reflects the increasing blend of fundamental and technological knowledge. In bridging the macroscopic, mesoscopic, and atomic-scale regimes, the authors present the state-of-the-art in growing, processing, and characterizing electronic junctions. As such, *Contacts to Semiconductors* is intended for both *(a)* technologists who require and up-to-date assessment of methods for contact formation, processing, and characterization, and *(b)* solid state researchers who desire the latest developments in understanding the basic mechanisms of interface physics and chemistry.

The first four chapters of the book focus primarily on the achievements of device properties and highlight the importance of microscopic, chemical and morphological, geometric parameters. In Chapter 1, Eric Marshall and Masanori Murakami introduce the fundamental mechanisms underlying current transport, barrier formation, and the limitation on device performance available with conventional methods of contact formation. They establish the role of interfacial microstructure in contact performance and reliability. In addition, they provide advanced growth and metallurgical methods for achieving highly ohmic contacts to III-V compounds while minimizing metallurgical disruption to a depth scale of only nanometers. Their chapter describes the state-of-the-art, high performance ohmic contacts now achievable and highlights the fundamental issues limiting further improvements in electrical control, process compatibility, thermal stability, and miniaturization.

In Chapter 2, Chris Palmstrøm and Tim Sands introduce state-of-the-art growth techniques for achieving contact metallizations to III-V compounds with both *(i)* high performance, uniformity, and stability, but also *(ii)* with more active device roles *via* novel multilayer metal-semiconductor structures. These advanced metallization structures enable three-dimensional integrated circuits for a variety of electronic, opto-electronic, and magneto-optic applications. Just as significantly, the achievement of Schottky barriers with near-ideal chemical and structural properties affords major opportunities to isolate and identify extrinsic versus intrinsic factors governing interface charge transfer. Palmstrøm and Sands provide the thermodynamic and lattice structural basis for developing such junctions and demonstrate the pronounced influence of controlled interface atomic structure on junction electronic properties.

In Chapter 3, Ray Tung addresses the electronic an structural properties of metal/Si and silicide/Si contacts. Beginning with a rigorous treatment of Schottky barrier properties and their measurement, this chapter establishes the importance of inhomogeneity on a micro-scale. Tung reviews the fabrication of Schottky diodes and ohmic contacts to Si—the wide variety of reactions between metals and silicon, the processing technologies available for producing contacts, the barriers measured for both epitaxial and non-epitaxial metals and silicides, and the barrier dependencies on processing. Because contacts to Si have been investigated extensively and structural properties can be controlled via epitaxial techniques, it is possible to analyze apparent anomalies in Schottky barrier behavior in terms of electrical inhomogeneities on a submicron scale. Tung provides a full review of electronic and structural results for both epitaxial and polycrystalline Si contact structures. As with the compound semiconductor junctions described earlier, the elemental semiconductor contacts covered here illustrate how control of interface atomic structure is being employed to elucidate mechanisms of charge transfer. Likewise, the advanced state of materials control revealed for Si technology provides new concepts and strategies for fundamental understanding and practical control.

In Chapter 4, Harry Wieder reviews the state of insulator-semiconductor contacts, complementing the overviews of semiconductor-metal junctions contained in prior chapters. Beginning with a rigorous introduction to the electrical properties of metal-insulator-semiconductor (MIS) and metal-oxide semiconductor (MOS) junctions, he establishes the basis for characterizing insulator bulk, surface, and interface states. Wieder provides an authoritative and comprehensive review of insulator-semiconductor fabrication and characterization for both elemental and compound semiconductor junctions, as well as heterojunction quasi-insulator junctions. As in pervious chapters, Chapter 4 highlights the role of interface chemistry and atomic structure in determining junction electronic properties. Wieder discusses the progress in reducing insulator interface state densities in terms of the same physical mechanisms introduced for metal-semiconductor junctions. Again, the dramatic improvements in device structure characteristics achieved via chemical techniques provide the basis for more refined fundamental understanding.

In Chapter 5, Len Brillson reviews and categorizes the various physical mechanisms which can give rise to interface states. These include both states intrinsic to the bulk materials forming the contact as well as those arising out of features extrinsic to either contact material. Those extrinsic states arising out of the growth and processing techniques described in earlier chapters are examined according to their macroscopic, conventional or microscopic, interface-specific origin. Included here is an overview of the ultrahigh vacuum (UHV) techniques now available to monitor the electronic, chemical, and geometrical structure of interfaces on an atomic scale. These techniques provide evidence for the existence of interface states arising from multiple physical mechanisms. A general, extended semiconductor interface picture emerges from the close correlation between atomic-scale chemical and electronic charges. Indeed, this correlation reveals a variety of microscopic avenues for controlling Schottky barrier formation, many of which have now been demonstrated.

In Chapter 6, Zuzanna Liliental-Weber, Eicke Weber, and Nate Newman provide an expanded description of the role of defects, epitaxy and morphology in forming Schottky barriers at metal/III-V compound junctions. Here transmission electron microscopy (TEM) measurements of atomic structure and morphology are joined with conventional electrical measurements for a broad array of metal/GaAs junctions. Morphology, stoichiometry, and epitaxial relationships are shown to play a major role in barrier formation and contact stability. Changes in structural and electrical properties of these junctions with thermal and electrical stress reveal evidence for the role of lattice defects associated with deviations in local stoichiometry. Besides establishing clear links between atomic-scale structure and electronic properties, this chapter highlights the importance of surface preparation of the semiconductor before metal deposition on the resultant interface morphology grain size, and orientation of the overlayers. These results are reviewed in the context of physical models for Schottky barrier formation, complementary experimental results, and recent theoretical calculations. They provide an impetus for even more refined structural analysis of electrically-active interface defects, impurities, and other crystalline imperfections.

In Chapter 7, Celso Aldao and John Weaver explore the role of interface chemistry at metal-semiconductor interfaces in greater depth. Drawing upon soft x-ray photoemission spectroscopy (SXPS) and scanning tunnelling microscopy (STM) results in conjunction with advanced deposition techniques, they describe in detail the formation and morphology of metal-semiconductor interfaces during the initial stages of Schottky barrier formation. Included here are interfacial phenomena such as atom-induced substrate disruption, chemical reaction, out-diffusion of one or more substrate species, and surface segregation. In turn, these phenomena contribute in varying degree according to the mode of overlayer growth, which can be altered via cluster deposition, ion deposition, and low temperature growth. Hence a picture emerges of the metal-semiconductor junction on an atomic scale which is far from abrupt, whose atom distribution and morphology is governed by both kinetic and thermodynamic parameters, and whose chemical and electronic properties can be altered by advanced growth techniques.

In Chapter 8, Charlie Duke and Bob Dandrea review recent advances in the theoretical description of semiconductor interface properties. This chapter provides a broad overview of the development of theoretical models and concepts in semiconductor interface science. It is intended to provide an introduction and general understanding of topics in interface science. The authors examine both metal-semiconductor as well as semiconductor-semiconductor interfaces from a fundamental perspective, with a view to synthesizing key theoretical and experimental results in the extensive literature. They also address the complications introduced by the non-ideal chemical and electrical nature of the metal-semiconductor contact. The discussion presented here provides a useful framework on which to address the electronic properties of advanced semiconductor junctions. Finally, in Chapter 9, Jim McKinley and Giorgio Margaritondo describe the atomic-scale control of heterojunction band lineups. Following on the advanced experimental and theoretical techniques described in previous chapters, they describe progress in manipulating the band lineups and discontinuities via controlled introduction of new chemical species. They show how such results permit a deeper understanding of the physical mechanisms governing semiconductor interfaces in general.

Overall, we have assembled a broad array of the latest semiconductor interface science and technology, ranging from advanced ohmic, Schottky, and heterojunction contacts to the refined perspectives of microscopic junctions gleaned from ultrahigh vacuum surface science techniques. Not only has considerable progress been made in these areas over the last few years, but further advances in understanding and control are clearly indicated.

Webster, New York
May 1, 1993

Leonard J. Brillson

# Contributors

**Celso M. Aldao**
Department of Chemical Engineering
and Materials Science
University of Minnesota
Minneapolis, MN

**Leonard J. Brillson**
Xerox Corporation
Joseph C. Wilson Center
for Technology
Webster, NY

**Robert G. Dandrea**
Xerox Corporation
Xerox Webster Research Center
Webster, NY

**Charles B. Duke**
Xerox Corporation
Xerox Webster Research Center
Webster, NY

**Zuzanna Liliental-Weber**
Lawrence Berkeley Laboratory
University of California
Berkeley, CA

**Giorgio Margaritondo**
Institut de Physique Appliquée
Ecole Polytechnique Fédérale
Lausanne, Switzerland

**Eric D. Marshall**
IBM, Thomas J. Watson
Research Center
Yorktown Heights, NY

**James C. McKinley**
Department of Physics
and Astronomy
Vanderbilt University
Nashville, TN

**Masanori Murakami**
Department of Metal Science
and Technology
Kyoto University
Kyoto, Japan

**Nathan Newman**
Lawrence Berkeley Laboratory
University of California
Berkeley, CA

**Chris J. Palmstrøm**
Bell Communications Research
Red Bank, NJ

**Timothy D. Sands**
Bell Communications Research
Red Bank, NJ

**Raymond T. Tung**
Bell Telephone Laboratories
Murray Hill, NJ

**John H. Weaver**
Department of Chemical Engineering
and Materials Science
University of Minnesota
Minneapolis, MN

**Eicke R. Weber**
Lawrence Berkeley Laboratory
University of California
Berkeley, CA

**Harry H. Wieder**
Department of Electrical and
Computer Engineering
University of California, San Diego
LaJolla, CA

# Contents

# 1

# Ohmic Contacts to GaAs and Other III-V Compounds: Correlation of Microstructure with Electrical Properties

*Eric D. Marshall and Masanori Murakami*

## 1.0 ELECTRICAL PROPERTIES OF OHMIC CONTACTS

### 1.1 Introduction

Gallium arsenide (GaAs) and other III-V compound semiconductors are continuing to undergo an intensive research and development effort as a material for heterostructure-based optoelectronic devices, discrete microwave electronic devices, and high-speed very large scale integrated (VLSI) electronics. The technology is now at a point where its potential for commercial success is being tested.[1]-[4] The motivation for these efforts stems both from intrinsic properties of GaAs and from technologies which have been developed around GaAs. Advantages of GaAs over Si include higher electron mobility and a direct bandgap. Hence, faster electronic devices which dissipate less power and efficient optical devices are possible. Advanced technologies, such as molecular beam epitaxy (MBE), metal organic chemical vapor deposition (MOCVD), and dry etching techniques allow for compositional and structural control on a nearly atomic scale. This can be translated into great control over material electronic and optical properties.

In order to successfully utilize the advantages of GaAs devices, connections to the devices must be made which do not add noticeable parasitic effects. Information is most commonly transferred electrically through wires and metal lines. The material and geometry of the interconnection metallization can contribute to the parasitic impedance, but will not be discussed in this review. The physical element which is of interest here is the point at which electrons flow between the metal and the semiconductor. The major point to be addressed is the relationship between the metallurgical properties of the metal-semiconductor interface and the impedance of the interface to current flow. While for some applications, a predictable interfacial impedance is desirable, for the purpose of making an ohmic contact, it is undesirable.

## 1.2 Ohmic Contacts Defined

In this study, the word *contact* refers to metallization which is applied to a semiconductor device or specimen, and processed, which makes good electrical and physical contact to the semiconductor. In the strict definition, *ohmic* implies that the resistance, R, across such a contact to a semiconductor is independent of applied voltage, V, and current, I. That is, the resistance follows the linear relation: R = V/I = constant. Under normal conditions, uniform bulk metals and semiconductors follow this relation. At the interface between two unlike materials, however, the resistance to electron transport across the interface may become a function of applied voltage. Since current flows from a metal through the interface and into the semiconductor, the current/voltage, or I/V, characteristics will appear linear if the interface is truly ohmic or if the voltage drop across the interface is negligible compared to that in the bulk of the metal and the semiconductor. From a practical perspective, Rhoderick suggests that: *The important feature of such contacts is that the voltage drop across them must be negligible compared with the voltage drop across the device or specimen, so that the contacts do not affect the I/V characteristics* [5].

By Rhoderick's definition, the choice of metallization used for an ohmic contact will depend very much upon the characteristics of the device to which it will be applied. High resistance devices can tolerate higher contact resistances than low resistance devices. Also, as devices shrink in size, the specific contact resistivity, $\rho_c$ = R x area, must decrease in order to retain the same contact resistance. Thus, moves to smaller or lower-resistance semiconductor devices provide incentive for seeking ohmic contacts with lower contact resistivity.

Improved ohmic contacts are also desired since ohmic contacts constitute interfaces between different materials, and a thermodynamic driving force may exist for reaction between the ohmic contact metallization and the semiconductor. Indeed, formation of ohmic characteristics often requires heat treatments which encourage this reaction. The issue then becomes optimization and control over such a reaction. Certain reactions require narrow time-temperature process windows under which good contacts are formed. From a manufacturing standpoint, this is highly undesirable. Moreover, reaction intimates that a certain amount of semiconductor material is consumed during the reaction. The extent and morphology of reaction can be detrimental to device behavior if the active regions of the device are located close enough to the contact to be involved in the reaction.

### 1.3 Electron Energy Band Diagrams

Electronic properties of a semiconductor can often be ascertained from analysis of its energy band diagram.[6] In order to show spatial variations in a material which may be encountered by an electron, one plots the highest valence band edge and lowest conduction band edge as a function of position. In this manner, spatial changes in band structure and electron energy gradients, or forces, may be denoted. The forces may be due to externally applied fields or internally generated fields, such as those due to variations in doping density or composition. In a metal-semiconductor system, the only feature of the band structure of the metal which is plotted is the Fermi energy level.

The energy band diagram for a metal in contact with an n-type semiconductor with applied bias is plotted in Fig. 1. $E_c$ is the conduction band edge, $E_F$ is the Fermi level, and $E_v$, which is not shown, can be plotted by drawing a line at $E_c - E_g$, where $E_g$ is the bandgap of the semiconductor. $V_n$ is the difference $E_c - E_F$ which can be determined from knowledge of the free carrier concentration and Fermi-Dirac statistics. The band-bending in the semiconductor can be determined by solving Poisson's equation while using knowledge of the applied bias, qV, the uncorrected barrier height, $q\phi_{B0}$, (see Sec. 1.5) the image force lowering, $q\Delta\phi$ , and the electrically active impurity spatial distribution. The effective potential barrier height to a thermally excited electron flux is $\phi_B = \phi_{B0} - \Delta\phi$. When $\phi_B > (V_n + V)$, the surface region of the semiconductor will be depleted of free charge to a depth, w, termed the depletion width. For a uniformly doped n-type semiconductor, according to the depletion approximation:

Eq. (1) $$w = \sqrt{\left[\frac{2\varepsilon_s}{qN_D}\right]\left[\phi_{B0} - V_n - V - \frac{kT}{q}\right]}$$

where $\varepsilon_s$ is the static dielectric constant for the semiconductor and $N_D$ is the donor concentration. Thus, the depletion width decreases with decreasing $\phi_{B0}$ and increasing $N_D$. This point is critical in the design of ohmic contacts; it will be shown that it directly affects tunneling current.

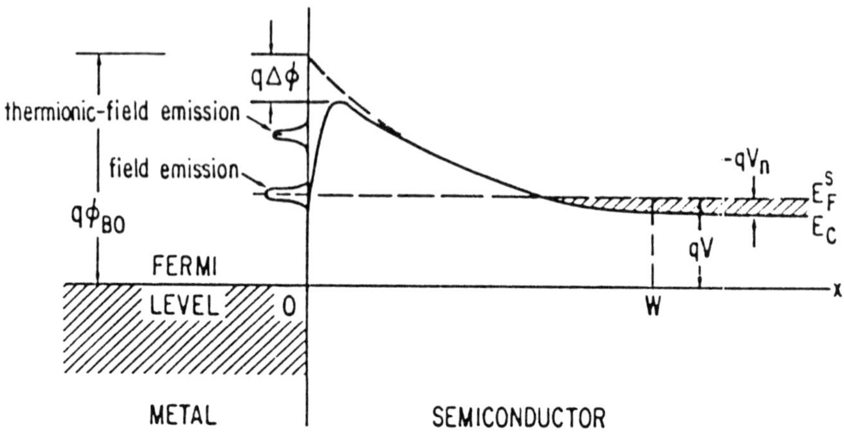


**Figure 1.** Schematic energy band diagram of a metal-semiconductor interface under forward bias. $E_F{}^s$ is the Fermi level of the semiconductor displaced relative to that of the metal by applied potential V. *(Adapted from Refs. 6 and 16.)*

## 1.4 Current Transport Mechanisms and Calculation of Contact Resistivity

Electron transport across a metal-semiconductor interface may be calculated if two quantities are known as a function of electron energy, E. First, the supply function, N(E), must be known (Eq. 2). It is a measure of the number of electrons per second per unit area having energy associated with motion normal to the junction within a range dE which are incident on the barrier. It is given by:[7]

Eq. (2) $$N(E)\,dE = \left(\frac{A^*T}{k}\right)\left[\ln\left(1+e^{\frac{-(E-E_F)}{kT}}\right) - \ln\left(1+e^{\frac{-(E-E_F+qV)}{kT}}\right)\right]dE$$

where $E_F$ is the Fermi level in the metal, V is the applied bias, and the Richardson constant is

Eq. (3)
$$A^* = \frac{4\pi^2 q k^2 m^*}{h^3}$$

where h is Planck's constant. The effective mass, m*, is a function of effective mass components in directions perpendicular and parallel to current flow. For non-degenerate semiconductors, Chang, et al.[8] make the approximation:

Eq. (4)
$$N(E)\,dE = \left(\frac{A^*T}{k}\right)\left[\frac{1}{1+\exp\left(\frac{E-E_F}{kT}\right)} - \frac{1}{1+\exp\left(\frac{E-E_F+qV}{kT}\right)}\right]dE$$

Under Maxwell-Boltzmann statistics, further simplification results:

Eq. (5)
$$N(E)\,dE = \left(\frac{A^*T}{k}\right)\exp\left[\frac{-(E-E_F)}{kT}\right]\left(1-\exp\frac{-qV}{kT}\right)dE$$

Typically, approximations are made for the purpose of developing tractable solutions at the expense of generalization.[9]-[10] As can be seen from Eqs. 2 - 5, the supply function is primarily a function of the Fermi level, effective mass in the semiconductor, bias, and temperature. The Fermi level, in turn, is dependent upon the doping level, band structure of the semiconductor, and the temperature.

The second quantity which must be known in order to calculate the current density is the transmission probability for electrons as a function of energy. It is a strong function of the details of the band-bending at the interface. If the potential, $\phi(x) = E_c(x) - E_F$, is known as a function of position near the interface, the transmission probability may be calculated by solving Schrödinger's equation.[8] The generalized Wenzel-Kramers-Brillouin (WKB) approximation is commonly used where

Eq. (6)
$$P(E) = \{1+\exp[2i\int_{x_1}^{x_2} k(x)\,dx]\}^{-1}$$

for tunneling through the barrier which is often approximated to be:

Eq. (7) $$P(E) = \exp\left[-2i\int_{x_1}^{x_2} k(x)\,dx\right]$$

and P(E) ~ 1 for emission over the barrier. The limits of integration, $x_1$ and $x_2$, are the classical turning points where the integrand becomes zero. The integrand is:[9]

Eq. (8) $$k(x) = -i\sqrt{\frac{8\pi^2 m^*}{h^2}\left[q\phi(x) - E\right]}$$

Boudville and McGill[11] have pointed out that a two-band model should be used where

Eq. (9) $$k(x) = -i\sqrt{\frac{8\pi^2 m^*}{h^3 E_g}\left[q\phi(x) - E\right]\left[E_g + E - q\phi(x)\right]}$$

This becomes most important when tunneling electrons have energy near the valence band edge while in the bandgap. For a uniformly doped semiconductor and simple parabolic band bending, the analytic solution for P(E) is[9]

Eq. (10) $$P(E) = \exp\left[\frac{-E_b y(\alpha)}{E_{00}}\right]$$

where $\alpha = E/E_b$, $E_b = q(\phi_B - V - V_n)$,

Eq. (11) $$y(\alpha) = \sqrt{1-\alpha} - \ln\left(\frac{1+\sqrt{1-\alpha}}{\sqrt{\alpha}}\right)$$

and

Eq. (12) $$E_{00} = \frac{qh}{4\pi}\sqrt{\frac{N_D}{m_r \varepsilon_r}} = 18.5\times 10^{-12}\left(\frac{N_D}{m_r \varepsilon_r}\right) \mathrm{eV}$$

where $m_r$ is the tunneling effective mass measured in units of the free electron mass, $N_D$ the impurity concentration in units of $cm^{-3}$, and $\varepsilon_r$ the static dielectric constant. Chang, Fang, and Sze[12]-[13] make the approximation:

Eq. (13) $$P(E) \approx \exp\left(\frac{E_b}{E_{00}}\right)$$

This approximation only holds for $E \approx 0$. A slightly better approximation[14] is

Eq.(14) $$P(E) \approx \exp\left[-\frac{(E_b - E)}{E_{00}}\right]$$

which becomes worse as E becomes greater than zero. As can be seen from Eqs. 6 - 14, the transmission probability is dependent upon semiconductor effective mass and bandgap, bias, and the exact nature of the potential barrier at the interface. Calculation of current density follows by solution of:

Eq. (15) $$J = \int_0^\infty P(E)\, N(E)\, dE$$

The specific contact resistivity, $\rho_c$, follows as:

Eq. (16) $$\rho_c = \left[\frac{\delta J}{\delta V}\right]_{V=0}^{-1}$$

For all practical purposes, Eq. 15 can be broken down into two components. One is due to thermionic emission (TE) of electrons over the barrier. The other is due to tunneling of electrons through the barrier. Tunneling which takes place at an energy near the Fermi level is termed field emission (FE). Thermally assisted tunneling, above the Fermi level, is called thermionic field emission (TFE, see Fig. 1). Approximate solution of Eq. 15 for the thermionic emission component is:[6]

Eq. (17) $$J_{TE} = A^* T^2 \exp\left[\frac{-q(\phi_{B0} - \Delta\phi)}{kT}\right]\left[\exp\left(\frac{qV}{kT}\right) - 1\right]$$

The field emission and thermionic field emission component is then:

Eq. (18) $$J_{FE\,and\,TFE} = \int_0^{\phi_{B0} - \Delta\phi} P(E)\, N(E)\, dE$$

Equation 18 is not very amenable to analytic solution. Thus, either series solution approximations[9]-[10][15]-[17] or numerical solutions[8][11]-[12] are attempted.

The series solutions are valid only under very specific conditions which have been assumed in order to make the approximations. They have proved successful in incorporating thermionic field emission into thermionic-emission-type equations.[9] The approximations, however, may fail in the case of low resistance ohmic contacts. Their utility for ohmic contacts lies in their simplified form which easily leads to an intuition as to which contact physical parameters predominantly affect contact resistance. The important quantity which comes out of the series solution is $E_{00}$ (Eq. 12). It is a parameter which is mainly determined by the band-bending at the interface. When $kT \sim E_{00}$, thermionic field emission is dominant. Under certain conditions as specified by Padovani,[13] specific contact resistivity in the thermionic field emission (TFE) regime is approximated to be:[17]

Eq. (19)

$$\rho_c = \left[\frac{kT}{qA^*T^2}\right]\left[kT\cosh\left(\frac{E_{00}}{kT}\right)\right]\sqrt{\frac{\coth\left(\frac{E_{00}}{kT}\right)}{\pi(\phi_B - V_n)E_{00}}}\exp\left[\frac{\phi_B - V_n}{E_0} + \frac{V_n}{kT}\right]$$

where

Eq. (20)
$$E_0 = E_{00}\coth\left(\frac{E_{00}}{kT}\right)$$

It must be stressed that different physical parameters may require approximations to be made other than those which lead to the ones stated here.[13] When $kT \ll E_{00}$, field emission is dominant. In the field emission(FE) regime, contact resistivity is approximated to be:

Eq. (21)
$$\rho_c = \left[\frac{A^*T^2\pi q\exp\left(\frac{-\phi_B}{E_{00}}\right)}{kT\sin(\pi c_1 kT)} - \frac{A^*T^2 c_1 q\exp\left(-\frac{\phi_B}{E_{00}} + c_1 V_n\right)}{(c_1 kT)^2}\right]^{-1}$$

where $c_1$ is mainly a function of $E_{00}$, $\phi_B$, and $V_n$. When $\phi_B \gg V_n$, Eqs. 19 and 21 for conduction in the TFE and FE regimes reduce to

Eq. (22) $$\rho_c = A \exp\left(\frac{\phi_B}{E_{00}}\right)$$

where A is a function of $E_{00}$, $\phi_B$, $V_n$, and T. When kT » $E_{00}$, thermionic emission is the predominant current transport mechanism. In the thermionic emission (TE) regime, contact resistivity is given by:

Eq. (23) $$\rho_c = \frac{kT}{qA^*T^2} \exp\left(\frac{\phi_B}{kT}\right)$$

The practical consequence of this analysis is that the design of low resistance ohmic contacts requires low barrier height, $\phi_B$, and high doping, since from Eq. 12, $N_D$ is inversely proportional to $E_{00}^2$. In Si technology, low barrier heights and high doping levels can be routinely realized. In GaAs technology, barrier heights are typically constrained to high values and doping levels are limited. Moreover, the metallurgy is usually more complicated as reactions involve three or more elements while Si ohmic contacts utilize self-limiting silicide reactions. Hence, the development of ohmic contacts to GaAs has been delayed relative to that for contacts to Si.

## 1.5 Barrier Height

In the Schottky-Mott theory, barrier height, $\phi_{B0}$, is equal to the difference between metal work function and semiconductor electronegativity. In the Bardeen and similar models, the barrier height is determined by interfacial electronic charge, and hence may be independent of metal work function. In practice, the degree of dependence of barrier height on choice of metal is found to be very dependent upon the semiconductor, the nature of its surface before contact formation, and the interfacial chemistry.

For GaAs, barrier heights do not usually follow the Schottky theory. The barrier heights for metals deposited under ultra-high vacuum (UHV) conditions onto oxide-free n-type liquid-encapsulated-Czochralski (LEC)-grown GaAs wafers are plotted in Fig. 2.[18] No simple correlation was found between $\phi_B$ and metal work function, nor between $\phi_B$ and the chemical reactivity at the metal-GaAs interface. Moreover, barrier heights were constrained to within a range of ~0.2 eV while metal work functions varied by ~0.8 eV. This result is similar to those usually obtained for contacts to chemically cleaned GaAs which also display only ~0.2 eV variation in $\phi_B$. Therefore, a great deal of latitude in barrier height adjustment is not afforded by simply choosing an appropriate metal. On the other hand, Viturro, et al.[19] have shown that, under certain conditions, a variation of ~0.9 eV in $\phi_B$ is possible while $\phi_M$ varies by ~1.6 eV for metals deposited at low temperatures onto molecular beam epitaxially (MBE) grown GaAs(100) layers (Fig. 2). This suggests that the semiconductor bulk properties, as well as semiconductor surface properties, determine the degree of dependence of $\phi_B$ on $\phi_M$.

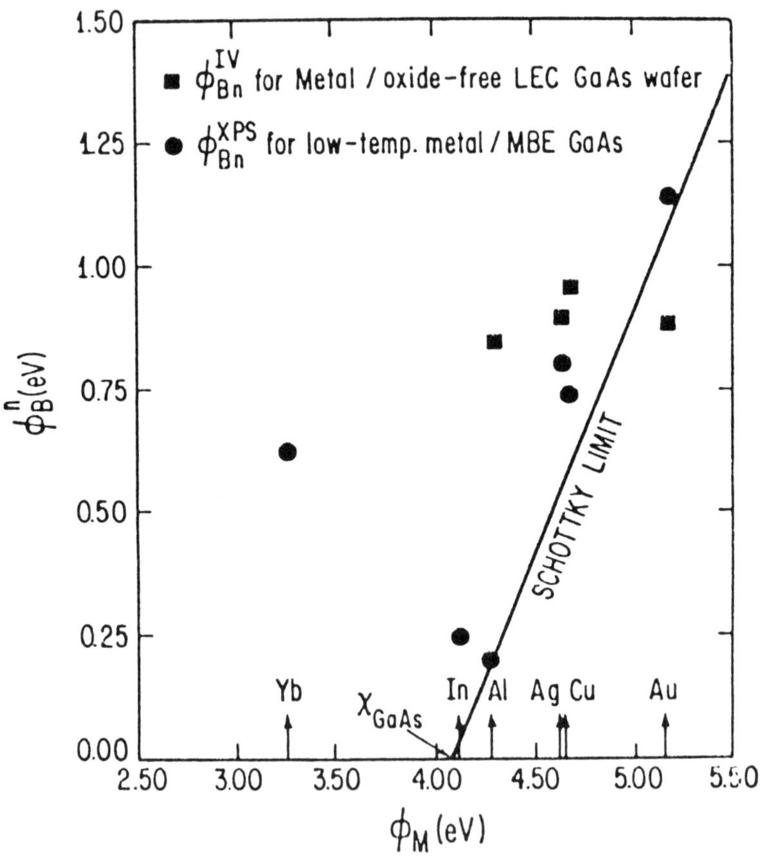


**Figure 2.** Barrier height, $\phi_B$, as a function of metal work function, $\phi_M$, for n-type GaAs(100). Squares are from current-voltage measurements on metal/oxide-free LEC-grown GaAs wafers. Circles are from XPS measurements on low-temperature deposited metal/MBE-grown GaAs layers. Under most conditions, barrier heights are constrained to ~0.7 - 0.9 eV *(squares)*. The recent XPS results *(circles)* indicate that ideal Schottky behavior *(solid line)* may be approached under specific material growth and deposition conditions. *(Adapted from Refs. 18 and 19.)*

## 1.6 Doping Limitations

Besides barrier height, the other material-dependent quantities which affect the energy band diagram for metal-semiconductor interfaces (Fig. 1) are implicit in $V_n$ and w. Both of these quantities are dependent upon the concentration of electrically active donors and acceptors. Typically, the lower limit for these values is set by unintentionally incorporated dopants and defects—both of which are somewhat controllable by varying the semiconductor growth technique. For n-type GaAs, there are at least three types of upper limits on donors. Firstly, the most commonly used donor species for GaAs are group IV elements, such as Si. They are amphoteric dopants. The proportion of impurity atoms which sit substitutionally on Ga as opposed to As sites is, in part, controlled by the relative proportion of Ga to As vacancies during growth or activation anneals. The charged vacancy concentrations are, in turn, a function of the Fermi level, which is itself dependent upon dopant activation. Depending on growth parameters, the upper limit of donor activation in MBE-grown material is usually in the $10^{18}$ $cm^{-3}$ range.[20] If the amphoteric limit is surmounted, then DX centers in GaAs create resonant states in the conduction band which should limit free electron concentration to ~$2 \times 10^{19}$ $cm^{-3}$.[21] DX centers create donor-associated energy levels which are labeled $E_d$ in Fig. 3. For GaAs the DX energy level, $E_d$ (dashed line), lies above the direct gap conduction band minimum, $E^{\Gamma}$, and represents an upper limit on $E^F$. This limit corresponds to ~$2 \times 10^{19}$ $cm^{-3}$ free electron concentration. Note that, although the free carrier concentration is limited by the DX energy level, the ionized donor concentration within the depletion region may not be limited by the same mechanism. An exception to these limitations has been reported for Ge-doping when a free electron concentration of $2 \times 10^{20}$ $cm^{-3}$ was achieved under certain MBE growth conditions.[23] Above this level, dopant precipitation and phase formation may occur due to solid solubility limitations. Higher doping levels are more easily achievable in p-type GaAs than in n-type GaAs since dopants which are not amphoteric are typically used and DX centers are associated with only donors. Thus, ohmic contacts to p-GaAs fabricated by high doping levels often lead to satisfactory values of contact resistance.

## 1.7 Other Current Transport Mechanisms

Mechanisms other than thermionic and field emission may result in more current flow than expected for a given junction bias. If a significant concentration of electronic states is formed within the semiconductor bandgap at the interface, defect-assisted tunneling, interface state thermally-assisted tunneling, or enhanced generation-recombination current may be expected. The current-voltage characteristics will be dependent upon the concentration and both the spatial and energy distribution of the midgap states. This, in turn, may be dependent upon the perfection and purity of any interfacial layer.

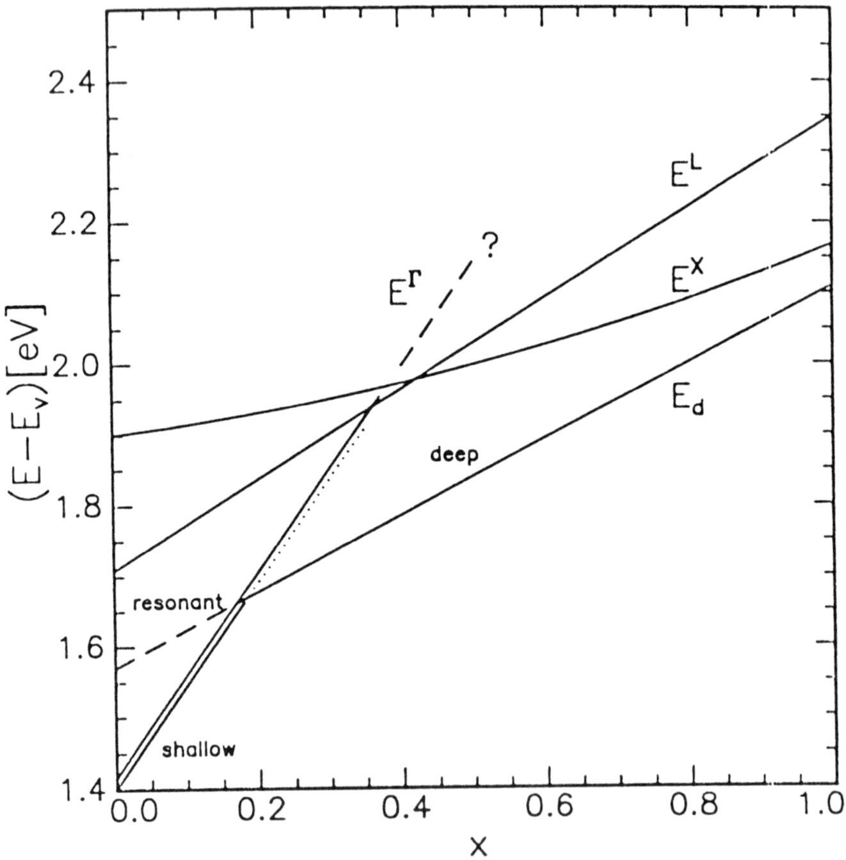


**Figure 3.** Variation of the Γ-, L-, and X-band energy minima in $Al_xGa_{1-x}As$ as a function of AlAs mole fraction, x ($E^\Gamma$, $E^L$, and $E^X$, respectively). $E_d$ corresponds to the DX center energy level. It is a dominant deep donor for x > 0.22. For x < 0.22, the DX level lies above the lowest conduction band minimum, $E^\Gamma$, is resonant with energy levels within the conduction band, and will limit $E_F$ movement in the conduction band and hence limit the maximum free electron concentration. The dominant donor level at these low x values is a shallow level. The dependence of $E^\Gamma$ on x was taken from Ref. 24 and temperature corrected using Varshni temperature coefficients from Ref. 25. *(Adapted from Refs. 21 and 22).*

## 2.0 IN-SITU EPITAXIAL GROWTH OF INTERLAYERS FOR OHMIC CONTACTS TO n-GaAs

### 2.1 Highly Doped Surface Layers for Effective Barrier Height Reduction

Given the limited range of metal/n-GaAs barrier heights and the doping limitations, alternative schemes for ohmic contact formation have been developed. Schemes using well-controlled means typically involve modification of the thin interfacial layer between the metal and the n-GaAs. If the fixed charge concentration in the surface depletion region is limited to the same value as the free carrier levels, low resistance contacts would not be expected. Yet, good ohmic contacts have been formed by growing highly doped surface layers by MBE (Fig. 4).[26]-[28] Barnes and Cho achieved a contact resistivity of $2 \times 10^{-6}$ $\Omega$-cm$^2$ using Sn doping with $|N_D - N_A| = 6 \times 10^{19}$ cm$^{-3}$.[26] The depleted surface region may contain a greater ionized donor concentration than the free carrier concentration in the bulk.[27][20] Kirchner et al. reported $\rho \sim 10^{-6}$ $\Omega$-cm$^2$ for highly doped Si-doped MBE layers and deduced a surface active donor concentration of $\sim 10^{20}$ cm$^{-3}$.[27] The placement of a plane of donor atoms very near to the interface (delta doping) has resulted in a high enough electric field at the interface to yield low values of contact resistivity ($\sim 10^{-6}$ $\Omega$-cm$^2$).[28] This is due to the effectiveness of high concentration of donors, and possibly deep levels, close to the interface. A number of calculations of the effect of a uniformly doped $n^+$ layer on contact resistivity have been developed.[29]-[32] They are predominantly based upon the equations reported in Sec. 1. The result has much to do with the thickness of the $n^+$ surface layer. If it is much thinner than the depletion width, then the contact behaves like a Schottky diode with a reduced effective barrier height.[33] If the thickness is much larger than the depletion width, then the contact behaves as an ohmic contact to a semiconductor with the doping of the $n^+$ layer in series with a high-low junction between the $n^+$ region and the n-substrate. In order to obtain the lowest contact resistivity, the surface should be doped as high as possible and its thickness should be approximately the same or greater than the depletion width.

### 2.2 Compositionally Different Surface Layers for Real Barrier Height Reduction

Another useful set of techniques involves the growth of an intermediate layer between the metal and the GaAs. In the case of a compositionally

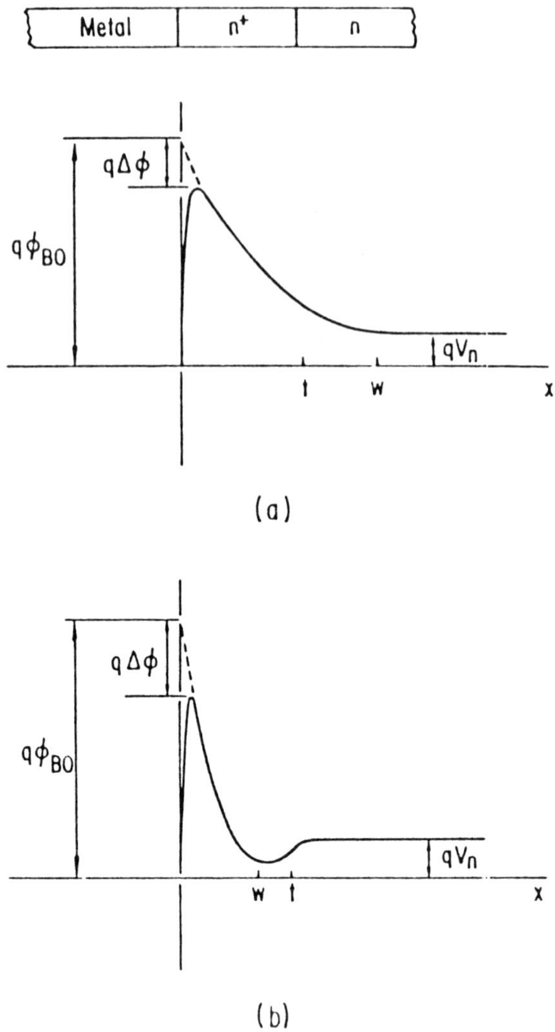


**Figure 4.** Schematic energy band diagram of a metal-semiconductor interface with a highly doped $n^+$ surface layer of thickness t under zero applied bias. *(a)* Surface layer is contained in the space charge region, $t < w$. *(b)* Surface layer extends beyond the space charge region, $t > w$. *(Adapted from Ref. 30.)*

homogeneous interlayer, the 0.8 eV barrier can be exchanged for two lower barriers in series. One successful MBE-grown contact involves the growth of a thin $n^+$ Ge layer on the GaAs (Fig. 5a).[34] The conduction band discontinuity between the GaAs and the Ge is low (0.08 eV[35] to 0.3 eV[36]). Since $E_F$ lies above $E_c$ in the Ge, the barrier at the Ge-GaAs heterojunction is reduced. Thus, this interface presents little resistance to electron flow. At the metal/$n^+$ Ge interface, the interfacial resistance is low because the barrier height for metals to n-Ge is lower than that for n-GaAs (~0.5 - 0.6 eV)[37] and Ge can be doped heavily n-type ($\geq 10^{20}$ $cm^{-3}$). Hence, tunneling at this interface can be significant. These two interfaces in series offer very little resistance to electron flow ($\leq 10^{-7}$ $\Omega$-$cm^2$).

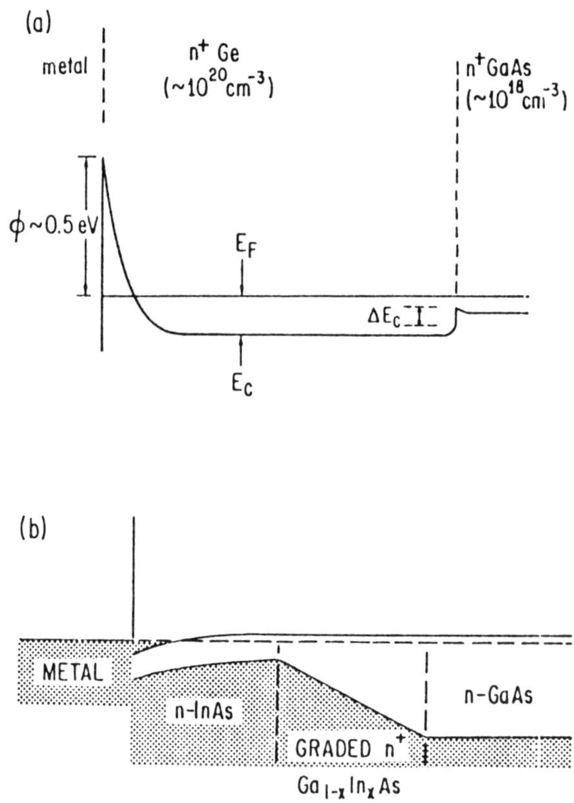


**Figure 5.** *(a)* Conduction band diagram for the MBE-grown metal/n+Ge/GaAs ohmic contact.[34] *(b)* Conduction band diagram for the MBE-grown metal/n+InAs/graded $Ga_{1-x}In_xAs$/GaAs ohmic contact.[38]

If the interlayer is compositionally graded and the barrier between the metal and the surface of the interlayer is negligible, then a contact with no barrier to current flow can be formed. This has been successfully accomplished with a graded n-$In_xGa_{1-x}As$ interlayer (Fig. 5b).[38] The Fermi level is pinned in the conduction band for metal contacts on normal n-InGaAs surfaces with $x > 0.8$.[39]-[40] The grading from $In_xGa_{1-x}As$ to GaAs effectively compensates the electric field which would exist at an abrupt InAs/GaAs interface. If done properly, an interface results with no barrier to electron flow ($\leq 10^{-7}$ Ω-$cm^2$).[41]

Overall, the preceding two sections demonstrate that a sufficient level of understanding is currently at hand to allow formation of ohmic contacts to n-GaAs using highly controlled fabrication techniques. These techniques, however, usually place a number of demanding limitations on the technology (i.e., costly or complex process, high process temperatures, difficult to pattern). Thus, more viable means have been sought.

## 3.0 CORRELATIONAL STUDIES OF INTERFACIAL MICROSTRUCTURE AND AuGeNi OHMIC CONTACT RESISTANCE

### 3.1 Brief History of AuGeNi Ohmic Contact Development

AuGeNi contacts were invented in 1967 by Braslau et al.[42] and have been extensively used as ohmic contacts for advanced GaAs devices over a quarter century. This ternary AuGeNi system was originally used in Gunn diodes due to simple fabrication of ohmic contacts rather than good electrical properties. The development of this contact is briefly described below. Binary AuGe contacts were first used by Gunn in 1964 in his diodes.[43] The contacts were very attractive for the diodes which did not require extremely low resistances. The contacts were prepared by evaporating AuGe pellets in a conventional vacuum system. Relatively low contact resistances were obtained by annealing contacts with eutectic composition (Au-13wt%Ge) at relatively low temperature of ~360°C. At that time this simple fabrication process was more attractive than low resistances. However, the AuGe contacts had a rough surface and the diffusion (vertical diffusion) depth into GaAs was deep.[44]-[50] In addition, lateral surface diffusion of Au was observed,[50][52] which limited applicability of these contacts to devices with small dimensions. Several improvements in the morphology were achieved: the surface morphology was significantly improved by decreasing the annealing temperature[53][54] or reducing the annealing time down to microseconds by laser annealing.[55]-[60] However, the thermal instability of the contact resistance during subsequent annealing after contact formation[46] and the large spread of the contact resistances over a given wafer could not be significantly improved. Slight improvement of the thermal stability was achieved by decreasing the amount of Ge added to the AuGe from the eutectic composition.[61]

Significant improvement of the binary AuGe contacts was made by adding Ni to the AuGe.[42][62] This vital Ni constituent for the AuGe contacts was first obtained from the contamination of the tungsten evaporator source. The big effect of the Ni addition to the AuGe contacts was reduction of surface roughness,[42][63]-[65] although the surface was still rough. In addition, several other improvements in the AuGeNi contacts were observed: improvement of adhesion[64][66][67] by depositing a thin first layer of Ni, improvement of reproducibility[63][68][69] and improvement of device reliability.[68]

The reasons why the AuGeNi contacts have been used over a quarter century are summarized as follow: *(a)* the contacts provide low resistance by annealing at relatively low temperatures, *(b)* the contacts have excellent run-to-run reproducibility, *(c)* the contacts are prepared by conventional evaporation and lift-off techniques, and *(d)* the contacts have a wide process window and can be prepared by various annealing methods.

## 3.2 Contact Resistance and Microstructure of Contacts Annealed at Various Temperatures

**Annealing Temperature Dependence of $\rho_c$.** The contact resistances were usually measured by the Transmission Line Method (TLM) after annealing at various temperatures for a few minutes. A typical temperature dependence of contact resistances of AuGeNi contacts on GaAs doped with Si at 1 x $10^{18}$ $cm^{-3}$ is shown in Fig. 6.[70]

The AuGeNi contacts show Schottky behavior when annealed at temperatures below 300°C. Ohmic-like behavior is first observed at temperatures above 360°C which corresponds to the eutectic temperature of AuGe. At 420°C, fairly high resistance ohmic contacts were observed and the spread in the $\rho_c$ values was extremely large (five orders of magnitude). Increasing the temperature to 440°C reduced the $\rho_c$ values by a factor of ~$10^4$ and a lowest value of mid-$10^{-7}$ $\Omega$-$cm^2$ was obtained. The (mean) resistances did not change with further increase in the annealing temperature up to 550°C, while at 600°C $\rho_c$ increased twofold. The $\rho_c$ values and the spread of each $\rho_c$ value were found to be reduced by depositing a thin Ni layer while the total Au, Ge and Ni layer thicknesses were kept the same.

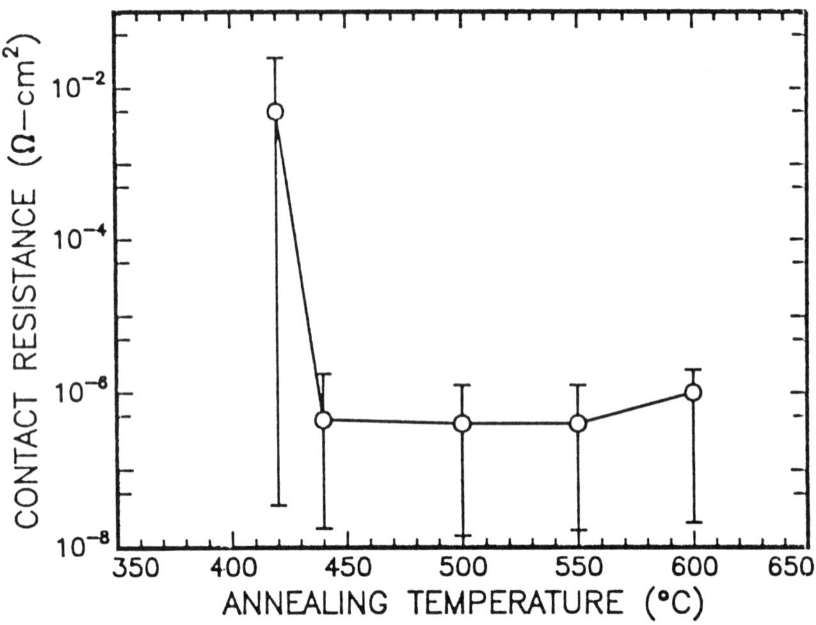


**Figure 6.** Contact resistance of AuGeNi contacts annealed at various temperatures.[70]

This U-shape dependence of contact resistances on annealing temperatures was seen in AuGeNi contacts with different layer thicknesses and deposition sequences. However, the optimum temperatures (which provide the lowest contact resistances) and the lowest $\rho_c$ values are influenced by the GaAs substrate surface cleaning procedure,[66][70][71] the deposition sequence/layer thickness,[72]-[79] and the annealing method/temperatures.[55][74]-[85] There are many combinations of process parameters to fabricate low resistance ohmic contacts and it is impossible to exercise all combinations. In order to find a process window which routinely yields low contact resistance, the interfacial microstructure should be correlated with the electrical properties. The microstructures of AuGeNi contacts annealed at various temperatures are given in next section.

**Annealing Temperature Dependence of Microstructure.** The interfacial microstructure for the AuGeNi contacts was examined by x-ray diffraction (XRD), Auger electron spectroscopy (AES) and transmission electron microscopy (TEM)[68][86]-[88] and correlated with the electrical properties. The kinetics of compound formation between the AuGeNi films and the GaAs substrates was studied by analyzing the same samples used for the $\rho_c$ measurements of Fig. 6 where the samples were annealed at temperatures between 300 and 700°C in intervals of 50°C. Before annealing, x-ray diffraction peaks corresponding to Au, Ni and Ge were observed, and no compound formation or substantial mixing between the layers occurred in the as-deposited sample, suggesting that interdiffusion between layers was negligible at room temperature. (Note that when Ge alone was deposited onto GaAs, the Ge layer usually had an amorphous structure. However, when Ge was premixed with Au and co-evaporated with it, the Ge layer had polycrystalline structure.) After annealing at higher temperatures, the Au, Ni and Ge started to react with GaAs and new compounds were formed. The relative volume fractions of these compounds which were calculated from the measured x-ray diffraction intensities are plotted in Fig. 7, where Figs. 7a and 7b show the transformation processes of the Au and Ni phases, respectively.[87] The Au was stable at temperatures below 450°C, formed a hexagonal close-packed (hcp) β-AuGa compound[89] above 450°C, and then transformed to the α-AuGa phase with a face-centered-cubic structure[89] at 600°C. The Ni and Ge formed $Ni_3Ge$ compounds at temperatures below 350°C, whereas an hcp NiAs(Ge) compound[90] was formed at 450°C and transformed to a NiGe compound with orthorhombic structure[91] at 600°C. [The NiAs(Ge) compound was determined by AES to be NiAs containing a small amount of Ge.] The results indicate that upon heating the Ni and Ge layers first interdiffused to form the $Ni_3Ge$ compound, and then Ga and As atoms diffused out from the substrate to form compounds with Au, Ni and Ge at 450°C. The β-AuGa and NiAs(Ge) compounds were not thermally stable above 600°C.

Comparing Figs. 6 and 7, the formation of the NiAs(Ge) and β-AuGa compounds was essential for low contact resistance. However, the x-ray diffraction technique did not provide information for location of these phases which was essential to correlate the interfacial structure and the electrical properties. Intensive studies by cross-sectional TEM were carried out by Kuan et al.[88] and Shih et al.[86] The microstructure of the AuGeNi contact annealed at 420°C (which had $\rho_c$ of mid-$10^{-3}$ $\Omega$-$cm^2$) showed no compound formation between the Au-Ge layer and GaAs. A small amount of Ga out-diffusion from the GaAs substrate to the Au layers was detected by x-ray energy dispersive spectroscopy (EDS) microanalysis. Large amounts of Ge diffused from the Au-Ge layer to the Ni layer to form $Ni_3Ge$. After annealing at 440°C, significant reaction was observed and the NiAs(Ge) and β-AuGa phases were formed as seen in the cross-sectional TEM micrograph of Fig. 8. The NiAs(Ge) grains grew to a depth as large as 250 nm and a width of ~1 $\mu$m. The β-AuGa grains were formed in a relatively uniform layer on top of the NiAs(Ge) and the GaAs substrate. Note that the interface is rough.

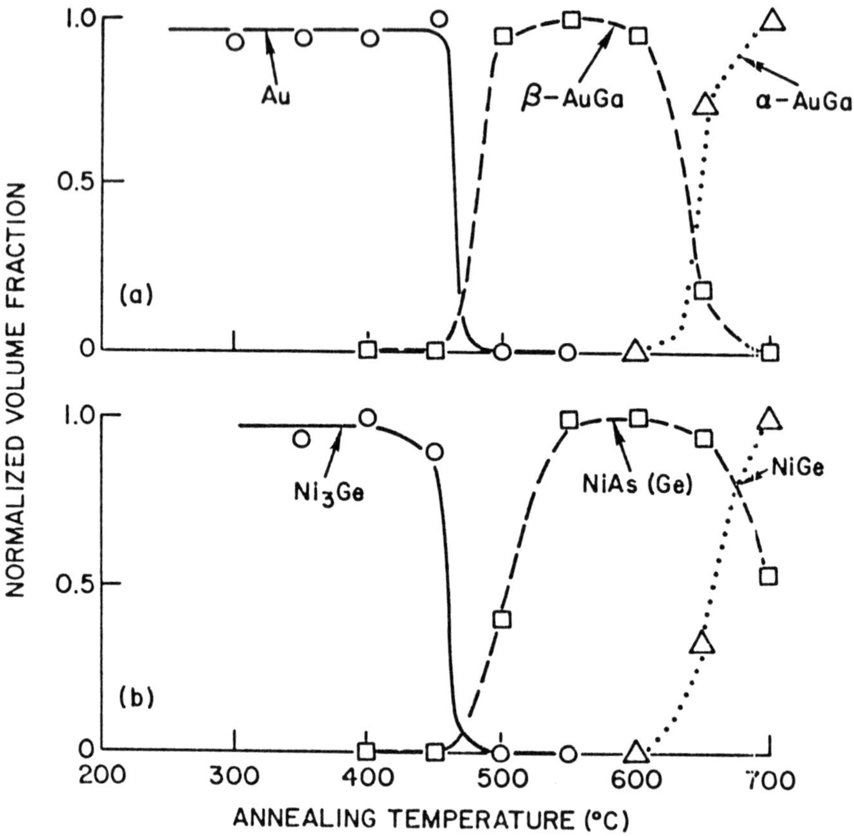


**Figure 7.** Normalized volume fractions of *(a)* Au and *(b)* Ni phases formed in AuGeNi contacts after annealing at various temperatures.[87]

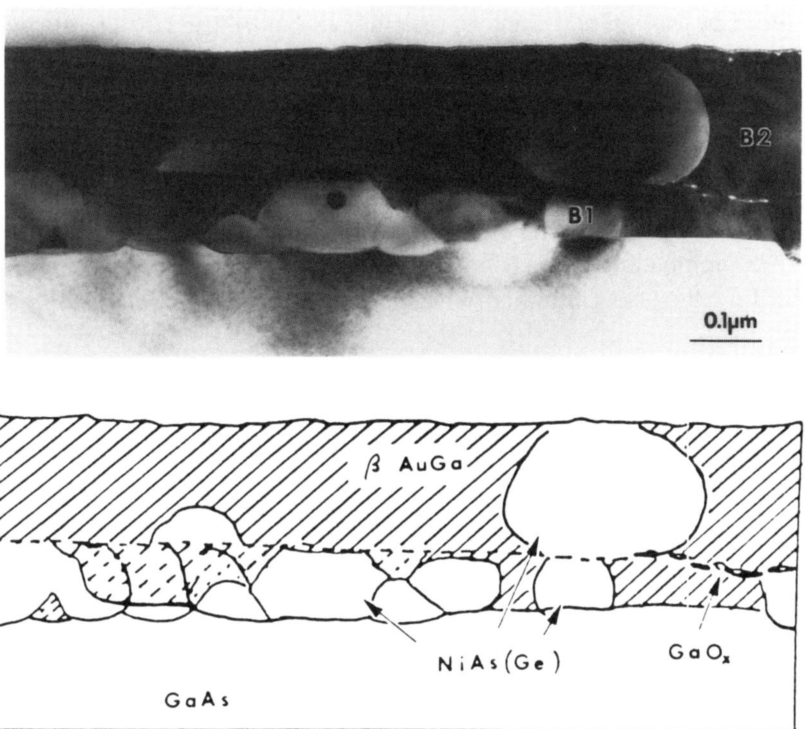


**Figure 8.** Bright-field image of cross-sectional view of AuGeNi contact after annealing at 440°C for 2 min. The diagram below the image indicates the phases observed in the image area.[86]

### 3.3 Roles of Interfacial Microstructure on $\rho_c$ and Thermal Stability

XRD and TEM examinations of samples with various $\rho_c$ values suggested that the NiAs(Ge), β-AuGa, and other compounds formed at the interface of the contacts which provide low contact resistances. The roles of each compound are discussed below.

**Role of NiAs(Ge) Compounds on $\rho_c$.** In order to understand the role of the NiAs(Ge) compounds on $\rho_c$, the $\rho_c$ value of the AuGeNi contacts are plotted in Fig. 9 as a function of the percentage of the GaAs interface covered by the NiAs(Ge) compounds. The percentages were measured from TEM micrographs of the AuGeNi contacts where GaAs substrates were doped with Si at a dose of 1 x $10^{18}$ $cm^{-3}$.[86][88] It is seen that the $\rho_c$ values decrease monotonically with increasing total area of GaAs interface covered by NiAs(Ge). Contact resistances below 1 x $10^{-6}$ Ω-$cm^2$ were obtained if more than 60% of the GaAs interface was covered by the NiAs(Ge) phase. A similar conclusion was also obtained by Bruce et al.[92] for the AuGeNi contacts to GaAs which was doped with Se at a level of 1 x $10^{18}$ $cm^{-3}$. The importance of the NiAs(Ge) phase in direct contact with the GaAs interface was demonstrated by several other authors.[78][93][94] Therefore, an ideal structure for low contact resistance is the GaAs surface covered completely by the NiAs(Ge) phase and the β-AuGa grains formed away from the GaAs interface, which is schematically illustrated in Fig. 10.

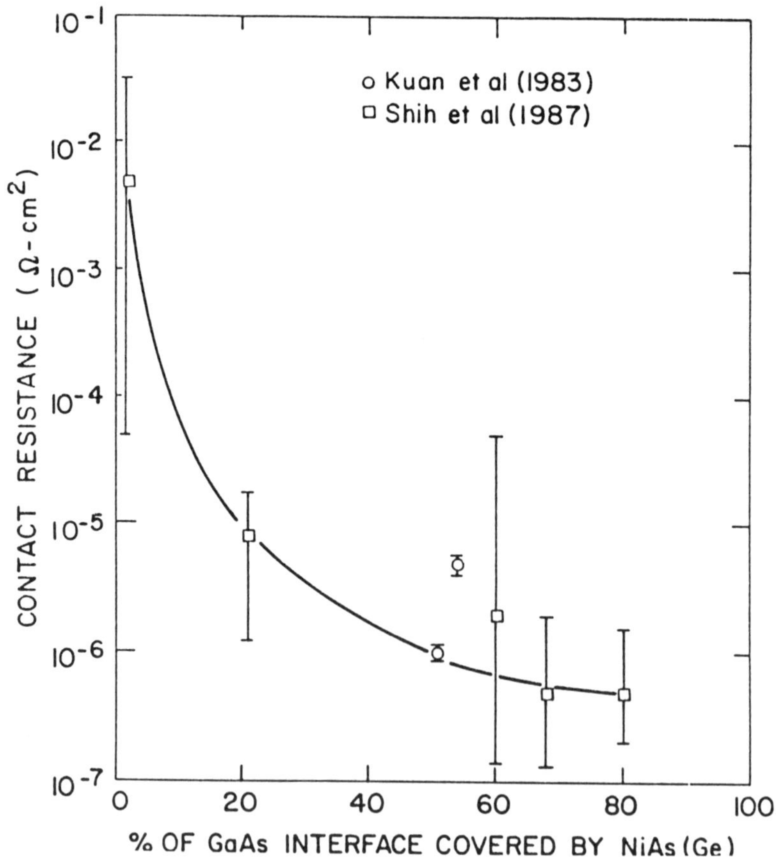


**Figure 9.** Contact resistance of AuGeNi contacts as a function of percentage of GaAs interface covered by NiAs(Ge) phases.

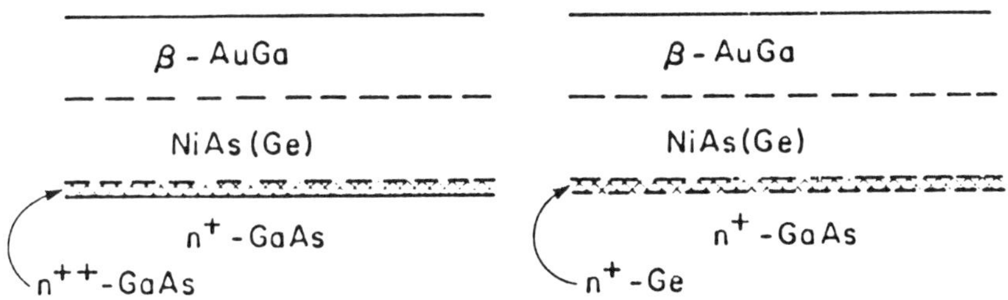


**Figure 10.** Cross-section of low-resistance AuGeNi contacts with heavy doping (left) or a thin Ge layer (right) at the GaAs interface.

The exact current transport mechanism of this contact is not clear at the moment. Two different models are established based on the interfacial microstructures shown in Fig. 10. The most widely accepted model is that the low contact resistance is due to heavy Ge doping at the metal/GaAs interfaces as shown in the left of Fig. 10. During contact annealing Ga atoms diffuse out from the GaAs substrate to the contact metal (forming β-AuGa phases), creating Ga vacancies, while Ge diffuses from the NiAs(Ge) phase to the Ga vacancy sites, providing donors close to the metal/GaAs interface at a level higher than $5 \times 10^{18}$ $cm^{-3}$. The depletion region at the metal/GaAs interface becomes narrow and allows tunneling of electrons through the barrier.[72][75][77][79][93][95]-[97] This model seems very reasonable, but such heavy doping in Ge at the metal/GaAs interface has not been experimentally detected.

Another model to explain the low $\rho_c$ values is that the $\rho_c$ reduction was due to the reduction of the barrier height at the metal/GaAs interfaces by forming a thin Ge layer. Waldrop and Grant[98][99] deposited a thin (~1 nm) Ge layer on a clean GaAs substrate at temperatures in the range of 200 - 325°C under conditions where As incorporation occurred, which made a thin n-type Ge layer as shown in the right of Fig. 10. The barrier height of this thin Ge(As) layer was measured by x-ray photoemission spectroscopy to be in the 0.2 - 0.4 eV range. The low band bending state was preserved upon deposition of a thin NiAs layer. In order to simulate the interfacial structure close to that shown in the right of Fig. 10, they prepared Ge(As)/NiAs/Au contacts and measured the $\phi_b$ value to be 0.3 - 0.4 eV by the I-V method. However, the barrier height of Ge(As)/Au contacts without a NiAs layer was as high as 0.7 eV. These experiments suggested that for annealed AuGeNi contacts, the NiAs(Ge) phases might be separated from the GaAs by an extremely thin Ge layer with a low $\phi_b$ to GaAs. If AuGa phases are in direct contact to these Ge layer regions, high $\phi_b$ regimes are formed. The $\rho_c$ decreased with increasing total area of GaAs interface covered by NiAs(Ge) phase.

However, thin Ge layers were not detected in annealed AuGeNi contacts. Only one experiment which may support this low-barrier-height model is a recent secondary ion mass spectroscopy (SIMS) analysis from the back side of a GaAs wafer.[100] The largest reduction in $\rho_c$ occurred before the period when Ge moves into the GaAs substrate, indicating that n-doping might be less important in the reduction of $\rho_c$ than formation of NiAs(Ge) and the Ge layer. To distinguish the above two models, high-resolution TEM experiments and electrical measurements are required.

**Roles of β-AuGa Compounds on Thermal Stability.** The β-AuGa compounds have two roles on AuGeNi contacts: one is reduction of the contact resistance and the other is thermal instability after contact formation. As explained in Fig. 10, if the dominant electron transport mechanism is tunneling of carriers through the metal/semiconductor interface, Ga vacancies are needed in the GaAs substrate close to the surface. These vacancies are created by forming β-AuGa compounds at the interface. Therefore, the β-AuGa compounds are essential for reduction of the contact resistance and have a positive role for the electrical properties of AuGeNi ohmic contacts.

However, the β-AuGa compounds have a negative effect on the thermal stability of AuGeNi contacts after contact formation. The melting point of the bulk β-AuGa compounds is reported to be about 360°C. Direct evidence of melting of AuGeNi contacts was obtained by XRD and differential scanning calorimetry (DSC). Kim and Chung[101] determined the melting point of Au-12%Ge-4%Ni contact to be 363°C by XRD. Increasing the thickness of the Ni layer increased the melting point during subsequent heating to 413°C, which was explained to be due to dissolution of Ni into the β-AuGa compounds.

Ball[103] also obtained evidence of melting of AuGeNi contacts by DSC. He observed an endothermic peak at a temperature just above 400°C, which was explained to be due to melting of the β-AuGa compounds.

The formation of low melting point β-AuGa compounds at the interface deteriorates the electrical properties and surface morphology of the AuGeNi contacts.

The AuGeNi contacts were thermally unstable at temperatures above 350°C even though the contacts were prepared at temperatures higher than 400°C.[51][69]-[71][92][103] The $\rho_c$ values of the AuGeNi contacts, which were prepared by annealing at 440°C for 2 min. and were subsequently annealed at 400°C, are shown in Fig. 11.[70] It is seen that the $\rho_c$ values increase after annealing for 2 hrs. The increase in the $\rho_c$ value is believed to be due to non-ideal interfacial morphology at the metal/semiconductor contacts. Since the AuGeNi contacts were annealed at a given temperature for a short time, the metal/GaAs reaction was not completed. (This is called a metastable state.) The reaction continued during subsequent annealing even though the temperature was lower than the ohmic contact formation temperature. When the contacts were annealed at lower temperature for a relatively long time or annealed at higher temperature for a short time, a small amount of unreacted Au was frequently detected in the contacts due to incompletion of the Au/GaAs reaction. During subsequent annealing, Au continued to react with GaAs, forming β-AuGa phases. This reaction resulted in an increase of the total area of β-AuGa/GaAs interfaces and the percentage of the GaAs interface covered by NiAs(Ge) decreased. This caused an increase of the $\rho_c$ values as predicted from Fig. 9. For contacts which were annealed at higher temperature, almost all of the Au reacted with GaAs and no unreacted Au was detected. Even in such contacts, grain growth of the β-AuGa phase was observed during subsequent annealing at temperatures higher than 350°C, causing an increase in the $\rho_c$ values for the reason described above. Since the melting point of the β-AuGa phase is ~360°C, melting occurs above this temperature.

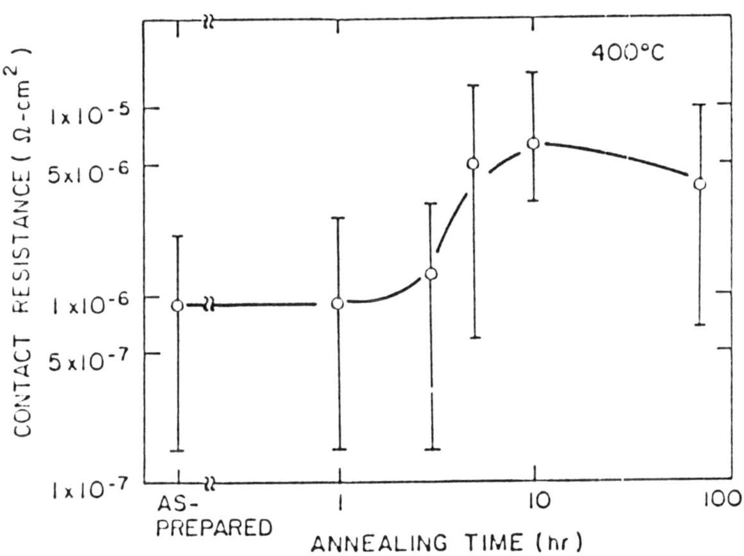


**Figure 11.** Change of the contact resistances of AuGeNi contact during isothermal annealing.[70]

In addition, the edge profiles of the contacts deteriorated due to the melting of β-AuGa compounds. Lateral diffusion of the contacts was observed in as-prepared AuGeNi contacts.[86][104] Cross-sectional TEM micrographs at the contact edges are shown in Fig. 12. The edges were sharply defined before contact annealing as shown by a dashed line, but graded toward the base GaAs surface after annealing at 440°C.[65][86] The typical distance of edge deterioration after annealing at 440°C measured from the original lithographic boundary was about 0.19 μm (Fig. 12a). The worst case of edge deterioration among ten observed specimens is shown in Fig. 12b and was ~0.5 μm. Kamada et al.[105] observed that the contact resistances measured by TLM depended on the angle between the current flow direction and the (110) crystalline direction of the substrate. This orientation dependence was explained to be due to variation of spacings between the contact pads caused by anisotropic lateral diffusion of the contact materials. Also, the contact materials diffused into the GaAs substrate to a measured depth of about 0.2 μm. These large vertical and horizontal diffusions were due to melting of the contact metals and dissolution of the GaAs during annealing at temperatures above 360°C. Therefore, these contacts are conventionally called "alloyed" contacts. Dissolution leads to the possibility of regrowth upon cooling. However, regrown areas would be local and it is difficult to observe them by conventional TEM. High resolution TEM study is required to confirm the presence of regrown layers.

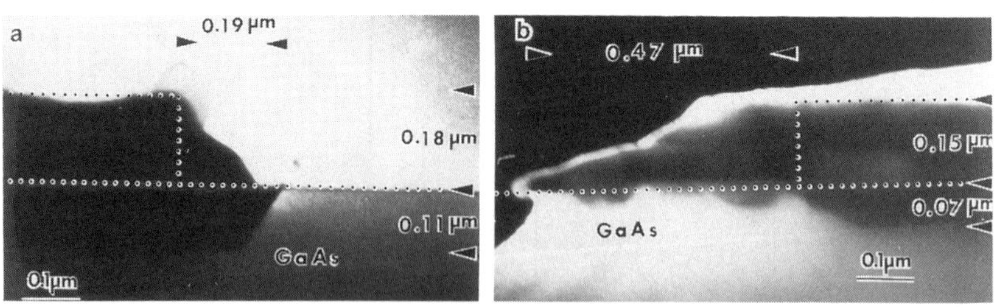


**Figure 12.** Bright-field images of AuGeNi ohmic contacts annealed at 440°C for 2 min: *(a)* typical edge deterioration and *(b)* worst edge deterioration.[86]

### 3.4 Improvement of Surface Morphology and Thermal Stability of Contacts

As described above, the weak points of AuGeNi contacts are a rough surface morphology and thermal instability during isothermal annealing at elevated temperatures above 350°C after contact formation. The improvements made for the AuGeNi contacts are described as follows.

**Surface Morphology.** The surface roughness was reduced partially by covering the contact materials by top cap layers. A smooth surface was obtained after annealing AuGeNi contacts covered with $SiO_2$ passivating layers at ~400°C.[106][107] However, since β-AuGa phases existed in the contacts, it was almost impossible to make a "mirror-like" surface, because the contact materials melted partially during contact annealing as described above. Note that the surface roughness did not correlate with the thermal stability during subsequent annealing. Improvements in thermal stability by reducing the amount of Au added to the AuGeNi contacts are described in the next section.

**Thermal Stability.** Improvement of the thermal stability of the conventional AuGeNi contacts was made by minimizing the amount of Au added to the contacts.

Fricke et al.[108] prepared Ge/Au/Ni contacts where the Ge layer was directly deposited on the n-GaAs and a very thin Au layer (2.5 nm thick) was deposited between the Ge and Ni layer. The Ge/Au/Ni contacts were covered by a $WSi_2$ layer prior to the top Au layer deposition. The contacts provided a relatively low resistance of ~5 x $10^{-6}$ Ω-$cm^2$ after annealing at 590°C. They fabricated metal-semiconductor field-effect transistor (MESFET) devices with this new ohmic contact and tested the thermal stability. The devices did not deteriorate after annealing at 300°C for 1000 hrs. and a small deterioration was reported after annealing at 400°C for 100 hrs.

Lustig et al.[109] also succeeded in improving the thermal stability of the conventional AuGeNi contacts by reducing the amounts of Au used in the contacts. They prepared Ni/Au/Ge/Ni/W contacts where a 6 nm thick Au layer was added to the NiGeW contacts. The contacts were annealed by rapid thermal annealing (RTA) for short times (5 - 30 sec). The $\rho_c$ values were observed to be very sensitive to the annealing temperatures and the lowest $\rho_c$ value of ~1.3 x $10^{-6}$ Ω-$cm^2$ was obtained at 650°C. This value is comparable with that obtained routinely in the conventional AuGeNi contacts.

The contacts were isothermally annealed at 400°C for long times and changes in the $\rho_c$ values are shown in Fig. 13. The contacts were relatively stable at this temperature and less than 50% increase in the $\rho_c$ values was observed after annealing at 400°C for 20 hrs.

The microstructures of contacts with small Au addition was analyzed by XRD. The β-AuGa compounds formed in these NiGeAu contacts was about a thousand times less than that observed in the conventional AuGeNi contacts. Therefore, it is believed that suppression of the low melting point β-AuGa compound formation led to the improvement of the thermal stability.

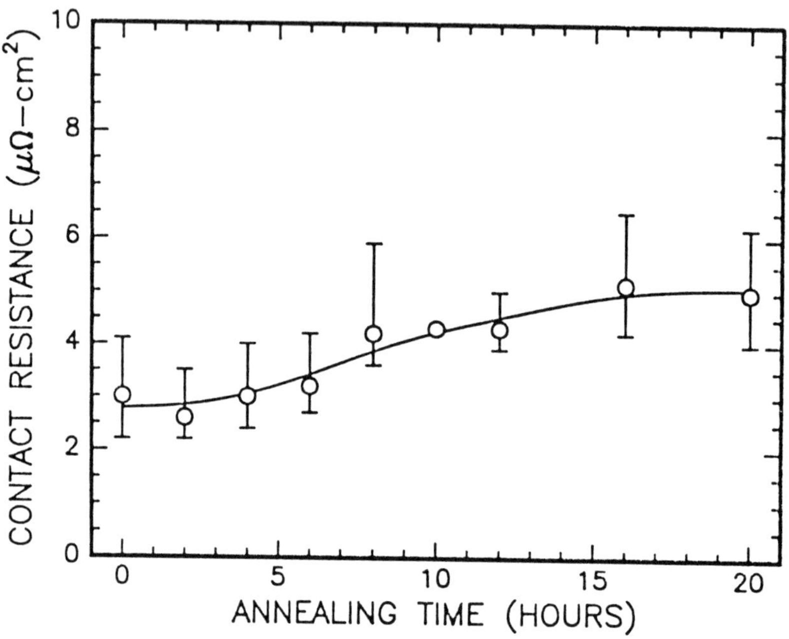


**Figure 13.** Changes of contact resistances during isothermal annealing at 400°C for NiGe(Au)W contacts with 6 nm Au.[110]

The most thermally stable contacts were prepared by Kolawa et al.[110] by eliminating Au addition in the AuGeNi contacts. They prepared NiGe contacts by using sputter-deposition techniques where Ni was directly deposited onto the GaAs substrates. To reduce the contact metal sheet resistance, a WN layer was deposited on the Ge layer followed by Au deposition. The WN barrier layer blocked Au diffusion toward the GaAs substrate, thus Au did not influence the contact resistance. The Ni/Ge/WN/Au contact annealed at 500°C for 10 min. provided a $\rho_c$ value of $2 \times 10^{-6}$ Ω-cm$^2$. This contact was stable at 500°C for 1 hr. Since the Ni layers were deposited directly onto GaAs, the contact formation process was different from that of the previous NiGe contacts in which the Ge layers were directly deposited onto the GaAs substrates. Although detailed microstructure analyses of the NiGe contacts have not been carried out, the NiGe/GaAs reaction is believed to be similar to that of PdGe which was proposed by Marshall et al.[111]-[114] However, the reproducibility of this NiGe contacts was extremely poor, indicating that the process window for low contact resistances is very narrow.

## 4.0 INTERFACIAL EPITAXIAL LAYER FORMATION BY LIMITED REACTIONS FOR OHMIC CONTACT FORMATION TO n-GaAs

### 4.1 Solid-Phase Consumption and Regrowth of an Interfacial Epitaxial Phase in the (Ge or Si)/(Pd or Ni)/n-GaAs System

The spatial non-uniformity and thermal instability of the AuGeNi contact provide impetus for development of a less reactive and more controllable planar ohmic contact. This drive becomes particularly strong as ohmic contacts are placed within a few thousand angstroms of other contacts or device active regions within the semiconductor. In order to overcome the problems of the AuGeNi contact, and to circumvent the complexities of MBE-grown ohmic contacts, controllable limited solid-phase reacted contacts have been investigated. Some early attempts have employed annealed metal/Ge bilayer films on n-GaAs.[115]-[118] Sinha et al. deposited a bilayer structure of Pd/Ge/GaAs and annealed it at a temperature (350 - 500°C) which allowed only solid-phase reactions.[115] The thicknesses of the layers were such that some of the Pd would react with most of the Ge to form PdGe and the remainder of the Pd would react with GaAs to form Pd-GaAs intermetallic compounds. Grinolds and Robinson[116] correlated the onset of ohmicity in this contact with the presence of Ge at the GaAs surface. Anderson et al. examined the effect of annealing a Ni film deposited on an epitaxial Ge film on n-GaAs.[118] An ohmic contact was formed when the thicknesses of the layers were such that some of the Ni would react with most of the Ge to form NiGe and the remainder of the Ni would react with GaAs to form Ni-GaAs intermetallic compounds after annealing at 450 - 650°C. Auger depth profiling analysis suggested that Ge existed deeper in the structure than Ni. Contact resistivity was measured in all of these studies[115][116][118] to be above $10^{-5}$ Ω-cm$^2$ for GaAs doping levels of $10^{16}$ and $10^{17}$ cm$^{-3}$. Tiwari et al. reduced contact resistivity to the $10^{-6}$ Ω-cm$^2$ range by using layers of Ge and a refractory metal, Mo, and high annealing temperatures (750°C) under an arsenic overpressure provided by InAs decomposition.[117] The presence of In, though, was found to influence the contact electrical behavior.[119]

In order to improve upon these contacts and to investigate the ohmic contact formation mechanism in depth, a concentrated effort on the (Ge or Si)/(Pd or Ni)/n-GaAs systems was undertaken. The initial goal of the study was to reproduce the structure of the MBE-grown metal/n$^+$Ge/n-GaAs ohmic contact (Fig. 5a) using simple processing techniques which did not require the ultra-high vacuum deposition environment of MBE.

**Interfacial Microstructure and Correlation with Contact Resistivity.** ***Ge/Pd/GaAs System.*** The approach chosen to form such a structure was solid-phase epitaxy (SPE) through a transport medium. A Ge/GaAs heterostructure results after the solid state reaction of a layered structure deposited in a conventional electron-beam evaporation system.[111]-[113][120]-[122] The technique involves deposition of a metallic transport medium, Pd, onto which a layer of amorphous Ge is then deposited without breaking vacuum (Fig. 14). The thickness of the Ge and Pd layers are chosen such that, upon annealing, the entire layer of Pd is consumed in the formation of a palladium germanide layer. The Pd reacts with the Ge to first form $Pd_2Ge$ and finally PdGe[37] with normalized thicknesses consumed and produced according to: 1.00 Å Pd + 1.54 Å Ge → 2.13 Å PdGe. The remaining amorphous Ge is then transported through the germanide to grow epitaxially on the GaAs substrate due to the lower free energy of the crystalline semiconductor compared to that in its amorphous state (Fig. 15c). The interface between the contact and the GaAs is planar and abrupt to within atomic dimensions. Consequently, the surface morphology of the annealed Ge/Pd/GaAs is virtually featureless.

It was found that in order to obtain low contact resistance using the Ge-Pd system on n-GaAs (~$10^{-6}$ Ω-cm$^2$ for $N_d \sim 10^{18}$ cm$^{-3}$): *(i)* the presence of Ge is necessary in the structure, *(ii)* excess Ge over that required for PdGe formation leads to lower specific contact resistivity than that obtained with excess Pd, and *(iii)* Pd must be initially in contact with the GaAs.

The observed benefit of the placement of Pd in direct contact with GaAs and closer examination of Fig. 14 points to an effect other than the epitaxy of Ge on GaAs which plays a role in the ohmic contact formation. A very limited uniform reaction occurs at the Pd-GaAs interface in the as-deposited state or in the initial stages of annealing. This results in a very thin (≤ 100 Å) epitaxial $Pd_4GaAs$ layer. This continuous layer present in the as-deposited sample (Fig. 14) is not seen at the grown Ge-GaAs interface (Fig. 15c). X-ray diffraction studies show that the onset of ohmic behavior can be correlated with the decomposition of this phase.[120]

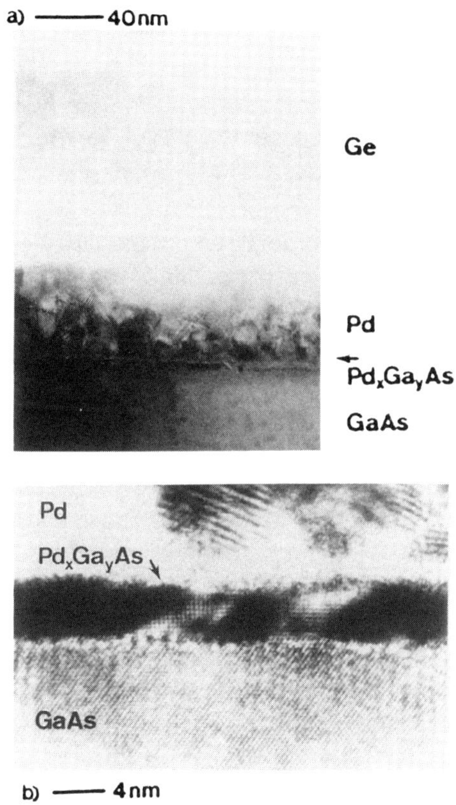


**Figure 14.** *(a)* Cross-sectional view of an as-deposited Ge/Pd/GaAs sample. *(b)* Lattice image of the interface between the GaAs and the as-deposited Pd layer. An epitaxial layer appears at the interface.[112]

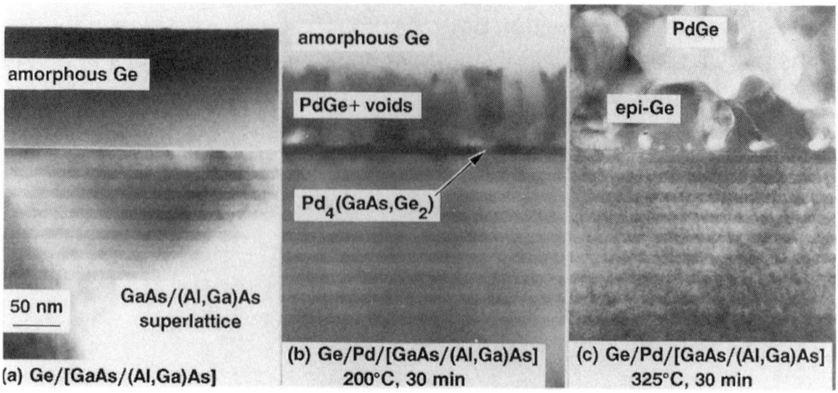


**Figure 15.** Cross-sectional TEM images of *(a)* as-deposited Ge/[GaAs/(Al,Ga)As], *(b)* Ge/Pd/[GaAs/(Al,Ga)As] after annealing at 200°C for 30 min., and *(c)* Ge/Pd/[GaAs/(Al,Ga)As] after annealing at 325°C for 30 min.[122]

**Si/Pd/GaAs System.** In order to determine the relative importance of semiconductor heteroepitaxy and of the role of the initial Pd-GaAs reaction in the formation of an ohmic contact in the Ge/Pd/GaAs system, the analogous Si/Pd/GaAs system was investigated.[123][124] Pd reacts with Si to form $Pd_2Si$ at temperatures greater than 100 - 300°C and less than 800°C. Solid-phase transport and epitaxy of excess Si is not expected in this system due to the large lattice mismatch between Si and GaAs (~4%) and a much higher transport temperature for Si through $Pd_2Si$ (~550°C).

It was found that good ohmic characteristics to n-GaAs resulted (~5 x $10^{-6}$ $\Omega$-cm$^2$ for $N_d$ ~$10^{18}$ cm$^{-3}$) without any evidence of macroscopic transport and epitaxy of Si to the GaAs interface. Moreover, the presence of excess Si over that necessary for $Pd_2Si$ formation was found to be critical for ohmic contact formation. The excess Si was also found to stabilize the GaAs against dissociation at higher temperatures (> 400°C). The major reactions affected are those between Pd and GaAs, which begin in the as-deposited state:

Eq. (24) $$Pd + \tfrac{1}{4}GaAs \leftrightarrow \tfrac{1}{4}Pd_4GaAs$$

and between Pd and Si, which begins after annealing at temperatures greater than 100°C:

Eq. (25) $$Pd + \tfrac{1}{2}Si \leftrightarrow \tfrac{1}{2}Pd_2Si$$

If these two reactions are considered to be competing in the consumption of Pd, Eqs. 24 and 25 can be reduced by canceling the common component, Pd, through subtraction to yield the thermodynamic equation:

Eq. (26) $$2Si + Pd_4GaAs \leftrightarrow 2Pd_2Si + GaAs$$

This reaction has associated with it a Gibbs' free energy change which is not known at present since the heat of formation of the $Pd_4GaAs$ phase is not known. The effect of excess Si and the thermodynamic stability of $Pd_2Si$ drive Eq. 26 to the right.[123] High-resolution TEM showed that in samples with excess Si, after annealing at temperatures at which the contact became ohmic, the thin epitaxial $Pd_4GaAs$ layer, which was seen at the interface of an as-deposited sample, was replaced by a thin epitaxial $Pd_2Si$ layer. The excess Si effectively reduced the $Pd_4GaAs$ to $Pd_2Si$ and GaAs. Lattice imaging showed dislocation cores located approximately 100 Å below the epitaxial phase which were speculated to demarcate an interface between the GaAs substrate and solid-phase regrown homoepitaxial GaAs which resulted from $Pd_4GaAs$ dissociation.

**Si/Ni/GaAs System.** The analogous Si/Ni/GaAs system results in similar reactions as found in the Si/Pd/GaAs system, except with much thicker, and hence more easily observable, reacted and regrown layers.[125] Approximately 1000 Å of GaAs was definitively observed to be initially consumed by reaction with Ni to form $Ni_xGaAs$ and subsequently epitaxially regrown. The regrown GaAs formed as a product of the reaction

Eq. (27) $$xSi + Ni_xGaAs \leftrightarrow xNiSi + GaAs$$

which is driven to the right in the presence of excess Si.

***High Resolution Interfacial Analysis in the Ge/Pd/n-GaAs System.*** While the initial motivation behind the Ge/Pd/GaAs contact was to form a low barrier Ge/GaAs heterojunction, investigation of the analogous Si/Pd/GaAs and Si/Ni/GaAs contact systems pointed towards GaAs regrowth as an important factor in the formation of ohmic characteristics. In order to determine whether regrowth was indeed occurring in the Ge/Pd/GaAs system, high-resolution analysis techniques were applied. Most analysis techniques capable of looking for heterojunctions or interfacial doped layers are two-dimensional in nature. Even though they may only sample a thin region or are capable of high-resolution depth profiling, they average over the probe cross-sectional area. Hence, in order to obtain any useful information, the contact studied should be laterally homogeneous at each point in depth over the cross-sectional area of the analysis probe. This is one of the reasons that the laterally non-homogeneous alloyed Au-Ge-Ni contact has evaded understanding for so long. Since the Ge/Pd/GaAs contact undergoes limited solid-phase reactions and retains interfacial planarity on a nearly atomic scale, it presents an opportunity to correlate high-resolution interfacial studies with the contact electrical behavior.

The use of a substrate which terminates in a GaAs/(Al,Ga)As superlattice facilitates the analysis (Fig. 15).[121][122] The uppermost GaAs layer is thin enough (100 Å) that its consumption and regrowth can be monitored by counting atomic layers using TEM lattice imaging and using the underlying (Al,Ga)As as a marker layer (Fig. 16). The incorporation of contact elements (i.e., Ge) into the interfacial region of the GaAs can be observed by selective removal of the substrate and SIMS depth profiling from the backside (Fig. 17). This allows for great depth resolution as the single crystalline substrate remains planar during sputtering, the ion beam mixing (knock on) of the contact elements do not reduce sensitivity, and the superlattice can be used as a calibration standard for sputter rate.

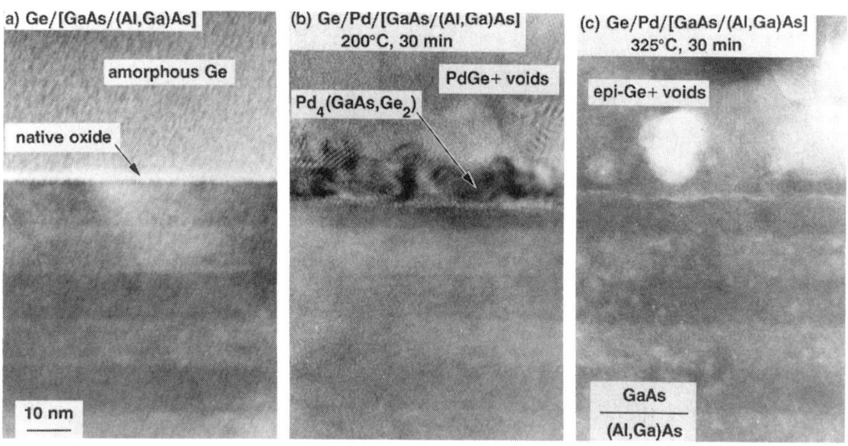


**Figure 16.** Higher magnification cross-sectional TEM images of the interfacial regions in Fig. 15. The initial condition of the uppermost GaAs layer may be ascertained from *(a)*. $Pd_4(GaAs,Ge_2)$ growth has consumed GaAs as seen in *(b)*. Decomposition of the $Pd_4(GaAs,Ge_2)$ leads to a thicker GaAs surface layer in *(c)*. This is indicative of solid-phase consumption and regrowth of a thin GaAs surface layer.[122]

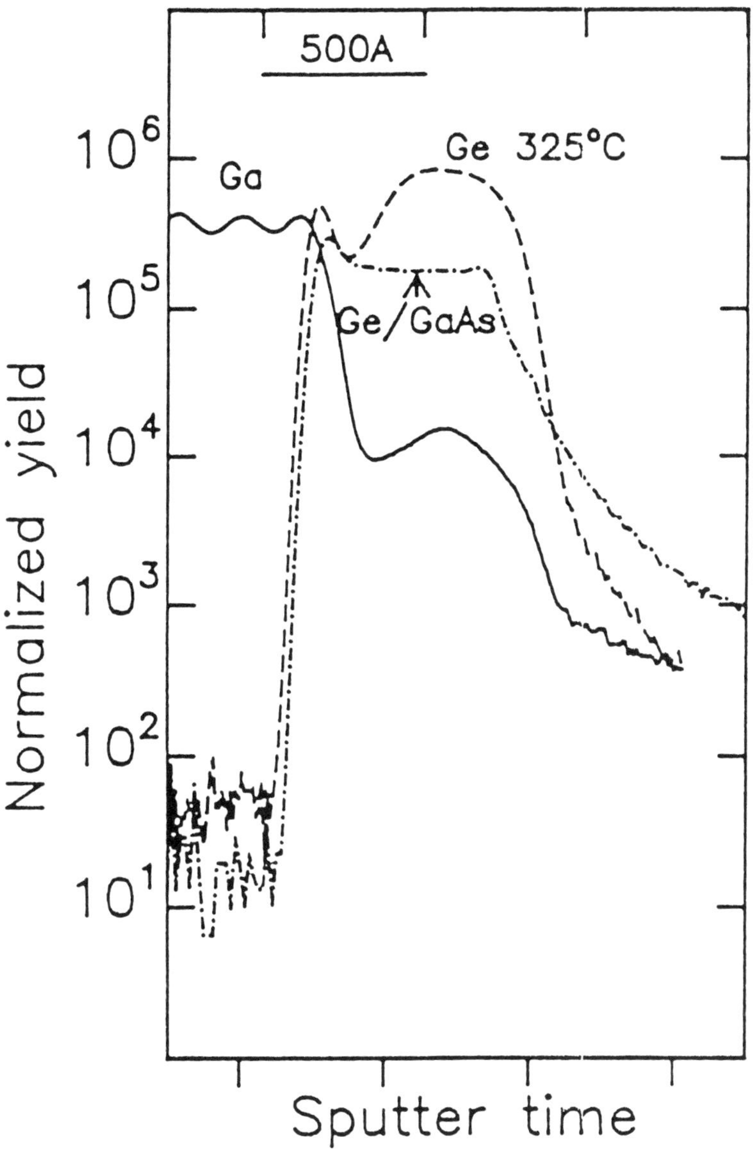


**Figure 17.** Backside SIMS sputter depth profiles of Ge in an unannealed Ge/(type II substrate) (labeled *Ge/GaAs*) and a 325°C 30 min. annealed Ge/Pd/[GaAs/(Al,Ga)As] (labeled *Ge 325°C)* sample using a 2.3 keV $O_2^+$ ion beam. The Ga profile from the 325°C annealed Ge/Pd/[GaAs/(Al,Ga)As] is also plotted. The depth scale marker is for GaAs only.[122]

**Electrical Transport Mechanism.** ***Contact Formation Mechanism.*** An understanding of the ohmic contact formation mechanism develops from correlation of the contact electrical and metallurgical behavior. This is shown schematically in Fig. 18. Upon deposition or after heat treatment of less than 100°C, Pd moves through and disperses the oxide to form a thin epitaxial $Pd_4GaAs$ layer. The thickness of this phase is limited to 100 Å. The major reaction is between Pd and Ge to first form $Pd_2Ge$ and then PdGe. After the Ge reaches the $Pd_4GaAs$, some Ge may be incorporated into this layer [$Pd_4(GaAs,Ge_2)$]. While this interfacial epitaxial

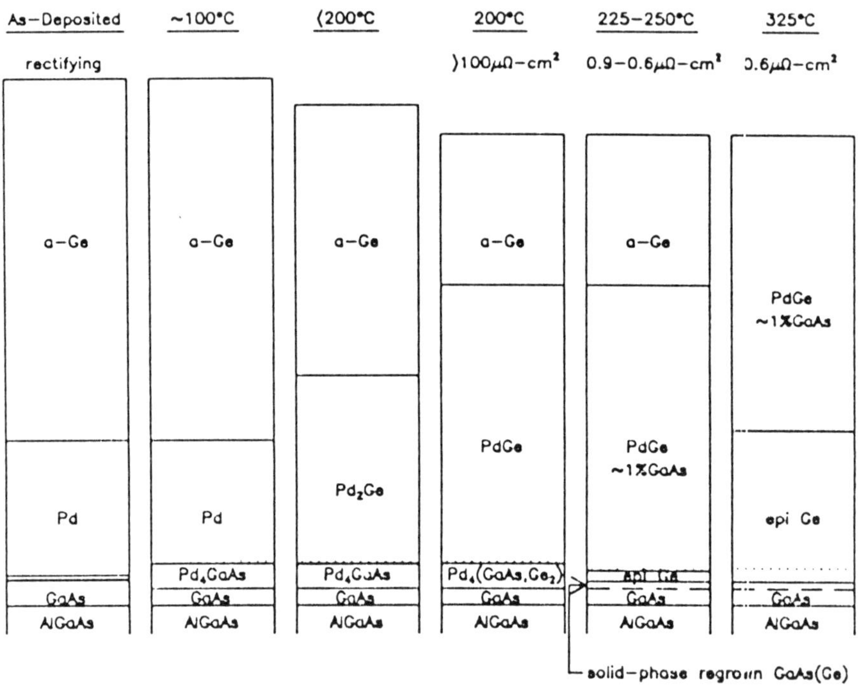


**Figure 18.** Hypothetical reaction sequence as a function of anneal for the Ge/Pd/[GaAs/(Al,Ga)As] structure. The closely spaced lines above the GaAs in the as-deposited sample represent the native oxide layer. The dots in the subsequent figures represent the position of the oxide layer which has been dispersed by the initial Pd-GaAs reaction and any voids which may have formed as a result of the reaction. It serves as a marker to indicate the position of the original interface. The specific contact resistivity ($\mu\Omega$-$cm^2$) is quoted for the different annealing conditions.[122]

phase is present, as in the 200°C sample, the contact does not demonstrate good ohmic behavior. After annealing at temperatures ≥ 225°C, the presence of excess Ge drives the decomposition of the epitaxial $Pd_4(GaAs,Ge_2)$ layer. The result of this decomposition is the solid-phase regrowth of Ge-incorporated GaAs, followed by the growth of a thin epitaxial Ge layer. Deep (> 100 Å) penetration of high levels (> $5 \times 10^{18}$ $cm^{-3}$) of Ge is not found below the contact. At this point, the contact displays good ohmic behavior with contact resistivity in the $10^{-7}$ $\Omega$-$cm^2$ range. Annealing at temperatures of ~325°C leads to solid-phase transport and epitaxy of the excess amorphous Ge. The contact resistivity, however, does not change. Hence, the onset of good ohmic behavior is correlated with the decomposition of the intermediate epitaxial $Pd_4(GaAs,Ge_2)$ phase and solid-phase regrowth of Ge-incorporated GaAs followed by the growth of a thin Ge epitaxial layer as

Eq. (28) $$4Ge + Pd_4GaAs \leftrightarrow 4PdGe + GaAs$$

is driven to the right in the presence of excess Ge. By using backside SIMS and TEM with thin marker layers, it is seen that all GaAs is accounted for in the reaction. About 35 Å of GaAs is consumed during $Pd_4GaAs$ formation. About 25 Å of GaAs is seen to regrow. The 1% GaAs in the PdGe layer corresponds to 10 Å GaAs. Hence,

Eq. (29) 35 Å *(consumed GaAs)* = 10 Å *(in contact)* + 25 Å *(regrown GaAs)*

***Contact Resistivity as a Function of Substrate Doping and Temperature.*** The dependence of specific contact resistivity on n-type substrate doping level at room temperature is shown in Fig. 19. A least squares fit yields the relation $\rho_c \propto N_D^{-1/2}$. For comparison, the line for Au-Ge-Ni contacts based on the results of Braslau[126] is also included in Fig. 19. The temperature dependence of contact resistivity, $\rho_c$, as a function of substrate doping concentration can be found in Fig. 20. The line drawn through the contact resistivity points is a non-linear least squares fit to the phenomenological equation based on thermionic field emission where

Eq. (30) $$\rho_c = C \exp\left[\phi_b / (E_{00} \coth \frac{E_{00}}{kT})\right]$$

The value of $\phi_b$ gives an estimate of effective barrier height in the high-

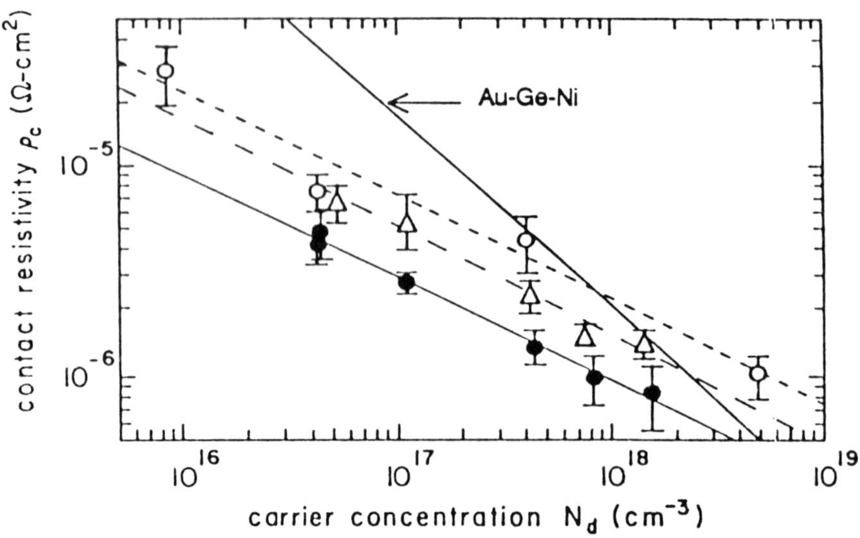


**Figure 19.** Specific contact resistivity as a function of substrate doping level for samples annealed at 325°C for 30 min. The different symbols correspond to three repetitions of the experiment. Least squares log-log fits to the averages yield the relation $\rho_c \propto N_D^{-1/2}$ for each repetition. The line calculated for Au-Ge-Ni contacts by Braslau[126] is included for comparison.[112]

temperature Arrhenius plot regime. The value of $E_{00}$ is a measure of the temperature at which electronic conduction becomes independent of temperature and does not necessarily have the same meaning as that given in Eq. 12. These numbers are more meaningful as the low-temperature and high-temperature linear regimes become better defined. For a degenerate $10^{18}$ cm$^{-3}$ doped sample, contact resistivity is only weakly dependent upon measurement temperature. Samples which are not degenerate at room temperature display much more pronounced temperature dependence. The transition temperature at which electronic transport becomes temperature independent, as evidenced by $E_{00}$, is similar for the three samples. The equivalence of $E_{00}$ values is due to one of two points. If it has the same meaning as given in Eqs. 12, then either the surface doping level is the same in all samples at ~3 x $10^{17}$ cm$^{-3}$, or, more generally, the shape of the band bending is the same for all samples. The alternative explanation lies in the fact that, functionally, $E_{00}/k$ is the temperature near the knee of the $\rho_c$ versus 1000/T curve at which contact resistivity becomes independent of tempera-

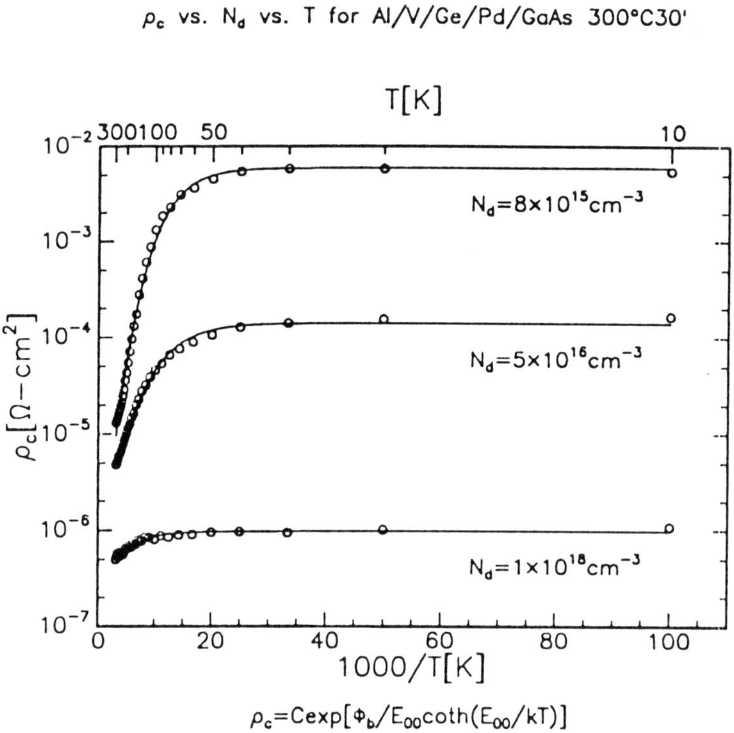


| $N_d[cm^{-3}]$ | $\Phi_b$[meV] | $E_{00}$[meV] | $E_{00}/k$[K] | log(C) |
|---|---|---|---|---|
| $8\times10^{15}$ | 104±4 | 10.2±0.3 | 118±3 | −6.62±0.08 |
| $5\times10^{16}$ | 43.7±1.4 | 8.8±0.2 | 102±2 | −6.00±0.03 |
| $1\times10^{18}$ | 12.6±1.8 | 11.9±2.6 | 138±30 | −6.47±0.03 |

**Figure 20.** Temperature-dependent contact resistivity for 300°C 30 min. annealed Al/V/Ge/Pd/GaAs ohmic contacts as a function of donor concentration. The line drawn through the contact resistivity points is a non-linear least squares fit to the phenomenological equation based on thermionic field emission (Eq. 30). The value of $\phi_b$ gives an estimate of effective barrier height in the high-temperature Arrhenius plot regime. The value of $E_{00}$ is a measure of the temperature at which electronic conduction becomes independent of temperature. These numbers are more meaningful as the low-temperature and high-temperature linear regimes become better defined.[111]

ture. This can occur if field emission is possible. Field emission, in turn, can occur if the Fermi level lies in the conduction band of the semiconductor. Indeed, calculations of $(E_F - E_c)$ based on the Hall effect measurements, which were taken simultaneously with contact resistivity measurements (see Fig. 21), show that $E_F$ approaches $E_c$ at approximately $E_{00} / k$. The calculation of $(E_F - E_c) / kT$ is derived from the measured approximate free electron carrier concentration by use of the approximations to the Fermi Dirac integral where[127]

Eq. (31) $$\frac{E_F - E_c}{kT} = -\ln\left(\frac{N_c}{n} - 0.27\right)$$

for $(E_F - E_c) / kT < 1.1$, and

Eq. (32) $$\frac{E_F - E_c}{kT} = \sqrt{\left(\frac{3\sqrt{\pi}}{4}\right)\left(\frac{n}{N_c}\right)^{\frac{4}{3}}}$$

for $(E_F - E_c) / kT > 1.1$, where

Eq. (33) $$N_c = (4.7 \times 10^{17}\ \text{cm}^{-3})(T / 300)^{3/2}$$

and T is the temperature in Kelvin. The effective barrier height, on the other hand, is a function of the GaAs doping level. Note that the effective barrier height is close to the value of $V_n$ at room temperature which is approximately one order of magnitude less than the usual Schottky barrier height to n-type GaAs of 0.8 eV. Since the temperature dependence of $\rho_c$ is dependent upon doping level, the dependence of $\rho_c$ on $N_d$ will be a function of temperature. While $\rho_c \propto N_d^{-1/2}$ at room temperature, $\rho_c \propto N_d^{-3/2}$ at less than 100 K.

***Implications of Results on Ohmic Contact Formation Models.*** The temperature dependence of contact resistivity as a function of doping and composition is dependent upon the exact nature of the metallurgical changes at the interface. For instance, the fit of contact resistivity as a function of substrate doping (Fig. 20) to Eq. 34 yields doping-independent values of $E_{00}$ and differing values of effective barrier height, $\phi_b$. If the fact that $E_{00}$ is equal for the three samples is not simply due to the temperature at which the GaAs becomes degenerate, then this fact implies that the interfacial band bending is the same for the three samples with very

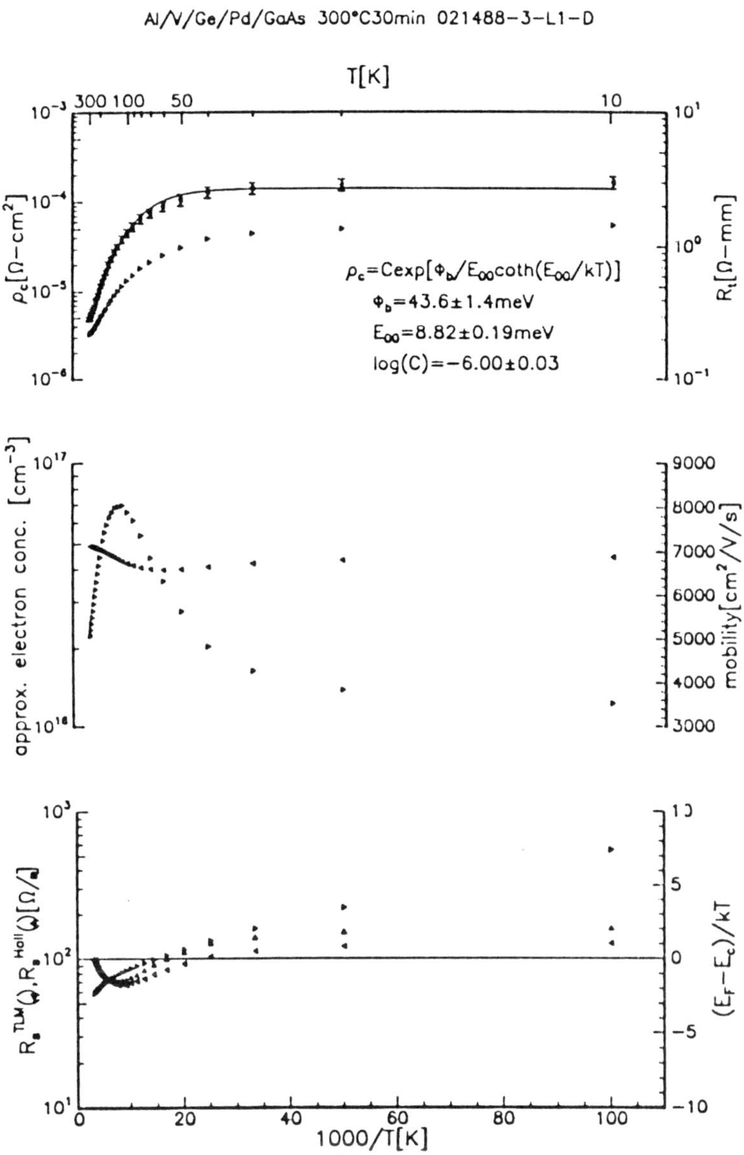


**Figure 21.** Results from variable temperature transmission line model (TLM) measurements and Hall effect measurements for 300°C annealed Al/V/Ge/Pd/GaAs ohmic contacts where the GaAs is doped with Si to ~5 x $10^{16}$ $cm^{-3}$.[111]

different doping levels. This could occur if the GaAs surface region was doped heavily enough to bring the conduction band edge down to the level of the effective barrier height within the thin modified interfacial region. The extracted values of effective barrier height, $\phi_b$, in Fig. 20 are all different and are all quite a bit lower than the normal barrier height of contacts to n-type GaAs of 0.8 eV. They are similar to the room temperature values of $V_n$ (= $E_c - E_F$). Thus, conduction takes place at energy levels near $E_c$. This could happen if the real barrier height was lowered to ≤ 0.1 eV. Alternatively, the potential barrier could be thin enough at ≤ 0.1 eV to allow electron tunneling.

Barrier thinning could take place if the surface region was highly doped. High-resolution backside SIMS and cross-sectional TEM with marker layers have shown that a 25 ± 10 Å regrown layer exists at the interface which may have significant Ge incorporation. Analytic solutions for band-bending in the presence of a highly doped surface region may lead to insight into the type of surface modification necessary to result in the observed contact resistivity behavior. The band-bending, $V(x) = E_c(x) - E_c(\infty)$, was calculated by:[128]

Eq. (34) $$V(x) = \phi_b - V_n + \left(\frac{N_{d1}}{\varepsilon}\right)\left(\frac{x^2}{2} - tx\right) + \left(\frac{N_{d0}}{\varepsilon}\right)\left(\frac{x^2}{2} - wx\right);\ 0 \le x \le t$$

Eq. (35) $$V(x) = \left(\frac{N_{d0}}{2\varepsilon}\right)(w - x)^2;\ t \le x \le w$$

where

Eq. (36) $$w = \sqrt{\left(\frac{2\varepsilon}{N_{d0}}\right)\left(\phi_b - V_n - \frac{N_{d1}t^2}{2\varepsilon}\right)}$$

$N_{d0}$ is the substrate doping level, $N_{d1}$ is the highly doped surface layer doping level, t is the thickness of the highly doped surface layer, and w is the depletion layer thickness (Fig. 4). Eqs. 34 - 36 are only true if w > t. If w ≤ t, then

Eq. (37) $$V(x) = \left(\frac{N_{d1}}{2\varepsilon}\right)(w - x)^2;\ 0 \le x \le w$$

where

Eq. (38) $$w = \sqrt{\left(\frac{2\varepsilon}{N_{d1}}\right)\left(\phi_b - V_n|_{N_{d1}}\right)}$$

If Eqs. 34 - 36 are not valid and the depletion width as calculated by Eq. 38 is greater than t, then w is set equal to t and Eq. 37 is used for V(x); this is only an approximation, whereas a full solution requires numerical solution of Poisson's equation. Image force lowering effects are included as a perturbation on band-bending by recalculating:

Eq. (39) $$V'(x) = V(x) - 1/(16\pi\varepsilon x)$$

If the barrier height is assumed to be unmodified and is the same as for most metal/GaAs junctions, 0.8 eV, the band-bending that results with a thin, highly doped surface layer can be found in Fig. 22. If the layer is 25 Å thick, then a doping level of $2 \times 10^{20}$ cm$^{-3}$ is just sufficient to cause ohmic behavior (Fig. 22b). Doping at less than $2 \times 10^{20}$ cm$^{-3}$ would lead to rectifying behavior (Fig. 22a) and doping more than $2 \times 10^{20}$ cm$^{-3}$ would be more than sufficient for ohmic behavior (Fig. 22c). Likewise, if the layer were doped to $2 \times 10^{20}$ cm$^{-3}$, a 15 Å layer would lead to a contact which is rectifying and a 35 Å layer is more than sufficient for ohmic behavior. As discussed in Sec. 1.6, $2 \times 10^{20}$ cm$^{-3}$ doping is about the highest value ever achieved and is not easily attainable.

An alternative mechanism which could lead to similar band-bending as calculated for Fig. 22 would be the presence of a sufficient quantity of electrically active defects at the regrown GaAs/GaAs substrate interface which could effectively pin the Fermi level near the conduction band edge. Since the conduction band edge would be brought from $\phi_b$ to $E_F$ within ~25 Å, a tunneling ohmic contact could result. This mechanism is akin to that of delta doping as discussed in Sec. 2.1. Verification of this proposed mechanism requires investigation of the electrical and metallurgical nature of the defects at the interface.

Lower surface doping levels would also be required if the actual barrier height were lowered. If the barrier height were 0.3 eV, the upper limit on the Ge-GaAs conduction band discontinuity, only ~$5 \times 10^{19}$ cm$^{-3}$ doping is required. If the barrier height were 0.1 eV, the lower limit on the Ge-GaAs conduction band discontinuity, then only $\geq 1 \times 10^{19}$ cm$^{-3}$ doping is required. A 0.1 eV barrier without a highly doped surface layer is unlikely as it would lead to a different dependence of contact resistivity on temperature than observed. From a plot of band-bending versus temperature, it can be seen

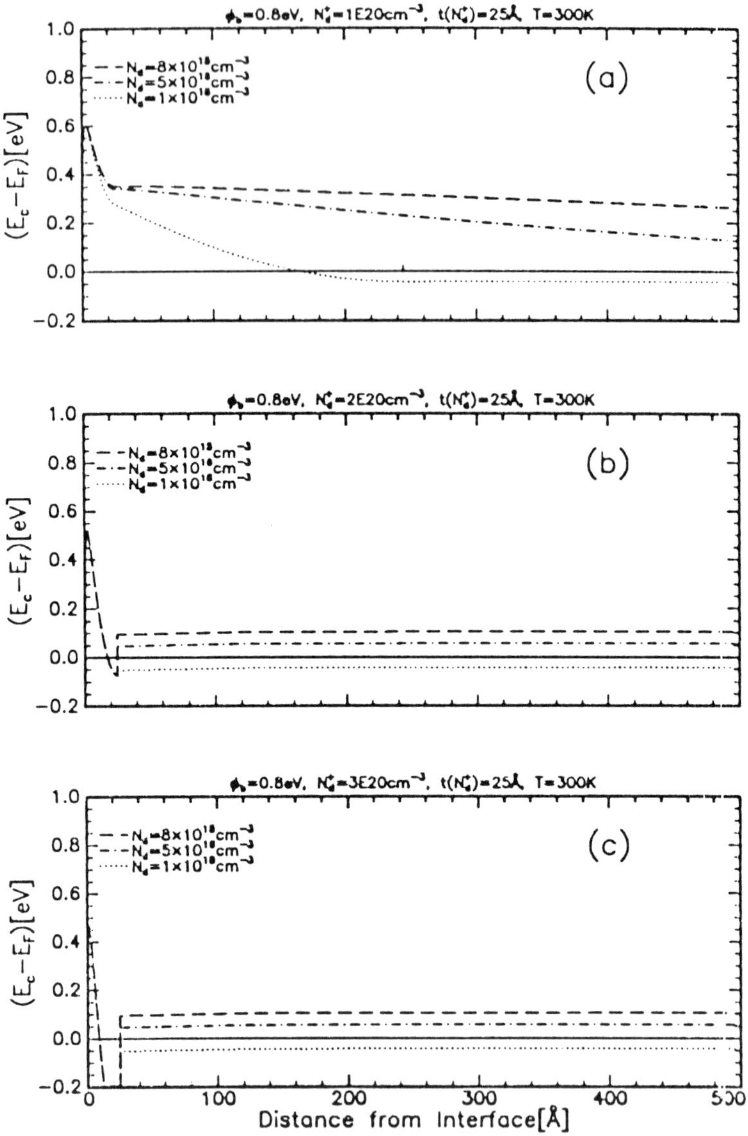


**Figure 22.** Band-bending as calculated by Eqs. 34 - 39 for GaAs doped to three different levels *(a)*, *(b)* and *(c)* as a function of the doping level of a highly doped surface layer. For a 25 Å thickness and a 0.8 eV surface barrier, a ~2 x $10^{20}$ $cm^{-3}$ doping is required for sufficient band bending for ohmic contact formation at 300 K.[111]

that as the temperature is lowered, the Fermi level approaches the conduction band edge in the 8 x $10^{15}$ $cm^{-3}$ and 5 x $10^{16}$ $cm^{-3}$ doped substrates. In all cases, the surface band-bending is predominantly controlled by the highly doped surface layer. In this manner, temperature-independent field emission is expected for all three samples at low temperatures. Indeed, this has been observed (see, for example, Fig. 20). A problem arises if the barrier height is 0.1 eV and the surface is not highly doped. At 50 K, the band-bending is distributed over a distance so large that tunneling is not likely. Hence temperature-dependent thermally activated transport would be expected. This is in contrast with observed contact behavior.

Another possible mechanism which may control contact resistivity is the presence of a high-low junction within the GaAs.[31][32] This would be due to a junction between a thick surface $n^+$ region and the GaAs substrate with lower doping. This mechanism is not likely since no strong evidence for a thick $n^+$ region has been found. Theories which invoke spreading resistance due to laterally inhomogeneous contacts[126] or a high-low barrier due to a thick $n^+$ surface region ($n^+$ layer thickness » depletion width)[32] predict that $\rho_c \propto N_D^{-1}$. Thus, the resistances of these contacts are most likely not limited by lateral inhomogeneity or by the creation of a thick $n^+$ region.

Another alternative mechanism which cannot be ruled out is the presence of a compositionally graded region between the Ge and the GaAs. In fact, metastable $(Ge_2)_x(GaAs)_{1-x}$ layers have been fabricated previously.[129] The effect of the grading distance on barrier height can be calculated by a simple theory.[130] After making approximations, such as $\Delta E_c \gg kT$ and using Maxwell-Boltzmann statistics, an expression for the temperature-dependent grading length, $W_c$, necessary for reduction of the barrier height from the value of the conduction band discontinuity, $\Delta E_c$, to kT is

Eq (40) $$W_c = \left(\frac{\Delta E_c}{kT}\right) L_D$$

where the extrinsic Debye length, $L_D$ is given by

Eq. (41) $$L_D = \sqrt{\frac{\varepsilon kT}{e^2 N_d}}$$

Application of the above equations shows that if the grading were to take place over the extent of the regrown region, ~25 Å, then a surface doping of >$10^{19}$ $cm^{-3}$ would be required if $\Delta E_c$ was 0.08 eV. A surface doping of >$10^{20}$ $cm^{-3}$ would be required if $\Delta E_c$ was 0.3 eV. Thus, without considering the effect of tunneling through the reduced barrier, grading over a thin region would require high doping levels in the graded region as well. Taking tunneling through the barrier into effect, the required grading length will lessen. For instance, if the barrier need only be reduced to ~0.05 eV, then at 300 K the required doping levels would be in the $10^{18}$ $cm^{-3}$ range for $\Delta E_c$ = 0.08 eV and in the $10^{19}$ $cm^{-3}$ range for $\Delta E_c$=0.3 eV.

The final possible mechanism to be mentioned for ohmic behavior relies on the fact that a solid-phase regrown layer may be imperfect. If it contains electrically active defects, it could either affect band-bending at the interface, as discussed previously, or provide midgap electronic states which might enhance tunneling current through the barrier. As discussed by Sebestyen,[131] this could take place if a transitional region existed at the interface between a disordered top surface and the crystalline substrate. A bandgap full of deep trap states due to the disorder could lead to trap-assisted tunneling. This type of contact should lead to an ohmic contact regardless of the dopant type in the semiconductor. In the case of the Ge/Pd/GaAs contact, when the contact becomes ohmic to n-type GaAs, it becomes more rectifying to p-type GaAs. Hence, the ohmic nature is not due to defect states distributed uniformly in energy throughout the bandgap. It could, however, be due to defect states with a distribution located near the conduction band edge.

***Effect of Intentionally Deposited Impurities within the Contact Structure.*** Solid-phase epitaxially grown layers can be doped or compositionally altered by the deposition of thin elemental layers within the contact structure.[132] A significant amount of the added element can be incorporated into the regrown substrate layer. This effect can be useful for the alteration of the effective barrier height in which In could be used to form an ohmic contact (forming interfacial $In_xGa_{1-x}As$) and Al can be used to form a contact with a higher effective barrier height (forming interfacial $Al_xGa_{1-x}As$). The element may be incorporated into the substrate as a dopant. This can be useful for forming p-type ohmic contacts utilizing high substrate doping levels. Addition of these thin layers, however, has been found to sometimes offer kinetic impedance to solid phase reactions (depending on species and quantity) which, in turn, affects controllability.

## 4.2 Interfacial InGaAs Formation for Ohmic Contacts to n-GaAs

**In and GaAs Reaction at Contacts.** In order to understand the electrical and thermal properties of In-based ohmic contacts, the microstructures at the metal/GaAs interfaces were examined. Since the reaction between multi-component contact-metal (containing In) and GaAs is complicated, the reaction between (pure) In and GaAs will be described first and the effects of other elements added to the In contacts will be described later.

The In and GaAs interaction at high temperatures had been predicted by phase diagrams before experiments were carried out by XRD and TEM. Ternary InGaAs phase diagrams were calculated[133][134] using an equation given by Antypas,[135] where the activity coefficient in the liquid phase was calculated using Darken's quadratic equation[136] and the solid solution in equilibrium with the ternary liquid was assumed to be regular. An example of a phase diagram calculated at 700°C is shown in Fig. 23.[133] The dashed lines indicate tie-lines which connect two thermal equilibrium phases at 700°C. At this temperature the solid $In_xGa_{1-x}As$ phase and the liquid In(Ga,As) phase (indicated by points *a* and *b*, respectively) coexist when the "effective" average composition is at the point *c* which is within the two-phase region.

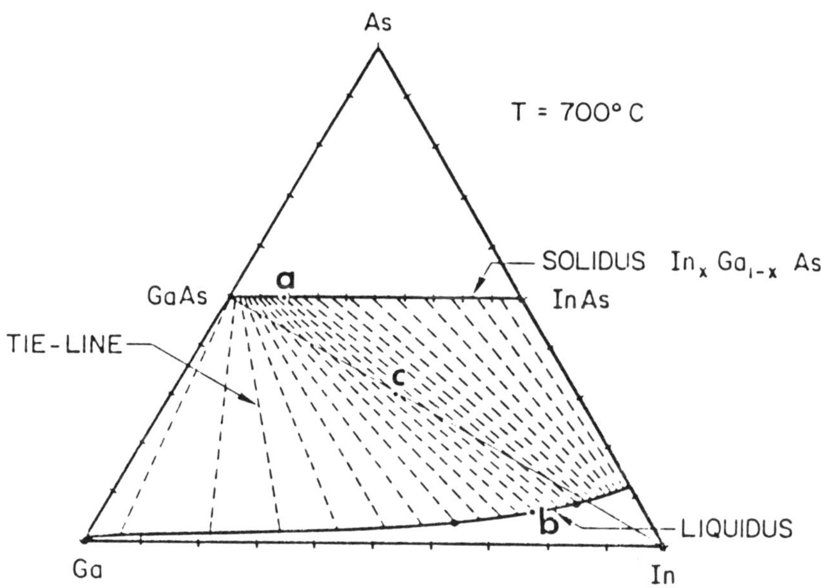


**Figure 23.** Ternary InGaAs phase diagram calculated at 700°C.[133]

The coexistence of two phases after annealing In/GaAs contacts was confirmed by TEM observation. The microstructures of In deposited on GaAs were observed by Ding et al.[137] and Kim et al.[133] As an example, a high-resolution cross-sectional TEM micrograph of the In/GaAs contact is shown in Fig. 24 in which an In layer of 100 nm thickness was deposited onto the GaAs substrate and annealed at 700°C for a short time.[133] No interaction between In and GaAs was observed in an as-deposited sample. A thin oxide layer observed at the interface prevented the In and GaAs reaction. After annealing at 700°C, the $In_xGa_{1-x}As$ phases were observed to be epitaxially grown locally on the GaAs substrate. No misfit dislocations were observed at the $In_xGa_{1-x}As$/GaAs interfaces. This indicates that the In concentration, x, in the $In_xGa_{1-x}As$ layer was small. The selected area diffraction (SAD) analysis of the $In_xGa_{1-x}As$ layer also indicated that the composition of the $In_xGa_{1-x}As$ layer was close to GaAs. The In concentrations (x) of the $In_xGa_{1-x}As$ layers were determined by EDX for the contacts annealed at various temperatures, and their x values were smaller than 0.4. This experimentally determined x value agreed very well with that calculated from the InGaAs phase diagram.

The roles of $In_xGa_{1-x}As$ and In(Ga,As) phases on the In-based ohmic contacts are described below.

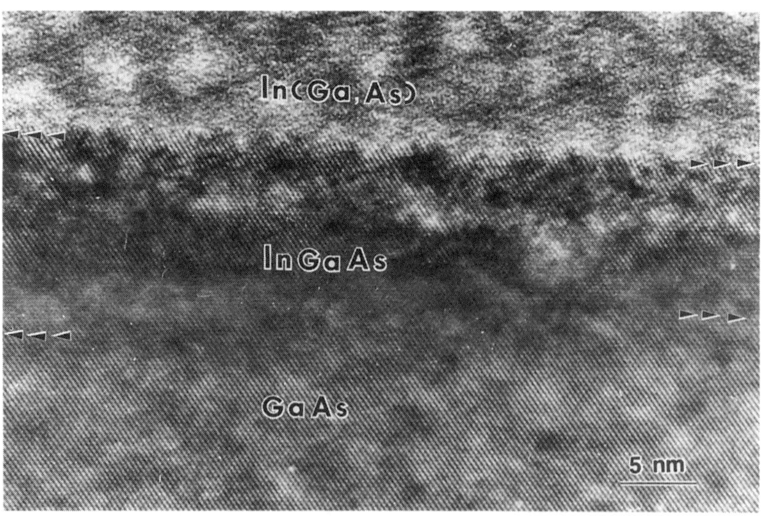


**Figure 24.** High-resolution TEM micrograph showing $In_xGa_{1-x}As$/GaAs interface of InW contact annealed at 700°C.[133]

**Role of $In_xGa_{1-x}As$ Phases on the Electrical Properties.** From TEM observation $In_xGa_{1-x}As$ phases were found to grow epitaxially on the GaAs substrate after heating In/GaAs contacts. Kajiyama et al.[39] measured the energy gap of bulk $In_xGa_{1-x}As$ and the barrier height ($\phi_a$) at the metal/n-$In_xGa_{1-x}As$ contacts. The barrier height is observed to decrease with increasing In concentration (x) in the $In_xGa_{1-x}As$ layer, and $\phi_a$ becomes negative for the metal/$In_xGa_{1-x}As$ with $x > 0.7$.[39]

The feasible energy band diagram at the metal/$In_xGa_{1-x}As$/GaAs interface constructed based on the energy gap measurement of $In_xGa_{1-x}As$ is schematically shown in Fig. 25.[138] There are two energy barriers for electron transport through this interface; one is the energy barrier ($\phi_a$) between metal and $In_xGa_{1-x}As$ and the other is the energy barrier ($\phi_b$) between $In_xGa_{1-x}As$ and GaAs. When a metal contacts directly to GaAs, the value of $\phi_a$ is measured to be close to 0.8 eV. The $\rho_c$ value at the GaAs/$In_xGa_{1-x}As$/metal interface was determined by the series resistances through these two barriers. As described above, the $\phi_a$ value decreases with increasing x value in the $In_xGa_{1-x}As$ phases. However, the lattice parameter of the ternary phases decreases with increasing In concentration and the lattice mismatch at the $In_xGa_{1-x}As$/GaAs interfaces increases. The barrier $\phi_b$ is associated with band gap discontinuities and the interface states are related to lattice mismatch. The $\phi_b$ value increases with increasing In concentration in the $In_xGa_{1-x}As$ phases, which leads to an increase in the $\rho_c$ value. Therefore there is an optimum In concentration in the intermediate $In_xGa_{1-x}As$ phases of Fig. 25.

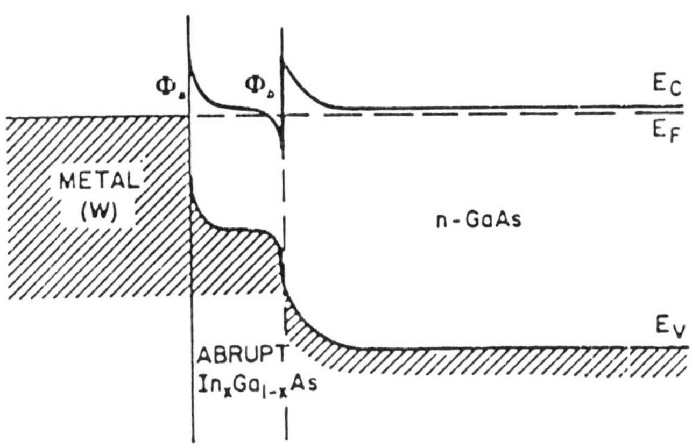


**Figure 25.** Energy band diagram at metal/$In_xGa_{1-x}As$/n-GaAs interface.[138]

Transition of Schottky to ohmic behavior by formation of intermediate $In_xGa_{1-x}As$ phases was observed experimentally. I-V curves of as-deposited In/W contacts showed Schottky behavior in which In was deposited directly on the GaAs substrate and the W layer was primarily used as a cap layer. Ohmic behavior was observed after annealing at temperatures higher than 400°C, which was found to be due to formation of the $In_xGa_{1-x}As$ phases at the In/GaAs interface. The same conclusion was obtained in the AgInGe contacts.[139] Although Cox and Strack[140] originally thought that In addition to AgGe contacts was only to lower the liquid surface tension and promote adhesion to GaAs, In played a primary role in the reduction of contact resistances by forming $In_xGa_{1-x}As$ phases. (Ge segregation to the GaAs interface forming an $n^+$ layer was found by McGuire et al.,[141] which also contributed to lower the resistance.)

In order to confirm that the $In_xGa_{1-x}As$ phases have the primary effect on the contact resistances, the In-based contacts with various metallizations were prepared and annealed at temperatures in the range of 700 to 900°C. The $\rho_c$ values of In-based contacts are plotted in Fig. 26 as a function of the percentage of the GaAs interface covered by the $In_xGa_{1-x}As$ phases. The percentages were measured from TEM micrographs of various In-based ohmic contacts prepared by different annealing methods. The $\rho_c$ values on the vertical axis are those in which the contribution of the contact metal sheet resistances ($R_s$) were subtracted from the measured $\rho_c$ values using an equation given by Marlow and Das[142] except one indicated by a filled circle which was an experimental value of the NiInW contact with low metal sheet resistance.[143] It is noted that the $\rho_c$ values decrease with increasing percentage of GaAs interface covered by $In_xGa_{1-x}As$. This result supports the above conclusion that the $In_xGa_{1-x}As$ phases had the primary effect on the electrical properties of these contacts. From Fig. 26, it is noted that the elements added with In to the contacts had a strong role in the $In_xGa_{1-x}As$ coverage, but the resultant phases did not have a strong effect on the electrical properties. Therefore, to prepare low-resistance contacts the total area of GaAs interface covered by $In_xGa_{1-x}As$ phases should be as large as possible.[144]

The In concentration (x) in the $In_xGa_{1-x}As$ phases of these contacts annealed at temperatures in the range of 700 to 900°C was determined to be about 0.4.[133][145] The measured x values agreed well with those extrapolated from the InGaAs phase diagrams. Therefore, the corresponding $\phi_b$ value was constant (close to 0.4 eV)[39] and thus the dependence of $\phi_b$ on the contact resistance cannot be studied for the contacts shown in Fig. 26.

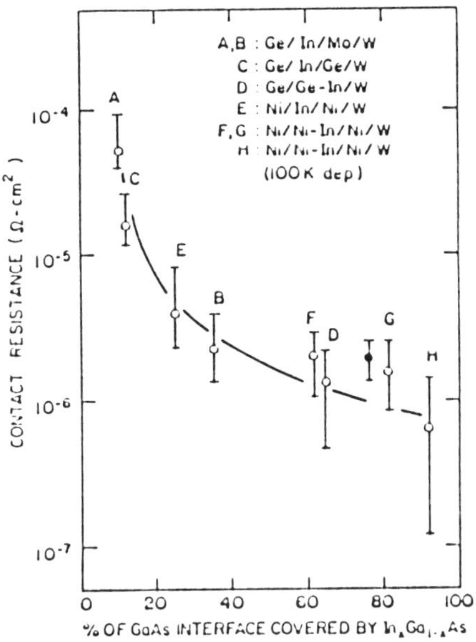


**Figure 26.** Contact resistance of In-based contacts as a function of percentage of GaAs interface covered by $In_xGa_{1-x}As$ phases.[138]

In order to prepare the contacts with high In concentration in the $In_xGa_{1-x}As$ layers, NiInAsW contacts were prepared by sputter-depositing InAs layers in an Ar atmosphere using an InAs target, with base pressure of $8 \times 10^{-5}$ Pa.[146] The deposition rate was about 0.1 nm/sec and the resulting In to As ratio in the film was close to unity. The Ni and W layers were deposited in an e-beam evaporation system, which was pumped down to about $8 \times 10^{-6}$ Pa before metal deposition. InAs/Ni/W and Ni/InAs/Ni/W contacts were thus prepared. The typical thicknesses of Ni, InAs, and W layers were 10, 30, and 40 nm, respectively. Samples were then capped by 50 nm thick $Si_3N_4$ layers, followed by rapid thermal annealing.

The electrical properties of the InAs/Ni/W and Ni/InAs/Ni/W contacts were measured after annealing the contacts at various temperatures. The current-voltage (I-V) curves of the as-deposited contacts showed Schottky behavior. For the InAs/Ni/W contacts, the ohmic behavior was first observed at 700°C. The $\rho_c$ values decreased with increasing annealing temperature and a $\rho_c$ of about $6 \times 10^{-6}$ Ω-cm² was obtained in the temperature range of 750 - 850°C as shown in Fig. 27. By annealing these contacts at higher temperatures, an increase in the $\rho_c$ value was observed and a typical U-shaped dependence of $\rho_c$ on T was obtained. A similar U-shaped curve in the $\rho_c$ vs. T plots was observed in the Ni/InAs/Ni/W contacts, but the $\rho_c$ values had less annealing temperature dependence. These contacts were isothermally annealed at 400°C and the $\rho_c$ values of these contacts did not deteriorate even after annealing for more than 100 hrs.

The microstructures of the InAs/Ni/W contacts annealed at 300°C for 30 min. and at 800°C for 1 sec. were examined by cross-sectional TEM and XRD. In the contact annealed at 300°C, large polycrystalline InAs grains were observed. Some grains grew epitaxially on the GaAs substrate, but most grains had random crystal orientations. The average grain size of the InAs grains was measured to be approximately 20 nm. Ni reacted with the InAs layer at this temperature and a small volume fraction of $NiAs_2$ phases which have cubic structure were observed within the InAs grains. Indium islands, precipitated out from the InAs layer after Ni/InAs reaction, were observed close to the GaAs surfaces. The micrograph of the contact annealed at 800°C is shown in Fig. 28, where $In_xGa_{1-x}As$ phases are observed to cover about 80% of the GaAs surface. The In concentration (x) in the $In_xGa_{1-x}As$ phases was uniform and relatively high ($x > 0.7$). Misfit dislocations and twins which accommodated the lattice mismatch of about 5% (at $x = 0.7$) between $In_xGa_{1-x}As$ and GaAs were observed at the interfaces. The presence of a strain contrast due to misfit dislocations indicates the presence of an abrupt interface. The $NiAs_2$ compounds were found to transform to NiAs compounds with hexagonal structure and no "pure" In was detected in this contact.

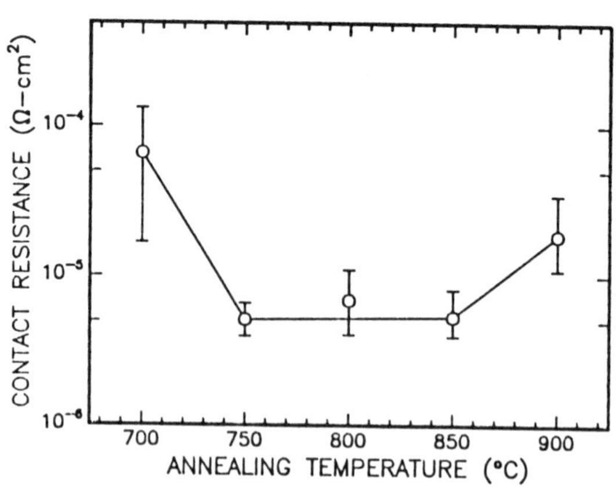


**Figure 27.** Contact resistance of NiInAsW contacts annealed at various temperatures.[146]

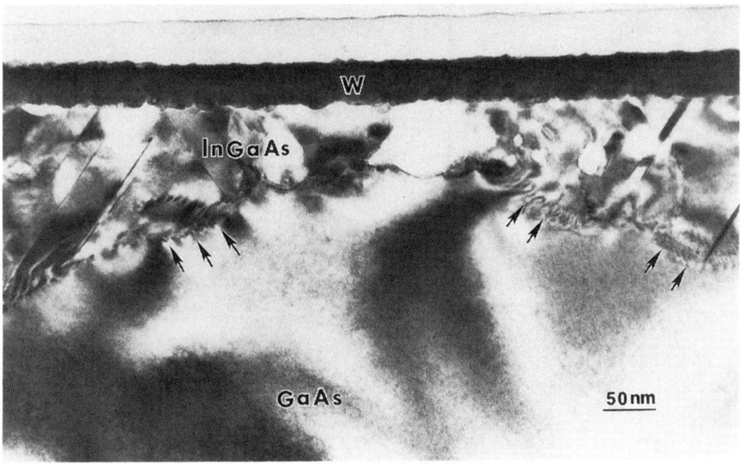


**Figure 28.** TEM micrograph of NiInAsW contact annealed at 800°C.[146]

The conventional NiInW contacts which form an $In_xGa_{1-x}As$ layer with x being about 0.4 at the GaAs interface, routinely provide $\rho_c$ values of about $3 \times 10^{-6}$ Ω-cm². Although, the NiInAsW contacts form an $In_xGa_{1-x}As$ layer with x = 0.7, the $\rho_c$ values did not reduce as seen in Fig. 27. The reason is believed to be formation of a high density of misfit dislocations at the $In_xGa_{1-x}As$/GaAs interface as observed by TEM. Woodall et al.[147] found that the pinning of the GaAs Fermi level at the interface between the heteroepitaxial layer and GaAs arose from the formation of misfit dislocations. Therefore, although "zero" barrier is expected at the metal/$In_xGa_{1-x}As$ interface, the barrier height at the $In_xGa_{1-x}As$/GaAs interface was high and the $\rho_c$ values of the metal/$In_xGa_{1-x}As$/GaAs contacts were high. The expected energy band diagram is shown in Fig. 29. (Note that low resistance contacts were prepared by annealing InAs/GaAs contacts grown by MBE[148] where the InAs layer was heavily doped. The difference between MBE-grown and sputter-deposited ohmic contacts is believed to be due to doping levels in the InAs layer and the location of dislocations. However the exact reason is not known at the moment.) The only way to reduce both barriers ($\phi_a$, $\phi_b$) is to prepare a compositionally graded $In_xGa_{1-x}As$ layer as an intermediate phase as demonstrated using MBE by Woodall et al.[147]

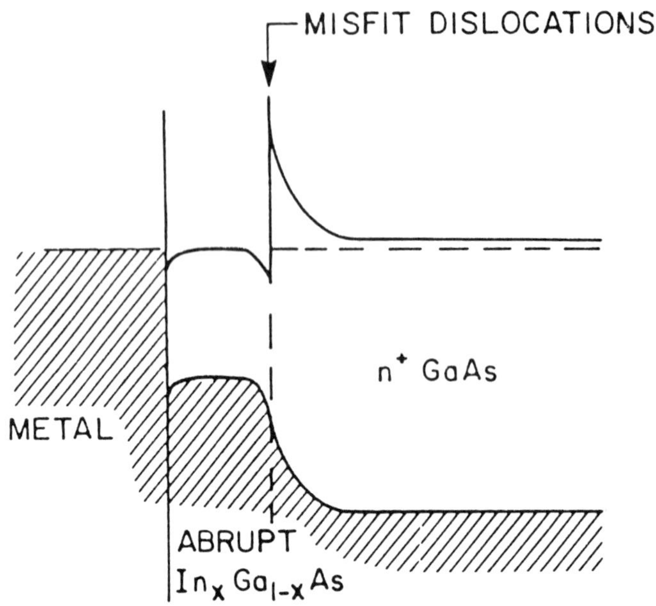


**Figure 29.** Energy band diagram of metal/$In_xGa_{1-x}As$/n-GaAs with misfit dislocations at the $In_xGa_{1-x}As$/GaAs interface.

**Role of In(Ga,As) Phases on Thermal Stability.** The In contacts with thick In layers had low resistance comparable to those routinely obtained in AuGeNi contacts, but had rough contact surface and poor thermal stability during annealing at 400°C after contact formation. These poor properties were due to the formation of In-rich In(Ga,As) phases, which have low melting points ($T_m$), as observed by TEM in Fig. 24. The $T_m$ of the In(Ga,As) phases depends on concentrations of Ga and As in the phases and decreases down to 156°C with decreasing the concentrations. To improve the thermal stability, the volume fraction of the In(Ga,As) phases should be minimized. The effects of the In layer thickness and addition of other elements to In affected the thermal stability and these effects are described in this section.

When the amount of In added to the contact metal is large, the In composition, averaged within In and GaAs reaction region, is high and the volume fraction of the In(Ga,As) phases in the contacts becomes large as predicted from the InGaAs phase diagram of Fig. 23. Therefore, the thickness of the deposited In layer strongly influences the thermal stability of the contacts after contact formation.

In/W contacts with In thickness of 3, 10 or 100 nm were prepared by depositing at -170°C and annealing at 900°C for a short time by RTA.[133] These contacts provided low resistances. Changes in the $\rho_c$ values of these contacts during subsequent isothermal annealing at 400°C are shown in Fig. 30. The $\rho_c$ values of the contact with a 100 nm thick In layer increased upon annealing at 400°C, but those of contacts with a 3 nm thick In layer were relatively stable and increased only slightly after annealing for 20 hrs. TEM observation of these contacts revealed that the volume fraction of the In(Ga,As) phases was significantly reduced by reducing the In thickness. Therefore, the improved stability of the In contact with reduced In thickness was attributed to the reduced volume fraction of the In(Ga,As) phases after contact formation.

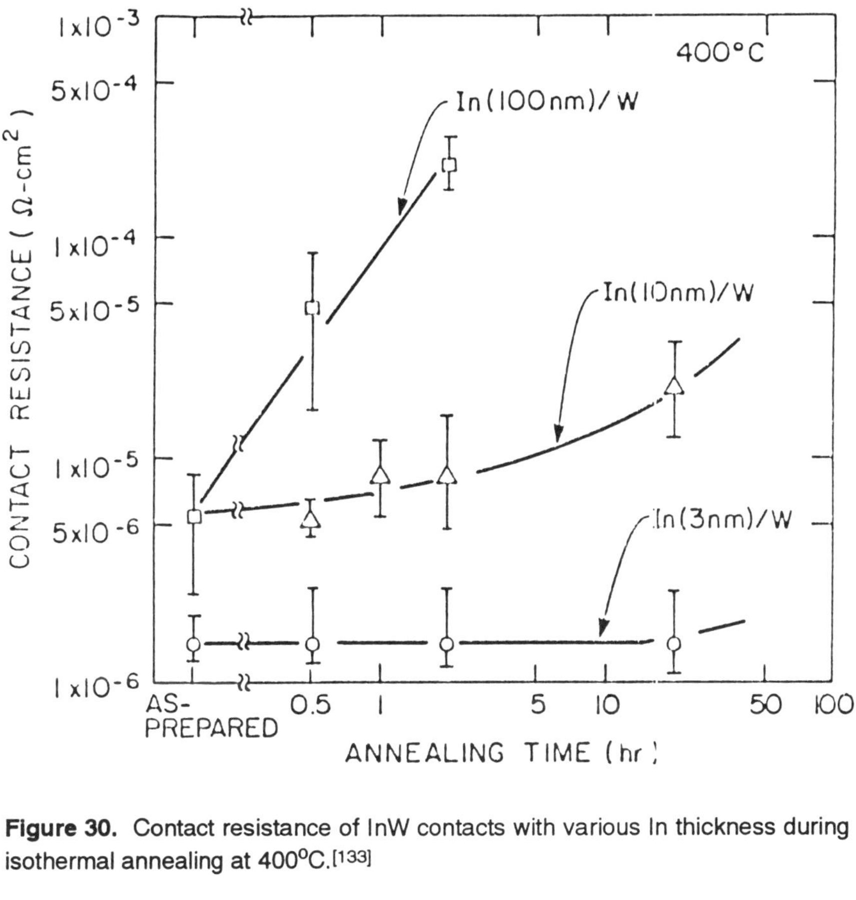


**Figure 30.** Contact resistance of InW contacts with various In thickness during isothermal annealing at 400°C.[133]

Although the reduction of the In layer thickness improved significantly the thermal stability after ohmic contact formation, the InGaAs phase diagram predicts that the In(Ga,As) phases coexist with the $In_xGa_{1-x}As$ phases even though the volume fraction is extremely small. To eliminate the In(Ga,As) phases, addition of an element which forms high-melting-point ($T_m$) intermetallic compounds with In is essential.[138] The effect of the additional elements on the thermal stability was studied for GeInW and NiInW contact systems. According to binary phase diagrams, In and Ge are immiscible and no intermetallic compounds are formed.[150] Changes in the $\rho_c$ values of the GeInW and NiInW contacts were measured during isothermal annealing at 400 and 500°C.[138] The $\rho_c$ values of the GeInW contacts increased during isothermal annealing at 400°C, but those of the NiInW contacts stayed constant even after annealing for 100 hrs. TEM observation showed that the In(Ga,As) phases formed $Ni_3In$ compounds which have their melting point higher than 900°C.[145] Therefore, the addition of elements such as Co, Pd, Pt, or Mn will improve the thermal stability by forming high-$T_m$ compounds with In(Ga,As).[151] The ideal interfacial microstructure which provides thermally stable low resistance is shown in Fig. 31, where M indicates a metal which forms an intermetallic compound with In.

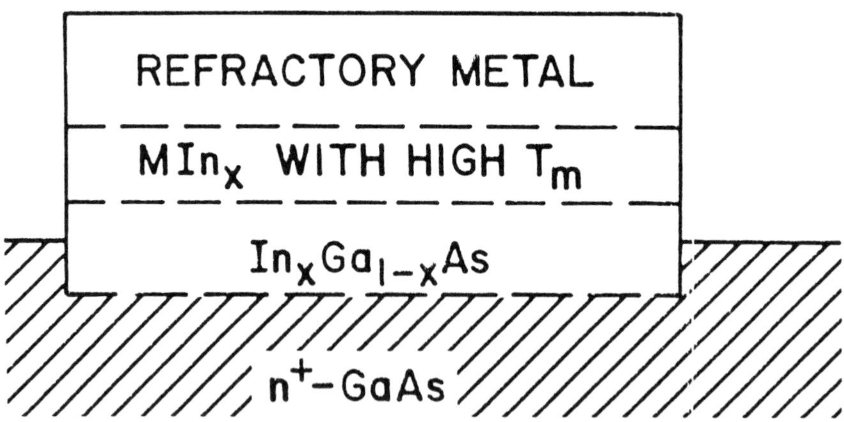


**Figure 31.** Cross-section of ideal In-based ohmic contact.

## 5.0 OHMIC CONTACTS TO p-GaAs

Ohmic contacts to p-type GaAs are much easier to fabricate than to n-type GaAs. Thus, they have received less critical attention. Even though contacts to p-GaAs have similar barrier height limitations as those to n-GaAs (Sec. 1.5, as $\phi_B{}^p = E_g - \phi_B{}^n$), the doping limitations are not similar (as described in Sec. 1.6).

Many of the techniques developed for contacts to n-type GaAs can be applied to p-type GaAs through appropriate choice of materials and processing conditions. In-situ growth of interlayers can be effective in achieving good ohmic contacts. Contact resistivity of less than $10^{-8}$ Ω-cm$^2$ has been achieved using MBE growth with carbon doping levels of up to $10^{21}$ cm$^{-3}$.[152] Low contact resistivity ($3 \times 10^{-7}$ Ω-cm$^2$) results from use of GaSb/GaAs strained superlattices for barrier height reduction.[153] Alloyed ohmic contacts are routinely used which usually incorporate Au and a non-amphoteric p-type dopant species such as Zn, Mn, Mg, or Be. These contacts form an interfacial GaAs layer which is sufficiently highly doped to allow tunneling to take place. Inclusion of Pd in the contact structure and optimization of these contacts can result in contact resistivities of less than $10^{-7}$ Ω-cm$^2$ for $5 \times 10^{18}$ cm$^{-3}$ doped GaAs.[154] Solid-phase reactions can be employed to form more controlled ohmic contacts by means of dopant diffusion[155] or dopant incorporation into regrown layers.[156]

## 6.0 OHMIC CONTACTS TO OTHER III-V COMPOUNDS

Ohmic contacts to other III-V compound semiconductors have been explored far less than those to GaAs. In general, however, the preceding discussion of ohmic contacts to GaAs applies to other III-V compound semiconductors. Factors which must be taken into account when comparing different semiconductors include metal/III-V barrier heights and the degree of dependence of the barrier height on choice of metal, maximum achievable doping levels, and the characteristics of interaction between metallization and semiconductor elements.

One of the more commonly used III-V compounds is $Al_xGa_{1-x}As$. The standard alloyed AuGeNi ohmic contact can be applied to n-$Al_xGa_{1-x}As$. The density of protrusions of the contact into the $Al_xGa_{1-x}As$, however, is seen to increase with the AlAs mole fraction, x.[157] Oxidation of $Al_xGa_{1-x}As$ in air at room temperature is a factor which must be considered as the oxide will present an insulating barrier to current flow and usually leads to laterally inhomogeneous and impeded alloying. In order to overcome this problem, an interlayer such as Al[158] or an epitaxial GaAs cap layer is applied. Planar ohmic contacts have been formed on n-$Al_xGa_{1-x}As$ utilizing the Ge/Pd/ system.[120] The ohmic contact formation mechanism is found to be independent of x and $\rho_c$ increases with x ($0 \leq x \leq 0.3$). A 20 nm undoped GaAs cap decreases $\rho_c$ by one order of magnitude for x = 0.2 or 0.3. DX center compensation is extremely important at these AlAs mole fractions. For x = 0.3, while the Si donor concentration is $10^{18}$ $cm^{-3}$, the room temperature free electron concentration is less than $10^{17}$ $cm^{-3}$ and it decreases exponentially with temperature. Hence, $\rho_c$ also increases rapidly with decreasing temperature.

InP-based compounds present factors which require special attention. Ohmic contacts are easier to attain on n-InP than p-InP as the typical barrier height of metals to n-InP is much less than that to p-InP. The Fermi level of air-exposed InAs is usually pinned in the conduction band. Hence, true ohmic contacts to n-InAs are possible with any metal. Ohmic contacts to p-InAs are not difficult, though, as its bandgap is relatively small and InAs can be highly doped. The reaction behavior of contacts on InP and InAs is different than that on GaAs. The melting points of InP and InAs are 200 - 300°C less than GaAs, hence the elements tend to be more volatile and reaction kinetics are quickened. In fact, Pd-InP initial reactions lead to a ternary $Pd_xInP$ layer similar to $Pd_4GaAs$ except that it is initially amorphous and then becomes epitaxial with further annealing.[159][160]

Ohmic contacts to III-V compounds other than those in the $(Al,Ga,In)_x(As,P)_{1-x}$ family continue to remain relatively unexplored.[161][162]

## 7.0 FUTURE DIRECTIONS

Future applications require greater understanding and control of ohmic contact formation behavior. Correlation of high-resolution metallurgical analysis and contact electrical behavior will allow engineered interfaces through appropriate choice of metallurgy. Control, process compatibility, miniaturization, and thermal stability may be afforded by a process which allows patterned epitaxial growth of ohmic contacts on previously processed semiconductor wafers.[163]

## REFERENCES


1. R. C. Eden, *Proc. IEEE,* 76:756 (1988)
2. C. G. Kirkpatrick, *Proc. IEEE,* 76:792 (1988)
3. J. H. Magerlein, D. J. Webb, A. Callegari, J. D. Feder, T. Fryxell, H. C. Guthrie, P. D. Hoh, J. W. Mitchell, A. T. S. Pomerene, S. Scontras, G. D. Spiers, and J. H. Greiner, *J. Appl. Phys.,* 61:3080 (1987)
4. T. Ikoma, "Very High Speed Integrated Circuits: Gallium Arsenide LSI", and "Very High Speed Integrated Circuits: Heterostructure", Vol. 29, in *Semiconductors and Semimetals,* (R. K. Willardson and A. C. Beer, eds.), Academic Press, Boston (1990)
5. E. H. Rhoderick, *Metal-Semiconductor Contacts,* p.14, Clarendon, Oxford (1978)
6. S. M. Sze, *Physics of Semiconductor Devices,* Wiley, New York (1981)
7. S. J. Fonash, *Solid- State Electron.,* 15:783 (1972)
8. C. Y. Chang and S. M. Sze, *Solid-State Electron.,* 13:727 (1970)
9. C. R. Crowell and V. L. Rideout, *Solid-State Electron.,* 12:89 (1969)
10. V. L. Rideout and C. R. Crowell, *Solid-State Electron.,* 13:993 (1970)
11. W. J. Boudville and T. C. McGill, *J. Vac. Sci. Technol.,* B3:1192 (1985)
12. C. Y. Chang, Y. K. Fang, and S. M. Sze, *Solid-State Electron.,* 14:541 (1971)
13. F. A. Padovani, *Semiconductors and Semimetals,* (R. K. Willardson and A. C. Beer, eds.), Vol. 7, Ch. 2, Academic Press, New York (1971)
14. L. S. Yu, personal communication
15. E. L. Murphy and R. H. Good, Jr., *Phys. Rev.,* 102:1464 (1956)
16. F. A. Padovani and R. Stratton, *Solid-State Electron.,* 9:695 (1966)
17. A. Y. C. Yu, *Solid-State Electron.,* 13:239 (1970)
18. J. R. Waldrop, *Appl. Phys. Lett.,* 44:1002 (1984)
19. R. E. Viturro, S. Chang, J. L. Shaw, C. Mailhiot, L. J. Brillson, A. Terrasi, Y. Hwu, G. Margaritondo, P. D. Kirchner, and J. M. Woodall, *J. Vac. Sci. Technol.,* B7:1007 (1989)
20. C. E. C. Wood, *The Technology and Physics of Molecular Beam Epitaxy,* (E. H. C. Parker, ed.), p. 67, Plenum, New York (1985)
21. T. N. Theis, P. M. Mooney, and S. L. Wright, *Phys. Rev., Lett.* 60:361 (1988)

22. N. Chand, T. Henderson, J. Klem, W. T. Masselink, R. Fischer, Y.-C. Chang, and H. Morkoc, *Phys. Rev.,* B30:4481 (1984)

23. A.-Z. Li, A. G. Milnes, Z. Y. Chen, Y. F. Shao, and S. B. Wang, *J. Vac. Sci. Technol.,* B3:629 (1985)

24. T. F. Kuech, D. J. Wolford, R. Potemski, J. A. Bradley, K. H. Kelleher, D. Yan, J. P. Farrell, P. M. S. Lesser, and F. H. Pollack, *Appl. Phys. Lett.,* 51:505 (1987)

25. D. E. Aspnes, *Phys. Rev.,* B14:5331 (1976)

26. P. A. Barnes and A. Y. Cho, *Appl. Phys. Lett.,* 33:651 (1978)

27. P. D. Kirchner, T. N. Jackson, G. D. Pettit, and J. M. Woodall, *Appl. Phys. Lett.,* 47:26 (1985)

28. E. F. Shubert, J. E. Cunningham, W. T. Tsang, and T. H. Chiu, *Appl. Phys. Lett.,* 49:292 (1986)

29. R. S. Popovic, *Solid- State Electron.,* 21:1133 (1978)

30. G. Brezeanu, C. Cabuz, D. Dascalu, and P. A. Dan, *Solid-State Electron.,* 30:527 (1987)

31. R. P. Gupta and W. S. Khokle, *IEEE Electron. Dev. Lett.,* EDL-6:300 (1985)

32. W. Dingfen, W. Dening, and K. Heime, *Solid-State Electron.,* 29:489 (1986)

33. J. M. Shannon, *Solid-State Electron.,* 19:537 (1976)

34. R. Stall, C. E. C. Wood, K. Board, and L. F. Eastman, *Electron. Lett.,* 24:800 (1979)

35. J. M. Ballingall, C. E. C. Wood, and L. F. Eastman, *J. Vac. Sci. Technol.,* B1:675 (1983)

36. P. Chiaradia, A. D. Katnani, H. W. Sang, Jr., and R. S. Bauer, *Phys. Rev. Lett.,* 52:1246 (1984)

37. E. D. Marshall, C. S. Wu, C. S. Pai, D. M. Scott, and S. S. Lau, in: *Thin Films: The Relationship of Structure to Properties,* (C. R. Aita and K. S. Sree Harsha, eds.), p. 161, Materials Research Society, Pittsburgh (1985)

38. J. M. Woodall, J. L. Freeouf, G. D. Pettit, T. Jackson, and P. Kirchner, *J. Vac. Sci. Technol.,* 19:626 (1981)

39. K. Kajiyama, Y. Mizushima, and S. Sakata, *Appl. Phys. Lett.,* 23:458 (1973)

40. H. H. Wieder, *Appl. Phys. Lett.,* 38:60 (1981)

41. T. Nittono, H. Ito, O. Nakajima, and T. Ishibashi, *Jap. J. Appl. Phys.,* 25:L865 (1986)

42. N. Braslau, J. B. Gunn, and J. L. Staples, *Solid State Electron.,* 10:381 (1967)
43. J. B. Gunn, *IEEE Trans. Elect. Dev.,* ED-23:705 (1976)
44. T. Kim and D. D. L. Chung, *J. Vac. Sci. Technol.,* B4:762 (1986)
45. P. Auvray, A. Guivarc'h, H. l'Haridon, and J. P. Mercier, *Thin Solid Films,* 127:39 (1985)
46. J. Gyulai, J. W. Mayer, V. Rodriguez, A. Y. C. Yu, and H. J. Gopen, *J. Appl. Phys.,* 42:3578 (1971)
47. L. Y. Zee and Z. A. Munir, *J. Mater. Sci.* 10:1929 (1975)
48. W. Katz, G. Smith, O. Aina, B. J. Baliga, and K. Rose, *Appl. Surf. Sci.,* 9:122 (1981)
49. T. Jung and E. Nebauer, *Phys. Stat. Sol.,* 77a:K203 (1983)
50. M. Procop and B. Sandow, *Phys. Stat. Sol.,* 95a:K211 (1986)
51. G. S. Marlow, M. B. Das, and L. Tongson, *Solid State Electron.,* 26:259 (1983)
52. A. A. Lakhani, R. C. Potter and D. M. Beyea, *Semicond. Sci. Technol.,* 3:605 (1988)
53. O. Aina, W. Katz, B. J. Baliga, and K. Rose, *J. Appl. Phys.,* 53: 777 (1982)
54. J. G. Werthen, and D. R. Scifres, *J. Appl. Phys.,* 52:1127 (1981)
55. P. Sircar, *Can J. Phys.,* 61:1218 (1983)
56. R. B. Gold, R. A. Powell, and J. F. Gibbons, *AIP Conf. Proc.,* 50:635 (1978)
57. S. Margalit, D. Fekete, D. M. Pepper, C. P. Lee, and A. Yariv, *Appl. Phys. Lett.,* 33:346 (1978)
58. G. Badertscher, W. Luthy, R. P. Salathe, K. Affolter, and M. Wittmer, *Rapport de la Session de Printemps de la Societe Suisse de Physique* 52:359 (1979)
59. O. Aina, J. Norton, W. Katz, G. Smith, and K. Rose, *Laser and Electron Beam Interactions with Solids,* Materials Research Society, Pittsburgh (1981)
60. P. Sircar, *Phys. Stat. Sol.,* 97a:K69 (1986)
61. R. Binkis, G. Pyatraukas, and A. Sakalas, *Sov. Phys. Semicond.,* 21:1163 (1987)
62. J. L. Staples, U. S. Patent 3,386.867 (1968)
63. W. D. Edwards, W. A. Hartman, and A. B. Torrens, *Solid State Electron.,* 15:387 (1972)

64. G. Y. Robinson, *Solid State Electron.,* 18:331 (1975)

65. M. Otsubo, H. Kumabe, and H. Miki, *Solid State Electron.,* 20:617 (1977)

66. K. Heime, U. Konig, E. Kohn, and W. Workmann, *Solid State Electron.,* 17:835 (1974)

67. R. P. Gupta and J. Freyer, *Intern. J. Electron.,* 47:459 (1979)

68. M. Ogawa, *J. Appl. Phys.,* 51:406 (1980)

69. A. Iliadis and K. E. Singer, *Solid State Commun.,* 49:99 (1984)

70. A. Callegari, E. T.-S. Pan, and M. Murakami, *Appl. Phys. Lett.,* 46:1141 (1985)

71. A. Callegari, D. Lacey, and E. T.-S. Pan, *Solid State Electron.,* 29:523 (1986)

72. J. Dell., H. L. Hartnagel, and A. G. Nassibian, *J. Phys.,* D16:L243 (1983)

73. T. S. Kalkur, J. Dell, and A. G. Nassibian, *J. Intern. Electron.,* 57:729 (1984)

74. A. G. Nassibian and T. S. Kalkur, *Appl. Surf. Sci.,* 22/23:1019 (1985)

75. F. Lonnum and J. S. Johannessen, *Electron. Lett.,* 22:632 (1986)

76. W. Patrick, W. S. Mackie, S. P. Beaumont, and C. D. W. Wilkinson, *Appl. Phys. Lett.,* 48:986 (1986)

77. T. Lalinsky, S. Chromik, M. Porges, and D. Gregusova, *Elektrotech. Cas.,* 37:354 (1986)

78. M. Procop, B. Sandow, H. Raidt, and L. D. Son, *Phys. Stat. Sol.,* 104:903 (1987)

79. Y. P. Chang and W. N. Choon, *Trans. Korean Inst. Electr. Eng.,* 36:888 (1987)

80. S. Tsou, W. Wang, C. Lin, and G. Xia, *Proc. 14th Inter. Conf. on Ion Implantation,* p.13 (1983)

81. H. G. Henry, D. E. Dawson, Z. J. Lemnios, and H. Kim, *IEEE Trans. Electron Devices,* ED-31:1100 (1984)

82. D. D. Cohen, T. S. Kalkur, G. J. Sutherland, and A. G. Nassibian, *J. Appl. Phys.,* 60:3100 (1986)

83. S. Imanaga, H. Kawai, K. Kajiwara, K. Kaneko, and N. Watanabe, *J. Appl. Phys.,* 62:2381 (1987)

84. T. S. Kalkur and A. G. Nassibian, *Solid State Electron.,* 30:619 (1987)

85. C. Lin, *Nucl. Tech.,* 10:1 (1987)

86. Y. C. Shih, M. Murakami, E. L. Wilkie, and A. Callegari, *J. Appl. Phys.*, 62:582 (1987)
87. M. Murakami, K. D. Childs, J. M. Baker, and A. Callegari, *J. Vac. Sci. Technol.*, B4:903 (1986)
88. T. S. Kuan, P. E. Batson, T. N. Jackson, H. Rupprecht, and E. L. Wilkie, *J. Appl. Phys.*, 54:6952 (1983)
89. C. J. Cooke and W. Hume-Rothery, *J. Less-Common Met.*, 10:42 (1966)
90. D. F. Hewitt, *Econ. Geol.*, 43:408 (1948)
91. H. Pfisterer and K. Schubert, *Z. Metallkd.*, 41:358 (1950)
92. R. A. Bruce and G. R. Piercy, *Solid State Electron.*, 30:729 (1987)
93. K. Taeil, D. D. L. Chung, and S. Mahajan, *Interfaces, Superlattices, and Thin Films Symposium* (1987)
94. W. O. Barnard and J. B. Malherbe, *Appl. Surf. Sci.*, 31:437 (1988)
95. M. Heiblum, M. I. Nathan, and C. A. Chang, *Solid State Electron.*, 25:185 (1982)
96. R. P. Gupta and W. S. Khokle, *Solid State Electron.*, 28:823 (1985)
97. A. P. Botha and E. Relling, in: *Thin Films-Interfaces and Phenomena*, Materials Research Society, Pittsburgh (1985)
98. J. R. Waldrop and R. W. Grant, *Appl. Phys. Lett.*, 50:250 (1987)
99. R. W. Grant and J. R. Waldrop, *J. Vac. Sci. Technol.*, B 5:1015 (1987)
100. R. A. Bruce, W. T. Moore, T. Lester, D. A. Clark, and A. J. Springthorpe, *Microscopy of Semiconducting Materials* (1989)
101. T. Kim and D. D. L. Chung, *Thin Solid Films*, 147:177 (1987)
102. R. K. Ball, *Thin Solid Films*, 176:55 (1989)
103. A. Callegari, M. Murakami, J. M. Baker, Y. C. Shih, and D. Lacey, *Proc. 17th European Solid State Devices Research Conference*, ESSDERC 87:601 (1988)
104. T. G. Finstad, *Thin Solid Films*, 47:279 (1977)
105. M. Kamada, T. Suzuki, F. Nakamura, Y. Mori, and M. Arai, *Appl. Phys. Lett.*, 49:1263 (1986)
106. F. Vidimari, *Electron. Lett.*, 15:675 (1979)
107. A. A. Lakhani, *J. Appl. Phys.*, 59:2082 (1986)
108. K. Fricke, H. L. Hartnagel, R. Schutz, G. Schweeger, and J. Wurfl, *IEEE Elect. Dev. Lett.*, EDL-10:577 (1989)

109. N. Lustig, M. Murakami, M. Norcott, and K. McGann, *Appl. Phys. Lett.,* 58:2093 (1991)

110. E. Kolawa, W. Flick, C. W. Nieh, J. M. Molarius, M. A. Nicolet, J. L. Tandon, J. H. Madok, and F. C. T. So, *IEEE Trans. Electron Devices,* ED-36:1223 (1989)

111. E. D. Marshall, "Ohmic Contacts to n-Type Aluminum Gallium Arsenide Utilizing Limited Solid-Phase Reactions", Ph.D. dissertation, University of California, San Diego (1989)

112. E. D. Marshall, B. Zhang, L. C. Wang, P. F. Jiao, W. X. Chen, T. Sawada, S. S. Lau, K. L. Kavanagh, and T. F. Kuech, *J. Appl. Phys.,* 62:942 (1987)

113. E. D. Marshall, W. X. Chen, C. S. Wu, S. S. Lau, and T. F. Kuech, *Appl. Phys. Lett.,* 47:298 (1985)

114. A. Paccagnella, C. Canali, G. Donzelli, E. Zanoni, L. C. Wang, and S. S. Lau, *Electron. Lett.,* 24:708 (1988)

115. A. K. Sinha, T. E. Smith, and M. J. Levenstein, *IEEE Trans. Electron. Dev.,* ED-22:218 (1975)

116. H. H. Grinolds, and G. Y. Robinson, *Solid State Electron.,* 23:973 (1980)

117. S. Tiwari, T.-S. Kuan, and E. Tierney, *International Electron Device Meeting Technical Digest,* p.115 (1993)

118. W. T. Anderson, Jr., A. Christou, and J. E. Davies, *IEEE J. Solid State Circuits,* SC-13:430 (1978)

119. M. Murakami, W. H. Price, Y-C. Shih, K. D. Childs, *J. Appl. Phys.,* 62:3288 (1987)

120. E. D. Marshall, L. S. Yu, S. S. Lau, T. F. Kuech, and K. L. Kavanagh, *Appl. Phys. Lett.,* 54:721 (1989)

121. C. J. Palmstrøm, S. A. Schwarz, E. D. Marshall, E. Yablanovitch, J. P. Harbison, C. L. Schwartz, L. Florez, T. J. Gmitter, L. C. Wang, and S. S. Lau, *Advanced Surface Processes for Optoelectronics,* p.283, Materials Research Society, Pittsburgh (1988)

122. E. D. Marshall, S. S. Lau, C. J. Palmstrøm, T. Sands, S. A. Schwarz, J. P. Harbison, C. L. Schwartz, and L. T. Florez, *Chemistry in Semiconductor Heterostructures,* p.148, Materials Research Society, Pittsburgh, (1989)

123. E. D. Marshall, B. Zhang, L. C. Wang, F. Fang, S. S. Lau, T. Sands, and T. F. Kuech, "Solid-Phase Reactions in the Ge/Pd/GaAs and Si/Pd/GaAs Systems for Planar Ohmic Contacts to n-GaAs", at 1987 Electrochem. Soc. Symp. on Contacts to Compound Semiconductors

124. L. C. Wang, B. Zhang, F. Fang, E. D. Marshall, S. S. Lau, T. Sands, and T. F. Kuech, *J. Mater. Res.*, 23:922 (1988)

125. T. Sands, E. D. Marshall, and L. C. Wang, *J. Mater. Res.*, 3:915 (1988)

126. N. Braslau, *J. Vac. Sci. Technol.*, 19:803 (1981)

127. J. S. Blakemore, *Semiconductor Statistics*, Dover, New York (1987)

128. C.-Y. Wu, *J. Appl. Phys.*, 251:4919 (1980)

129. I. Banerjee, D. W. Chung, and H. Kroemer, *Appl. Phys. Lett.*, 46:94 (1985)

130. R. M. Raymond and R. E. Hayes, *J. Appl. Phys.*, 48:1359 (1977)

131. T. Sebestyen, *Solid-State Electron.*, 25:543 (1982)

132. S. S. Lau, C. Canali, Z. L. Liau, K. Nakamura, M.-A. Nicolet, J. W. Mayer, R. J. Blattner, and C. A. Evans, Jr., *Appl. Phys. Lett.*, 28:148 (1976)

133. H. J. Kim, M. Murakami, W. H. Price, and M. Norcott, *J. Appl. Phys.*, 67:4183 (1990)

134. M. Murakami, Y. C. Shih, H. J. Kim, and W. H. Price, *Proc. 20th Inter. Conf. on Solid State Device Materials*, D-2-3:283 (1988)

135. G. A. Antypas, *J. Electrochem. Soc.*, 117:1393 (1970)

136. L. S. Darken, *Trans. Met. Soc.*, AIME 239:80 (1967)

137. J. Ding, J. Washburn, T. Sands, and V. G. Keramidas, *Appl. Phys. Lett.*, 49:818 (1986)

138. M. Murakami, Y.C. Shih, W. H. Price, and E. L. Wilkie, *J. Appl. Phys.*, 64:1974 (1988)

139. H. L. Hartnagel, K. Tomizawa, L. H. Herron, and B. L. Weiss, *Thin Solid Films*, 36:393 (1976)

140. H. Cox and H. Strack, *Solid State Electron.*, 10:1213 (1967)

141. G. E. McGuire, W. R. Wisseman, R. D. Ragle, and J. H. Tregilgas, *J. Vac. Sci. Technol.*, 16:141 (1979)

142. G. S. Marlow and M. B. Das, *Solid State Electron.*, 25:91 (1982)

143. M. Murakami, H. J. Kim, W. H. Price, M. Norcott, and Y. C. Shih, *Mater, Res. Soc. Symp. Proc.*, Vol. 148 (1989)

144. M. Murakami, Y. C. Shih, W. H. Price, N. Braslau, K. D. Childs, and C. C. Parks, *Inst. Phys. Conf. Ser.*, 91:55 (1988)

145. Y. C. Shih, M. Murakami, and W. H. Price, *J. Appl. Phys.*, 65:3539 (1989)

146. H. J. Kim, M. Murakami, S. L. Wright, M. Norcott, W. H. Price, and D. LaTulipe, *J. Appl. Phys.*, 68:2475 (1990)

147. J. M. Woodall, H. Rupprecht, R. J. Chicotka, and G. Wicks, *Appl. Phys. Lett.*, 38:639 (1981)

148. S. L. Wright, R. F. Marks, S. Tiwari, T. N. Jackson, and H. Baratte, *Appl. Phys. Lett.*, 49:1545 (1986)

149. J. M. Woodall, G. D. Pettit, T. N. Jackson, and C. Lanza, *Phys. Rev. Lett.*, 51:1783 (1983)

150. M. Hansen and K. Anderko, *Constitution of Binary Alloys,* McGraw-Hill, New York (1958)

151. L. C. Wang, X. Z. Wang, S. N. Hsu, S. S. Lau, P. S. D. Lin, T. Sands, S. A. Schwarz, D. L. Plumton, and T. F. Kuech, *J. Appl. Phys.,* 69:4364 (1991)

152. T. Usagawa, M. Kobayashi, T. Mishima, P. D. Rabinzohn, A. Ihara, M. Kawata, T. Yamada, E. Tokumitsu, M. Konagai, K. Takahashi, *J. Appl. Phys.,* 69:8227 (1991)

153. J. I. Chyi, J. Chen, N. S. Kumar, C. Kiely, C. K. Peng, A. Rockett, H. Morkoc, *Appl. Phys. Lett.,* 55:570 (1989)

154. R. Bruce, D. Clark, S. Eicher, *J. Electron. Mater.,* 19:225 (1990)

155. H. P. Kattelus, J. L. Tandon, M.-A., Nicolet, *Solid-St. Electron.,* 29:903 (1986)

156. C. C. Han, X. Z. Wang, L. C. Wang, E. D. Marshall, S. S. Lau, S. A. Schwarz, C. J. Palmstrøm, J. P. Harbison, L. T. Florez, R. M. Potemski, M. A. Tischler, T. F. Kuech, *J. Appl. Phys.,* 68:5714 (1990)

157. Z. Liliental and R. W. Carpenter, *Ultramicroscopy,* 14:135 (1984)

158. V. G. Keramidas, *Thin Solid Films,* 96:347 (1982)

159. R. Caron-Popowich, J. Washburn, T. Sands, E. D. Marshall, *Chemistry in Semiconductor Heterostructures,* p.53, Materials Research Society, Pittsburgh, (1989)

160. D. G. Ivey, P. Jian, L. Wan, R. Bruce, S. Eicher, C. Blaauw, *J. Electron. Mater.,* 20:237 (1991)

161. Y. K. Su, F. S. Juang, K. J. Gan, *Jpn. J. Appl. Phys.,* 30:914 (1991)

162. G. Y. Robinson, *Physics and Chemistry of III-V Compound Semiconductor Interfaces* (C. W. Wilmsen, ed.) Plenum Press, New York (1985)

163. A. Palevski, P. Solomon, T. F. Kuech, M. A. Tischler, *Appl. Phys. Lett.,* 56:171 (1990)

# 2

# Stable and Epitaxial Contacts to III-V Compound Semiconductors

*Chris J. Palmstrøm and Timothy D. Sands*

## 1.0 INTRODUCTION

Thin-film metallizations serve the essential functions of making direct contact with the semiconductor material and connecting semiconductor devices to one another on an integrated circuit. These metallizations are, therefore, classified as either *contact* or *interconnect* metallizations. In this chapter we focus on contact metallizations for III-V semiconductors. Rather than reviewing the evolution and drawbacks of the present generation of empirically-derived contact metallizations, we adopt the approach of defining the "ideal" metallization. We discuss recent work toward achieving this ideal. The approach is not just an academic exercise. In the process of striving for such a goal, one acquires a deeper understanding of the factors that control contact performance, uniformity and stability. This insight is key to modifying and, in some cases, revolutionizing contact metallizations used in today's III-V devices and circuits to enhance circuit yield, performance and reliability.

Today's contact metallizations fall into two categories; low resistance ohmic contacts that pass current into and out of the semiconductor and rectifying (Schottky) contacts. An ideal ohmic contact should pass current through the metal/semiconductor interface with negligible resistance. In a field-effect transistor (FET) the rectifying contact controls the flow of current in an underlying semiconductor *channel*. An ideal rectifying contact acting as the gate electrode in an FET should modulate the flow of current in the channel without drawing current through the metal/semiconductor interface. Furthermore, this performance should be achieved at a reasonable cost (e.g., simple, reproducible processing steps compatible with the processing of the semiconductor device) and maintained through the lifetime of the device. On a more detailed level, these requirements necessitate metallizations that are microstructurally and electrically uniform, thermally stable and environmentally robust. Since grain boundaries are fast diffusion paths facilitating interface roughening, reactions, contamination and electromigration,[1] the ideal contact should be *monocrystalline and epitaxial.* The ultimate in contact stability can be achieved by choosing a metallization that has a *high melting point* and is *thermodynamically stable* in contact with the semiconductor. Finally, environmental robustness demands that the metallization, if exposed to the ambient during processing or operation, be *oxidation and corrosion resistant.*

Although these criteria seem overly ambitious for such a simple device as an FET, modern III-V epitaxial growth techniques have yielded a wide variety of sophisticated device heterostructures from which devices with remarkable properties and performance have been demonstrated. These artificially structured semiconductors may have doping that is confined to a few atomic planes, semiconductor layers of different composition than the substrate and with thicknesses of only a few nanometers, and strained heteroepitaxial layers for enhanced mobility or improved laser performance. These complex heterostructures put added demands on contact performance and reliability, particularly as lateral device dimensions become considerably smaller than one micron.

With the control of metallizations comes the ability to utilize metals in more *active* device roles. For example, epitaxial nanometer-thick metal layers that are compatible with III-V growth conditions could, in principle, serve as buried gates as in the permeable base transistor, or as the base in a metal base transistor. More generally, buried metal structures form a necessary element in three-dimensional integrated circuits. Epitaxy also permits the exploitation of the anisotropic properties of certain metals. For example, thin epitaxial metals can behave as metallic quantum wells with rather striking transport properties[2]-[5] and the magnetic, optical and magneto-optical properties of non-cubic metals can be controlled and enhanced via epitaxy.[6] Furthermore, these properties can be used to modulate charge transport or propagating light in the surrounding semiconductor.[7]

From a scientific point-of-view, studies of structurally and chemically ideal Schottky barriers promise to yield meaningful progress toward a better understanding of the intrinsic and extrinsic factors that govern the measured barrier height at a metal/semiconductor interface, a topic of considerable debate over the past decades.[8]-[14] The key issues are the origins of Fermi level pinning and whether or not there is an *intrinsic* Fermi level pinning position at the semiconductor surface in metal/III-V semiconductor Schottky barriers. A number of models have been proposed for Schottky barrier formation involving metal induced gap states (MIGS),[13] virtual induced gap states (VIGS),[15] defect states[11][16][17] (including the *advanced unified defect model*[12]), and *effective work functions*.[9] These studies have involved trying to measure the dependence (or lack of dependence) of the barrier height with the metal work function.[14] The two different surfaces which have been extensively studied are the (110) cleaved surface and the technologically important (100) surface. Apart from studies involving arsenic capped (100) molecular beam epitaxy (MBE) grown layers, the Schottky barrier height for GaAs appears to be pinned in the range ~0.7 - 0.9 eV[14] independent of GaAs orientation and only weakly dependent upon the metal work function. In the case of arsenic capped MBE grown (100) layers, a larger variation in Schottky barrier height with metal work function was observed.[18] However, these studies have, in general, involved the use of reactive and polycrystalline metals, which are not in thermodynamic equilibrium on the semiconductor surface. This suggests that metal/semiconductor reactions will be taking place. Until recently, the most ideal metal/semiconductor interfaces formed have been the thermodynamically stable epitaxial transition metal disilicide/Si interfaces. Epitaxial metal silicide/Si Schottky barrier interfaces have been extensively studied in the past decade.[8][19][28] The $NiSi_2$/Si Schottky barrier was found for Si(111) to depend upon whether the $NiSi_2$ was *type A* (unrotated) or *type B* (180° rotated) with respect to the Si(111) substrate,[19][20][27][28] and it was also found to depend on the orientation of the Si substrate itself.[21][22] The difference in barrier height between type A and type B $NiSi_2$/Si(111) has been predicted theoretically and arises from a difference in the interface dipole due to the different bonding configuration.[29] Such an effect has not been observed for metal/GaAs interfaces.

The ability to grow thermodynamically stable epitaxial contacts to III-V semiconductors allows the study of metal/III-V semiconductor interfaces which are substantially more ideal than those studied previously and thus, may lead to a much better understanding of Schottky barrier formation, as has been the case for the thermodynamically stable epitaxial $CoSi_2$/Si and $NiSi_2$/Si contacts. In Sec. 6 we address the effects of mismatch, crystallographic orientation and deposition conditions for thermodynamically stable epitaxial $Sc_{1-x}Er_xAs$/GaAs Schottky barriers. The effect of modifying the interfacial semiconductor layers is also addressed for CoGa, CoAl and NiAl epitaxial contacts. These studies are performed on probably the most 'ideal' metal/III-V semiconductor systems to date.

We begin this chapter by discussing the criteria that must be met in achieving epitaxy and thermal stability (Secs. 2.1 and 2.2). The generic issues related to the growth of metal/III-V heterostructures are covered in Sec. 2.3, followed by discussions of specific elemental metal/III-V systems in Sec. 3. The majority of the stable and epitaxial metal systems fall into one of two categories, the rare earth monopnictides (RE-V, e.g., ErAs [30]-[48]) and the transition-metal group III intermetallics (TM-III, e.g., NiAl [49]-[61]). We describe the growth of these RE-V/III-V and TM-III/III-V heterostructures in Sec. 4. In Sec. 5 the growth of other epitaxial intermetallic compounds is discussed. The transport, magnetic and optical properties of these heterostructures are summarized in Sec. 6 and the fundamentals of Schottky barrier formation in Sec. 7. The chapter is concluded with an overview of the device potential of stable and epitaxial metallizations.

## 2.0 CRITERIA FOR THE METAL LAYER: HOW DOES ONE CHOOSE THE IDEAL METALLIZATION?

### 2.1 Epitaxy

The first criterion for the metal layer is for the metallization to be epitaxial on the III-V semiconductor. The most stringent criterion would then be for the metal to be cubic with the same lattice parameter as the III-V semiconductor. However, there are other conceivable criteria: either the metal lattice parameter is an integer multiple of the III-V semiconductor or vice versa. For example, the metal lattice parameter may be half that of the semiconductor and one could still expect epitaxy. These criteria will be referred to as cube-on-cube as shown in Fig. 1 (a,b). The lattices could also be rotated with respect to each other as shown in Fig. 1 (c,d). A 45° rotation would require the lattice parameter of the metallic layer to be $1/\sqrt{2}\,a_{0sc}$ or $\sqrt{2}a_{0sc}$. These are the most likely orientation relationships for epitaxial growth, but clearly other rotations are also possible. The important alignment is in the interface plane, which relaxes the cubic criterion since the lattice parameter normal to the interface plane is not as important (except for steps). Layers with tetragonal symmetry would be possible candidates. Orthorhombic structures may also be candidates, but the structures would have to have combinations of *a*, *b* and *c* lattice parameters that would fit the above criteria. On {111} oriented III-V systems, hexagonal metallic layers could be good choices. Clearly, there are a large number of possibilities for potential epitaxy. However, the simplest criteria would be of the forms shown in Figs. 1a and 1b.

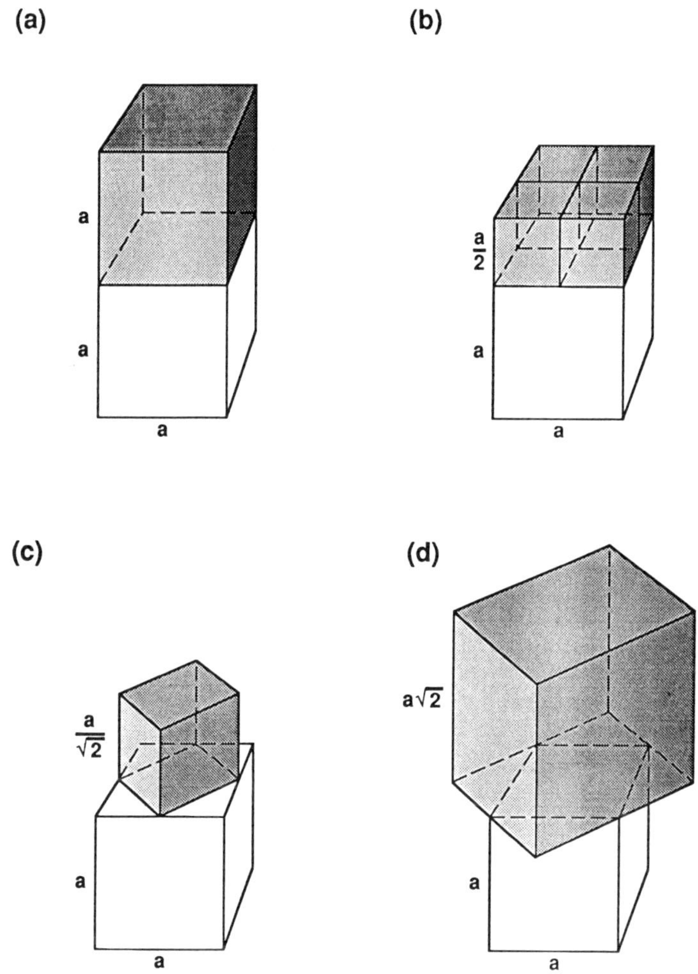


**Figure 1.** Schematic of simplest potential epitaxial alignments for a cubic metal on a cubic semiconductor: *(a)* and *(b)* are referred to as "cube-on-cube" with *(a)* $a_M = a_{0sc}$ and *(b)* $a_M = 0.5a_{0sc}$, *(c)* and *(d)* are referred to as 45° rotated alignment with *(c)* $a_M = 1/\sqrt{2}\,a_{0GaAs}$ and *(d)* $a_M = \sqrt{2}a_{0GaAs}$. Clearly, there are a number of other possible epitaxial alignments with different lattice parameters and rotations.

The lattice parameters for cubic metals are shown in Fig. 2 for both *fcc* and *bcc* structures. The number of elements with lattice parameter close to the III-V semiconductors is limited. In general, *fcc* metals can have a close lattice match to GaAs by a 45° rotation and $1/\sqrt{2}a_{0GaAs}$ lattice parameter (Fig. 2a with alignment as in Fig. 1c) and *bcc* metals by a cube-on-cube alignment with $0.5a_{0GaAs}$ (Fig. 2b with alignment as in Fig. 1b). As will be shown later, the metallic elements are generally not thermodynamically stable on III-V semiconductors. There are, however, a large number of metallic compounds with cubic structure. Compounds with the $CaF_2$, the CsCl, the NaCl and the $BiF_3$ structures have been of particular interest for potential epitaxial metallization. Some of these structures are shown in Fig. 3. Figure 4 shows the lattice parameters of transition metal-group III (TM-III) compounds with the CsCl structure for a number of transition metals. From this figure, it is seen that CoAl, CoGa, NiAl and NiGa have lattice parameters close to half that of GaAs. However, all the TM-III compounds have lattice parameters slightly larger than half that of GaAs. Most rare-earth monopnictides (RE-V) have the NaCl structure and the lattice parameters of these are plotted in Fig. 5. The monopnictides of Sc and Y have been included due to Sc and Y's close chemical similarity to the rare-earth elements. The actinide monopnictides, which have the same structure and similar lattice parameters to the rare-earth monopnictides, have been omitted in this plot for clarity. A very large range in lattice parameters, from 4.505 Å for ScN to 6.580 Å for LaBi[62], is available. This suggests that ternaries, formed by mixing rare-earth elements ($RE^a_{1-x}RE^b_x-V$) or group V elements ($RE-V^c_{1-y}V^d_y$), or quaternary systems, formed by mixing both ($RE^a_{1-x}RE^b_x-V^c_{1-y}V^d_y$), can be lattice matched to almost all group IV, III-V and II-VI semiconductors. Palmstrøm et al.[34][49] have used the former approach to grow lattice matched $Sc_{1-x}Er_xAs$ on GaAs. Guivarc'h and co-workers have used both approaches; they have grown lattice matched $Sc_{1-x}Yb_xAs$[47] and $ErP_{0.6}As_{0.4}$[35][42]-[45] to GaAs and $ErP_{0.47}Sb_{0.53}$[35] to InP. A number of rare-earth monochalcogenides (RE-VI) also have the NaCl structure. The lattice parameters vary from 5.192 Å (ScS) to 6.595 Å (SmTe)[62]. The rare-earth monochalcogenides are metals or semiconductors.[63] The possibility of making solid solutions of RE-V and RE-VI compounds such as $RE^a_{1-x}RE^b_x-V^c_{1-z}VI^d_z$ opens up further possibilities of tailoring the contact metallization.[34] In fact, a number of $RE-V_{1-z}VI_z$ solid solutions have been fabricated in the bulk.[64]

For Si the transition metal disilicides with the $CaF_2$ structure have been used extensively as epitaxial metallizations. A good analogy would be to use a compound with a similar structure to the metal silicides (such as the anti-$CaF_2$). Guivarc'h et al.[65] identified α-$Rh_2As$ as a potential candidate. It has a lattice parameter of 5.6753 Å and has, therefore, only a very small mismatch to GaAs (~0.4%). In practice, however, α-$Rh_2As$ is a poor choice due to problems of thermodynamic stability, which are discussed below.

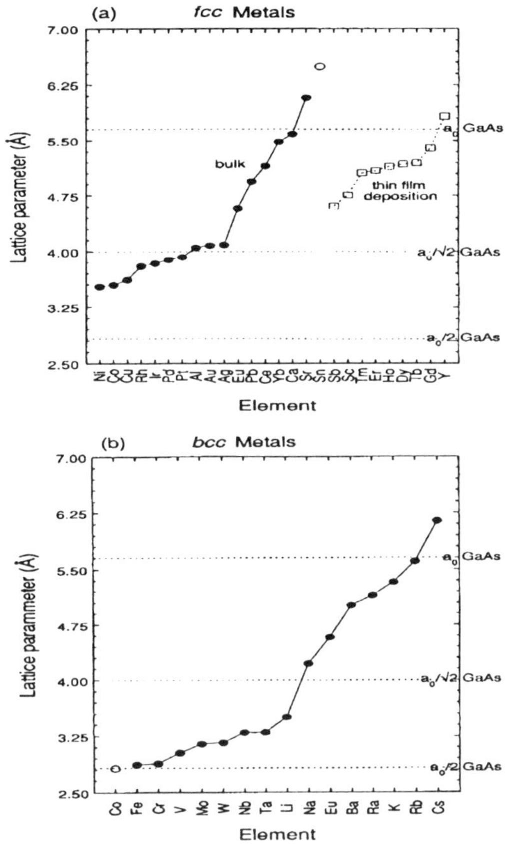


**Figure 2.** *(a)* Plot of lattice parameter vs. element for *fcc* metals (the filled circles are for the *fcc* metals, the open circle is for α-Sn, diamond structure, and the open squares are for *fcc* stabilized deposited thin films). Most *fcc* metals with potential lattice matching to GaAs can be achieved with the metal lattice parameter, $a_M = 1/\sqrt{2}\,a_{0GaAs}$ (e.g., Rh, Ir, Pd, Pt, Al, Au, Ag), resulting in the expected alignment as shown in Fig. 1c. Only a few (e.g., Yb and Ca) have the potential of the cube-on-cube alignment (Fig. 1a). *(b)* Plot of lattice parameter versus element for *bcc* metals (the open circle is for metastable *bcc* Co grown on GaAs).[231] The expected epitaxial alignment for *bcc* metals is the cube-on-cube with $a_M = 0.5a_{0sc}$ (Fig. 1b) for Co, Fe, Cr and V, and $a_M = a_{0sc}$ (Fig. 1a) for Rb. (The lattice parameters were taken from Ref. 232)

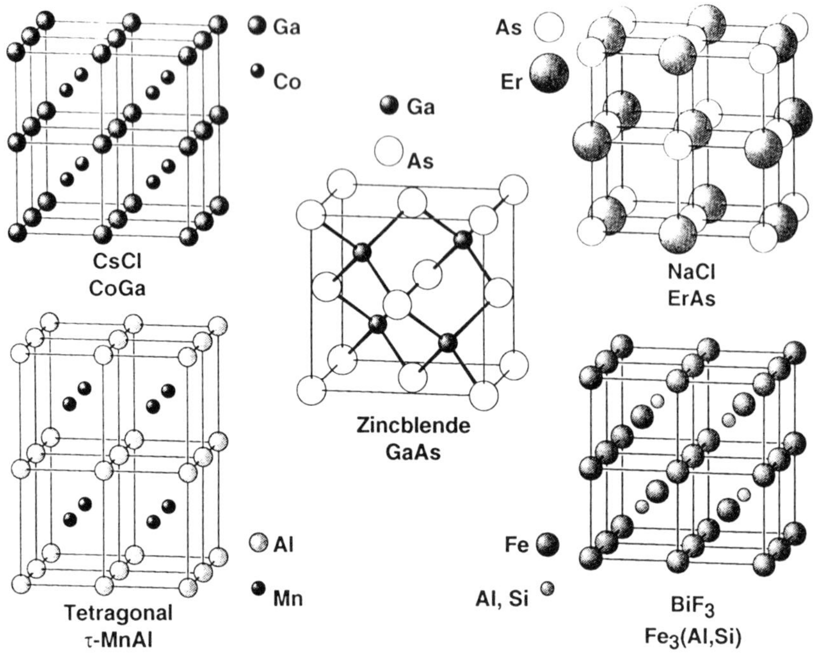


**Figure 3.** Comparison of different crystal structures with potential epitaxial alignment: CoGa structure (CsCl type), τ-MnAl (tetragonal), ErAs structure (NaCl type), $Fe_3$(Al,Si) structure ($BiF_3$ type) and GaAs (zincblende type). Eight unit cells of idealized CoGa (or NiAl) structure and a unit cell of ErAs structure and $Fe_3$(Al,Si) structure occupy approximately the same volume as a single unit cell of GaAs. The basal plane of τ-MnAl unit cell is nearly lattice matched to the {100} plane of GaAs.

Two other types of epitaxial metallic compounds have been grown on GaAs. Hong et al.[66] used a solid solution of $Fe_3Si$ and $Fe_3Al$ to form $Fe_3Al_{1-y}Si_y$ lattice matched to GaAs. This compound has the $BiF_3$ structure (Fig. 3). Also, the ferromagnetic metastable phase, τ-MnAl (Fig. 3), has been grown epitaxially on AlAs.[67] It has a tetragonally distorted CsCl structure and is nearly lattice matched to half the lattice constant of AlAs when the *c*-axis is oriented perpendicular to the AlAs (100) surface.

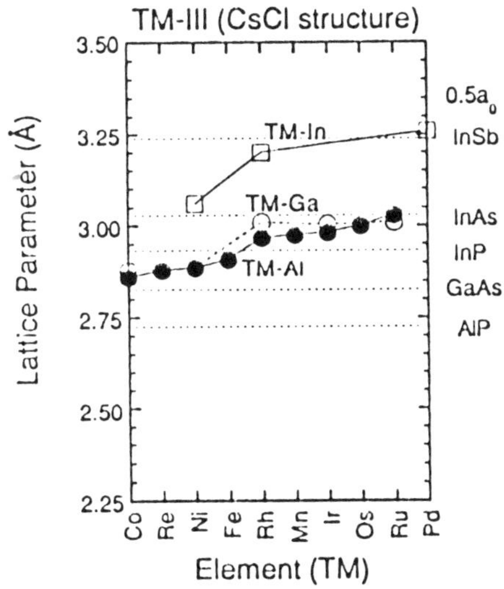


**Figure 4.** Lattice parameters of TM-III intermetallics with CsCl structure (data obtained from Ref. 62). The predicted alignment is the cube-on-cube with $a_M = 0.5a_{0sc}$ (Fig. 1b). Note that all the TM-III compounds have $a_M > 0.5a_{GaAs}$.

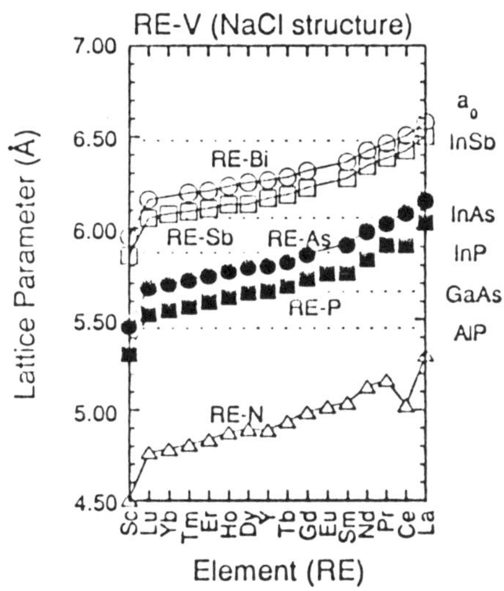


**Figure 5.** Lattice parameters of RE-V compounds with NaCl structure (data obtained from Ref. 62). The RE-V compounds span a large range of lattice parameters ranging from 4.505 Å (ScN) to 6.580 Å (LaBi), which suggests that ternaries, formed by mixing rare-earth elements ($RE^a_{1-x}RE^b_x - V$) or group V elements ($RE\text{-}V^c_{1-y}V^d_y$), or quaternary systems, formed by mixing both ($RE^a_{1-x}RE^b_x - V^c_{1-y}V^d_y$), can be lattice matched to almost all group IV, III-V and II-VI semiconductors.

## 2.2 Thermodynamic Stability

Thermodynamic stability implies that once the metallic layer has been deposited it does not react with the underlying III-V semiconductor during subsequent annealing and device operation. Although a contact is epitaxial, it does not necessarily follow that it is thermodynamically stable. For example, *bcc* Fe can be grown epitaxially on GaAs[68]-[73], but during deposition and subsequent annealing the Fe reacts with GaAs.[74] A stable contact may be formed by depositing a metal-III or metal-V compound on the III-V semiconductor. The choice of compound is critical for thermodynamic stability. The best example for this may be the case of $\alpha$-$Rh_2As$, a phase with an anti-$CaF_2$ structure and a lattice mismatch of ~0.4% to GaAs. Guivarc'h et al.[65] found that they could not grow $\alpha$-$Rh_2As$ with the cube-on-cube orientation (Fig. 1a, discussed below) and that the $\alpha$-$Rh_2As$ reacted with the GaAs above 400°C to form RhGa and $RhAs_2$. This behavior can be understood from knowledge of the ternary Rh-Ga-As phase diagram, which was determined by Guérin et al.[75] In such a diagram, single-phase regions are separated from one another by two-phase regions consisting of "bundles" of tie-lines. If the average composition of the system is within a two-phase region, the compositions of the two equilibrium phases are given by the endpoints of the tie-line that passes through the average system composition. The relative amounts of the phases are given by the Lever rule. The two-phase regions bound three-phase tie-triangles. For a system with its average composition in a three-phase region, the compositions of the three phases in equilibrium are given by the corners of the triangle. The Gibbs phase rule dictates that no more than three phases can be in equilibrium at a given temperature for a three-component system, except at special points (e.g., a ternary eutectic). The isothermal section for Rh-Ga-As at 1000°C is shown in Fig. 6. A tie-line between two phases indicates thermodynamic stability. Tie-lines go between GaAs, RhGa, and $RhAs_2$ showing that both RhGa and $RhAs_2$ are stable in contact with GaAs. However, there is no tie-line between $Rh_2As$ and GaAs indicating that $Rh_2As$ is not thermodynamically stable on GaAs. In this diagram one would expect to see tie-lines between $Rh_{10}Ga_{17}$, $RhGa_3$, $Rh_2Ga_9$ and $RhAs_3$, and GaAs. However, the phases may not be stable at the temperature at which this isothermal section of the equilibrium phase diagram was determined. In general, to determine the phases that are thermodynamically stable on a III-V semiconductor, one needs to know the M-III-V ternary phase diagrams. Unfortunately, the number of ternary phase diagrams that have been measured is limited to Ni-Ga-As,[76] Ni-Ga-Al-As,[77] Co-Ga-As,[78]

Pd-Ga-As,[79] Ag-Ga-As,[80] Au-III-V,[81][82] Nb-Ga-As,[83] Cr-Ga-As,[84] Rh-Ga-As,[75] Ir-Ga-As,[85] Pt-Ga-As,[86] Cu-Ga-As,[87] Co-In-P,[88] and Pd-In-Sb.[89] However, a number have been predicted[90]-[93] either through metal/III-V reactions or theory.

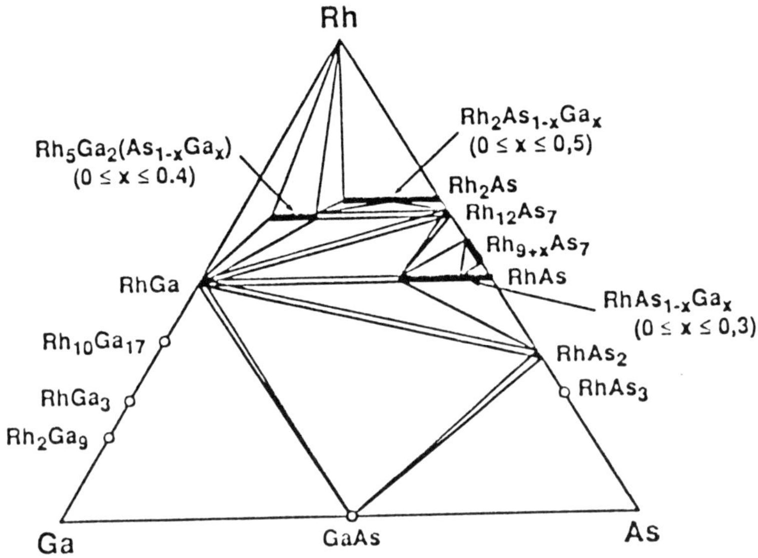


**Figure 6.** Isothermal section of the Rh-Ga-As phase diagram at 1000°C (adapted from Ref. 75). Tie-lines go between GaAs, RhGa, and $RhAs_2$ showing that both RhGa and $RhAs_2$ are stable in contact with GaAs. However, there is no tie-line between $Rh_2As$ and GaAs indicating that $Rh_2As$ is not thermodynamically stable on GaAs. Since tie-lines between the phases $Rh_{10}Ga_{17}$, $RhGa_3$, $Rh_2Ga_9$ and $RhAs_3$, and GaAs are not observed, these phases may not be stable at the temperature at which this isothermal section of the equilibrium phase diagram was determined.

The complete M-III-V phase diagram is not always necessary to determine stable phases in contact with a III-V semiconductor. Metal/III-V semiconductor thin film reactions studies allow one to make predictions of the ternary phase diagram and determine the thermodynamically stable phases in contact with the III-V semiconductor.[92][94] This approach can be understood by considering the schematic isothermal section of the equilibrium phase diagram of a hypothetical closed system consisting of M, III and V in Fig. 7. Assuming that the supply of M is limited, the average composition of the M/III-V system will be within the three-phase triangle defined by the phases M-III, M-$V_2$, and the III-V semiconductor. Initially, the system starts with the phases M and III-V. Clearly there is no tie line between M and III-V indicating that the system is not in thermodynamic equilibrium. During the reaction, the average composition of the metal plus the reaction products will follow the dashed line from pure M to the M-III-M-$V_2$ tie-line that bounds the M-III-M-$V_2$-III-V three-phase region. This dashed line is not to be confused with the "diffusion path"[78][95] which describes the actual sequence of phases formed during the reaction. This sequence is determined primarily by the relative diffusivities of the three species in each phase, information that is not contained within the bulk phase diagram. Thus by driving the thin film M/III-V reaction to completion one will end up with the thermodynamically stable M-III and M-V phases in contact with the III-V semiconductor, M-III and M-$V_2$ in this case. The deposition of either M-III or M-$V_2$ (or both) would result in thermodynamically stable contacts which would not react with the III-V semiconductor.

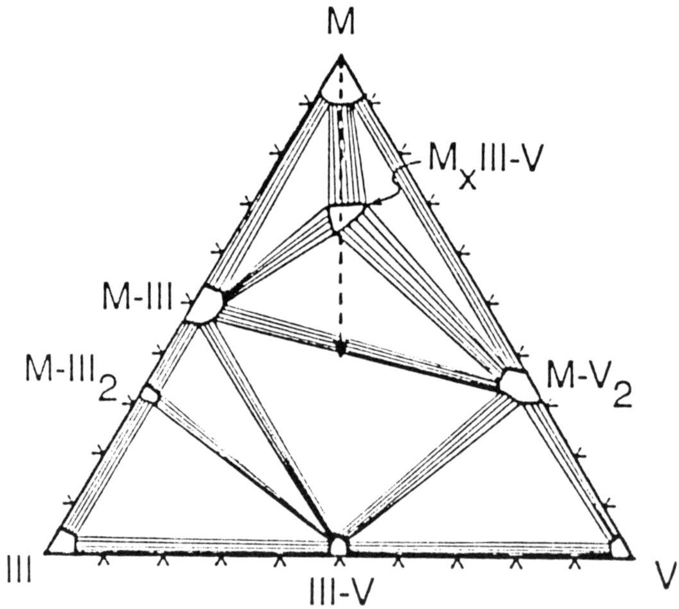


**Figure 7.** Schematic isothermal section of the equilibrium phase diagram of a hypothetical closed system consisting of the three components M, III and V, where M is a metal. The dashed line indicates the average composition of the subsystem consisting of the reaction products and the unreacted metal film during a reaction between a film of M on the III-V substrate. The reaction stops when M is consumed and the composition of the subsystem reaches a region of the phase diagram which includes the III-V as one of the stable phases. For the system depicted here, the final phases are M-III, M-$V_2$, and III-V. The phase M-$III_2$ is also stable in contact with the III-V substrate.

In practice the system is not closed, particularly in an ultra-high vacuum molecular beam epitaxy (UHV-MBE) system. The group-V species have high vapor pressures so that group-V loss from the III-V semiconductor is a problem at elevated temperature. In the case of MBE growth of III-V semiconductors, this preferential group-V loss is compensated for by the use of excess group-V flux as compared with the group-III flux. The equilibrium phase diagrams serve as guides for the choice of metallizations. The transition metal-group-III compounds (TM-III) and the rare-earth monopnictides (RE-V) are promising systems for epitaxial metallizations on III-V semiconductors. Schematic isothermal sections of the Ni-Ga-As and Er-Ga-As ternary phase diagrams are shown in Fig. 8. The former was determined using bulk samples[76] and the latter predicted from studies of ErAs and $ErGa_{2.6}$ thin films on GaAs.[93] Recently the phase diagram for Er-Ga-As[96] has been determined and it is essentially as that shown in Fig. 8 except that there appears to be some Ga solubility in ErAs. The Ni-Ga-As phase diagram shows tie-lines between GaAs and $Ni_2Ga_3$, NiGa and NiAs, indicating that NiGa, $Ni_2Ga_3$ and NiAs are thermodynamically stable on GaAs. For Er-Ga-As there is no tie-line between any Er-Ga phase and GaAs. Hence, only ErAs is stable on GaAs. These phase diagrams may be considered typical for the TM-III-V and RE-III-V. In some RE-V and RE-VI cases, however, phases richer in group V or group VI elements exist and control of the group V or group VI to rare-earth element flux ratio may be necessary. In growing Co-As phases on GaAs[97] with low Co to $As_4$ flux ratio, CoAs was found to form even though the phases $CoAs_2$ and $CoAs_3$ exist. This suggests that control of the group V or group VI to rare-earth element flux ratio may not be critical even in cases where phases richer in group V or group VI elements than the rare-earth monopnictides or monochalcogenides exist. For Sc and Er arsenides the most arsenic-rich phase is the rare-earth monoarsenide, which eliminates this problem. In summary, from a thermodynamic standpoint both the TM-III and RE-V compounds appear to be ideal candidates for epitaxial metallization systems. A summary of RE-As properties is shown in Table 1.

For III-V semiconductor overgrowth, the stability under group-V flux is important since III-V MBE growth is always performed under excess group-V flux. Note that there is no tie-line between NiGa and As, but there is between ErAs and As. This indicates that NiGa is not stable and ErAs will be stable in an appropriate $As_4$ or $As_2$ flux. From the phase diagram (Fig. 8a) it may deduced that NiGa would react with arsenic to form NiAs and GaAs. This is what thermodynamics dictates; however, kinetics play an important role. For example, if GaAs can be grown on top of the NiGa before arsenic starts to react with the NiGa, thermodynamic stability is again achieved. A technique to ensure that the group-V flux does not react with the TM-III layer using kinetics and a separate supply of the group-III element is described in Sec. 4.

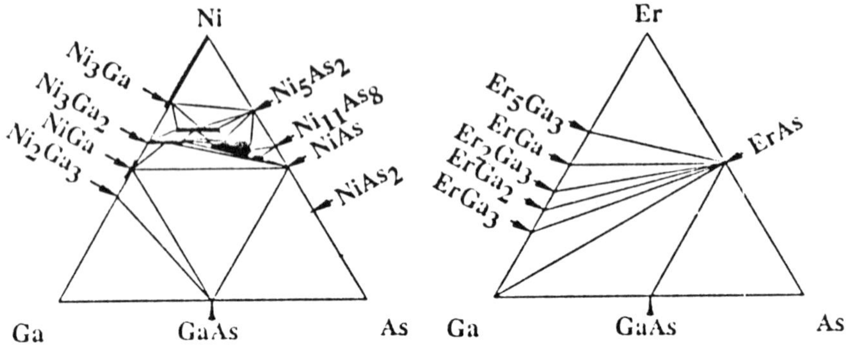


**Figure 8.** Schematic isothermal sections of the Ni-Ga-As (adapted from Ref. 76) and the Er-Ga-As phase diagrams. These are representative TM-III-V and RE-III-V phase diagrams. For Ni-Ga-As the stable phases in contact with GaAs are NiGa, NiAs, and $Ni_2Ga_3$. (Since a tie-line between $NiAs_2$ and GaAs is not observed, $NiAs_2$ may not be stable at the temperature at which the phase diagram was determined.) There is no tie-line between NiGa and As indicating that NiGa will react with an As-flux to form GaAs and NiAs. There are no Er-Ga phases stable in contact with GaAs. However, ErAs is thermodynamically stable in contact with both GaAs and As, which indicates that ErAs may be grown under excess As conditions and that GaAs growth on top of ErAs should be possible without interfacial reactions.

Thus far, only the thermodynamic stability of ternary systems have been considered. A quaternary system is substantially more complicated, but can add additional degrees of freedom such as lattice parameter tailoring. Since, the additional element is most commonly from the same group in the periodic table, that is from either the group of M, III or V designated M′, III′, and V′, and the chemistry for elements in the same group is very similar, then the quaternary system may be considered to be a perturbation on the ternary system. This suggests that if M-III and M-V phases are stable, the single-phase alloys of the form (M,M′)-III, (M,M′)-V, M-(III,III′), and M-(V,V′) may then also be potentially stable phases. For example, the substitution of isovalent Al for Ga in NiGa or CoGa results in a phase with the same structure and similar lattice parameter.[98] Similarly,

P or Sb may be substituted for As in many cubic RE-As phases without changing structure.[35] Two RE elements can also be codeposited with a group-V element to grow single-phase (M,M′)-V film with the NaCl structure, for example, (Sc,Er)As[34] and (Sc,Yb)As[47][96].

**Table 1.** Properties of RE-As compounds with NaCl structure

| Compound | Lattice parameter (Å)[62] | Mismatch to GaAs (%) | Melting point (°C)[158][159] | Heats of formation (kcal/mole)[230] | Other arsenides |
|---|---|---|---|---|---|
| ScAs | 5.464 | -3.35 | | -65.2 | $Sc_3As_2$,$Sc_5As_3$,$Sc_{15}As_6$ |
| YAs | 5.796 | 2.53 | | -77.4 | |
| LaAs | 6.1504 | 8.80 | | -73.0 | $LaAs_2$,$La_4As_3$ |
| CeAs | 6.086 | 7.66 | | -68.9 | $CeAs_2$,$Ce_4As_3$ |
| PrAs | 6.0274 | 6.62 | | -73.4 | $PrAs_2$,$Pr_3As_4$ |
| NdAs | 5.986 | 5.89 | 2220 | -72.7 | $NdAs_2$,$Nd_3As$ |
| SmAs | 5.9131 | 4.60 | | -72.0 | |
| EuAs | hexagonal | | ~1050 | | many from $EuAs_3$ to $Eu_5As_3$ |
| GdAs | 5.860 | 3.66 | | -74.4 | |
| TbAs | 5.8203 | 2.96 | | -75.0 | |
| DyAs | 5.7894 | 2.41 | | -78.1 | |
| HoAs | 5.769 | 2.04 | | -72.3 | |
| ErAs | 5.7427 | 1.58 | >2500 | -75.6 | |
| TmAs | 5.7188 | 1.16 | | -72.8 | |
| YbAs | 5.6979 | 0.79 | | -61.8 | $Yb_4As_3$,$Yb_5As_3$ |
| LuAs | 5.6761 | 0.41 | | -75.2 | |

Thermodynamic stability is affected by these isovalent substitutions, particularly if III′ or V′ is substituted for III or V, respectively. The M-III/III-V and M-III′/III′-V systems may be thermodynamically stable but, in general, the system M-(III′,III)/III-V will not. If the latter system is allowed to approach equilibrium, the exchange of III′ and III across the interface would be expected. Since III′ and III are chemically similar, the thermodynamic driving force for this interdiffusion will be relatively small. This type of an isovalent exchange reaction has been observed in the case of NiAl films on GaAs.[51][99][100] Annealing at high temperatures results in the formation of Ni(Al,Ga) alloy and an interfacial semiconductor layer of (Ga,Al)As. The interdiffusion does not alter the structure of either the metal or the semiconductor. Yet, the electronic properties that are determined by the interface chemistry such as the Schottky barrier height are significantly changed.[99]-[103] To eliminate the driving force for interdiffusion at the metal/semiconductor interface, an interfacial layer of AlAs has been inserted.[52] The unstable interface then becomes the AlAs/GaAs interface, a problem that can be ignored at low and moderate temperatures because of the low diffusivities of Ga and Al in monocrystalline (Al,Ga)As. In the case of M-(V,V′)/III-V interfaces the same arguments apply. However, an additional complication results during the actual growth of the M-(V,V′) film. At typical growth temperatures the semiconductor surface must always have an impinging group-V flux to compensate for the high volatility for the group-V component and growth usually occurs under excess group-V conditions. The semiconductor (III-V) may, therefore, have impeding V and V′ fluxes before the M-(V,V′) growth starts, which can result in surface exchange reactions resulting in III-(V,V′) semiconductor formation prior to the M-(V,V′) growth.

## 2.3 Growth Modes

Epitaxial growth proceeds in different ways depending upon surface, interface and strain energies of the system. From a thermodynamic standpoint, if the surface energy per unit area of the metal, $\gamma_M$, plus the interface energy per unit area, $\gamma_{int}$, is lower than the semiconductor surface energy per unit area, $\gamma_{III\text{-}V}$, then the metal will wet the semiconductor ($\gamma_M + \gamma_{int} < \gamma_{III\text{-}V}$) and growth should proceed in a layer-by-layer manner. This growth mode is referred to as Frank-van de Merwe.[104] It follows that if the metal wets the semiconductor surface then the semiconductor will not wet the metal surface ($\gamma_{III\text{-}V} + \gamma_{int} > \gamma_M + 2\gamma_{int}$), and growth will be by a three dimensional island growth mode known as Volmer-Weber.[104] If the semiconductor does not wet the metal it does not immediately follow that the metal will wet the semiconductor. This will only occur if $\gamma_M - \gamma_{III\text{-}V} < \gamma_{int} < \gamma_{III\text{-}V} - \gamma_M$. From a thermodynamics argument the growth of III-V/M/III-V structures will be difficult since ideally the metal should wet the III-V and the III-V the metal, which appears to be impossible. The only way to achieve this would be to modify the surface or interface energies during the growth. For example, if the metal wets the semiconductor then the use of a surfactant to either increase the metal surface energy prior to the III-V overgrowth or to decrease the III-V surface energy of the growing layer could result in the III-V wetting the metal. This approach has been used to suppress island formation during Ge growth on Si.[105] In order for island formation to occur, the growing layer needs sufficient atomic mobility for the atoms to coalesce into islands before more material covers the surface.

Therefore, if growth is done at low enough temperatures and/or high enough deposition rates, then the surface mobility may not be sufficient to allow dewetting and a "quasi-two-dimensional" growth mode may be achieved. In this growth mode the surface may be atomically rough, but smooth on a larger scale. There is of course a tradeoff; if the temperature is too low or the deposition rate too high, then epitaxy will not occur because the depositing atoms will not have sufficient time and mobility to find lattice sites. From a practical standpoint, one must look for a temperature window within which epitaxy occurs without sufficient surface diffusion for agglomeration to occur.

Achieving epitaxial III-V/M/III-V structures is also facilitated if the melting point of the metal is comparable to, or higher than, that of the semiconductor (surface and bulk diffusivities generally decrease for increasing melting point). If the melting point of the metal is considerably lower, the metal film may agglomerate during semiconductor overgrowth at 400 - 600°C. This problem may be exacerbated if the metal layer is extremely thin (less than about 100 Å). At such thicknesses, it is likely that extended defects in the metal film (e.g., dislocations, stacking faults, second-phase inclusions or pinholes) will traverse the film thickness, thereby providing favorable sites for the initiation of film agglomeration. At the other extreme, if the metal's melting point is extremely high, the smooth growth temperature for the metal may be above the decomposition temperature of the semiconductor (about 550°C for GaAs without an $As_4$ or $As_2$ flux). For smooth growth of a metal, experiments[106] and theory[107] suggest that the substrate must be brought to a temperature of at least ~$3T_m/8$, where $T_m$ is the melting point. Taking $3T_m/8$ as a conservative estimate for the smooth growth temperature of the metal (certain metallic films such as the rare earth arsenides actually have lower relative smooth growth temperatures), the metal's melting point would have to be higher than about 2200°C to make GaAs decomposition a concern. In practice then, one should choose the metal with the highest melting point possible. Note that most of the elemental metals that can be grown epitaxially on III-Vs and that are relatively unreactive with GaAs (e.g., Au and Al) also have low melting points (1065 and 660°C, respectively).

Once a uniform layer covers the surface there will be an activation barrier for island nucleation due to the increase in film surface area needed to form a pinhole. The substrate temperature may then be raised or the deposition rate lowered so that true layer-by-layer growth occurs. For smooth growth of a metal the substrate must be brought to a temperature of at least ~$3T_m/8$. To an extreme, the initial layer may be deposited in the amorphous state and solid phase epitaxy can be used to recrystallize the layer, which now is used as a template for further epitaxy. The use of templates will be discussed further later in this section. From these arguments, the most likely way to achieve III-V/M/III-V structures is to have as small a difference in surface energies (i.e., $\gamma_M \sim \gamma_{III\text{-}V}$) as possible and to perform the growth at low temperature. This approach is described in Sec. 4.

The effect of lattice mismatch on heteroepitaxial growth has been widely studied for semiconductor heteroepitaxial growth.[108] The lattice mismatch introduces an additional term to the excess interfacial energy term. Molecular dynamics simulations that consider the interfacial energy term associated with the mismatch only, have shown that three-dimensional island growth is preferred over two-dimensional layer-by-layer growth for any finite lattice mismatch.[109] In practice, Stranski-Krastanov[104] growth is observed, where initially layer-by-layer growth is observed up to a critical thickness above which three-dimensional island growth occurs. In this growth mode there is always a finite two-dimensional layer covering the substrate surface. Experimentally, growth of most metals at relatively low deposition rates and moderate substrate temperatures occurs by the Volmer-Weber mode or Stranski-Krastanov mode.[110] We show in Sec. 4 that a growth mode which is neither strictly Stranski-Krastanov or Volmer-Weber can occur for GaAs growth on $Sc_{1-x}Er_xAs$.

## 2.4 Inherent Symmetry and Defects in the Semiconductor Overgrowth

The growth of a defect-free zincblende semiconductor layer seeded entirely on a metal surface is very difficult. Such a high level of structural perfection would require that the metal meet criteria more stringent than those that must be satisfied for nominally defect-free growth of the metal on the semiconductor. The primary cause in III-V overlayer defect nucleation is the difference in crystal symmetry between the metal and the semiconductor. Generally, cubic metals have higher symmetry than the zincblende semiconductors. In other words, operations such as rotation, inversion and mirror reflection that leave a cubic metal's crystal structure invariant (a symmetry operation), may not leave the structure of the semiconductor invariant. In particular, metals and intermetallics with the *bcc*, *fcc*, NaCl and CsCl structures possess the inversion symmetry element. The zincblende structure, however, is noncentrosymmetric (no center of inversion symmetry). Consequently, the semiconductor growing on the metal does not "know" which way is up. The resulting two-dimensional defect is known as an inversion domain boundary since it separates two structural domains of the crystal. Such defects are also common to III-V semiconductor layers grown on Si (the diamond structure possesses a center of inversion symmetry). However, as in the case of III-V semiconductor growth on Si,[111] vicinal surfaces can be used to reduce the symmetry of the metal surface.

A second symmetry-related type of defect is the translation domain boundary. Translation domains arise when a translation symmetry operation of the metal is not a symmetry operation of the semiconductor. For example, metals and intermetallics with the *bcc* and CsCl structures generally have lattice parameters that are about half those of the III-V semiconductors. Thus, there are two equivalent positions that an epitaxial III-V nucleus can adopt on the metal surface. Impingement of III-V grains growing from nuclei of different variants leads to the formation of a translation domain boundary.[49] Growth over surface steps can also nucleate such a boundary if the vector that describes the step is not a translation symmetry element of the semiconductor.

Lattice mismatch between the metal and the semiconductor will also generate defects in the semiconductor. These defects include {111} stacking faults and threading dislocations that terminate misfit dislocations in the interface. This type of defect can be minimized by matching the lattice constant of the metal to the semiconductor by alloying or by growing a metal film that is thin enough to remain coherent with the semiconductor overlayer. In either case, the symmetry-related defects discussed above are still problematic.

## 2.5 Morphological Stability of Buried Layers

Assuming the metal/III-V interface has a finite positive excess energy per unit area, a metal film buried in semiconductor will always be metastable with respect to agglomeration.[49] Once such a heterostructure has been fabricated (using growth techniques described above), the increase in the interface area needed to nucleate a pinhole provides an activation barrier for agglomeration. In practice, however, variations in interface energy with interface orientation or extended defects in the metal film may lower this barrier. Considerations of interface fluctuations and pinhole nucleation are more severe for very thin metal films. Agglomeration of thick buried metal films is not expected, particularly if they have already endured a high temperature III-V overgrowth step without agglomerating.

## 3.0 GROWTH OF III-V/M/III-V STRUCTURES

Simple inspection of the structures and lattice parameters of a metal and a semiconductor can lead to expectations for a specific orientation relationship during epitaxial growth, especially if the metal and semiconductor are both cubic and their lattice parameters are commensurate. For example, the elemental metals, Ag and Al, both have *fcc* lattices while their lattice parameters are approximately equal to $1/\sqrt{2}a_{0GaAs}$ (Fig. 2a). One might then reasonably expect that these metals should grow on {100} GaAs with the orientation relationship $\{100\}_{Al,Ag}||\{100\}_{GaAs}$ and $\langle 010\rangle_{Al,Ag}||\langle 011\rangle_{GaAs}$ as shown in Fig. 9a. Experiments, however, have revealed much more complicated phenomena. While epitaxy does occur in both cases, the orientation relationships are hardly straightforward. Instead, they depend on substrate temperature and GaAs surface termination (reconstruction/composition).[110][112]-[115] In the case of films of Al on {100}GaAs, they nucleate in three different orientations, only one of which is easily rationalized on the basis of lattice matching (Fig. 9). The dominant variant is determined by the GaAs substrate surface reconstruction.[110][112][113] In the case of Ag, the expected orientation relationship (Fig. 10a) is generally not observed. For growth temperatures less than about 200°C, two orientation variants with $\{110\}_{Ag}||\{100\}_{GaAs}$ (not strictly identical because of the noncentrosymmetric structure of GaAs) are observed[110][114] (Fig. 10b). At higher growth temperatures, the {100} plane of Ag is parallel to the {100} GaAs surface, but the unit cell of Ag is unrotated (Fig. 10c) resulting in an apparent lattice mismatch of ~30%! Recent experiments, however, have shown that a 6 monolayer (ML) thick epitaxial film of *bcc* Fe deposited prior to the Ag[116][117] (Fig. 10d) can lead to the expected orientational relationship of Fig. 10a. This phenomenon is not yet fully understood. It does demonstrate, though, that the application of the template approach to heteroepitaxy can lead to the achievement of heterostructures that cannot be obtained by single-step continuous growth. This approach is discussed in Sec. 4 for the growth of TM-III/III-V heterostructures.

It is clear that the growth of high quality single crystal Al layers is difficult on GaAs. This together with the low melting point of Al makes Al a poor choice for the growth of GaAs/metal/GaAs heterostructures. Nevertheless, a few attempts have been made.[118]-[121] The Al grows by the Volmer-Weber mode and, therefore, layer thicknesses >100Å are required before a smooth film is obtained. For a growth temperature of ~40°C the

Al films were predominantly {110} on GaAs(100).[118][119] Growth at ~0°C resulted in {111} textured films.[121] Regardless of the Al orientation, the GaAs overgrowth was similar. Growth at 400°C resulted in a completely reacted Al film and no GaAs/Al/GaAs heterostructure. The GaAs overgrowth must be performed at ~200°C to ensure minimal Al/Ga exchange reaction and Al reaction with the impinging As-flux. Conventional MBE was found to yield polycrystalline layers.[118] Crystalline layers were obtained by using migration enhanced epitaxy (MEE),[119]-[121] but were of poor crystalline quality. These results are a clear indication of the importance of *epitaxy, thermodynamic stability and melting point* discussed in Sec. 2 for the choice of the metallic layer.

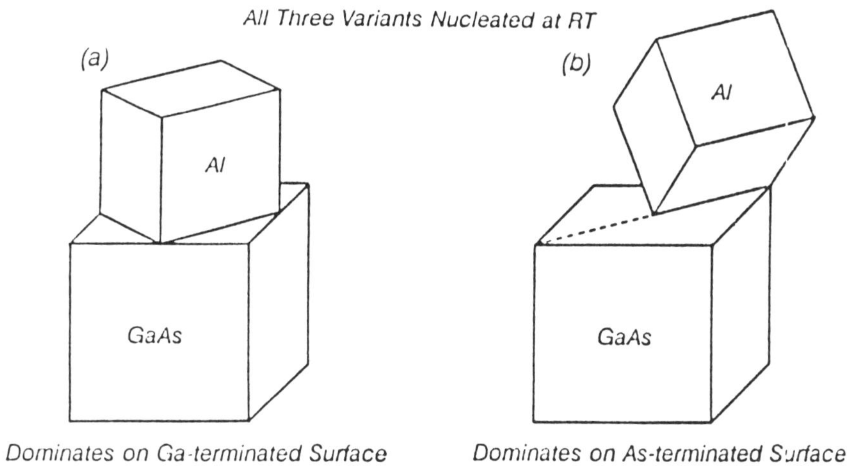


**Figure 9.** Schematic illustrations of the orientation relationships observed for the epitaxial growth of Al on {100}GaAs. Based on the ratio of lattice parameters, approximately equal to $\sqrt{2}$, the expected orientation relationship is shown in (a). Two additional orientation relationships are observed with {110}Al || {100}GaAs. One is depicted in (b). The dominant orientation relationship is determined by the initial GaAs surface reconstruction. The unit cells are depicted as simple cubes for clarity (see references in text).

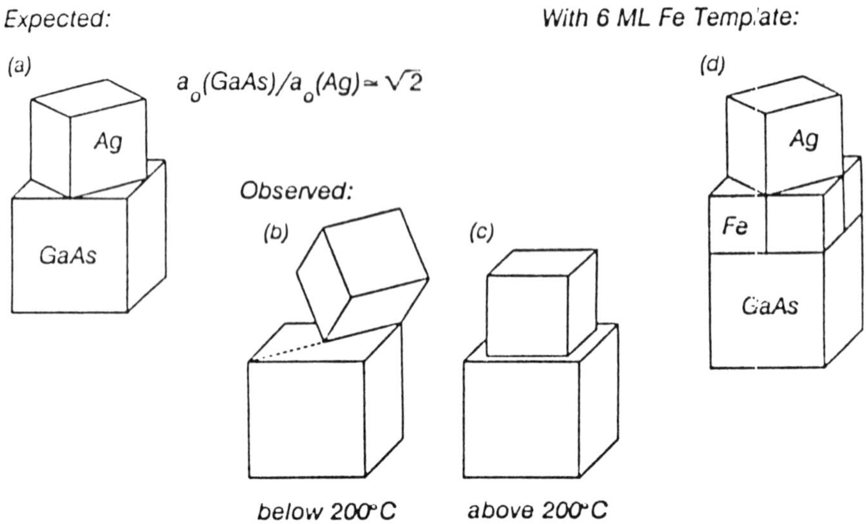


**Figure 10.** Schematic illustrations of expected and observed orientation relationships for Ag on {100}GaAs. Like Al, the lattice parameter of Ag is approximately $1/\sqrt{2}\,a_{0GaAs}$. The expected orientation relationship *(a)*, with the unit cell of Ag rotated 45° about the surface normal, is not observed. Instead, two orientation relationships with {110}Ag || {100}GaAs *(b)* and the unrotated orientation relationship *(c)* are observed, depending on the substrate temperature during deposition. Farrow and coworkers have shown that deposition of a thin template of Fe causes the Ag overlayer to adopt the 45° rotated orientation relationship *(d)* (see references in text).

## 4.0 GROWTH OF THE THERMODYNAMICALLY STABLE METALLIC COMPOUNDS

The growth of RE-V compounds on III-V semiconductors has been explored by three groups: Bellcore (USA), Fraunhofer (Germany) and CNET (France). The following RE-V/III-V systems have been grown: ErAs/GaAs,[30][32][39][40] LuAs/GaAs,[32] YbAs/GaAs,[31] $Sc_{1-x}Er_xAs$/GaAs,[34] $Sc_{1-x}Yb_xAs$,[47][96] $ErP_{1-y}As_y$/GaAs[42]-[45] and $ErP_{1-y}Sb_y$/InP.[35] The growth mechanisms are very similar and are described below.

A conventional MBE system intended for III-V compound semiconductor growth can be used to grow the RE-As/III-V heterostructures. However, for the growth of RE-P or RE-$P_{1-y}As_y$ compounds, a gas source MBE system or an MBE system with valved cracker cells is preferable due to difficulties in handling phosphorous from a conventional effusion cell. The rare-earth metals can be evaporated using effusion cells or an electron gun evaporation source. The main difficulties in using electron guns are deposition rate control and the tendency for the rare-earth elements to sublime. Effusion cells are easier to use and more compatible with a conventional MBE system. Conventional effusion cells with a pyrolytic boron nitride (PBN) crucible can be used for Er and some other rare-earth elements. Since the temperatures required for reasonable evaporation rates are often close to the maximum operation temperature for PBN and conventional cells, the use of high temperature effusion cells with tungsten crucibles is recommended. They are commercially available and allow operating temperatures up to ~2000°C. Organometallic chemical vapor deposition (OMCVD), organometallic molecular beam epitaxy (OMMBE) or vapor phase epitaxy (VPE) techniques could be used with the right choice of gases.

## 4.1 RE-V Growth on GaAs(100)

Figure 11 shows a typical sequence for the MBE growth of TM-III/III-V and RE-V/III-V heterostructures. The III-V semiconductor is grown with the normal III-V MBE growth conditions (i.e. for (100)GaAs: a substrate temperature ~580 - 600°C, an arsenic flux greater than the gallium flux, and a growth rate ~0.1 - 1.0 $\mu$m/hr). Before the RE-V is grown the substrate temperature may be lowered, keeping the group V flux on to inhibit surface decomposition due to evaporation of the group V element. The typical substrate temperature ranges from ~350 - 600°C ($\leq$ 0.22 - 0.31 $T_m$) depending upon film thickness. The RE-V films are always grown under excess group V flux to inhibit RE/III-V interfacial reactions. On {100}GaAs surfaces single variant {100}RE-V growth is always observed. Figure 12 shows a RHEED sequence during the growth of a GaAs/ErAs/GaAs structure. The starting GaAs surface shows the expected c(4x4) As-stabilized reconstruction expected at 450°C with the $As_4$ flux on. When the ErAs is deposited on the GaAs the reconstruction changes to a (1x3) after ~0.2 ML. After 1 ML of deposition the RHEED pattern changes to a (1x1), which is also observed for thicker films.

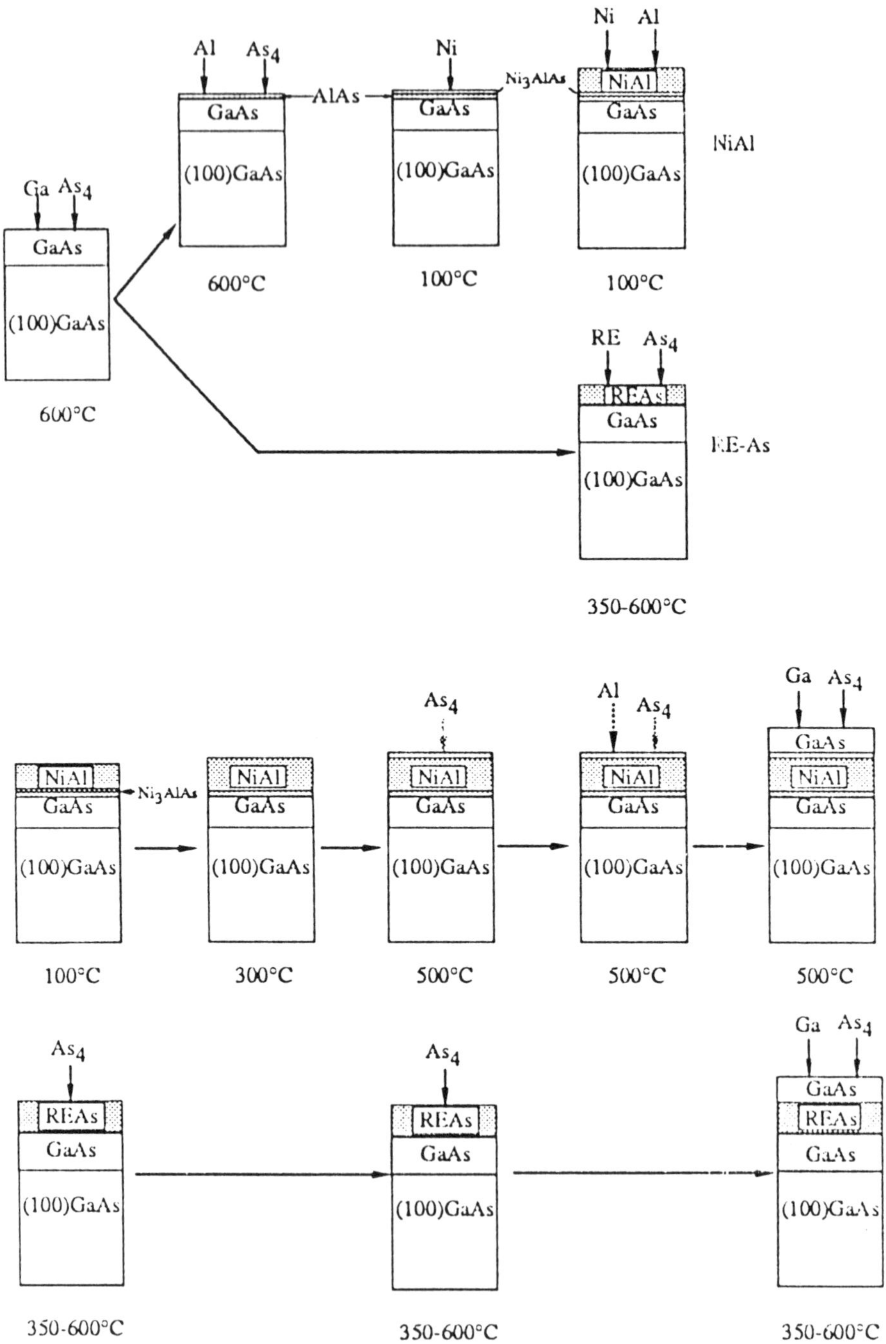
Ga As4
GaAs
(100)GaAs
600°C
Al As4
AlAs
GaAs
(100)GaAs
600°C
Ni
Ni3AlAs
GaAs
(100)GaAs
100°C
Ni Al
NiAl
GaAs
(100)GaAs
100°C
NiAl
RE As4
REAs
GaAs
(100)GaAs
350-600°C
RE-As
NiAl
Ni3AlAs
GaAs
(100)GaAs
100°C
300°C
As4
500°C
Al As4
500°C
Ga As4
GaAs
500°C
As4
REAs
GaAs
(100)GaAs
350-600°C
As4
350-600°C
Ga As4
GaAs
REAs
350-600°C

**Figure 11.** Schematic of NiAl and RE-As growth on GaAs and subsequent GaAs overgrowth. Prior to NiAl and RE-As growth a buffer layer of GaAs is grown at 600°C. In order to minimize Al-Ga exchange reactions, a thin AlAs layer is grown before the NiAl deposition. After the AlAs layer is grown, the substrate heater is turned off and the sample is allowed to cool to ~100°C. The As-flux is turned off, since NiAl is not thermodynamically stable in contact with As. About one monolayer of Ni is deposited prior to the growth of NiAl. This monolayer of Ni, which is believed to form a thin $Ni_3AlAs$ template layer, is necessary to ensure {100} NiAl growth on GaAs(100) and the low substrate temperature is necessary to reduce agglomeration of the NiAl for ultrathin layers. Since RE-As is thermodynamically stable in contact with both GaAs and As, there is no need to grow an AlAs layer or reduce the As-flux prior to growing the RE-As. The substrate temperature may be higher than that during NiAl growth due to the higher melting point of RE-As. The GaAs overgrowth procedure is also different for growth on NiAl and RE-As. For NiAl the substrate is heated to 500°C for the GaAs overgrowth and the As-cell temperature must be raised to achieve the desired $As_4$-flux for GaAs growth. During the heating process Ni from the $Ni_3AlAs$ layer is believed to react with excess Al in the NiAl, which results in more NiAl forming and the regrowth of a thin AlAs layer. Arsenic reaction with the NiAl is prevented by supplying an Al flux from the Al-cell instead of the NiAl resulting in AlAs growth on NiAl. The correct amount of Al is determined by observing the RHEED pattern (see details in text). This procedure is used to grow a thin AlAs layer prior to growing the GaAs overlayer. Since RE-As is stable in contact with GaAs and As, the procedure for growing GaAs on top of the RE-As layer only involves opening the Ga-shutter.

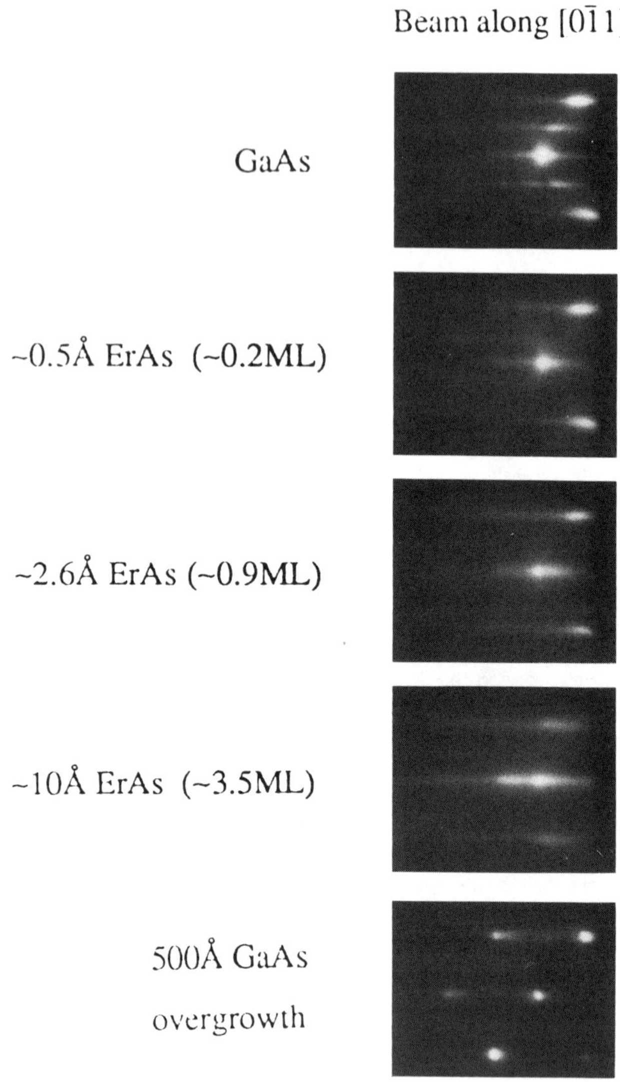


**Figure 12.** RHEED sequence of GaAs/ErAs/GaAs(100) growth at 450°C. The starting GaAs surface has the c(4x4) As-rich structure. After 0.2 ML of ErAs deposition the diffraction pattern changes to a (1x3). For thicknesses ≥ 0.9 ML a (1x1) pattern is observed. The patterns are streaky indicating a fairly flat surface. When GaAs is grown on top of the ErAs the diffraction pattern becomes spotty after 1 ML of GaAs deposition and remains spotty even after 500 Å of GaAs growth. This indicates that the GaAs overgrowth is 3-D like, resulting in a rough surface.

RHEED allows the real time study of the growth of both the RE-As and GaAs layers. Oscillations are observed in the specular spot intensity during $Sc_{1-x}Er_xAs$ growth (Fig. 13). From Rutherford backscattering (RBS) measurements of the film thickness post growth it has been determined that each oscillation corresponds to 1 ML $Sc_{1-x}Er_xAs$ growth, indicating that the growth occurs in a monolayer-by-monolayer fashion. When the Sc and Er shutters are first opened, the specular spot intensity drops dramatically and the diffraction pattern also becomes faint. This drop arises from increased surface roughness and possibly disorder until the $Sc_{1-x}Er_xAs$ (NaCl) structure is established. The specular spot intensity increases after about 1 ML of $Sc_{1-x}Er_xAs$ growth. A maximum in the RHEED intensity is observed for about 4 ML of deposition, the intensity then slowly decreases for thicker films (Fig. 13) due to an increase in surface roughening. Figure 14 shows high-resolution cross-sectional TEM micrographs of GaAs/ErAs/GaAs with ErAs thickness of 1, 2 and 3 ML. It is clear from Fig. 14 that the first monolayer does not go down uniformly, but instead appears as islands 2 - 3 ML thick. The quality of the overgrown GaAs is very good for ErAs layers which are 1 and 2 ML thick, but substantially worse for the 3 ML thick layer. This is confirmed from the backscattering minimum yield versus ErAs film thickness shown in Fig. 15. For ErAs layers ≤ 2 ML, both the ErAs and overgrown GaAs have relatively good minimum yields of about ≤ 24% and ≤ 13%, respectively. For ErAs layers ≥ 3 ML, there is an abrupt increase in the minimum yields for both the overgrown GaAs (~40%) and the ErAs. The GaAs minimum yield remains approximately constant while the ErAs minimum yield decreases with increasing ErAs film thickness above 3 ML in a similar manner to that observed for samples with either Al or amorphous Si capping layers.[122] The results in Fig. 15 may be explained by the 1 and 2 ML thick ErAs layers having sufficient number of pinholes enabling the overgrown GaAs to seed on GaAs in the pinholes, resulting in good quality epitaxial GaAs. This is consistent with electrical measurements which showed the 1 ML film electrically discontinuous and the films ≥ 2 ML to be electrically continuous. After 2 ML of deposition the regions between the islands start to fill in, and after 3 ML a uniform layer is seen, as discussed below. Hence, for ErAs films ≥ 3 ML the pinhole density is not sufficient to allow seeding in pinholes to dominate the GaAs epitaxial overgrowth mechanism and growth proceeds by GaAs nucleation on the ErAs surface, which results in a decrease in the crystal quality.

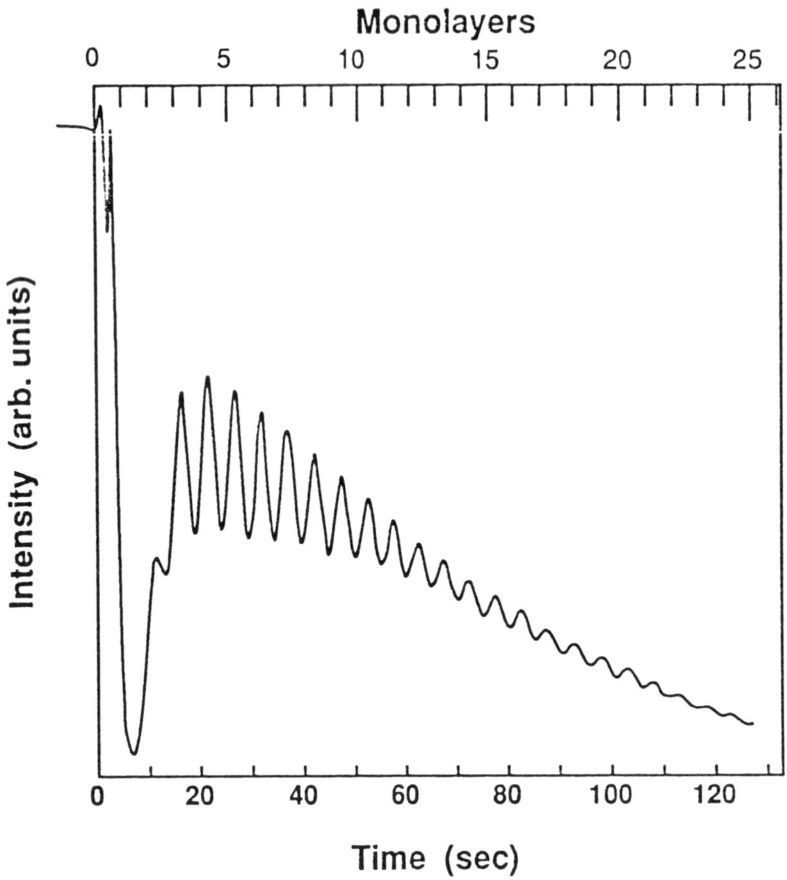


**Figure 13.** RHEED intensity oscillations for $Sc_{0.32}Er_{0.68}As$ growth on GaAs(100) at 400°C. The intensity of the specular beam decreases dramatically during the first ML of $Sc_{0.32}Er_{0.68}As$ growth and then it starts to increase. RHEED intensity oscillations are observed after 2 ML of deposition. The maximum in the RHEED intensity is found after about 4 ML of deposition. The RHEED intensity and the amplitude of the oscillations decrease with further $Sc_{0.32}Er_{0.68}As$ deposition. This behavior is attributed to the $Sc_{0.32}Er_{0.68}As$ growing as small islands during the first 1-2 ML of growth and the islands coalescing after 3 ML of growth. The smoothest films are found for ~4 ML layers. Further deposition results in an increase in the surface roughness.

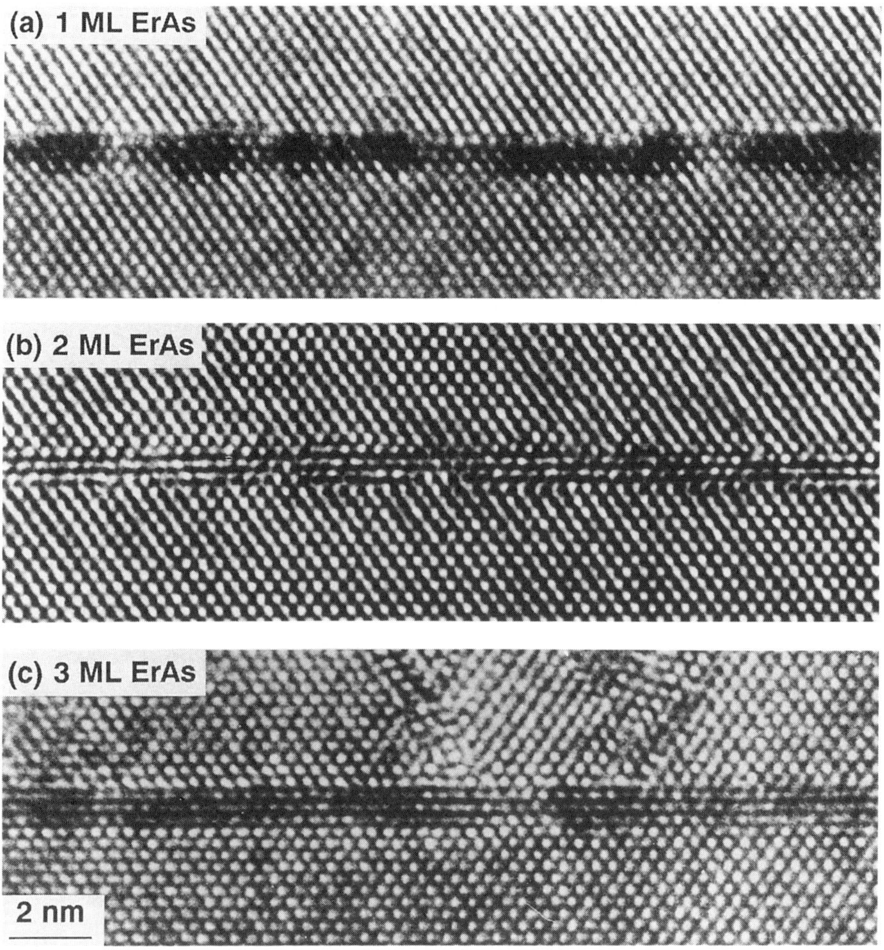


**Figure 14.** High resolution cross-section TEM micrographs of GaAs/ErAs/GaAs(100) structures; *(a)* 1 ML ErAs, *(b)* 2 ML ErAs and *(c)* 3 ML ErAs as determined by deposition conditions and RBS measurements. Distinct ErAs islands, which may be estimated to be ~2-3 ML thick, can easily be seen in *(a)*. This film is not electrically continuous. For the 2 ML film the micrograph *(b)* indicates a more continuous film and it is electrically continuous. For the micrograph of the 3 ML thick sample a relatively uniform film structure is evident. Note that the quality of the overgrown GaAs is substantially worse for the 3 ML sample. This suggests that there is a high density of pinholes in the 1 and 2 ML ErAs films, which allows the overgrown GaAs to seed on the GaAs substrate and not the ErAs.

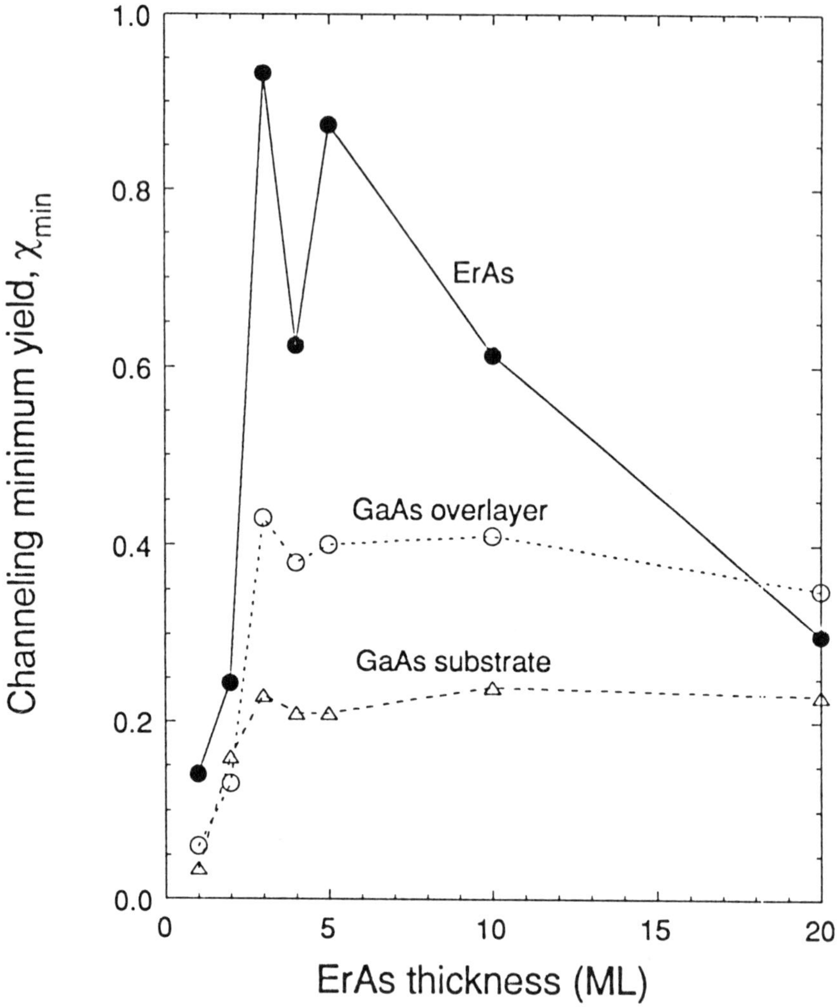


**Figure 15.** Ion channeling minimum yield, $\chi_{min}$, vs. ErAs thickness for GaAs(500Å)/ErAs/GaAs(100) structures. The large increase in the minimum yields for the overgrown GaAs, ErAs and GaAs substrate for ErAs films ≥ 3 ML arises from a dramatic decrease in the overgrown GaAs crystalline quality due to a substantial decrease in pinholes in the ErAs films (see text).

A more quantitative way of measuring the ErAs film thickness fluctuations can be done using x-ray extended range specular reflectivity.[123][124] Figure 16a shows x-ray reflectivity for a 2 ML thick ErAs film buried beneath 500 Å of GaAs. Because of the small quantity of buried ErAs the low angle reflectivity does not directly measure the thin layer, in fact, this portion of the reflectivity could be equally well modeled by two GaAs slabs separated by a gap. In contrast, the GaAs scattering is much reduced at the {002} position so that the ErAs layer directly produces a broad feature there, the width of which is related to the ErAs thickness. At the {004} position the GaAs scattering again dominates. Miceli et al.[123,124] have done model calculations of the data by summing the contributions to the scattering from each atomic layer in the system. The solid curve in Fig. 16a is a least squares fit of the model to the data. This fit was obtained with individual probabilities, $P_i$, of having layers of each thickness (i.e., $P_n$ for the layer to be n monolayers thick), yielding $P_1 = 0.0 \pm 0.02$, $P_2 = 0.04 \pm 0.08$, $P_3 = 0.41 \pm 0.07$, $P_4 = 0.19 \pm 0.07$, $P_5 = 0.07 \pm 0.04$, and $P_n = 0$ for $n \geq 6$. The average coverage is the sum of the probabilities times the layer thickness for all the layers. This yields 2.4 ML ErAs average coverage, which is consistent with the growth conditions. The results show, however, that the film is predominantly 3 ML to 4 ML thick, which suggests that there are regions of no coverage (i.e. pinholes in the ErAs). Since the probabilities are normalized, the uncovered fraction is obtained directly $P_0 = 0.3 \pm 0.2$. This is a measurement of the pinhole density. Figure 16b shows data for a 5 ML thick ErAs film. Three subsidiary maxima are clearly visible, indicating that the ErAs layer is very uniform. A least squares fit to the data yields (to within $\pm 0.03$) $P_3 = 0.01$, $P_4 = 0.41$, $P_5 = 0.55$, $P_6 = 0.03$, and all other $P_n = 0$. Thus the average coverage is 4.6 ML with nearly equal amounts of 4 and 5 ML regions. Note that no pinholes are present and the film is uniform with single atomic steps due to the 0.5 ML terminated growth.

The traditional picture of the Frank-van der Merwe growth mechanism is that initially two dimensional islands one monolayer thick nucleate on the surface and grow laterally until a complete layer is formed. This process continues for the following layers, resulting in layer-by-layer growth. Since $Sc_{1-x}Er_xAs$ initially nucleates as 3 ML thick islands, the argument may be made that $Sc_{1-x}Er_xAs$ does not wet the GaAs(100) surface and the growth is better described by a Volmer-Weber mode. However, since $Sc_{1-x}Er_xAs$ has a completely different crystal structure, bulk cohesive energy and bonding from GaAs, the argument may be made that a critical thickness greater than 1 ML is required for the formation of $Sc_{1-x}Er_xAs$. Thus the initial 3 ML islands may then be interpreted as a two dimensional nucleation stage of $Sc_{1-x}Er_xAs$ rather than from dewetting, and the $Sc_{1-x}Er_xAs$ appears to wet the GaAs(100) surface. From the thermodynamic arguments in Sec. 2, together with this assumption, it is expected that GaAs should not wet $Sc_{1-x}Er_xAs$. The overgrowth of GaAs in RE-As(100) films is, therefore, expected to be substantially more difficult than the growth of RE-As on GaAs(100) and will be discussed in Sec. 4.2.

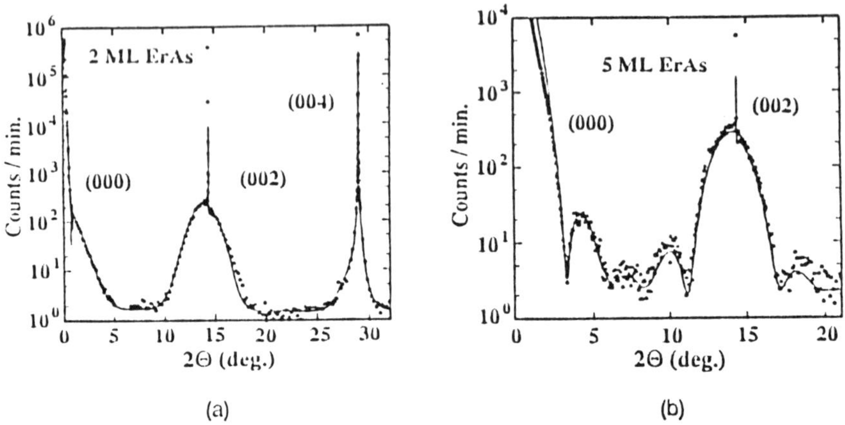


**Figure 16.** X-ray extended reflectivity from GaAs/ErAs/GaAs(100) structures with ErAs layer thickness of *(a)* 2 ML and *(b)* 5 ML. Solid curves are least squares fit of a discrete atomic layer model to the data, yielding ErAs atomic layer thickness probabilities $P_0 = 0.3 \pm 0.2$, $P_1 = 0.0 \pm 0.02$, $P_2 = 0.04 \pm 0.08$, $P_3 = 0.41 \pm 0.07$, $P_4 = 0.19 \pm 0.07$, $P_5 = 0.07 \pm 0.04$, and $P_n = 0$ for $n \geq 6$ for *(a)* and $P_3 = 0.01$, $P_4 = 0.41$, $P_5 = 0.55$, $P_6 = 0.03$ (to within $\pm$ 0.03), and all other $P_n = 0$ for *(b)*. This indicates that for *(a)* the ErAs film consists primarily of 3 and 4 ML thick regions and that pinholes in the ErAs correspond to about 30% coverage. For *(b)* the ErAs film consists primarily of 3 and 4 ML thick regions with no pinholes. *(Courtesy of P. F. Miceli, Bellcore)*

It is interesting to note that Yang and Flynn[125] observed RHEED oscillations during the growth of (100) alkali-halide with the NaCl structure for substrate temperatures as low as ~$0.1T_m$. This growth temperature is substantially lower than that required for metals (~$3T_m/8$)[106] and semiconductors (~$0.5T_m$). Yang and Flynn postulated that this occurs due to the weak binding of alkali-halide molecules to terraces and strong bonding at ledges in these ionic crystals. The growth of RE-As appears similar to that of the alkali-halides (≤ $0.22T_m$[125]). The lowest temperature for which RHEED oscillations can be observed has not been determined. However, for growth temperatures >450°C RHEED oscillations were not observed, indicating that the growth mode changes to step flow or more 3-D type growth.

The dampening of the RHEED oscillations during $Sc_{1-x}Er_xAs$ growth indicates an increase in surface roughness with thickness. The ErAs layer thickness fluctuations have been determined from x-ray Bragg reflectivity as a function of film thickness.[123,124] Figure 17 shows x-ray reflectivity data and a model fit for 35 and 140 Å ErAs films. The ErAs/GaAs interface roughness was 1 ML and the ErAs layer thickness fluctuations were modeled as Gaussian, giving 1 ML for the 35 Å and 2.5 ML for the 140 Å ErAs films. For thicker films the ErAs films roughened further and the ErAs film thickness fluctuations were better described by an exponential rather than a Gaussian.[124] This roughening is also evident from RHEED during growth, which also indicates that some faceting may be occurring.

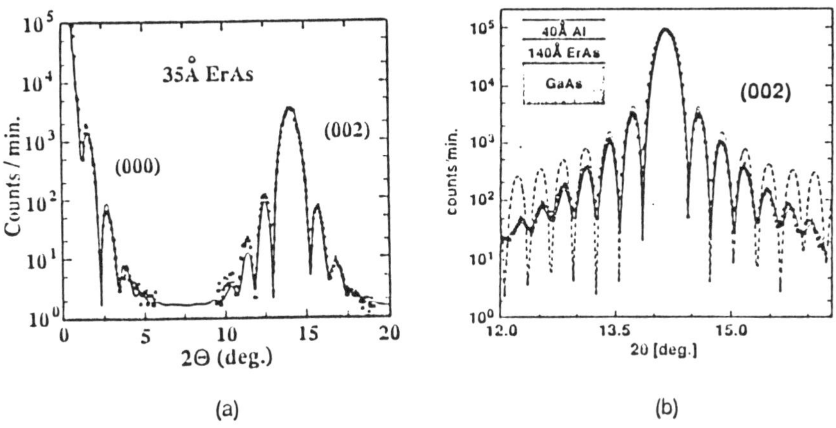


**Figure 17.** The measured {002} extended reflectivity (points) and fit to the data (solid curve) for *(a)* α-Si/ErAs(35Å)/GaAs(100) and *(b)* Al(40Å)/ErAs(140Å)/GaAs(100). Calculations assuming with thickness fluctuations (solid curve) and without thickness fluctuations (dashed curve) are shown in *(b)* (see text for details). *(Courtesy of P. F. Miceli, Bellcore)*

To model the growth mechanism and the electronic structure of GaAs/$Sc_{1-x}Er_xAs$/GaAs heterostructures the detailed interface atomic structure must be determined. Stoffel et al.[126] have used ion channeling measurements to determine the interface lattice registry of ultrathin buried ErAs layers with respect to the overgrown GaAs. In their study, the difference in symmetry between the overgrown GaAs and the ErAs is used to identify the Er lattice registry. To obtain reliable data, the crystalline quality of all the layers had to be excellent. Hence, an ultrathin 0.5 ML ErAs layer was used. If the lattice registry is as that shown in Fig. 18, that is with the As-sublattice continuous across the interface, the Ga and As atoms in the GaAs overlayer will shadow the Er atoms beneath when channeling is performed in a <100> direction. However, when channeling is performed in a <110> direction the Er atoms are not shadowed, as is shown in Fig. 19a. Note that the Er atom is not centered in the channel, therefore, the Ga atoms on the right should shadow the Er atoms more than the As atoms on the left during a horizontal (azimuthal Fig. 19a) angular scan. Such a scan is shown in Fig. 19b. It is clear that the Er yield peak is not centered on the Ga and As yield minimum, indicating that the Er is not centered in the channel. In fact, from comparisons with detailed Monte Carlo simulations, the Er atom was determined to be in the position shown in Fig. 18 within 0.1 Å.[126]

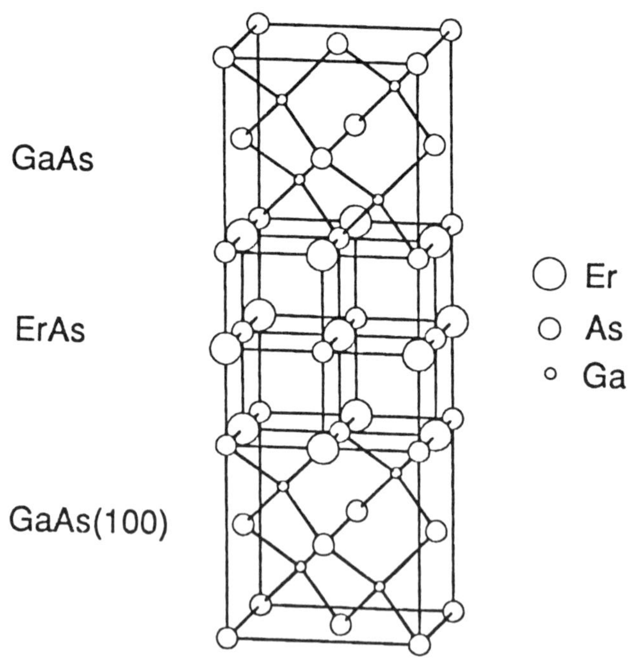


**Figure 18.** Proposed GaAs/ErAs/GaAs(100) lattice alignment. The As-sublattice is continuous across the interfaces.

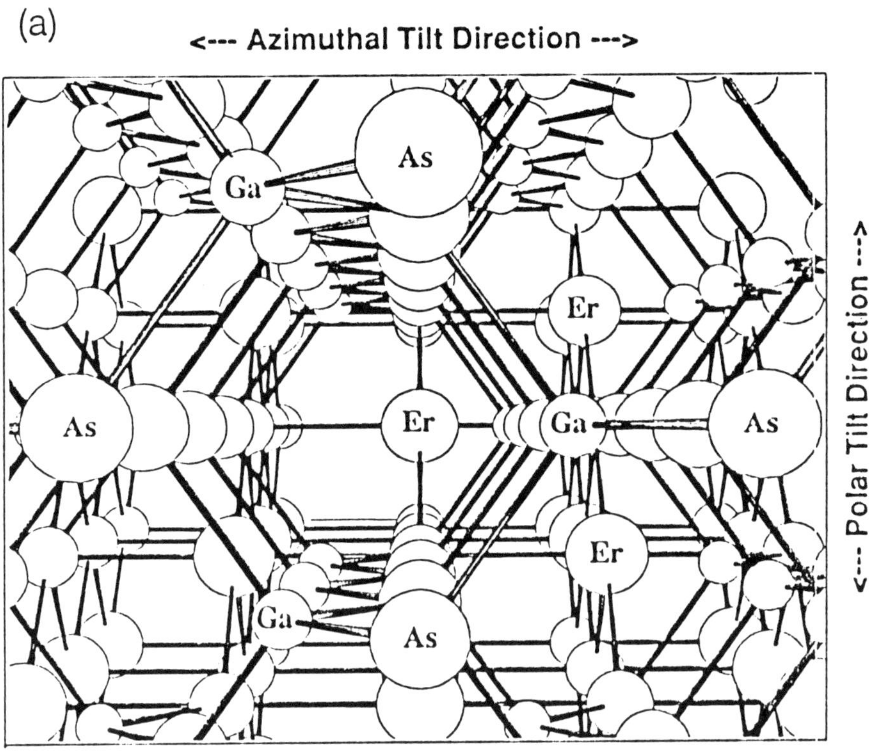


**Figure 19** ***(a).*** A perspective view along the [011] direction for a slab of GaAs(001) on top of an ErAs(001) slab. The registry of the two layers is such that the As-sublattice joins continuously. Note that the Er atom is not centered in the [011] channel. The Ga-atomic row on the right will shadow the Er atoms more than the As-atomic row on the left.

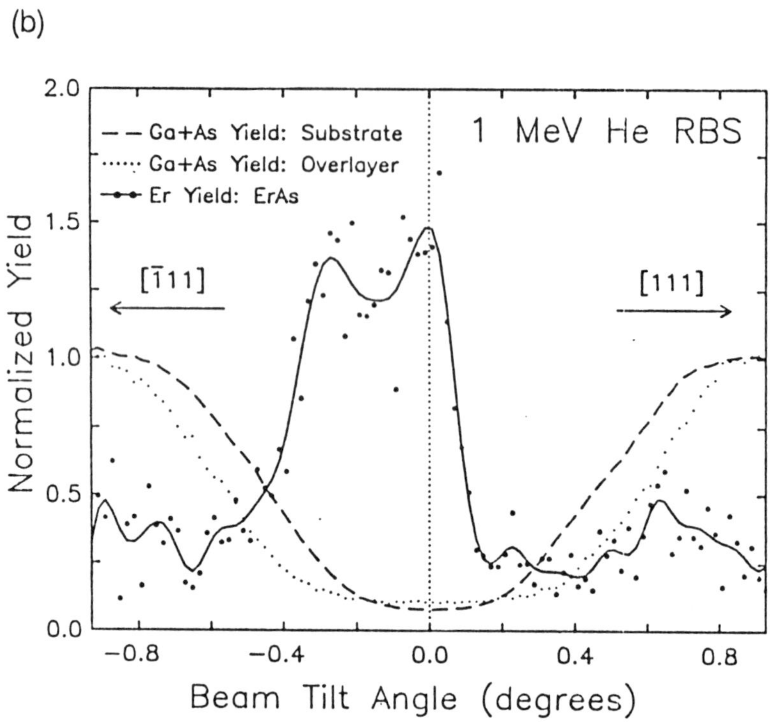


**Figure 19** ***(b)***. RBS yield as a function of azimuthal angle for a scan of the ion beam through the [011] axis. The positive tilt direction is taken to be towards the [111] axis. The sample has an ErAs layer of 0.5 ML average thickness with a 200 Å GaAs cap layer. The Er yield is not centered at the minimum of the Ga plus As yields. The double peak in the Er yield results from focusing of the incident $He^+$-ion beam onto the Er atom in the channel by the As and Ga atomic rows. This rocking curve is consistent with the lattice alignment shown in *(a)* and Monte Carlo simulations of the channeling are consistent with the Er atom being in the position as shown in *(a)* to within 0.1Å. (From Ref. 126.)

Since ErAs is not lattice matched to GaAs, one expects to see lattice relaxation for films thicker than the critical thickness. For films thicker than ~70 Å the ErAs lattice relaxes with film thickness, exhibiting a change in curvature around ~300 Å and eventually reaching the bulk value at ~500 Å (Fig. 20). Thus the critical film thickness is ≤ 70 Å. The low growth temperatures used (≤ $0.22T_m$) explain the slow relaxation kinetics. No misfit dislocations are observable by TEM for films 23 Å[127] and 37 Å,[128] but they are easily detected in films 300 Å thick.[127]–[129] Most of the misfit dislocations are characteristic of the NaCl structure. They are essentially along [010] and [001] directions with Burgers vectors *a*/2<101> or *a*/2<10$\bar{1}$>.[127] The average dislocation spacing (~350 Å) is about twice that anticipated for a completely relaxed film (~180 Å), indicating that the film is only partially relaxed. Thicker films result in increased dislocation densities and more irregular dislocations.[127] During the relaxation the ErAs film tilts with respect to the GaAs. Ion channeling has shown no tilt for 150 Å films, but ~0.26° tilts for 300 Å thick films.[122] X-ray diffraction also substantiates this. Ralston et al.[39] reported tilting for very thick ~3000 Å ErAs films from x-ray diffraction data. Ion channeling[122] and x-ray Bragg reflectivity[123][130] indicate that the relaxation is not uniform through the ErAs film. Ion channeling shows, for example, that the tilting was less at the ErAs/GaAs interface than at the surface.[122] This may result from the increase in the dislocation irregularity with thickness.

Associated with the lattice relaxation is an increase in the film mosaic. Figure 20b shows the film thickness dependence of the in-plane mosaic, $\Delta\phi$, defined as the full width to half maximum (FWHM) peak width obtained at the (222) reflection by rotating the sample about an axis normal to the film plane. Below 70 Å the peak is extremely sharp but the width increases rapidly between 70 and 300 Å with $\Delta\phi$ saturating to approximately 0.8°. $\Delta\phi$ will depend upon the dislocation density[131], $\rho$, so that Fig. 20b is indicative of $\rho$. It is interesting to note that the $\Delta\phi$ has attained 90% of maximum around 300 Å, which implies that the dislocation density causing the in-plane rotations is saturating and lattice relaxation above 300 Å must involve a process which does not couple to the in-plane rotations. The out-of-plane mosaic is obtained at the {002} reflection by rotating the sample about the axis normal to the diffraction plane. In contrast to the in-plane mosaic which yielded the expected Gaussian line shape, the out-of-plane line shape displays two contributions, a sharp specular component (~0.002° FWHM) and a much broader, diffuse part (~0.8° FWHM), as shown in Fig. 21. Longitudinal scans perpendicular to the plane of the film for both the sharp and broad components exhibit the same intensity profile indicating that each arises from the entire film thickness. The sharp component is resolution limited, implying flatness over a correlation range of at least several microns, whereas the diffuse part arises from short range orientation order associated with the misfit dislocations. As the film thickness increases, the specular component remains resolution limited, however, the intensity in this component decreases (Fig. 21). The integrated specular intensity divided by the square of the film thickness is plotted in Fig. 20c which shows that the specular scattering falls rapidly towards 300 Å. It appears, therefore, that the ErAs lattice planes "roughen". This change in the film microstructure may be responsible for the change in surface roughness described above.

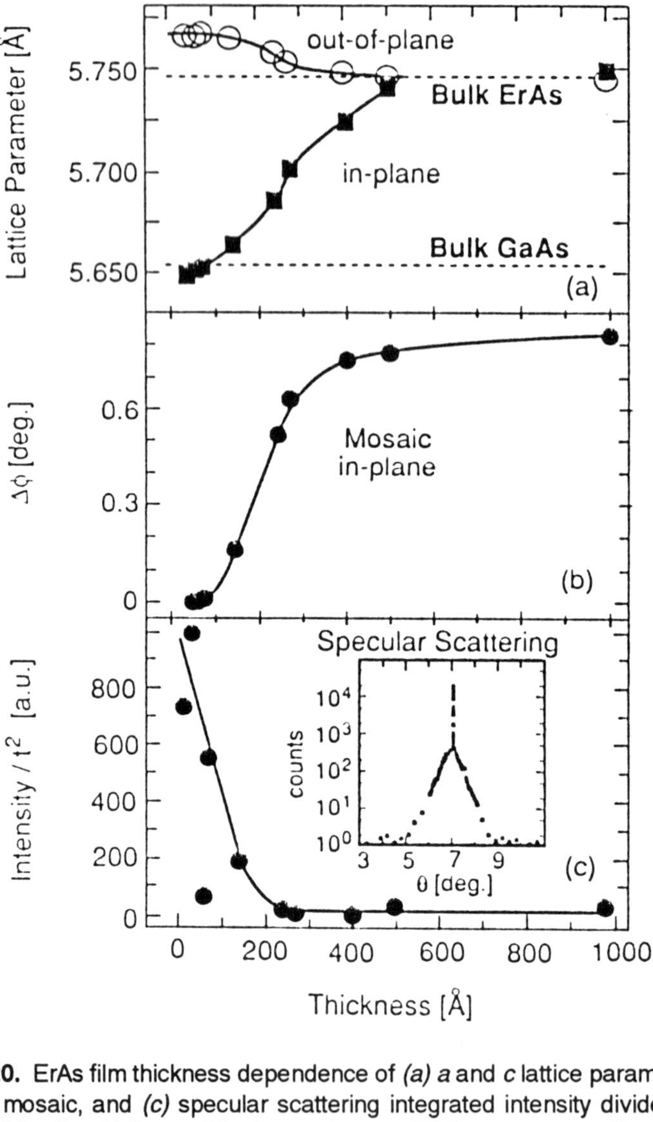


**Figure 20.** ErAs film thickness dependence of *(a) a* and *c* lattice parameters, *(b)* in-plane mosaic, and *(c)* specular scattering integrated intensity divided by the square of the film thickness. The insert shows a (002) transverse reflectivity scan for a 72 Å ErAs film. The in-plane mosaic reaches 90% of its final value for a 300 Å film while the in-plane lattice parameter indicates only 50% relaxation of the film. The change in lattice parameter with thickness shows an inflection point for this thickness. This suggests that the relaxation mechanism changes for films ≥ 300Å. The (002) transverse reflectivity scan indicates that the out-of-plane mosaic contains two components: a sharp specular component (~0.002° FWHM), and a much broader, diffuse part (~0.8° FWHM). *(Courtesy of P. F. Miceli, Bellcore.)*

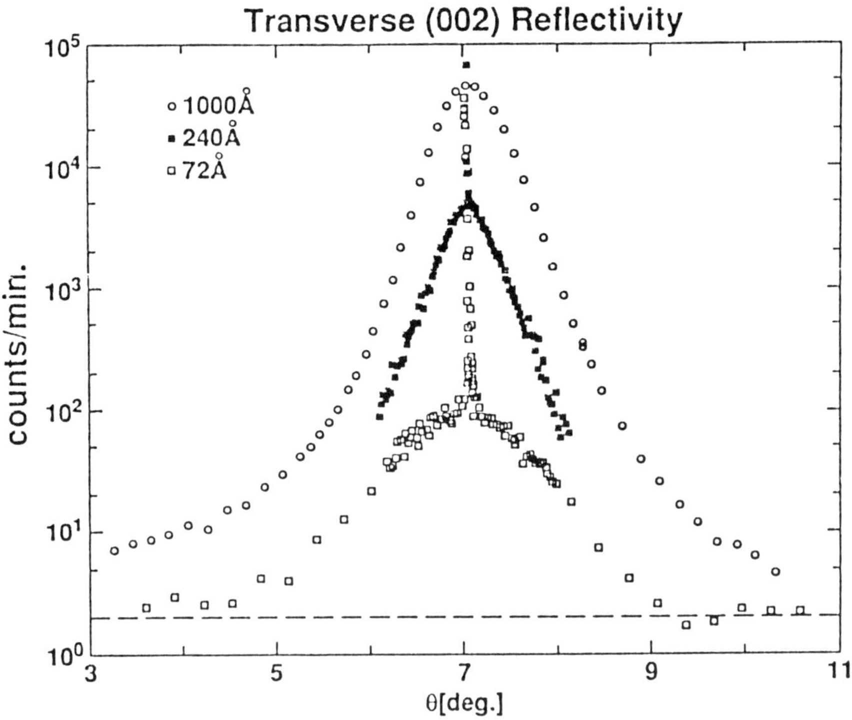


**Figure 21.** Transverse scans taken at the center of the ErAs reflectivity for different film thickness. A sharp component (~0.002° FWHM) and a broad diffuse component (~0.8° FWHM) are observed. The sharp component is resolution limited, implying flatness over a correlation range of at least several microns, whereas the diffuse part arises from short range orientation order associated with the misfit dislocations. As the film thickness increases, the specular component remains resolution limited, however, the intensity in this component decreases and the intensity in the broad diffuse component increases. *(Courtesy of P. F. Miceli, Bellcore.)*

Improvements in film quality should be possible if lattice matched growth is achieved. The period of the RHEED oscillations can be used for determining the rare-earth metal flux (as one does for the group III flux in III-V semiconductor growth). The oscillations for the lattice matched $Sc_{0.32}Er_{0.68}As$ growth appear stronger than for the mismatched ErAs (compression) and ScAs (tension) films, which is indicative of a superior quality film. From x-ray diffraction measurements the mismatch is estimated to be ≤ 0.1%.[34] Lattice matching can also been achieved with $Sc_{0.2}Yb_{0.8}As$[47][96] and $ErP_{0.6}As_{0.4}$[42]-[45] on GaAs and $ErP_{1-y}Sb_y$[35]) on InP.

The mechanical properties of the metallic layers are important for estimating stress and critical thickness for mismatched layers, which directly relates to device reliability. The ratio of the elastic constants, $C_{12}/C_{11}$, of ErAs has been determined from the tetragonal distortion obtained from the data in Fig. 20a.[132][133] A $C_{12}/C_{11}$ ratio of 0.126 is small compared to compound III-V semiconductors which typically exhibit values near 0.5. However, it is similar to a value $C_{12}/C_{11}$ = 0.11 for PrSb[132][133] which has been extracted from the work of Mullen et al.[134] By measuring the lattice parameter of a thin (72 Å) ErAs layer on GaAs[133] as a function of temperature the thermal expansion coefficient for the ErAs layer has been determined to be ~12 x $10^{-6}$ °$K^{-1}$, which is about twice as large as for GaAs.

## 4.2 Semiconductor Overgrowth on RE-V/III-V(100)

The initial stages of GaAs growth on $Sc_{1-x}Er_xAs$(100) have been studied in detail in situ using RHEED, LEED and Auger electron spectroscopy[135][136] and ex situ using TEM.[137][138] The RHEED diffraction patterns obtained for various GaAs thicknesses deposited at 480°C is shown in Fig. 22. The diffraction patterns obtained from the 1 ML thick GaAs shows no 3-D diffraction features. RHEED and LEED indicate a (1x3)/(3x1) reconstruction for this surface and Auger spectroscopy measurements indicate wetting of the $Sc_{1-x}Er_xAs$. Once more than 1 ML of GaAs is deposited, distinct 3-D diffraction features become visible and the Auger suggests incomplete GaAs coverage on the $Sc_{1-x}Er_xAs$ surface. The RHEED patterns show curved lines when the electron beam is incident along <010> direction, but straight lines when incident along <011> and <0$\bar{1}$1> directions. The curved lines are interpreted as resulting from diffraction from needle shaped islands which are oriented along the <011> and <0$\bar{1}$1> directions. The length of the curved lines observed depends upon the growth temperature. At ~330°C they are not as long and well developed. TEM studies confirm this interpretation. Plan-view TEM images of GaAs islands after 20 ML GaAs growth at 330°C and 480°C are shown in Fig. 23. As expected, higher growth temperatures result in a lower density of larger islands and more needle like islands.

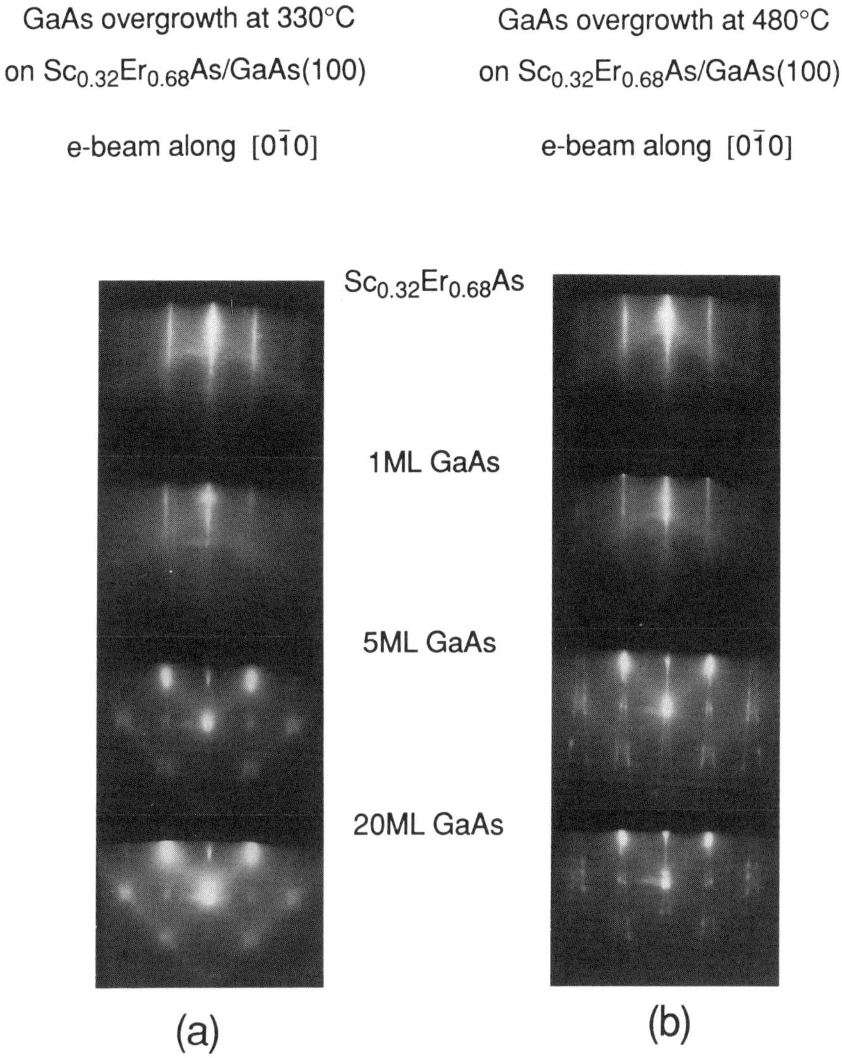


**Figure 22.** RHEED diffraction patterns with the electron beam along $[0\bar{1}0]$ direction obtained from $GaAs/Sc_{1x}Er_xAs(100)$ for GaAs grown at *(a)* 330°C and *(b)* 480°C with GaAs thickness of 0 ML, 1 ML, 5 ML and 20 ML. 3-D like diffraction features and curved streaks appear for GaAs films > 1 ML. The curved streaks are observed with the electron beam along $[0\bar{1}0]$ and not with the beam along the <011> and <0$\bar{1}$1> directions. These streaks result from diffraction from needle shaped islands in <011> and <0$\bar{1}$1> directions. The films grown at the higher temperature have sharper and more pronounced curved streaks, indicating larger and more needle shaped islands.

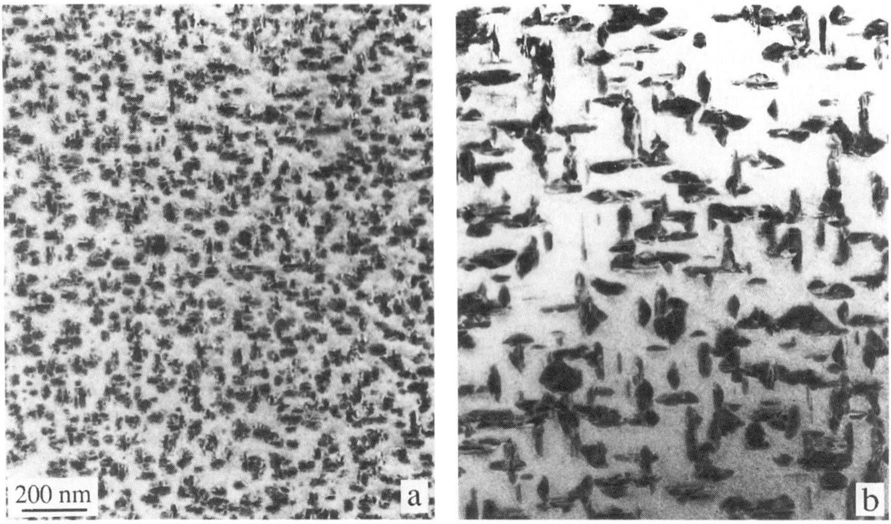


**Figure 23.** Plan-view bright field TEM images showing GaAs island formation resulting from 20 ML GaAs growth on top of $Sc_{1-x}Er_xAs(100)$. The GaAs growth was at (a) 330°C and (b) 480°C. The needle shape and island size increases, and the density decreases with growth temperature. *(Courtesy of J. G. Zhu and C. B. Carter, Cornell University.)*

For growth >500°C, no GaAs wetting of the $Sc_{1-x}Er_xAs$ occurred and the growth mode can easily be described as Volmer-Weber. The growth at <500°C does not, however, correspond to any current simple theory of epitaxial growth. A schematic of the growth sequence is shown in Fig. 24. Stranski-Krastanov growth would predict that the wetting layer remains on the surface after the islands start to form. This is not what is observed. The simplest explanation is that the first ML of GaAs is a metastable phase. Finstad et al.[135][136] propose that the bonding in this layer is mostly within the layer, with no dangling bonds out of the surface and only a few bonds to the $Sc_{1-x}Er_xAs$, which act as anchor points for the orientation relationship to the substrate. The type of bonding within the layer could be $sp^2$-like. The 3-D island growth sets in when nuclei with GaAs covalent $sp^3$-bonds form. When the metastable (1x3)/(3x1)GaAs surface layer has been built, the nuclei can form by increasing the super-saturation of Ga and As atoms on top of the (1x3)/(3x1) layer.

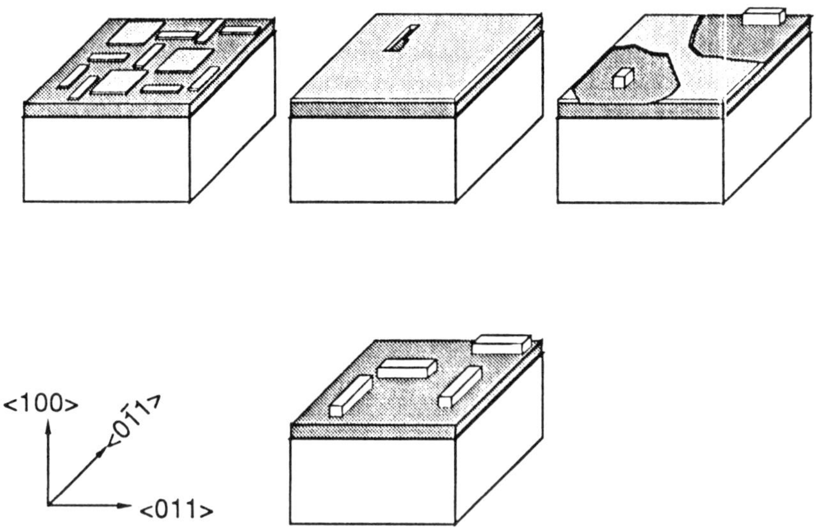


**Figure 24.** Schematic of GaAs growth on $Sc_{1-x}Er_xAs(100)$ at growth temperatures <500°C. The first monolayer of GaAs wets the $Sc_{1-x}Er_xAs$ surface. Once islands are formed this layer de-wets the surface and is incorporated in the islands. The islands are needle shaped growing in $\langle 011 \rangle$ and $\langle 0\bar{1}1 \rangle$ directions.

Cross-sectional TEM micrographs of the islands are shown in Fig. 25. The islands formed for the 480°C growth (Fig. 25b) show distinct faceting, with {111} and {110} facets. These surfaces have lower energy than {100}. This suggests that the islands are minimizing the GaAs surface energy. It is interesting to note that there is substantial twinning within the GaAs island. The island in Fig. 25a contains well aligned material in the center and singly and doubly twinned material on both the left and right. At the interface between the twinned GaAs and $Sc_{1-x}Er_xAs$, a different phase boundary has been found, {122} GaAs on (100) $Sc_{1-x}Er_xAs$ with $\langle 01\bar{1} \rangle$ GaAs parallel to $\langle 01\bar{1} \rangle$ $Sc_{1-x}Er_xAs$ and $\langle \bar{4}11 \rangle$ GaAs parallel to $\langle 011 \rangle$ $Sc_{1-x}Er_xAs$, which has the same misfit as (100) GaAs on (100) $Sc_{1-x}Er_xAs$. For growth temperatures ≥ ~400°C, {111} oriented GaAs grains have also been observed.[128] The cross-sectional image of a GaAs island grown on $Sc_{1-x}Er_xAs(100)$ at 480°C is shown in Fig. 25b. On the left hand side of the island, {122} GaAs has grown on (100) $Sc_{1-x}Er_xAs$ with the same orientation relationship described for the 330°C growth. The GaAs on the right hand side of the island has epitaxial relationship of {111} GaAs on (100) $Sc_{1-x}Er_xAs$ with $\langle \bar{2}11 \rangle$ GaAs parallel to $\langle 011 \rangle$ $Sc_{1-x}Er_xAs$ and $\langle 01\bar{1} \rangle$ GaAs parallel to $\langle 01\bar{1} \rangle$ $Sc_{1-x}Er_xAs$. The {111} planes are bent gradually and the orientation relationship becomes {122} GaAs/(100)$Sc_{1-x}Er_xAs$ in the area close to the center. The angle between <122> and <111> is 15.8°. The {111} planes, therefore, curve through ~16° across ~340Å, which indicates that small islands can tolerate a very large amount of strain. Only a very small part of this island shows the expected {100} GaAs on (100) $Sc_{1-x}Er_xAs$ and it is close to the center of the island.

It is interesting to speculate how these {122} twinned crystallites are formed. An explanation of the formation of {122} crystallites is that they are created by twins on {111} faceted islands during growth. One can also speculate that the (1x3)/(3x1) reconstructed layer could either be a precursor or play a role in their formation. The {122} GaAs and (100) $Sc_{1-x}Er_xAs$ have a coincidence lattice with a periodicity of 3x1 on the (100) $Sc_{1-x}Er_xAs$ plane.[135][137][138] All islands studied were found to contain {122} oriented grains even for only 5 ML of GaAs deposition.[139] This together with the strong tendency for the formation of the lowest energy {111} and {110} facets (Fig. 25b) suggests that the island energy is reduced by twinning to allow minimal surface area and the formation of low energy surface facets.

As growth proceeds, the {122} oriented grains twin back to (100) epitaxial material by twinning on {111} planes. This occurs by ~200 Å of GaAs growth[128] (Fig. 26). The {111} oriented grains cannot twin back to (100) and therefore remain {111} during growth. The growth of GaAs on (100) $Sc_{1-x}Er_xAs$ tends to result in a rough surface morphology due to the island growth mode. The smoothest surface is obtained by growth at lower temperature, where the island size is small and density large. For ultrathin ErAs layers ~1 - 2 ML thick, the surface of the overgrown GaAs is very flat. The major defects in the overgrown GaAs is stacking faults. A stacking-fault pyramid density of about $5 \times 10^9$ $cm^{-2}$ was observed for a 1 ML ErAs film.[137] This change in GaAs growth is an indication of seeding through pinholes in the ultrathin ErAs film. The sudden onset of twins for growth on ≥ 3 ML $Sc_{1-x}Er_xAs$ films is consistent with a sudden increase in the channeling minimum yield for GaAs/ErAs/GaAs(100) structures for ErAs films ≥ 3 ML shown in Fig. 15.

In Sec. 2 it was suggested that the use of vicinal surfaces will reduce the symmetry of the metal surface and, therefore, result in improved III-V semiconductor overgrowth. By using (100) offcut by 6° towards (111) the overgrown GaAs had a substantial reduction in {122} oriented grains[139] as compared with (100) surfaces. This has also been observed for overgrown GaAs on $Sc_{1-x}Yb_xAs$ on (100)-4° vicinal samples.[47] In general, however, the poor GaAs wetting of the RE-V(100) surface results in poor crystal quality of the GaAs grown on RE-V(100) surfaces.

(a)

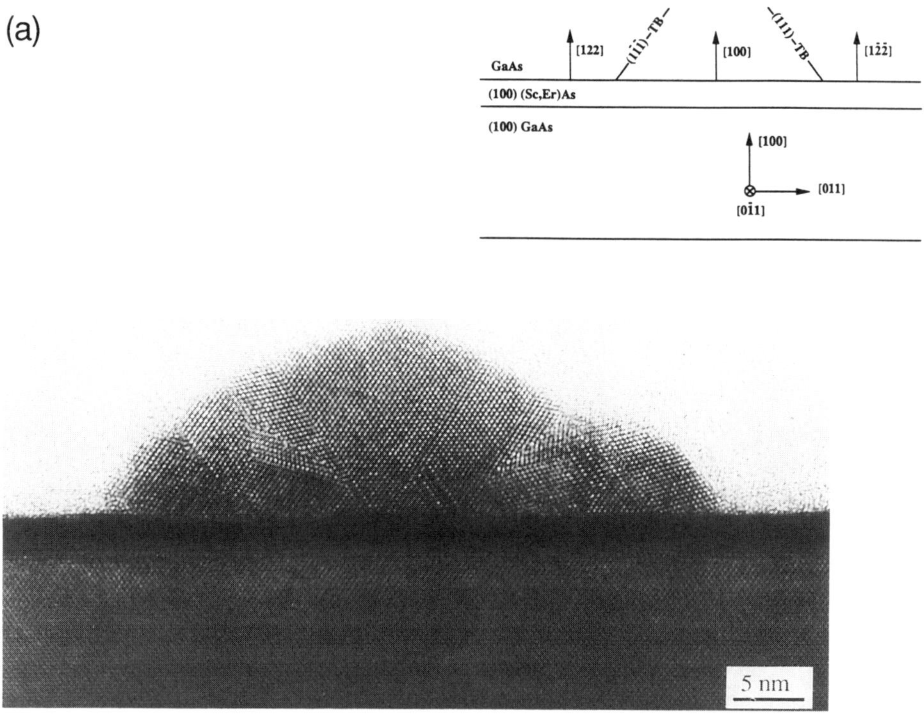


**Figure 25.** $\{0\bar{1}\bar{1}\}$ cross-sectional high-resolution images of GaAs islands formed after 20 ML of GaAs growth on $Sc_{1-x}Er_xAs(100)$ at (a) 350°C and (b) 480°C. Clearly 20 ML of GaAs does not wet the $Sc_{1-x}Er_xAs(100)$ surface. The islands are twinned as is evident from the schematic insert for each figure. The insert in (a) shows the twin relation between (100) and {122}. The insert in (b) shows the schematic of the island, which consists of four twin related grains (A-D). Region X of grain C has the epitaxial relationship $(1\bar{1}\bar{1})$ GaAs on (100) $Sc_{1-x}Er_xAs$ with $[211]_{GaAs} \parallel [01\bar{1}]_{Sc_{1-x}Er_xAs}$ (or $[01\bar{1}]_{GaAs} \parallel [01\bar{1}]_{Sc_{1-x}Er_xAs}$). The $(1\bar{1}\bar{1})$ lattice planes in grain C bend gradually from $[1\bar{1}\bar{1}]$-oriented region above X to the $[1\bar{2}\bar{2}]$-oriented region above Y. The angle between $[1\bar{2}\bar{2}]$ and $[1\bar{1}\bar{1}]$ is 15.8°. The $(1\bar{1}\bar{1})$ planes therefore curve through almost 16° across 340 Å, which indicates that a small island can tolerate a large amount of strain. The {111} twin boundaries are labeled with T. There is only a small amount of (100) GaAs, grain D, in this island. The surface of the island is faceted on {111} and {110} lattice planes, which are low energy surfaces. The lower temperature growth results in {100} and {122} oriented GaAs on $Sc_{1-x}Er_xAs(100)$. Growth at ≥ 450°C results in {111} oriented GaAs in addition. The islands grown at higher temperature are more faceted than the ones grown at lower temperature. *(Courtesy J. G. Zhu and C. B. Carter, Cornell University.)*

(b)

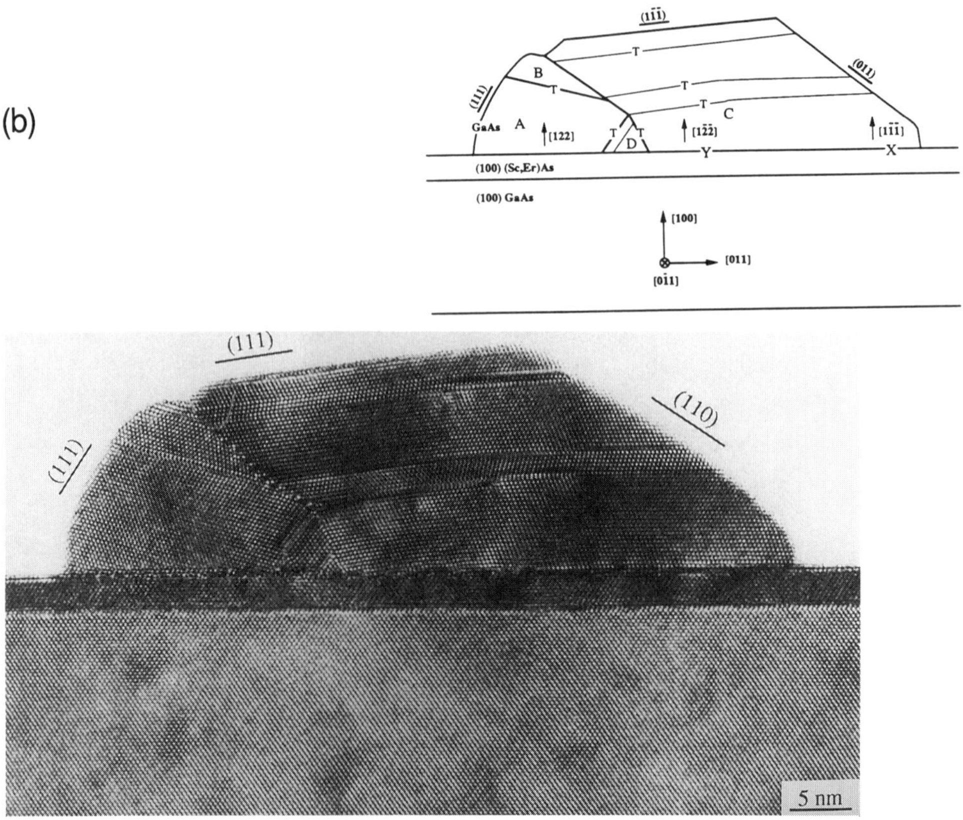


**Figure 25.** (Continued)

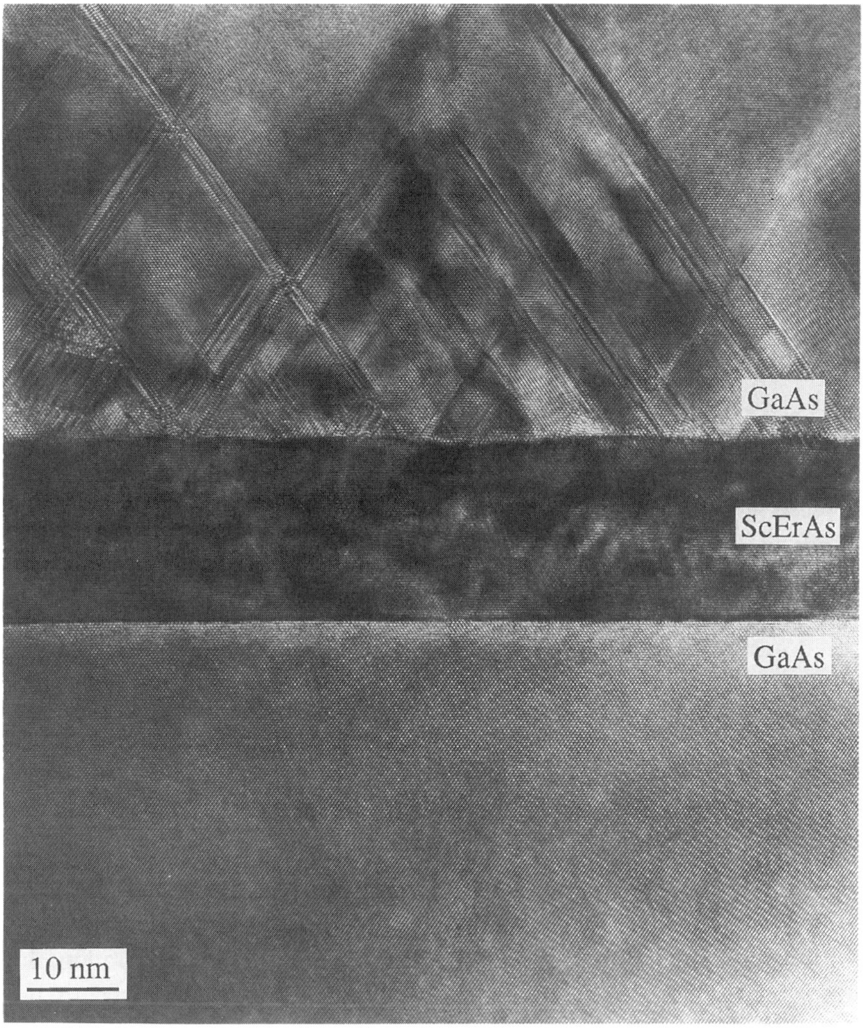


**Figure 26.** Cross-sectional TEM image of GaAs/$Sc_{0.32}Er_{0.68}As$(200Å)/GaAs(100). The bottom $Sc_{1-x}Er_xAs$/GaAs interface is atomically abrupt. The $Sc_{1-x}Er_xAs$ surface is somewhat rough. Stacking faults and microtwins are clearly evident in the overgrown GaAs. The {122} grains twin back to {100}-oriented growth within the first ~200 Å of GaAs growth, resulting in stacking faults as the major defect further away from the interface. *(Courtesy J. G. Zhu and C. B. Carter, Cornell University.)*

## 4.3 Growth of III-V/RE-As/III-V on {h11} III-V Orientations

It is expected from the thermodynamic arguments in Sec. 2.3 that GaAs wetting on $Sc_{1-x}Er_xAs(100)$ would be a problem. From these arguments, the wetting can only be improved by either increasing the $Sc_{1-x}Er_xAs$ surface energy or decreasing the GaAs surface energy. An analogy can be made with NaCl, whose nonpolar (100) surface has a free energy of 150 ergs/cm$^2$[140] and the polar (111) surface 872 ergs/cm$^2$,[140] which indicates that the $Sc_{1-x}Er_xAs$ {111} surfaces can be expected to have a much higher surface free energy than the (100). Another approach is to look at the details of the interface structures. Figure 27 shows schematic diagrams for possible GaAs/$Sc_{1x}Er_xAs$ interfaces for (100) and {111} orientations. The stacking sequence for $Sc_{1-x}Er_xAs(100)$ is layers of $Sc_{1-x}Er_xAs$, each of which is non-polar. The GaAs stacking sequence is alternating layers of Ga and As, resulting in a polar surface. Hence, with this orientation one is trying to grow a polar material on a non-polar substrate. This is difficult since there must be a charge imbalance at the interface. Further inspection also indicates a difficult bonding configuration. The GaAs being primarily covalently bonded and the $Sc_{1-x}Er_xAs$ ionically bonded adds another interface problem. The bonding configuration drawn schematically in Fig. 27a suggests that the interface As atoms are 7-fold coordinated and that they are covalently bonded to Ga in the GaAs and ionically bonded to As in the $Sc_{1-x}Er_xAs$. This type of bonding configuration is not desirable.

In the {111} case, however, the situation is much better. The stacking sequence for $Sc_{1-x}Er_xAs$ consists of alternating planes of $Sc_{1-x}Er_x$ and As, which will result in polar surfaces. Similarly, for GaAs the stacking sequence is alternating planes of Ga and As, again yielding polar surfaces. The interface structure proposed in Fig. 27b shows the interface As atoms 6-fold coordinated with a much more desirable bonding configuration. In this figure the GaAs has been drawn as a type-B (180° rotated) twin on the $Sc_{1-x}Er_xAs(111)$. This gives a more reasonable bonding configuration than for the type-A orientation. Although the bonding of the first Ga layer to the As in $Sc_{1-x}Er_xAs$ is not known, the proposed bonding configuration is close to $sp^3$, which is the covalent bonding in GaAs. Hence, the next As layer is probably $sp^3$ bonded. From this simplistic approach, one would expect the interface bonding to be substantially more inducive for GaAs to wet $Sc_{1-x}Er_xAs$ in the {111} than the (100) case.

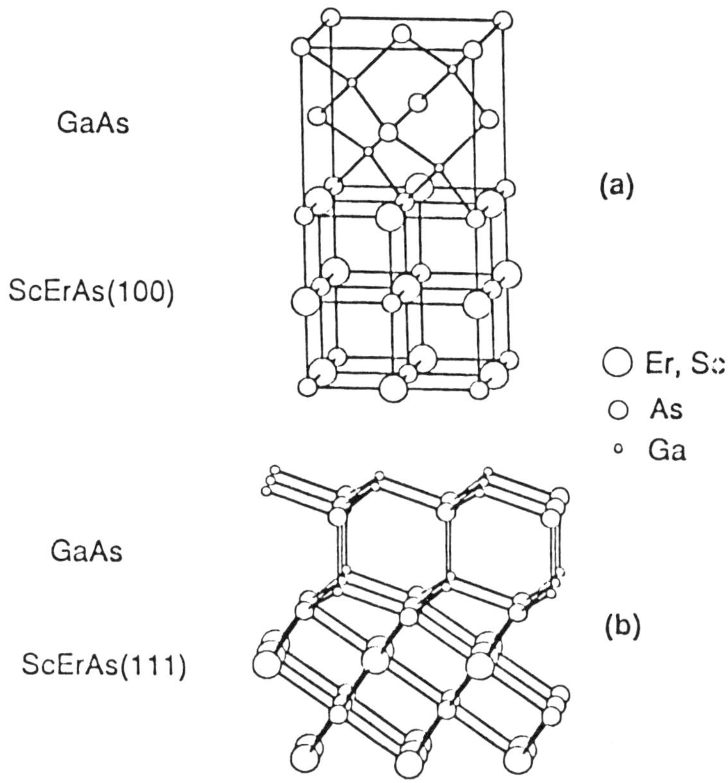


**Figure 27.** Schematic diagrams of possible GaAs/$Sc_{1-x}Er_xAs$ interfaces:

*(a)* GaAs/$Sc_{1-x}Er_xAs$(100). The stacking sequence for $Sc_{1-x}Er_xAs$(100) consists of layers of $Sc_{1-x}Er_xAs$, each of which is non-polar. The GaAs stacking sequence is alternating layers of Ga and As, resulting in a polar surface. Hence, with this orientation the growth is of a polar material on a non-polar substrate. The bonding configuration drawn schematically in (a) suggests that the interface As atoms are 7-fold coordinated and that they are covalently bonded to Ga in the GaAs and ionically bonded to As in the $Sc_{1-x}Er_xAs$. This type of bonding configuration is not desirable.

*(b)* GaAs/$Sc_{1-x}Er_xAs$(111). In this case, the interface charge imbalance is much better. The stacking sequence for $Sc_{1-x}Er_xAs$ consists of alternating planes of $Sc_{1-x}Er_x$ and As, which will result in polar surfaces. Similarly, for GaAs the stacking sequence is alternating planes of Ga and As, again yielding polar surfaces. The interface structure proposed in (b) shows the interface As atoms 6-fold coordinated with a much more desirable bonding configuration than in (a). The GaAs has been drawn as a 'type-B' (180° rotated) twin on the $Sc_{1-x}Er_xAs$(111). This gives a more reasonable bonding configuration than for the 'type-A' orientation.

Although in practice considerable improvement in the GaAs wetting of the $Sc_{1-x}Er_xAs$ is found for {111} surfaces, this is unfortunately the twinning plane for GaAs and $Sc_{1-x}Er_xAs$ and substantial twinning during growth is observed. This is discussed in more detail below. However, by using surfaces which have both {111} and {100} character, a tradeoff between wetting and twinning may be achieved. Figure 28 shows schematic diagrams of bulk termination for a number of different surfaces. Surfaces of type {h11} consist of terraces and ledges of {100} and {111} character (Fig. 28b). These diagrams assume bulk termination, which is unlikely to be the case for semiconductors. These surfaces will most likely reconstruct, the reconstruction depending upon temperature, As/Ga flux ratio and orientation.

The optimum growth conditions for GaAs and the $Sc_{1-x}Er_xAs$ depend upon the substrate orientation. For (100), (100)-2°, (100)-6°, $(\bar{7}\bar{1}\bar{1})$, $(\bar{5}\bar{1}\bar{1})$, and $(\bar{3}\bar{1}\bar{1})$ substrates, a substrate temperature of 580-600°C was used for GaAs growth. Growth of $(\bar{1}\bar{1}\bar{1})$ GaAs has been studied recently by a number of groups.[141]-[147] In general, optimum growth occurs with low As/Ga flux ratio. The use of vicinal substrates also improves the growth.[141][142][144] This suggests that the preferable growth mode is by step flow rather than by layer-by-layer growth. In this study nominal $(\bar{1}\bar{1}\bar{1})$ GaAs substrates were used. It was found that by growing at high substrate temperatures (~700°C), twin-free material could be grown. A similar growth temperature was found to work for $(\bar{2}\bar{1}\bar{1})$. For (110) growth a substrate temperature of ~510°C, a 0.5 $\mu$m/hr growth rate and As/Ga flux ratio ~6 were used. These conditions are similar to the optimum conditions reported by Pfeiffer et al.[148] RHEED oscillations were observed for $Sc_{1-x}Er_xAs$ growth on (100) surfaces for substrate temperatures ≤ 450°C. For higher substrate temperatures and on other substrate orientations, RHEED oscillations were not observed. This suggests that the growth mode changes to step flow or more 3-D type growth. As in the case of GaAs growth, $Sc_{1-x}Er_xAs$ growth appeared from RHEED to be smoother at higher substrate temperatures for $(\bar{1}\bar{1}\bar{1})$ and $(\bar{2}\bar{1}\bar{1})$.

In all cases the films were epitaxial as determined by RHEED and LEED, and also substantiated by TEM and RBS. The crystalline quality of the lattice matched $Sc_{1-x}Er_xAs$ was dependent upon the substrate orientation. In the case of $(\bar{1}\bar{1}\bar{1})$ growth, the $Sc_{1-x}Er_xAs$ films tended to be heavily twinned as did the overgrown GaAs cap layer. A reduction in twinning may occur on (111) GaAs.[45]

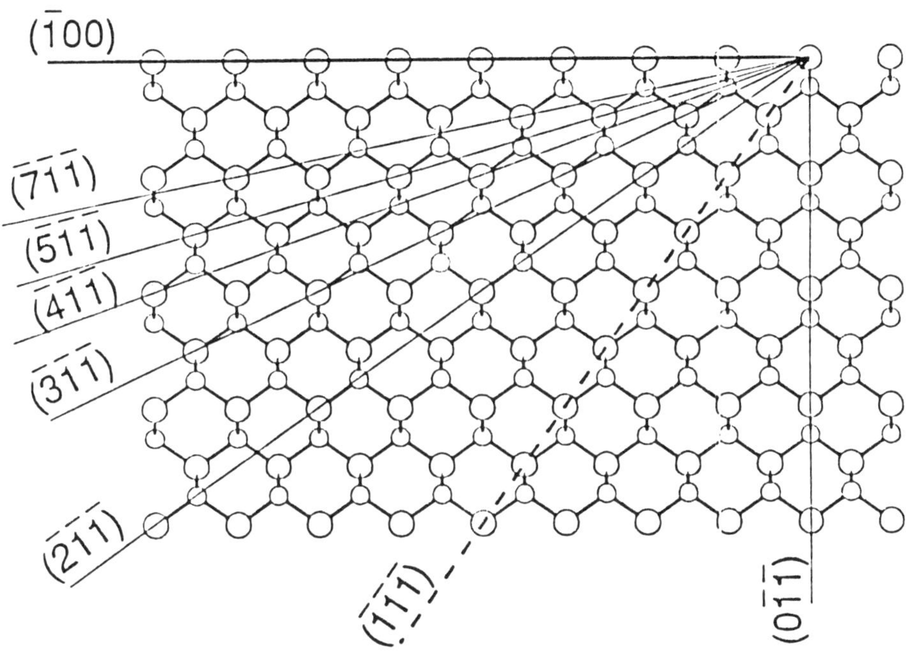


(a)

**Figure 28.** *(a)* {011} cross-sectional view of GaAs. *(b)* {h11} bulk terminated surfaces showing {100} and {111} steps and ledges. The solid lines indicate {100} surfaces and the dashed lines {111} surfaces. By using {h11} surfaces, the degree of {100} and {111} character can be adjusted.

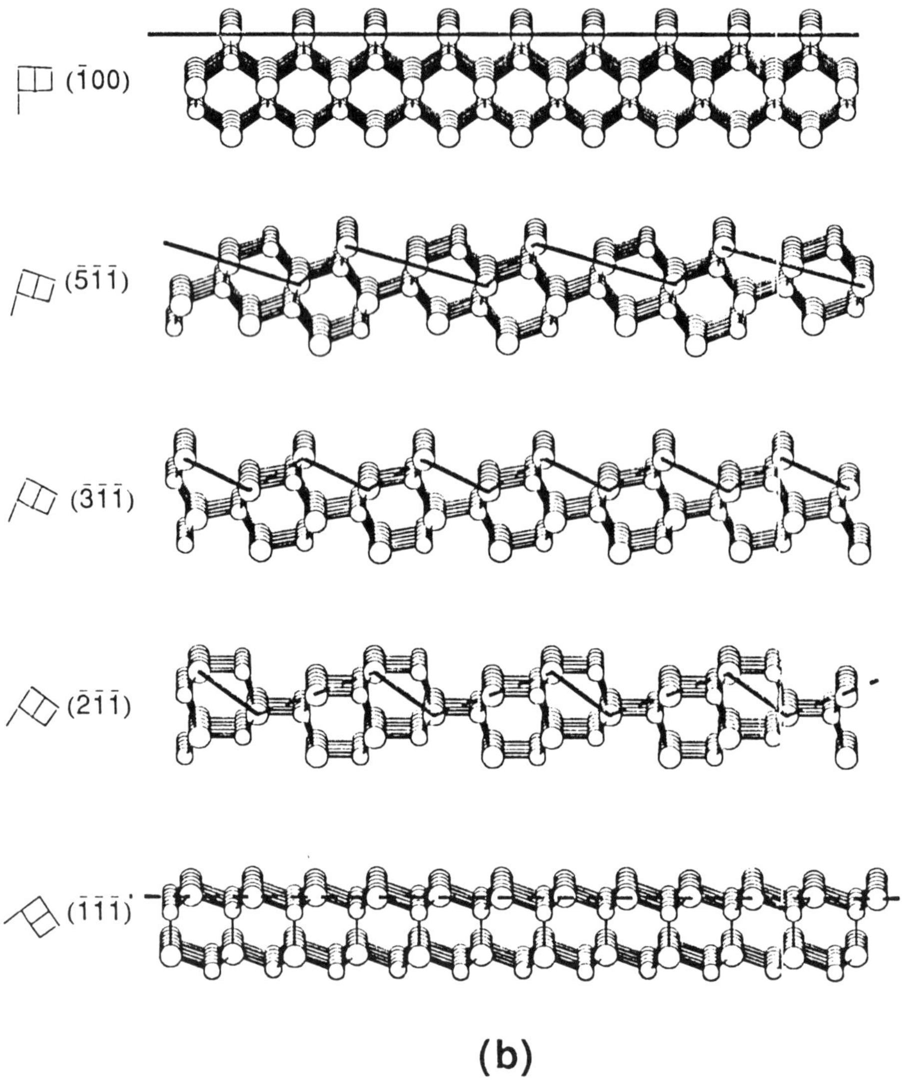


**Figure 28** (Continued)

Figure 29 shows RHEED patterns taken after GaAs and $Sc_{1-x}Er_xAs$ growth for a number of different substrate orientations. The GaAs surfaces tend to be reconstructed. The diffraction patterns for the $Sc_{1-x}Er_xAs$ clearly show a stepped surface, as would be expected for bulk termination. During the initial growth of $Sc_{1-x}Er_xAs$ on {h11} surfaces, the $Sc_{1-x}Er_xAs$ growth tends to be 3-D and RHEED studies suggest that substantial faceting occurs. In general, {100} facets appear to form. Upon further deposition the islands coalesce to produce flat films as are observed by the RHEED patterns in Fig. 29. The degree of faceting and 3-D growth was found to depend upon orientation and growth temperature. For $(\bar{1}\,\bar{1}\,\bar{1})$ and $(\bar{2}\,\bar{1}\,\bar{1})$ the smoothest films were grown at high temperature (~600°C). For the $(\bar{1}\,\bar{1}\,\bar{1})$ surfaces the films tended to be more twinned when grown at high temperature. It has been suggested that this is due to the problems of growing on a ($\sqrt{19}$x$\sqrt{19}$ R23. 4°) Ga-rich surface rather than on the (2x2) As-rich surface[149]. For films grown at ~600°C severe twinning occurred, which results in an interface with mixed twinned and untwinned (180° rotated) $Sc_{1-x}Er_xAs$/GaAs $(\bar{1}\,\bar{1}\,\bar{1})$ analogous to mixed 'type-A' and 'type-B' $NiSi_2$/Si(111). For (100) the c(4x4) As-rich surface is the starting GaAs surface for growths at T<500°C. When $Sc_{1-x}Er_xAs$ is initiated on this surface, the RHEED pattern becomes weak during the first monolayer of growth. By one monolayer of growth the pattern has changed to a (1x1) characteristic of $Sc_{1-x}Er_xAs$. For growth at 580°C the GaAs starting surface has the typical As-rich (2x4)/c(2x8) pattern prior to $Sc_{1-x}Er_xAs$ growth. Once the Sc and Er shutters are opened, the surface reconstruction changes to a (4x2) or c(8x2), which is characteristic of the Ga-rich GaAs surface. After about one monolayer of $Sc_{1-x}Er_xAs$ growth, the characteristic (1x1) pattern for $Sc_{1x}Er_xAs$ is observed. This suggests that at high temperature the deposition of $Sc_{1-x}Er_xAs$ destabilizes the As-rich surface resulting in the loss of As-dimers and the formation of Ga-dimers. Another way of looking at this is to consider the deposition of small coverages of $Sc_{1-x}Er_xAs$ as changing the relative stability of the (2x4)/c(2x8) compared with the c(8x2). For GaAs(100) the transition temperature is ~630°C for the As-rich (2x4)/c(2x8) to convert to the Ga-rich c(8x2). Small quantities of $Sc_{1-x}Er_xAs$ may lower this temperature to below 580°C.

It is difficult to determine at this time how the growth mode and GaAs surface reconstruction prior and during the initial $Sc_{1-x}Er_xAs$ deposition influences the interface structure. Figure 30 shows high resolution cross-sectional TEM micrographs of GaAs/$Sc_{1-x}Er_xAs$/GaAs structures for (100) and (311) GaAs substrates. The (100) interface appears more abrupt than the (311). This may arise from the inherent steps on the surface on the (311) surface. However, in both cases the interface appears abrupt to about 2 atomic layers. Recently, Nötzel et al.[150] suggested from RHEED studies that regular

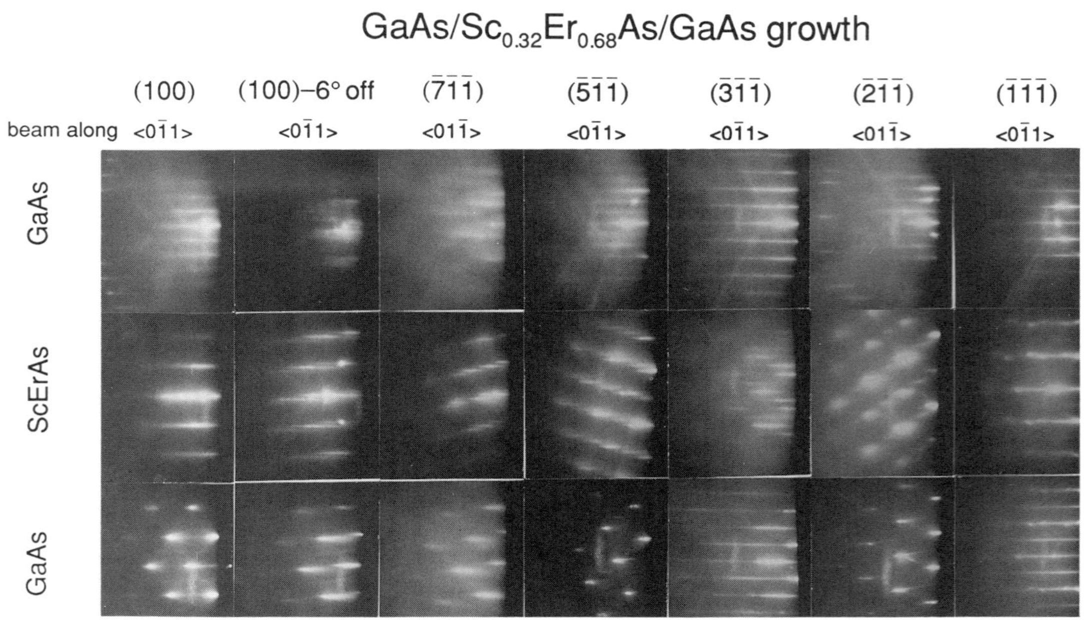


**Figure 29.** RHEED patterns from the as-grown GaAs{h11} surfaces, after $Sc_{1-x}Er_xAs$ deposition and after GaAs overgrowth. The diffraction patterns from the GaAs surfaces prior to $Sc_{1-x}Er_xAs$ suggest surface reconstructions and stepped surfaces. The RHEED patterns from the $Sc_{1-x}Er_xAs$ indicate bulk terminated surfaces with steps and ledges as would be expected from Fig. 28. Generally, the diffraction patterns from the overgrown GaAs show 3-D like features. (100) vicinal surfaces were somewhat smoother than (100) surfaces. Unfortunately, the growth conditions were not optimized for each orientation and smoother surfaces may be expected for $(\bar{5}\bar{1}\bar{1})$ and $(\bar{2}\bar{1}\bar{1})$ orientations. However, the streaks observed for the $(\bar{3}\bar{1}\bar{1})$ overgrown GaAs indicates a relatively smooth overgrown layer and that the GaAs wets the $Sc_{1-x}Er_xAs\{311\}$ surface.

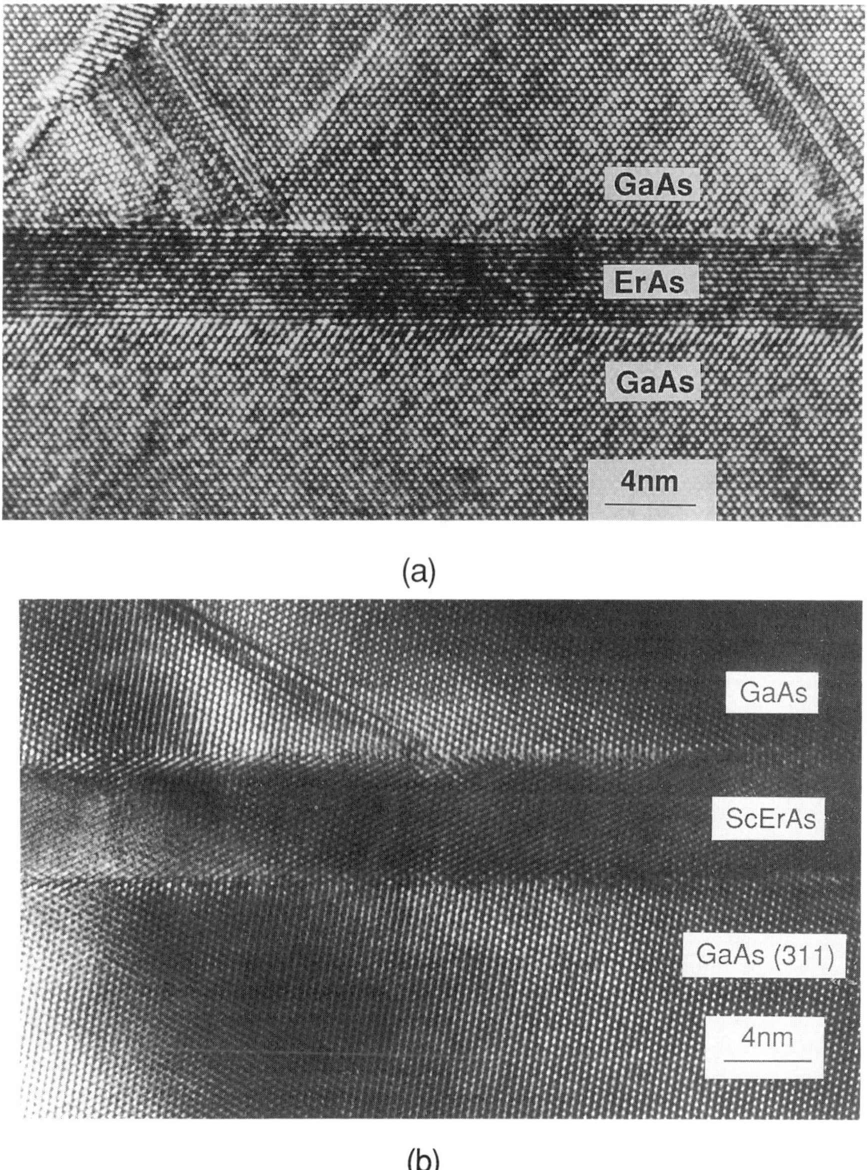


**Figure 30.** High resolution TEM cross-sectional micrographs of GaAs/$Sc_{1-x}Er_xAs$/GaAs interfaces a) (100) (40Å ErAs) and b) (311) (50Å $Sc_{0.32}Er_{0.68}As$). The $Sc_{1-x}Er_xAs$/GaAs interface appears flatter in the (100) than (311) case. However, the overgrown GaAs is of substantially better quality for the (311) growth. Stacking faults are the only defect seen in the overgrown GaAs in the (311) case, but both microtwins and stacking faults are observed in the (100) case. *(Courtesy J. G. Zhu and C. B. Carter, Cornell University.)*

macrosteps run in <$\bar{2}$33> direction, i.e. perpendicular to this <011> cross-sectional view on the GaAs (311) surface. This would also result in a less abrupt interface in this cross-sectional view. These regular macrosteps are not observed for $(\bar{3}\bar{1}\bar{1})$ surfaces. Figure 31 shows RBS-channeling measurements of GaAs $(\bar{3}\bar{1}\bar{1})$ and GaAs/$Sc_{1-x}Er_xAs$/GaAs $(\bar{3}\bar{1}\bar{1})$ structures. For GaAs $(\bar{3}\bar{1}\bar{1})$, the minimum yield, $\chi_{min}$, is ~3% for <111> and <100> channeling and ~7.5% for <311> channeling. The higher minimum yield in the <311> channeling results from the channel being less open in this direction than in <111> and <100> directions. For GaAs/$Sc_{1-x}Er_xAs$/GaAs $(\bar{3}\bar{1}\bar{1})$ structures, the GaAs substrate <111> and <311> channeling is identical to the GaAs $(\bar{3}\bar{1}\bar{1})$, indicating excellent crystalline quality. However, the <100> channeling is slightly poorer, which is evidence of some defects which cause dechanneling in this direction. The presence of a some stacking faults in the overgrown GaAs is evident from the cross-section TEM micrograph in Fig. 30b and these may explain the increased dechanneling in the <100> direction. The quality of the overgrown GaAs on $Sc_{1-x}Er_xAs$\{311\} is substantially better than when grown on $Sc_{1-x}Er_xAs$(100).

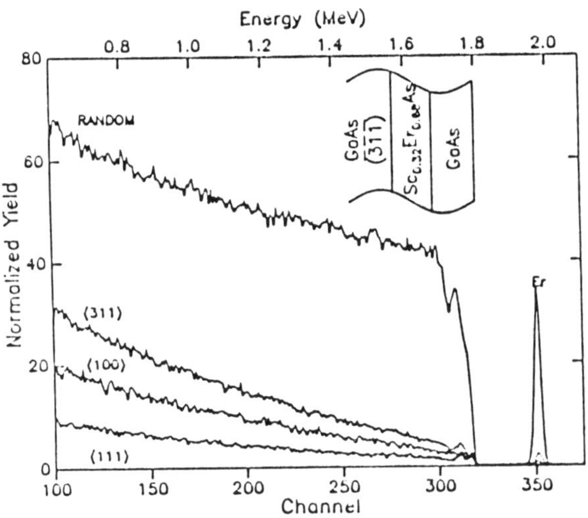


**Figure 31.** Random and aligned RBS spectra obtained from GaAs/$Sc_{1-x}Er_xAs$/GaAs $(\bar{3}\bar{1}\bar{1})$. The channeling minimum yields ($\chi_{min} \leq 3\%$ <111> and <100>, and 6% for <311>) and the dechanneling rates are the same for the GaAs/$Sc_{1-x}Er_xAs$/GaAs $(\bar{3}\bar{1}\bar{1})$ and bulk GaAs $(\bar{3}\bar{1}\bar{1})$ for channeling in the <311> and <111> directions. The dechanneling rate in the GaAs substrate is slightly higher for <100> alignment in the GaAs/$Sc_{1-x}Er_xAs$/GaAs $(\bar{3}\bar{1}\bar{1})$ structure than bulk GaAs $(\bar{3}\bar{1}\bar{1})$ (which is similar to the <111> alignment). The increased dechanneling in the <100> GaAs/$Sc_{1-x}Er_xAs$/GaAs $(\bar{3}\bar{1}\bar{1})$ aligned spectrum may result from the stacking faults in the overgrown GaAs. *(Courtesy B. J. Wilkens, Bellcore.)*

## 4.4 Growth of TM-III Intermetallics on III-V Surfaces: A Template Approach

The first attempts at growing epitaxial metallic compounds on GaAs involved the TM-III intermetallics.[33][50][51][97][99][151]-[153] Because of the significant differences in structure and chemistry between the TM-III intermetallics and the III-V semiconductors, the growth conditions for the metal and the semiconductor are substantially different. Furthermore, the first few monolayers of growth determine the morphology and crystallographic orientation of the metal overlayer. In some cases, the phase that nucleates is also determined by the initial growth conditions. Generally, this template stage of growth must be considered as distinct from the subsequent continuous growth of the metal film under nearly homoepitaxial growth conditions.

Some of the important variables that impact the formation of the template are the *substrate surface reconstruction, initial deposition temperature, deposition sequence, template composition, template thickness, and template annealing temperature.* As in the case of RE-V growth, a MBE system can be used to grow the III-V semiconductor and the TM-III compounds. In general, an electron beam evaporator mounted in the MBE growth chamber has been used to deposit the transition metals Fe, Co and Ni. These metals require higher temperatures for substantial deposition rates than Er and Sc. Although the recent high temperature effusion cells can easily operate at the desired temperatures, the most critical component is the crucible material. For substantial deposition rates the Ni and Co must be raised above their melting points. Unfortunately, both graphite and tungsten form eutectics with both Ni and Co,[154] which makes them poor choices as crucible material. Although alumina works for Ni, there is some problem with O contamination during deposition.[155] The best choice so far seems to be magnesia, which has been found to work for Ni.[155][156] As in the case of Al in PBN, the heating rate must be very slow through the melting point to ensure that the crucible does not break. The higher vapor pressure of Mn allows evaporation from a standard effusion cell at moderate temperatures ($\leq 1000°C$). Unlike the growth of the RE-As compounds where growth is done under excess $As_4$ or $As_2$ flux since any excess As will re-evaporate from the surface, the growth of TM-III compounds requires careful control of both the TM and group-III fluxes. Fortunately, most of the TM-III compounds have large phase fields.[154][157]-[159] For example, the bulk β′-NiAl phase field extends from 40 to 55 atomic % (at%) Al.[154] On the Ni-rich side, non-stoichiometry is accommodated by Ni atoms on Al sites. On the Al-rich side, non-stoichiometry is accommodated by Ni vacancies.[160][161] The large phase field relaxes the compositional control to some degree. However, from a thermodynamic standpoint the group-III rich composition will probably be the most stable, as has been shown for CoGa.[162] The lattice parameter also changes with composition, and for CoGa the lattice parameter is smallest for the Ga-rich compositions.[163][164] The flux of TM provided from effusion cells can be determined using a quartz crystal thickness monitor or as a function of effusion cell temperature by ex situ film thickness measurements (e.g., by RBS). The group-III flux is most easily calibrated by using RHEED oscillations during growth of a III-V semiconductor on a reference sample.

A cursory comparison of the CsCl (e.g., NiAl and CoAl) and zincblende crystal structures leads to the presumption that the favorable configuration for epitaxy would be with the cube axes parallel. Thin-film TM/GaAs reaction studies, however, revealed three possible orientation relationships, the simple "cube-on-cube"[165][166] and two orientation relationships with $\{110\}_{TM\text{-}Ga} \parallel \{100\}_{GaAs}$.[166] TM-Ga phases with lattice parameters smaller than ~2.94 Å exhibited primarily the cube-on-cube orientation relationship, whereas TM-Ga phases with larger lattice parameters (e.g., RhGa and IrGa) tended to exhibit only the latter two relationships. This difference was rationalized by noting the equivalence of the $\{300\}_{TM\text{-}III}$ and $\{044\}_{GaAs}$ plane spacings for a TM-III lattice parameter of 3.0 Å (6.1% greater than $0.5a_{0GaAs}$).[167] In the case of NiAl/GaAs, the effective lattice mismatch is 2.1%. In studies of reacted Ni/Al/Ni/GaAs structures, the NiAl was found to grow in the cube-on-cube orientation.[98] When NiAl films were grown directly on the c(4x4) As-stabilized {100} surface of AlAs, however, a mixture of the three variants was found[52][61] (Fig. 32). To gain control of the NiAl film orientation, the template stage was isolated from the continuous growth stage. This approach enabled exploration of the effect of compositional deviations during the first monolayers of growth. Al/Ni multilayers with Ni in initial direct contact with GaAs were found in earlier work to yield cube-on-cube NiAl after annealing.[51][98] This result led to the hypothesis that a starting monolayer of Ni could have the same effect.[52] Indeed, the first ML of Ni was observed by RHEED to have reacted with the AlAs to form a surface structure similar to that of $Ni_xGaAs$, the first product phase of the Ni/GaAs reaction. $Ni_xGaAs$ has a hexagonal unit cell derived from that of NiAs and $Ni_3Ga_2$.[168][169] Reacted $Ni_xGaAs$ layers on {100}GaAs grow as epitaxial films with four variants corresponding to the nominal orientation relationship with {0001} approximately parallel to the four inclined {111} planes of GaAs and the {0$\bar{1}$1} planes of $Ni_xGaAs$ approximately parallel to the {100}GaAs surface. Since $c_0/a_0$ is nearly equal to $\sqrt{3}/\sqrt{2}$ (the "pseudo-cubic" criterion),[170] the {0$\bar{1}$1} planes present a nearly square surface template. X-ray photoelectron diffraction and LEED studies support a cubic-like structure for this reacted layer.[171] The data gave a good fit to a CsCl structure, but no attempt was made to fit to the pseudo-cubic structure.[172] Subsequent growth of NiAl on this template resulted in single variant films with the cube-on-cube orientation (Fig. 32b). It is possible that this template is *structurally* more favorable for the growth of cube-on-cube NiAl than the original c(4x4)AlAs surface.

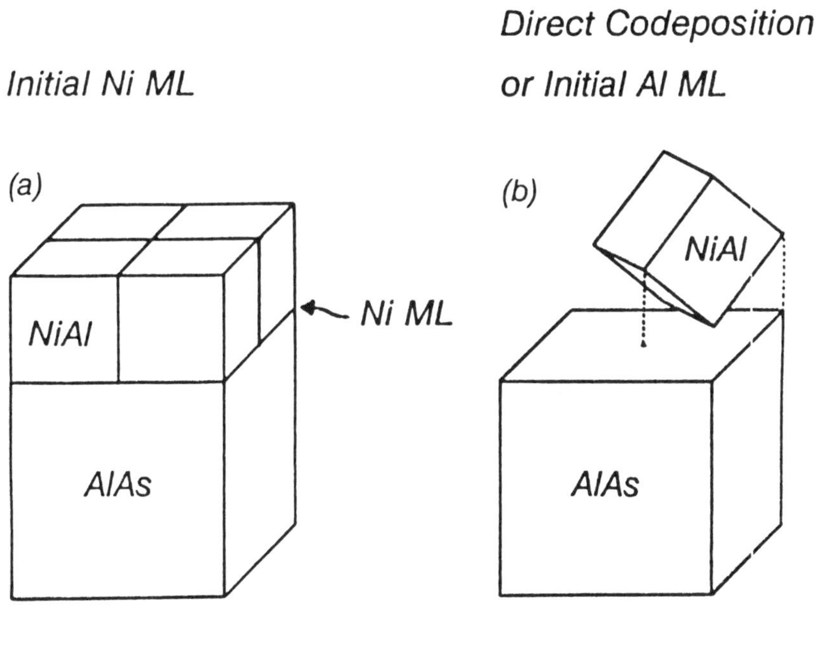


**Figure 32.** Schematic illustrations of the orientation relationships observed for NiAl films grown on the c(4x4){100}AlAs surface at temperatures below 200°C. If a monolayer of Ni is deposited first, the cube-on-cube orientation relationship shown in *(a)* is adopted. If Al is deposited first or if both Ni and Al shutters are opened simultaneously, two orientation relationships with {110}NiAl || {100}AlAs are observed. One of these variants is shown in *(b)*.

A second hypothesis is that Al or Al-rich phases are the first to nucleate on the c(4x4)AlAs surface during Ni and Al codeposition at low temperatures (<200°C). We know from previous work on the epitaxy of Al on As-terminated GaAs that two orientational variants with $\{110\}_{Al}||\{100\}_{GaAs}$ are observed along with one variant with $\{100\}_{Al}||\{100\}_{GaAs}$ (unit cell rotated 45° about the surface normal).[110][112][113] The multiple variants of NiAl observed during direct codeposition could be related to these multiple-variant Al grains. The deposition of a ML of Al prior to NiAl codeposition was also found to result in the formation of multiple NiAl variants[61], although direct experimental evidence for the nucleation of *fcc* Al has not been obtained.

A third possible explanation comes from preliminary embedded atom method (EAM) calculations of the energetics of various Ni-Al configurations on the {100} surface. Lacking an appropriate method for treating arsenides within the embedded atom scheme, the substrate for these calculations is an artificial Al crystal with the diamond structure and the AlAs lattice parameter. These calculations show that the addition of a Ni-rich surface layer results in a lower energy for the system than does the addition of an equiatomic NiAl layer.[173][174] Furthermore, molecular dynamics simulations of growth indicate that alternating layers of Ni and Al form along the growth direction when growth is initiated on the Ni-rich template, whereas growth directly on the pseudo-AlAs leads to compositional disorder within the layers. Indeed, bulk NiAl is compositionally ordered along the <100> directions and of mixed composition on the {110} planes (see Fig. 3). Although these calculations are based on some unrealistic assumptions and simplifications, the appealing results thus far suggest that further improvements in the simulation capabilities will lead to a much more thorough understanding of the initial stages of heteroepitaxy.

Although CoGa and NiAl have identical structures and nearly the same lattice parameter, one cannot simply translate the growth procedure from one material to the other. The orientation of CoGa films grown on GaAs is also controlled by the surface reconstruction and initial deposition, but in a different manner than observed for the growth of NiAl on AlAs. Palmstrøm et al.,[97] in growing epitaxial CoGa, have observed that a group III termination on (Al,Ga)As surfaces ensures the cube-on-cube orientation relationship for the CoGa overlayer. Failure to control the surface termination can lead to growth of {110} oriented films. More recently, Kuo et al.[175] and Tanaka et al.[176] have reported that a similar growth procedure to that used for NiAl (i.e., TM predeposition) also worked for CoGa and CoAl, respectively. At higher growth temperatures (400 - 600°C), the results of Kamigaki et al.[177] on NiAl suggest that direct codeposition is sufficient to ensure the growth of the correct orientational variant.

## 4.5 Growth of III-V Semiconductors on Epitaxial TM-III Films

The epitaxial growth of a III-V semiconductor on a TM-III intermetallic substrate presents a set of problems and constraints that are similar in difficulty to those encountered during the epitaxy of the TM-III film. Again, the primary challenge is in establishing the proper template. Under typical GaAs growth conditions, the substrate temperature is near 600°C and the growth occurs under an arsenic overpressure with the growth rate determined by the flux of Ga (assuming a sticking coefficient of about one and negligible GaAs evaporation). After growth, the epitaxial NiAl surface tends to exhibit a 50 at% Ni- 50 at% Al surface composition.[178][179] Exposing a TM-III surface to an arsenic ambient at 600°C, however, leads to substantial reaction with the group III component. Continued exposure eventually results in the formation of a TM arsenide. Furthermore, ultrathin films (<~10 nm) will agglomerate at temperatures above 500°C. Growth of the III-V layer is best initiated at lower temperatures, providing just enough As background pressure to react with the deposited group III atoms.

The most thoroughly studied system is GaAs/AlAs/NiAl. Harbison, Sands and coworkers[49][51][52][61] found that the AlAs interlayer was necessary to suppress the Ga-Al exchange reaction. When the {100}NiAl surface is exposed to a low background pressure of arsenic, the RHEED pattern shows a gradual evolution toward an {100}AlAs-like RHEED pattern. In order to prevent excessive reaction with As, Al is deposited one monolayer at a time. After each monolayer, the Al is allowed to "soak" in the As background until the RHEED pattern takes on the characteristics of {100}AlAs. After about 20 Å of AlAs has been formed, the As cell temperature is ramped up (or the As cracker valve is opened) and continuous growth of the III-V overlayer is begun. At this stage, the growth conditions are identical to those employed during homoepitaxy except that the substrate temperature is generally kept below 500°C to prevent agglomeration of the thin NiAl film. Identical growth conditions were found to be appropriate for GaAs/AlAs structures grown on CoAl as well.[49][61]

Using this growth procedure, epitaxial GaAs/AlAs/NiAl/AlAs/GaAs heterostructures containing electrically continuous NiAl films as thin as 10 Å have been fabricated.[49] The surface morphology of the III-V overlayer can be quite smooth, appearing specular to the eye. The fact that both NiAl growth on GaAs and GaAs growth on NiAl can lead to relatively smooth epitaxial films suggests that the surface energies of NiAl and GaAs (or AlAs) are comparable.

The quality of the GaAs overlayers is limited by the same factors that limit the quality of GaAs layers grown on Si: mismatches in thermal coefficient of expansion, lattice constant and symmetry. In the case of III-V/TM-III systems, the primary problem appears to be the symmetry difference,[180] as discussed in Sec. 2.4. Since the TM-III phases with the CsCl structure have lattice constants that are approximately one half the lattice constant of the III-V, translation domain boundaries (sometimes known as antiphase domain boundaries) can be expected in the III-V overlayer. Inversion domain boundaries (also known, inaccurately, as antiphase boundaries) are a problem as well since the CsCl structure is centrosymmetric while the zincblende structure is not. Both types of defects are inherent; they can only be eliminated by devising clever techniques to lower the symmetry of the TM-III surface. In the absence of any such recipe, III-V layers grown on TM-III films contain a high density of planar defects and dislocations. Near-surface defect densities decrease as the III-V layer is grown thicker, primarily as a result of defect interactions on impinging {111} glide planes.

Although the surface layer of a III-V overgrowth may be adequate for majority carrier devices, minority carrier electronic (e.g., heterojunction bipolar transistors) and optoelectronic (e.g., lasers) devices are not likely to perform satisfactorily. Despite this prognosis, devices that employ majority carrier transport across the upper GaAs/AlAs/NiAl interface have been fabricated. Comparison of the upper and lower Schottky barriers in these double heterostructures reveals nearly ideal thermionic emission characteristics for the lower diode (barrier height, $\phi_{bn}$ = 0.98 eV and ideality factor, n =1.1) and somewhat poorer results from the upper diode ($\phi_{bn}$ = 0.78 eV, n = 1.7).[180] The higher ideality factor and lower effective barrier height are attributable to excess leakage (shunting) currents associated with the high defect density in the interfacial region. A similar conclusion was reached in the analysis of double barrier resonant tunneling structures with 3 - 5 nm thick NiAl films. In these devices, larger peak-to-valley negative differential resistance regions (NDRR) were always observed when the upper $n^+$-GaAs layer was biased positive with respect to the substrate.[4]

## 5.0 GROWTH OF OTHER EPITAXIAL INTERMETALLIC COMPOUNDS

### 5.1 Control of Phase and Orientation; MnAl/AlAs

There are no general predictive tools as yet for the behavior of epitaxial films grown on structurally dissimilar substrates. This view has been strengthened by observations during studies of the growth of MnAl on AlAs.[6][67][181] τ-MnAl is a ferromagnetic phase with a high degree of uniaxial magnetocrystalline anisotropy. The easy magnetization direction is along the *c* axis of the tetragonal unit cell. Such a material, if grown on GaAs with the easy magnetization axis parallel to the substrate surface normal, could form the basis for a magnetic or magneto-optic information storage medium integrated with compound semiconductor electronic and photonic devices.[6] At first glance, τ-MnAl appears to be structurally and chemically compatible with AlAs. Like the cubic TM-Al intermetallic phases, NiAl, CoAl and FeAl, τ-MnAl is nearly 50 at% Al, suggesting a minimal driving force for reaction with AlAs. Although the structure of τ-MnAl is not cubic, it can be viewed simply as a tetragonally distorted derivative of the CsCl structure of NiAl, CoAl, and FeAl (Fig. 3). Furthermore, the basal plane lattice parameter of τ-MnAl is approximately half that of AlAs. Thus, epitaxy on AlAs with the desired orientation of the easy magnetic axis could be expected. Unfortunately, however, τ-MnAl is not a bulk equilibrium phase in the Mn-Al system. Near the equiatomic composition, a rhombohedral Al-rich phase and β-Mn (saturated with Al) are the stable phases at room temperature. τ-MnAl is formed in the bulk only by quenching the non-magnetic high-temperature hexagonal ε-MnAl phase at a composition of about 55 at% Mn.[182] Since vapor deposition can be viewed as a form of quenching, it is not surprising that the τ phase has also been observed in sputtered Mn-Al films.[183]

As shown in Fig. 33, initial attempts to grow single-phase τ-MnAl films by direct codeposition onto arsenic terminated {100}AlAs led to films consisting primarily of the high-temperature hexagonal ε phase. These initial results suggested that a template approach could circumvent this problem.[6][67][181] A thin (3.5 - 11 Å) amorphous layer of MnAl, deposited at ~20°C, will crystallize during in situ annealing at a temperature in the range of 150 - 300°C, yielding a RHEED pattern characteristic of the τ phase in the desired orientation. Because RHEED only samples the uppermost monolayer or two, one is not able to record the actual sequence of the nucleation process. However, the appearance of the epitaxial template suggests that the τ phase nucleates at the interface between the amorphous MnAl and the AlAs. This thin epitaxial film can then be used as a template for the growth of τ-MnAl by codeposition. Depending on the substrate temperature during this stage of growth, the τ-MnAl film will grow until either an amorphous phase or the ε phase takes over. Magnetoresistance measurements of the τ-MnAl films grown by this template technique reveal a strong extraordinary Hall effect with hysteresis characteristic of perpendicular magnetization.[6][181][184]

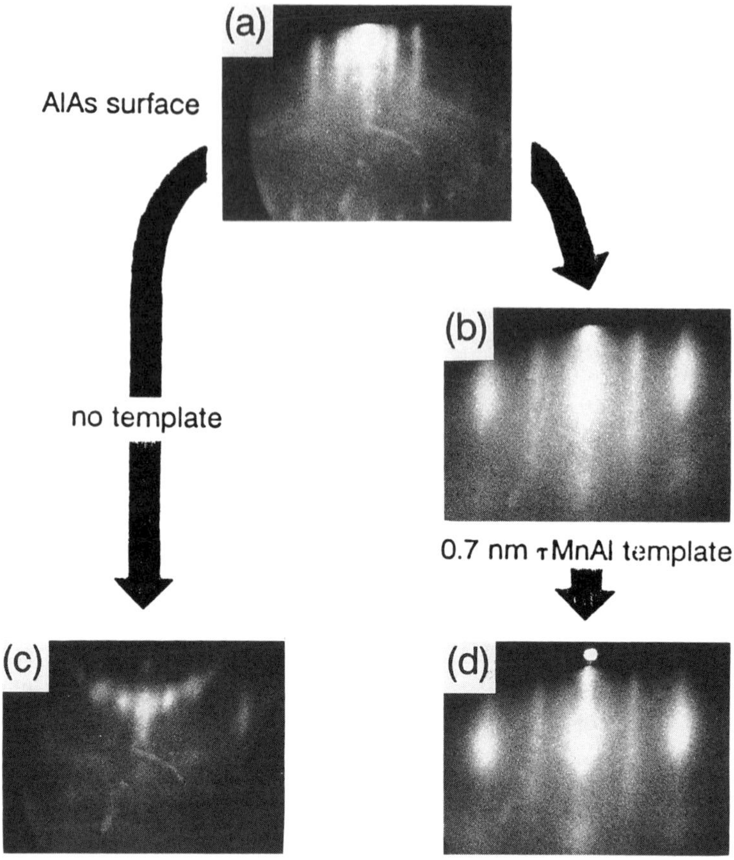


**Figure 33.** <110> azimuth (relative to substrate) RHEED patterns illustrating the importance of a template in the nucleation of the desired τ-MnAl phase. The pattern in *(a)* is typical of the initial c(4x4){100}AlAs surface. On one such surface an amorphous MnAl film of thickness 7 Å was deposited at room temperature. Annealing induced solid phase crystallization of this film to form an epitaxial τ-MnAl template. *(b)* shows the crystallized film at 115°C during heating to 300°C. The pattern in *(d)* indicates that the τ-MnAl phase remained the dominant phase after growth of 17 Å on this template at 300°C. Under identical growth conditions without the template, a film consisting primarily of the non-magnetic ε-MnAl phase is formed. The characteristic RHEED pattern is shown in *(c)*.

Because of the eventual appearance of an amorphous phase or the non-magnetic ε phase, we have experimented with alternative methods to form thicker single-phase films of τ-MnAl. Since the formation of the thin τ-MnAl template occurs in the solid state, the obvious choice was to try thicker amorphous templates. Unfortunately, excessively high temperatures (>500°C in some cases) were required to fully crystallize templates thicker than about ten nanometers. Furthermore, no evidence of epitaxial τ-MnAl was obtained by RHEED. A bilayer reaction technique to form thick templates (30 - 100 Å) was explored. In Fig. 34 two variants of this bilayer technique are compared. These two methods differ only in the first monolayer deposited on the AlAs surface at room temperature. If a ML of Mn is deposited, followed by all of the Al required to create the desired final MnAl film thickness, followed by the necessary remaining Mn, then the film reacts during annealing to form an epitaxial layer of τ-MnAl with the *c* axis perpendicular to the film surface. The extraordinary Hall resistance measurement for a 5 nm film measured at room temperature is shown in Fig. 34c. Note that the hysteresis loop is remarkably square, indicative of uniform film microstructure and perpendicular magnetization. Figure 34g shows an extraordinary Hall effect measurement from a film of the same thickness and prepared in the identical manner except that the initial Mn monolayer was omitted (it was "added" to the top Mn layer instead to maintain the proper Mn:Al ratio). The hysteresis loop is sheared suggesting structural or compositional nonuniformity and/or oblique remnant magnetization.

Comparison of the RHEED data for samples prepared with and without the initial Mn ML reveal direct evidence for a dramatic difference in τ-MnAl orientation. Without the Mn ML, the Al layer was epitaxial with multiple orientations as expected. The Mn overlayer was found to be fine-grain polycrystalline. During annealing at 400°C, the τ-MnAl appeared to take its orientational clues from the epitaxial Al. The diffraction pattern observed at this stage was consistent with {110} and {101} MnAl approximately parallel to the film surface. Thus, the magnetically easy axis was in the plane or at an angle to the substrate normal. In contrast, the Al film deposited on the Mn ML appeared to be amorphous up to a thickness of about 1.5 nm, after which faint indications of {111} textured polycrystalline Al began to appear. The disorder imparted in the first Al layers by the Mn ML is not surprising given the propensity for the formation of low symmetry phases (most notably, the quasicrystals) in Al-rich Mn-Al. Deposition of the upper Mn layer followed by annealing at 400°C resulted in the appearance of a RHEED pattern consistent with τ-MnAl in the "perpendicular" orientation. This example dramatically illustrates the effect of a single monolayer, not only on the film structure, but on the magnetic properties of an epitaxial film as well.

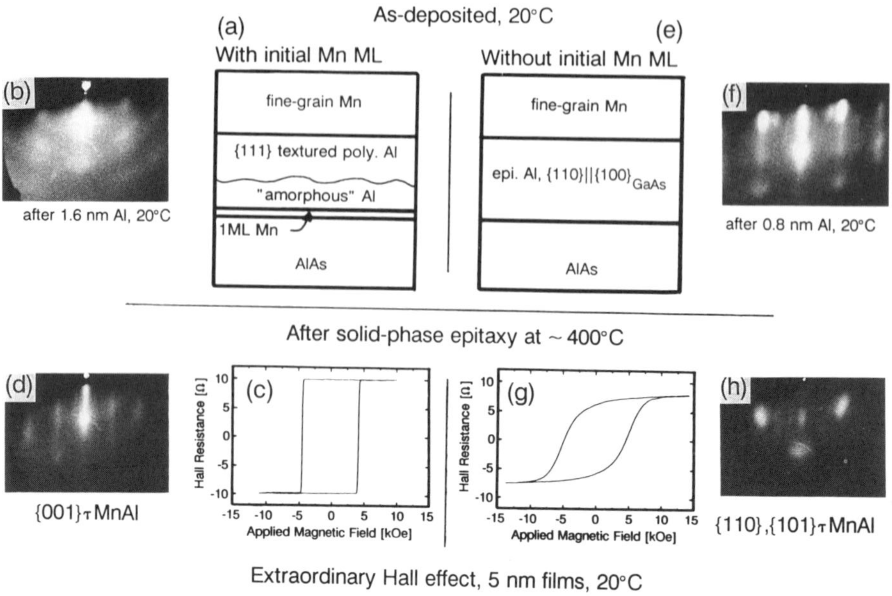


**Figure 34.** Comparison of RHEED patterns, as-deposited film structure, and extraordinary Hall effect (EHE) measured in τ-MnAl films prepared by reaction between elemental layers. The only difference in the preparation of these films is that the as-deposited structure shown on the left side *(a)* contains a monolayer of Mn at the interface between the Al and the AlAs. The as-deposited structure on the right *(e)* was prepared by direct deposition of Al onto the AlAs surface. Without the initial Mn monolayer, the Al film is epitaxial (two variants) *(f)*. During annealing at 400°C, the τ-MnAl adopts orientation relationships suggesting that τ-MnAl nucleates on the epitaxial Al *(h)*. The resulting oblique orientations of the magnetically easy axis give rise to a sheared EHE hysteresis loop *(g)*. With the initial Mn monolayer, the Al film is polycrystalline (b). Annealing at 400°C yields an epitaxial τ-MnAl film with a magnetically easy axis parallel to the film normal as indicated by the RHEED *(d)* and the nearly ideal EHE hysteresis loop (details in text).

## 5.2 Growth of $Fe_3Al_{1-y}Si_y$

Another class of compounds with a $BiF_3$ structure (Fig. 3), which is closely related to the CsCl structure, have recently been studied as a potential epitaxial metallization on GaAs. One such compound is $Fe_3Al_{1-y}Si_y$.[66] By varying the Al/Si-ratio the lattice parameter can be adjusted from 5.64Å ($Fe_3Si$) to 5.74Å ($Fe_3Al$). Fortunately, the $Fe_3Al_{1-y}Si_y$ phase extends over a finite range of composition (as is observed for the TM-III compounds with CsCl structure), which allows for some deviation from the ideal Fe/(Al,Si)-ratio of three. For a lattice match to GaAs a composition of

$Fe_3Al_{0.13}Si_{0.87}$ is required. Hong et al.[66] have grown $Fe_3Al_{1-y}Si_y$ epitaxially on GaAs. They used e-guns for evaporating Fe and Si and an effusion cell for Al. When the $Fe_3Al_{1-y}Si_y$ was grown on Ga-stabilized GaAs(100) surfaces at 300 - 450°C, a (100)/[011] $Fe_3Al_{1-y}Si_y$ || (100)/[011] GaAs orientation relationship was observed. This is the same orientation relationship found for CoGa growth on Ga-stabilized GaAs(100).[33][97][152]

## 6.0 PROPERTIES OF EPI-METALLIC FILMS

### 6.1 Properties of RE-V Compounds

In the bulk, the RE-V compounds have not been as well studied as the TM-III due to the difficulties in obtaining high-purity RE-metals, the high reactivity of the RE metals (particularly with oxygen), and the high equilibrium vapor pressures of pnigogens above the RE-V compounds at high temperatures. Studies of the RE-V and RE-VI compounds performed in the 1960s and 1970s have been reviewed by Hulliger[63] (RE-V) and Flahaut[185] (RE-VI). Unlike the TM-III phases, the RE-V phases are generally considered to be line compounds (i.e., they have narrow phase fields). Taylor et al.[186] measured the non-stoichiometry of bulk RE-As compounds fabricated using a variety of techniques and found that they tended to be slightly arsenic deficient. In general, the arsenic content could be reduced to RE-$As_{\sim 0.95}$ without much change in lattice parameter. More significant arsenic deficiencies were observed for SmAs and PrAs. A composition of $PrAs_{\sim 0.8}$ was measured after annealing at 1611°C, yet no evidence of ordering of point defects was observed by x-ray diffraction. Taylor et al.[186] also detected RE deficiency in some samples which they attributed to the formation of RE vacancies. It should be noted, however, that the RE elements used in their experiments contained high levels of impurities including carbon, oxygen, hydrogen, fluorine and chlorine. The RE-As compounds tended to contain high levels of these impurities as well, although the authors suggest that the impurity levels were insufficient to explain the observed deviation from stoichiometry.

The high degree of localization of the RE 4f electrons has permitted the unambiguous determination of the valence of the RE ion in RE-V compounds. The RE ions are trivalent in all the NaCl-structure RE-V compounds with the exceptions of CeN and CeP.[63] The electronic structure of the rare-earth monopnictides has been a controversial issue.[63][187][188] Early results suggested both metallic[189]-[191] and semiconducting[192]-[194]

behavior. Optical measurements on thin films suggested a small band gap (~0.7 - 1.2 eV) consistent with semiconductor behavior.[63][193][194] Early theoretical work predicted semiconductor behavior.[195] However, later band structure calculations by Hasegawa and Yanase[196] for the Gd monopnictides suggested that GdN was a semiconductor and that the other monopnictides were all semi-metals.

Detailed optical characterization of ErAs grown on GaAs has been reported by Ralston et al.[39] For films ~1500Å thick they have observed the crystal-field split 1.54 $\mu$m $^4I_{15/2}$-$^4I_{13/2}$ transitions of $Er^{3+}$ ($4f^{11}$) both in Fourier transform infrared absorption and cathodoluminescence. Their absorption and x-ray data are consistent with unstrained ErAs layers. The 1.54 $\mu$m $Er^{3+}$ emission is not observed in photoluminescence do to the high free carrier absorption.

## 6.2 Properties of TM-III Intermetallic Phases

The TM-III phases with the CsCl structure (Fig. 3) are characterized by their relatively high melting points (e.g., >1600°C for NiAl and CoAl), high temperature strength and corrosion resistance. Their ordered structures can accommodate considerable non-stoichiometry, a property that has garnered much attention among researchers studying the basic properties of intermetallic phases. For example, the NiAl phase with the CsCl structure is stable over the range of 40 - 55 at% Al. The minimum in electrical resistivity and the maximum in lattice parameter occur at about the equiatomic composition. These properties are reviewed in more detail in Ref. 49. For the purposes of this chapter, we focus on the transport properties of the TM-III phases since these properties are most relevant for device applications.

The TM-III intermetallic phases with the CsCl structure have resistivities that are intermediate between those of the pure metals (~1 - 5 $\mu\Omega$-cm) and the resistivities typical of disordered metals (~100 - 200$\mu\Omega$-cm). The TM-III phase with the lowest room temperature resistivity is NiAl at 9.8 $\mu\Omega$-cm[197]. In general, the resistivity increases as the TM moves to lower atomic number (Co and Fe) and with the substitution of Ga for Al. This trend is consistent with the variation in Fermi level within the density of states. In particular, in moving from NiAl toward FeAl, the Fermi level moves into the peak in the density of states associated with the TM d-bands. Concomitant with this change comes an increase in the scattering rate for point defects.[197]-[200]

## 6.3 Lateral Magnetotransport in RE-V Epitaxial Layers

Lateral magnetotransport measurements can be used to probe the electronic structure of metallic layers. Detailed magnetotransport measurements on $Sc_{1-x}Er_xAs$ films grown on GaAs(100) and capped by a GaAs capping layer have been reported recently.[201]-[204] A very useful guide for the understanding of the lateral transport measurements described below comes from the band-structure calculations of Hasegawa and Yanase for Gd-monopnictides.[196] The calculations show GdAs to be a semi-metal with overlap of the conduction and valence bands yielding equal electron and hole concentrations, ~$3.6 \times 10^{20}$ cm$^{-3}$, with the holes being at $\Gamma$ and electrons at X (Fig. 35). Figure 36a shows the magnetoresistance for a $Sc_{1-x}Er_xAs$ sample as a function of magnetic field with temperature as a parameter. The most striking feature is the curvature and sign change of the Hall resistance from "electron"-like to "hole"-like as the temperature is reduced from room temperature to liquid-helium temperatures. This is clear evidence that the transport is circumscribed by a compensated metal with nearly equal numbers of electrons and holes. Using the Hasegawa and Yanase[196] band structure for GdAs as a guide one may assume that $Sc_{1-x}$ $Er_xAs$ has an isotropic hole Fermi surface centered at $\Gamma$ with hole density $n_h$=n (where n is the carrier density) and mobility $\mu_h$, and three equivalent ellipsoidal electron Fermi surfaces at X with electron density $n_e$=1/3n each and anisotropic electron mobilities $\mu_{e,1}$ and $\mu_{e,2}$, which characterize the electron mobility along and perpendicular to the axis of the ellipsoid of revolution (see Fig. 35). Figure 36b shows the temperature dependence of the hole and electron densities and mobilities extracted from a least squares fit of the magnetotransport data in Fig. 36a.[203] It is interesting to note that the electron and hole densities do not change significantly with temperature. The change in the electron and hole mobilities with temperature is responsible for the change from "electron"-like to "hole"-like behavior with temperature. More recent magnetotransport measurements on $Sc_{1-x}Er_xAs$ layers with pulsed magnetic fields up to 46 Tesla[204] indicate that the model assuming equal electron and hole concentration breaks down for measurements at high magnetic fields (>20 T). A better fit to the measurements is obtained with a total electron density of $3.0 \pm 0.3 \times 10^{20}$ cm$^{-3}$ and hole density of $2.8 \pm 0.3 \times 10^{20}$ cm$^{3}$.

Shubnikov-de Haas (SdH) oscillations are another measure of the film quality. For fields <9 Tesla, SdH oscillations, resulting from the electrons, have been observed in $Sc_{1-x}Er_xAs$ films, but not in pure ErAs films grown under similar conditions, despite transport mobilities that are not much

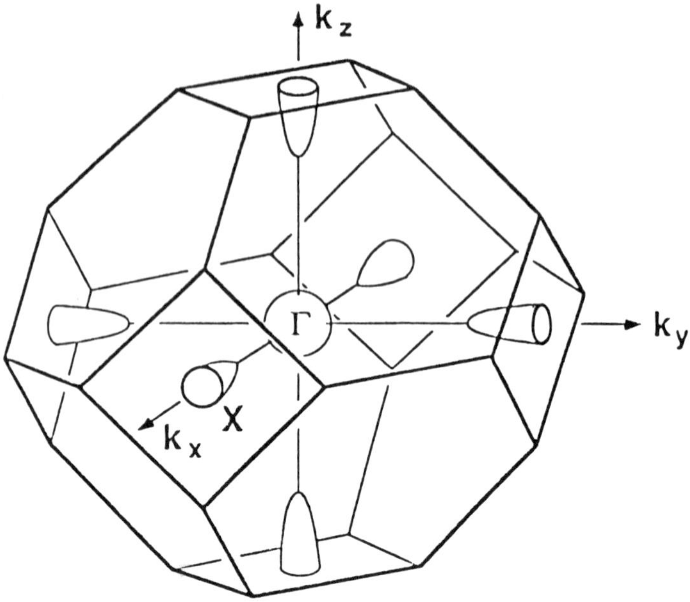


**Figure 35.** Schematic of the Brillouin zone and the hole and electron Fermi surfaces for $Sc_{1-x}Er_xAs$ that are used to model the magnetotransport data. $Sc_{1-x}Er_xAs$ has an isotropic hole Fermi surface centered at Γ with hole density $n_h$=n (where n is the carrier density) and mobility $\mu_h$, and three equivalent ellipsoidal electron Fermi surfaces at X with electron density $n_e$=1/3n each and anisotropic electron mobilities $\mu_{e,1}$ and $\mu_{e,2}$, which characterize the electron mobility along and perpendicular to the axis of the ellipsoid of revolution. (from Ref. 203.)

higher in $Sc_{1-x}Er_xAs$ than the pure ErAs films.[203] It is possible, however, that by achieving a measure of lattice match between the $Sc_{1-x}Er_xAs$ and underlying GaAs substrate, the "grain" size of the epitaxial material is substantially larger and small-angle scattering from the boundaries is significantly reduced. Boundary scattering will have little effect on the transport mobilities but any scattering event affects the coherent cyclotron motion and suppresses quantum oscillations like the SdH oscillations described here. This is consistent with the x-ray Bragg reflectivity, RBS-channeling and TEM measurements described in Sec. 4. They have clearly shown that the $Sc_{1-x}Er_xAs$ films are of superior crystal quality compared to pure ErAs, particularly for films greater than the ErAs critical thickness. SdH oscillations have recently been observed in ultrathin (~10 - 20 ML) ErAs films with pulsed magnetic fields up to 46

Tesla.[204] Fig. 37 shows SdH obtained from a 200 Å $Sc_{1-x}Er_xAs$ sample. The fractional changes in resistance produced by the oscillations are quite small when compared with those seen in the two-dimensional electron gas in semiconductors. The Fourier transform of the oscillations is shown in Fig. 37c. The beating that appears in the data is reflected in the appearance of two peaks in the Fourier transform at 330 ± 5 and 385 ± 5 Tesla. There is a strong angular dependence of the SdH oscillation suggesting an elliptical or cylindrical piece of Fermi surface. A fit to the angular dependence to a three-dimensional model based on an ellipsoid of revolution suggests that the surface in question is the electron surface that emerges from the band structure theory of Hasegawa and Yanase[196] and is located at the X points. However, there is little to distinguish between a fit to a cylinder and an ellipsoid of revolution in the range of angles over which the SdH oscillations are observed. The amplitude of the SdH oscillations should rapidly diminish as the magnetic field is rotated away from the surface normal due to the collisions of the orbiting electrons with the surfaces of the thin epitaxial film. The critical field required to fit a circular electron orbit inside the epitaxial layer is shown in Fig. 38 together with the experimental points for the smallest magnetic field at which SdH oscillations are seen. The large discrepancy between the observation of the oscillations and the prediction that the three-dimensional orbit will collide with the surface is a demonstration that the electron moves in a size-quantized orbit, constrained to move in the plane of the epitaxial layer. The angular dependence reflects the fact that only the normal component of the magnetic field controls the period of the SdH oscillations and the oscillatory transport in Fig. 37 is *quasi-two-dimensional.* Studies with higher magnetic field (up to 46 T) have shown the quantum confinement of holes as well as electrons.[204] This type of quantum confinement has also been observed for relatively thick (~1000Å) layers of gray tin.[205]

Allen et al [203] fit the amplitude of the SdH oscillations as a function of temperature and magnetic field to extract the effective mass and the scattering rate. The mass of 0.17 $m_0$ for the pocket of electrons directed towards X is close to the mass calculated by Hasegawa and Yanase, 0.1 $m_0$.

The strong beating between the quantum oscillations has been interpreted to be as a result of a splitting of the conduction band caused by the exchange interaction of the fully polarized 4*f* spin on the $Er^{3+}$ with the conduction electrons.[203] The splitting is given by the difference in the Fermi levels for the spin up, $\mu\uparrow$, and spin down, $\mu\downarrow$, measured from the spin-up and spin-down conduction band bottoms and is 38 ± 7 meV for the sample in Figs. 37. The splitting was found to scale with the Er concentration. For pure ErAs the Fermi energy splitting is 64 ± 7 meV.[204]

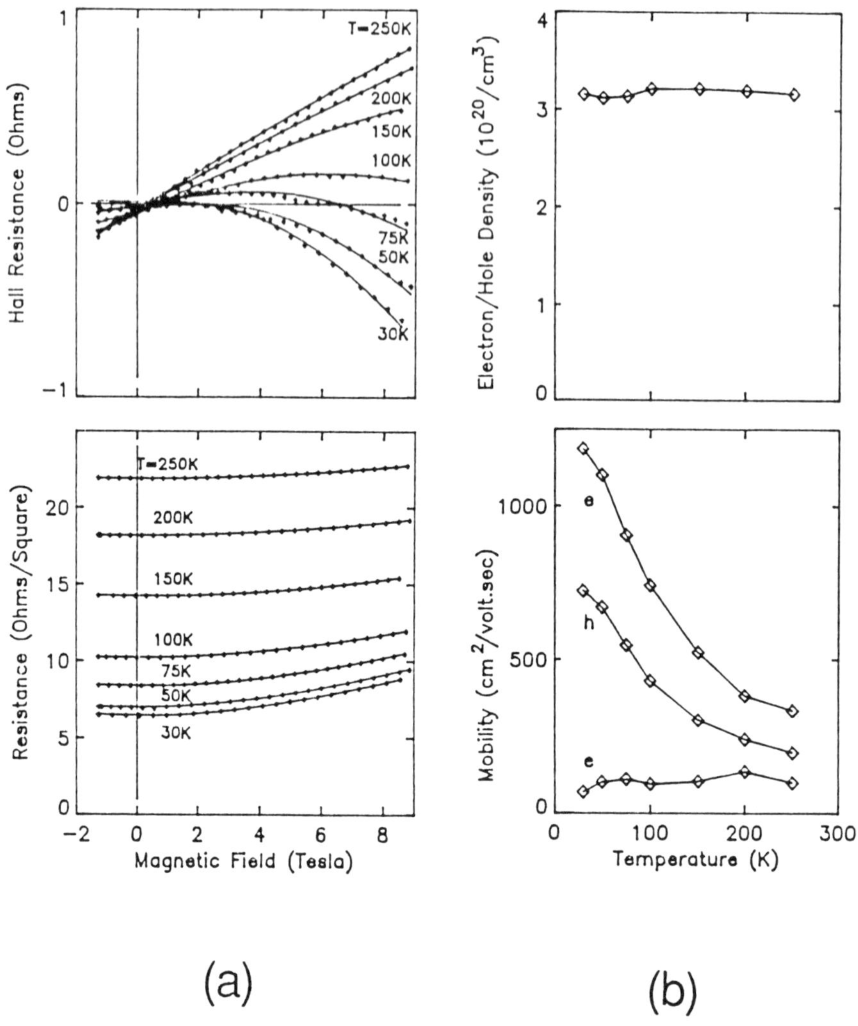


**Figure 36.** *(a)* Hall resistance and longitudinal magnetoresistance for a 200 Å thick $Sc_{0.43}Er_{0.57}As$ layer with temperature as a parameter. The solid lines are the result of a least-squares fit to the parameters shown in *(b)*. *(b)* Electron or hole density and mobilities extracted from the magnetoresistance data of *(a)*. The change in sign of the Hall resistance with temperature is direct evidence of the presence of both electron and hole conduction. It is interesting to note that the electron and hole densities do not change significantly with temperature. The change in the electron and hole mobilities with temperature is responsible for the change from electron-like to hole-like behavior with temperature. (From Ref. 203.)

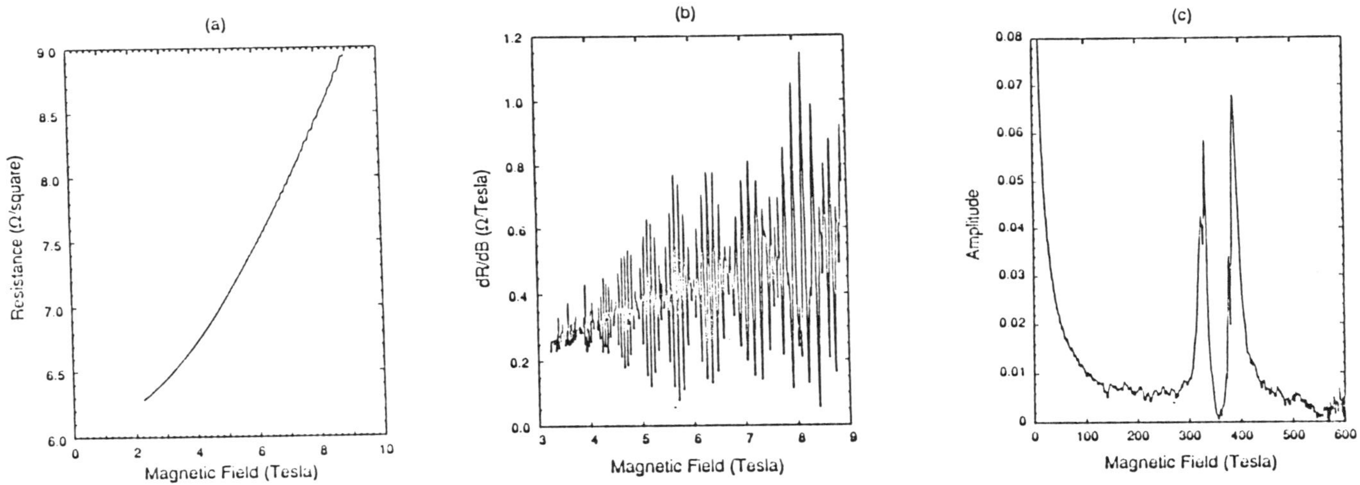


**Figure 37.** Longitudinal magnetoresistance of a 200Å thick $Sc_{0.43}Er_{0.57}As$ layer buried in GaAs at 1.5K *(a)*, the first derivative *(b)* displaying Shubnikov-de Haas oscillations and *(c)* Fourier transform of the first derivative of the Shubnikov-de Haas oscillations shown in *(b)*. The appearance of Shubnikov-de Haas oscillations is indicative of a high quality film with little charge carrier scattering. The fractional changes in resistance produced by the oscillations are quite small when compared with those seen in the two-dimensional electron gas in semiconductors. The beating that appears in the data in *(b)* is reflected in the appearance of two peaks in the Fourier transform at 330 ± 5 and 385 ± 5 Tesla in *(c)*. A fit to the angular dependence of the Shubnikov-de Haas oscillations to a three-dimensional model based on an ellipsoid of revolution suggests that the Fermi surface in question is the electron surface. (Adapted from 203.)

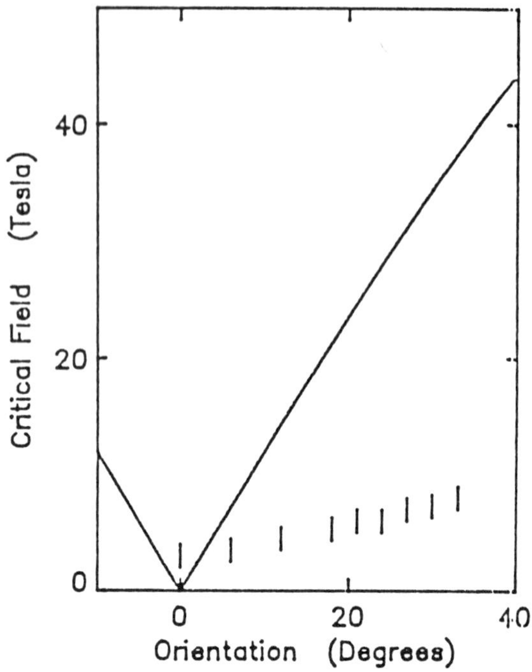


**Figure 38.** Calculated (solid line) minimum magnetic field required to just fit an electron orbit inside a 200 Å thick $Sc_{0.43}Er_{0.57}As$ epitaxial film at a tip angle $\theta$. Experimental points are the smallest magnetic field at which Shubnikov-de Haas oscillations are seen. There is a large discrepancy between the observation of the oscillations and the prediction that the three-dimensional orbit will collide with the film surface, demonstrating that the electron moves in a size-quantized orbit, constrained to move in the plane of the epitaxial layer. The angular dependence reflects that only the normal component of the magnetic field controls the period of the Shubnikov-de Haas oscillations and that the oscillatory transport in Fig. 37 is *quasi-two-dimensional.* (From Ref. 203.)

Since the conduction electrons can scatter elastically from the $4f$ $Er^{3+}$ spin and at low temperature ErAs orders antiferromagnetically, one expects to observe the antiferromagnet transition in magnetotransport measurements. Figure 39(a) shows the resistivity of an epitaxial 150 Å ErAs layer buried in GaAs(100) without an applied magnetic field. The anomaly at low temperature results from the antiferromagnetic-paramagnetic transition. This was verified by measuring the resistivity as a function of temperature with a 8.9 T applied magnetic field (Fig. 39b). Without an applied field the resistivity increases rapidly between 2 and 5 K. This is not observed when the

4*f* $Er^{3+}$ spins are aligned with the applied magnetic field. From the change in resistivity without an applied magnetic field, the Neél temperature is ~4.5 K.

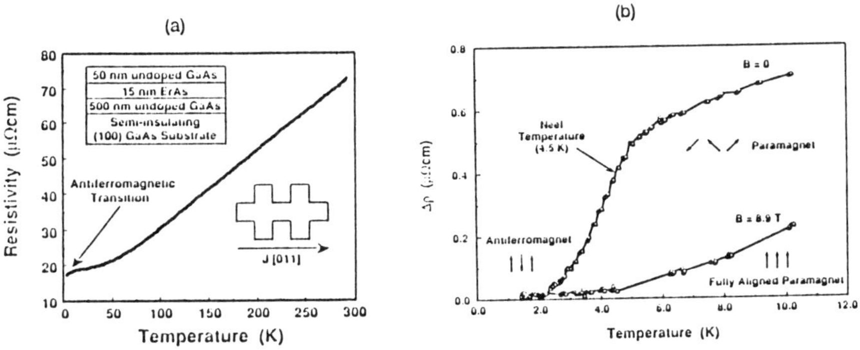


**Figure 39.** Plots of resistivity versus temperature *(a)* and change in resistivity versus temperature with and without an applied magnetic field *(b)* for a 150 Å ErAs epitaxial layer buried in GaAs(100). The conduction electrons can scatter elastically from the 4*f* $Er^{3+}$ spin. There is an anomaly at low temperature without an applied magnetic field, which results from the antiferromagnetic-paramagnetic transition. When the 4*f* $Er^{3+}$ spins are aligned with the applied magnetic field no anomaly is observed. From the change in resistivity without an applied magnetic field the Neél temperature is ~4.5 K. *(Courtesy of S. J. Allen, Jr., Bellcore.)*

Figure 40 shows the (1/2 1/2 1/2) neutron diffraction intensity versus temperature. This diffraction condition corresponds to the scattering spins ordering on {111} planes with periodicity of double the {111} lattice spacing. The antiferromagnetic order consists of ferromagnetically aligned sheets perpendicular to a <111> direction with neighboring sheets aligned antiferromagnetically with respect to each other. This is the same as that observed in ErP and ErSb.[206] The ordering temperature determined from Fig. 40 is ~4.5 K for both the 1000 Å and 200 Å thick films. This is in close agreement with magnetotransport measurements discussed above and the peak in susceptibility on a different 250 Å thick film found at $T_N = 4.8 \pm 0.2$ K. However, the ordering temperature measured for these epitaxial ErAs films is substantially higher than that reported for bulk ErAs, ~2.9 K[207] and 3.5 K.[208] The reason for this difference has not been resolved. It may arise from improved purity of these later films (fabricated by MBE in UHV, using possibly higher purity Er and capped with GaAs), stress in the ErAs films due to the mismatch with GaAs and/or different thermal expansion coefficients, or possibly due to the GaAs substrate changing the mechanical properties of the films, which inhibits the lattice distortion which occurs at the paramagnetic to antiferromagnetic transition.

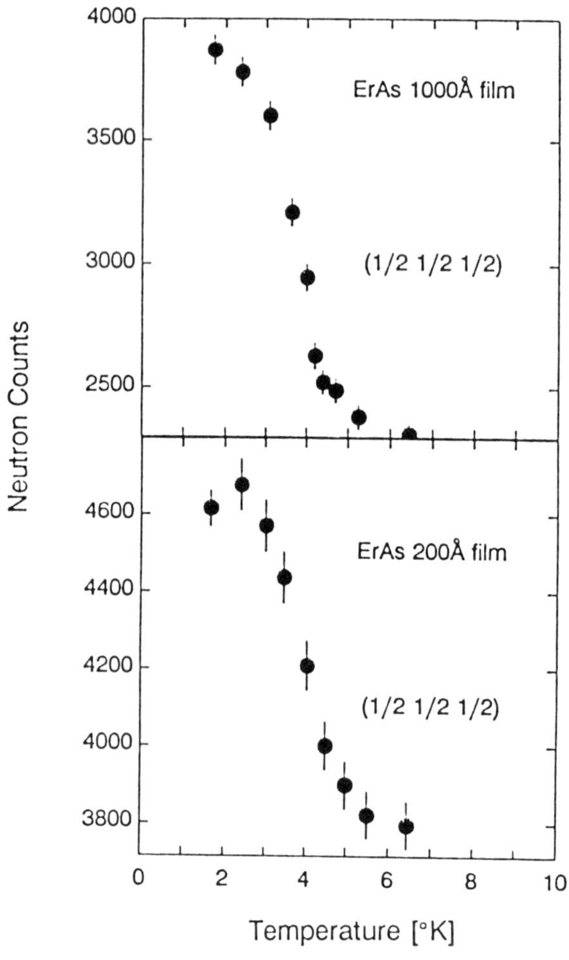


**Figure 40.** Plots of the neutron diffraction intensity of the (1/2 1/2 1/2) diffraction versus temperature for a 1000 Å and a 200 Å buried ErAs epitaxial film on GaAs(100). This diffraction condition corresponds to the scattering spins ordering on {111} planes with periodicity of double the {111} lattice spacing. The antiferromagnetic order consists of ferromagnetically aligned sheets perpendicular to a <111> direction with neighboring sheets aligned antiferromagnetically with respect to each other. The ordering temperature determined from this figure is ~4.5 K for both the 1000 Å and 200 Å thick films. *(Courtesy P. F. Miceli, Bellcore.)*

In magnetoresistance measurements a cusp-like anomaly is caused by spin-disorder scattering at the antiferromagnetic-paramagnetic phase boundary. By measuring position of the cusp-like anomaly with temperature,[201] the low temperature magnetic phase diagram has been deduced and is shown in Fig. 41. Similar measurements have been made on ErP, $ErAs_{0.4}P_{0.6}$ and ErSb.[46]

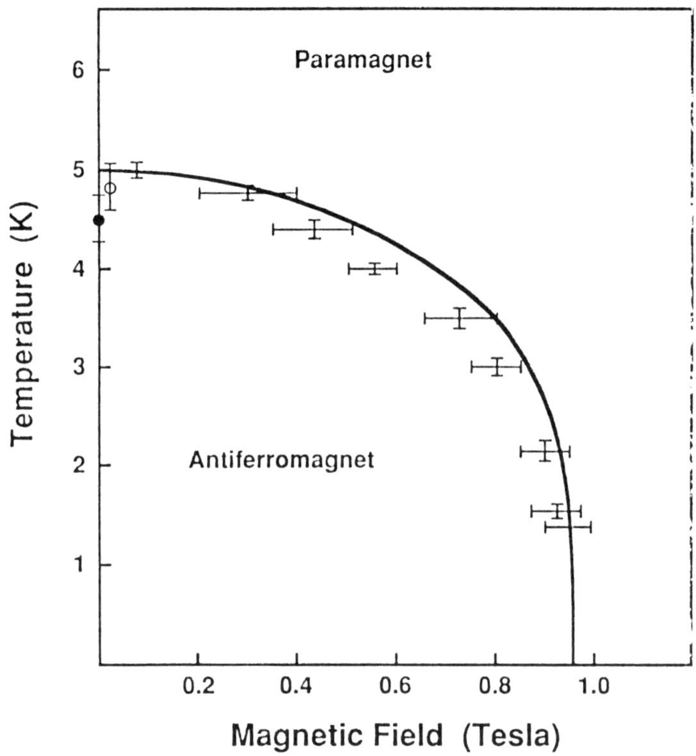


**Figure 41.** The low-temperature ErAs magnetic phase diagram deduced from the temperature dependence of the resistance *(solid circle),* magnetoresistance, *(crosses),* and susceptibility *(open circle).* The *solid curve* is a calculated phase boundary. (From Ref. 201.)

The magnetic ordering in bulk samples has been extensively studied.[63][206] In general, most rare-earth monopnictides are antiferromagnets at low temperature, $T_N$ ~1-25 K.[63] A summary of the transport properties of RE-V epitaxial films grown on III-V semiconductors is shown in Table 2.

**Table 2.** Electrical properties of RE-V films grown on III-V semiconductors

| Compound | Mismatch to GaAs (%) | Resistivity ($\mu\Omega$·cm) $\rho_{RT}$ | $\rho_0$ | Thickness (Å) | $\phi_{bn}$ (eV) | $\phi_{bp}$ (eV) | $T_N$ (K) | Reference |
|---|---|---|---|---|---|---|---|---|
| ErAs | 1.6 | 60-70 | 16-17 | 150 | ~0.88* | | 4.5 | 32,39,45,201 |
| $Sc_{0.32}Er_{0.68}As$ | <0.1 | 50 | 13 | 200 | ~0.85* | | | 36,48,203 |
| ScAs | -3.4 | | | | ~0.81* | | | 36,47,48 |
| LuAs | 0.4 | 90 | 45 | 65 | | | | 32 |
| $Sc_{0.2}Yb_{0.8}As$ | <0.02 | | | | 0.84 | | | 47 |
| YbAs | 0.9 | | | | 0.89 | | | 47 |
| $ErP_{0.6}As_{0.4}$ | <0.025 | 80 | 26 | | 0.88 | 0.48 | 3.1 | 44,46 |
| ErP | | 150 | 85 | | | | 4.8 | 35,44,46 |
| ErSb | | 46-80 | 19 | | | | 4.4 | 35,44,46 |
| $ErP_{0.47}Sb_{0.53}$ | 0(InP) | | | | | | | 35 |

* Depends on GaAs orientation and deposition conditions (see text).[36][48]

## 6.4 Transport in Ultrathin RE-V Epitaxial Films

As a semi-metal layer thickness is reduced, one may expect that size quantization would result in a semi-metal to semiconductor transition.[202] Figure 42a shows the predicted thickness dependence of the energy gap assuming a free electron value for the electron and hole mass and a bulk electron and hole concentration of $3 \times 10^{20}$ cm$^{-3}$. At three monolayers the hole and electron states should be separated by as much as 1 eV by the quantum confinement. Figure 42b shows the reciprocal Hall constant, which determines an effective sheet carrier density for the layers, as a function of the thickness in monolayers. The shaded area is an extrapolation from a 150 Å thick sample. The thinnest sample differs from the bulk extrapolation by only a factor of two. Hence, it is an over simplification to treat a 3 ML film by an effective mass "particle in a box" approximation. Xia et al.[209] have done detailed band structure calculations for GdAs/GaAs superlattices and show that no semi-metal to semiconductor transition is expected, which is in agreement with these results. However, for ultrathin (10 and 20 ML thick) ErAs films the magnetotransport measurements suggest that the films tend to be more "electron" like as the thickness decreases.[204]

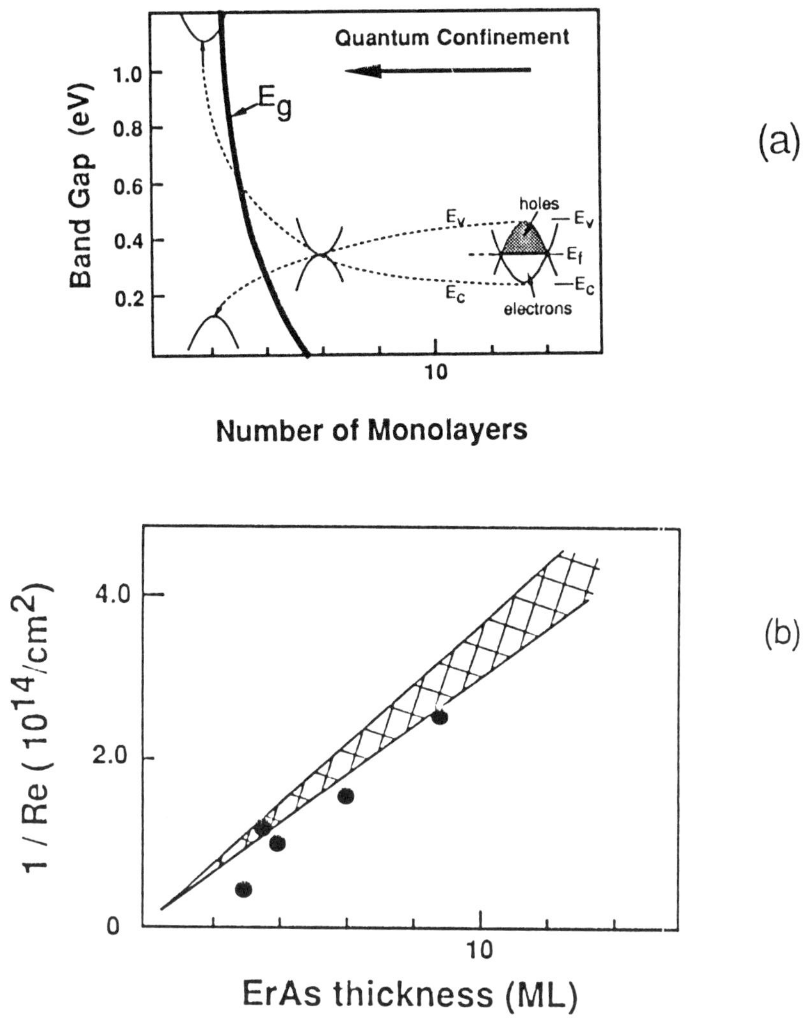


**Figure 42.** *(a)* Predicted energy gap produced by size quantization versus layer thickness in monolayers for ErAs. A schematic of the band alignment predicted from quantum confinement is also shown. For thick layers the conduction and valence band overlap. As the ErAs thickness decreases the model predicts a separation of the conduction and valence bands, which results in a 3 ML thick film having a band gap of 1 eV. *(b)* Plot of the reciprocal Hall coefficient versus layer thickness in monolayers for ErAs epitaxial layers buried in GaAs(100). The shaded area is an extrapolation from a 150 Å thick sample. The thinnest sample differs from the bulk extrapolation by only a factor of two. This shows that it is an over simplification to treat a 3 ML film by an effective mass "particle in a box" approximation. (Adapted from Ref. 202.)

The reciprocal temperature dependence of the conductance for different film thicknesses, with a constant magnetic field to align the 4*f* $Er^{3+}$ spins, is plotted in Fig. 43. Carrier-carrier scattering dominates for the 2 ML film. As expected for uniform films, the slope of the resistance with temperature (dR/dT) at high temperature is proportional to the reciprocal of thickness. The log of the low temperature conductance has been plotted against the log of the film thickness in Fig. 44. Note that the slope of the conductance goes as the cube of thickness as is expected from the theories of Trivedi and Ashcroft[210] and Tesanovic.[211] The best fit to their theory assuming no quantum confinement is with interface roughness of ~3 monolayers. However, including quantum confinement effects would decrease the effective interface roughness. A more reasonable interface roughness may then be ~1 - 2 monolayers, which is more consistent with the x-ray Bragg reflectivity measurements described in Sec. 4.

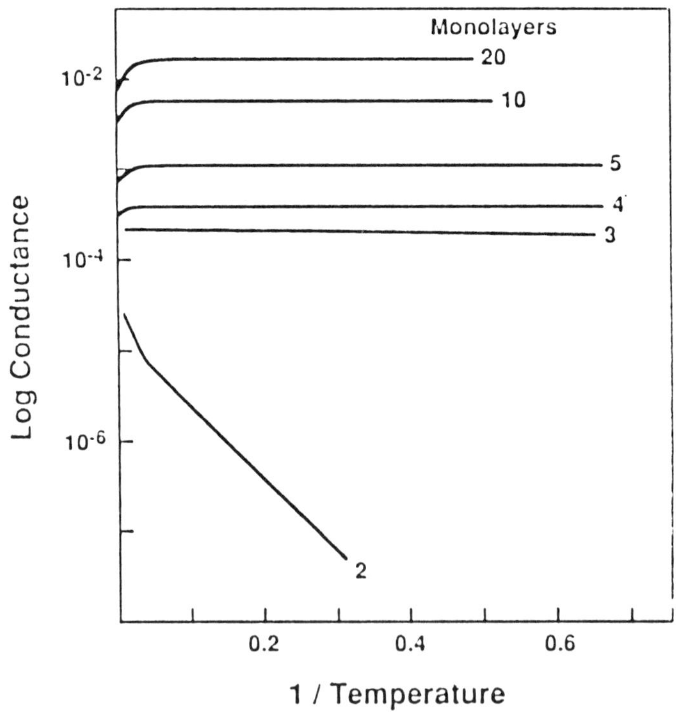


**Figure 43.** Plot of log conductance versus reciprocal temperature with film thickness as a parameter for ErAs epitaxial layers in GaAs(100). Carrier-carrier scattering dominates for the 2 ML film. (A constant magnetic field was applied to align the 4*f* spins.) *(Courtesy of S. J. Allen, Jr., Bellcore.)*

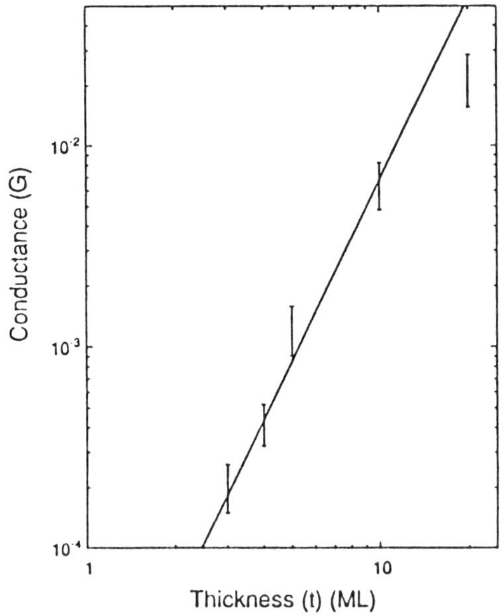


**Figure 44.** Plot of the low temperature conductance versus film thickness for ErAs epitaxial films grown in GaAs(100). The data shows a good fit to a $t^3$ dependence *(solid line). (Courtesy of S. J. Allen, Jr., Bellcore.)*

The antiferromagnet ordering is of the face-centered second kind, type II, which implies that next-nearest-neighbor interactions dominate. One would expect, therefore, that as the ErAs epitaxial layer is made thinner the ordering temperature may change. Figure 45 shows $T_N$ variation with ErAs layer thickness determined from magnetotransport measurements. It is interesting to note that the transition temperature does not decrease until the films are thinner than 10 ML. For a 3 ML thick film the transition temperature drops to ~2.5 K. In Fig. 46 the sheet conductance of another ~3 ML thick ErAs film has been plotted as a logarithmic function of temperature. With no magnetic field applied the conductance drops precipitously at low temperature. Remarkably a modest magnetic field of the order of 3 Tesla applied in any direction restores the conductance values that are obtained from the high temperature extrapolation. A possible explanation for this large negative magneto-resistance is the existence of bound magnetic polarons or localization by the carrier 4*f* spin interaction. The reduced dimensionality of the thinnest layers suppresses the antiferromagnetic order that is established below 5 K in the bulk layers, permitting a large amount of 4*f* spin susceptibility. This could lead to localization due to the spin disorder itself or self-trapping by bound magnetic polarons as the spin of the carrier polarizes the 4*f* spins in its immediate environment.[212] Either mechanism can lead to localization that would be quenched in magnetic fields that completely polarize the 4*f* spins. The exact nature of this localization by the carrier 4*f* spin interaction can only be elucidated by more experiments and detailed modeling.

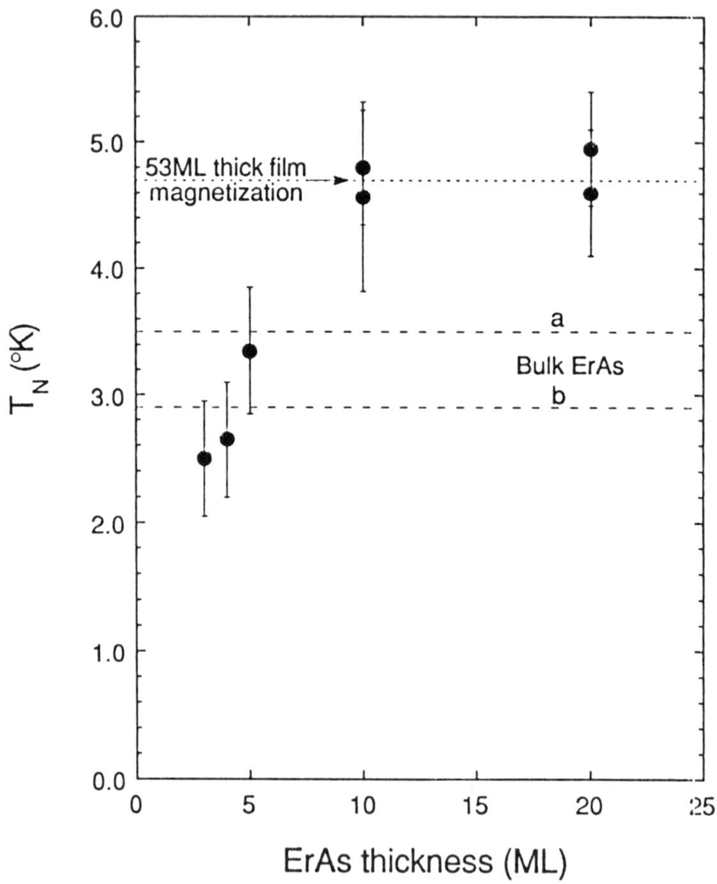


**Figure 45.** Plot of the Neél temperature versus layer thickness for ultrathin buried ErAs epitaxial layers grown on GaAs(100). The Neél temperature for bulk ErAs has also been plotted, *(a)* is from Ref. 208 and *(b)* from Ref. 207. For epitaxial films ≥ 10 ML grown on GaAs the Neél temperature is substantially higher than the bulk value. The Neél temperature decreases for layers less than 10 ML thick. *(Courtesy of S. J. Allen, Jr., Bellcore.)*

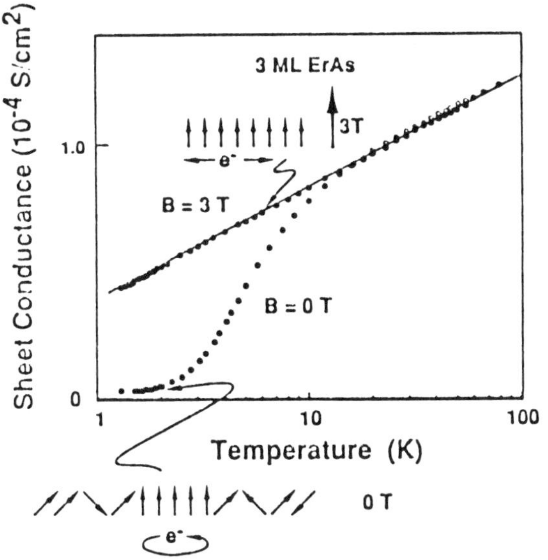


**Figure 46.** Plot of the temperature dependent conductance with magnetic field as a parameter for a 3 ML thick ErAs buried in GaAs(100). The schematic suggests that the low temperature conductivity without an applied magnetic field results from a bound magnetic polaron (see text for details). *(Courtesy of S. J. Allen, Jr., Bellcore)*

## 6.4 Transport in Ultrathin TM-III Epitaxial Films

The high melting points of the TM-III phases combined with their stability and epitaxy on III-V substrates permits the growth of ultrathin, electrically continuous thin films that can withstand the temperatures required for III-V overgrowth and III-V device processing. Measurements of film resistivity versus film thickness are shown in Fig. 47. These films behave similarly and the resistivities scale with bulk resistivities. For each system studied (NiAl,[5][49] CoAl,[213] NiGa,[55] and CoGa,[214]) the resistivity increases rapidly as the film thickness is reduced. Qualitatively, this behavior is consistent with Fuchs' model[215] which includes non-specular scattering from interfaces. Detailed comparisons[5][214] reveal many discrepancies between the data and this simple model. One interesting aspect to these data is that the 10 Å NiAl and CoAl films show lower resistivities than slightly thicker films. These are also the only NiAl and CoAl films that are thinner than the estimated critical thicknesses for misfit dislocation introduction. It is interesting to speculate that the lower resistivities may arise from reduced interface scattering due to coherent epitaxy. An alternative explanation could be that the 10 Å films became partially agglomerated during the semiconductor overgrowth step at ~400°C. Note that the only other measurement of a 10 Å TM-III film is that of CoGa. This film was not capped with epitaxial III-V material, however. The high resistivity (750 $\mu\Omega\cdot$cm) may reflect the loss of material due to oxidation of the exposed surface.

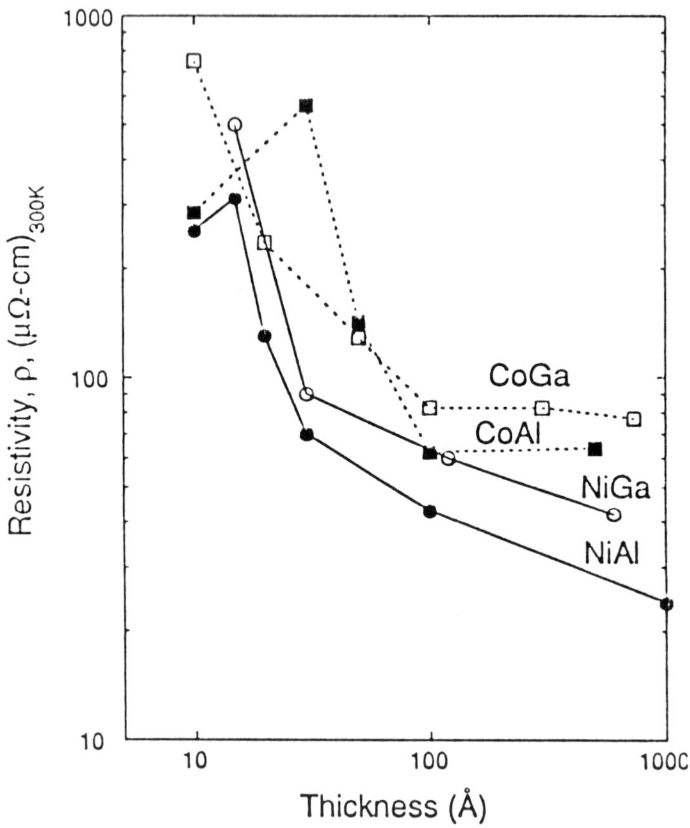


**Figure 47.** Plot of resistivity versus film thickness of epitaxial TM-III intermetallic films grown on $Ga_{1-x}Al_xAs(100)$. The data for NiAl were taken from Refs. 5 and 49, NiGa from Ref. 55, CoAl from Ref. 213 and CoGa from Ref. 214. The behavior is nearly identical for all the TM-III intermetallics. Only the thinnest NiAl and CoAl films are thinner than the critical thickness. The improved crystal quality of pseudomorphic films may explain the decrease in resistivity for the thinnest NiAl and CoAl as compared with slightly thicker films.

Epitaxial thin metal films provide interesting vehicles for the study of quantum confinement in metals. Early work by Jaklevic et al.[2][3] on tunneling in metal/insulator/metal structures with textured metal films revealed "standing wave" phenomena consistent with the band structure of the metal. Later work on much thinner (30 - 50 Å) epitaxial NiAl films by Tabatabaie et al.[4] demonstrated negative differential resistance regions at room temperature in the vertical I-V characteristics. These data were interpreted as tunneling through subbands in a metallic quantum well. The optical properties of metallic quantum wells have also been modeled.[216] The primary conclusion of this work is that screening effects dominate in metals and the subbands, although spaced in energy by as much as an eV, are not likely to be resolved by optical absorption techniques. Nevertheless, the optical and electronic properties of metallic quantum wells remains an area of experimental physics that is relatively unexplored. Given the recent advances in heteroepitaxy, thorough investigations are now possible.

Transport in ferromagnetic thin films has been the topic of numerous investigations, particularly in recent years with the discovery of giant magnetoresistance effects in multilayers. These effects are seen in the *longitudinal* magnetoresistance. If one measures the *transverse* voltage generated as current passes through a ferromagnetic film in the presence of a magnetic field oriented normal to the substrate, a signal two or three orders of magnitude greater than the ordinary Hall effect can be measured. This effect, known as the extraordinary or anomalous Hall effect, arises from asymmetric scattering of carriers from the individual magnetic ions. In the case where the film is perpendicularly magnetized in zero field, this effect can be used, in principle, to fabricate a nonvolatile memory element with the "up" and "down" states corresponding to binary "0" and "1." As discussed previously in Sec. 5, epitaxial MnAl films grown on AlAs exhibit this hysteretic behavior in the extraordinary Hall effect. Voltage differences as high as 20 mV between the two states have been measured at room temperatures for a 1 mA sense current passing through a 50 Å thick epitaxial MnAl film.[217] Furthermore, these films are thermally stable up to about 400°C making them compatible with III-V device processing.

## 7.0 FUNDAMENTALS OF SCHOTTKY BARRIER FORMATION

The Fermi level position determined from Schottky barrier height measurements of CoGa on $Ga_{1-x}Al_xAs$(100) are summarized in Fig. 48. The Fermi level appears to be pinned close to mid gap in GaAs and at positions in $Ga_{1-x}Al_xAs$ which are in good agreement with most Schottky barrier models.[97] This figure illustrates that by adjusting the composition of a $Ga_{1-x}Al_xAs$ layer beneath the CoGa, the effective Schottky barrier height to GaAs can be tailored. These results also show the importance of any interfacial exchange reaction to the electronic properties of the metal/III-V semiconductor interface. Clearly, if CoAl is deposited on GaAs, then a Ga-Al exchange reaction will result in the formation of a $Ga_{1-x}Al_xAs$ beneath the contact and the barrier height would depend strongly on the Al content. As already mentioned, an AlAs layer is generally grown between the TM-Al contact metallization and GaAs to suppress this interaction. The effective Schottky barrier height, however, is found to depend upon the thickness of the AlAs layer[49][180][218] (Fig. 49). The observed barrier height depends critically on the measurement technique. For internal photoemission the barrier height appears to be sensitive only to the first ~10 Å of semiconductor, yielding an approximately constant barrier height for all AlAs thicknesses. Current-voltage measurements on the other hand are very sensitive to the AlAs thickness. The CoAl effective barrier height can be varied from 0.76 eV to 1.1 eV by increasing the AlAs from 0 to 100 Å. This results from the change in electron tunneling probability with AlAs thickness.

The results from TM-III/$Ga_{1-x}Al_xAs$(100) barrier height measurements have not yielded new insight into the mechanisms involved in Schottky barrier formation. This may be partially due to the limited number of studies and the use of AlAs interlayers in the case of TM-III compounds. The RE-V/III-V compound semiconductor systems offer additional parameters which may be varied, such as lattice mismatch, deposition and annealing conditions, and substrate orientation.[36][48] Figure 50 shows a plot of the Schottky barrier height versus ErAs mole fraction (x in $Sc_{1x}Er_xAs$) for layers grown at temperatures ~350 - 400°C on (100)GaAs.[36][48] Using current-voltage measurements, the Schottky barrier height varies approximately linearly from ~0.81 eV for ScAs (-3.3% mismatch) to ~0.88 eV for ErAs (+1.6% mismatch). These films were about 200 Å thick, which is above the critical thickness for ErAs[132][133] and ScAs. Hence, misfit dislocations in the $Sc_{1-x}Er_xAs$ films do not appear to influence strongly the Schottky barrier height. This is in agreement with lack of thickness dependence observed for epitaxial silicide/Si Schottky barriers.[20] Figure 50 also shows results from in situ annealing under an $As_4$-flux. The barrier height increases for lattice matched $Sc_{0.32}Er_{0.68}As$ layers from ~0.85 eV for layers grown at 350°C to ~1.03 eV for layers annealed at T ≥ 500°C in an $As_4$ flux post deposition. A similar increase is observed for ScAs. Internal photoemission measurements verified the 1.03 eV barrier height.

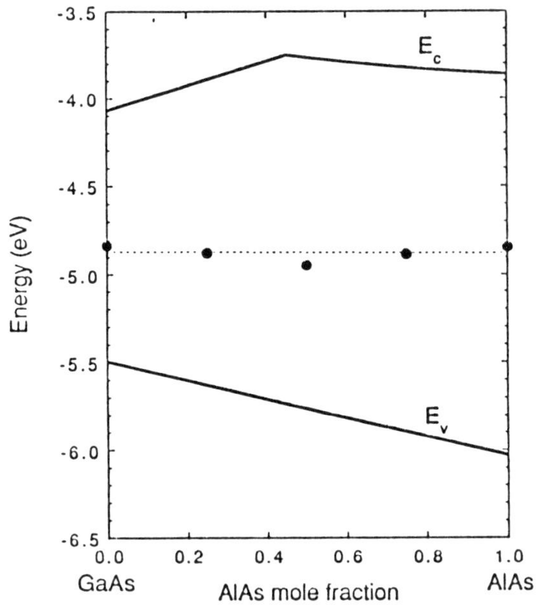


**Figure 48.** Fermi level pinning position for epitaxial CoGa/n-$Ga_{1-x}Al_xAs$(100) Schottky barriers plotted on an absolute energy scale. The conduction band and valence band energies are also shown. Note that the Fermi level is approximately constant. (Adapted from Ref. 97.)

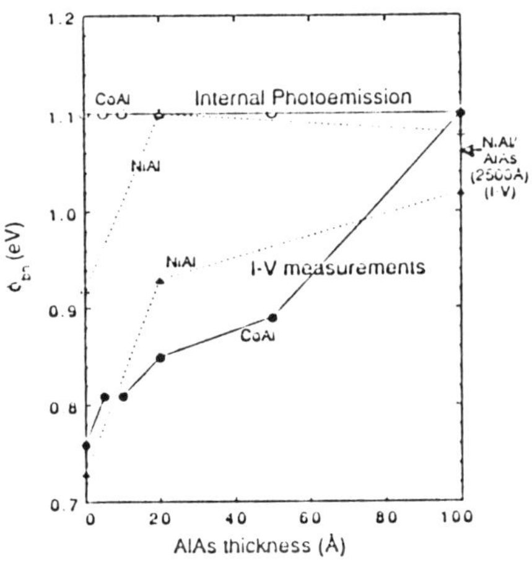


**Figure 49.** Schottky barrier heights for epitaxial NiAl/AlAs/n-GaAs(100) and CoAl/AlAs/n-GaAs(100) plotted against AlAs interlayer thickness. The internal photoemission determination of the Schottky barrier height is nearly insensitive to the AlAs thickness whereas the current-voltage measurements are very sensitive to the AlAs thickness. (Adapted from Refs. 49, 180, 218.)

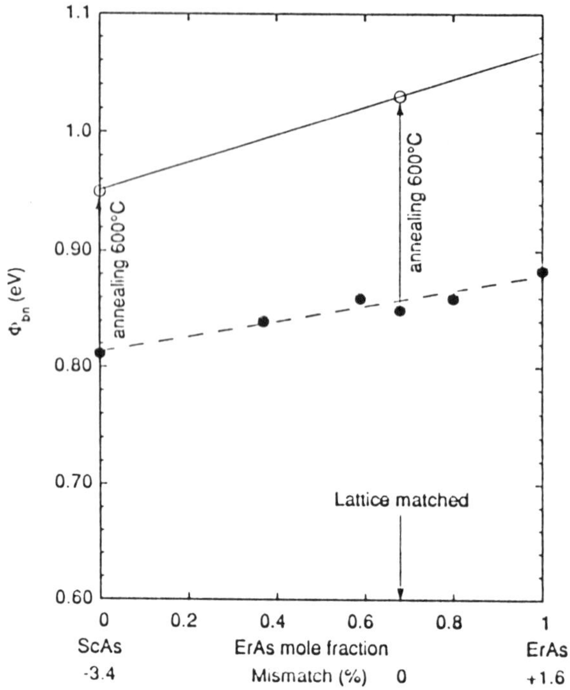


**Figure 50.** Schottky barrier heights plotted against the mole fraction of ErAs for $Sc_{1-x}Er_xAs$/n-GaAs(100) diodes, as-deposited and with subsequent annealing. The $Sc_{1-x}Er_xAs$ was grown at ≤400°C. The barrier height does not depend strongly on the composition and strain. Since the ScAs and ErAs films are thicker than the critical thickness, the barrier height does not appear to be sensitive to misfit dislocations in the $Sc_{1-x}Er_xAs$ layers. The barrier height does, however, increase by nearly 0.2 eV with annealing.

Tung et al.[219] have observed large variations in the $NiSi_2$/Si(100) Schottky barrier height depending upon deposition conditions. From plan-view TEM dark field images they were able to conclude that the intrinsic barrier height was low on (100)Si and that higher barrier heights resulted from parallel high barrier height regions formed by facet bars with {111} facets. These high barrier height regions tended to pinch-off the lower barrier height regions. Similar TEM plan-view images of $Sc_{1-x}Er_xAs$/GaAs(100) structures as deposited at 380°C and annealed at 580°C are shown in Fig. 51. Small white lines can be seen in <011> type directions. The origin of the contrast in these images is not clear. The white lines could possibly be similar to the facet bars observed for $NiSi_2$/Si(100), but on a much smaller scale. The difference in the images from the unannealed and annealed samples is subtle, if any, and is substantially less than that reported for the $NiSi_2$/Si(100) structures. RBS-channeling measurements were also unable to detect any difference in the unannealed and annealed films. However, it is interesting to note that at ≥ 500°C a slight increase in the sharpness of both LEED and RHEED diffraction patterns has been observed for $Sc_{1-x}Er_xAs$ films. This may suggest an improved interface structure due to removal of point defects by annealing. A similar result has been reported by Hricovini et al.[220] for epitaxial Pb/Si. They attribute the increase in the Schottky barrier height as a result of annealing to a rearrangement of the Pb atoms on the surface.

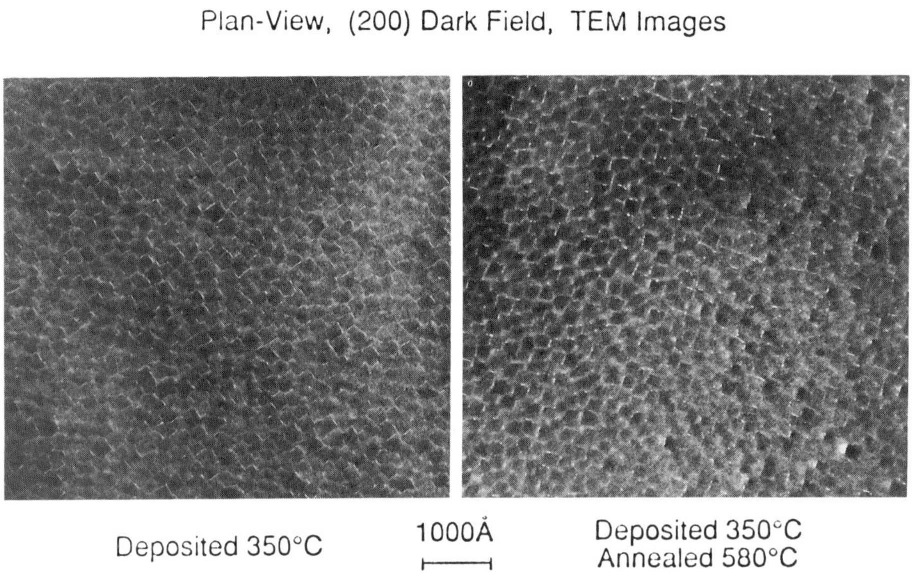


**Figure 51.** Plan view, (200) dark field, TEM images of α-Si/$Sc_{1-x}Er_xAs$/GaAs(100) structures as deposited at 350°C and after annealing at 580°C. Small white lines can be seen in <011> type directions. The origin of the contrast in these images is not clear. The difference in the film microstructure, as deposited and after annealing at 580°C is subtle if any. *(Courtesy of J. G. Zhu and C. B. Carter, Cornell University.)*

The advanced unified defect model[12] would predict that as the As-concentration decreases at the interface, the barrier height to n-GaAs should increase. Since, growth at 350°C occurs on the very As-rich c(4x4) GaAs(100) surface, which corresponds to ~1.75 ML of As coverage[221], it may be postulated that excess arsenic is trapped at the $Sc_{1-x}Er_xAs$/GaAs(100) interface during deposition and that annealing results in As outdiffusion resulting in a less As-rich interface. Thus the advanced unified defect model[12] could possibly explain the annealing results. However, associating the increase in the barrier height to a reduction in interface As concentration alone is questionable since the films deposited directly at 580°C yield barrier heights comparable to the ones deposited at 350°C (Figs. 50 and 52). Here the starting surface is an As-rich (2x4)/c(2x8) (~0.75 ML As), which RHEED suggests converts to the Ga-rich c(8x2) (~0.75 ML Ga) during the initial deposition of $Sc_{1-x}Er_xAs$. Thus the advanced unified defect model would not explain the high temperature deposition results. However, one interesting point to note is that the GaAs(100) work function is strongly dependent upon the surface reconstruction.[222]-[225] A plot of change in work function versus annealing is shown in Fig. 53. A work function difference as large as ~0.75 eV is observed depending upon surface preparation and reconstruction.[223] As the As-coverage decreases, the work function increases sharply in going from the c(4x4)-As-rich surface to the (2x4)/c(2x8)-$\gamma$ As-rich surface. The (2x4)/c(2x8)-$\beta$ surface has the highest work function and the (2x4)/c(2x8)-$\alpha$ a somewhat lower value. The work function then decreases rapidly in going to the more Ga rich (3x1) and the (1x6) surfaces, reaching a minimum close to the (1x6)/c(8x2) phase boundary. The value starts to increase again for the c(8x2) Ga-rich surface towards the value for the c(4x4) As-rich surface.[223] The differences in work function can arise from changes in band bending and changes in the surface dipoles, which result in changes in the effective electron affinity. Svensson et al.[226] found for in situ MBE prepared surfaces that the Fermi level was pinned close to mid gap (within ± 0.1 eV) for c(4x4), (2x4)/c(2x8) and (4x6) surfaces. This is in agreement with recent results on decapped GaAs(100) surfaces.[224][225] Since these studies suggest that the Fermi level is pinned at or close to band center it appears reasonable to suggest that the change in work function may be due to differences in surface dipole layers.[223][224] *It is, therefore, important to point out that the effective electron affinity cannot be considered to be a constant for a specific semiconductor, but will depend upon the detailed surface structure.* The $Sc_{1-x}Er_xAs$ semi-metal would screen some of the dipole charge and, therefore, may reduce the extreme change in barrier height. During the deposition of $Sc_{1-x}Er_xAs$ on the c(4x4) As-rich surface at 350°C the GaAs c(4x4) surface reconstruction is lost immediately. In general, GaAs surface reconstructions are very sensitive to the surface charge distribution (electron counting) and, therefore, if some $Sc_{1-x}Er_xAs$ is

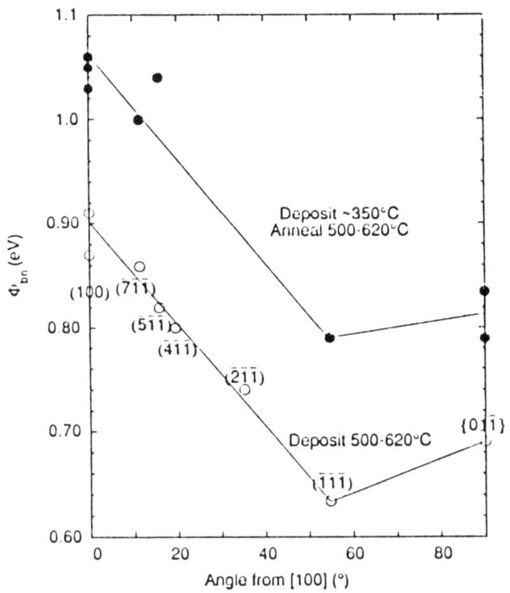


**Figure 52.** Plot of Schottky barrier height versus substrate orientation for lattice matched $Sc_{1-x}Er_xAs$/n-GaAs {h$\bar{1}\bar{1}$} diodes. The barrier height is highest for (100) and it decreases approximately linearly with angle towards ($\bar{1}\bar{1}\bar{1}$) for $Sc_{0.32}Er_{0.68}As$ layers grown at 500 - 600°C. The barrier height is higher for samples with the $Sc_{1-x}Er_xAs$ grown at ~350°C and annealed at 500 - 600°C in situ in an $As_4$ flux than ones grown at 500-620°C. The barrier height variation ranges from ~1.03 eV for (100) deposited at 350°C and annealed at ~580°C to ~0.63 eV for ($\bar{1}\bar{1}\bar{1}$) deposited at ~600°C, a total variation of ~0.4 eV for the same metallization on GaAs.

deposited on the surface, the change in the surface electron distribution may completely destabilize the GaAs surface reconstruction. This is what was observed during the $Sc_{1-x}Er_xAs$ deposition at ~580°C, where the (2x4)/c(2x8) reconstruction changed to the c(8x2). All of the reconstructions consist of either As- or Ga-dimers. The destabilization can be thought of as 'unzipping' of the As-dimers on the (2x4)/c(2x8) and evaporation of $As_2$ resulting in a Ga-dimerized surface. When $Sc_{1-x}Er_xAs$ is deposited on the c(4x4) surface, which consists of ~1.75 ML of As,[221] the As-dimers may start to unzip in the top and possibly the lower As-layer. Although RBS-channeling measurements of ultrathin buried ErAs layers indicate that the As-sublattice is continuous across the interface and consistent with the alignment shown in Fig. 54a,[126] there may still be some excess As trapped at the interface. One could think of the 350°C deposited samples as corresponding to a fairly high GaAs work function surface [close to the (2x4)/c(2x8)] which approaches a more Ga-rich interface during annealing [more like the (1x6) surface], resulting in a lower work function and therefore higher barrier height. Deposition at high temperature could result in the formation of an interface even more Ga-rich (closer to the c(8x2) surface reconstruction), which can have a fairly high work function and thus lower barrier height.

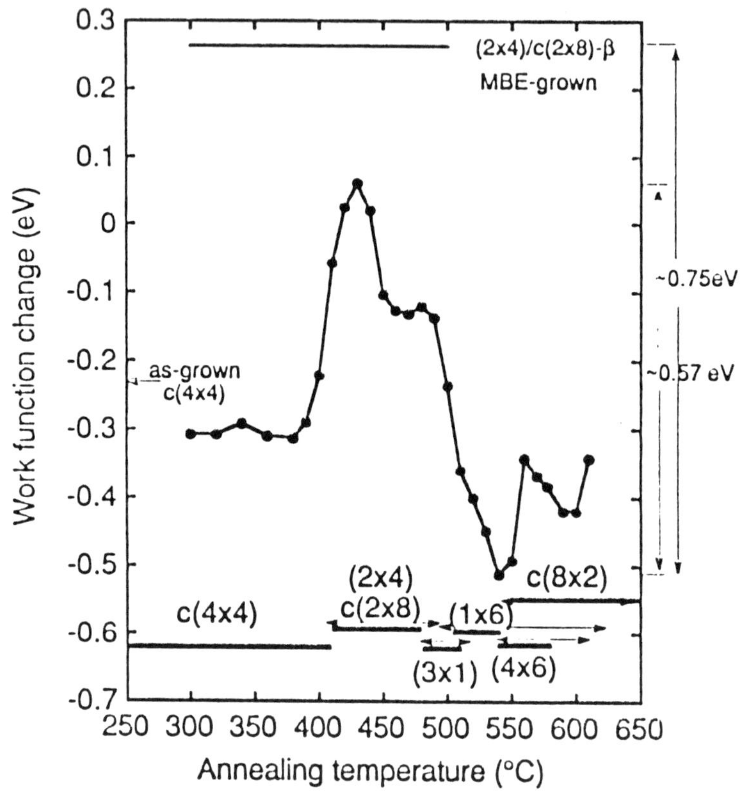


**Figure 53.** Plot of the work function change as a result of annealing in situ fabricated MBE-grown GaAs(100) c(4x4) As-rich surfaces. The work function for a (2x4)/c(2x8)-β surface fabricated in situ by annealing under an $As_4$-flux is also shown. The surface reconstructions are shown at the bottom of the figure. The work function of GaAs is not constant, but depends strongly on the surface reconstruction. A variation of ~0.75 eV in GaAs work function is observed. The variation in work function is believed to result from changes in the GaAs electron affinity and not band bending. (Adapted from Ref. 223.)

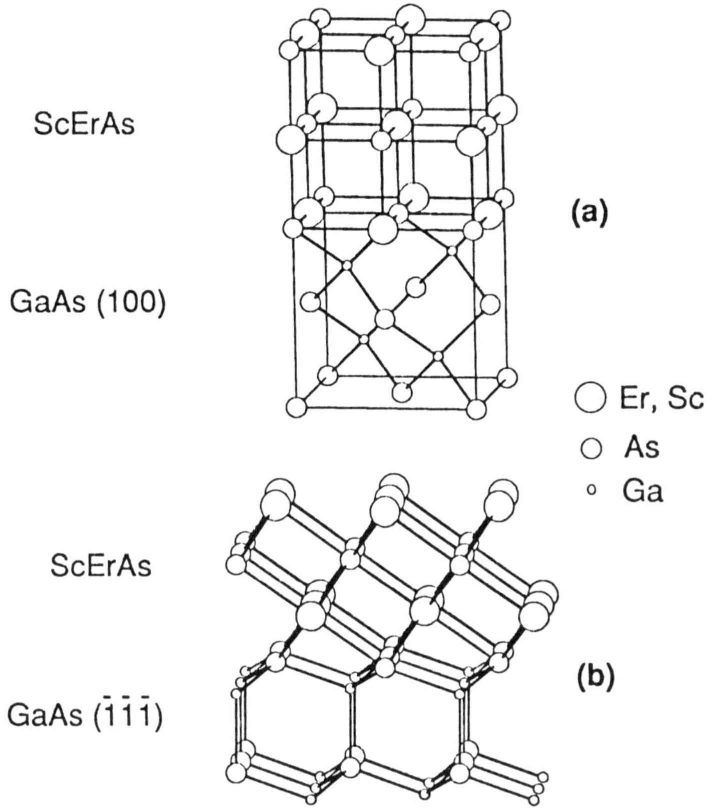


**Figure 54.** Schematic diagrams of possible $Sc_{1-x}Er_xAs$/GaAs interfaces: *(a)* $Sc_{1-x}Er_xAs$/GaAs(100) and *(b)* $Sc_{1-x}Er_xAs$/GaAs($\bar{1}\,\bar{1}\,\bar{1}$). The stacking sequence and interface bonding arrangements are very different in the two cases. In *(a)* the GaAs consists of alternating layers of Ga and As (polar-layers) and the $Sc_{1-x}Er_xAs$ consists of layers of $Sc_{1-x}Er_xAs$ (i.e. mixed layers which are non-polar). In *(b)* the GaAs consists of alternating layers of Ga and As (polar-layers) and the $Sc_{1-x}Er_xAs$ consists of layers of $Sc_{1-x}Er_x$ and As (polar-layers). This difference in stacking sequence between *(a)* and *(b)* must yield different interface structures. In going from GaAs to $Sc_{1-x}Er_xAs$ the stacking is made up of layers of As, Ga, As, Ga and then a mixed $Sc_{1-x}Er_xAs$ layer in *(a)*, and the stacking is made up of layers of Ga, As, Ga, As and then a $Sc_{1-x}Er_x$ layer in *(b)*. The interface bonding between the primarily covalently bonded GaAs and the primarily ionically bonded $Sc_{1-x}Er_xAs$ must be different in the two cases, which will result in different interface dipoles. The change in interface dipole is believed to be responsible for the variation in Schottky barrier height with orientation (Fig. 52).

Figure 52 shows the Schottky barrier height as a function of orientation for lattice matched layers grown at 500 - 600°C, and samples grown at 350 - 400°C and annealed at temperatures ≥ 500°C.[36][48] The barrier height is highest for (100) and it decreases approximately linearly with angle towards $(\bar{1}\bar{1}\bar{1})$ for $Sc_{0.32}Er_{0.68}As$ layers grown at 500 - 600°C.[48] As was observed for (100), deposition at ~350°C and annealing at ~500 - 600°C results in higher barrier heights than deposition at ~500 - 600°C for all orientations. The barrier height variation ranges from ~1.03 eV for (100) deposited at 350°C and annealed at ~580°C to ~0.63 eV for $(\bar{1}\bar{1}\bar{1})$ deposited at ~600°C, a total variation of ~0.4 eV for the same metallization on GaAs. It is difficult to see how subsurface defects in the GaAs, such as an antisite, can explain this difference since their energy levels should not be orientation dependent. Similarly, one may not expect MIGS to be strongly orientation dependent. Invoking surface defects to explain the annealing and orientation dependence of barrier height is more difficult to rule out, but considering the amount of arsenic at the interface may be too simplistic. Clearly, the interface bonding can be essential in that a change in bonding configuration will lead to a change in the interface dipole. It is interesting to note that bulk terminated surfaces of type {h11}, h ≥ 1, consist of ledges and steps of {100} and {111} character (Fig. 28). For example, the {311} surface has equal {100} and {111} character. Thus, if the Schottky barrier height is dependent upon substrate orientation and the surface $(\bar{1}\bar{1}\bar{1})$ yields the lower barrier height, one may expect to see a barrier height variation of that shown in Fig. 52 due to the change in the amount of {100} and {111} character.

For GaAs/ErAs/GaAs(100) interfaces, RBS-channeling rocking curve measurements[126] are consistent with the As sublattice being continuous across the interface and the last layer of As in the GaAs being incorporated in the first layer of $Sc_{1-x}Er_xAs$, as shown in Fig. 54a. Similar RBS-channeling rocking curve measurements on GaAs/$Sc_{1-x}Er_xAs$/GaAs$(\bar{3}\bar{1}\bar{1})$ were also consistent with a continuous As-sublattice across the interfaces.[227] The structure shown in Fig. 54b may be one of many possible structures for the $(\bar{1}\bar{1}\bar{1})$ case. Due to the high degree of twinning of the $Sc_{1-x}Er_xAs$ and GaAs overgrown layers, RBS-channeling rocking curve measurements could not be used for lattice site locations for the GaAs/$Sc_{1-x}Er_xAs$/GaAs{111} interface structure. However, without knowing the details of the interface structure it is clear that in the case of {100}, the $Sc_{1-x}Er_xAs$ stacking sequence consists of layers of $Sc_{1-x}Er_xAs$ (i.e., mixed layers which are non-polar) and the GaAs of alternating layers of Ga and As (polar), as is shown in Fig. 54a. In the {111} case, the $Sc_{1-x}Er_xAs$ stacking consists of alternating layers of $Sc_{1-x}Er_x$ and As (i.e., separated polar layers) and the GaAs of alternating layers of Ga and As (Fig. 54b). The $Sc_{1-x}Er_xAs$/GaAs interfaces should, therefore, be substantially different, both chemically and electronically depending on the crystal orientation. It is suggestive that the change in barrier height with orientation results from this difference in interface structure. Similar arguments have been used by Tung and co-workers[19][21][22] for the changes in the Schottky barrier heights in epitaxial $NiSi_2$/Si and $CoSi_2$/Si with Si and silicide orientation. In the specific case of $NiSi_2$/Si(111) theoretical calculations indicate that the difference in barrier height between type-A (unrotated) and type-B (rotated 180°) results from differences in the interface dipole due to the different bonding configuration.[29] This suggests that many of the concepts which have been applied to epitaxial Schottky barriers on Si are also applicable to epitaxial Schottky barriers on III-V compound semiconductors.

## 8.0 CONCLUSIONS

As an "ideal" metallization, the transition-metal group III intermetallics have the advantage of slightly lower resistivity than the RE-V compounds. They have, however, the disadvantage of having a greater potential for interface exchange reactions, which motivated the use of AlAs cladding layers on each side of buried NiAl layers. The lack of interface exchange reactions for RE-As on III-As semiconductors allows for tailoring the Schottky barrier height at each semiconductor/RE-As interface. For example, in metal-base transistor applications it is desirable to have a high barrier height for the emitter and a lower barrier height for the collector. This asymmetry would improve the electron transmission probability through the metal base. Such a structure could be realized by having a $Ga_{1-x}Al_xAs$/RE-As interface for the emitter and a RE-As/GaAs interface for the collector. This feature of tailoring the barrier height at each interface is not possible in epitaxial metal silicide/Si structures without the introduction of stress (i.e. with the use of Si-Ge alloys). For Si/$CoSi_2$/Si(111) metal base transistor structures, there is an additional complication of a very high quantum mechanical reflection coefficient at the Si/$CoSi_2$ interface. This has been shown theoretically to result from the lack of kinematically allowed bands in $CoSi_2$ near the Si conduction band minimum for the electrons to tunnel into.[228] Calculations for $NiSi_2$/Si(111) also suggest poor transmission and a strong dependence on the interface structure.[229] The situation may be better for other orientations. Similar calculations have not been performed on the GaAs/RE-As/GaAs structures. However, the fact that GaAs/RE-As/

GaAs heterostructures have been grown on numerous orientations suggests that by adjusting the barrier heights, RE-V compounds and crystal orientation, structures with minimum quantum mechanical reflection coefficients may be realized. Unfortunately, the quality of the overgrown III-V semiconductor may not yet be of sufficient quality to enable high performance metal base transistors to be a reality.

An additional advantage of RE-V compounds may arise from the fact that they are semi-metals with fairly low carrier density (electron and hole densities ~3 x $10^{20}/cm^3$), which suggests that the free carrier absorption should be significantly lower than for metals and that they may be useful as semi-transparent metallizations.

In summary, a worldwide effort to achieve the "ideal" III-V metallization was triggered in the mid-1980's by the realization that conventional metallizations, which were morphologically and electrically non-uniform as well as thermally unstable, would soon limit III-V device and circuit technology. By combining the phase diagram approach with advances in heteroepitaxy, the rare-earth monopnictides (e.g., ErAs) and the transition metal-group III metal intermetallics (e.g., NiAl) have been shown to meet the stability and epitaxy criteria that constitute, in part, the definition of an ideal metallization. Electrically continuous, thermally robust and epitaxial, films of these metallic phases as thin as three atomic planes have been grown sandwiched between epitaxial III-V semiconductor layers. These demonstrations make possible new classes of electronic, optoelectronic and magnetic devices that employ buried ultrathin metallic films as interconnects, semi-transparent electrodes, metallic quantum wells, magnetic storage and non-reciprocal optical (magneto-optic) elements. The experimental advances required to achieve such heterostructures have furthered our understanding of the epitaxy of dissimilar materials. These nearly ideal metal/III-V interfaces have also provided model systems for addressing the fundamental electronic behavior of metal/semiconductor interfaces, including the origins of Fermi level pinning.

The work on stable and epitaxial metal/III-V systems has implications as well for more conventional applications of contact metallizations, such as the source, drain and gate contacts in an FET. The identification of TM-III and RE-V phases as thermodynamically stable on III-V semiconductors forms the basis for the design of stable and shallow ohmic contacts and Schottky barriers fabricated by conventional deposition and lift-off techniques.[94] As device dimensions are reduced, the enhanced stability and uniformity of these metallizations will prove to be an important advantage in achieving high yield and reliability. In the more distant future, stable and epitaxial electrodes, interconnects and contacts may also be implemented to eliminate the morphological non-uniformities and grain-boundary-enhanced degradation mechanisms present in polycrystalline, low-melting-point Al and Au-based metallizations.

## ACKNOWLEDGMENTS

We gratefully acknowledge the many contributions of our present and former colleagues at Bellcore, Cornell University, University of Oslo, Rutgers University, University of Trondheim and Ecole National Supérieure d'Electronique et de Radioélectricité. In particular, the willingness of S. J. Allen, Jr. (UCSB), C. B. Carter (Univ. Minnesota), T. L. Cheeks, R. Duszak, T. G. Finstad (Oslo U.), J. P. Harbison, P. F. Miceli, N. G. Stoffel, N. Tabatabaie (Univ. Wisconsin), B. J. Wilkens and J. G. Zhu (NREL) to use some of their unpublished data. The many technical interactions with V. G. Keramidas (Bellcore), A. Guivarc'h (CNET, Lannion), S. A. Chambers (Pacific Northwest Laboratory), J. D. Ralston (Fraunhofer), R. Guérin (U. Rennes), J. Lee (Rutgers U.), G.-C. Joo (Rutgers U.), T. Tsakalakos (Rutgers U.), R. S. Williams (UCLA) and Y. A. Chang (U. Wisconsin-Madison) are much appreciated. The assistance of R. A. Hamilton in the manuscript preparation and P. Rickman in preparation of figures is also appreciated.

## REFERENCES

1. C. J. Palmstrøm and D. V. Morgan, in *Gallium Arsenide Materials, Devices, and Circuits*, (M. J. Howes and D. V. Morgan, eds.), p. 195, John Wiley & Sons Ltd., Chichester (1985)
2. R. C. Jaklevic, J. Lambe, M. Mikkor, and W. C. Vassell, *Phys. Rev. Lett.,* 26:88 (1971)
3. R. C. Jaklevic and J. Lambe, *Phys. Rev.,* B12:4146 (1975)
4. N. Tabatabaie, T. Sands, J. P. Harbison, H. L. Gilchrist, and V. G. Keramidas, *Appl. Phys. Lett.,* 53:2528 (1988)
5. N. Tabatabaie, T. Sands, J. P. Harbison, H. L. Gilchrist, L. T. Florez, and V. G. Keramidas, *Appl. Phys. Lett.,* 54:2112 (1989)
6. T. Sands, J. P. Harbison, M. L. Leadbeater, S. J. Allen, Jr., G. W. Hull, R. Ramesh, and V. G. Keramidas, *Appl. Phys. Lett.,* 57:2609 (1990)
7. G. A. Prinz, *Science,* 250:1092 (1991)

8. L. J. Brillson, *Surface Science Reports,* 2:123 (1982)
9. J. L. Freeouf and J. M. Woodall, *Appl. Phys. Lett.,* 39:727 (1981)
10. L. J. Brillson, R. E. Viturro, S. Chang, J. L. Shaw, C. Mailhiot, R. Zanoni, Y. Hwu, G. Margaritondo, P. Kirchner, and J. M. Woodall, *Mat. Res. Soc. Symp. Proc.,* 148:103 (1989)
11. W. E. Spicer, I. Lindau, P. R. Skeath, C. Y. Su, and P. Thye, *Phys. Rev. Lett.,* 44:420 (1980)
12. W. E. Spicer, Z. Liliental-Weber, E. Weber, N. Newman, T. Kendelewicz, R. Cao, C. McCants, P. Mahowald, K. Miyano, and I. Lindau, *J. Vac. Sci. Technol.,* B6:1245 (1988)
13. J. Tersoff, *Surf. Sci.,* 168:275 (1986)
14. E. H. Rhoderick and R. H. Williams, *Metal-Semiconductor Contacts,* Ch. 2 and references therein, Oxford University Press, Oxford, (1988)
15. W. Mönch, *Phys. Rev.,* B37:7129 (1988)
16. R. E. Allen and J. D. Dow, *Phys. Rev.,* B25:1423 (1982)
17. M. S. Daw and D. L. Smith, *Appl. Phys. Lett.,* 36:690 (1980)
18. R. E. Viturro, S. Chang, J. L. Shaw, C. Mailhiot, L. J. Brillson, A. Terrasi, Y. Hwu, G. Margaritondo, P. D. Kirchner, and J. M. Woodall, *J. Vac. Sci. Technol.,* B7:1007 (1989)
19. R. T. Tung, *Phys. Rev. Lett.,* 52:461 (1984)
20. R. T. Tung, *J. Vac. Sci. Technol.,* B2:465 (1984)
21. R. T. Tung and J. M. Gibson, *J. Vac. Sci. Technol.,* A3:987 (1985)
22. R. T. Tung, *J. Vac. Sci. Technol.* A, 7:598 (1989)
23. X. Wu, H. L. Evans, E. S. Yang, M. Liehr and P. S. Ho, *J. Vac. Sci. Technol.,* B3:1151 (1985)
24. M. Liehr, P. E. Schmidt, F. K. LeGoues, and P. S. Ho, *Phys. Rev. Lett.,* 54:2139 (1985)
25. M. Liehr, P. E. Schmid, F. K. LeGoues and P. S. Ho, *J. Vac. Sci. Technol.,* B3:1190 (1985)
26. J. G. Clabes, G. Meyer, H. von Löbbecke, and C.-D. Simm, *J. Vac. Sci. Technol.,* A3:931 (1985)
27. R. J. Hauenstein, T. E. Schlesinger, T. C. McGill, B. D. Hunt, and L. J. Schowalter, *Appl. Phys. Lett.,* 47:853 (1985)
28. M. Ospelt, J. Henz, L. Flepp, and H. von Känel, *Appl. Phys. Lett.,* 52:227 (1988)
29. G. P. Das, P. Blöchl, O. K. Andersen, N. E. Christensen, and O. Gunnarsson, *Phys. Rev. Lett.,* 63:1168 (1989)

30. C. J. Palmstrøm, N. Tabatabaie, and S. J. Allen, Jr., *Appl. Phys. Lett.,* 53:2608 (1988)
31. H. J. Richter, R. S. Smith, N. Herres, M. Seelman-Eggebert, and P. Wennekers, *Appl. Phys. Lett.,* 53:99 (1988)
32. C. J. Palmstrøm, K. C. Garrison, S. Mounier, T. Sands, C. L. Schwartz, N. Tabatabaie, S. J. Allen, Jr., H. L. Gilchrist, and P. F. Miceli, *J. Vac. Sci. Technol. B,* 7:747 (1989)
33. C. J. Palmstrøm, K. C. Garrison, B.-O. Fimland, T. Sands, and R. A. Bartynski, *Mat. Res. Soc. Symp. Proc.,* 144:583 (1989)
34. C. J. Palmstrøm, S. Mounier, T. G. Finstad, and P. F. Miceli, *Appl. Phys. Lett.,* 56:382 (1990)
35. A. Guivarc'h, J. Caulet, and A. LeCorre, *Electron. Lett.,* 25:1050 (1989)
36. C. J. Palmstrøm, T. L. Cheeks, H. L. Gilchrist, J. G. Zhu, C. B. Carter, and R. E. Nahory, *Mat. Res. Soc. Ext. Abstr.,* EA-21: 63 (1990)
37. C. J. Palmstrøm, J. P. Harbison, T. Sands, R. Ramesh, T. G. Finstad, S. Mounier, J. G. Zhu, C. B. Carter, L. T. Florez, and V. G. Keramidas, *Mat. Res. Soc. Symp. Proc.,* 198:153 (1990)
38. C. J. Palmstrøm, T. Sands, J. P. Harbison, T. G. Finstad, S. Mounier, L. T. Florez, V. G. Keramidas, J. G. Zhu, and C. B. Carter, *SPIE Proceedings,* 1285:85 (1990)
39. J. D. Ralston, H. Ennen, P. Wennekers, P. Hiesinger, N. Herres, J. Schneider, H. D. Muller, W. Rothemund, F. Fuchs, J. Schmalzlin, and K. Thonke, *J. Electron. Mater.,* 19:555 (1990)
40. J. D. Ralston, F. Fuchs, J. Schneider, and J. Schmälzlin, *J. Appl. Phys.,* 68:2176 (1990)
41. J. D. Ralston, J. Wagner, F. Fuchs, and P. Hiesinger, *J. Cryst. Growth,* 111:989 (1991)
42. A. Le Corre, J. Caulet, and A. Guivarc'h, *Appl. Phys. Lett.,* 55:2298 (1989)
43. J. Caulet, A. Le Corre, B. Guenais, G. Ropars, and A. Guivarc'h, *J. Crystal Growth.,* 102:309 (1990)
44. A. Guivarc'h, A. Le Corre, J. Caulet, B. Guenais, M. Minier, G. Ropars, P. A. Badoz, and J. Y. Duboz, *Mat. Res. Soc. Symp. Proc.,* 160:331 (1990)
45. A. Le Corre, B. Guenais, A. Guivarc'h, D. Lecrosnier, J. Caulet, M. Minier, G. Ropars, P. A. Badoz, and J. Y. Duboz, *J. Crystal Growth,* 105:234 (1990)

46. J. A. Chroboczek, P. A. Badoz, F. Arnaud d'Avitaya, A. Guivarc'h, A. Le Corre, W. Joss, A. Briggs, S. Auffret, and J. Pierre, *Proc. 20th Int. Conf. on the Physics of Semiconductors,* Thessalonicki (1990)

47. V. Durel, J. Caulet, Y. Ballini, M. Minier, B. Guenais, G. Dupas, and A. Guivarc'h, *Ext. Abstr. EUROMBE-91* (1991)

48. C. J. Palmstrøm, T. L. Cheeks, H. L. Gilchrist, J. G. Zhu, C. B. Carter, B. J. Wilkens, and R. Martin, *J. Vac. Sci. Technol. A,* 10:1946 (1992)

49. T. Sands, C. J. Palmstrøm, J. P. Harbison, V. G. Keramidas, N. Tabatabaie, T. L. Cheeks, R. Ramesh, and Y. Silberberg, *Mat. Sci. Rep.,* 5:99 (1990)

50. A. Guivarc'h, R. Guérin, and M. Secoué, *Electron. Lett.,* 23:1004 (1987)

51. T. Sands, J. P. Harbison, W. K. Chan, S. A. Schwarz, C. C. Chang, C. J. Palmstrøm, and V. G. Keramidas, *Appl. Phys. Lett.,* 52:1216 (1988)

52. J. P. Harbison, T. Sands, N. Tabatabaie, W. K. Chan, L. T. Florez, and V. G. Keramidas, *Appl. Phys. Lett.,* 53:1717 (1988)

53. A. Guivarc'h, J. Caulet, B. Guenais, Y. Ballini, R. Guérin, A. Poudoulec, and A. Regreny, *J. Cryst. Growth,* 95:427 (1989)

54. S. A. Chambers, *J. Vac. Sci. Technol. B,* 7:737 (1989)

55. V. Durel, B. Guenais, Y. Ballini, J. Caulet, A. Chomette, G. Dupas, G. Ropars, M. Minier, A. Guivarc'h, and A. Regreny, *Inst. Phys. Conf. Ser. No,* 112:129 (1990)

56. V. Durel, B. Guenais, Y. Ballini, J. Caulet, A. Chomette, G. Dupas, G. Ropars, M. Minier, A. Guivarc'h, and A. Regreny, *Le Vide, les Couches Minces* N°251, 106 (1990)

57. B. Guenais, A. Poudoulec, J. Caulet, and A. Guivarc'h, *J. Crystal Growth,* 102:925 (1990)

58. J. P. Harbison, T. Sands, R. Ramesh, N. Tabatabaie, H. L. Gilchrist, L. T. Florez, and V. G. Keramidas, *J. Vac. Sci. Technol. B,* 8:242 (1990)

59. T. Sands, J. P. Harbison, N. Tabatabaie, W. K. Chan, H. L. Gilchrist, S. A. Schwarz, C. L. Schwartz, L. T. Florez, and V. G. Keramidas, *Mat. Res. Soc. Symp. Proc.,* 144:571 (1989)

60. T. Sands, J. P. Harbison, N. Tabatabaie, W. K. Chan, H. L. Gilchrist, T. L. Cheeks, L. T. Florez, and V. G. Keramidas, *Surf. Sci.,* 228:1 (1990)

61. T. Sands, J. P. Harbison, R. Ramesh, C. J. Palmstrøm, L. T. Florez, and V. G. Keramidas, *Mater. Sci. and Eng. B,* 6:147 (1990)

62. P. Villars and L. D. Calvert, *Pearson's Handbook of Crystallographic Data for Intermetallic Phases, Vols. 2&3,* American Society for Metals, Metals Park, Ohio (1985)

63. F. Hulliger, in *Handbook on the Physics and Chemistry of Rare Earths, Vol. 4-Non-Metallic Compounds II*, (K. A. Gschneider, Jr. and L. Eyring, eds.), p. 153, North-Holland, New York (1979)

64. O. R. Vogt and K. Mattenberger, *J. Less-Comm. Metals,* 133:53 (1987)

65. A. Guivarc'h, M. Secoué, B. Guenais, Y. Ballini, P. A. Badoz, and E. Rosencher, *J. Appl. Phys.,* 64:683 (1988)

66. M. Hong, H. S. Chen, J. Kwo, A. R. Kortan, J. P. Mannaerts, B. E. Weir, and L. C. Feldman, *J. Cryst. Growth,* 111:984 (1991)

67. J. P. Harbison, T. Sands, R. Ramesh, L. T. Florez, B. J. Wilkens, and V. G. Keramidas, *J. Cryst. Growth,* 111:978 (1991)

68. J. R. Waldrop and R. W. Grant, *Appl. Phys. Lett.,* 34:630 (1979)

69. G. A. Prinz and J. J. Krebs, *Appl. Phys. Lett.,* 39:397 (1981)

70. S. A. Chambers, F. Xu, H. W. Chen, I. M. Vitomirov, S. B. Anderson, and J. H. Weaver, *Phys Rev. B,* 34:6605 (1986)

71. J. J. Krebs, B. T. Jonker, and G. A. Prinz, *J. Appl. Phys.,* 61:2596 (1987)

72. N. C. Koon, B. T. Jonker, F. A. Volkening, J. J. Krebs, and G. A. Prinz, *Phys. Rev. Lett.,* 59:2463 (1987)

73. M. Rubenstein, F. J. Rachford, W. W. Fuller, and G. A. Prinz, *Phys. Rev. B,* 37:8689 (1988)

74. M. W. Ruckman, J. J. Joyce, and J. H. Weaver, *Phys. Rev. B,* 33:7029 (1987)

75. R. Guérin, A. Guivarc'h, Y. Ballini, and M. Secoué, *Revue Phys. Appl.,* 25:411 (1990)

76. R. Guérin and A. Guivarc'h, *J. Appl. Phys.,* 66:2122 (1989)

77. R. Guérin, S. Deputier, J. Caulet, M. Minier, A. Poudoulec, Y. Ballini, V. Durel, G. Dupas, and A. Guivarc'h, *Mater. Res. Soc. Symp. Proc.,* 160:319 (1990)

78. J.-C. Lin, K.-C. Hsieh, K. Schulz, and Y. A. Chang, *J. Mater. Res.,* 3:148 (1988)

79. M. El-Boragy and K. Schubert, *Z. Metallkd.,* 72:279 (1981)

80. M. B. Panish, *J. Electochem. Soc.,* 114:516 (1967)

81. C. T. Tsai and R. S. Williams, *J. Mater. Res.,* 1:352 (1986)

82. R. S. Williams, J. R. Lince, T. C. Tsai, and J. H. Pugh, *Mater. Res. Soc. Symp. Proc.,* 54:335 (1986)

83. K. J. Schulz, X.-Y. Zheng, and Y. A. Chang, *J. Mater. Res.,* 4:1462 (1989)

84. M. D. Deal, R. A. Gasser, and D. A. Stevenson, *J. Phys. Chem. Solids,* 46:859 (1985)

85. K. J. Schulz, O. A. Musbah, and Y. A. Chang, *J. Appl. Phys.,* 67:6798 (1990)

86. X.-Y. Zheng, K. J. Schulz, J.-C. Lin, and Y. A. Chang, *J. Less-Common Met.,* 146:63 (1989)

87. M. B. Panish, *J. Electrochem. Soc.,* 114:516 (1967)

88. D. Swenson and Y. A. Chang, *Mater. Sci. and Eng. B,* 8:225 (1991)

89. M. El-Boragy and K. Schubert, *Z. Metallkde.,* 63:667 (1971)

90. R. Beyers, K. B. Kim, and R. Sinclair, *J. Appl. Phys.,* 61:2195 (1987)

91. R. Schmid-Fetzer, *J. Electronic Mater.,* 17:193 (1988)

92. T. Sands, *J. Metals,* 38:31 (1986)

93. C. J. Palmstrøm, K. C. Garrison, B.-O. Fimland, T. Sands, and R. A. Bartynski, paper presented at the Fall 1988 MRS meeting (1988)

94. T. Sands, *Mater. Sci. and Engin. B,* 1:289 (1989)

95. J.-C. Lin and Y. A. Chang, *Mater. Res. Soc. Symp. Proc.,* 148:3 (1989)

96. A. Guivarc'h, paper presented at the Electronic Materials Conference, Boulder, CO (1991)

97. C. J. Palmstrøm, B.-O. Fimland, T. Sands, K. C. Garrison, and R. A. Bartynski, *J. Appl. Phys.,* 65:4753 (1989)

98. T. Sands, *Appl. Phys. Lett.,* 52:197 (1988)

99. T. Sands, W. K. Chan, C. C. Chang, E. W. Chase, and V. G. Keramidas, *Appl. Phys. Lett.,* 52:1338 (1988)

100. S. A. Chambers, V. A. Loebs, and D. H. Doyle, *J. Vac. Sci. Technol.,* B8:985 (1990)

101. S. A. Chambers and V. A. Loebs, *J. Vac. Sci. Technol.,* B8:724 (1990)

102. S. A. Chambers and V. A. Loebs, *J. Vac. Sci. Technol.,* A 8:2074 (1990)

103. S. A. Chambers and V. A. Loebs, *J. Vac. Sci. Technol.,* A10:1940 (1992)

104. E. Bauer, *Z. Kristallogr.,* 110:372 (1958)

105. M. Copel, M. C. Reuter, M. Horn von Hoegen, and R. M. Tromp, *Phys. Rev.,* B42:11682 (1990)

106. J. E. Cunningham, J. E. Dura, and C. P. Flynn, in *Metallic Multilayers and Epitaxy,* (M. Hong, S. Wolf, and D. C. Gubser, eds.), p. 77, The Metallurgical Soc., Warrendale, PA (1988)

107. C. P. Flynn, *J. Phys. F: Met. Phys.,* 18:L195 (1988)

108. E. G. Bauer, B. W. Dodson, D. J. Ehrlich, L. C. Feldman, C. P. Flynn, M. W. Geis, J. P. Harbison, R. J. Matyi, P. S. Peercy, P. M. Petroff, J. M. Phillips, G. B. Stringfellow, and A. Zangwill, *J. Mater. Res.,* 5:852 (1990)

109. M. H. Grabow and G. H. Gilmer, *Surface Sci.,* 194:333 (1988)

110. R. Ludeke, *J. Vac. Sci. Technol.* B2:400 (1984)

111. H. Kroemer, *Mat. Res. Soc. Symp. Proc.,* 67:3 (1986)

112. R. Ludeke, L. L. Chang, and L. Esaki, *Appl. Phys. Lett.,* 23:201 (1973)

113. A. Y. Cho and P. D. Dernier, *J. Appl. Phys.,* 49:3328 (1978)

114. J. Massies, P. Delescluse, P. Etienne, and N. T. Linh, *Thin Solid Films,* 90:113 (1982)

115. S. K. Donner, K. P. Caffey, and N. Winograd, *J. Vac. Sci. Technol.,* B7:742 (1989)

116. R. F. C. Farrow, V. S. Speriosu, S. S. P. Parkin, C. Chien, J. C. Bravman, R. F. Marks, P. D. Kirchner, G. A. Prinz, and B. T. Zonker, *Thin Films: Stresses and Mechanical Properties,* 130:281 (1989)

117. C. J. Chien, J. C. Bravman, and R. F. C. Farrow, *J. Appl. Phys.,* 68:4343 (1990)

118. K. Okamoto, C. E. C. Wood, L. Rathbun, and L. F. Eastman, *J. Appl. Phys.,* 53:1532 (1982)

119. B. Tadayon, S. Tadayon, M. G. Spencer, G. L. Harris, L. Rathbun, J. T. Bradshaw, W. J. Schaff, P. J. Tasker, and L. F. Eastman, *Appl. Phys. Lett.,* 53:2664 (1988)

120. J. E. Oh, P. K. Battacharya, J. Singh, W. Dos Passos, R. Clarke, N. Mestres, R. Merlin, K. H. Chang, and R. Gibala, *Surf. Sci.,* 228:16 (1990)

121. P. Bhattacharya, J. E. Oh, J. Singh, D. Biswas, R. Clarke, W. Dos Passos, R. Merlin, N. Mestres, K. H. Chang, and R. Gibala, *J. Appl. Phys.,* 67:3700 (1990)

122. B. J. Wilkens, J. A. Martinez, S. Mounier, and C. J. Palmstrøm, *J. Appl. Phys.,* 70:4890 (1991)

123. P. F. Miceli, C. J. Palmstrøm, and K. W. Moyers, in *Surface X-ray and Neutron Scattering,* (H. Zabel and I. K. Robinson, eds.), p. 203, Springer Proceedings in Physics, Vol. 61 (1992)

124. P. F. Miceli, C. J. Palmstrøm, and K. W. Moyers, Appl. Phys. Lett., 61:2060 (1992)

125. M. H. Yang and C. P. Flynn, *Phys. Rev. Lett.,* 62:2476 (1989)

126. N. G. Stoffel, C. J. Palmstrøm, and B. J. Wilkens, *Nucl. Inst. and Meth.*, B56/57:792 (1991)

127. J. G. Zhu, C. J. Palmstrøm, S. Mounier, and C. B. Carter, *Mater. Res. Soc. Symp. Proc.*, 160:325 (1990)

128. J. G. Zhu, C. B. Carter, C. J. Palmstrøm, and S. Mounier, *Appl. Phys. Lett.*, 56:1323 (1990)

129. C. B. Carter, J. G. Zhu, and C. J. Palmstrøm, in *Proceedings 47th Ann. Meet. Electron Micros. Soc. Amer.*, (G. W. Bailey, ed.), p. 446, San Francisco Press, San Francisco (1989)

130. P. F. Miceli, unpublished

131. J. W. Matthews, in *Epitaxial Growth, Part B*, (J. W. Matthews, ed.), p. 559, Academic Press, New York (1975)

132. P. F. Miceli, C. J. Palmstrøm, and K. W. Moyers, *Appl. Phys. Lett.*, 58:1602 (1991)

133. P. F. Miceli, K. W. Moyers, and C. J. Palmstrøm, *Mat. Res. Soc. Symp. Proc.*, 202:579 (1991)

134. M. E. Mullen, B. Luthi, P. S. Wang, E. Bucher, L. D. Longinotti, and J. P. Maita, *Phys. Rev.*, B10:186 (1974)

135. T. G. Finstad, C. J. Palmstrøm, S. Mounier, V. G. Keramidas, J. G. Zhu, and C. B. Carter, *Mat. Res. Soc. Symp. Proc.*, 202:413 (1991)

136. T. G. Finstad, C. J. Palmstrøm, S. Mounier, and V. G. Keramidas, to be published

137. J. G. Zhu, C. J. Palmstrøm, and C. B. Carter, *Mat. Res. Soc. Symp. Proc.*, 198:177 (1990)

138. J. G. Zhu, C. J. Palmstrøm, and C. B. Carter, *Inst. Phys. Conf. Ser. No.*, 117:419 (1991)

139. J. G. Zhu, "Growth and Structural Characterization of Novel Semiconductor Heterostructures," Ph.D. Thesis, Cornell University (1991)

140. S. Z. Bokshtein, in *Diffusion and Structure of Metals*, 1:123, Metallurgiya, Moscow (1973)

141. T. Hayakawa, M. Kondo, T. Suyama, K. Takahashi, S. Yamamoto, and T. Hijikata, *Jap. J. Appl. Phys.*, 26:L302 (1987)

142. T. Hayakawa, M. Kondo, T. Morita, K. Takahashi, T. Suyama, S. Yamamoto, and T. Hijikata, *Appl. Phys. Lett.*, 51:1705 (1987)

143. K. Elcess, J.-L. Liévin, and C. G. Fonstad, *J. Vac. Sci. Technol.*, B6:638 (1988)

144. A. M. Dabiran, R. R. Keller, J. E. Angelo, W. W. Gerberich, and P. I. Cohen, to be published in J. Appl. Phys.

145. M. Y. Yen and T. W. Haas, *J. Vac. Sci. Technol.,* B8:308 (1990)

146. M. Y. Yen and T. W. Haas, *Appl. Phys. Lett.,* 56:2533 (1990)

147. K. C. Rajkumar, P. Chen, and A. Madhukar, *J. Appl. Phys.,* 69:2219 (1991)

148. L. Pfeiffer, K. W. West, H. L. Stormer, J. P. Eisenstein, K. W. Baldwin, D. Gershoni, and J. Spector, *Appl. Phys. Lett.,* 56:1697 (1990)

149. H. H. Farrell and C. J. Palmstrøm, to be published,

150. R. Nötzel, N. N. Ledentsov, L. Däweritz, M. Hohenstein, and K. Ploog, *Phys. Rev. Lett.,* 67:3812 (1991)

151. A. Guivarc'h, M. Secoué, and B. Guenais, *Appl. Phys. Lett.,* 52:948 (1988)

152. K. C. Garrison, C. J. Palmstrøm, and R. A. Bartynski, *Mat. Res. Soc. Symp. Proc.,* 144:613 (1989)

153. A. M. Wowchak, J. N. Kuznia, and P. I. Cohen, *J. Vac. Sci. Technol.,* B7:733 (1989)

154. M. Hansen, in *Constitution of Binary Alloys*, McGraw-Hill, New York (1958)

155. S. A. Chambers, Private Communication (1991)

156. J. P. Harbison, Private Communication (1991)

157. R. P. Elliot, in *Constitution of Binary Alloys, First Suppl.*, McGraw-Hill, New York (1965)

158. F. A. Shunk, in *Constitution of Binary Alloys, Second Supplement*, McGraw-Hill, New York (1969)

159. W. G. Moffatt, in *The Handbook of Binary Phase Diagrams*, Genium, Schenectady, New York (1984)

160. A. J. Bradley and A. Taylor, *Proc. Royal Soc.,* A159:50 (1937)

161. M. J. Cooper, *Phil. Mag.,* 8: 805 (1963)

162. D. A. Baugh, A. A. Talin, R. S. Williams, T.-C. Kuo, and K. L. Wang, *J. Vac. Sci. Technol.,* B9:2154 (1991)

163. K. Schubert, H. L. Lukas, H.-G. Meissner, and S. Bhan, *Z. Metallkde.,* 50:534 (1959)

164. K. M. Wunsch and E. Wachtel, *Z. Metallkde.,* 73:311 (1982)

165. C. J. Palmstrøm, C. C. Chang, A. Yu, G. J. Galvin, and J. W. Mayer, *J. Appl. Phys.,* 62:3755 (1987)

166. T. Sands, V. G. Keramidas, K. M. Yu, J. Washburn, and K. M. Krishnan, *J. Appl. Phys.,* 62:2070 (1987)

167. T. Sands, in *Proc. 45th Ann. Meet. of the Electron Micros. Soc. of Amer.*, (G. W. Bailey, ed.), p. 322, San Francisco Press, San Francisco (1987)

168. T. Sands, V. G. Keramidas, A. J. Yu, K.-M. Yu, R. Gronsky, and J. Washburn, *J. Mater. Res.,* 2:262 (1987)

169. S. H. Chen, C. B. Carter, and C. J. Palmstrøm, *J. Mater. Res.,* 3:1385 (1988)

170. A. Lahav, M. Eizenberg, and Y. Komem, *J. Appl. Phys,* 60:991 (1986)

171. S. A. Chambers and V. A. Loebs, *Appl. Phys. Lett.,* 60:38 (1992)

172. S. A. Chambers, Private Communication (1992)

173. G. C. Joo, T. Tsakalakos, and S. P. Chen, *Phil. Mag. Lett.,* 63:249 (1991)

174. G. C. Joo, Z. Kalman, T. Tsakalakos, S. P. Chen, T. Sands, J. P. Harbison, and V. G. Keramidas, to be published in Scripta Met.

175. T. C. Kuo, T. W. Kang, and K. L. Wang, *J. Cryst. Growth,* 111:996 (1991)

176. M. Tanaka, H. Sakakibara, and T. Nishinaga, *Appl. Phys. Lett.,* 59:3115 (1991)

177. K. Kamigaki, S. Yuda, H. Kato, M. Ishida, H. Terauchi, N. Sano, and S. Hiyamizu, *J. Appl. Phys.,* 69:2196 (1991)

178. S. A. Chambers, *J. Vac. Sci. Technol.,* A8:2062 (1990)

179. S. A. Chambers, *Phys. Rev.,* B42:10865 (1990)

180. T. L. Cheeks, T. Sands, R. E. Nahory, J. P. Harbison, N. Tabatabaie, H. L. Gilchrist, B. J. Wilkens, and V. G. Keramidas, *Appl. Phys. Lett.,* 56:1043 (1990)

181. T. Sands, J. P. Harbison, S. J. Allen, Jr., M. L. Leadbeater, T. L. Cheeks, M. J. S. P. Brasil, C. C. Chang, R. Ramesh, L. T. Florez, F. DeRosa, and V. G. Keramidas, *Mat. Res. Soc. Symp. Proc.,* 231:341 (1992)

182. H. Kono, *J. Phys. Soc. Japan,* 13:1444 (1958)

183. A. Morisako, M. Matsumoto, and M. Naoe, *J. Appl. Phys.,* 61:4281 (1987)

184. M. L. Leadbeater, S. J. Allen , Jr., F. DeRosa, J. P. Harbison, T. Sands, R. Ramesh, L. T. Florez, and V. G. Keramidas, *J. Appl. Phys.,* 69:4689 (1991)

185. J. Flahaut, in *Handbook on the Physics and Chemistry of Rare Earths, Vol. 4-Non-Metallic Compounds II,* (K. A. Gschneider, Jr. and L. Eyring, eds.), p. 1, North-Holland, New York (1979)

186. J. B. Taylor, L. D. Calvert, J. G. Despault, E. J. Gabe, and J. J. Murray, *J. Less-Common Metals,* 37:217 (1974)

187. G. Güntherodt, G. E. Kaldis, and P. Wachter, *Solid State Commun.,* 15:1435 (1974)

188. J. M. Honig, *J. Solid State Chem.,* 1:19 (1969)

189. E. Kaldis, G. von Schulthess, and P. Wachter, *Solid State Communications,* 17:1401 (1975)

190. E. V. Goncharova, V. P. Zhuze, V. V. Zhdanova, T. B. Zhukova, I. A. Smirnov, and E. V. Shadrichev, *Soviet Phys.-Sol. State,* 10:1052 (1968)

191. N. H. Andersen, P. E. Lindelof, H. Smith, O. Splittorff, and O. Vogt, *Phys. Rev. Lett.,* 37:46 (1976)

192. L. H. Brixner, *J. Inorg. & Nucl. Chem.,* 16:199 (1960)

193. W. M. Yim, E. J. Stofko, and R. T. Smith, *J. Appl. Phys.,* 43:254 (1972)

194. S. E. R. Hiscocks, and J. B. Mullin, *J. Mater. Sci.,* 4:962 (1969)

195. N. Sclar, *J. Appl. Phys.,* 33:2999 (1962)

196. A. Hasegawa and A. Yanase, *J. Phys. Soc. Jap.,* 42:492 (1977)

197. G. R. Caskey, J. M. Franz, and D. J. Sellmyer, *J. Phys. Chem. Solids,* 34:1179 (1973)

198. J. J. Rechtien, C. R. Kannewurf, and J. O. Brittain, *J. Appl. Phys.,* 38:3045 (1967)

199. Y. Yamaguchi, D. A. Kiewit, T. Aoki, and J. O. Brittain, *J. Appl. Phys.,* 39:231 (1968)

200. S. R. Butler, J. E. Hanlon, and R. J. Wasilewski, *J. Phys. Chem. Solids,* 30:1929 (1969)

201. S. J. Allen, Jr., N. Tabatabaie, C. J. Palmstrøm, G. W. Hull, T. Sands, F. DeRosa, H. L. Gilchrist, and K. C. Garrison, *Phys. Rev. Lett.,* 62:2309 (1989)

202. S. J. Allen, Jr., N. Tabatabaie, C. J. Palmstrøm, G. W. Hull, T. Sands, F. DeRosa, H. L. Gilchrist, and K. C. Garrison, *Surf. Sci.,* 228:13 (1990)

203. S. J. Allen, Jr., F. DeRosa, C. J. Palmstrøm, and A. Zrenner, *Phys. Rev.,* B43:9599 (1991)

204. R. Bogaerts, L. V. Bockstal, F. Herlach, F. M. Peeters, F. DeRosa, C. J. Palmstrøm, and S. J. Allen, Jr., *Physica,* B177:425 (1992)

205. S. N. Song, X. J. Yi, J. Q. Zheng, Z. Zhao, L. W. Tu, G. K. Wong, and J. B. Ketterson, *Phys. Rev. Lett.,* 65:227 (1990)

206. H. R. Child, M. K. Wilkinson, J. W. Cable, W. C. Koehler, and E. O. Wollan, *Phys. Rev.,* 131:922 (1963)
207. F. Hulliger and B. Natterer, *Solid State Comm.,* 13:221 (1973)
208. P. Junod, A. Menth, and O. Vogt, *Phys. Kondens. Mat.,* 8:323 (1969)
209. J.-B. Xia, S.-F. Ren, and Y.-C. Chang, *Phy. Rev.,* B43:1692 (1991)
210. N. Trivedi and N. W. Ashcroft, *Phys. Rev.,* B38:12298 (1988)
211. Z. Tesanovic, M. V. Jaric, and S. Maekawa, *Phys. Rev. Lett.,* 57:2760 (1986)
212. S. von Molnar and T. Penney, in *Localization and Metal Insulator Transitions,* (H. Fritzche and D. Adler, ed.), p. 183, Plenum, New York (1985)
213. S. J. Allen, Jr., T. Sands, and J. P. Harbison, unpublished
214. T. C. Kuo and K. L. Wang, *Appl. Phys. Lett.,* 59:3399 (1991)
215. K. Fuchs, *Proc. Cambridge Philos. Soc.,* 34:100 (1938)
216. Y. Silberberg and T. Sands, to be published in J. Quantum Electronics
217. T. Sands, J. P. Harbison, S. J. Allen, Jr., M. Leadbeater, T. L. Cheeks, M. J. S. P. Brasil, C. C. Chang, R. Ramesh, L. T. Florez, F. DeRosa, and V. G. Keramidas, *Mater. Res. Soc. Symp. Proc.,* 231:341 (1992)
218. T. L. Cheeks, T. Sands, R. E. Nahory, J. P. Harbison, H. L. Gilchrist, and V. G. Keramidas, *J. Electronic Mater.,* 20:881 (1991)
219. R. T. Tung, A. F. J. Levi, J. P. Sullivan, and F. Schrey, *Phys. Rev. Lett.,* 66:72 (1991)
220. K. Hricovini, G. L. Lay, A. Kahn, A. Taleb-Ibrahimi, J. E. Bonnet, L. Lassabatère, and M. Dumas, *Surf. Sci.,* 251/252:424 (1991)
221. D. K. Biegelsen, R. D. Bringans, J. E. Northrup, and L.-E. Swartz, *Phys. Rev.,* B41:5701 (1990)
222. J. Massies, P. Devoldere, and N. T. Linh, *J. Vac. Sci. Technol.,* 16:1244 (1979)
223. R. Duszak, C. J. Palmstrøm, L. T. Florez, Y.-N. Yang, and J. H. Weaver, *J. Vac. Sci. Technol.,* B10:1891 (1992)
224. W. Chen, M. Dumas, D. Mao and A. Kahn, *J. Vac. Sci. Technol.,* B10:1886 (1992)
225. I. M. Vitomirov, A. D. Raisanen, A. C. Finnefrock, R. E. Viturro, L. J. Brillson, P. D. Kirchner, G. D. Pettit, and J. M. Woodall, *J. Vac. Sci. Technol.,* B10:1898 (1992)
226. S. P. Svensson, J. Kanski, T. G. Andersson, and P.-O. Nilsson, *J. Vac. Sci. Technol.,* B2:235 (1984)

227. N. G. Stoffel, C. J. Palmstrøm, and B. J. Wilkens, unpublished,

228. L. F. Mattheiss and D. R. Hamann, *Phys. Rev.,* B37:10623 (1988)

229. M. D. Stiles and D. R. Hamann, *Phys. Rev.,* B40:1349 (1989)

230. R. Hanks and M. M. Faktor, *Trans. Faraday Soc.,* 63:1130 (1967)

231. G. A. Prinz, *Mat. Res. Soc. Symp. Proc.,* 56:139 (1986)

232. P. Villars and L. D. Calvert, *Pearson's Handbook of Crystallographic Data for Intermetallic Phases, Second Edition,* ASM International, The Materials Information Society, Materials Park, OH, (1991)

233. T. L. Cheeks, R. E. Nahory, T. Sands, J. P. Harbison, H. L. Gilchrist, and V. G. Keramidas, in paper presented at the March 1990 Meeting of the Amer. Phys. Soc. (1990)

# 3

# Schottky Barriers and Ohmic Contacts to Silicon

*Raymond T. Tung*

## 1.0 INTRODUCTION

Virtually all semiconductor electronic and opto-electronic devices receive power and signals from, and often transmit output to, metallic contacts which connect them to external circuitry. The fabrication and the performance of metal-semiconductor (MS) structures are of vital importance to all electronic devices and circuits. Our knowledge of the physical, electronic, and optical properties of semiconductors has progressed tremendously in recent years. So has our ability to process and manufacture high-precision nanostructures and devices on semiconductors. In contrast, our understanding of the electronic properties of MS interfaces has advanced little in the past several decades. Despite much progress in our understanding of the chemistry and metallurgy at MS interfaces, little is firmly understood or experimentally proven concerning the formation mechanism of the Schottky barrier (SB).[1]-[4] Until recently, the absence of a strong dependence of the experimentally observed Schottky barrier heights (SBH) on the metal work function was thought to be due to the presence of a high density of interface states which pin the Fermi-level (FL).[4][5] Electronic states due to defects[6]-[9] and metal induced gap states (MIGS)[5][10]-[12] were conjectured to be likely origins of FL pinning. According to such FL pinning mechanisms, the SBH of a MS system does not have a first-order dependence on the atomic structure of the interface. In this review, it will be pointed out that such a view is not consistent with the majority of experimental results.

It seems intuitively obvious that the structure of a MS interface should influence the decision on the SBH. The role played by the structure on the formation of the SB is one of the most pressing questions which experimental and theoretical studies must address. However, the relationship between the SBH and the interface structure has thus far not been assessed thoroughly because of two difficulties. The first has to do with the fact that polycrystallinity and other crystalline imperfections have made the atomic structures at an ordinary MS interface too complicated to study by experimental means. Without information on the structure of real MS interfaces, no first principles calculation of the interface electronic properties may be carried out. Existing SB models invariably resort to mechanisms not directly linked to the interface structure, not because such a connection does not exist, but because this relationship is too complicated to study. The second difficulty is the general ignorance on the issue of SBH uniformity, or the lack thereof. This issue is important because the possible spatial variation of the SBH at MS interfaces directly influences the identification of the main formation mechanism of the SB. For example, if the SB mechanism depends on structure, then the SBH at an ordinary MS interface should vary spatially, because the atomic structure varies at such an interface. One may add that if the FL were pinned, the SBH should be uniform within one SB diode. Hence, a knowledge about the homogeneity of the SBH can certainly clarify the role played by interface structure.

Recently, there have been developments in both of these areas of difficulty. Several nearly perfect *epitaxial* MS structures have been fabricated and their electronic properties investigated. Since these structures are single-crystalline, there is but one atomic structure for the entire MS interface. State of the art microscopic and spectroscopic techniques have been used to determine the atomic structures at epitaxial MS interfaces, which, in turn, have been used in direct calculations of the interface electronic structures. Epitaxial MS interfaces provide an unprecedented opportunity to test the validity of various SB models from first principles. Results from epitaxial MS interfaces show the importance of interface structure on the formation of the SB. The second development was a recent break-through in our understanding of the SBH uniformity issue. The reason that little evidence for SBH inhomogeneity has been reported is that, until recently, one did not know how to recognize it. One notes that the majority of SB investigations automatically assumes a uniform SBH. Many abnormal phenomena have routinely been observed from MS interfaces and have been explained, rather empirically, by a wide variety of mechanisms, often with questionable assumptions. It was recently pointed out,[13] however, that these anomalous behaviors are all in very good agreement with the presence of SBH inhomogeneity. Hence, SBH inhomogeneities are likely a common feature of most MS contacts and need to be characterized experimentally and addressed theoretically.

Fortunately, an imprecise knowledge about how SB's are formed has not impeded the satisfactory application of MS interfaces in advanced microelectronic devices. Because of the highly-developed purification processes for bulk crystals, attractive oxide characteristics, and unparalleled fabrication technologies, silicon has maintained its role as the leading semiconductor substrate on which the majority of microelectronic devices are fabricated. As the dimensions of ultra large scale integration (ULSI) devices continue to decrease, much more stringent requirements are being placed on the material, processing, and electrical performance of low-resistance MS contacts to Si. The front-end metallization of current ULSI processing is faced with many problems and will, undoubtedly, become even more difficult in the future. Understanding the basic concept of ohmic contacts and the practical materials issues in microelectronic processing is the only way to ensure that the advance of the contact technology can keep up with the demands of the Si ULSI industry.

In this paper, recent developments and progress in the physics and applications of Schottky barriers and ohmic contacts to Si are reviewed. The early development of SB theories and the formulation of the basic potential and transport equations at MS interfaces had drawn heavily from works involving SB's fabricated on Si. The various behaviors of Si SB diodes played a very important role in shaping the basic concepts with which we view the SB in general. It is the purpose of this paper to re-examine some classic experimental data and the associated basic concepts in the light of recent discoveries in this field. We also want to examine the present status of Si contact technologies, to anticipate the needs of next generation devices, and to comment on possible future contact schemes and strategies. The organization of this paper is as follows. We will first examine the traditional view of interface states and how they are thought to lead to FL pinning. This is followed by a slight diversion to discuss the newly developed treatment of inhomogeneous SB's, before returning to the discussion of the measurement of the SBH and the fabrication of SB diodes on Si. The experimentally observed SBH's from silicide and metal SB's are discussed, followed by careful analyses of the various "anomalous" behaviors, where it will be argued that the SBH's of ordinary polycrystalline SB's are frequently inhomogeneous. This leads to the discussion of the fabrication and the

SBH's of epitaxial silicide SB diodes, where it is shown that the atomic structure of the MS interface plays an important role on the formation of the SBH. Finally, we end with a discussion of the strategies and technologies of ohmic contacts to Si in the submicron device era. At the end of this chapter a glossary is provided for the convenience of the reader. It contains a listing of the symbols and notations used in this chapter.

## 2.0 SCHOTTKY BARRIER FUNDAMENTALS

### 2.1 Basic Band Diagram

The band diagram at the interface of a metal and a non-degenerate n-type semiconductor, at zero bias, is traditionally sketched in the fashion shown in Fig. 1(a). The difference between the conduction band minimum (CBM) and the FL is the flat-band SBH for n-type semiconductors, $\Phi_{Bn,o}$.

Eq. (1) $$\Phi_{Bn,o} \equiv E_C - E_F$$

The difference between the FL and the valence band maximum (VBM) is the flat-band SBH for p-type semiconductors, $\Phi_{Bp,o}$, as shown in Fig. 1(b). It is assumed that some fundamental mechanism at the MS interface has already led to a decision on the FL position (see discussion in the next section). Suffice it to mention, for the discussion at hand, that the interface FL position of any MS system is generally regarded as independent of the doping type and the doping level of the semicondutor. It follows that the sum of the two flat-band SBH's is simply the bandgap.

Eq. (2) $$\Phi_{Bn,o} + \Phi_{Bp,o} = E_g$$

The use of the expression *flat-band SBH* implies an intrinsic semiconductor is in contact with the metal. Generalized, it refers to the situation when SBH lowering mechanisms are absent. Since in SBH studies, the semiconductor is always doped, corrections to the apparent SBH are sometimes deemed necessary in order to deduce the true interface FL position.

The electric potential at the MS interface is governed by Poisson's equation. For a uniformly doped n-type semiconductor, the potential of the CBM with respect to the FL of the metal may be written, within the depletion approximation, as [see Fig. 1(c)]

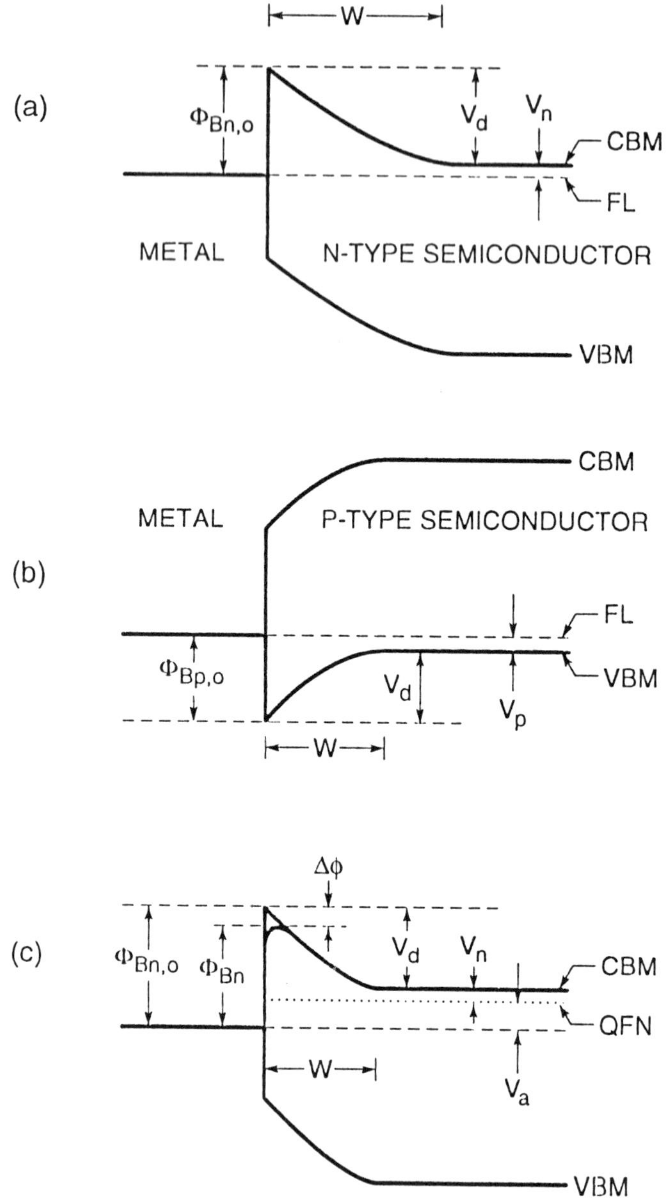


**Figure 1.** Band diagram at a MS interface, for *(a)* n-type semiconductor and *(b)* p-type semiconductor, in the absence of an applied bias. Under an applied forward bias, the quasi-Fermi level position and the band diagram for an n-type Schottky barrier is shown in *(c)*. SBH lowering is included only in *(c)*.

Eq. (3) $$V(z) = V_d(1 - z/W)^2 + V_n + V_a \qquad (0 \le z \le W)$$

where $V_a$ is the applied bias, $V_n$ is the difference between CBM and the FL in the neutral semiconductor, $V_d$ is the total band-bending (the diffusion potential), given by

Eq. (4) $$V_d = \Phi_{Bn,o} - V_n - V_a$$

and W is the depletion region width, given by

Eq. (5) $$W \approx \sqrt{\frac{2\varepsilon_s V_d}{qN_D}} = \sqrt{2\eta V_d}$$

where $\eta \equiv \varepsilon_s/(qN_D) = \beta\lambda_D^2$ and $\lambda_D$ is the Debye length of the semiconductor. Other symbols have their usual meanings. The maximum electric field near the MS interface is given by

Eq. (6) $$E_{max} \approx \sqrt{\frac{2qN_D V_d}{\varepsilon_s}} = \sqrt{\frac{2V_d}{\eta}}$$

All the equations and discussions presented so far, and those about to be presented, also apply to the metal/p-type semiconductor interfaces, with an appropriate change in the relevant subscripts.

## 2.2 FL Pinning and Interface States

Perhaps a more correct way of describing the SBH is to say that it is determined by how the semiconductor band edges at the interface have been positioned with respect to the FL of the metal. But more often, the situation at the MS interface is described using the bandgap of the semiconductor as a fixed reference point. For example, the determination of the SBH is often referred to as a measure of the interface FL position with respect to the semiconductor bands. As an example of this type of usage, one notes that the common phrase *FL pinning* usually describes the condition that the SBH, and therefore "the FL" is determined by some strong mechanism which makes the magnitude of the SBH relatively insensitive to the choice of the metal, the structure of the interface, the processing conditions, etc. It is noted that the term *FL pinning* has occasionally been used in the literature to indicate a SBH which is independent of the applied bias. The two meanings of FL pinning, although attributable to a single mechanism within certain SBH models, should be carefully distinguished. In general, a SBH which is independent of bias does not need to be formed by FL pinning. Occasionally, FL pinning is concluded, rather curiously, for some MS system simply because Eq. (7) below is not satisfied! One notes that the word *pinning* implies the presence of some special feature, such as a discrete interface state level or a charge-neutrality level, at the energy where the interface FL is established.

According to Schottky's original description,[14] the SBH between a metal with a work function of $\phi_m$ and a n-type semiconductor with an electron affinity of $\chi$ should be

Eq. (7) $$\Phi_{Bn,o} = \phi_m - \chi$$

as schematically shown in Fig. 2(a). However, experimentally observed SBH's are less dependent on the metal work function than predicted by Eq. (7).[15] This led to proposals of FL pinning by interface states. Present models of interface states can be approximately divided into two groups, namely, those which involve an interfacial insulating layer and those which do not. Bardeen was the first to attribute the lack of a metal dependence of the observed SBH to the presence of electronic states.[4] His proposal called for pinning by states on the semiconductor surface prior to the semiconductor being brought into contact with the metal. When the contact is made, there is simply a lineup of the two FL's, leading to a SBH which is independent of the metal, as schematically shown in Fig. 2(b). Bardeen's model involves no interaction at the interface and is, strictly speaking, only applicable when a thin dielectric layer separates the metal and the semiconductor. Under the assumption of a uniform distribution of interface states, $D_{ss}$, and a dielectric layer of a fixed thickness, d, the expected SBH may be expressed as a function of the metal work function:[16]-[18]

Eq.(8) $$\Phi_{Bn,o} = \gamma_{is}(\phi_m - \chi) + (1 - \gamma_{is})(E_g - \phi_o)$$

where the parameter $\gamma_{is}$, defined by

Eq. (9) $$\gamma_{is} = \left(1 + \frac{D_{ss}d}{\varepsilon_i}\right)^{-1}$$

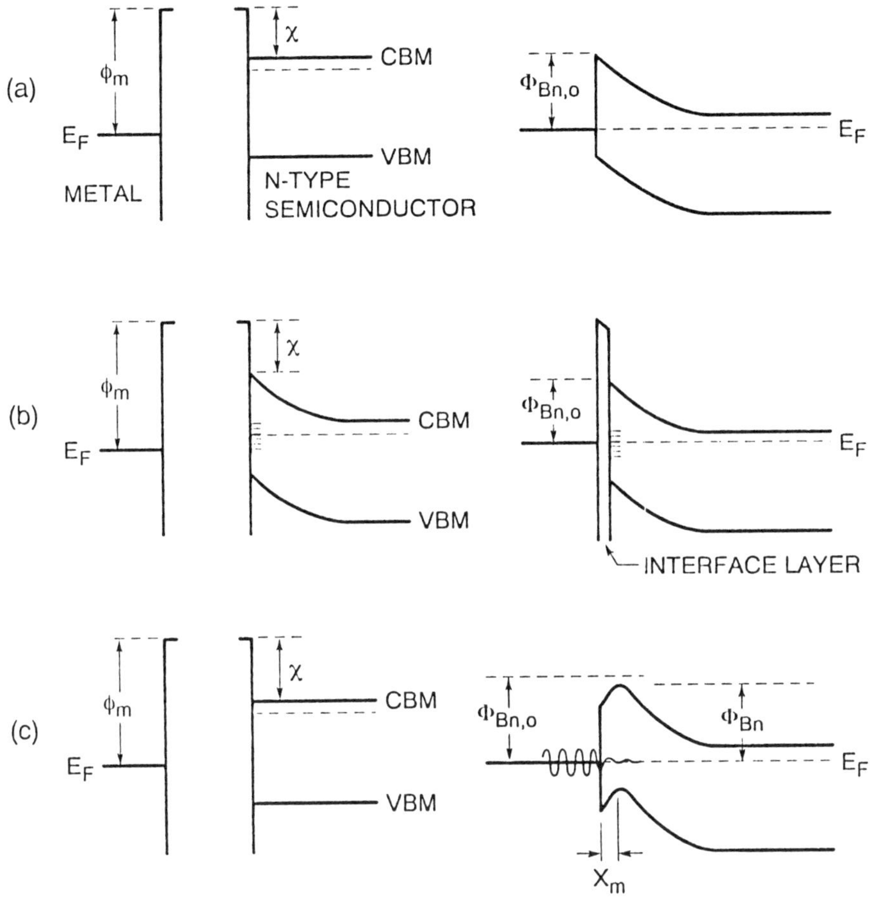


**Figure 2.** Band diagrams before *(left panel)* and after *(right panel)* the metal and the n-type semiconductor are brought into contact. *(a)* corresponds to the non-interacting Schottky model. *(b)* represents the situation when the semiconductor FL is pinned by surface states before metallization. A dielectric layer is assumed to be present at the subsequent MS interface to take up the difference in the two FL's. *(c)* is an intimate MS interface, with short-range band bending due to penetration of MIGS into the semiconductor. In *(c)*, the effective electrical interface (position of maximum electric potential) is located in the semiconductor, at some distance from the metallurgical MS interface. *(c)* is sometimes referred to as the negative charge model. Image force lowering is not included in these drawings [the upward band bending at the interface in *(c)* is not due to image-force lowering].

is related to the famous S-parameter[19][20] frequently discussed, and $\phi_o$ is the charge neutrality level of the interface states. Since interface states are assumed to be independent of the metal, Eq. (8) predicts a dependence of the SBH on the applied bias similar to that on the metal work function. As can be seen from Eq. (9), a critical parameter in such a model of the interface states is the product $D_{ss}d$. Clearly, the Bardeen limit corresponds to $\gamma_{is} = 0$ ( $D_{ss} = \infty$), and the Schottky limit, Eq. (7), is reached when $\gamma_{is} = 1$ or, equivalently, when $D_{ss} = 0$. Levine[21] and Crowell[22] pointed out that if the charge of the interface states follows an exponential distribution,

Eq. (10) $$Q_{ss} = Q_f\left[\exp\left(-\frac{\phi_b - \phi_o}{E_o}\right) - 1\right]$$

where $Q_f$ and $E_o$ are constants, then the $T_o$ anomaly observed in current-voltage (I-V) measurements of many SB's may be explained. Although not specifically discussed,[21] it is clear that the presence of a dielectric layer is also required in such a model.[22]

Subsequent to Bardeen's proposal, Heine[5] pointed out that for intimate MS contacts, electronic interaction between the metal and the semiconductor is unavoidable, even in the absence of any metallurgical reaction. Metal wave functions penetrate into the semiconductor and lead to MIGS and the formation of an interface dipole. For an analysis of the interface electronic properties, an exponential decay of the charge into the semiconductor, with a single decay length $\lambda_s \sim 5$ - 20 Å, is usually assumed. For Si, the depth of the metal wave function tail was estimated to be ~8 Å, with a density of ~$3 \times 10^{14}$ $cm^{-2}$ $eV^{-1}$.[5] It is understood that the (negative) charge extending into the semiconductor is offset by an equal (positive) charge residing in the metal. However, since the charge is assumed to penetrate a negligible distance into the metal compared to its spatial extension into the semiconductor, potential variation inside the metal is usually ignored. Therefore, once $\lambda_s$ and $D_{ss}$ are known, the band-bending at an intimate MS interface may be uniquely determined, as shown in Fig. 2(c).[23][24] An expression for the SBH at intimate MS interfaces almost identical to Eq. (8) may be deduced[25] with these assumptions of the MIGS model. Since the flat-band SBH is usually defined as the difference between the majority carrier band edge and the FL position at the "interface", it is interesting to note that, for a MIGS-derived band diagram, the "interface" is well inside the semiconductor, at some distance from the metallurgical interface (Fig. 2c). The short-range band bending due to MIGS, considered part of the formation mechanism of the SBH, supposedly varies with the metal in a fashion which makes the SBH's insensitive to the metal work function. Tersoff's[12] proposal that the FL position is determined by the charge neutrality condition of the MIGS is analogous to the strong-pinning (Bardeen) limit for interfaces with a dielectric layer. The mechanism proposed by Spicer et al,[6] i.e., interface FL position pinning by defect states, is independent of whether a dielectric layer is present at the interface. However, in the absence of a dielectric layer, a fixed separation of the defect states from the metal, of the order of 5 - 10 Å, is usually assumed,[26][27] for an estimation of the defect density.

## 2.3 Weaknesses of Interface States Models

While there is hardly any question that electronic states particular to the MS interface are present, the question is whether they lead to the formation of the SBH in the fashion proposed by existing models. Because of various experimental difficulties, it has not been possible to unequivocally deduce the distribution of interface states by experimental means. Techniques such as photoemission,[28] junction capacitance,[29]-[31] photoelectric spectroscopy,[32] and optical absorption[33] measurements, thus far employed for this purpose, all involve arbitrary assumptions, the validity of which is difficult to evaluate. The most popular of these techniques, the capacitance-voltage (C-V) technique at forward bias, assumes the existence of an insulating layer or some other mechanism which leads to a fixed separation of charges on the semiconductor side from that on the metal side of the MS interface.[34] Under an applied bias, the charge on the semiconductor side is assumed to be controlled by the quasi-FL position. Such an assumption, which is actually inconsistent with the band diagrams for intimate SB interfaces, has to be made because without it, interface states cannot lead to excess capacitances.[35] There is no experimental evidence or scientific basis for the assumption of a fixed separation between the interface states and the metal at intimate MS interfaces. The assumption that the population of interface states is governed by the quasi-FL is also unfounded. Besides these obvious difficulties of attributing excess capacitance at intimate SB junctions to interface states, one notes that minority carriers[36] can lead to large junction capacitances. Hence, the issue of interface states is far from being resolved experimentally. Even though experimental evidence for a causal relationship between interface gap states and the SBH is generally lacking, FL pinning by interface states seems to be a concept which has already been, mistakenly, accepted. This likely has to do with a (mis)conception that interface states are able to account for both a lack of dependence of the SBH on metal work function and, at the same time, a host of non-ideal behaviors routinely observed from SBH experiments. Actually, the fact that interface state models have implications on both the SBH's dependence on metal work function and its dependence on other parameters (such as bias, $N_D$, and T) raises doubts concerning the validity of these models, rather than providing support. For example, it has been pointed out that a dilemma is created if the interface states responsible for FL pinning are also the ones responsible for the ideality factor.[37] Discussions below will show that the formation mechanism of the SBH may be irrelevant to the apparent dependencies of the SBH on bias and other parameters.

Present models of interface states involve many assumptions. An obvious assumption of interface state models is that the positive charge and the negative charge giving rise to the interface dipole may be separated physically and controlled by different quasi-FL positions. Another simplification of existing models is the assumed independence of the states on the interface structure. As a result, the distribution of interface states is assumed to be laterally homogeneous. According to Eq. (9), both the density and the effective depth of the interface states have to remain constant for the local SBH to remain laterally uniform. At real MS interfaces, it seems unlikely that these conditions are met. For example, the native oxide layer, suspected of being present at some MS interfaces, is most certainly non-uniform in its thickness and/or its composition. Structural defects are certainly not, as present models invariably assume, uniformly distributed at a constant distance from the MS interface. As will be discussed, many consequences of an inhomogeneous SBH cannot be derived from transport equations involving homogeneous SB's. It appears that the very assumptions which have made the interface state models easy to analyze may have made them too simplified to describe real experimental situations.

The existence of charge neutrality[10][12] is obviously an important reason for the observed insensitivity of SBH on metal work function. However, one notes that the neutrality of the entire MS interface region, including the space charge region, is always satisfied. It is not clear that the requirement of charge neutrality, stemming from the minimization of energy, should be exercised on a local level. Furthermore, it is interesting to ponder where one should impose this additional charge neutrality rule.[12] As was noted, within the MIGS picture, the flat-band SBH is determined by the asymptotic FL position at the tail-end of the MIGS distribution, where approximate charge neutrality is already guaranteed due to the rapid decay of MIGS away from the MS interface. However, Tersoff's charge neutrality criterion seems to apply to the location of maximum MIGS concentration, i.e., the MS interface. It further implies that the short-range band-bending near the MS interface is missing. Since the interaction between metal wave functions and semiconductor wave functions is strongest at the interface, it is not clear how accurately the charge neutrality level can be estimated using the band structure of the semiconductor alone. It seems inappropriate to concentrate only on the energetic distribution of the charge, since it is the *spatial* distribution of all the charge at the MS interface region which accounts for the interface dipole and, hence, the SBH.

What has been generally overlooked in interface states models so far is that the distribution of states with energies lying within the semiconductor bandgap represents only a small portion of the total redistribution of charge at the MS interface. After all, interface dipole may arise not only from a change in the population of states but also from a change in the electron wave functions. Recent theoretical calculations[38][39] based on the atomic structures at real MS interfaces show that the interaction of the conduction band of the metal and the valence band of the semiconductor plays a more important role in establishing the dipole at the MS interface than do states in the bandgap. Almost all the essential assumptions of the interface gap state models, including *(i)* a larger penetration of interface charge into the semiconductor than into the metal, *(ii)* a unilateral transfer of charge across the MS interface, *(iii)* an independence of the MIGS distribution on the interface atomic structure, and *(iv)* a net charge transfer in the direction of the metal (the negative charge model), are found to be violated at realistic MS interfaces, as shown in Figs. 27 and 28 below. In fact, it has not been possible to identify the role played by MIGS in the formation of the SBH. It thus appears that, in order to understand the formation mechanism of the SBH, one should not concentrate exclusively on MIGS and ignore the role played by bonding electrons.

## 2.4 SBH Measurement Techniques

There are many techniques with which to determine the SBH of a MS interface. I-V, C-V, photoresponse/internal photoemission (PR), and surface photoemission spectroscopy (PES) are the most commonly used techniques. The principles and practices of these techniques have been discussed at length elsewhere in this volume and will not be reiterated.

However to facilitate subsequent discussions, we write down the equation for the junction current

Eq. (11) $$I(V_a) = I_s\left[\exp\left(\frac{qV_a}{nk_BT}\right) - 1\right]$$

where the saturation current, $I_s$, is given by the thermionic emission theory as

Eq. (12) $$I_s = A^*AT^2\exp\left(-\frac{q\Phi_{Bn}}{k_BT}\right) = A^*AT^2\exp(-\beta\Phi_{Bn})$$

where A* is the Richardson constant, A is the area of the diode, and $\beta \equiv q/(k_BT)$. The ideality factor, n in Eq. (11), is a fit to the slope of an experimentally obtained (semi-logarithmic) I-V curve. An ideality factor of 1 is predicted by the thermionic emission theory. It is clear that if formulas like Eqs. (11) and (12) are used to analyze any I-V trace which has a semi-logarithmic relationship, a single parameter $\Phi_{Bn}$ is obtained. When the SB is homogeneous, $\Phi_{Bn}$ may be identified as the SBH for n-type semiconductors ($\Phi_{Bp}$ for p-type semiconductors). It is generally assumed that the interface FL may be deduced (or defined) by adding to the $\Phi_{Bn}$ measured from I-V and PR experiments, a correction term, $\delta\Phi$, due to various SBH lowering mechanisms, such as image-force lowering, negative interface charge, and thermionic-field emission (TFE),

Eq. (13) $$\Phi_{Bn,o} = \Phi_{Bn} + \delta\Phi$$

as schematically shown in Fig. 1(c). On the other hand, the C-V method gives the flat-band SBH's, $\Phi_{Bn,o}$, because it is not influenced by SBH lowering mechanisms. In Table 1, some important numbers relevant to the measurement of SB's on silicon are listed.

Investigations of the lateral uniformity of the SBH have occasionally been reported. Electron beam induced current (EBIC)[40] and voltage (EBIV)[41] make use of a possible variation of the local depletion width due to SBH inhomogeneity, and typically have a lateral resolution on the order of the larger of the beam excitation size or the depletion width. The lateral resolution of scanning cathodoluminescence and photoresponse are typically much larger than the depletion length. Recently, exciting prospects of studying SBH variations on an atomic scale were offered by the invention of ballistic electron emission microscopy (BEEM).[42] Electrons which tunnel from a scanning tunneling microscope (STM) tip at adjustable energies and traverse the very thin metal layer ballistically are used as a probe of the local SBH. The transmission of these electrons through the MS interface can then be used to deduce valuable information on the electronic structure of the interface. However, since electrons have to reach the neutral semiconductor, the resolution of BEEM is also limited by potential pinch-off in a fashion similar to any I-V study, as recently pointed out.[13][43] Furthermore, effects due to band structures can also make the interpretation of BEEM difficult. The true advantage of BEEM is its unique ability to correlate local effective SBH's to the structures concurrently studied.

**Table 1.** Relevant Parameters of Si Schottky Barriers

| Parameter | Value at 300K | |
|---|---|---|
| | n-type | p-type |
| Effective Density of States ($cm^{-3}$) | $N_C = 3.22 \times 10^{19}$ | $N_A = 1.83 \times 10^{19}$ |
| Richardson Constant (Amp $K^{-2}$ $cm^{-2}$) | 112 | 32 |
| Resistivity, $1 \times 10^{15}$ $cm^{-3}$ doping | 4.5Ω-cm | 14Ω-cm |
| Resistivity, $1 \times 10^{16}$ $cm^{-3}$ doping | 0.5Ω-cm | 1.5Ω-cm |
| Resistivity, $1 \times 10^{17}$ $cm^{-3}$ doping | 0.08Ω-cm | 0.2Ω-cm |
| Band Gap | 1.12 eV | |
| Dielectric Constant | 11.8 | |
| Electron Affinity (111) | 4.85 eV | |

## 2.5 Potential of Inhomogeneous SB's

Since the very earliest of SBH studies, there has been an automatic assumption about the uniformity of the FL position at a MS interface in the vast majority of the investigations. A priori, there is not any more reason to assume the SBH to be homogeneous than there is to assume that it is inhomogeneous. However, because this tacit assumption of homogeneity has never been challenged, possible variation in the SBH has not been treated as a major issue. SB diodes often exhibit many non-ideal behaviors which have, until recently, been attributed to interface states. That these abnormal behaviors may simply be explained on the basis of SBH inhomogeneity was recently pointed out.[13] Philosophically, one may argue that it makes more sense to start the treatment of SB junctions with a general theory of inhomogeneous SB's than it is to immediately assume a uniform SBH. After all, a theory of inhomogeneous SB's may be easily applied to treat the homogeneous SB's as a special case. But historically, the logic has worked in reverse, namely, one usually assumes that the behavior of an inhomogenous SB may be accounted for by a simple summation of behaviors observed at homogeneous SB's with different SBH's. To understand why and how such a view is incorrect, a knowledge of the potential of an inhomogeneous SB is required.

When the SBH varies locally at a MS interface, the potential also varies from region to region. The solution to such a problem is usually obtained by solving Poisson's equation, with the SBH contours supplied as the boundary condition,

Eq. (14) $$V(x,y,0) = \Phi_{Bn,o}(x,y) = \Phi_{mean} + \delta(x,y)$$

where $\delta$ is the difference between the local SBH and a "mean" SBH, $\Phi_{mean}$. Numerical solutions to this boundary value problem in two convenient geometries all showed the interesting phenomenon of potential pinch-off.[44]-[48] *Pinch-off* is a terminology often used to describe the operation of a field effect transistor. In the present context, an area (of the MS interface) is said to be pinched-off if majority carriers originating from outside the space charge region need to go over a potential barrier, higher than the local majority band-edge position at the MS interface, in order to reach it. The two geometries studied are the most interesting forms of possible SBH inhomogeneities consisting of small areas of the MS interface with a low SBH, embedded in an interface with an, otherwise uniform, higher SBH. One is in the form of a circular patch[44][45][48] with a small radius, $R_o$, the other consists of a strip[46]-[48] with a width of $L_o$, placed at the origin. The SBH is $\Phi_{mean}$ everywhere except for the patch/strip which has a lower but constant SBH of $\Phi_{mean} - \Delta$.

Potential pinch-off has been demonstrated in an extensive numerical study of inhomogeneous SB's with a multi-purpose device simulator.[48] When pinch-off occurs, there is a *saddle-point*, in front of every low-SBH patch/strip, whose potential is a local maximum along the *z*-axis, as shown in Fig. 3, but is a local minimum along the lateral axis, as shown in Fig. 4. Obviously, the height and the distribution of the potential at the saddle-points are of vital importance to understanding the physics at inhomogeneous SB's, because they control the electron transport to and from the low-SBH patches/strips. However, the complicated dependencies of the potential distribution near the saddle-points on the patch/strip characteristics, the applied bias, the doping level, the temperature, etc., make a conceptual grasp of the pinch-off phenomenon difficult through numerical simulations.[44][47] In addition, it is impractical to solve the boundary value problems for a large number of parameters in order to explain the electrical data from inhomogeneous SB's.

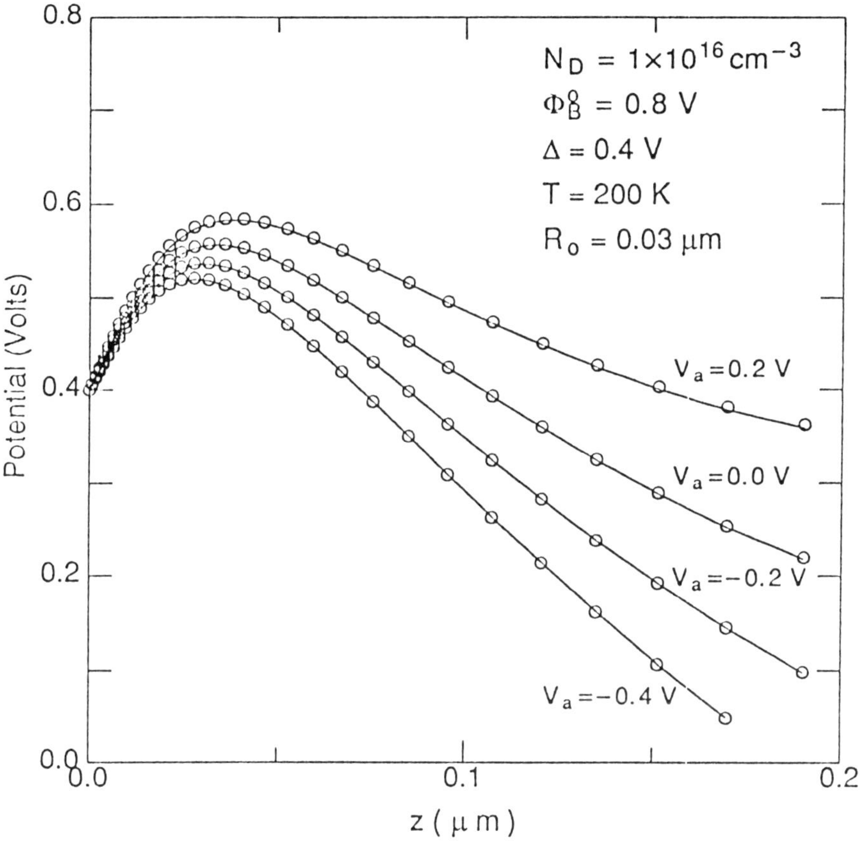


**Figure 3.** Potential of the CBM at different applied biases, in front of a low-SBH (0.4 V) circular patch which is embedded in a high-SBH (0.8 V) background. Circles represent a numerical solution to Poisson's equation.[48] Solid curves are calculated from the analytic expression[13] for a dipole layer, Eq. (15). The saddle point potential, at z ~ 0.03 - 0.04 $\mu$m, varies with the applied bias.

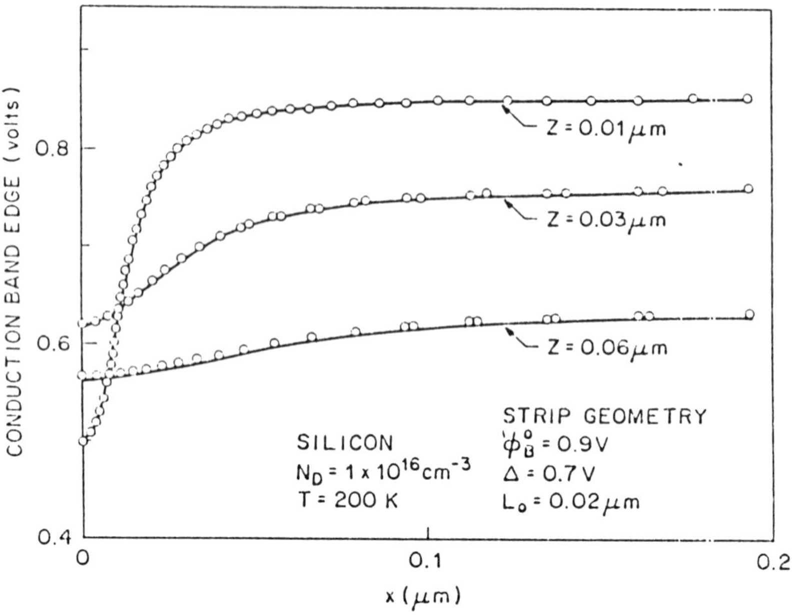


**Figure 4.** CBM potential at zero bias, referenced to the FL, in front of a low-SBH region in the shape of a narrow strip. The saddle point is close to z = 0.03 μm. Circles are from a numerical solution to Poisson's equation.[48] Curves are calculated from an analytic theory,[13] Eq. (16).

Fortunately, an analytic solution to the potential (and the electron transport) at inhomogeneous SB's was recently obtained.[13] For instance, the potential for the patch geometry mentioned above may be expressed in the analytic form[13]

Eq. (15)
$$V_{patch}(0,0,z) = V_d\left(1-\frac{z}{W}\right)^2 + V_n + V_a - \Delta\left[1 - \frac{z}{\sqrt{z^2+R_o^2}}\right]$$

For the strip geometry, the potential is[13]

Eq. (16)

$$V_{strip}(x,y,z) = V_d\left(1-\frac{z}{W}\right)^2 + V_n + V_a - \frac{\Delta}{\pi}\tan^{-1}\frac{|x|+\frac{L_o}{2}}{z} + \frac{\Delta}{\pi}\tan^{-1}\frac{|x|-\frac{L_o}{2}}{z}$$

In Eqs. (15) and (16), the first term is the potential due to a SBH of $\Phi_{mean}$ ($V_d$ and $W$ are parameters corresponding to the average SBH, $\Phi_{mean}$, namely, $V_d = \Phi_{mean} - V_n - V_a$), and the last term(s) is due to the low-SBH patch/strip. The near perfect agreement between these analytic expressions and numerical solutions generated by computer simulations, e.g., Figs. 3 and 4, has been extensively demonstrated.[48]

## 2.6 Electron Transport of Inhomogeneous SB's

Traditionally, electron transport at inhomogeneous MS junctions has been treated by a parallel conduction model, namely, the current is assumed to be a sum of currents flowing in all the individual patches ($I_i$), each with its own area ($A_i$) and SBH ($\Phi_i$):

Eq. (17) $$I(V_a) = \sum_i I_i(V_a) = A^*T^2\left[\exp(\beta V_a) - 1\right]\sum_i A_i \exp(-\beta\Phi_i)$$

Such a concept was so intuitively obvious that this model had been applied many times[49]-[51] before it was formally discussed by Ohdomari and Tu.[52] Note that the $\Phi_i$'s in Eq. (17) are the local SBH's at the MS interface. From previous discussions of the potential at inhomogeneous SB's, it is clear that when pinch-off occurs, the minimum potential barrier between a low-SBH area at the MS interface and the neutral semiconductor is the *height* of the potential at the saddle point, situated some distance away from the MS interface plane. The saddle-point potential may be significantly higher than that at the MS interface. Therefore, when SBH inhomogeneity occurs on a lateral scale less than the depletion length, Eq. (17) is no longer accurate.

Numerical simulations of electron transport at inhomogeneous SB's, taking into account the pinch-off effect, have existed for some time.[45][47] However, it was not until recently that the importance of the saddle-point potential has been recognized and that the concept of assigning an effective SBH to each low-SBH area has been developed.[13] Numerical simulations[48] gave excellent support to expressing the total current as a modified sum of currents flowing in each patch:[13].

Eq. (18) $$I(V_a) = A^*T^2\left[\exp(\beta V_a) - 1\right]\sum_i A_{i,eff} \exp(-\beta\Phi_{i,eff})$$

where the effective SBH, $\Phi_{i,eff}$, is simply the height of the saddle-point in front of the i-th low-SBH patch. The effective area of a patch, $A_{i,eff}$, is related to the rate of lateral ascent of the potential near the saddle-point. Note that the difference between Eqs. (17) and (18) is that $A_{i,eff}$'s and $\Phi_{i,eff}$'s depend on the applied bias, the doping level, the temperature, etc., while $A_i$'s and $\Phi_i$'s do not. Since the analytic theory describes the potential near the saddle-point extremely well, it is not surprising that the numerically simulated currents[48] showed quantitative agreement with the predictions of the analytic theory using only saddle-point potentials, i.e., Eq. (18). This discovery obviates the need for complicated computer calculations and makes physical interpretations of the experimental data transparent. Results from the analytic theory of inhomogeneous SB's are summarized in Table 2. Basically, the potential at a low-SBH patch (strip) may be uniquely described by a region parameter $\gamma(\omega)$ which measures the strength of a low-SBH patch (strip). The dimensionless parameter $\Gamma(\Omega)$, on the other hand, is a parameter for the space charge region, which measures the current-carrying capability of a patch (strip). It is clear that since the saddle point potential increases with forward bias (Fig. 3), the component of current flowing to a low-SBH area has an ideality factor greater than 1. Simulated ideality factors as functions of the region parameter and the doping level are shown in Fig. 5, illustrating the overall success of the analytic equations[13] in Table 2 in describing the electron transport of inhomogeneous SB's. The effect of series resistance on the current transport of a low-SBH patch may be approximated by using the spreading resistance[53] for a disc with an area equal to $A_{eff}$.

**Table 2.** Electron transport at an isolated low-SBH region with a local SBH of $\phi_{mean} - \Delta$, surrounded by regions with a SBH of $\phi_{mean}$.

| Geometry | Circular Patch | Semi-infinite Strip |
|---|---|---|
| Dimension | Radius = $R_o$ | Width = $L_o$, Length = $L_{strip} >> L_o$ |
| Region Parameter | $\gamma = 3(\Delta R_o^2/4)^{1/3}$ | $\omega = 2(\sqrt{2}\,L_o\Delta/\pi)^{1/2}$ |
| Space Charge Parameter | $\Gamma = (R_o^2\Delta/4W^2V_d)^{1/3}$ | $\Omega = (L_o\Delta/2\pi W V_d)^{1/2}$ |
| Effective Area | $A_{eff} = (4\pi\gamma/3\beta)(\eta/V_d)^{2/3}$<br>$= (4/3)\pi\lambda_D^2\Gamma$ | $A_{eff} = (\pi\omega/\beta)^{1/2}(\eta/V_d)^{3/8}L_{strip}/2$<br>$= \sqrt{\pi\Omega}\,\lambda_D L_{strip}$ |
| Effective SBH | $\Phi_{eff} = \Phi_{mean} - 3\Gamma V_d$<br>$= \Phi_{mean} - \gamma(V_d/\eta)^{1/3}$ | $\Phi_{eff} = \Phi_{mean} - 4\Omega V_d$<br>$= \Phi_{mean} - \omega(V_d/\eta)^{1/4}$ |
| Ideality Factor | $n \approx 1 + \Gamma$<br>$= 1 + \gamma\eta^{-1/3}V_d^{-2/3}/3$ | $n \approx 1 + \Omega$<br>$= 1 + \omega\eta^{-1/4}V_d^{-3/4}/4$ |

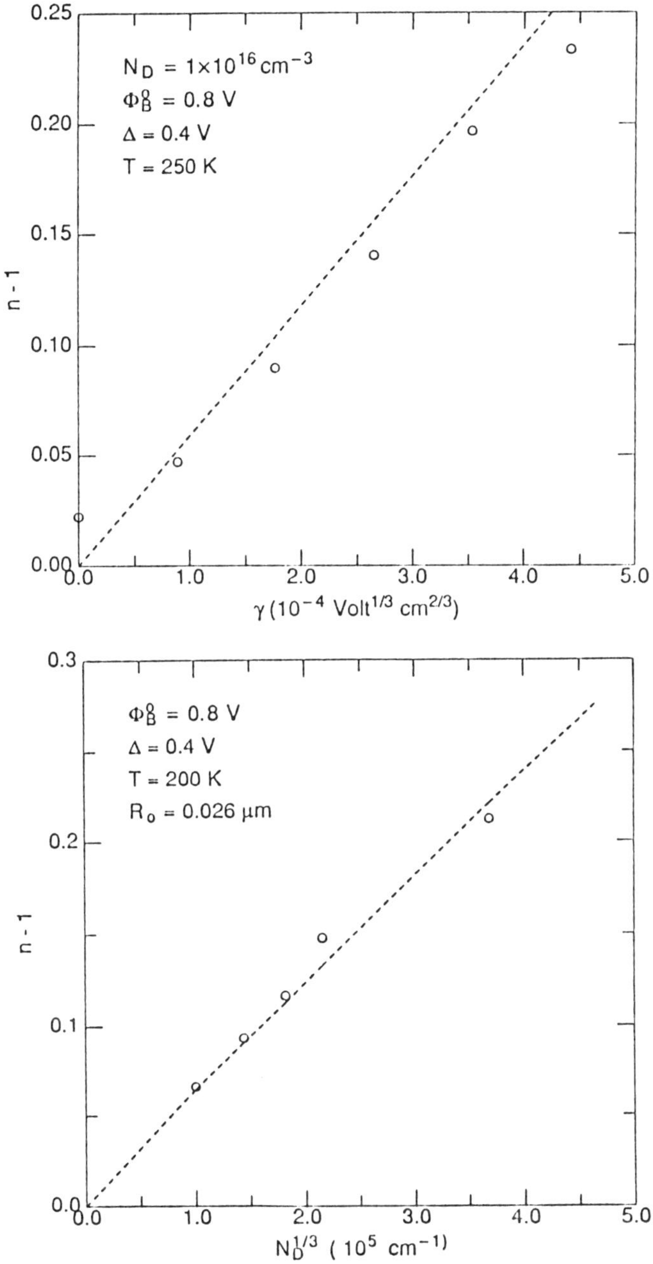


**Figure 5.** Computer simulated ideality factors[48] of circular low-SBH patches, as a function of *(a)* the parameter $\gamma$ and *(b)* the doping level. The dashed lines are predictions from theory,[13] i.e., Table 2.

At a real MS interface, SBH variations may occur at a variety of shapes and spatial frequencies. The observed I-V behavior from an inhomogeneous diode depends very much on the characteristics of SBH variation. When the junction current is dominated by patches with a narrowly distributed $\gamma$'s, it may appear, from Table 2, that the ideality factor is independent of temperature. However, experimental I-V traces are usually recorded at a constant current range, leading to a decrease in the effective bias range with temperature, and, hence, a decrease in the measured ideality factor with temperature. This phenomenon is illustrated in Fig. 6. A random variation of the SBH at a MS interface may be simulated by a statistical distribution of the patch/strip characteristics. For example, one may assume a Gaussian distribution of the patches

Eq. (19) $$N(\gamma) = c_1/\sigma \ \exp(-\gamma^2/2\sigma^2) \qquad (\gamma > 0)$$

where $N(\gamma)d\gamma$ is the density of circular patches with a $\gamma$ between $\gamma$ and $\gamma + d\gamma$, $c_1$ is a constant, and $\sigma$ is the standard deviation of the normal distribution. It has been shown[13] that the current flowing in such a composite inhomogeneous SB diode is made up of two components. One component is the current over the entire diode, which is governed by a uniform SBH of $\Phi_{mean}$, and the other is an additional current due to the presence of the low-SBH patches. The combined effect of all the low-SBH patches is as if there were a big low-SBH region in the diode with an effective SBH of

Eq. (20) $$\Phi_{EFF} = \Phi_{mean} - \frac{\beta\sigma^2 V_d^{2/3}}{2\eta^{2/3}}$$

and an "overall ideality factor" of

Eq. (21) $$n_{EFF} \approx 1 + \frac{\beta\sigma^2}{3\eta^{2/3} V_d^{1/3}}$$

Eqs. (20) and (21) are fascinating. They show that, even though the effective SBH and the ideality factor of each individual patch are roughly temperature-independent, put together in the same diode, they may be represented by an effective SBH and an ideality factor which depend on temperature explicitly! Obviously, the I-V characteristic of a SB diode depends on its exact "composition," and is not limited to the above examples. Furthermore, series resistance may also significantly affect the appearance of I-V traces. More on the junction currents of inhomogeneous SB diodes are discussed in Sec. 5.

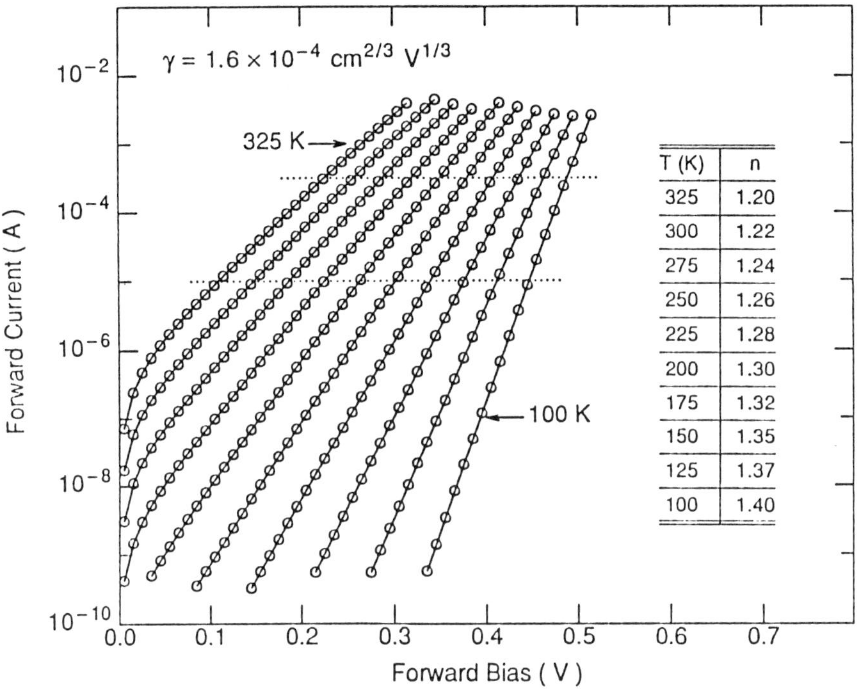


**Figure 6.** I-V traces, calculated with analytic expressions of Table 2, of a SB diode which contains a large number of low-SBH patches with a unique $\gamma$. ($\Phi_b{}^o = 0.8$ V, $\gamma_o = 1.6 \times 10^{-4}$ cm$^{2/3}$ V$^{1/3}$, and $N_D = 1 \times 10^{17}$ cm$^{-3}$). The SBH distribution is assumed to be independent of temperature. Series resistance, which is included in the calculation, has a negligible effect on the I-V characteristic. Ideality factors are deduced in a current range of $1 \times 10^{-5}$ to $3 \times 10^{-4}$ ampere.

## 3.0 FABRICATION OF SCHOTTKY DIODES ON SILICON

### 3.1 Processing Technologies for Si SB Diodes

Silicon SB diodes enjoy a few major material advantages over diodes fabricated on other semiconductors, in terms of providing high quality electrical characteristics for fundamental studies of the SB mechanism. First of all, the low density of traps and the high purity of bulk-grown single crystals have put a wide range of useful doping levels at our disposal. The inert interfaces of native and grown silicon oxides provide a convenient and reliable means of isolation and passivation, thus helping to avoid problems associated with surface recombination, which plague many compound semiconductors. In addition, the maturity of technologies for processing device structures on silicon has made it possible to design and manufacture special SB structures for dedicated investigations and applications. In the literature, one finds many examples of significant improvements in the quality of the SB diodes as a direct result of more careful processing. To avoid spurious effects and to investigate genuine fundamental phenomena at MS interfaces, the importance of careful processing cannot be overemphasized. In the following, common processing steps involved in the fabrication of SB diodes on Si are discussed.

The definition of SB diodes may be carried out before or after the metal deposition by patterning, or during the metal deposition by the use of evaporation masks. Sometimes, a combination of these techniques is necessary. For predeposition patterning, oxide windows (sometimes referred to as *vias,* in VLSI terminology) have been the most popular structure, although lift-off of metal has also been used occasionally. When prepatterned thick oxides are used, one should be aware of possible changes in the surface doping profile and the generation of point defects during the oxidation process. In this respect, deposited oxides are preferred over thermally-grown oxides. Silicon dioxide layers may also be used as implantation/diffusion masks, as self-alignment definition for silicide formation, or even as a remote gate during electrical measurements.[54] *Self-alignment* denotes the transfer of a pattern, of oxide in this case, into the next layer, metal/silicide in this case, without the use of an additional photolithographic step. It may be accomplished through the preferential formation of silicide in vias and the selective etching of unreacted metal from on top of the oxide.[55] It may also be accomplished by the selective deposition of certain metals only in Si windows, and not on the oxide.[56][57] Both of these processes may be used in front-end metallization, e.g., self-aligned silicide (salicide) in the source/drain and gate of a metal-oxide-semiconductor field-effect transistor (MOSFET), and selective metal fillings (trench-plugs) of the vias. To eliminate effects due to edges, guard rings based on either p-n junctions[58][59] or high-SBH metals[60] have been shown to be very effective. Other schemes have also been proposed to reduce effects due to edges.[54][61] Contacts to the metal layers may be accomplished by probing or wire bonding. Remote contacting (on top of oxide), which may be facilitated by further metallization and/or patterning, is preferred over direct pressuring the MS interface.

Existing technologies of ohmic contacting are largely based on heavily doping the semiconductor. The stringent requirements and practices of small-area ohmic contacts to ULSI devices on Si will be discussed in a later section. In comparison, contacts to Si for SBH studies are usually large and, therefore, require little sophistication. To avoid heating the SB structures to high temperatures, surface doping, by diffusion or implanation and activation of dopants, for ohmic contacts are usually accomplished before the fabrication of SB diodes. High quality ohmic contacting after the fabrication of SB diodes is also possible by laser melting of deposited dopants. Eutectic amalgamates, such as InGa, may also be used, although the quality of such contacts is less reliable. The high currents flowing in SB's sometimes make the unwanted voltage drop due to series resistance a significant part of the circuit. The use of epitaxially-grown lightly-doped layers (epilayers) on heavily-doped substrates alleviates the series resistance problem and, at the same time, provides easy ohmic back contacts. To further reduce series resistance, or for optical measurements involving back surface entrance, front ohmic contacts may be used.[62] Device geometries based on "diffusions" (diffused doped tubs) and implantation are also convenient for multi-terminal characterizations of the SB diodes, e.g., using both n- and p-type contacts to study effects due to minority carriers.[63][64][65] The use of a very thin (< 100 Å), undoped, MBE-grown, epitaxial Si layer on a heavily-doped Si substrate allows the tunneling current to be studied.[66] A few of the many structures of SB diodes which have been fabricated on Si are shown schematically in Fig. 7. It is somewhat ironic that early works on Si SB's often involved very careful processing of the SB diodes, including the use of epilayers and guard rings, while the more recent works rarely do.

### 3.2 Reaction Between Metals and Si

Almost all metal-Si interfaces involve some degree of interdiffusion and/or compound formation. Because of the stability of their structures and electrical properties, silicides have already secured an important place in microelectronics fabrications. However, in terms of the fundamental understanding of the SBH phenomenon, the reaction between metals and Si to form various phases of silicide have both desirable and undesirable elements. On the positive side, silicide reaction is an important mechanism for the elimination of the native oxide layer at the Si surface, which leads to

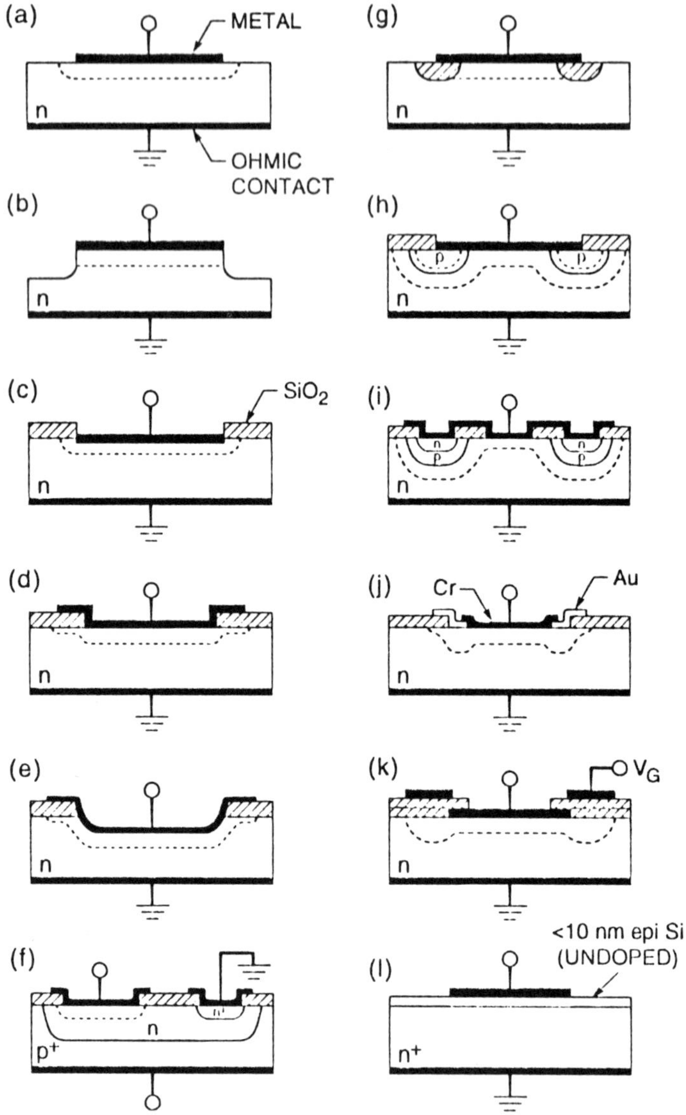


**Figure 7.** Assorted SBH diode structures fabricated on Si. Dark areas are metals or silicides, hatched areas are $SiO_2$, and dashed lines indicate edges of the depletion regions. *(a)* Lift-off or shadow-masked metal, *(b)* etched mesa, *(c)* self-aligned silicide or selectively deposited metal/silicide, *(d)* overlapped metal, *(e)* moat-etched Si,[61] *(f)* diffusion tub,[65] *(g)* LOCOS isolation, *(h)* diffused guard ring, *(i)* double diffused guard ring, *(j)* high-SBH guard ring,[60] *(k)* gate-controlled structure,[187] and *(l)* tunnel diode.[66]

intimate MS contacts with more reliable and predictable characteristics. Many high quality silicide-Si interfaces have been manufactured by well controlled reaction and growth techniques. Single crystal silicide-Si interfaces are the best understood and the most perfect MS interfaces presently available. The richness of possible compound structures which one may form on Si leaves plenty of opportunities to study systematics of the SBH's. However, on the negative side, silicide reactions may lead to extended and point defects, metal impurities, stress, dopant diffusion, and other unwanted changes to the semiconductor material. Moreover, inhomogeneities in silicide diffusion and reaction often lead to incomplete dissolution of the native oxide layer, rough morphologies, pinholes, edge-specific structures, multiple compounds/phases, and generally more complicated interface structures/stoichiometries. The electronic properties of a metal-Si junction is often incomprehensible without a knowledge of the structure of the silicide film. Only by understanding the various aspects of silicide reaction may we turn the many structures of the intimate silicide interface to our advantage in investigating the SBH formation mechanism.

The preparation of the Si surface may significantly affect the growth of silicides. Among non-UHV silicide investigations, etching Si in a solution containing HF to remove the native oxide on the surface, is frequently the last step of chemical cleaning. Even though the Si is usually loaded into a deposition chamber immediately afterwards, chemically cleaned surfaces contain significant amounts of oxygen and carbon. Under high vacuum conditions carbon impurities are more difficult to remove than oxygen. Therefore, the most common practices of the silicon surface science and molecular beam epitaxy (MBE) community is to intentionally grow a protective oxide layer on the surface as the last step of chemical cleaning.[67] This thin oxide layer may be removed easily in a UHV chamber by direct heating, Si beam cleaning,[68] or sputtering and annealing. However, it should be noted that heating of Si at temperature in excess of $1000^0$C is known to allow the unavoidable boron contamination on the original silicon surface to diffuse into the Si and change the doping characteristics.[69][70] Therefore, prolonged heating at high temperatures should be avoided when SBH studies are carried out on lightly n-type doped Si.[71] Atomically clean surfaces may be prepared by the UHV techniques described earlier or by simply cleaving a Si crystal inside a vacuum chamber, which exposes the (111) surface. Since atomically clean Si surfaces are obviously preferred over chemically etched surfaces in terms of consistency of the electrical properties, searches for possible treatments which could allow these clean surfaces to be achieved without the laborious UHV practices have been active. Chemical techniques, in controlled environments, which allow near perfect passivation by hydrogen[72]-[74] or other surfactants[75] have been demonstrated. Apparently, the implementation of these techniques in routine clean room environment has already been demonstrated and shown to lead to MS interfaces with amazing quality and stability.[74] It is expected that the control of contamination will always remain at the forefront of semiconductor processing research, and for good reasons.

The most common methods of metal deposition are e-beam evaporation and sputtering. Thermal evaporation, from a heated filament or crucible, is limited to low-temperature metals. Occasionally, the silicon surface has been maintained at an elevated temperature during metal deposition. Considerable advances have been made recently in the application of ion beams in the deposition of metals. Electrical results from high energy implantation[76] and ionized cluster beam deposition[77] are scarce, though it is already clear that damages in the process of ion implantation will strongly influence the electrical properties of the junction thus formed. Chemical vapor deposition (CVD) and electroless deposition may also be used for either blanket or selective deposition of metals. It should be noted that the selective CVD process usually requires the consumption of a thin layer of silicon from the substrate surface.[56][78] In the vast majority of SBH studies, only one metal is deposited. However, for certain applications, co-deposition of alloys and silicides, or sequential depositions of multi-layers of metals and silicon may be desired.

The growth of silicides involves two thermally activated processes, nucleation and diffusion. Starting from a metal layer deposited on Si, annealing usually leads to the nucleation and growth of a metal-rich silicide phase, e.g., $Pd_2Si$ and $Ni_2Si$. Upon the completion of the growth of the first silicide phase, and often upon annealing at a higher temperature, a second silicide phase, richer in silicon, is nucleated from the interface and grown. Simultaneous growth of more than one silicide phase is possible, but not frequently encountered in thin film studies. This process continues as the temperature is raised, until the final, most silicon-rich phase, e.g., PdSi and $NiSi_2$, is grown on the silicon. The morphology of the silicide layer is often related to an important aspect of the silicide reactions, namely, the diffusing species. This is particularly true when the silicide reaction occurs in a non-planar geometry. For example, it is known that Si is the more mobile species in the initial Ti silicide reaction. As a result, reaction of Ti with Si contact areas inside oxide windows usually spreads over the sidewall, and even the top, of the oxide layer. This is less desirable than a silicide process which involves the diffusion of metal atoms. It is also interesting to point out the possible complexity of the diffusing species issue in some silicide reactions.

For example, the reaction of $CoSi_2$ involves two intermediate silicide phases, $Co_2Si$ and CoSi. Cobalt is believed to be the moving species in $Co_2Si$ and $CoSi_2$ reactions, while Si has been shown to be the more dominant moving species in the CoSi reaction. Fortunately for the $CoSi_2$ technology, such diffusion tendencies are not in conflict with the self-alignment process.

Silicide reaction is often strongly influenced by impurities in the layers and in the ambient.[79] Continued, or higher-temperature, annealing of the final silicide phase may lead to grain growth, epitaxy, islanding, and other orientational and morphological changes, which may or may not be desirable. The issue of epitaxial formation of silicides is addressed in Sec. 7. Since a small concentration of most metals may dissolve in Si at high temperatures, prolonged anneals should be avoided for most applications. Reducing the thermal budget also helps to minimize modification of the doping profile. The influence of dopants on the reaction of silicides and the redistribution of dopants during silicide reactions are well-known phenomena which have direct bearing on the electrical properties of a SB junction. For certain metals, the silicide reaction rejects dopants originally contained in the silicon, which leads to an accumulation of excess electrically active dopants near the silicide interface (the *snow-plow* effect)[80][81][82] and markedly changes the electrical characteristic from that expected based on the original Si doping level. An example of the snow-plow effect is shown in Fig. 8. Occasionally, the dependence of the observed SBH on silicide film thickness may be traced to the snow-plow effect.[83] Obviously, the reaction of a silicide with a dopant depends on the ternary phase diagram of the metal-Si-dopant system. Therefore, the degree of snow-plow varies with the metal and the dopant species. For example, snow-plow is of less importance[84] for silicides such as $NiSi_2$, $CoSi_2$ and $TiSi_2$ than it is for the near-noble $Pd_2Si$ and PtSi.

The morphology and the structure of a silicide layer is obviously related to all the processes which have occurred on that substrate, prior to and including the final anneal. Inhomogeneities in surface oxide, deposition, nucleation, diffusion, and localized defects all lead to morphological nonuniformity. To improve the uniformity of a silicide layer, various methods of deposition and annealing have been exploited. Under optimized conditions, two-step furnace annealing,[85][86] rapid thermal annealing (RTA),[84][87] and electron-beam and laser annealing[88] have all shown improved uniformity over routine furnace annealing. Rapid thermal processes have the advantage of better maintaining the doping profile and preventing dopant loss. Ion beam mixing through the MS interface, although clearly capable of producing very uniform silicide layers, may lead to damages and metal impurities

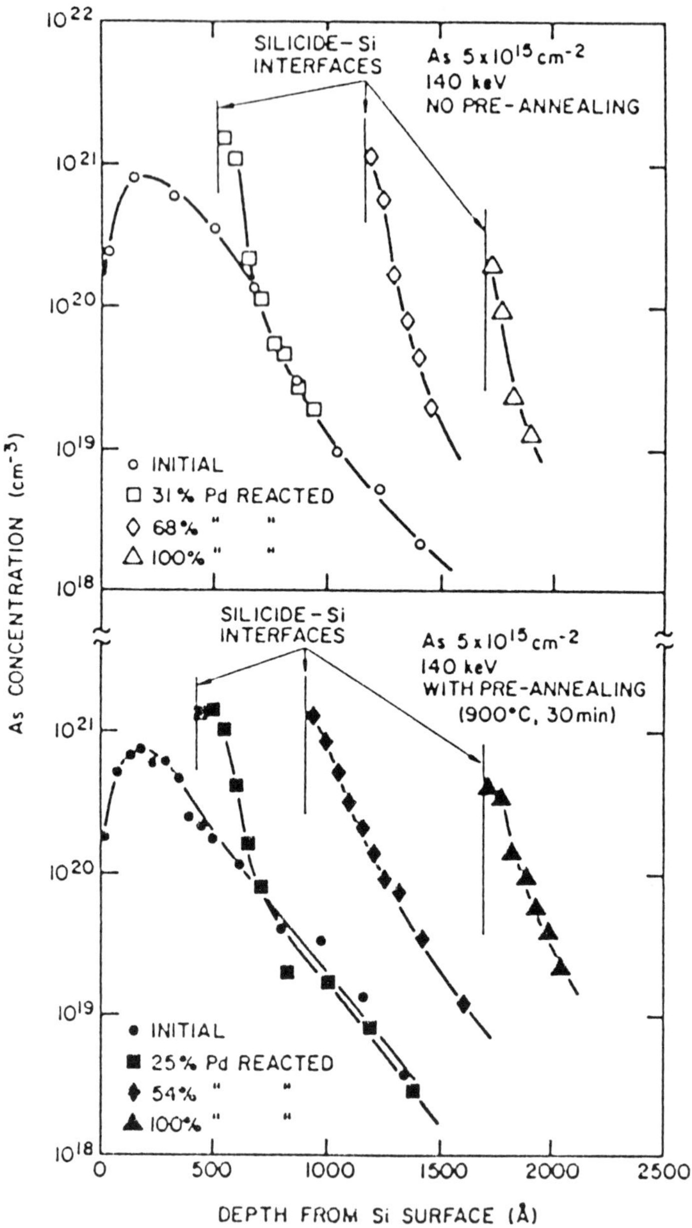


**Figure 8.** Depth profiles of As concentration for samples without *(open symbols)* and with *(solid symbols)* a 900°C, 30-min, dopant-activating, anneal prior to Pd deposition. The *squares, diamonds,* and *triangles* represent the distribution after isothermal annealing at 250°C for 17, 70, and 375 min, respectively. The shift in the As profile with the $Pd_2Si$ reaction, the "snow plow" effect, is clearly demonstrated. (After Ohdomari, Tu, Suguro, Akiyama, Kimura, and Yoneda.)[82]

in silicon. To achieve shallower contacts without sacrificing sheet conductance, co-deposition of silicides or layered deposition of both metals and silicon has been reported.[89]-[93] The deposited silicon limits the reaction of the deposits with the silicon substrate, and therefore the depth of the junction is reduced. Such a concept of achieving shallow contacts through limited reaction was also exploited by diluting the metal deposits with a higher-temperature metal.[94]-[96] Since only one metal from such an alloy, e.g., Pt from a Pt-W alloy,[94] reacts with silicon at the chosen annealing temperatures, more uniform shallow junctions may be achieved. However, the growth of the lower-temperature silicides may be affected by the presence of the alloy, as changes in both the growth temperature and the phase sequence have been observed.[96][97] The unreacted higher-temperature metal layer, on the silicide layer, may also serve a beneficial purpose, as a diffusion barrier.[94][95] In the case of deposited titanium, the nitride reaction with a nitrogen ambient has been utilized to limit the depth of the silicide junction, as well as providing a diffusion barrier and limiting lateral diffusion over oxide. The latter is crucial in the success of the self-aligned $TiSi_2$ technology. Despite the success achieved by various strategies using common Si processing technologies, it should be pointed out that the most uniform and shallow silicide layers are those grown by UHV methods. Because of the clean conditions, growth of silicides in UHV is inherently more homogeneous and there are more options in tailoring the deposition schedule to minimize inhomogeneities. Examples of UHV processing will be given in our discussion of epitaxial silicide formation.

## 4.0 SBH'S OF NON-EPITAXIAL METALS AND SILICIDES

The SBH's of various metals and silicides on Si have previously been reviewed.[1][98][99] Extensive literature on the SBH's of the most common metals and silicides has recently been compiled by Ng.[100] The usual approach to this subject is to concentrate on the magnitudes of the observed SBH's from different metals/silicides and, from the systematics, make comments on possible SB models. To avoid duplication, and simply because this author is not convinced that one should concentrate only on the magnitudes of the experimentally observed SBH, the present discussion will take a different approach. We will focus more on the material aspects of the SB's and also on the various observed variations of the SBH within one MS system. In particular, it will be shown that evidence for SBH inhomogeneity is abundant in almost every metal/Si or silicide/Si system that people have studied. The absolute magnitude of the apparent SBH is but a small part of the entire electronic structure of any MS interface. Therefore, understanding the variation of the SBH at even one MS interface provides insight into the SBH formation mechanism, more so than concentrating on experimental trends involving many MS interfaces which are themselves inhomogeneous.

## 4.1 Deposited Metals: Dependence on Surface Preparation

The preparation of the Si surface prior to metal deposition is an important step of SB fabrication. Different surface preparations lead to a difference in the amounts of chemical impurities, oxides, and structural defects which often reside within a few monolayers of the silicon surface, and, hence, greatly influence the electronic properties of the eventual MS interface. Processes such as sputtering of the surface, implantation, chemical etching, and hydrogenation change the properties of the silicon. It is not difficult to see that surface treatments may influence the electronic structure of the interfaces between the as-deposited metals and silicon, since the surface essentially becomes part of the interface. (The SBH's of the annealed silicide/Si interfaces may also be affected, because the formation of the silicide structure is often influenced by surface treatments.)

Before the development of UHV processing techniques now considered routine, silicon surfaces were essentially prepared by two methods: chemical etching, which leaves native oxides, hydrides, and carbonaceous compounds on the surface, and cleaving, which exposes the atomically clean surfaces of silicon albeit with some structural damage. One of the earliest discoveries concerning the influence of surface treatment on SBH was the observation of *aging* (variation of SBH characteristics with time) on chemically etched surfaces,[101] and not on surfaces cleaved in vacuum. Such an effect was at first attributed to ionic activity within the native oxide at the MS interface.[101] Apparently not all etched surfaces behaved the same way, as aging was found to be totally absent in many studies.[102][103] It has been proposed that the slow variation of the SBH after metal deposition is due to oxygen from the air diffusing to the MS interface.[104] The role played by hydrogen in varying the interface state density and, hence, changing the observed SBH has also been pointed out.[105] It is known that certain metals, such as Ti, Al, and Ag, exhibit a lower SBH on n-type Si surfaces with a native oxide layer, than on clean Si surfaces.[106]-[108] The existence of interface states at the oxide interface has been offered as an explanation of this effect. Because sputtering has been the method of metal deposition in some of these studies, surface damage and donor-like traps have also been invoked to explain such a behavior of the observed SBH's.[107]-[109] Prolonged sputtering of the Si is known to lead to positively charged defects in the near surface region.[110][111] It has been reported that etching the Si substrate prior to metal deposition may bring about a decrease in the n-type SBH's (an increase in the p-type SBH's),[112][113] which has also been attributed to donor-like defects near the surface. There is no reason to immediately attribute all the variations of the SBH's thus far discussed to one mechanism, but the similarities should be noted. Results from a careful UHV and surface spectroscopies study,[114] using e-beam evaporated Ti and both n- and p-type Si, showed that such a variation of the SBH arose from the surface oxide and not necessarily from the introduction of any structural defects in Si. It is interesting to note that a lower SBH for titanium oxide than for Ti on n-type Si (the reverse is true on p-type Si) is not consistent with the trend in the work functions of these two conductors.[114] In reality, a thin amorphous silicide layer, present at the as-deposited Ti/Si interface, is likely responsible for the SBH.[115][116] One notes that the suggestion[117][118] that the $SiO_2$ interface may be in accumulation for n-type Si, and hence contributes significantly to the junction current when oxide windows are used to define SB diodes, was largely based on results obtained from Ti SB diodes. From the present discussion, it is clear that these edge-related currents[117][118] could easily have come from an oxide-related, lower, local, n-type SBH near the edges.[114]

High temperature annealing of chemically cleaned n-type Si in vacuum was shown to lead to higher SBH's for subsequently deposited metals.[112] This effect is likely due to the now familiar boron-contamination problem, made famous by the dispute over $NiSi_2$ SBH's.[71][119] Besides the many dependencies of SBH's which may be clearly attributed to certain surface treatments, there are other dependencies and inconsistencies which may also be related to surface treatments. For example, as-deposited Ni shows SBH's ranging from ~0.7 eV[101][120][121] to ~0.5 eV on n-type Si.[112][122] The SBH's of deposited metals sometimes show dependencies on other properties of the Si crystal, such as the orientation[123] and the doping level.[124][125] However, the experimentally observed dependencies of (raw) SBH's on the doping level were usually analyzed with some assumed models of the interfaces to allow a determination of the density of the interface states.[124][125] Not surprisingly, the result of such a "reverse analysis" is usually a conclusion that the true SBH's do not depend on the doping level.[124][125] Deposited metals often show non-ideal I-V behaviors and sometimes display the famous $T_o$ effect,[60][126] as shown in Fig. 9.

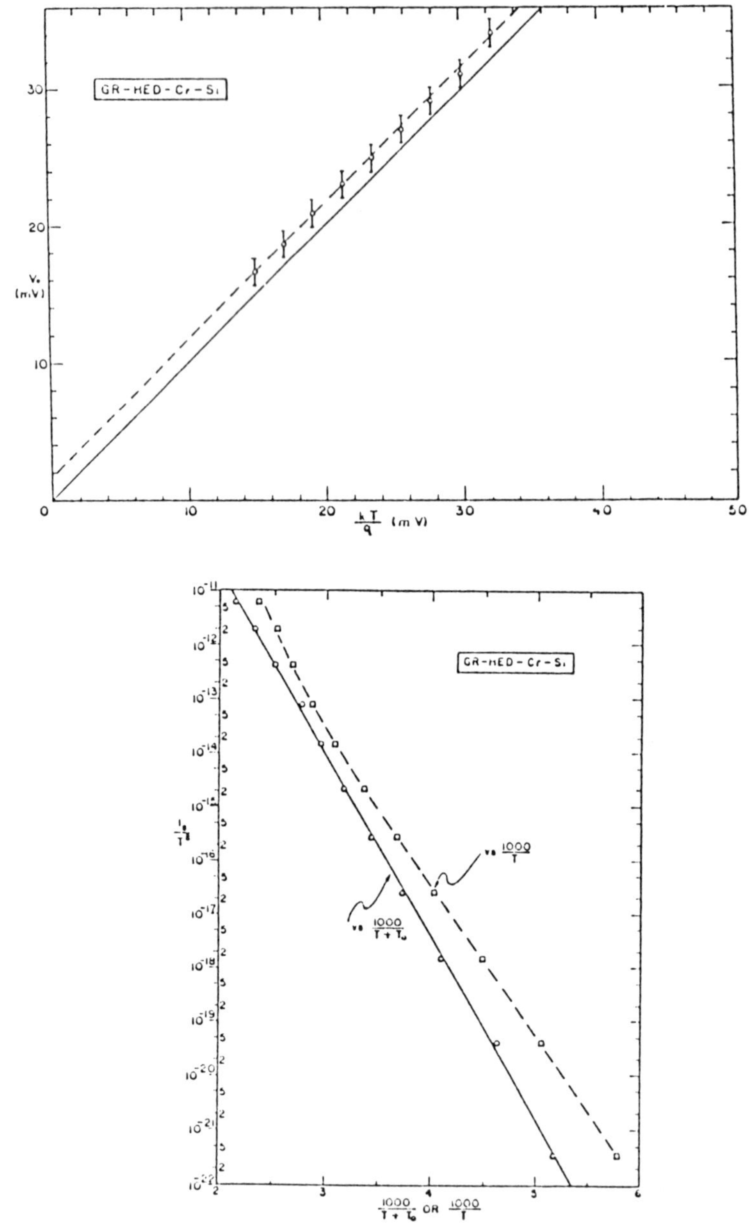


**Figure 9.** *(a)* The inverse slope of a guarded Cr SB diode on n-type Si, plotted against measurement temperature. *(b)* Regular (vs. 1/T) and modified (vs. 1/nT) Richardson plots of the saturation current, illustrating the $T_0$ effect. Note that the regular Richardson plot does not fit a straight line, whereas the modified plot does. (After A. N. Saxena.)[60]

Cleaving a Si crystal in vacuum has also been a popular technique of Si surface preparation. A lot of early attention has been devoted to explaining the different SBH results obtained on cleaved surfaces from those obtained on etched surfaces.[1] Some results seemed to suggest that the observed SBH's on cleaved surfaces are independent of the metal species.[17][101] However, significant variations in the SBH's were observed in other experiments on cleaved surfaces.[127]-[129] It has been demonstrated that the quality of cleaved surfaces depends much on the particular cleave, and may even vary considerably from region to region on one cleaved surface. This fact has led to large fluctuations and inconsistencies in the observed SBH's.[129]-[131] The famous discrepancy, of typically 0.1 - 0.35 eV, observed on cleaved surfaces between the SBH obtained by the C-V method and that deduced by I-V methods[127] was briefly discussed in terms of interface states. Other studies, however, show that the observed SBH's for many metals on cleaved Si(111) do not significantly depend on the measurement technique.[130][132]

## 4.2 Annealed Intimate SB's: Various Dependencies

Annealing of deposited metals generally leads to intimate MS interfaces and more reproducible electrical characteristics. For most metals, this is accomplished through a change in the depth of the MS junction due to silicide reaction. The consistency of the SBH usually improves upon annealing, even in the absence of silicide reaction, such as for aluminum and gold, which do not form silicides, and for W and Ta, which have high silicide reaction temperatures. The aging effect discussed in the last section, which may be accelerated through annealing,[101] is an example of this improved consistency. Because of its technological importance, annealed Al-Si interfaces have been the most studied of all non-silicide contacts. Only rarely have very high quality characteristics been observed from Al diodes.[133] The famous *spiking* problem[50] and the steady increase of the (n-type) SBH[134] or contact resistivity[135] with annealing, extensively studied in the 1970's, led to a general consensus that Al is not a suitable contact material for devices requiring shallow junctions. Since Si has a significant solubility in Al at temperatures higher than ~400°C, silicon at the interface is etched by Al during annealing. This reaction is often laterally inhomogeneous due to an uneven distribution of interfacial oxide and a crystallographic dependence of the dissolution process, resulting in a rough Al-Si interface which consists of deep Al pits penetrating into the Si. Upon cooling, the solubility decreases leading to the precipitation of Al-doped Si at the MS interface and other areas in contact with the Al (including the free surface). The precipitation process apparently depends on the distribution of temperature across the entire sample, and may at times depend on the cooling rate.[136] It was initially speculated[134] and later repeatedly confirmed[137][138] that the increase of both the apparent n-type SBH and the ideality factor upon annealing of Al-Si contacts is due to the precipitation of p-type, heavily (~5 x $10^{18}$ $cm^{-3}$) Al-doped Si at the interface. The problem of spiking with Al is more severe near edges of an oxide window. Since the amount of (additional) Si dissolved in Al depends on the thickness and the existing Si concentration of the Al, spiking may be considerably reduced by using thin layers of Al, by using Al already doped with Si, or by reducing the annealing time and temperature. The use of epitaxial Al grown by deposition processes assisted with an ion beam may alleviate the symptoms of Al spiking.[139] Presumably, Al deposited under very clean processing conditions should also exhibit higher stability.[74] Like Al, tungsten is a metal which does not form silicide upon annealing at moderate temperatures. Tungsten is becoming increasingly important in ULSI applications because of various advantages.[56][140] Both highly ideal characteristics[141] and some not-so-ideal behaviors[142] have been observed from W SB diodes.

Since there is usually more than one stable silicide compound for a metal, whether different silicides have different SBH's and how silicide SBH's compare with those of the unreacted metals have often been investigated. For almost all the metals studied, including Ni,[120]-[122][143] Pd,[92][143]-[145] Pt,[86][143][146]-[148] Ir,[51][149] Co,[93][150] Ti,[62][108][118][151] W,[142][152] Cr,[154] and some rare earth metals,[155]-[157] it is generally observed that various silicides of one metal have similar SBH's and, furthermore, SBH's of silicides are similar to that of the as-deposited metals. However, some minor variations in the magnitude of the SBH and in other junction characteristics, such as the ideality factor, the difference between I-V and C-V measurements, the reverse currents, etc., are almost always present. Frequently, more than one component of current were observed in the I-V characteristic and were attributed to generation-recombination and/or edge related conduction. For example, the ohmic behavior is occasionally simply ascribed to edges.[121] An interesting microscopy study[146] revealed the physical spiking of silicide near the edges of PtSi diodes and increased currents which correlate with the diode peripheral length. However, the edge-related current in this case had an ideality factors[146] which was lower than the current flowing in the planar portion of the diode! Occasionally, the SBH is found to depend on the orientation of the Si substrate,[51][147][158] the thickness of the silicide layer,[146]-[148][159][160] the degree of epitaxy,[161] the degree of crystallinity,[116] and the technique of measurement.[162] Some, but clearly not all, of the observed dependencies of the SBH on layer thickness may be attributed to trivial explanations such as dopant pileup[83] and pinholes in the silicide.[49] Hydrogen is known to affect the SBH of silicide interfaces,[105][163]-[165] which is usually attributed to the passivation of interface states. The observed dependencies of the SBH (and the ideality factor) on the doping level and the magnitude of the reverse currents are almost always larger than those predicted by image-force lowering and TFE.[147][151][166][167] Interface states have been the mechanism most often mentioned[166] in connection with these discrepancies, although mechanisms such as multi-step tunneling[151][167] and diffusion[147] have also been proposed. However, as explicitly shown recently,[13][48] these behaviors of the SBH are more consistent with SBH inhomogeneities.

The ideality factors of some intimate silicide diodes were observed to be close to unity, independent of the measurement temperature.[58][133][149][168] But much more often, there is an increase in the ideality factor of the silicide diodes when the measurement temperature is lowered.[49][118][142][151][152][167] This increase in the ideality factor frequently, but not always, follows the expected behavior of the $T_o$ effect. A variety of other dependencies of the ideality factor on temperature have also been observed from silicide SB's. Some ideality factors were found to remain constant yet significantly higher than unity,[151] as shown in Fig. 10. Occasionally, the ideality factor is found to decrease when the temperature is lowered, as shown in Fig. 11, contradicting the prediction of all existing models of the ideality factor based on interface states or tunneling. Another interesting phenomenon which has repeatedly been noticed is a correlation of the ideality factor with the SBH.[51] For silicides of one metal, higher ideality factors in I-V measurements were often found to accompany lower observed SBH's,[51][143] as shown in Fig. 12. The rise in the ideality factor, during silicide reaction, has occasionally been observed to correlate with a state of maximum confusion, such as prior to the completion of silicide reactions.[160] Various approaches to fabricate more uniform silicide layers, such as by better cleaning of the Si surface,[169] better choice of the deposited material,[90][92] and better annealing schedule,[86] have resulted in not much changes in the measured SBH's, but significant improvements in their ideality factors.

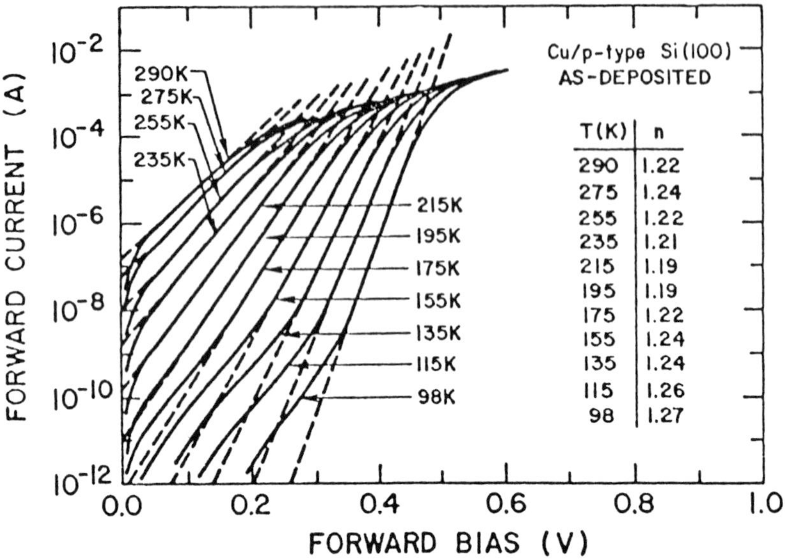


**Figure 10.** Forward current-voltage characteristics of as-deposited Cu on p-type Si(100) as a function of temperature. (After Aboelfotoh, Cros, Svensson, and Tu.)[167]

## 4.3 SB's Involving Two or More Metals

The compatibility of metals/silicides with other processing requirements is of primary concern in choosing a suitable contact scheme for any application. Since some additional annealing has to take place after the initial MS junctions are formed, during the processing of dielectrics and interconnections, it is important that the electrical characteristics do not degrade during these anneals. The stability in the electrical characteristics of the junction depends on, but is much more demanding than, the observed stabilities in the metallurgical and physical properties of the layer/interface. The annealing behavior of silicide SB's in contact with common interconnection material such as Al (and W) have been extensively studied.[170] Stacks of Si/silicide/Al or Si/silicide/barrier/Al were often annealed at different temperatures and for different periods of time to determine the maximum allowed thermal budget without junction failure. Very similar behaviors were observed for silicides such as PtSi,[171] $Pd_2Si$,[172] NiSi,[173] $CoSi_2$[174] and $TiSi_2$[175] in terms of how the apparent SBH and ideality factors of the junction vary with increasing Al diffusion. A typical example, involving the

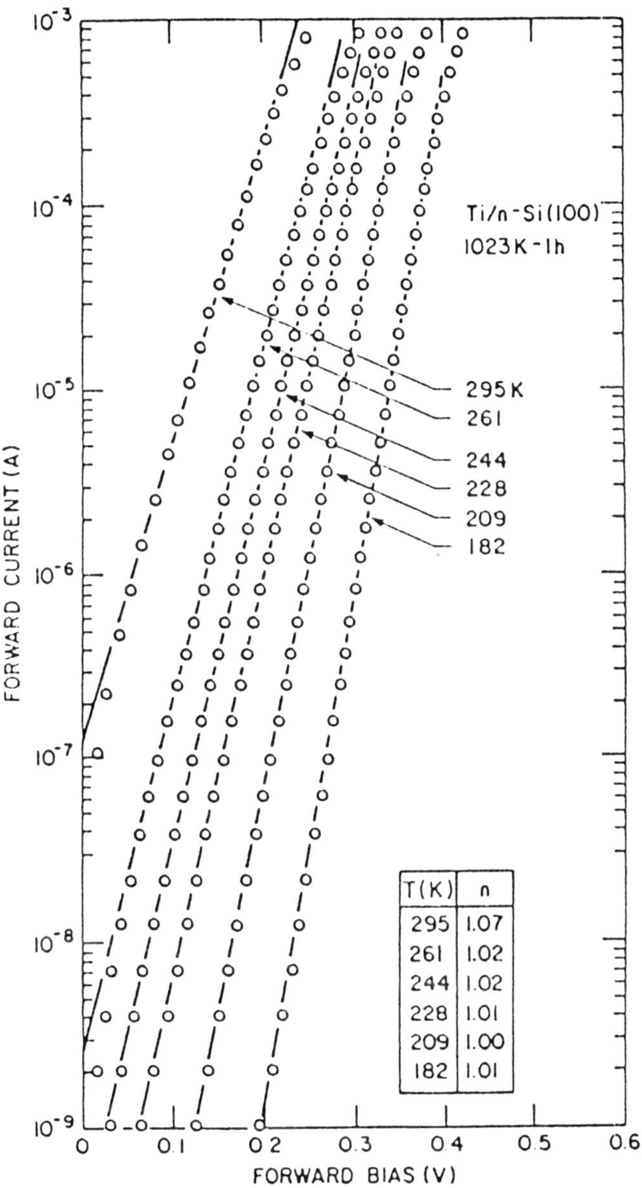


**Figure 11.** Forward current-voltage characteristics of Ti on n-type Si(100) as a function of measurement temperature for samples annealed at 1023 K for 1 h. Diode diameter is 129 $\mu$m. (After Aboelfotoh and Tu.)[118]

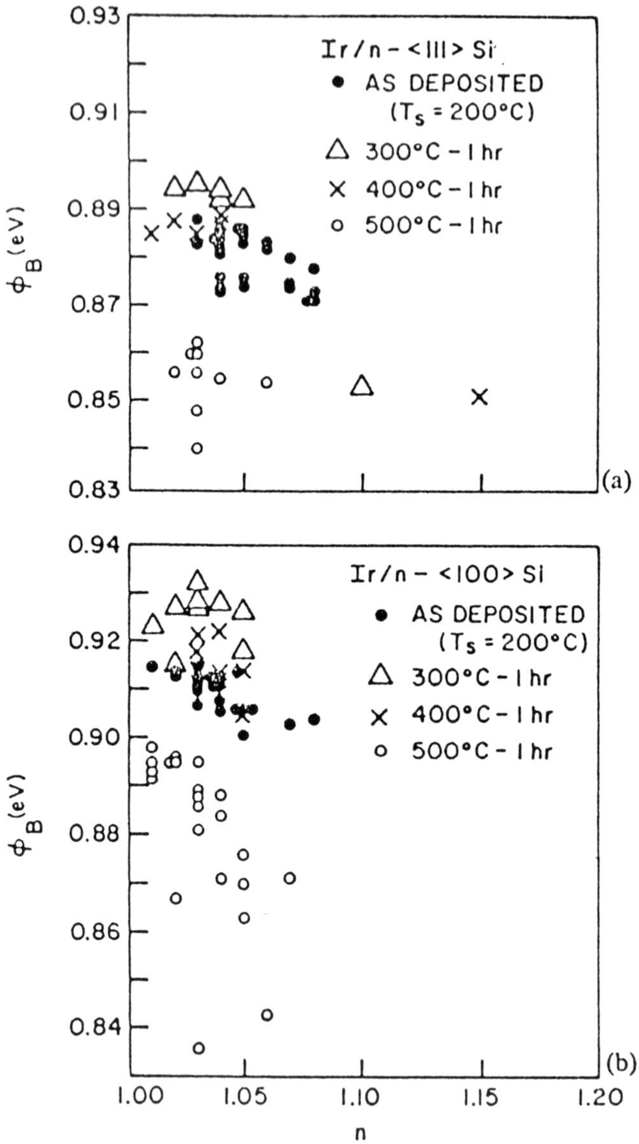


**Figure 12.** The distribution of $\Phi_{Bn}$ versus n of Ir/Si diodes as-deposited and annealed at 300, 400, and 500°C for 1 h. Each symbol represents one measurement. *(a)* Ir on n-type Si(111); *(b)* Ir on n-type Si(100). (After Ohdomari, Kuan, and Tu.)[51]

annealing of n-Si/$Pd_2Si$/Al,[172] is shown in Fig. 13. The behavior of SBH's and ideality factors from I-V measurements may be roughly divided into three regions. Before any Al reaches the silicide interface, the SBH is characteristic of the silicide and the ideality factor is usually not far from unity. When Al first diffuses to the interface, the SBH decreases to a value characteristic of intimate Al SB's and the ideality factor increases significantly. For $CoSi_2$ and $TiSi_2$, which have initial SBH's slightly lower than the Al SBH, there is no decrease in the SBH, just an increase in the ideality factor. Some studies show that the ideality factor reaches its maximum just before the SBH reaches a minimum.[171] Others notice the more pronounced variation in the ideality factor than in the measured SBH.[95] The SBH measured by the C-V technique becomes noticeably higher than that measured by I-V techniques when Al reaches the interfaces. Finally, after annealing proceeds further, the SBH increases[171][95][172] and the difference between the SBH's measured by C-V and I-V techniques also increases.

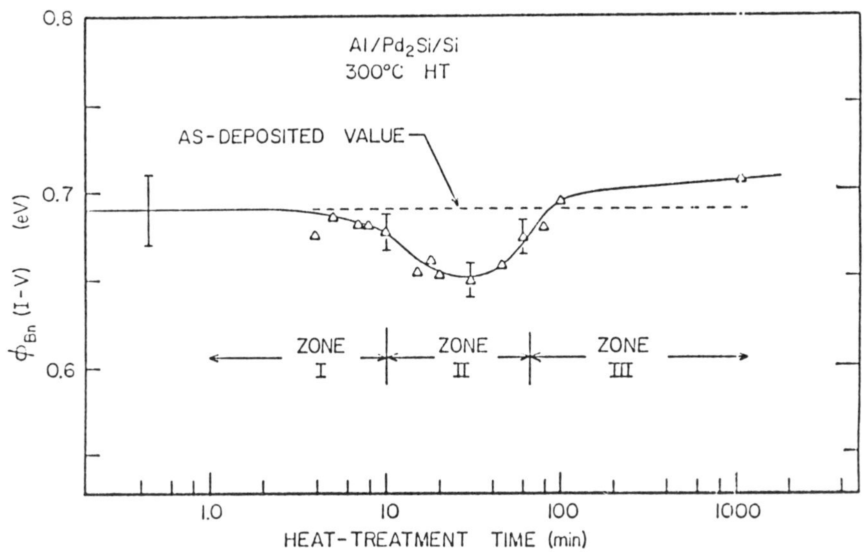


**Figure 13.** I-V determined $\Phi_{Bn}$ as a function of heat-treatment time at 300°C for Al/$Pd_2Si$/Si Schottky diodes with as-deposited Al film thickness of 125 nm. Zone I displays the as-grown $Pd_2Si$ SB diode characteristics. The transition to Zone II, which corresponds to a change in the composition of the sample as observed by Auger profiling, sees a decrease of the SBH to that usually observed for Al SB diodes and an increase in the ideality factor. The increase of the SBH in Zone III is likely due to Al doping at the interface. (After Grinolds and Robinson.)[172]

Although the increase of the SBH upon extended annealing has occasionally been attributed to intermetallic compound formation,[173][176] the inescapable similarities to behaviors of annealed pure Al SB's would suggest that Al-doped Si at the MS interface must play an important role.[138] The increase of the SBH was more pronounced when Si-doped Al or thicker layers of Al were used.[177] A relevant question to the study of SBH formation is what happens to the interface electronic structure when Al just reaches the silicide/Si interface. The increase in the ideality factor, the decrease of the I-V determined SBH, and the difference between I-V and C-V SBH's all suggest the presence of inhomogeneity in the SBH, as will be discussed. Since the diffusion of material through thin films is dominated by grain boundaries and defects, lateral inhomogeneity in the Al profile is expected. With better control over the grain size of the silicide layers, impurity diffusion may even serve as a means to artificially fabricate electrically inhomogeneous SB's. The diffusion of Ti to silicide interfaces has also been shown to lead to a rise in the ideality factor and a drop in the SBH.[148] Various materials such as TiW,[146][175][177]-[179] TiN,[170][180][181] Cr,[95][182] and W[176] have been proposed as diffusion barriers between the silicides and the Al metallization. Generally, diffusion barriers help maintain the integrity of silicide contacts up to about 500 - 550°C. Above these temperatures, failure modes similar to that observed in the absence of the diffusion barrier are observed. Occasionally, diffusion barrier material may cause edge-related problems on its own.[179] Contacts formed by co-deposition of more than one metal,[94][183][184] or by the sequential deposition of more than one layer of metal,[185] have actually not been subjected to detailed analyses like these Al-diffused junctions just described.

## 5.0 SBH ANOMALIES: INTERFACE STATES OR SBH INHOMOGENEITY?

The usual descriptions of the currents and capacitances at SBH's have all been more or less based on idealized MS interfaces. Experimentally observed electrical characteristics from real SB's, on the other hand, often show considerable departures from ideal behavior. In this section, some of the more common non-ideal, or anomalous, phenomena and their likely explanations are discussed.

### 5.1 Leakages and Edge-Related Currents

Experimentally observed I-V curves have frequently been analyzed[186] and shown to be comprised of two or more components of current. At small biases, the forward current is sometimes dominated by a *soft,* or *leaky,* component which leads to both a curvature in the I-V curve and tangential slopes corresponding to ideality factors much in excess of 1. As bias increases, the I-V relationship becomes semi-logarithmic with an ideality factor not far from unity (although not necessarily very close to unity), before turning over at large bias, due to series resistance. The existence of multiple current components is more discernible at lower temperatures, as shown[167] in Fig. 10. It is customary to attribute the linear portion of the I-V curve to the main conduction mechanism, e.g., thermionic emission over the SB, and the leakage current at low biases to a different mechanism. Generation and recombination in the space charge region[187] and edge-related conduction[58][188][189] are the mechanisms most frequently thought to lead to the leakage currents. Because edge-related currents scale with the peripheral length of the diode, and not with the area of the diode, they may be unambiguously identified by using diodes with different sizes.[189][190] Until recently, the additional current component associated with edges was thought to be due to a larger electric field at the diode edges which leads to increased tunneling and/or increased generation-recombination.[187] Because the oxide-silicon interface is usually unpinned, the increase in the electric field near the edges is not significant enough to explain the magnitude of the observed edge current by tunneling. In addition, the edge corrections to the total volume of the depletion region are negligible for usual diode sizes and doping levels. Sometimes, detailed analyses showed that edge-related currents have ideality factors similar to,[189] or smaller than,[146][190] that associated with the center portions of the diodes. Sometimes, edge-related currents were attributed to recombination even when the activation energy of such a current component was shown to differ significantly from one half of the Si bandgap.[149][151] Clearly, the indiscriminate attribution of all edge-related currents to tunneling and generation-recombination is questionable. When the leakage current is not correlated with edges, generation-recombination is still the explanation most commonly invoked. However, the fact that leakage currents were observed to be clearly dominating in some diodes and completely missing in other diodes on the same sample[190] seems inconsistent with a uniform distribution of generation-recombination centers.

From the earlier discussion of electron transport at inhomogeneous SB's, it is clear that experimentally observed leakage currents are consistent with SBH inhomogeneity. The presence of a few large low-SBH regions (with their large $\gamma$'s and, hence, large ideality factors) in the SB diode can certainly lead to the appearance of a leaky component in the junction current, as illustrated in Fig. 14. The large ideality factor of this current and effects due to series resistance limit the predominance of this leakage component to small forward biases. Because of a lower effective SBH, the presence of even a single low-SBH region can lead to the observation of leakage current. Experimentally observed diode-to-diode variations of the leakage current are in much better agreement with isolated *leakage spots* due to low local SBH than with a distribution of recombination centers in the space-charge region. The existence of a current component which is proportional to the perimeter of the diode is also in good agreement with SBH inhomogeneity. A low-SBH patch is less effectively pinched-off when it is situated in close proximity (on the order of the depletion width) to the edge than when it is in the central portion of a diode, as has been clearly demonstrated by computer simulations.[48] When a SB diode contains a uniform distribution of low-SBH patches, the number of patches which are found near the edges is proportional to the perimeter, giving rise to an edge-related current component. The fact that edge-related currents have either a smaller or a larger ideality factor than that of the central portion of the diode is also consistent with SBH inhomogeneity. When the planar portion of the diode is dominated by a roughly uniform SBH, the edge-related current has a larger ideality factor. When the current from central portion is already dominated by low-SBH patches, the edge-related current will have a smaller ideality factor because the patches near the edges are less pinched-off! However, series resistance usually causes edge-related currents to give the appearance of a large ideality factor.

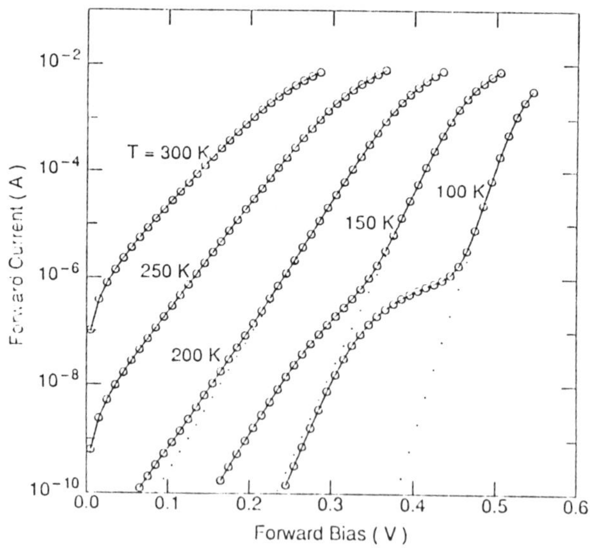


**Figure 14.** Calculated forward current-voltage characteristics from a SB diode ($\Phi_{mean}$ = 0.65 V and $N_D$ = 1 x $10^{15}$ $cm^{-3}$), which contains a low-SBH patch ($\gamma$ = 8 x $10^{-4}$ $cm^{2/3}$ $V^{1/3}$), showing that an inhomogeneous SB can give the appearance of a "leaky" component of current at low temperatures.

## 5.2 Greater-Than-Unity Ideality Factors

Experimentally observed I-V characteristics are almost always semi-logarithmic in some bias range, in agreement with the thermionic emission theory. However, the slopes of such traces often differ from the theoretical prediction, which necessitated the inclusion of the empirical ideality factor[191] in the description of the junction current, Eq. (11). An ideality factor greater than 1 has no direct explanation within the thermionic emission theory, and is generally attributed to a SBH which is bias-dependent. Image force lowering,[192] generation-recombination, interface states (negative charge),[18][193] and TFE[194][195] have all been discussed as possible mechanisms which could lead to a greater-than-unity ideality factor. Since the image-force lowering and the TFE may be calculated and the generation-recombination contribution can be distinguished experimentally, the maximum ideality factor due to these mechanisms may be estimated. Observed ideality factors often far exceed these estimates, prompting the proposal that interface states are a main origin of greater-than-unity ideality factors.

There are two entirely different proposals to explain the non-ideal behavior based on interface states: the interface layer (the tunnel MIS diode) approach and the intimate MIGS (negative charge) approach. In the presence of an interfacial dielectric layer (cf. Fig. 2b), the charge at the dielectric-semiconductor interface becomes more negative (or less positive) with applied forward bias, leading to an increase of the SBH with bias and, hence, an ideality factor greater than 1. The gross non-linearity of experimental I-V curves has been attributed to features in the distribution of interface states. The observation of leakage-like components in the I-V curves, such as that shown in Fig. 15a, has been attributed to this mechanism.[29] The flat portions of the curves shown in Fig. 15a have been ascribed to a high density of interface states at ~0.35 eV below the CBM.[29] However, such interpretations of the I-V curves have difficulties. For example, the current levels at the plateau region of the I-V curves should differ by roughly two orders of magnitude [the difference in the factor $\exp(-0.35\text{eV}/k_BT)$] for the two measurement temperatures, if the proposed interface state mechanism were correct. This is in disagreement with the experimental observations. The plateau-shaped I-V curves are consistent with SBH inhomogeneity and series resistance effects, as demonstrated in Fig. 15b. Generally speaking, the interpretation based on an interface layer is not consistent with both the bias dependence and the work function dependence of the SBH's, as already discussed.[37]

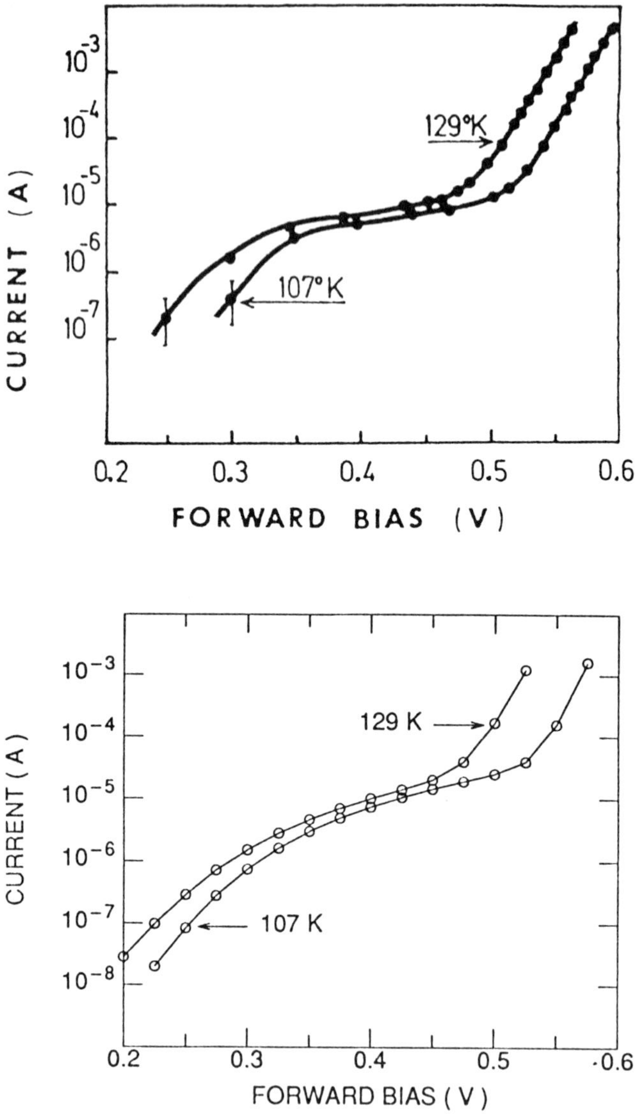


**Figure 15.** *(a)* Forward I-V characteristic of an Au SB diode on n-type Si. (After Barret and Muret.)[29] *(b)* Simulated I-V characteristics of a SB diode with a $\Phi_{mean}$ of 0.8 V, which contains a low-SBH patch ($\gamma = 7.4 \times 10^{-4}$ cm$^{2/3}$V$^{1/3}$ and $N_D = 1 \times 10^{16}$ cm$^{-3}$). The plateau-shaped characteristic results from a combination of SBH inhomogeneity and series resistance.

At intimate SB interfaces, interface states have also been blamed for large ideality factors. The upward bending of the semiconductor bands near the MS interface, due to the spatial extension of the (negative) charge, results in different turning points of the potential (different effective SBH's) for different electric fields. This dependence of the SBH on the electric field, shown in Fig. 2c, is the mechanism with which ideality factors are explained by interface states. However, there is a major consequence of such an explanation which is in conflict with experimental results. The short-range band bending at the MS interface (Fig. 2c) is independent of the semiconductor doping type, and, therefore, may be used to explain only the ideality factor on one type of semiconductor. Specifically, for a particular MS system, MIGS can lead to large ideality factors on only n-type (or p-type, if the charge is positive) semiconductors, and not on both types of semiconductors. Experimental results show large ideality factors on both n- and p-type semiconductors, in disagreement with the interface state mechanism. There are other experimental observations which are not consistent with interface states. For instance, the ideality factors are often found to vary significantly with processing, or from diode to diode on the same sample, while the SBH's are essentially the same. These results are difficult to explain with interface states, because the interface states are assumed to decide both the magnitude of the SBH and the ideality factor of a SB diode. Therefore, contrary to popular belief, interface states and/or interface layers cannot provide a consistent explanation of the various behaviors of the ideality factor.

This bias dependence of the effective SBH's (saddle-point potential) of an inhomogeneous SB can explain all the observed behavior of the ideality factors. Since the currents of SB diodes are often dominated by low-SBH patches, greater-than-unity ideality factors are routinely observed. When the doping level increases, the $\Phi_{eff}$ of a low-SBH patch with a fixed $\gamma$ decreases and its ideality factor increases, in good agreement with experimental observations. The correlation shown in Fig. 12 may be expected when the local SBH varies about the same mean SBH, but with random amplitude and/or length scale. The diode-to-diode variation of the ideality factor and the dependence on processing are also consistent with variations in the distribution of local SBH in the diodes under study. The observation of large ideality factors when the diode is in a state of maximum confusion is also in good agreement with the interpretation of ideality factors based on SBH inhomogeneity. One notes that greater-than-unity ideality factors were recently proposed to be related to SBH inhomogeneities,[196] even though the underlying reason, namely potential pinch-off, was not recognized.

## 5.3 The $T_o$ Anomaly

Many different temperature dependencies of the ideality factor have been observed experimentally. Most frequently, the ideality factor of a diode increases when the sample temperature is lowered. At many MS interfaces, the deduced SBH and ideality factors are found to vary with the measurement temperature in a fashion generally known as the *$T_o$ anomaly*.[60][167][197][198] Such a phenomenon has been observed from all types of SBs, on elemental semiconductors[60][126] and compound semiconductors[197][199] alike. A diode is said to display the $T_o$ effect if its junction current may be expressed as

Eq. (22) $$I = A^{**}AT^2 \exp\left(-\frac{q\Phi}{k_B(T+T_o)}\right)\left[\exp\left(\frac{qV_a}{k_B(T+T_o)}\right)-1\right]$$

where $T_o$ is a constant, typically 10 - 60 Kelvin. Demonstration of the $T_o$ effect is usually accomplished by plotting $nk_BT$ (the inverse slope of an I-V curve) against $k_BT$ and observing a straight line, with a slope of unity, which does not extrapolate through the origin. Concurrently, by changing the abscissa of the Richardson plot from 1/T to 1/nT, a straight line should be observed in cases displaying the $T_o$ anomaly. Examples of these two plots, reproduced from Saxena's work on guarded Cr-Si diodes,[60] are shown in Fig. 9. Presently, the $T_o$ anomaly has been ascribed to an exponential distribution of the density of interface states,[21][22] e.g., Eq. (11). However, such a model depends on the presence of an interface layer[22] and, hence, cannot explain the $T_o$ effect, which is frequently observed at intimate MS contacts.[151][142] Furthermore, there is no experimental evidence for an exponential distribution of states at MS interfaces. The fact that the measured $T_o$ varies significantly among similarly fabricated diodes[200] and the local non-uniformity of $T_o$ in large area diodes[201] are suggestive that the $T_o$ anomaly is not directly related to the formation mechanism of the SBH. Crowell[22] pointed out that the $T_o$ anomaly was consistent with band bending such as that arising from opposite-type doping at the interface.[202] However, deliberate doping modifications brought out changes[203] in the apparent $T_o$ exactly opposite to that expected from Crowell's proposal. The suggestion that the $T_o$ anomaly is a result of the temperature dependence of the work function[60] seems numerically off by at least an order of magnitude.

SBH inhomogeneities offer the only valid explanation of the $T_0$ anomaly. As the temperature is lowered, the current of an inhomogeneous SB displays two trends, either of which can lead to the $T_0$ effect. Firstly, as shown in Fig. 6, an increase in the bias results in an increase in the measured ideality factor. Secondly, the current of a random SB diode which contains low-SBH regions with a distribution of $\gamma$'s [cf. Eqs. (20) and (21)] may be expressed phenomenologically in a form identical to Eq. (22).[13] Therefore, when the temperature is lowered, the junction current is dominated by fewer low-SBH regions with lower effective SBH's and larger ideality factors. As the example of Fig. 16 shows, calculated junction current of a SB diode with low-SBH patches of just two distinct $\gamma$'s displays the $T_0$ effect. The empirical constant $T_0$ depends on the SBH distribution, the doping level, and the experimental procedure for ideality factor determination.[13] Since the fluctuation of SBH likely varies for different diodes, the inconsistency of the apparent $T_0$'s[200][201] and the doping dependence[203] are all naturally explained.

## 5.4 Other Temperature Dependencies of the Ideality Factor

It is usually assumed that a study of the dependence of the ideality factor on temperature can reveal the conduction mechanism of a particular SB diode. The $T_0$ phenomenon is only one of five distinct temperature dependencies according to the original categorization by Saxena,[60] as schematically shown in Fig. 17. A temperature-independent, large ideality factor, line 2 in Fig. 17, has often been observed experimentally. When the slope of the I-V trace approaches a constant at low temperature, as shown by line 4 in Fig. 17, the conduction mechanism is usually thought to be dominated by TFE.[194][195] Good quantitative agreements of experimental data with the TFE theory have been observed.[194] However, the occasional observations of data similar to line 4 in Fig. 17, under experimental conditions where tunneling should be negligible,[204] suggests that the interpretation of the conduction mechanism based on the temperature dependence of the ideality factor may not be unique. Frequently, the ideality factor of a SB diode is shown to follow different behaviors at different temperature ranges.[142] Furthermore, just like the magnitude of the ideality factor, the temperature dependence of the ideality factor of a diode also varies with processing. Both the diode-to-diode variation and the variation with processing are suggestive that the many distinctive dependencies of the ideality factor on temperature are not unrelated. It seems obvious that one should look for a single mechanism which may be compatible with all the observations.

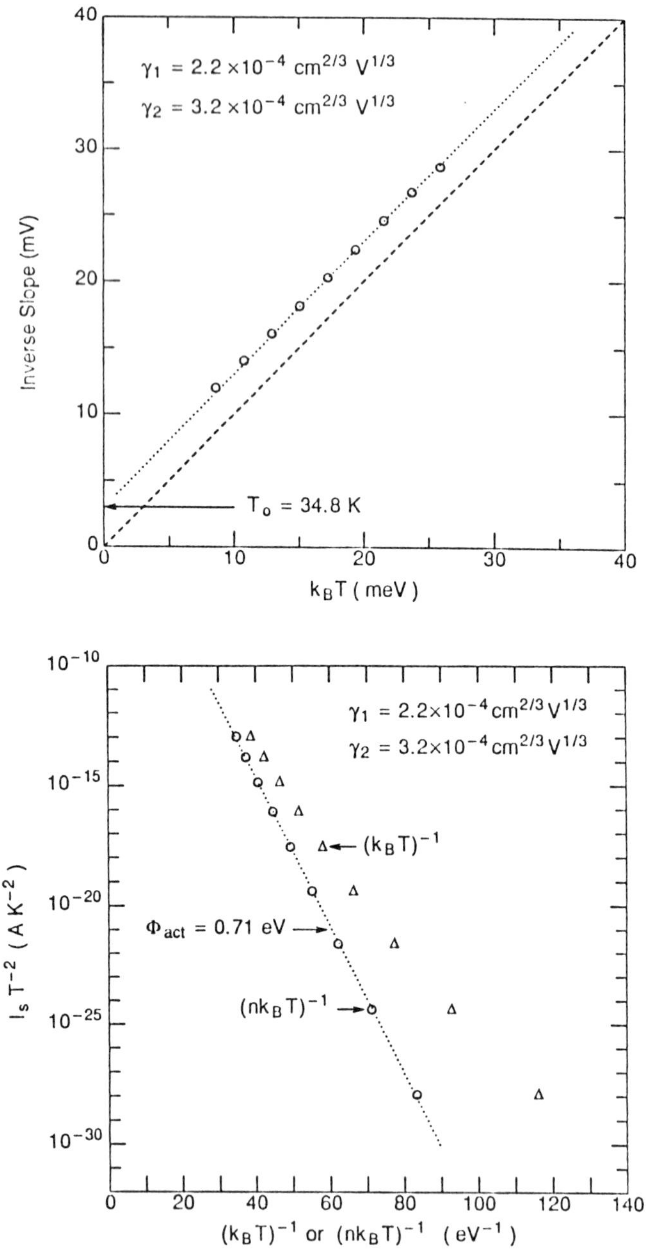


**Figure 16.** I-V characteristics, calculated with analytic expressions of Table 2, of an inhomogeneous SB diode, with a $\Phi_{mean}$ = 0.8 V and an area of $4 \times 10^{-3}$ cm$^2$. Low-SBH patches with $\gamma$'s of $3.2 \times 10^{-4}$ cm$^{2/3}$ V$^{1/3}$ (x 800) and $2.2 \times 10^{-4}$ cm$^{2/3}$ V$^{1/3}$ (x 800,000) are included. *(a)* plot of inverse slope (evaluated at ~$1 \times 10^{-6}$ ampere), and *(b)* regular and modified Richardson plot of the saturation current.

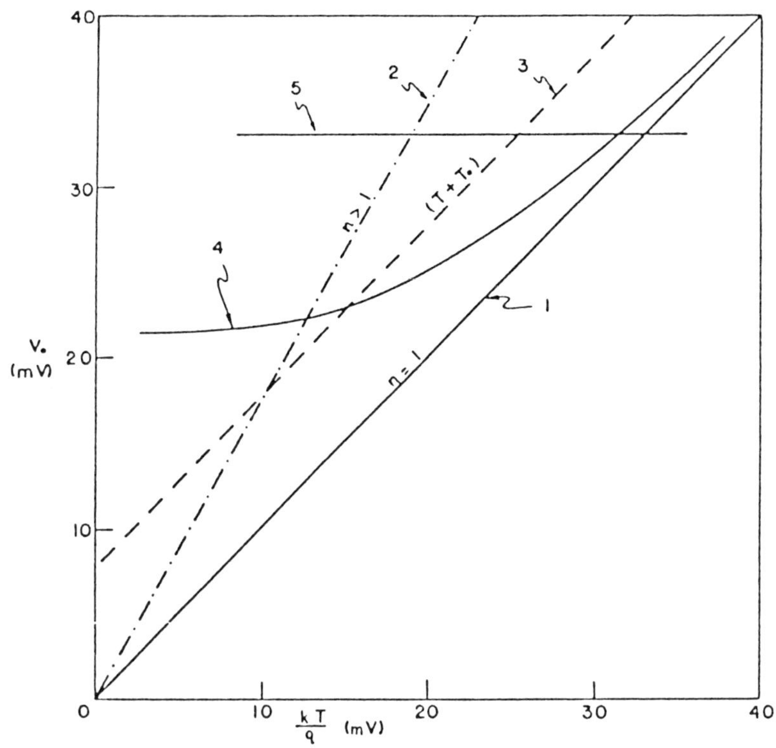


**Figure 17.** Plot of inverse slope ($V_0$) versus $k_B\,T/q$, showing the five basic categories of the temperature dependence of the ideality factor. Line *1* is an ideal SB which follows the prediction of thermionic emission theory. Line *2* shows a temperature-independent, greater-than-unity, ideality factor. Line *3* displays the $T_0$ effect. Lines *4* and *5* represent the behaviors when conduction is dominated by, respectively, TFE and FE. (After A. N. Saxena.)[60]

Recent simulations showed that almost all of the temperature dependencies of the ideality factor are consistent with SBH inhomogeneity.[48] Ideality factors from calculated I-V characteristic of a composite SBH diode are shown in Fig. 18, which qualitatively reproduces the behavior usually attributed to the TFE mechanism. A temperature-independent, greater-than-unity, ideality factor, such as that displayed in Fig. 10, is consistent with a SB diode whose current transport is dominated by low-SBH regions with narrowly distributed γ's, as shown in Fig. 19a. A decrease of the ideality factor with cooling, as shown in Fig. 11, is consistent with the presence of general SBH inhomogeneity about some mean SBH and, in addition, a small number of low-SBH regions which are large enough that they are not pinched-off, as shown in Fig. 19c. The apparent switch-overs, between different ideality factor categories[142] (switching between lines shown in Fig. 17), observed at different temperature ranges are also consistent with SBH inhomogeneity,[48] as shown in Fig. 19b. Since the ideality factor is simply a manifestation of the SBH uniformity, it is not surprising that it may be improved by improving the uniformity of the layer, which presumably also leads to a more uniform interface structure.[86][90][92][169] Nor should one find it odd that the ideality factor is the largest when the layer is the most non-uniform.[160] The perfect explanation of almost the entire spectrum of the observed temperature-dependencies with a single assumption of SBH inhomogeneity rules out interface states as being of any real significance to the ideality factor.

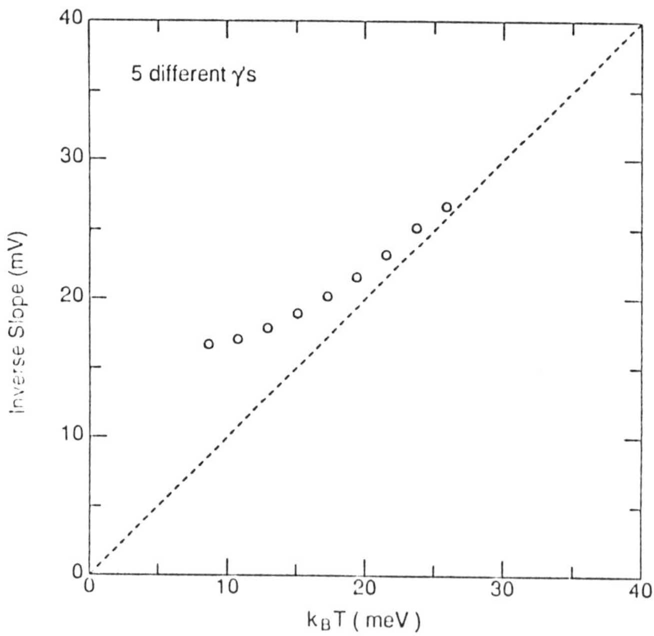


**Figure 18.** Inverse slope of the I-V curves from an inhomogeneous Si SB diode ($N_D$ = 1 x $10^{16}$ $cm^{-3}$) with a $\Phi_{mean}$ of 0.7 V and an area of 4 x $10^{-3}$ $cm^2$. The following low-SBH patches are included: $\gamma$ = 4 x $10^{-4}$ (x 1), 3.5 x $10^{-4}$ (x 20), 3 x $10^{-4}$ (x 400), 2.5 x $10^{-4}$ (x 8,000), and $\gamma$ = 2 x $10^{-4}$ $cm^{2/3}$ $V^{1/3}$ (x 160,000). The displayed temperature dependence of the ideality factor is similar to that usually ascribed to TFE.

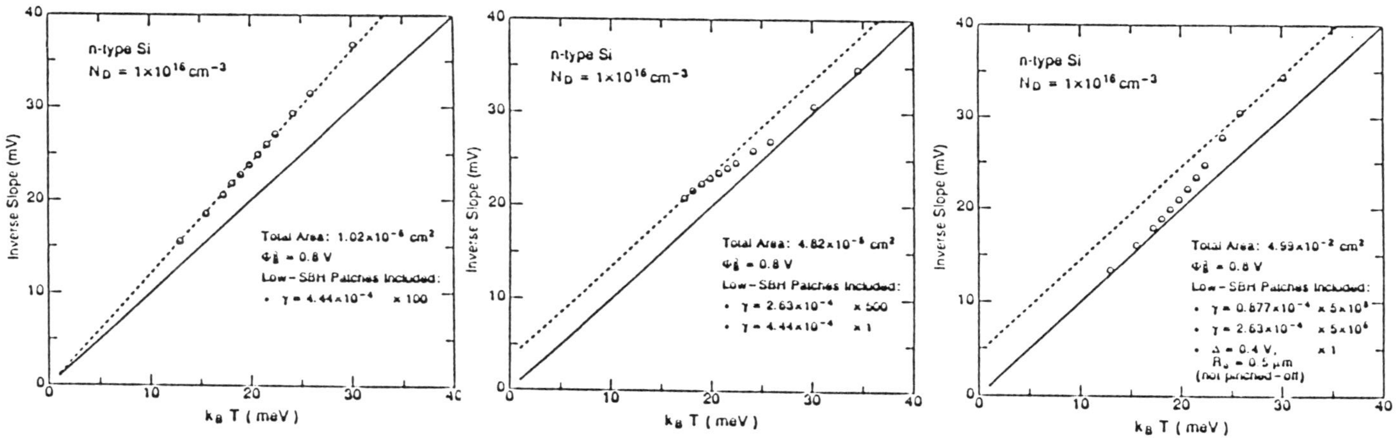


**Figure 19.** The inverse slope of simulated I-V plots as a function of $k_B$ T for inhomogeneous MS contacts composed of a distribution of low-SBH patches as described in the figures ($\gamma$ is in units of $V^{1/3} cm^{2/3}$): *(a)* plot showing a constant ideality factor in excess of unity, *(b)* plot showing the $T_o$ effect at low temperature and ideal behavior at high temperature, and *(c)* plot showing ideal behavior at low temperature and non-ideal behavior at high temperature. (After Sullivan, Tung, Pinto, and Graham.)[48]

## 5.5 *Soft* Reverse Characteristic

It is a universal observation that the current from any SB diode never truly saturates at large reverse bias. These *soft* reverse characteristics are observed even when the utmost care is taken to eliminate possible effects due to edges of the diode.[58] A linear relationship is often observed, at least for reverse biases not exceeding ~10V, by plotting the logarithm of the current against $V_d^{1/4}$,[102][133][187][205] in agreement with the functional dependence of the proposed SBH lowering mechanism from image-force. However, the observed slopes from such plots often far exceed that predicted by the image-force mechanism alone,[102][187][205] which led to the suggestion that the relevant permittivity was not the static permittivity of the semiconductor, $\varepsilon_s$. Essentially, it is rationalized that the experimentally observed slopes may be fitted to deduce an "appropriate" high-frequency permittivity, $\varepsilon_d$. But the question which immediately arose, namely, "Why does $\varepsilon_d$ show a strong dependence on temperature?" could not be answered. This dilemma was removed by Andrews and Lepselter[206] who did an extensive investigation of the reverse characteristics of SB's and proposed that, in addition to SBH lowering due to image-force, $\delta\Phi_{img}$, there is a SBH lowering, $\delta\Phi_{is}$, which is proportional to the electric field, i.e.

Eq. (23)
$$\delta\phi = \delta\Phi_{img} + \delta\Phi_{is} = \frac{1}{2}\sqrt{\frac{qE_{max}}{\pi\varepsilon_s}} + \alpha E_{max}$$

This additional SBH lowering is thought to arise from the upward bending of the semiconductor band due to MIGS,[5][23] identical to the mechanism proposed, and presently ruled out, to lead to greater-than-unity ideality factors. In Eq. (23), the constant $\alpha$ is thought to relate to the density and depth of the interface states.[23] Experimental results in agreement with the prediction of this model have been observed in many studies.[88][179][206]-[208] For example, the excellent fit of experimental data, over many decades, with the prediction of Eq. (23) is shown in Fig. 20(a). However, even though the functional form of the experimental reverse current may be explained by this model, other consequences of this model, due to interface states, have not been born out by experiments. For instance, it is not clear why very different $\alpha$'s are found for similar MS interfaces. Also, the proposed mechanism of SBH lowering is completely absent in some diodes.[192][133] But, as before, the most severe problem with the interface state model is that the proposed SBH lowering should be proportional to the electric field perpendicular to the

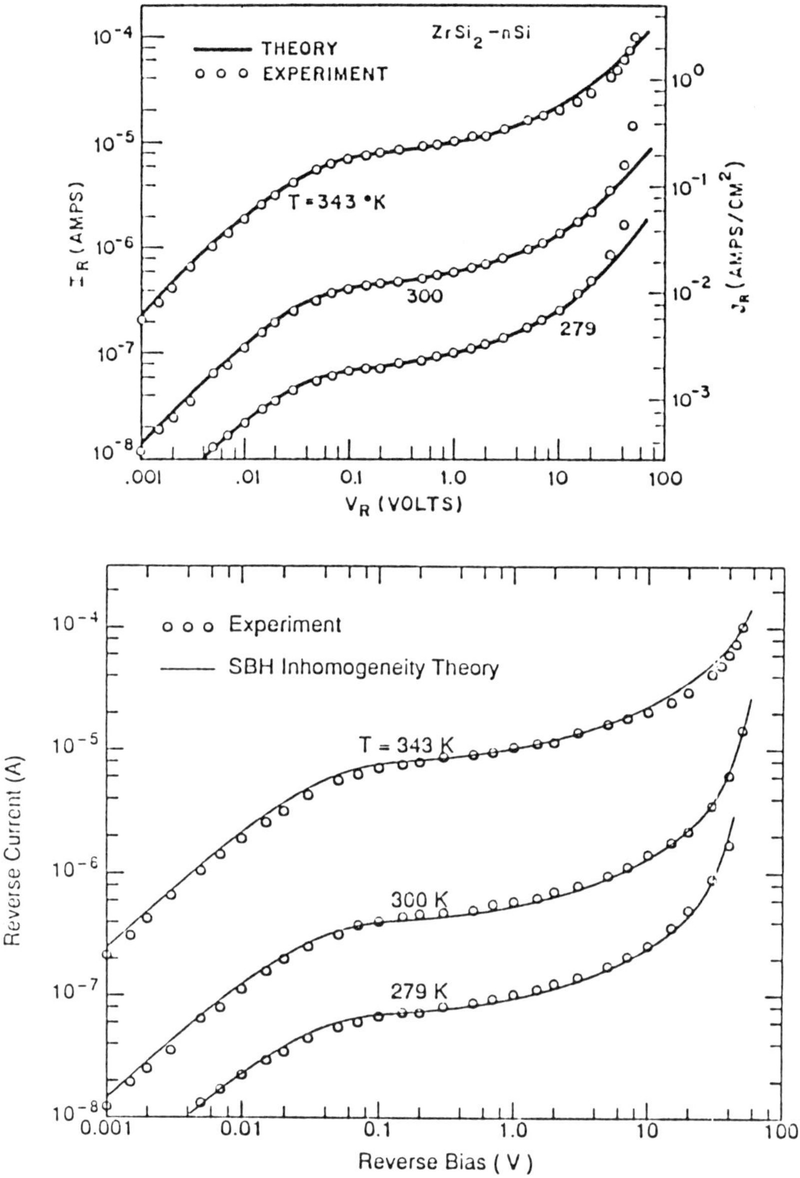


**Figure 20.** *(a)* Reverse characteristics of a $ZrSi_2$ Schottky diode formed on n-type silicon. Theoretical lines were generated using Eq. (23) with $\alpha$ = 10 Å (After Andrews and Lepselter.[206] *(b)* Experimental reverse currents, from Andrews and Lepselter,[206] compared with calculated currents of an inhomogeneous SB diode, which has a $\Phi_{mean}$ of 0.56 V and contains low-SBH patches with $\gamma$'s of 5.6 x $10^{-5}$ $cm^{2/3}$ $V^{1/3}$ (3 x $10^5$) and 4 x $10^{-5}$ $cm^{2/3}$ $V^{1/3}$ (1.7 x $10^6$).

interface, and NOT to its absolute value. In other words, for a particular MS system, the SBH lowering mechanism due to interface states, $\delta\Phi_{is}$, may be operative on either the n-type or the p-type semiconductor, but not on both. Experimental results,[117][179][207] including those shown by Andrews and Lepselter in their original paper,[206] indicated that the soft SB characteristics for many metals/silicides occurred with similar magnitudes on both types of Si substrate. Therefore, the SBH lowering proportional to $E_{max}$, in Eq. (23), is not due to interface states as originally thought.

Due to the limited range of available reverse bias, a reliable determination of the functional form of the current of any particular SB diode is difficult. In addition, the reverse currents of different diodes often show different behaviors. Therefore, even though Eq. (23) may account for the observed reverse current of some diodes, it is by no means the only lowering mechanism which may explain the experimental data. Any SBH lowering mechanism which varies with the electric field more rapidly than $E_{max}^{1/2}$, could lead to a very satisfactory fit with the experimentally observed reverse currents. As discussed, the reverse characteristic of an inhomogeneous SB depends critically on the actual variation of the SBH. A variety of dependencies of the SBH lowering on reverse bias, such as proportional to $E_{max}$ or $E_{max}^{4/3}$ are possible in inhomogeneous SB diodes,[13] which are certainly capable of explaining the bias dependence of experimentally observed reverse characteristics. One notes that Andrews and Lepselter's theory[206] tends to overestimate the current at very-large reverse biases at high temperatures and underestimate it at low temperatures (Fig. 20a). For large reverse biases, the experimentally observed currents are actually in slightly better agreement with that due to SBH inhomogeneity, as shown in Fig. 20b. The wide range of behaviors of reverse currents experimentally observed from various SB's suggests that the problem is complicated and that each specific diode may have its own individuality. Such a scenario is in complete accord with SBH inhomogeneity and is not easily reconciled with the interface state model.[206]

## 5.6 Dependence of SBH on Measurement Technique

The SBH measured by the I-V technique, $\Phi_{I\text{-}V}$, often decreases with increasing doping level, while the SBH measured by the C-V method, $\Phi_{C\text{-}V}$, remains constant. Frequently, the SBH's depend on the technique of measurement, namely, $\Phi_{C\text{-}V}$ sometimes significantly exceeds $\Phi_{I\text{-}V}$ and the SBH derived from PR techniques, $\Phi_{PR}$.[127][209][210] Identical to the proposed explanations of the ideality factor, lowering of the SBH by image-force, interface states, and TFE have frequently been invoked to explain the doping-level-dependence of $\Phi_{I-V}$'s.[124]-[166] While SBH investigations, especially those where the doping-dependence of the SBH have been studied, have generally concentrated on n-type semiconductors, it is known that SBH lowerings on p-type semiconductors also routinely exceed that predicted by image-force alone. The argument invoked in the last few sections concerning the sign of SBH lowering due to MIGS also applies to the present case. Thus, it is clear that the observed doping-dependence of moderately-doped, intimate, SB contacts may not be attributed to interface states as previously thought. Rather, it is in agreement with the presence of SBH inhomogeneities.[44][52][211][212] Current transport at inhomogeneous SB's is dominated by low-SBH patches, leading to the deduction by I-V and PR techniques of apparent SBH's which are lower than the arithmetic average of the entire diode. Since under usual circumstances, the C-V technique yields an average SBH for the whole diode,[13] the experimentally observed dependence of SBH on the technique of measurement is likely due to SBH inhomogeneity. Because effects due to SBH inhomogeneity are more pronounced on more heavily doped semiconductors, it is not surprising that the classic examples of the dependence of SBH on the measurement technique[127] (and examples of the $T_0$ effect)[60][126] were all obtained on Si with $N_D > 10^{16}$ cm$^{-3}$. To artificially reduce the difference between $\Phi_{I-V}$ and $\Phi_{C-V}$, it has been proposed that the true I-V SBH is the parameter $\Phi$ in Eq. (22), and not the parameter $\Phi_{Bn}$ of Eq. (12).[213] This procedure, which is equivalent to defining the SBH as $n\Phi_{Bn}$, has often been followed, even in cases where the diode I-V behavior is not in accordance with the $T_0$ phenomenon.[213][214] At the heart of the issue is the fact that a single parameter, either $\Phi_{Bn}$ or $\Phi$, is not sufficient to represent the entire junction characteristic in case of SBH inhomogeneity. However, it seems that $\Phi_{Bn}$, which is based on a physical model such as the thermionic emission theory, is still more meaningful than an arbitrary parameter such as $n\Phi_{Bn}$.

A common occurrence in C-V experiments is the deduction of an apparent SBH which is even higher than the true arithmetic average. This phenomenon is often associated with experimentation problems, namely, how to determine the true space charge capacitance from raw C-V data. There are many well-known phenomena which may lead to the measurement of apparent capacitances which are not the true junction capacitances, as previously discussed.[215] Besides possible explanations from deep levels,[216][217] doping variations,[218] edges,[219] interface states,[220] etc., one notes that a large series resistance may lead to such an observation.[215][221] This phenomenon is particularly noticeable at high modulation frequencies and when the magnitude of C is large.[215] When the SBH is uniform, or when the SBH fluctuates rapidly, the true junction capacitance may be deconvoluted, using a correct equivalent circuit diagram, and the measured value of the series resistance. But in the case of an inhomogeneous SB, the series resistance issue is difficult to handle because of the lateral inhomogeneity of the in-phase current. Because the series resistance is not a fixed parameter for an inhomogeneous diode, an appropriate equivalent circuit diagram cannot be drawn. It is difficult to utilize the C-V technique for SBH deduction under such conditions. Other C-V anomalies exist, e.g., the occasional observation of a downward curvature in $C^{1/2}$ plots has been explained in terms of an *excess capacitance,* ($C_o$), and attributed to the presence of interface states.[222]-[224] Excess capacitances observed at a SB in the forward bias have also been attributed to interface states.[29]-[31] However, as already noted, the explanation of excess capacitance in terms of interface states has many difficulties.[34][36] Excess capacitances are often found to correlate with large ideality factors,[31] a fact which, as discussed, is suggestive of a connection between capacitance anomalies and SBH inhomogeneities.

## 5.7 Other Inconsistencies

Since the very beginning of SBH experiments, it seemed that phenomena occurring at MS interfaces were always more complicated than proposed theories and mechanisms could explain. There have been a few examples of quantitative agreements between experiments and theories. However, departures from theoretical predictions were also often observed with regularity under almost identical experimental conditions. These inconsistencies in experimental results are suggestive of interface electronic properties which are more complicated than the simple models which have thus far been proposed. It has been a general practice for SBH investigators to attribute changes in the electronic properties of the MS interface to changes in parameters such as interface state densities and interfacial layer thicknesses. Furthermore, whatever changes due to various treatments are assumed to be uniform for the entire MS interface. The obvious reason for these assumptions is that one may apply these changes directly in the simple models one assumes for the interface states, such as those shown in Fig. 2. It should be kept in mind that the common attribution of SBH variations to interface states with a uniform lateral distribution originates from convenience, rather than through scientific deduction.

Direct evidence for the inhomogeneity in the SBH's has often been recognized.[225]-[227] It is obvious that common diode processing can lead to structures which are laterally inhomogeneous. For example, it was shown earlier that surface treatments bring about much variation in the observed SBH's. Specifics of the surface, such as oxide thickness, contamination particles, structural defects, morphologies, etc., are almost certainly non-uniform. Variations and inconsistencies in the observed SBH's are usually larger for as-deposited metals, or as-deposited silicides,[92] than for well-annealed junctions. This is sometimes rationalized as due to defects on the original surfaces which are subsequently annealed out. Since defects, impurities, etc. are nothing more than part of the structure of a surface, perhaps a more general way to view an as-deposited MS interface is to say that it has a less stable structure, which is reminiscent of the original surface and which may certainly fluctuate from region to region, than the structure of an annealed interface, which is more a result of thermodynamics. The chemistry and the metallurgical reactions at any intimate MS interface may be quite heterogeneous. Experimentally observed fluctuations and variations of the SBH data are in good agreement with the fact that almost all SBH's are inhomogeneous!

## 6.0 NON-EPITAXIAL SBH TRENDS AND MODELS

When the SBH's of a large number of metals and silicides are summarized onto a table or a plot, it is usually tempting to explain whatever correlations and systematics which one may find, despite large scatters in the experimental data. Guided by the Schottky-Mott relationship,[228] Eq. (7), early correlations have mostly been made to the electronegativity[19][229] and the work function[230] of the metal. Although the SBH's are usually higher for metals with larger work functions or electronegativities, the fitted slope of such a plot is far less than that predicted by Eq. (7).[15][19] This led to the suggestion of surface/interface states, as discussed earlier. Andrews and Phillips[231] noted that, with few exceptions, the silicide SBH's exhibit a linear relationship with the heat of formation of the silicides, $\Delta H_f$. This relationship, which is illustrated in Fig. 21, was argued to indicate the fact that the degree of hybridized bonding between transition metals and Si is linear in $\Delta H_f$. It has also been pointed out that these silicide SBH's may be correlated with the eutectic temperature of the metal-Si binary system,[232] which led to the proposal that an interfacial (amorphous) layer determines the SBH. The use of effective work functions, characteristics of an interfacial $MSi_4$ stoichiometry, have also produced a correlation with the observed silicide SBH's.[233] Recent experiments have shown the abruptness of most silicide interfaces, with no evidence for amorphous or additional interface layers. Tersoff[12] proposed that the SBH's are determined by the charge neutrality level of the MIGS, which implies that the SBH does not have a first order dependence on the metal. A refined treatment[234] considered the screening of the interface dipole by MIGS and derived a relationship of the SBH with the metal electronegativity. Schmid[99] plotted the silicide SBH's against the Miedema electronegativities of the metals and produced a relationship which is reminiscent of the theoretical predicted FL pinning behavior[26] by a finite density of interface states. This result is further expanded to stress the importance of considering effects due to both defects and MIGS.[25]

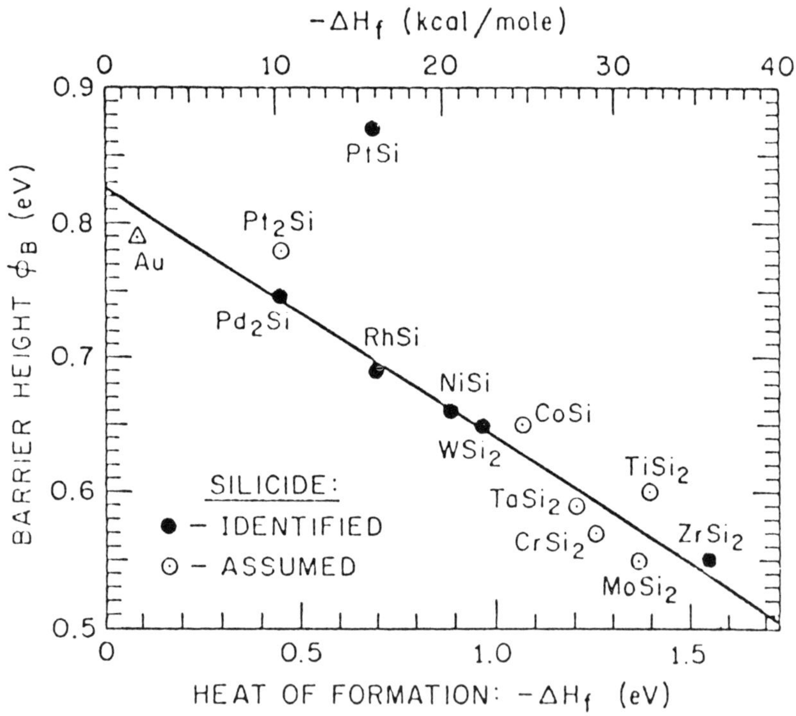


**Figure 21.** Barrier heights of transition metal silicide-Si interfaces plotted against heats of formation $\Delta H_f$ of the silicide. The correlation coefficient (excluding the PtSi datum) is 97%. Line indicates $\Phi_B = 0.83 + 0.18\,\Delta H_f$. (After Andrews and Phillips.)[231]

In the light of massive evidence for SBH inhomogeneities in polycrystalline Si SB diodes discussed earlier, the very approach of concentrating only on the magnitudes of the experimental SBH's, in order to assess the validity of a SB theory, seems inappropriate. If a MS interface has an inhomogeneous SBH, then the SBH experimentally obtained from this interface is just an averaged value of some weighted distribution of different FL positions. Such an average does not necessarily have physical significance and certainly should not be the only data used for a deduction of the SB mechanism. It is obvious that if it had been known that most of the MS interfaces had inhomogeneous SBH's, then FL pinning would not have been an attractive candidate to propose as the formation mechanism of the SB. Generally speaking, the simplicity and the rigidity of the FL pinning concept are difficult to reconcile with the diversity of phenomena observed at MS interfaces. The most direct explanation of inhomogeneous SBH's is that the SB mechanism depends on some local parameters of the MS interface. It is the variation of these local specifics which causes the local SBH to change. The true formation mechanism of the SB is, of course, the one which determines the local FL position based on these local specifics.

Discussions of the last section indicate that if one is willing to abandon the notion that all SBH's are spatially homogeneous then many abnormal behaviors cease to be abnormal. The issue of SBH inhomogeneity should no longer be ignored, because understanding why and how the SBH varies at a MS interface is probably tantamount to understanding the formation mechanism of the SB. It is hard to conceive that the observed variations of the SBH could be completely independent of the local structure of the MS interface. Actually, the dependencies of the SBH on the Si orientation,[51][147][158] the degree of epitaxy,[161] the degree of crystallinity,[116] and the silicide thickness[146][147][160] are, by themselves, quite suggestive of SBH changes which are mediated through changes in the interface structure. One may also view the effect of hydrogen on the SBH[164][165] as a change in the interface structure, since H becomes part of the MS interface. However, the possible relationship between interface structure and Schottky barrier height is not easily studied in MS systems discussed so far. The reason is that the structure at an ordinary, polycrystalline, MS interface is too complicated to allow a meaningful correlation with the observed electronic structure. Fortunately, high quality epitaxial MS interfaces have become available which have made the study of the correlation of structure with FL position considerably easier, as will be discussed in detail.

# 7.0 STUDY OF SB MECHANISMS THROUGH EPITAXIAL SILICIDES

## 7.1 Advantages of Epitaxy

Ordinary MS interfaces consist of a host of different structures due to multiple compound formation, polycrystallinity, defects and other disorder. The complexity of the interface structures and the many possible ways electrical data may be affected by SBH inhomogeneity, as discussed in the last section, can easily mar experimental data and prevent an explicit deduction of the SB formation mechanism. Therefore, even though the formation mechanism of the SBH at ordinary MS interfaces is of the most scientific and technological interest, its deduction almost certainly has to come from simpler "model" systems. Epitaxially fabricated MS systems offer the best opportunity to understand the dependence of the electronic properties on the structure of a MS interface. Not only can the atomic structure of a single crystal MS interface be obtained by state-of-the-art experimental techniques, but also the electronic structures may be calculated based on the observed atomic structures. A comparison of the experimental and theoretical SBH results represents the best hope of understanding the formation mechanism of the SB. On account of their unsurpassed crystalline quality among all available epitaxial MS systems, epitaxial silicide-Si interfaces[235] have become the premium vehicle for a controllable investigation of the SBH.

## 7.2 $NiSi_2$ on Si(111)

**Fabrication and Structures.** On Si(111), two epitaxial orientations are possible for $NiSi_2$ (and $CoSi_2$). The type A silicide has the same orientation as the silicon substrate, and the type B silicide shares the surface normal <111> axis with Si, but is rotated 180° about this axis with respect to the Si.[236] High quality single crystals of $NiSi_2$ may be grown with either type A or type B orientation by a proper choice of template growth condition.[237]-[240] The epitaxial orientation of thin $NiSi_2$ layers depends on the amount of deposited nickel. When ~16 - 20 Å nickel is deposited at room temperature, subsequent annealing leads to the growth of type A $NiSi_2$.[237][241][242] From the consideration of the crystal symmetry, type A $NiSi_2$/Si(111) is the only epitaxial MS system which may be, and has been, fabricated without dislocations. All other existing MS interfaces, such as type B $NiSi_2$/Si(111), NiAl on AlAs(100), etc., necessarily contain dislocations because of the change in the crystal symmetry across the interface and the presence of steps at the interface. Even though dislocations are absent in thin layers, steps, due to the accidental misorientation of the Si(111) wafer, are often present at type A $NiSi_2$ interfaces.

If a small amount (< 5 Å) of nickel is deposited on Si(111) at room temperature, annealing leads to the growth of $NiSi_2$ with majority type B orientation. In reality, type B $NiSi_2$ layers grown by this technique are likely to be non-uniform in thickness and, furthermore, they often contain a small fraction of type A grains. So in practice, growth techniques involving deposition of Si or co-deposition of $NiSi_2$ are more often employed for the growth of purely type B oriented $NiSi_2$.[242]-[244] For instance, co-deposition of $NiSi_2$ at elevated temperature (the MBE technique) on Si(111) leads to the growth of purely B-type layers.[245][246] However, $NiSi_2$ layers grown by MBE at elevated temperatures are usually nonuniform and contain a high density of dislocations. Recently, it was discovered that single crystal type B $NiSi_2$ layers can be grown at room temperature by (pre-)deposition of a suitable amount, ~2 Å, of Ni and co-deposition of $NiSi_2$.[243] A typical room-temperature-grown $NiSi_2$ layer is shown in Fig. 22a. Although single crystal growth is demonstrated at room temperature, the density of defects contained in the grown layers is high, evidenced by a fuzziness in their TEM images, as shown in Fig. 22a. Structural variation remnant of the original 7x7 surface, clusters of point defects in the film, disorder at the silicide interface are all possible defects which may be present. Upon annealing at above 350°C the defects in these $NiSi_2$ layers are reduced, evidenced by an improvement of the channeling characteristics and the observation of much more robust TEM images, as shown in Fig. 22b.

The atomic structure of both type A and type B $NiSi_2$/Si(111) interfaces have been studied by high resolution electron microscopy (HREM) and found to have the 7-fold structure.[247]-[249] A high resolution image of the type B $NiSi_2$ interface is shown in Fig. 23. The terminology for the structure of a silicide interface, e.g., 7-fold, is based on the number of nearest Si neighbor atoms to a metal atom at the interface. In a bulk disilicide lattice, each metal atom has a coordination number of 8. Structural models of the two 7-fold $NiSi_2$/Si(111) interfaces are schematically shown in Fig. 24. These early HREM results were later confirmed by x-ray standing wave (XSW),[250][251] medium energy ion scattering (MEIS),[252] and x-ray interference[253] investigations. Even though all the investigations unanimously deduce the $NiSi_2$ planar interface to have the 7-fold structure,[254] there is considerable disagreement as to the magnitude of the vertical contraction at the interfaces. A longer Si-Si bond length at a type A interface than that at a type B interface was suggested by one XSW study,[250] while a conflicting conclusion was drawn in another XSW study.[251] In terms of the structural model of the interface, this discrepancy may seem an unimportant side issue. However, since the SBH turns out to depend on the details of the atomic structure, relaxation at the interface may have a significant effect on the observed SBH. Hence, an expeditious resolution of this experimental disagreement is desired.

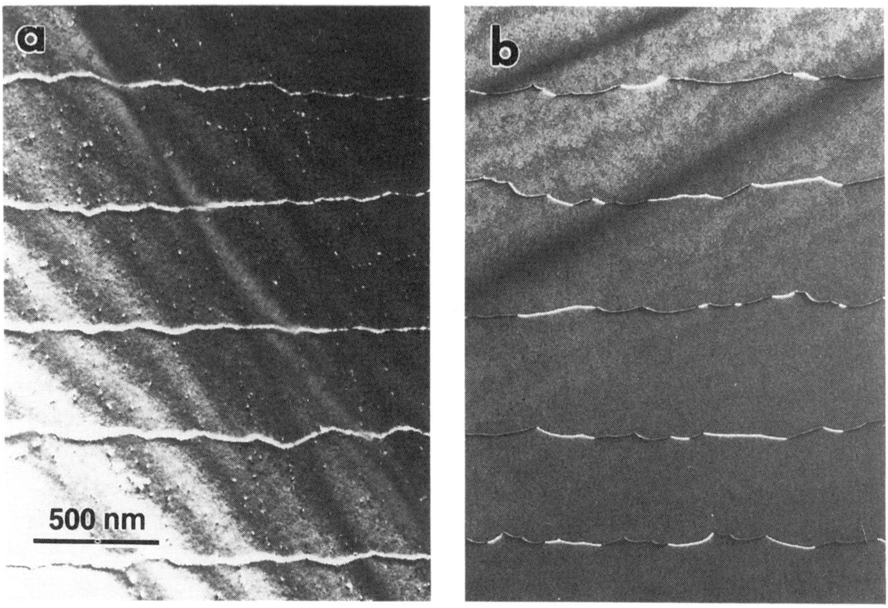


**Figure 22.** Planview, $(2\bar{2}0)$ weak beam, TEM images of ~50 Å thick type B $NiSi_2$ layers on Si(111). *(a)* A layer grown at room temperature by deposition of ~2 Å Ni and co-deposition, and *(b)* a layer grown by the same technique as *(a)*, but with an additional anneal at ~500°C.

**Electronic Properties.** An intriguing dependence of the SBH on the epitaxial orientation has been observed at the epitaxial $NiSi_2$/Si(111) interfaces: Type A and type B $NiSi_2$ have distinctively different SBH's.[211][255] Typical I-V characteristics from $NiSi_2$/Si(111) interfaces are shown in Fig. 25 to illustrate the significant difference between the two orientations. This dependence of SBH on the epitaxial orientation was briefly challenged.[119] However, very extensive studies by various groups[256]-[258] have since fully confirmed the original findings[255] that type B $NiSi_2$ has a SBH about

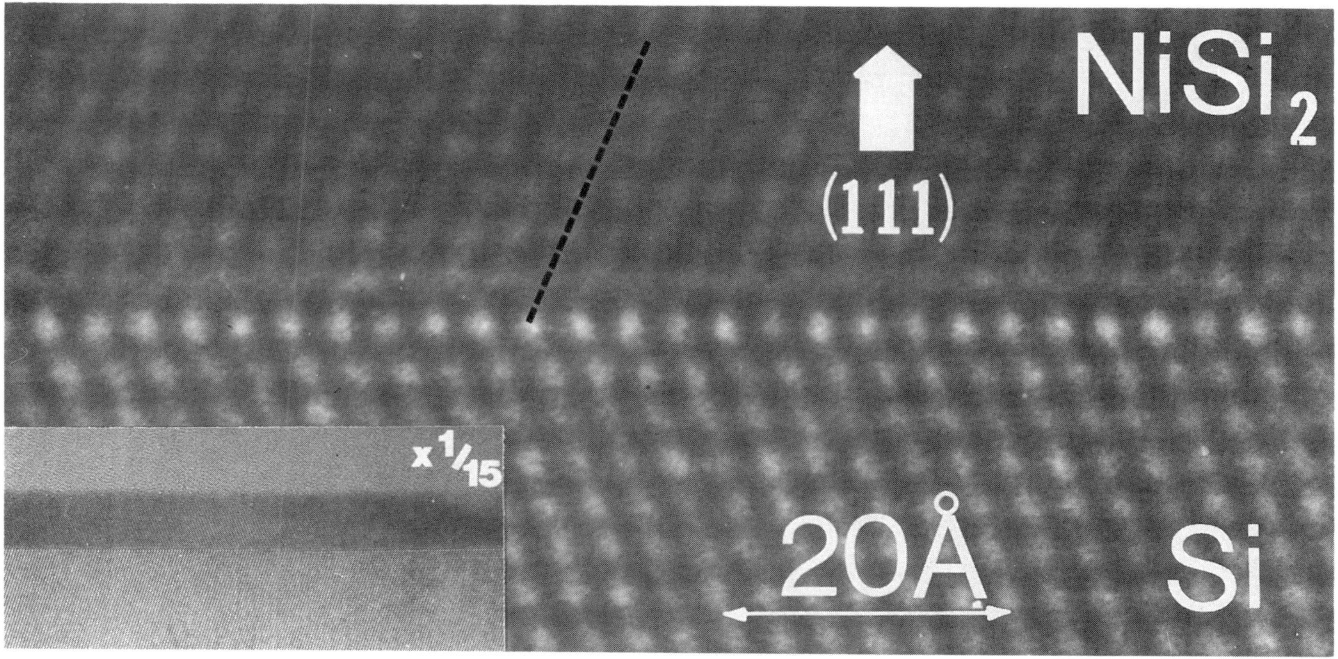


**Figure 23.** Cross-sectional HREM image of the lattice at a type B $NiSi_2$/Si(111) interface, viewed in the $[1\bar{1}0]$ direction. (After Gibson, Tung, and Poate.)[249]

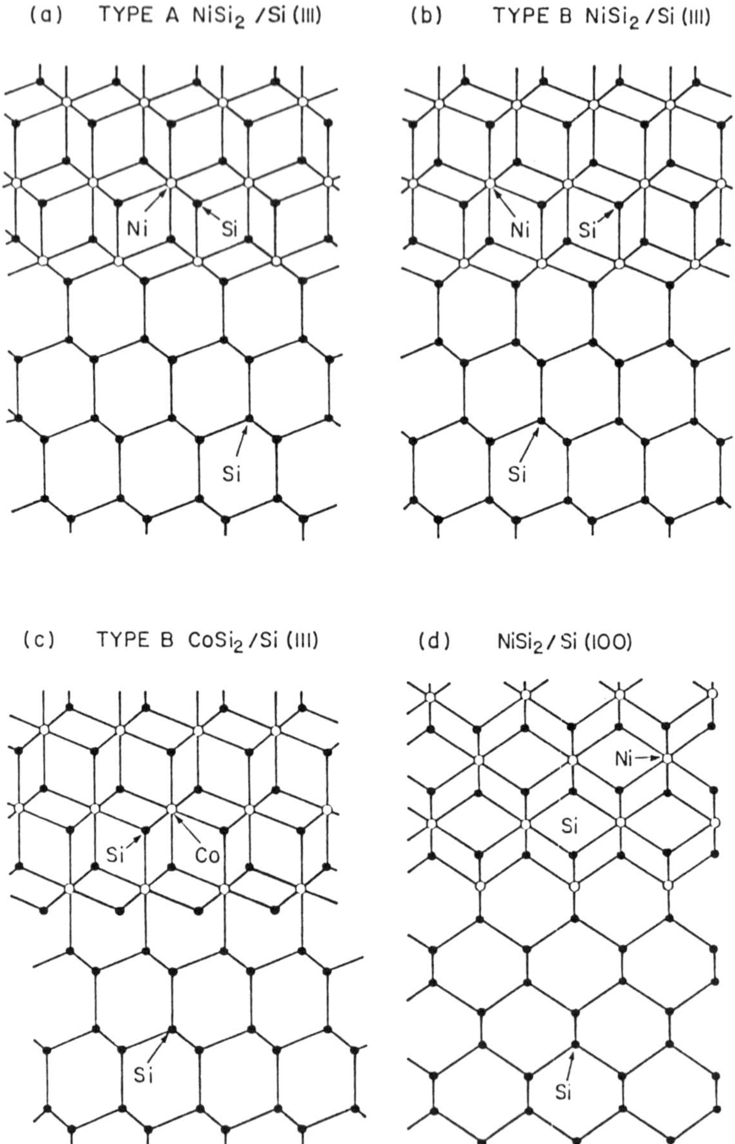


**Figure 24.** Ball and stick models of epitaxial silicide/Si interfaces, viewed in the direction. *(a)* the 7-fold type A $NiSi_2$/Si(111) interface, *(b)* the 7-fold type B $NiSi_2$/Si(111) interface, *(c)* the 8-fold type B $CoSi_2$/Si(111) interface, and *(d)* the 6-fold $NiSi_2$/Si(100) interface.

0.14 eV higher than type A $NiSi_2$ on n-type Si(111). The origin of the initial disagreement is now understood[71] to be a surface boron contamination problem associated with the experimental conditions employed in the study of Liehr et al.[119] Under careful procedures which avoid this boron problem, a low SBH, ~0.65 eV, for type A $NiSi_2$ on n-type Si(111) is very consistently measured by every laboratory. I-V studies of type A $NiSi_2$ diodes on n-type Si(111), such as that shown in Fig. 25a, indicate that the ideality factor for high quality type A diodes is almost exactly 1, independent of the measurement temperature. From the earlier discussion of inhomogeneous SBH's, this observation is consistent with a homogeneous SBH. Experimentally measured SBH's of epitaxial $NiSi_2$ are summarized in Table 3.

For the type B $NiSi_2$ interface, some variation of the SBH data has appeared in the literature, which suggests that the SBH may depend on growth, diode processing, and the method of SBH measurements.[256]-[259] The value of 0.79 eV, originally reported[255] for this interface, has since only been quantitatively reproduced by one group.[257] Other groups usually report a type B SBH which is somewhat lower.[256][258] In particular, it was claimed that the SBH of type B $NiSi_2$ showed a dependence on film thickness[260] which was attributed to a variation with respect to the dislocation density.[260][261] As shown earlier,[262] the dislocation density at a type B $NiSi_2$ interface is not a simple function of the film thickness, but rather is strongly dependent on the growth procedure. Furthermore, the samples employed in this study[260] were grown by co-deposition of $NiSi_2$ at ~400°C, which invariably leads to a very rough layer morphology.[262] Inclined facets of a type B $NiSi_2$ layer are incoherent twin boundaries with structures yet to be determined. It is very plausible that inclined facets have a lower local SBH than the 7-fold type B planar interface and are a major source of apparent SBH variation. (In contrast, inclined facets of a type A $NiSi_2$ layer are likely type A {111} interfaces themselves. Therefore, SBH homogeneity is expected at type A $NiSi_2$ interfaces, even those with non-uniform morphologies.) Even the most uniform type B $NiSi_2$ SB diodes often display a slightly larger ideality factor, as shown in Fig. 25b, than type A diodes grown on the same substrate, suggesting the presence of some minor SBH inhomogeneity. One may point out, in passing, that type B $NiSi_2$ interfaces represent an ideal system to study the effect of dislocations on the SBH,[263] because the density of dislocations may be controlled by the misorientation angle of the substrate.[264]

Type B $NiSi_2$ layers grown at room temperature by the deposition of Ni and the co-deposition of $NiSi_2$, such as the layer of Fig. 22a, show a SBH of ~0.67 eV.[262] It should be noted that the atomic structure of room-

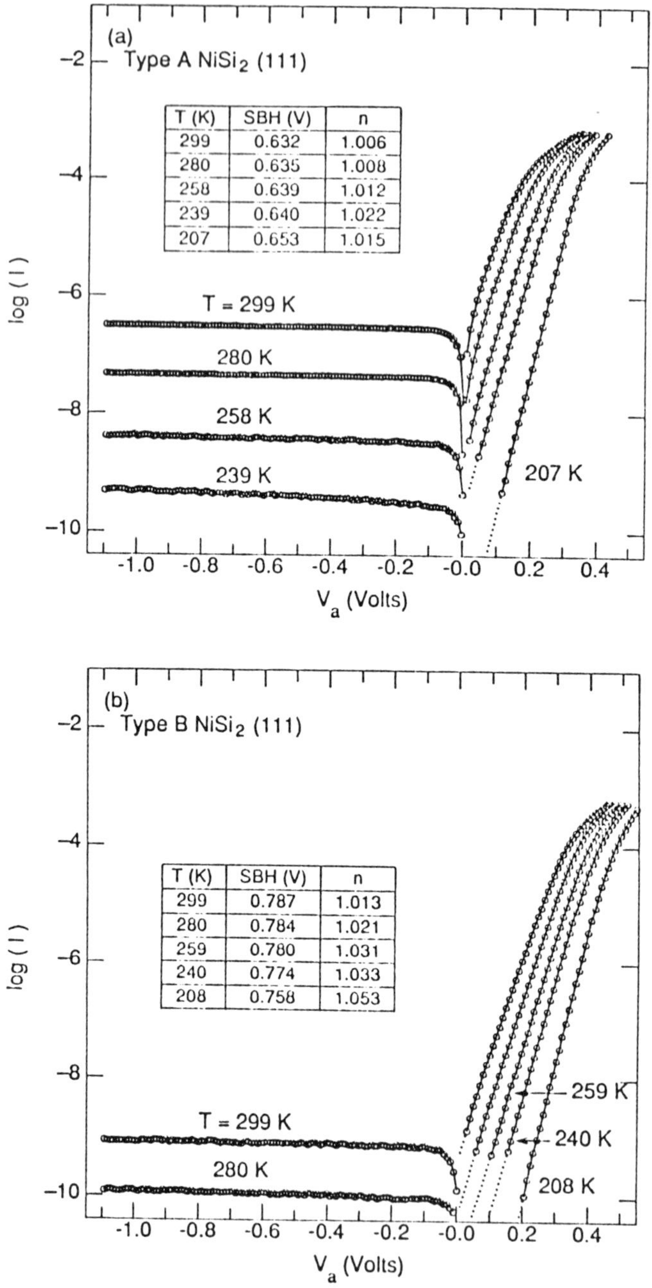


**Figure 25.** Current-voltage characteristics from epitaxial $NiSi_2$ diodes fabricated on n-type Si(111). *(a)* Type A orientation, *(b)* type B orientation.

**Table 3.** Schottky Barrier Heights of Epitaxial $NiSi_2$

| Silicide | Orientation | Substrate | Interface Structure | SBH (eV) | |
|---|---|---|---|---|---|
| | | | | n-type | p-type |
| $NiSi_2$ | Type A | Si(111) | 7-fold | 0.65 | 0.47 |
| $NiSi_2$ | Type B | Si(111) | 7-fold | 0.79 | 0.33 |
| $NiSi_2$ | (100) | Si(100) | 6-fold* | 0.40 | 0.73 |
| $NiSi_2$ | (100) | Si(100) | 7-fold {111}A | 0.65 | 0.45 |
| $NiSi_2$ | (110) | Si(110) | 7-fold {111}A | 0.65 | 0.45 |

* tentative

temperature-grown type B $NiSi_2$ interfaces has not been studied experimentally. Some degree of disorder is very likely. Annealing these layers at >350°C leads to a SBH of 0.79 eV, presumably through an ordering of the interface.[262] A soft x-ray PES experiment on monolayers of Ni deposited at room temperature on Si(111) and the subsequent formation of ordered type B $NiSi_2$ layer,[265] shown in Fig. 26, revealed a similar shift (~0.1 eV) in the SBH. The result of Fig. 26 is also suggestive that the establishment of a bulk-like SBH at a MS interface requires ≤ 7 Å of metal (the thickness of the $NiSi_2$ layer from the reaction of 2 Å Ni). Such an interpretation, in view of Fig. 28 below, is not unreasonable for the $NiSi_2$/Si system.

In order to understand the difference in SBH's of type A and type B $NiSi_2$, there have been a number of attempts to measure the density of interface electronic states experimentally. A forward bias capacitance technique has been used to study type A and type B $NiSi_2$/Si(111) interfaces.[30] However, the interpretation of such experiments is questionable, as already discussed. Deep level spectroscopy and hydrogenation studies of these epitaxial interfaces have shown the density of interface states to be low.[266] Infrared absorption measurements on thin layers of type A and type B $NiSi_2$ have seen no evidence of interface electronic states,[267] in sharp contrast to results from polycrystalline silicide films. These measurements set an upper limit of ~$10^{14}$ $cm^{-2}$ $eV^{-1}$ for the electronic state density at epitaxial $NiSi_2$/Si(111) interfaces. The dependence of the $NiSi_2$/Si(111) SBH on hydrostatic pressure has also been investigated[268] and been shown that the difference between type A and type B SBH's may not be simply attributed to a difference in the interface bond lengths.

However, such an interpretation of the SBH dependence on the hydrostatic pressure has recently been challenged.[269] BEEM studies[270][271] of the two $NiSi_2$ interfaces have generally not been conclusive about the electronic structures. By and large, the origin of the difference between A- and B-type SBH's has not been explained through direct experimental means.

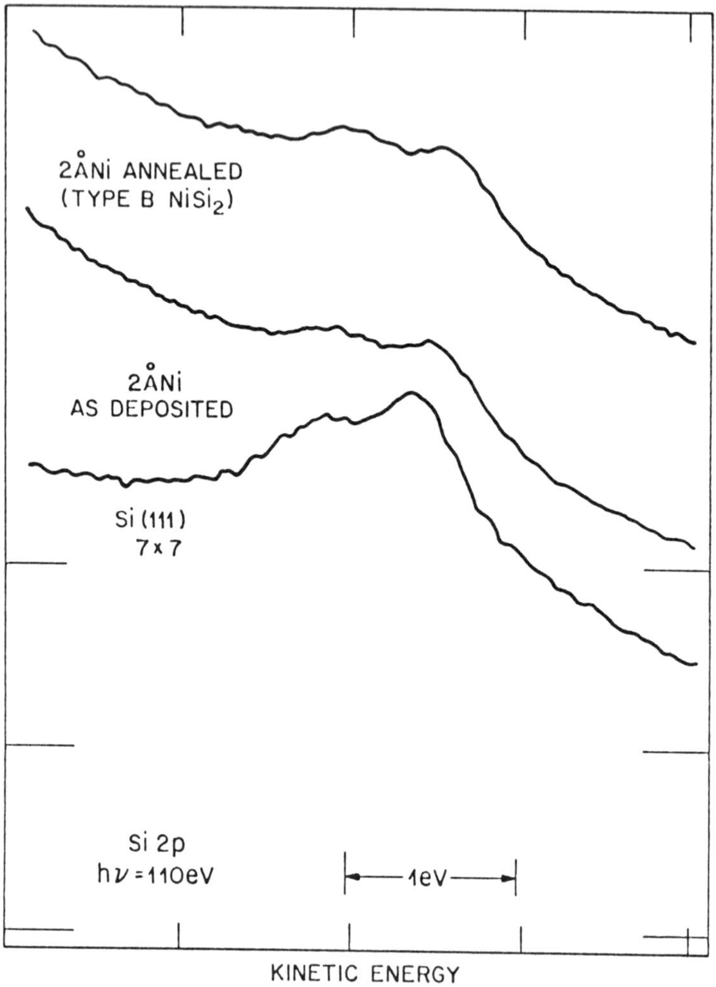


**Figure 26.** Soft x-ray photoemission spectra from a clean 7x7 Si(111) surface *(lower curve)*, after 2 Å Ni deposition *(middle curve)*, and after annealing at ~450°C to form type B $NiSi_2$ *(upper curve)*. Curves are displaced vertically for clarity. An upward shift by ~0.1 eV of the Si 2p core level, which corresponds to an increase in the SBH with the same magnitude, is observed when type B $NiSi_2$ is formed from deposited Ni. (After Kevan and Tung.)[265]

An important advantage of an epitaxial MS interface, in terms of SBH investigations, is the possibility of conducting theoretical calculations based on real atomic structures. Because of their high structural perfection and the intriguing variation of SBH with orientation, type A and type B $NiSi_2$ interfaces represent the best testing ground for a theoretical investigation of the formation mechanism of the SB. Not surprisingly, there have already been many calculations of the electronic structure and SBH properties at these interfaces. Calculations which concentrated on the possible role played by point defects[8][272] at the epitaxial interfaces were unable to explain the difference between the SBH's. The difference between the SBH's of A- and B-type interfaces has been empirically attributed to the "fault" at the type B interface in one study.[273] Apparently, such a proposal[273] is not capable of addressing the SBH problem at large. Calculations involving smaller supercells,[274]-[277] generally yielded similar SBH's for the A- and B-interfaces. It was suggested that the SBH at type B interface may be determined by a mechanism different from that for the type A interface.[25][275][278] Even though early calculations were unable to explain the difference between A and B type SBH's, they clearly showed that the SBH depends on interface structure.[274][276]

More recent calculations, involving very large supercells,[38][39][279] have all yielded SBH results which are in excellent agreement with experimental results.[255] An n-type SBH for the type B $NiSi_2$ interface which is 0.14 - 0.15 eV higher than the type A interface was obtained in two independent calculations,[38][39] to be compared with the 0.14 eV difference experimentally observed.[255] However, due to the known weakness of the local density approximation employed in these calculations, it seems more beneficial to examine the general picture presented by these calculations than to concentrate on features specific to the individual interfaces. Although the densities of MIGS are high at both interfaces,[39] as shown in Fig. 27, there is no easily identifiable feature in the energetic distribution of these states to suggest a pinning of the FL. Actually, the role played by MIGS in determining the FL position was not clearly revealed by these calculations.[38][39] One notes that there are significant differences in the distribution of MIGS at these two interfaces, as shown in Fig. 27. This dependence of MIGS on the interface structure indicates that the formalisms previously constructed for MIGS[5][23][12] are too simplified for real MS interfaces. Since MIGS's depend on interface structure, it does not appear that one may apply a charge neutrality criterion[12] based on the semiconductor band structure alone.

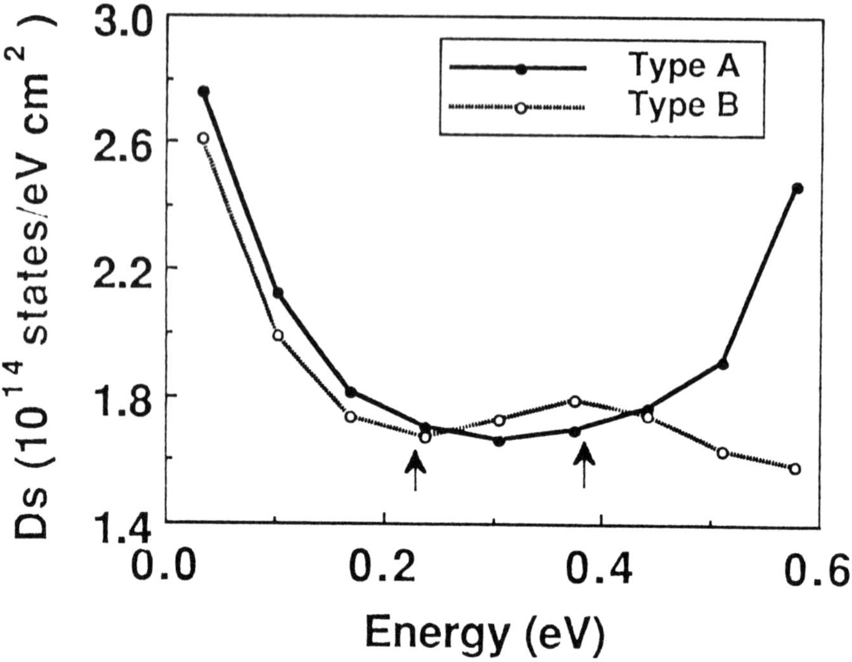


**Figure 27.** Calculated density of metal-induced gap states at single crystal $NiSi_2$/Si(111) interfaces. *Arrows* indicate the Fermi energies of the two types of interface. *Circles* show calculated energy points. By local density approximation, the Si thermal gap is depressed to about 0.6 eV. (After Fujitani and Asano.)[39]

The potential due to all charges at the two interfaces,[39] shown in Fig. 28, reveals several surprising results. First of all, the charge due to MIGS is only a small part of the total redistributed charge at the interface. Since the difference in the calculated SBH's originates from the difference in the interface dipoles for the two orientations, that this difference in dipoles is found to be due largely to states below the VBM is contradictory to the concept of FL pinning. Local structures and bonding geometries seem more important than MIGS in contributing to the difference in charge distributions at the $NiSi_2$/Si interface. Furthermore, the local charge oscillates rapidly at the MS interface and extends roughly an equal distance, ~6 - 9 Å, into both the Si and the $NiSi_2$ sides of the interface. Electric potential fluctuates with very large amplitudes at the MS interface, in disagreement with the assumption of a unilateral transfer of charge in MIGS and other interface/

surface states models. Even the direction of the net charge transfer is found to be opposite to that proposed in the (negative charge) MIGS model.[23] The potentials at these interfaces,[39] shown in Fig. 28, bear no resemblance to the monotonic and smooth band diagram from simple dipoles, usually assumed for MIGS,[5][23] shown in Fig. 2c. Also absent from the calculated potential of the $NiSi_2$/Si interfaces is the exponential decay of the interface charge which is the central argument for the attribution of SBH lowering on interface states.[5][23] The formation mechanism of the SBH at epitaxial $NiSi_2$ interfaces seems, in spirit, close to the charge transfer and chemical bonding mechanism of Andrews and Phillips[231] and others.[99] However, it is the specific charge redistribution due to the *interfacial* bonding, and not silicide bulk bonding, that seems to be more relevant. It is simply amazing that the minute difference in the interface structures between A and B interfaces is enough to produce a significant difference in the SBH's.

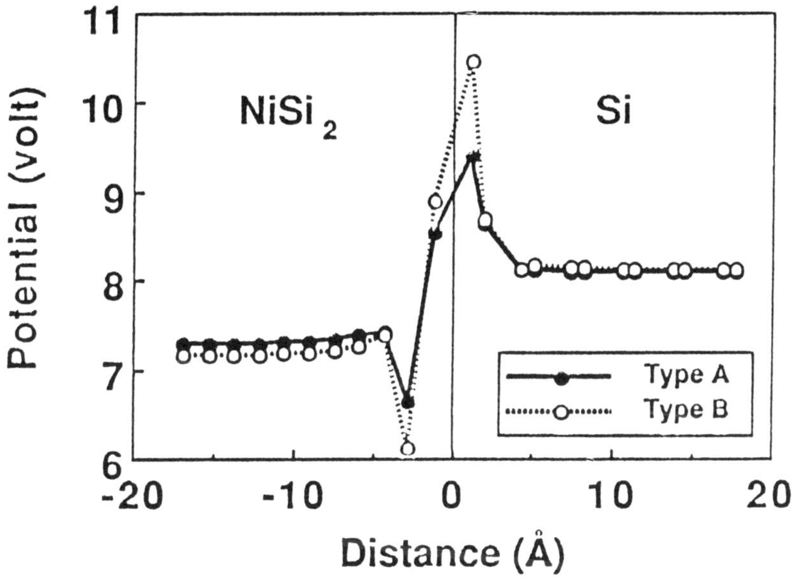


**Figure 28.** Potential in each Si site at the two $NiSi_2$/Si(111) interfaces. Potential was calculated by Madelung constants and charges in other atomic spheres. The center line is the interface. (After Fujitani and Asano.)[39]

## 7.3 $NiSi_2$ on Si(100)

**Fabrication and Structures.** Layer uniformity is the main issue facing the growth of $NiSi_2$ on Si(100). The nucleation of $NiSi_2$ on Si(100) requires ~350°C, but its homoepitaxial growth may take place at room temperature.[280] Epitaxial $NiSi_2$ layers grown at temperatures below 600°C often contain a high density, $10^8$ - $10^{10}$ cm$^{-2}$, of *facet bars* and therefore are very non-uniform in thickness.[281] A facet bar is a slender and straight $NiSi_2$ protrusion bound by two inclined <111> facets, usually ~100 Å in its depth and height and ~200 - 2000 Å in length.[281] The density and dimensions of facet bars are conveniently studied by planview TEM, where under (020)-type dark field these facet bars appear as short, bright streaks, as shown in Fig. 29b. In addition to facet bars, $NiSi_2$ layers contain a high density of line defects. Some defects are dislocations related to 1/4<111> and some are twin-related,[281][282] both of which are associated with steps at the interface. It was recently shown that high temperature (> 700°C) anneals significantly reduce the density of line defects and may completely eliminate facet bars in $NiSi_2$ layers, as shown in Fig. 29c. It was also discovered that the deposition of a thick layer of Ni on a thin $NiSi_2$ layer and annealing at ~400 - 600°C reproducibly lead to the fabrication of almost completely faceted interfaces, as shown in Fig. 29a. An examination of the three micrographs shown in Fig. 29 demonstrates that the morphology of the epitaxial $NiSi_2$/Si(100) interface, in particular the density of facet bars, may be controlled.[43]

The atomic structure of the planar $NiSi_2$/Si(100) interface has been studied experimentally by HREM.[249][283] Rigid shift measurements were used to conclude that the $NiSi_2$/Si(100) interface was 6-fold coordinated, as depicted in Fig. 24c. However, one should note that the 8-fold model considered in the early HREM studies consisted of four in-plane bonds for the interfacial Si.[283] The possibility of a more likely 8-fold model, with tetrahedral Si interfacial bonding, was not considered.[283] The measured rigid shifts from the tetrahedral 8-fold interface are not distinguishable from that of the 6-fold model. Therefore, the assignment of 6-fold coordination to the $NiSi_2$/Si(100) interface should be viewed as tentative.[249][283] Recent theoretical calculations are suggestive that the coordination number at this interface is higher than six.[284][285] In addition to the uncertainty in the 6-fold model, there is now new evidence for the presence of additional structures at this interface. Transmission electron diffraction shows streaky intensities at the < 0 ½ ½ >-related locations,[43] suggesting the $NiSi_2$/Si(100) interface may be partially 2x1 structured, similar to the well-known reconstruction at the (8-fold) $CoSi_2$/Si(100) interface.[286] The observation of this reconstruction suggests that the structure of even the most uniform, single crystal $NiSi_2$/Si(100) interfaces may be inhomogeneous.

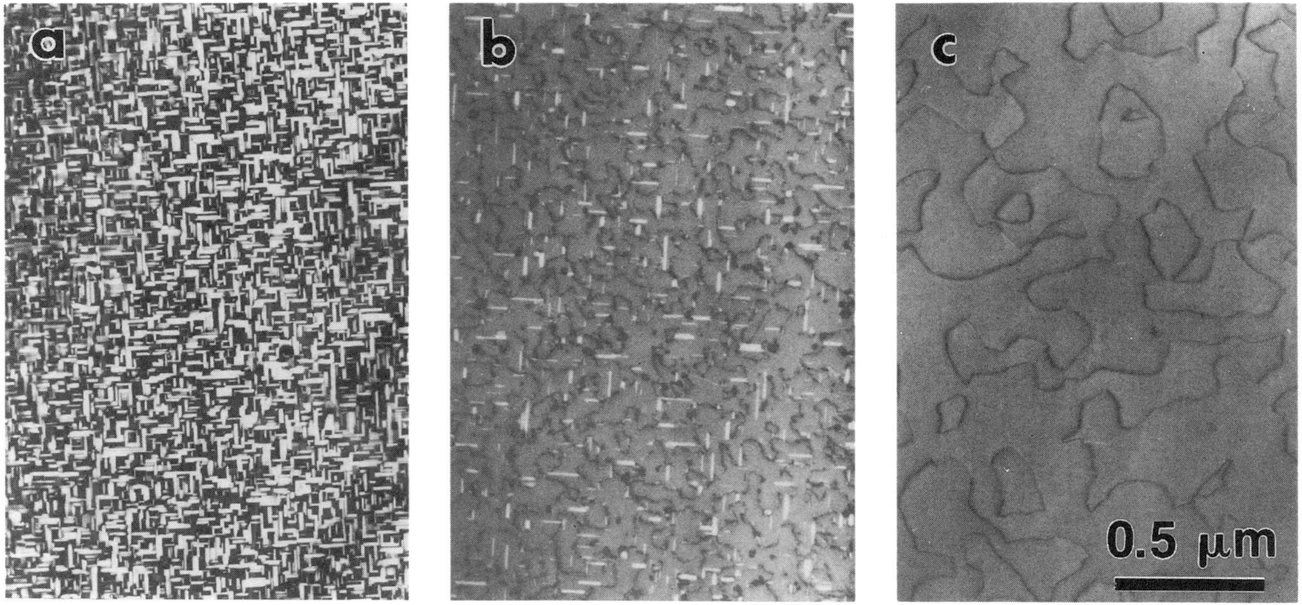


**Figure 29.** Planview, (002) dark field, TEM images of ~80 Å thick $NiSi_2$ layers grown on Si(100). A facet bar shows up as a bright streak under this imaging condition. Dark lines are defects with characters related to 1/4(111), which decorate steps with an odd number of atomic planes at the interface. *(a)* A layer which is nearly completely faceted; *(b)* a layer with mixed-morphology; (c) a uniform layer. (After Tung, Levi, Sullivan, and Schrey.)[43]

**Electronic Properties.** Early studies showed the SBH of the epitaxial $NiSi_2$/Si(100) interface is much lower (on n-type Si) than that at the $NiSi_2$/Si(111) interfaces.[287][288] However, results presented by Kikuchi et al[260] suggested that the SBH of $NiSi_2$/Si(100) may not be different from that observed at the type A $NiSi_2$/Si(111) interface. This discrepancy was resolved in a recent study,[43] where layers with a variety of morphologies were employed (cf. Fig. 29). It was demonstrated that the SBH measured at a $NiSi_2$/Si(100) interface depended very much on the observed morphology of the particular layer. Interfaces which are almost completely {111} faceted were shown to have SBH's similar to that found at a type A $NiSi_2$/Si(111) interface.[43] This result is consistent with the observation of Kikuchi et al,[260] whose growth procedures are known to lead to nearly completely faceted interfaces. Inclined facets at the $NiSi_2$/Si(100) interface are simply sections of a type A, 7-fold, $NiSi_2$/Si(111) interface. It is not surprising that the SBH found for a $NiSi_2$/Si(100) interface made up entirely of facets is identical to that found at a planar type A $NiSi_2$/Si(111) interface.[255]

Uniform $NiSi_2$/Si(100) interfaces show a SBH which is much lower on n-type Si than either of the two $NiSi_2$/Si(111) interfaces. Since the $NiSi_2$/Si(100) interface has an entirely different atomic structure from either of the two $NiSi_2$/Si(111) interfaces, it is perhaps not surprising that the SBH is also different. One notes that the low SBH of 0.4 eV measured from uniform $NiSi_2$ layers is also very different from the value of 0.6 - 0.7 eV usually observed for all phases of polycrystalline nickel silicides on Si.[122][143] On n-type substrates, ideality factors for I-V measurements are good ($n < 1.03$) and very consistent results were obtained from either I-V, C-V, or activation energy studies.[43] There is no observed dependence of measured SBH on the doping level.[43] However, on p-type Si(100), planar $NiSi_2$ diodes showed evidence for a slight inhomogeneity of the SBH, such as large ideality factors, $n \geq 1.08$, and discrepancies between SBH's measured by I-V and C-V techniques.[43] Therefore, even very uniform $NiSi_2$/Si(100) interfaces may in fact be electrically inhomogeneous. The origin of this SBH inhomogeneity is likely related to the 1x2 reconstruction which has been shown to likely be present at this interface.

The presence of a few facet bars at $NiSi_2$/Si(100) interfaces, which are otherwise flat, has little effect on the n-type SBH, but has a strong influence on the measured SBH on p-type Si. The I-V deduced p-type SBH decreases rapidly as the density of facet bars increases, while a slower, but noticeable, decrease of the C-V SBH is concurrently observed. As a result, the C-V measured SBH for any specific diode significantly exceeds that deduced from I-V. Mixed-morphology p-type diodes are *leaky*, having poor ideality factors ($n \geq 1.08$ for $N_A > 10^{16}cm^{-3}$) and displaying reverse currents which do not saturate. There is also a clear dependence of the electron transport on the substrate doping level. As an example, $NiSi_2$ layers with similar densities of facet bars, but grown on p-type Si with different doping levels, have been selected and their electrical properties[43] shown in Table 4. The C-V determined SBH's of these layers are similar, but the SBH deduced from I-V analysis decreases rapidly with increasing doping level. As discussed previously, these experimental results are a clear indication of an inhomogeneous SB. On a p-type substrate, almost all the current originates from the small, isolated facet bars which are partially pinched-off. Assuming a local $\Phi_{Bp}$ of 0.47 eV for areas occupied by facet bars and 0.72 eV for the flat areas, the saddle point potential for the facet bars was estimated using the analytic expressions[13] and an observed average width of larger facet bars, $L_o \approx 150$ Å. As shown in Table 4, such a (crude) estimate of the saddle-point potential semi-quantitatively reproduced the experimentally observed dependence on doping level.

**Table 4.** SBH of $NiSi_2$ layers on p-type Si(100) which have similar interface morphologies. Facets occupy ~10% of the total area. The dependence on doping level is in agreement with calculations based on saddle-point potential and it is not explained by the parallel conduction model of Ohdomari and Tu.[5]

| Doping | Experiments (eV) | | Calculated I-V $\Phi_{Bp,o}$ (eV) | |
|---|---|---|---|---|
| | $\Phi_{Bp,o}$ (C-V) | $\Phi_{Bp,o}$ (I-V) | Saddle-Point | Ohdomari and Tu[52] |
| $1\times10^{17}$ cm$^{-3}$ | 0.73 | 0.50 | 0.52 | 0.52 |
| $3.6\times10^{16}$ cm$^{-3}$ | 0.74 | 0.60 | 0.59 | 0.52 |
| $5\times10^{15}$ cm$^{-3}$ | 0.73 | 0.64 | 0.68 | 0.52 |

The epitaxial system of $NiSi_2$/Si(100) presents a very rare opportunity in the study of the SB formation mechanism, because it is the only demonstration of a SBH inhomogeneity, on a length scale smaller than the depletion width, which is artificially fabricated and studied. Results from this interface provided strong support for the transport theory of inhomogeneous SB's.[13] In addition, a dependence of the local SBH on the local atomic structure is strongly suggested by these experimental results. Because of the uncertainty concerning the atomic structure at planar $NiSi_2$/Si(100) interfaces, the electronic properties of this interface have not been conclusively studied by theoretical means. Very recent calculations by Fujitani and Asano[284] showed that the SBH at an 8-fold planar $NiSi_2$/Si(100) interface resembled that of the type A $NiSi_2$/Si(111) interface. However, very different redistributions of charge were found for these two interfaces.[284] At present, the relationship between the atomic structure and the SBH of the planar $NiSi_2$/Si(100) interface is not completely understood.

## 7.4 $NiSi_2$ on Si(110)

The growth of single crystal $NiSi_2$ on Si(110) is very easy, however interface faceting is an unavoidable problem.[289] Essentially uniform $NiSi_2$ layers may be grown by deposition of Ni and Si, or by co-deposition of $NiSi_2$, at room temperature and annealing at ~500°C. However, upon annealing to high temperature (>550°C), the interface breaks up into inclined <111> facets, as shown in Fig. 30. Since the $NiSi_2$/Si(110) interface is actually made up of <111> facets, it is not surprising that a SBH similar to that of type A $NiSi_2$/Si(111) interface, ~0.65 eV on n-type Si, is usually found.[287][290]

**Figure 30.** Planview, (002) dark-field, TEM image of a ~70 Å thick $NiSi_2$ layer on Si(110). Long streaks running along the direction indicate a near complete <111> faceting.

## 7.5 Other Epitaxial Metals and Silicides on Si

The SBH's of epitaxial $CoSi_2$/Si interfaces have also been investigated by a number of groups.[211][291][292] However, the SBH for well-annealed epitaxial $CoSi_2$ interfaces does not seem to vary significantly with the orientation of the Si. Preliminary results indicate a huge variation, ~0.4 eV, of the observed SBH for the type B $CoSi_2$/Si(111) interfaces prepared at < 500°C.[293] Due to a shear-induced phase transformation[294] and other line defects[293] at the type B $CoSi_2$/Si(111) interface, the atomic structure is not fully understood. Eight-fold structure is most commonly observed,[295]-[297] however, evidence for 7-fold structure has also been noted.[293] Much more work in the epitaxial $CoSi_2$ interfaces is needed to clarify the role played by the interface structure on the observed SBH.

Recently, a correlation of the observed SBH with the starting surface structure has been observed from the Pb/Si(111) system.[298] A SBH of ~0.7 eV was measured when the surface had a starting (7x7)-Pb structure, and 0.93 eV was measured when the $\left(\sqrt{3}\times\sqrt{3}\right)$R30°-Pb structure was present.[298] This dependence has been attributed to the difference in the Pb/Si interface structures, originating from the starting surfaces. This FL trend has also been observed by photoemission.[299] However, the preservation of these surface reconstructions at the eventual Pb/Si interface has been questioned.[300] In any event, the structures of the two Pb/Si interfaces, albeit possibly changed from those on the original surfaces, are still expected to be different. Hence, the observed difference in SBH may still be due to a structural difference. One notices that the high ideality factors and the difference between I-V and C-V measurements may be suggestive of some SBH inhomogeneity at one or both of these interfaces.[298]

## 7.6 Implications of SBH's of Epitaxial MS Interfaces

$NiSi_2$/Si interfaces have been studied extensively by various experimental techniques, and shown to be of very high structural perfection. A comparison of the SBH at $NiSi_2$/Si(100) interface with the type B $NiSi_2$/Si(111) interface shows that a difference of ~0.4 eV, or more than 1/3 of the Si bandgap, has been observed between the same metal and the same semiconductor. Such a difference is difficult to reconcile with FL pinning mechanisms. Experimental and theoretical results obtained from epitaxial $NiSi_2$, $CoSi_2$, and Pb interfaces strongly suggest that a causal relationship exists between the structural properties and the electronic properties of a MS interface.

For decades, the formation of the SBH was discussed in terms of FL pinning and interface states. It was clear that the goal of a SBH theory was to explain the magnitudes of the experimentally observed SBH's. Discussions in the last few sections point out that the concept of FL pinning is not consistent with the observed SBH inhomogeneity at polycrystalline MS interfaces and the observed SBH dependence on the atomic structure of epitaxial MS interfaces. The fact that the observed SBH's of inhomogeneous MS interfaces may be nothing more than some statistical average of a variety of FL positions is also pointed out. It seems more relevant to understand the local mechanism of SBH formation than to explain the observed average SBH's. Calculations show the complexity of the electronic structure at even perfect single crystal MS interfaces. In view of these difficulties, it is perhaps necessary to more carefully consider the question, What should be regarded as a satisfactory explanation of the formation mechanism of the SBH? With more knowledge on the complexity of the structures and the electronic properties of polycrystalline MS interfaces, the answer seems less obvious than before. For example, one may understand the formation mechanism of the SBH perfectly well, from studies of well-controlled MS interfaces, and still not be able to explain or predict SBH's of polycrystalline MS interfaces simply because the interface structures are not known or are too random. Perhaps it is convenient to divide the complete explanation of the SBH into a "physical", or core, mechanism which deals with the relationship between the structure and the dipole of the MS interface and a "chemical", or energetic, mechanism which addresses how and why certain MS interfaces are structured in certain ways. These two mechanisms are complicated and sometimes related. The physical mechanism has historically been regarded as the formation mechanism of the SBH, and perhaps should continue to be so regarded. As for the observed trends and systematics of polycrystalline SBH's, one really doesn't even know whether to attribute them to the chemical or the physical mechanism. It seems that the basic concepts of SBH formation, namely that it depends on the interface structure and bonding, are all in place. However, one's ability to quantitatively predict or explain the observed SBH can only improve when much better understanding of the interface structures and much better calculation schemes are realized.

## 8.0 TECHNOLOGICALLY IMPORTANT CONTACTS TO SILICON

### 8.1 Requirements of Ohmic Contacts

From the point of view of applications, there has not been much development in, nor are there any foreseeable demands for a significant improvement in, rectifying contacts to silicon. Rectifying SB contacts are potentially useful in metal-semiconductor field effect transistors (MESFET's) or power and microwave devices. However, the metallization technologies for these devices are mature. Therefore, our attention will be focused on low-resistance ohmic contacts which are an essential part of the vast majority of Si devices, such as the MOSFET's and bipolar devices. Whereas an absolute definition of an *ohmic* contact does not exist, the use of the word generally implies that the voltage drop at such a contact under ordinary operation conditions is an insignificant part of the total applied voltage.

Almost all the problems facing low-resistance contacts originate from the scaling of Si devices to submicron and deep submicron dimensions.[301][302] Ohmic contacts are often made to thin $p^+$ ( $n^+$ ) diffusion tubs in $n^-$ ( $p^-$ ) Si, and the integrity of the p-n junctions is as much an issue as the performance of the MS interfaces themselves. While there are continued discussions about how scaling of devices should be carried out,[303][304] there is little doubt that along with the reduction in the lateral dimensions of the device, a decrease in the vertical dimension of the junction and, possibly, an increase in the current density are necessary. Combined with the high material stability required by the sophisticated processing steps after the first metallization, the demand in the improvements of ohmic contact performance has, in many cases, surpassed what may be expected by refining existing technologies and has propelled research into looking at new possibilities. Even though the tolerable contact resistivity for a given ohmic contact is dependent on the specific device in question, a value less than 1 x $10^{-7}$ $\Omega cm^2$ is probably satisfactory for the anticipated current levels and $V_{DD}$'s of the next few generations of ULSI devices. It is also expected that if low-leakage shallow junctions may be fabricated on both n-type and p-type channel devices and with a depth no more than ~100 nm, including the thickness of the silicide layer, contacts should perform adequately for the next decade. Achieving these individual goals is not difficult. However, what make it difficult are other conditions attached to the silicide process, such as the compatibility with subsequent processes, and the stability of contacts, especially under high stressing conditions. For example, existing processing technologies require that an ohmic contact should sustain a high temperature (~850°C) anneal by itself, and should remain stable at ~550°C after further metallization is put into place. It is also desirable that the silicide has certain properties such as low film stress, inertness to etching and oxidation, etc. It may be pointed out that if a raised source-drain technology, e.g., by selective Si or poly Si deposition, is adopted, shallow junctions would cease to be an issue for the contact technology (although it would then become an issue for the Si deposition technologies). However, since we cannot speak of the future with any certainty, the subsequent discussions are still based on the assumption that shallow junction formation is an inseparable part of ohmic contact technologies.

The choice of contact technology is often not decided based on its own merits but is dictated by the most critical process in the entire fabrication process. Therefore, it is important for metallization technologies to be simple, inexpensive, applicable to gates and both n- and p-type sources/drains, and have a wide processing window. One needs to understand the performance of a process under as many conditions as possible, and not just under the optimum conditions. For this reason, process engineers have spent enormous amounts of time documenting the contact resistances as a function of various processing conditions. Unfortunately, the emphasis on empiricism also means that the explanation of the specific contact resistivities on a fundamental level has not been pursued with much enthusiasm. Large discrepancies between theory and experiments are often observed. The newly raised issue of SBH inhomogeneity may certainly influence the conduction mechanism at heavily doped ohmic contacts. However, it is too early to assess how much a bearing SBH inhomogeneity has and how one may take advantage of the naturally occurring inhomogeneity in designing better ohmic contact schemes. Clearly, these are good reasons why efforts toward the fundamental understanding of the formation mechanism of the SB and the ohmic contact should continue.

## 8.2 Strategies and Concepts

The most relevant parameter of a (heavily doped) MS interface is its specific contact resistivity, $\rho_c$. However, because $\rho_c$ depends critically on the doping level of the semiconductor, it is not as "fundamental" to a MS interface as $\Phi_{B,o}$ is to a Schottky junction. At a MS junction where the electron transmission is dominated by field emission (FE), i.e. $E_{oo} \gg k_BT$, the specific contact resistivity may be expressed as[194]

Eq. (24) $$\rho_c = \left[\frac{A^*\pi qT}{k_B \sin(\pi c k_b T)}\exp\left(\frac{-q\Phi_B}{E_{oo}}\right) - \frac{A^*q}{ck_B^2}\exp\left(\frac{-q\Phi_B}{E_{oo}} - cqV_n\right)\right]^{-1}$$

where $E_{oo}$ is a measure of the importance of tunneling, given by

Eq. (25)
$$E_{oo} = \frac{qh}{2}\sqrt{\frac{N_D}{m^*_{tun}\,\varepsilon_s}}$$

$m^*_{tun}$ is the tunneling effective mass,[305] and $c = (\frac{1}{2}E_{oo})\ln(4\Phi_B / V_n)$. Calculated $\rho_c$ for n-type Si(100),[306] based on appropriate values of the effective mass, is shown in Fig. 31 for different $\Phi_B$'s. Theoretical values of $\rho_c$. for holes and other orientations of Si have also been calculated.[306][307] Experimental results in good agreement with predictions of tunneling theories have been observed in many cases.[74][200][308][309] However, departures from these theories have also been frequently observed, especially at high doping levels.[310] As already pointed out, an inhomogeneous SBH is a common phenomenon for polycrystalline silicide/metal contacts, which certainly may have a strong influence on the measured $\rho_c$.

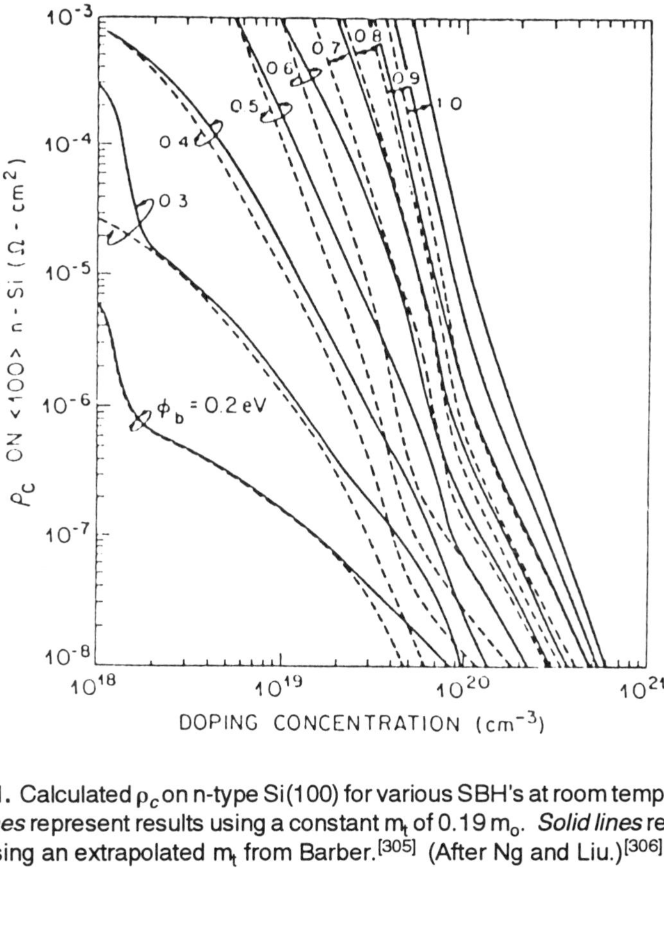


**Figure 31.** Calculated $\rho_c$ on n-type Si(100) for various SBH's at room temperature. *Dotted lines* represent results using a constant $m_t$ of 0.19 $m_o$. *Solid lines* represent results using an extrapolated $m_t$ from Barber.[305] (After Ng and Liu.)[306]

The choice of a metal with a low SBH naturally leads to a low resistance contact. However, this obvious approach for ohmic contacts has not been adopted in ULSI technology because of other overriding materials issues discussed in the previous section. Instead, metals with SBH's roughly one half the Si bandgap are preferred, because they are expected to yield satisfactory ohmic contacts to both n- and p-type Si. Since reliable contacts require the formation of intimate MS interfaces, proper cleaning of the Si surface prior to metallization is important. Furthermore, silicides are preferred over elemental metals, because the silicide reaction usually leads to intimate MS interfaces some distance away from the location of the original Si surface where much impurity resides. The use of self-aligned silicidation processes enjoys the maximum MS contact area without additional photolithographic steps. Properly designed, the salicide process may be used to significantly improve the conductivity of polycrystalline Si gates and interconnects, concurrent with the formation of metallized contacts. However, the use of silicides is not without its disadvantages. For example, the silicide reaction reduces the depth of existing p-n junctions, sometimes to an extent beyond that required by the natural silicidation process, because of significant diffusion of Si from the substrate. In addition, severe inhomogeneities in the thickness of the silicide layers and metal deep levels may have a devastating effect on the leakage of the contacts. In this regard, a selective deposition process (on Si but not on $SiO_2$) of metal or silicide contacts would be most welcome. The strategy of narrowing the bandgap by grading the composition of the semiconductor (e.g., with a $Si_xGe_{1-x}$ layer), demonstrated for compound semiconductor contacts, seems unnecessary for Si contacts. Ohmic contacts to Si are invariably formed by heavily doping the MS interface. At heavily doped MS junctions, FE and TFE processes become more dominant than thermionic emission, leading to high contact conductivities. Therefore, the technology of doping and the formation of shallow junctions weigh heavily on the choice of the metallization scheme. It is clear that the higher the doping concentration the lower the contact resistance. However, one should note that very high surface concentration of dopants naturally leads to deeper junctions because of concentration dependent diffusion. Furthermore, silicidation on heavily doped Si is often considerably more difficult than that on undoped Si. Therefore, the formation of a low-resistivity contact to very shallow diffusions cannot be implemented with just any metal indiscriminately. Ohmic contact technologies may be categorized into those which involve the formation of a p-n junction prior to metallization and those which form the p-n junction during, or after, metallization.

The traditional way of forming an ohmic contact is to form the p-n junction prior to metallization. The technologies of shallow p-n junction formation are generally considered mature, at least for junctions with depths over ~200 nm. However, a reduction of the junction depth below ~100 nm will bring into the scene issues such as the transient enhanced diffusion of dopants,[311][312] the removal of damage under limited thermal budget, and the performance of equipment for very low energy (< 1 kV) ion implantation. It is advantageous to cap the Si during implantation and the activating anneals. Other methods of dopant introduction, such as gas immersion laser doping (GILD)[313] and plasma implantation, have also received some attention recently. Since there are good reasons to expect that the problems with shallow junctions may be solved in time, we will only be concerned with the formation of the MS interface on these pre-formed shallow diffusions. Clearly the issues of the most concern are the uniformity, the thickness and the sheet resistance of the metal/silicide layer grown on heavily-doped Si, the quality of the MS interface, and the maintenance/modification of the doping profile during silicidation. To minimize the modification and loss of the doping profile, rapid thermal processing has been thought to be more effective.

It has become increasingly popular to introduce the dopants for shallow p-n junction formation after some metallization steps. For example, implantation of dopants into (or through) deposited metal layers, with or without some pre-annealing, followed by a dopant-activating annealing at high temperature has been proposed. Such an *implantation through metal* (ITM) technique[314][315][316] supposedly leads to the growth of more uniform layers of silicides due to the more uniform nucleation of silicides from ion beam mixing. Obviously, the issues of incomplete annealing of implantation damage, in both silicides and Si, and recoil implantation of metal atoms to the p-n junction region need to be dealt with. Dopants may also be shallow-implanted into preformed silicide layers and subsequently driven into the Si through an anneal.[317][318] The fact that dopants have much higher diffusivity in some silicides (or the grain boundaries of silicides) than that in Si may result in the formation of a p-n junction which follows the contour of the silicide-Si interface. This technique,[318] sometimes referred to as the *silicides as doping source* (SADS) technique, relaxes the requirement on the uniformity of silicide layers, and is still able to achieve shallow enough junctions to avoid short channel effects in MOSFET's. It is obvious that if the silicide reaction is known to have a strong tendency to snowplow, then shallow p-n junctions which follow the contour of the silicide interface may also be achieved without much difficulty. However, considerations based on the resistivity and the contact resistivity of silicides, the economy of the processes, and the stability of the contacts have already narrowed the choice of silicides for ULSI applications down to a handful of silicides, $TiSi_2$, $CoSi_2$ and, possibly, $WSi_2$. Junction formation by the snowplow effect is not a viable process with these silicides. One may point out that the SADS approach involves no implantation damage to the Si and, hence, does not require a high temperature anneal. However, the out-diffusion of dopants is only driven by thermodynamics when the dopant concentration exceeds the solid solubility in the silicide. This fact not only sets a lower limit on the implantation dose but also explains the failure of SADS for silicides which form stable compounds with the dopants.

## 8.3 Measurement of Specific Contact Resistivity

When the resistance of a contact is large, it may be precisely measured by various techniques such as the four point probe method,[319] the vertical method,[53][320] the potential profile method,[321] etc. However, errors involved in these techniques may be significant when the specific contact resistivity is low. The quantity which is actually measured in an experiment is the resistance of a contact, $R_c$, which depends not only on $\rho_c$ of the MS interface but also on the geometry of the contact and the sheet resistances of both the diffusion and the metal layer. Experimentally measured $R_c$'s need to be deconvoluted in some fitting procedures to deduce the specific contact resistivities. For instance, the series resistance of a source/drain,[322][323] as shown in Fig. 32, may be broken into the following components

Eq. (26) $$R_{total} = R_c + R_{sh} + R_{sp} + R_{ac}$$

where $R_{sh}$, $R_{sp}$, and $R_{ac}$ are, respectively, the sheet resistance of the diffusion ($R_{sh} = \rho_\square S/w$, where w is the width of the device and $\rho_\square$ is the sheet resistance per square), the spreading resistance, and the accumulation layer resistance.[324] Neglecting the sheet resistance of the metal, the contact resistance may be shown to be approximately given by

Eq. (27) $$Rc = \rho_\square L_t / [w \tanh(\ell/L_t)]$$

where $L_t$ is an important parameter known as the transfer length and is defined as

Eq. (28) $$L_t = (\rho_c/\rho_\square)^{1/2}$$

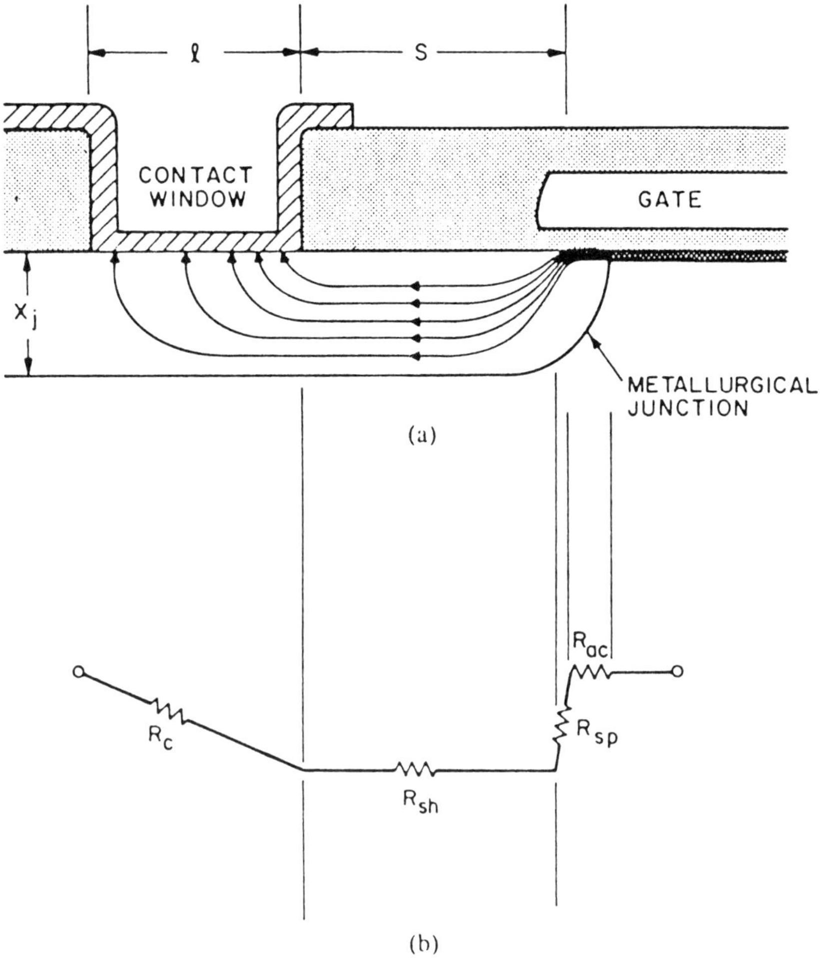


**Figure 32.** Schematic diagrams for *(a)* current flow pattern in the source/drain region, and *(b)* the associated resistance components. (After Ng and Lynch.)[322]

The contact resistance of a source/drain, Eq. (27), is identical to the "front resistance" of deliberately fabricated test structures for specific contact resistivity measurement and may be derived from the *transmission line model* (TLM).[325][326] Contact resistance determination by the TLM requires multiple electrical measurements and also requires a precise knowledge of the dimension of the tap resistor structures. Therefore, other horizontal test structures, such as the *cross bridge Kelvin resistor* (CBKR)[327] and the *contact end resistor* (CER), which allow the contact resistance to be measured in a single measurement, are preferred. The basic principle behind these measurement techniques is the same: a current I is passed from the diffusion into the metal through a contact window and the voltage drop across the contact, $V_c$, measured using two other terminals, gives the contact resistance as $R_c = V_c/I$. However, because of the difference in the location of the voltage probe, contact resistances are no longer given by Eq. (27). In the CBKR geometry, the voltage-probing terminal of the diffusion level makes a 90° angle with the current flow terminal (of the diffusion), and, therefore, there are good reasons to expect the specific contact resistivity to be simply the contact resistance multiplied by the area, $\rho_c = R_c w\ell$. The probe of the diffusion potential is placed at the backside of the current arm (180°) in the CER geometry, leading to an expression similar in appearance to Eq. (27), except that the hyperbolic tangent function is replaced by the hyperbolic sine function (sinh). However, it should be noted that, independent of whether a TLM, CER, or CBKR test structure is used, a difference between the width of the contact and that of the diffusion could lead to significant errors in these expressions for the contact resistance, which were all derived from a one-dimensional formalism. Lateral current crowding is particularly severe when the $\rho_c$ under study is small, as is usually the case for silicide contacts to Si. This effect leads to the deduction of a contact resistivity which is larger than the actual specific contact resistivity,[328] in a fashion shown in Fig. 33. In addition, significant errors in contact resistivity measurement may arise because of lateral misalignments of the contact window with respect to the diffused path. Misalignment also leads to an overestimation of the contact resistivity from the one-dimensional model. Other errors, such as that due to an inhomogeneous diffusion layer,[329] may also be present. In theory, the use of self-aligned contact processing should eliminate errors due to misalignment and current crowding. However, the lateral diffusion of implanted dopants may still lead to some current spreading which needs to be considered. Recent proposals of vertical test structures significantly simplify the deduction of a specific contact resistivity from an experimentally measured I-V relationship.[330][331] However, vertical structures require additional lithography and implantation steps and, therefore, are not conveniently implemented under the exact processing conditions of real MOS devices.

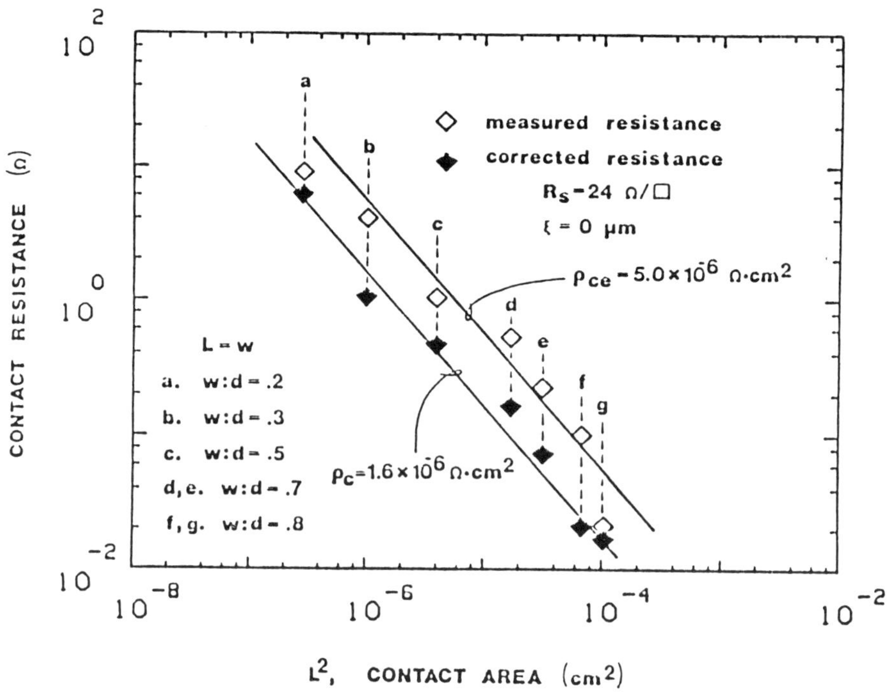


**Figure 33.** Experimental data taken on As-implanted Si for seven different Al contact sizes with five different w:d ratios. These CBKR test structures used, known as D-resistors, all have diffusion widths, w, identical to the widths, L, of the particular square contact. However, near the contact area, the diffusion is widened to a total width of d. *Open symbols* are measured data while *filled symbols* are the data corrected for current crowding. (After Gillenwater, Hafich, and Robinson.)[328]

## 8.4 Practical Ohmic Contacts

Before the 1980's, aluminum was the most popular ohmic contact metal to Si. Aluminum's low resistivity, good processability, and the fact that it was already widely in use for interconnects on VLSI devices all made it a logical choice for the first level metallization. However, the spiking of annealed Al contacts, as shown in Fig. 34, and the weak resistance to electromigration were two undesirable problems. Although the spiking problem, which, as already explained, is caused by the diffusion of Si into Al during annealing, may be alleviated by doping deposited Al with Si, other undesirable effects, such as the degradation of the n-type contact from precipitated Al-doped Si, occur. It does not seem likely that Al may be used for shallow contacts in the future. Recently, Al contacts with increased stability have been fabricated using novel processing techniques.[74] However, there are as yet no indications that the aluminum contact technology may be refined to meet the requirements of submicron ULSI processing.

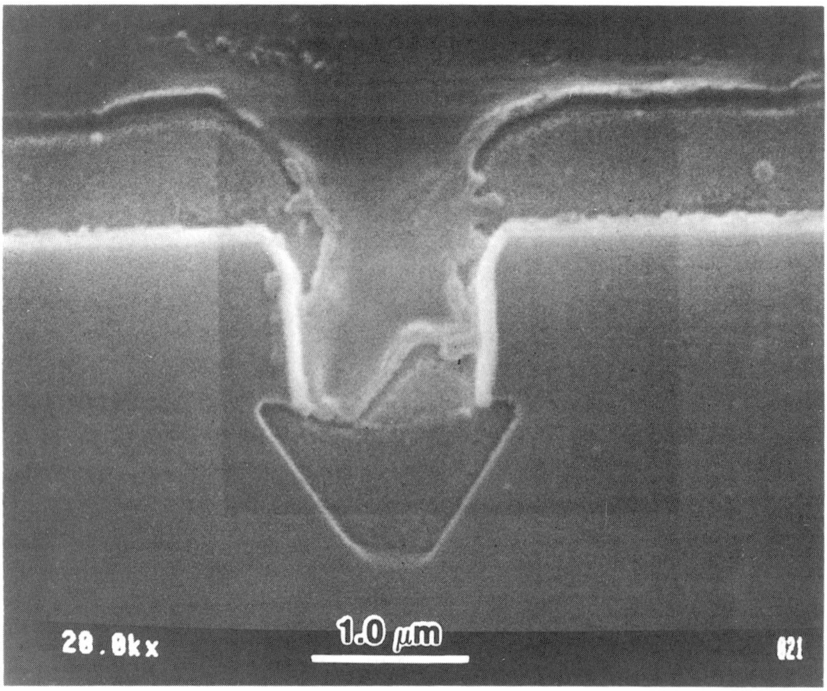


**Figure 34.** Scanning electron micrograph of the cross section of an Al contact which has been annealed at 550°C, displaying the spiking effect. (Courtesy of C. Case and R. Liu.)

Platinum silicide was one of the earliest silicides to find applications in Si IC's. The formation of ohmic PtSi contacts is relatively easy on both n- and p-type Si.[86][332] Self-aligned processing of PtSi has been demonstrated, although the selectivity in Pt etching is achieved through extrinsic means. Despite some early disagreement,[169] it is now generally accepted that a thin oxide layer grown on PtSi under usual processing conditions, protects the PtSi from being removed by aqua regia.[332] The PtSi silicidation process on heavily doped Si leads to clear snowplow of dopants, which has certain attractive features. For example, snowplow is an important reason for the formation of adequate contacts on n-type Si despite the known high SBH. PtSi reacts with Al at low temperatures and, hence, a diffusion barrier such as Ti-W or TiN is required to maintain the integrity of the contact at even moderate temperatures. The interest in PtSi as contact material has decreased due to its high contact resistivity on n-type Si, mediocre sheet resistance, high cost of processing, the poor thermal stability, reactivity with Al, and the absence of a dry etch for the metal. However, PtSi still has wide uses in Si devices as Schottky clamps, discrete power Schottky devices, and IR detectors.

Specific contact resistivities of technologically important contacts have been repeatedly studied. When the geometry is taken into consideration, a specific contact resistivity which is independent of the contact size is usually found. Errors introduced by current crowding and contact misalignment usually lead to a specific contact resistivity which increases slightly when the contact size is decreased. Occasionally, however, $\rho_c$'s which increase with contact size, such as those shown in Fig. 35, are observed on both n- and p-type diffusions.[333] Note that the trend shown in Fig. 35 cannot be explained by any measurement error (with the exception of an unlikely systematic error in the sizes of the contacts), but is consistent with a spatially inhomogeneous contact resistivity. Drawing from the previous discussion on SBH inhomogeneity, it is clear that the reduced pinch-off of low-SBH patches at edges of a MS interface may lead to the trend experimentally observed in Fig. 35. Whether this turns out to be case remains to be seen. Considerations of the stability, the sheet resistivity, the contact resistivity, the cost and the ease of processing, etc., have narrowed the choice of silicides to $CoSi_2$ and $TiSi_2$ as the prime candidates for contact materials of deep submicron ULSI applications. A brief description of these two silicide processes is given below. As already explained, the decision on which silicide to use is based more on the compatibility with other processing steps rather than on the silicide process itself. Therefore, only salient features of these technologies will be identified, and no attempt to determine a universal optimal silicide process will be made.

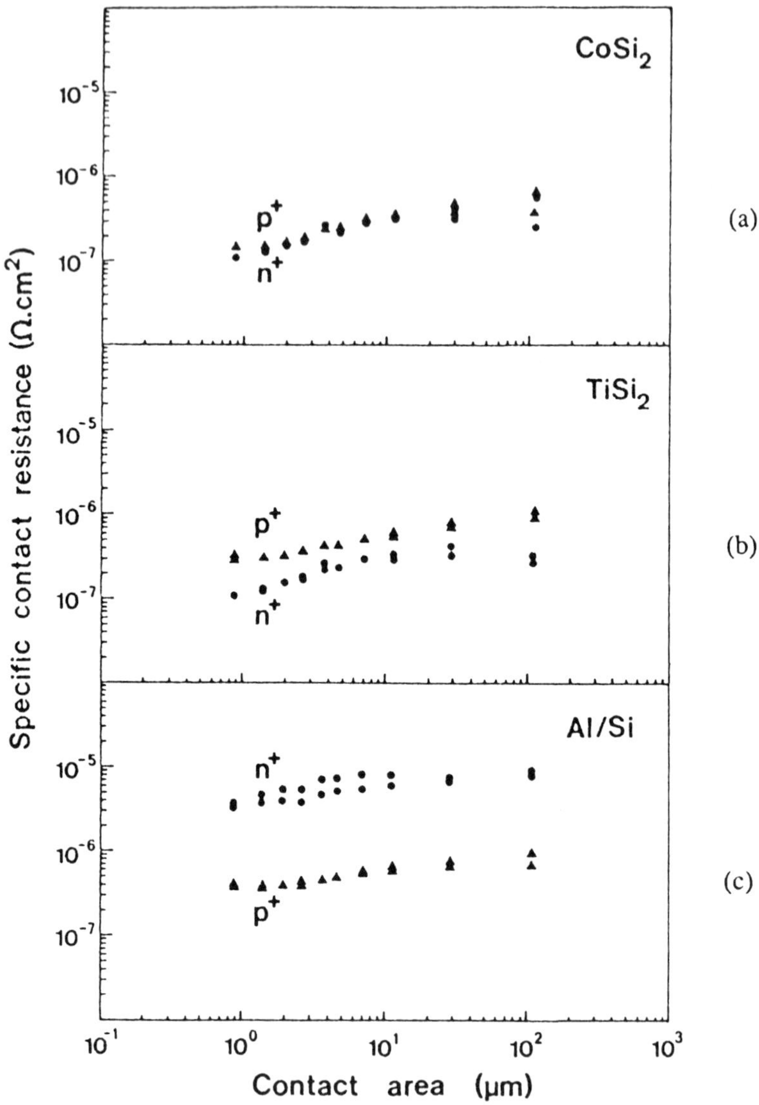


**Figure 35.** Specific contact resistivity versus contact area from *(a)* $CoSi_2$, *(b)* $TiSi_2$, and *(c)* Al/Si to $n^+$ and $p^+$ Si. The surface concentrations are about 1.5 x $10^{20}$ $cm^{-3}$ As and 6 x $10^{19}$ $cm^{-3}$ B, respectively. (After Van den Hove, Wolters, Maex, De Keersmaecker, and Declerck.)[333]

**$TiSi_2$ Technology**. It has long been realized that Ti might be a suitable material for stable contacts to both n- and p-type Si.[334]-[336] The silicide reaction of $TiSi_2$ is slightly more complicated than usual due to the existence of two $TiSi_2$ phases. The formation of the low temperature C49 $TiSi_2$ phase and the TiSi phase usually precedes the formation of the desired, high temperature, C54 $TiSi_2$ phase.[337] Since there is a sharp decrease in the resistivity when the transformation of the C54 $TiSi_2$ occurs (at ~800°C), the Ti-Si reaction may be monitored easily by sheet resistance measurement. Si is known to be the moving species in the Ti-Si reaction which, under usual vacuum annealing conditions, leads to the growth of silicide on sidewalls, the electrical shorting between gates and sources/drains (S/D), and the consumption of Si from the S/D region. However, self-aligned processing of $TiSi_2$ is still possible by using multiple step annealing and annealing in a nitrogen ambient.[338]-[340] In a nitrogen anneal, the rapid formation of TiN along the grain boundaries of Ti on $SiO_2$ and sidewalls impedes the lateral spreading of the $TiSi_2$ reaction by Si diffusion. However, since TiN is also formed in the intended contact areas, a reduction in the silicide thickness is unavoidable in a nitrogen anneal. When a layer of Ti deposited on Si is annealed in $N_2$, TiN reaction at the surface and $TiSi_2$ reaction at the interface proceed simultaneously. The final thicknesses of the TiN and the $TiSi_2$ layers depend on the kinetics of the two reactions. For this reason, factors which affect the silicide reaction, such as the surface cleanliness, the doping level of the Si, and the annealing condition, also significantly affect the thickness of the $TiSi_2$ layer. For example, the thickness of $TiSi_2$ layers grown by nitrogen anneal on heavily doped n-type Si is always less than that found on a comparably doped p-type Si.[341][342] Also, the cleaning of the contact hole prior to Ti deposition is known to have a significant effect on both the thickness and the morphology of the reacted $TiSi_2$ film. In self-aligned processing of $TiSi_2$, the first annealing temperature should be kept low, or the annealing time kept short,[340] to allow only the low temperature phase of the $TiSi_2$ to form and to minimize Ti/$SiO_2$ reaction, which may lead to electrical leakage. The TiN layer covering the whole surface and any unreacted Ti on top of the $SiO_2$ may be removed, leaving self-aligned $TiSi_2$ in the contact openings (and on top of polysilicon gates and runners if one chooses). With an additional lithography step, TiN may be used as convenient local interconnects.[343] Even though etch solutions which have very good selectivity for TiN are known, solutions which have a small etch rate for $TiSi_2$ are still preferred in order to remove the small amount of Ti silicide on $SiO_2$. Therefore, the etching condition needs to be carefully monitored to control the $TiSi_2$ layer thickness. Generally speaking, a loss in silicide thickness is unavoidable. The low-temperature silicide may be converted to the low-resistivity phase by a higher temperature anneal. It is important to control the conditions of this anneal to minimize dopant loss and/or redistribution[337][344] and the growth of oxide.[345] $TiSi_2$ is not stable above 900°C.

Experimentally determined $\rho_c$'s of common contacts on heavily-doped diffusions with typical concentration of ~5 - 10 x $10^{19}$ $cm^{-3}$ are shown in Fig. 35. It seems that the $\rho_c$'s for $TiSi_2$ are comparable to other silicides on n-type Si, but are slightly higher on p-type Si. Because the p-type diffusions have higher sheet resistances than n-type diffusions, an additional high contact resistivity to p-type Si is more of a nuisance than for n-type Si. Ion beam mixing of $TiSi_2$ has been studied[346]-[348] which lead to some improvement in the silicide morphology. However, the SADS approach to shallow junction formation is not compatible with the $TiSi_2$ technology because of the existence of a stable $TiB_2$ compound.[338][349] Shallow p-n junctions are usually formed before Ti deposition. The low temperature $TiSi_2$ reaction generally does not lead to much variation of the doping profile.[337] Also, the conversion to the C54 phase is reported to depend on the substrate and the dimension of the feature. The C49 to C54 conversion is impeded by heavy As doping. It is easier on blanket Si than on narrow lines. Even though silicidation of the gates and interconnects by $TiSi_2$ has been reported,[350][351] the high stresses of $TiSi_2$ polycide, especially due to mechanical pinning at edges of fine lines,[351] needs to be addressed. The thermal stability of $TiSi_2$/polysilicon stacks, although problematic at high temperatures, is much better than stacked $CoSi_2$ structures. The reaction between Ti and $SiO_2$ leads to a slight erosion of oxide even at typical first-step annealing temperatures. In addition, the poor resistance to HF-containing solutions and fluorine-based plasmas of the $TiSi_2$ (compared with e.g., $CoSi_2$) makes the etching and the cleaning of contact holes difficult. The deposition of a small amount of Ti in the contact holes, prior to the deposition of other barrier/plug metals, facilitates the metal-to-silicide contact. However, the depth of the silicide junction increases as a result. Therefore, the self-aligned $TiSi_2$ process is not without its problems. A process which would allow selective deposition of silicide contacts is, of course, highly desirable. For $TiSi_2$, a selective low pressure CVD deposition process, without the consumption of substrate Si, was recently reported.[352] There are still questions regarding the uniformity and the mechanism of this CVD process.[353] It remains to be seen whether such a technology could be implemented in a manufacturing environment.

**$CoSi_2$ Technology**. The Co-Si reaction is more straightforward than the Ti-Si reaction. The end-phase $CoSi_2$, which has the *fcc* structure, is the only disilicide phase of Co. Even though Co is the dominant diffusing species in the $CoSi_2$ reaction, the growth of the preceding CoSi phase involves the diffusion of Si, as already discussed. This may seem to pose a potential inconvenience for self-aligned processing of $CoSi_2$, but actually does not. This is because the Co diffuses into Si in the initial $Co_2Si$ reaction, thereby cutting off the connection between unreacted Co on top of $SiO_2$ and the contact metallization, except for very thick layers of deposited Co. (However, this phenomenon also has the undesirable consequence that the Co silicide layer thickness may depend on the contact hole geometry or it may be nonuniform across a contact, being thicker near the edges than at the center of the contact.) Self-alignment may be achieved by annealing at a lower temperature, stopping short of the $CoSi_2$ (or the CoSi) reaction, removing unreacted Co, and annealing (or RTA) at a higher temperature for the formation of $CoSi_2$. The cleanliness of the Si surface and the substrate doping do not affect the Co-Si reaction as much as the Ti-Si reaction, although $CoSi_2$ layers grown on heavily As-doped Si are usually less uniform than those formed on undoped or B-doped Si. Coarsening of $CoSi_2$ polycrystalline grains at high temperatures may lead to grains with sizes much larger than the film thickness, as shown in Fig. 36. Layer uniformity is one of the most important issues facing the $CoSi_2$ technology. This is a particularly difficult problem when very thin $CoSi_2$ layers are desired. When the $CoSi_2$ average thickness is less than ~400 Å, islanding of the silicide, in the form of both $CoSi_2$ and $Co_2Si$, occurs at even moderate temperatures.[354] Even though a nitrogen ambient is not required for the self-alignment of $CoSi_2$, it considerably improves the stability of the $CoSi_2$ layer in the final high temperature anneal. Snowplow and other changes of doping profile are not significant during usual $CoSi_2$ growth conditions. On preformed shallow junctions, a high temperature anneal of the $CoSi_2$ layer is necessary to reduce the leakage current.[318] However, modulation of the $CoSi_2$ layer thickness increases with annealing, and may be severe enough to cause shorting of the p-n junction. Fortunately, the SADS approach may be effectively utilized in conjunction with the $CoSi_2$ technology,[355] as shown in Fig. 36. The specific contact resistivity of $CoSi_2$, shown in Fig. 35, is very low on both n- and p-type diffusions. The selective formation of $CoSi_2$ on polysilicon gates and interconnects is also straightforward. $CoSi_2$ polycide structures are not under as much stress as $TiSi_2$ stacks. However, the thermal stability of $CoSi_2$/polysilicon stacked structures is poor. The diffusion of Co along the grain boundaries of polysilicon sometimes lead to lateral columnar structures.[356]

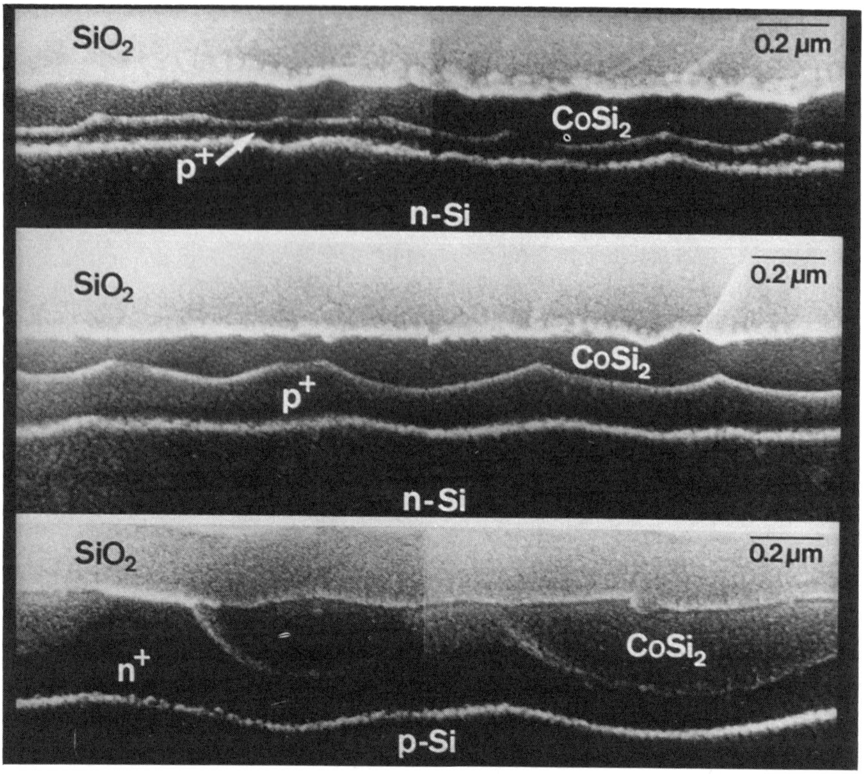


**Figure 36.** Cross-sectional electron micrographs showing the diffusion of dopants from the silicide. The $CoSi_2$ and highly doped regions (> $10^{18}cm^{-3}$) are delineated using an etching technique. *(Top panel)* B diffusion at 900°C/30 min, *(middle panel)* B diffusion at 1050°C/30 s and *(bottom panel)* As diffusion at 1100°C/60 s. (After Van den Hove, Maex, Hobbs, Lippens, De Keersmaecker, Probst, and Schaber.)[355]

Because of the good lattice match between $CoSi_2$ and Si, epitaxial growth of $CoSi_2$ layers has been demonstrated on all major orientations of Si.[357][358][359] High quality double epitaxial structures[357][360][361] and epitaxial superlattices[362][363] have also been demonstrated on Si(111) by MBE. Even though the epitaxial growth of stacked structures of metals and semiconductors brings up the possibility of many useful applications, the UHV conditions and very careful control of the processing parameters required for high quality epitaxy may limit their usefulness. Epitaxial silicide processing does not seem compatible with conventional Si clean room settings. Recently, high energy implantation of metal into Si, *mesotaxy*, was shown to be a viable way of forming buried single crystal metal layers.[364] High quality buried $CoSi_2$ layers have been fabricated on Si(111) and (100) using this technique.[364]-[367] However, the complexity and high cost of this method and the high density of defects in the mesotaxy layers make this an unlikely process to employ in device manufacturing. Despite the great advances which have been made in the fabrication of epitaxial $CoSi_2$/Si and Si/$CoSi_2$/Si structures, active applications, such as the frequently touted metal base transistor,[357][368] have yet to be demonstrated from these structures. Recently, it was demonstrated that the reaction of Ti-Co alloy and bimetallic layers leads to the growth of epitaxial $CoSi_2$ layers.[369][370] The simplicity of this process and the increased stability of single crystal silicide layers are attractive.

## 9.0 SUMMARY

Silicon has various advantages over other semiconductors in providing a testing ground for the investigation of the SB mechanism, which has thus far remained elusive. It is argued that a major obstacle to understanding the SB mechanism may be the ubiquitous assumption of the homogeneity of all SBH's. This intuitive assumption, which has invariably been made mainly out of convenience, has little scientific basis. Many non-ideal behaviors have been observed in routine SBH experiments and have been attributed to electronic states at the MS interface. In this review, popular models of interface states have been examined in detail and have been shown not to be consistent with all of the non-ideal behaviors experimentally observed. Instead, these common non-ideal behaviors are shown to be consistent with the presence of SBH inhomogeneity. It is pointed out that potential pinch-off at inhomogeneous MS interfaces provides the band-bending necessary for an explanation of the experimental results, which suggests that the majority of polycrystalline MS interfaces are likely to have inhomogeneous SBH's. The common occurrence of inhomogeneous SBH's has two major implications on the formation mechanism of the SBH. First of all, the FL position at the MS interface is not pinned as usually assumed, but is likely determined by local parameters which vary spatially at the interface. Secondly, the SBH's experimentally determined from routine measurements merely represent some averaged position of the FL with respect to the semiconductor bands and, therefore, do not necessarily reveal the trend of the local SBH mechanism. Following this logic, it seems that the study of polycrystalline SB's should be conducted at a local level, e.g., by BEEM, and the deduction of the SBH mechanism should be carried out in well-characterized MS interfaces, e.g., epitaxial silicide systems. Experimental and theoretical results from single-crystal epitaxial $NiSi_2/Si$ interfaces, discussed in this review, strongly support the close relationship between the SBH and the interface atomic structure. Furthermore, it is suggested that the distribution of the total charge at the interface, rather than just the charge redistribution in the semiconductor bandgap, should be used to address the SBH formation problem. Common ohmic contact technologies have also been reviewed. It is clear that the most important criteria for the choice of a low-resistivity contact stem from processing issues rather than from the electrical performance of the contact. Since there is no optimal contact technology per se which works for every processing design, we have opted to discuss the processing issues facing the two silicide technologies of most current interest.

## ACKNOWLEDGMENTS

I am grateful to Kwok Ng for making available to me his enormous, yet well-organized, collection of reprints on the subjects of Schottky barrier height and ohmic contact. By providing assistance and discussing many provocative ideas, John Sullivan has made working on this project a pleasure for me. I also thank Drs. M. O. Aboelfotoh, J. M. Andrews, H. Fujitani, R. Liu, K. Maex, I. Ohdomari, G. Y. Robinson and A. N. Saxena for kind permissions to use their results. Collaborations with J. P. Sullivan, F. Schrey, J. M. Gibson, A. F. J. Levi, S. D. Kevan, and M. Pinto on epitaxial silicides and computer simulations are also acknowledged.

## GLOSSARY OF NOTATION AND SYMBOLS

| Symbol | Description |
|---|---|
| $\alpha$ | A constant with the dimensionality of distance. |
| $\beta$ | $= q/(kBT)$ |
| $\Gamma$ | Dimensionless parameter for a low-SBH patch. |
| $\gamma$ | Characteristic parameter of a low-SBH patch. |
| $\gamma_{is}$ | Parameter related to interface states, defined in Eq. (9). |
| $\Lambda$ | Difference between the SBH of a low-SBH patch/strip and the average SBH for the entire diode. |
| $\delta(x,y)$ | Difference between the local SBH and the average SBH. |
| $\varepsilon_i$ | Permittivity of the interface layer. |
| $\varepsilon_d$ | Apparent high frequency permittivity deduced from reverse leakage current. |
| $\varepsilon_s$ | Semiconductor permittivity. |
| $\eta$ | $= \varepsilon_s/(qN_D)$ |
| $\lambda_D$ | Debye length of semiconductor. |
| $\lambda_s$ | Decay length of MIGS. |
| $\rho_c$ | Specific contact resistivity. |
| $\rho_\square$ | Sheet resistance per square of heavily doped semiconductor layer. |
| $\sigma$ | Standard deviation in $\gamma$ of the distribution of low-SBH patches in an inhomogeneous SB diode. |
| $\Phi_{Bn,o}$ | Flat-band SBH for n-type semiconductor. |
| $\Phi_{Bp,o}$ | Flat-band SBH for p-type semiconductor. |
| $\Phi_{Bn}$ | Apparent SBH on n-type doped semiconductor, from I-V and photoresponse measurements. |
| $\Phi_{Bp}$ | Apparent SBH on p-type doped semiconductor, from I-V and photoresponse measurements. |
| $\Phi_{mean}$ | Mean SBH for an inhomogeneous SB. |
| $\Phi$ | A constant with the dimensionality of volt. |
| $\Phi_i$ | The SBH of i-th patch. |
| $\Phi_{i,eff}$ | Effective SBH of the i-th patch. |

## Glossary (Cont'd)

| | |
|---|---|
| $\Phi_{EFF}$ | The overall effective SBH of all the low-SBH patches in an inhomogeneous SB diode. |
| $\delta\Phi$ | Total SBH lowering. |
| $\delta\Phi_{img}$ | Schottky lowering due to image force. |
| $\delta\Phi_{is}$ | SBH lowering due to interface states. |
| $\phi_m$ | Metal work function (volt). |
| $\phi_o$ | Charge neutrality level of the interface states. |
| $\chi$ | Semiconductor electron affinity (volt). |
| $\Omega$ | Dimensionless parameter for a low-SBH strip. |
| $\omega$ | Characteristic parameter for a low-SBH strip. |
| A | Area of diode. |
| A* | Richardson constant. |
| A** | Effective Richardson constant. |
| $A_i$ | Area of the i-th patch. |
| $A_{i,eff}$ | Effective area of the i-th patch. |
| c | $= (\frac{1}{2}E_{oo}) \ln(4\Phi_B/V_n)$. |
| $D_{ss}$ | Density of interface states. |
| d | Assumed separation between the interface states and the metal. |
| $E_{max}$ | Electric field at the MS interface. |
| $E_{oo}$ | Parameter, defined in Eq. (25), on the importance of tunneling in a semiconductor material. |
| $E_C$ | Conduction band minimum energy (volt). |
| $E_F$ | Fermi energy (volt). |
| $E_g$ | Semiconductor bandgap energy (volt). |
| h | Planck's constant. |
| $\Delta H_f$ | Silicide heat of formation. |
| I | Total electric current of SB diode. |
| $I_s$ | Saturation current of SB diode. |
| $k_B$ | Boltzmann constant. |
| $L_o$ | Width of a low-SBH strip. |

**Glossary (Cont'd)**

| | |
|---|---|
| $L_t$ | Transfer length, defined in Eq. (28). |
| $\ell$ | Length of contact. |
| $m_{tun}$ | Tunneling effective mass. |
| $N_A$ | Acceptor concentration. |
| $N_D$ | Donor concentration. |
| n | Ideality factor. |
| $n_{EFF}$ | The ideality factor of the sum of currents flowing in all low-SBH patches of an inhomogeneous SB diode. |
| $Q_f$ | A constant with the dimensionality of q $cm^{-2}$. |
| $Q_{ss}$ | Total charge of interface states. |
| q | Electronic charge. |
| $R_{ac}$ | Accumulation layer resistance. |
| $R_c$ | Contact resistance. |
| $R_o$ | Radius of a circular low-SBH patch. |
| $R_{sh}$ | Sheet resistance of heavily doped semiconductor layer. |
| $R_{sp}$ | Spreading resistance. |
| S | Length of heavily doped semiconductor diffusion. |
| T | Absolute temperature. |
| $T_o$ | A constant with the dimensionality of K (Kelvin). |
| V(x,y,z) | Semiconductor CBM potential, referenced to metal FL. |
| $V_a$ | Applied bias. |
| $V_c$ | Voltage drop across a contact. |
| $V_d$ | Diffusion potential (band bending) for the SB junction. In case of an inhomogeneous SBH, $V_d$ denotes the band bending corresponding to the mean SBH. |
| $V_n$ | Potential difference between CBM and the FL in a neutral n-type semiconductor. $V_n \approx \beta^{-1} \ln(N_C/N_D)$. |
| W | Depletion region width. In case of an inhomogeneous SBH, W denotes the depletion region width corresponding to a hypothetical SB with a uniform SBH equaling the average SBH of the inhomogeneous SB. |
| w | Width of heavily doped semiconductor diffusion. |

## REFERENCES


1. E. H. Rhoderick and R. H. Williams, *Metal-Semiconductor Contacts,* Clarendon Press, Oxford (1988)
2. L. J. Brillson, *Surf. Sci. Rep.*, 2:123 (1982)
3. M. S. Tyagi, in *Metal-semiconductor Schottky Barrier Junctions and Their Applications,* (B. L. Sharma, ed.), Plenum, New York (1984)
4. J. Bardeen, *Phys. Rev.,* 771:717 (1947)
5. V. Heine, *Phys. Rev.*, A138:1689 (1965)
6. W. E. Spicer, I. Lindau, P. Skeath, C. Y. Su, and P. W. Chye, *Phys. Rev. Lett.*, 44:420 (1980)
7. M. S. Daw and D. L. Smith, *Solid-St. Commun.*, 37:205 (1981)
8. O. F. Sankey, R. E. Allen, and J. D. Dow, *Solid St. Commun.*, 49:1 (1984)
9. H. Hasegawa, L. He, H. Ohno, T. Sawada, T. Haga, Y. Abe, and H. Takahashi, *J. Vac. Sci. Technol.,* B5:1097 (1987)
10. C. Tejedor, F. Flores, and E. Louis, *J. Phys., C*10:2163 (1977)
11. S. G. Louie, J. R. Chelikowsky, and M. L. Cohen, *Phys. Rev.*, B15: 2154 (1977)
12. J. Tersoff, *Phys. Rev. Lett.*, 52:465 (1984)
13. R. T. Tung, *App. Phys. Lett.,* 58:2821 (1991); and *Phys. Rev.,* B, 45:13509 (1992)
14. W. Schottky, *Z. Phys.*, 113:367 (1939)
15. A. Mead and W. G. Spitzer, *Phys. Rev.*, 134:A713 (1964)
16. A. M. Cowley and S. M. Sze, *J. Appl. Phys.*, 36:3212 (1965)
17. R. J. Archer and M. M. Atalla, *Ann. N. Y. Acad. Sci.,* 101:697 (1963)
18. C. R. Crowell and G. I. Roberts, *J. Appl. Phys.*, 40:3726 (1969)
19. S. Kurtin, T. C. McGill, and C. A. Mead, *Phys. Rev. Lett.,* 22:1433 (1970)
20. L. J. Brillson, *Phys. Rev. Lett.*, 40:260 (1978)
21. J. D. Levine, J. *Appl. Phys.,* 42:3991 (1971)
22. C. R. Crowell, *Solid-St. Electron.*, 20:171 (1977)
23. G. H. Parker, T. C. McGill, C. A. Mead, and D. Hoffman, *Solid-St. Electron.*, 11:201 (1968)
24. C. R. Crowell, *J. Vac. Sci. Technol,.* 11:951 (1974)
25. W. Mönch, *Phys. Rev. Lett.,* 58:1260 (1987)

26. A. Zur, T. C. McGill, and D. L. Smith, *Phys. Rev.,* B28:2060 (1983)
27. C. B. Duke and C. Mailhiot, *J. Vac. Sci. Technol.,* B3:1170 (1985)
28. G. W. Rubloff and P. S. Ho, *Thin Solid Films,* 93:19 (1982)
29. C. Barret and P. Muret, *Appl. Phys. Lett.,* 42:890 (1983)
30. P. S. Ho, E. S. Yang, H. L. Evans, and X. Wu, *Phys. Rev. Lett.,* 56:177 (1986)
31. P. Muret and D. Elguennouni, M. Missous, and E. H. Rhoderick, *Appl. Phys. Lett.*, 58:155 (1991)
32. A. Deneuville and B. K. Chakraverty, *Phys. Rev. Lett.,* 28:1258 (1972)
33. T. Flohr and M. Schulz, *Appl. Phys. Lett.,* 48:1534 (1986)
34. J. H. Werner, in *Metallization and and Metal-Semiconductor Interfaces*, (I. P. Batra, ed.), p. 35, Plenum, New York (1989)
35. For recent discussions, see X. Wu, E. S. Yang, and H. L. Evans, *J. Appl. Phys.,* 68:2845 (1990), and J. H. Werner, *ibid.* 70:1087 (1991)
36. J. H. Werner, A. F. J. Levi, R. T. Tung, M. Anzlowar, and M. Pinto, *Phys. Rev. Lett.,* 60:53 (1988)
37. J. L. Freeouf, *Appl. Phys. Lett.*, 41:285 (1982)
38. G. P. Das, P. Blöchl, O. K. Andersen, N. E. Christensen, and O. Gunnarsson, *Phys. Rev. Lett.,* 63:1168 (1989)
39. H. Fujitani and S. Asano, *Phys. Rev.*, B42:1696 (1990)
40. J. M. Gibson, D. C. Joy, R. T. Tung, J. L. Ellison, C. Pimentel, and A. F. J. Levi, *Mat. Res. Soc. Symp. Proc.*, 56:163 (1986)
41. H.-C. W. Huang, C. F. Aliotta, and P. S. Ho, *Appl. Phys. Lett.,* 41:54 (1982)
42. W. J. Kaiser and L. D. Bell, *Phys. Rev. Lett.,* 60:1406 (1988)
43. R. T. Tung, A. F. J. Levi, J. P. Sullivan, and F. Schrey, *Phys. Rev. Lett.,* 66:72 (1991); J. P. Sullivan, R. T. Tung, and F. Schrey, *J. Appl. Phys.,* 72:478 (1992)
44. J. L. Freeouf, T. N. Jackson, S. E. Laux, and J. M. Woodall, *Appl. Phys. Lett.*, 40:634 (1982)
45. J. L. Freeouf, T. N. Jackson, S. E. Laux, and J. M. Woodall, *J. Vac. Sci. Technol.*, 21:570 (1982)
46. I. Ohdomari and H. Aochi, *Phys. Rev.,* B35:682 (1987) However, a faulty boundary condition in this paper led to very significant errors in its estimation of the saddle-point potential.
47. A. I. Bastys, V. B. Bikbaev, J. J. Vaitkus, and S. C. Karpinskas, *Litovskii Fizicheskii Sbornik*, 28:191 (1988)

48. J. P. Sullivan, R. T. Tung, M. Pinto, and W. R. Graham, *J. Appl. Phys.*, 70:7403 (1991)

49. C. Canali, F. Catellani, S. Mantovani, and M. Prudenziati, *J. Phys.*, D10:2481 (1977)

50. R. Rosenberg, M. J. Sullivan, and J. K. Howard, in *Thin films Interdiffusion and Reactions*, (J. M. Poate, K. N. Tu, and J. W. Mayer, eds.), p. 13, Wiley, New York (1978)

51. I. Ohdomari, T. S. Kuan and K. N. Tu, *J. Appl. Phys.*, 50:7020 (1979)

52. I. Ohdomari and K. N. Tu, *J. Appl. Phys.*, 51:3735 (1980)

53. R. H. Cox and H. Strack, *Solid-St. Electron.*, 10:1213 (1967)

54. R. N. Sreenath, M. M. Chandra, and G. Suryan, *IEEE Proc.*, 131:63 (1984)

55. T. Shibata, H. Hieda, M. Sato, M. Konaka, R. L. M. Dang, and H. Izuka, *IEEE IEDM Tech. Dig.*, p.647 (1981)

56. R. A. Levy, M. L. Green, P. K. Gallagher, and Y. S. Ali, *J. Electrochem. Soc.*, 133:1905 (1986)

57. A. Bouteville, A. Royer, and J. C. Remy, *J. Electrochem. Soc.*, 134:2080 (1987)

58. M. P. Lepselter and S. M. Sze, *Bell Syst. Tech. J.*, 47:195 (1968)

59. R. A. Zettler and A. M. Cowley, *IEEE Trans. Elec. Dev.*, 16:58 (1969)

60. A. N. Saxena, *Surf. Sci.*, 13:151 (1969)

61. A. Rusu, C. Bulucea, and C. Postolache, *Solid-St. Electron.*, 20:499 (1977)

62. A. Kikuchi, *Jpn. J. Appl. Phys.* 25:L894 (1986)

63. D. L. Scharfetter, *Solid-St. Electron.*, 8:299 (1965)

64. A. Y. C. Yu and E. H. Snow, *Solid-St. Electron.*, 12:155 (1969)

65. M. J. Hargrove and R. L. Anderson, *Solid-St. Electron.*, 29:365 (1986)

66. W. Cabanski and M. Schulz, *Appl. Phys.*, A48:203 (1989)

67. A. Ishizaka and Y. Shiraki, *J. Electrochem. Soc.*, 133:666 (1986)

68. M. Tabe, *Jpn. J. Appl. Phys.*, 21:534 (1982)

69. F. G. Allen, T. M. Buck, and J. T. Law, *J. Appl. Phys.*, 31:979 (1960)

70. J. D. Mottram, A. Thanailakis, and D. C. Northrop, *J. Phys.*, D8:1316 (1975)

71. R. T. Tung, K. K. Ng, J. M. Gibson, and A. F. J. Levi, *Phys. Rev.*, B33:7077 (1986)

72. P. J. Grunthaner, F. J. Grunthaner, R. W. Fathauer, T. L. Lin, M. H. Hecht, L. D. Bell, W. J. Kaiser, F. D. Schowengerdt, and J. H. Mazur, *Thin Solid Films,* 183:197 (1989)

73. G. S. Higashi, R. S. Becker, Y. J. Chabal, and A. J. Becker, *Appl. Phys. Lett.,* 58:1656 (1991)

74. M. Miyawaki, S. Yoshitake, and T. Ohmi, *IEEE Elec. Dev. Lett.*, 11:10 (1990)

75. M. Copel, M. C. Reuter, E. Kaxiras, and R. M. Tromp, *Phys. Rev. Lett.*, 63:632 (1989)

76. A. E. White, K. T. Short, K. Maex, R. Hull, Y.-F. Hsieh, S. A. Audet, K. W. Goossen, D. C. Jacobson, and J. M. Poate, *Nucl. Instrum. & Methods B*, 59/60:693 (1991)

77. I. Yamada, H. Inokawa, T. Takagi, *J. Appl. Phys.,* 56:2746 (1984)

78. J. Lee and R. Reif, *J. Electron. Mater.*, 20:331 (1991)

79. C. A. Crider and J. M. Poate, *Appl. Phys. Lett.*, 36:417 (1980)

80. M. Wittmer and T. E. Seidel, *J. Appl. Phys.,* 49:5827 (1979)

81. M. Wittmer, C. Y. Ting, and K. N. Tu, *J. Appl. Phys.,* 54:699 (1983)

82. I. Ohdomari, K. N. Tu, K. Suguro, M. Akiyama, I. Kimura, and K. Yoneda, *Appl. Phys. Lett.,* 38:1015 (1981)

83. J. B. Bindell, W. M. Moller, and E. F. Labuda, *IEEE Trans. Elec. Dev.,* 27:420 (1980)

84. M. Tabasky, E. S. Bulat, B. M. Bulat, B. M. Ditchek, M. A. Sullivan, and S. C. Shatas, *IEEE Trans. Elec. Dev.,* 34:548 (1987)

85. L. J. Chen and T. T. Chang, *Thin Solid Films,* 104:183 (1983)

86. C.-A. Chang, *J. Appl. Phys.*, 59:3116 (1986)

87. C. S. Wei, J. van der Spiegel, J. J. Santiago, and L. E. Seiberling, *Appl. Phys. Lett.*, 45:527 (1984)

88. M. Wittmer, W. Lüthy, B. Studer, and H. Melchior, *Solid-St. Electron.,* 24:141 (1981)

89. J. C. Bean and J. M. Poate, *Appl. Phys. Lett.,* 37:643 (1980)

90. B.-Y. Tsaur, D. J. Silversmith, R. W. Mountain, L. S. Hung, S. S. Lau, and T. T. Sheng, *J. Appl. Phys.,* 51:5243 (1981)

91. M. Eizenberg, H. Föll, and K. N. Tu, *J. Appl. Phys.*, 52:861 (1981)

92. S. Kritzinger and K. N. Tu, *J. Appl. Phys.*, 52:305 (1981)

93. C. D. Lien, M. Finetti, and M.-A. Nicolet, *Appl. Phys.,* A35:47 (1984)

94. M. Eizenberg, G. Ottaviani, and K. N. Tu, *Appl. Phys. Lett.*, 37:87 (1980)

95. K. N. Tu, W. N. Hammer, and J. O. Olowolafe, *J. Appl. Phys.*, 51:1663 (1980)

96. G. A. Rozgonyi, J.-H. Lee, D. Knoesen, D. Adams, B. Patnaik, N. Parikh, A. S. M. Salih, and P. Balducci, *Appl. Phys. Lett.*, 58:729 (1991)

97. R. de Reus, H. C. Tissik, and F. W. Saris, *J. Mater. Res.*, 5:341 (1990)

98. S. M. Sze, *Physics of Semiconductor Devices*, Wiley, New York (1981)

99. P. E. Schmid, *Helvetica Physica Acta*, 58:371 (1985)

100. K. K. Ng, in *Properties of Silicon*, INSPEC, IEE, London (1988)

101. M. J. Turner and E. H. Rhoderick, *Solid-St. Electron.*, 11:291 (1968)

102. T. Arizumi and M. Hirose, *Jpn. J. Appl. Phys.*, 8:749 (1969)

103. M. Hirose, N. Altaf, and T. Arizumi, *Jpn. J. Appl. Phys.*, 9:260 (1970)

104. J. P. Ponpon and P. Siffert, *J. Appl. Phys.*, 49:6004 (1978)

105. A. Diligenti, M. Stagi, and V. Ciuti, *Solid State Commun.*, 45:347 (1983)

106. H. C. Card, *IEEE Trans. Elec. Dev.*, 23:538 (1976)

107. M. Finetti, I. Suni, M. Bartur, T. Banwell, M.-A. Nicolet, *Solid-St. Electron.*, 27:617 (1984)

108. W. De Bosscher, R. L. Meirhaeghe, A. De Laere, W. H. Laflere, and F. Cardon, *Solid-St. Electron.*, 31:945 (1988)

109. M. Wittmer, B. Studer, and H. Melchior, *J. Appl. Phys.*, 52:5722 (1981)

110. S. Ashok, T. P. Chow, and B. J. Baliga, *Appl. Phys. Lett.*, 42:687 (1983)

111. S. Ashok, H. Kräutle, and H. Beneking, *Appl. Phys. Lett.*, 45:431 (1984)

112. Y. Itoh, *Jpn. J. Appl. Phys.*, 9:926 (1970)

113. G. A. Adegboyega, A. Poggy, E. Susi, A. Castaldini, and A. Cavallini, *Appl. Phys.*, A48:391 (1989)

114. M. A. Taubenblatt, D. Thomson, and C. R. Helms, *Appl. Phys. Lett.*, 44:895 (1984)

115. I. J. M. M. Raaijmakers and K. B. Kim, *J. Appl. Phys.*, 67:6255 (1990)

116. S. Ogawa, T. Kouzaki, T. Yoshida, and R. Sinclair, *J. Appl. Phys.*, 70:827 (1991)

117. A. M. Cowley, *Solid-St. Electron.*, 12:403 (1970)

118. M. O. Aboelfotoh and K. N. Tu, *Phys. Rev.*, B34:2311 (1986)

119. M. Liehr, P. E. Schmidt, F. K. LeGoues and P. S. Ho, *Phys. Rev. Lett.*, 54:2139 (1985)
120. K. E. Sundström, S. Petersson, and P. A. Tove, *Phys. Stat. Sol.*, (a)20:653 (1973)
121. D. J. Coe and E. H. Rhoderick, *J. Phys.*, D9:965 (1976)
122. P. E. Schmid, P. S. Ho, H. Föll, and T. Y. Tan, *Phys. Rev.*, B28:4593 (1983)
123. P. Gutknecht and M. J. O. Strutt, *Appl. Phys. Lett.*, 21:405 (1972)
124. R. J. Archer and T. O. Yep, *J. Appl. Phys.*, 41:303 (1970)
125. D. Kahng, *Solid-St. Electron.*, 6:281 (1963)
126. F. A. Padovani, *J. Appl. Phys.*, 38:891 (1967)
127. A. Thanailakis and A. Rasul, *J. Phys.*, C9:337 (1976)
128. C. R. Crowell, W. G. Spitzer, L. E. Howarth, and E. E. LaBate, *Phys. Rev.*, 127:2006 (1962)
129. J. D. van Otterloo, *Surf. Sci.*, 104:L205 (1981)
130. J. D. van Otterloo and J. G. De Groot, *Surf. Sci.*, 57:93 (1976)
131. R. W. Soshea and R. C. Lucas, *Phys. Rev.*, A138:1182 (1965)
132. J. D. van Otterloo and L. J. Gerritsen, *J. Appl. Phys.*, 49:723 (1978)
133. A. Y. C. Yu and C. A. Mead, *Solid-St. Electron.*, 13:97 (1970)
134. K. Chino, *Solid-St. Electron.*, 16:119 (1973)
135. H. M. Naguib and L. H. Hobbs, *J. Electrochem. Soc.*, 124:573 (1977)
136. C. M. Wu and E. S. Yang, *J. Appl. Phys.*, 52:4700 (1981)
137. J. M. Shannon, *Appl. Phys. Lett.*, 25:75 (1974)
138. T. M. Reith and J. D. Schick, *Appl. Phys. Lett.*, 25:524 (1974)
139. T.-M. Lu, P. Bai, A. S. Yapsir, P.-H. Chang, and T. J. Shaffner, *Phys. Rev.*, B39:9584 (1989)
140. K. C. Saraswat, S. Swirhun, and J. P. McVittie, *MRS Proc. 409*
141. C. R. Crowell, J. C. Sarace, and S. M. Sze, *Trans. TMS/AIME*, 233:478 (1965)
142. M. O. Aboelfotoh, *J. Appl. Phys.*, 66:262 (1989)
143. G. Ottaviani, K. N. Tu, and J. W. Mayer, *Phys. Rev.*, B24:3354 (1981)
144. R. J. Purtell, P. S. Ho, G. W. Rubloff, and P. E. Schmid, *Physica*, 117/118B:834 (1983)
145. D. J. Fertig and G. Y. Robinson, *Solid-St. Electron.*, 19:407 (1976)

146. F. La Via, P. Lanza, O. Viscuso, G. Feria, and E. Rimini, *Thin Solid Films,* 161:13 (1988)

147. L. Stolt, P. A. Tove, and S. Petersson, *Phys. Scripta,* 18:410 (1978)

148. D. Moy, S. Basavaiah, C. T. Chuang, G. P. Li, E. Hackbarth, S. B. Brodsky, and M. R. Polcari, *Solid-St. Electron.*, 31:843 (1988)

149. M. Wittmer, *Phys. Rev.,* B42:5249 (1990)

150. G. J. van Gurp, *J. Appl. Phys.,* 46:4308 (1975)

151. M. O. Aboelfotoh, *J. Appl. Phys.*, 64:4046 (1988)

152. K. Shenai, *IEEE Elec. Dev. Lett.,* 12:169 (1991)

153. G. Kano, M. Inoue, J. Matsuno, and S. Takayanagi, *J. Appl. Phys.,* 37:37 (1966)

154. A. Martinez, D. Esteve, A. Guivarc'h, P. Auvray, P. Henoc, and G. Pelous, *Solid-St. Electron.*, 23:55 (1980)

155. K. N. Tu, R. D. Thompson, and B. Y. Tsaur, *Appl. Phys. Lett.,* 38:626 (1981)

156. H. Norde, J. de Sousa Pires, F. d'Heurle, F. Pesavento, S. Petersson, and P. A. Tove, *Appl. Phys. Lett.,* 38:865 (1981)

157. V. M. Koleshko, V. F. Belitsky, and A. A. Khodin, *Vacuum,* 36:669 (1986)

158. P. W. Pellegrini, C. E. Ludington, and M. M. Weeks, *J. Appl. Phys.,* 67:1417 (1990)

159. J. Silverman, P. Pellegrini, J. Comer, A. Golvbovic, M. Weeks, J. Mooney, J. Fitzgerald, *MRS Symp. Proc.,* 54:515 (1986)

160. N. Toyama, T. Takahashi, H. Murakami, and H. Horiyama, *Appl. Phys. Lett.,* 46:557 (1985)

161. R. C. McKee, *IEEE Trans. Elec. Dev.,* 31:968 (1984)

162. E. Calleia, J. Garrido, J. Piqueras, and A. Martinez, *Solid-St. Electron.,* 23:591 (1980)

163. T. T. Bardin, J. G. Pronko, R. C. Budhani, and R. F. Bunshah, *J. Vac. Sci. Technol.*, A4:3121 (1986)

164. D. Tripathi, P. C. Srivastava, and S. Chandra, *Phys. Rev.,* B39:13420 (1989)

165. B.-Y. Tsaur, J. P. Mattia, and C. K. Chen, *Appl. Phys. Lett.,* 57:1111 (1990)

166. R. F. Broom, *Solid-St. Electron.*, 14:1087 (1971)

167. M. O. Aboelfotoh, A. Cros, B. G. Svensson, and K. N. Tu, *Phys. Rev.,* B41:9819 (1990)

168. M. Wittmer, *Phys. Rev.*, B43:4385 (1991)

169. M. Severi, E. Gabilli, S. Guerri, G. Celloti, *J. Appl. Phys.*, 48:1998 (1977)

170. C. Y. Ting and M. Wittmer, *J. Appl. Phys.*, 54:937 (1983)

171. H. H. Hosack, *Appl. Phys. Lett.* 2:256 (1972); and *J. Appl. Phys.*, 44:3476 (1973)

172. H. Grinolds and G. Y. Robinson, *J. Vac. Sci. Technol.*, 14:75 (1977)

173. E. Hökelek and G. Y. Robinson, *Thin Solid Films*, 53:135 (1978)

174. G. J. van Gurp and W. M. Reukers, *J. Appl. Phys.*, 50:6923 (1979)

175. D. C. Chen, P. Merchant, and J. Amano, *J. Vac. Sci. Technol.*, A3:709 (1985)

176. M. Bartur and M.-A. Nicolet, *Appl. Phys. Lett.*, 39:822 (1981)

177. P. C. Parekh, R. C. Sirrine, and P. Lemieux, *Solid-St. Electron.*, 19:493 (1976)

178. C. Canali, F. Fantini, and E. Zanoni, *Thin Solid Films*, 97:325 (1982)

179. V. F. Drobny, *J. Elec. Mater.*, 14:283 (1985)

180. R. J. Schutz, *Thin Solid Films*, 104:89 (1983)

181. M. Finetti, I. Suni, and, M.-A. Nicolet, *J. Elec. Mater.*, 13:327 (1984)

182. M. Bartur and M.-A. Nicolet, *J. Electrochem. Soc.*, B131:1118 (1984)

183. Arizumi, M. Hirose, and N. Altaf, *Japan. J. Appl. Phys.*, 7:870 (1968)

184. M. Finetti, E. T. S. Pan, I. Suni, and M.-A. Nicolet, *Appl. Phys. Lett.*, 42:987 (1983)

185. C. Narayan, A. S. Karakashian, G. H. R. Kegel, and Z. Rivera, *Appl. Phys. Lett.*, 59:2541 (1991)

186. J. H. Werner, *Appl. Phys.*, A47:291 (1988)

187. A. Y. C. Yu and E. H. Snow, *J. Appl. Phys.*, 39:3008 (1968)

188. Z. Liliental-Weber, R. Gronsky, J. Washburn, N. Newman, W. E. Spicer, and E. R. Weber, *J. Vac. Sci. Technol.*, B4:912 (1986)

189. D. Dascalu, Gh. Brezeanu, P. A. Dan and C. Dima, *Solid-St. Electron.*, 24:897 (1981)

190. J. P. Sullivan, R. T. Tung, F. Schrey, and W. R. Graham, *J. Vac. Sci. Technol.*, A10:1959 (1992)

191. M. M. Atalla and R. W. Soshea, Scientific Report No. 1, Contract No. AF19 (628)-1637, Hewlett-Packard Associates (1962)

192. S. M. Sze, C. R. Crowell, and D. Kahng, *J. Appl. Phys.*, 35:2534 (1964)

193. H. C. Card and E. H. Rhoderick, *J. Phys.*, D4:1589 (1971)

194. F. A. Padovani and R. Stratton, *Solid-St. Electron.*, 9:695 (1966)
195. V. L. Rideout and C. R. Crowell, *Solid-St. Electron.*, 13:993 (1970)
196. J. H. Werner and H. H. Güttler, J*. Appl. Phys.*, 69:1522 (1991)
197. F. A. Padovani and G. G. Sumner, *J. Appl. Phys.*, 36:3744 (1965)
198. S. Ashok, J. M. Borrego, and R. J. Guttmann, *Solid-St. Electron.,* 22:621 (1979)
199. B. Tuck, G. Eftekhari, and D. M. de Cogan, *J. Phys.,* D15:457 (1982)
200. F. A. Padovani, in *Semiconductors and Semimetals,* Vol. 7A, (R. K. Willardson and A. C. Beer, eds.), Academic Press, New York (1971)
201. G. S. Visweswaran and R. Sharan, *Proc. IEEE,* 67:436 (1979)
202. J. Basterfield, J. M. Shannon, and A. Gill, *Solid-St. Electron.,* 18:290 (1975)
203. B. Studer, *Solid-St. Electron.,* 23:1181 (1980)
204. V. B. Bikbaev, S. C. Karpinskas, and J. J. Vaitkus, *Phys. Stat. Sol.,* (a)75:583 (1983)
205. V. W. L. Chin, J. W. V. Storey, and M. A. Green, *J. Appl. Phys.,* 68:4127 (1990)
206. J. M. Andrews and M. P. Lepselter, *Solid-St. Electron.,* 13:1011 (1970)
207. J. M. Andrews and F. B. Koch, *Solid-St. Electron.,* 14:901 (1971)
208. C. J. Kircher, *Solid-St. Electron.,* 14:507 (1971)
209. N. Newman, M. van Schilfgaarde, T. Kendelwicz, M. D. Williams, and W. E. Spicer, *Phys. Rev.,* B33:1146 (1986)
210. A. B. McLean and R. H. Williams, *J. Phys.,* C21:783 (1988)
211. R. T. Tung, *J. Vac. Sci. Technol.,* B2:465 (1984)
212. H. H. Güttler and J. H. Werner, *Appl. Phys. Lett.,* 56:1113 (1990)
213. R. Hackam and P. Harrop, *Solid State Commun.,* 11:669 (1972)
214. V. W. L. Chin, M. A. Green, and J. W. V. Storey, *J. Appl. Phys.,* 68:3470 (1990)
215. A. M. Goodman, *J. Appl. Phys.,* 34:329 (1963)
216. G. L. Miller, D. V. Lang, and L. C. Kimerling, *Ann. Rev. Mat. Sci.,* 7:377 (1977)
217. F. J. Bryant, J. M. Majid, C. G. Scott and D. Shaw, *Solid State Commun.,* 63:9 (1987)
218. H. Grinolds and G. Y. Robinson, *J. Vac. Sci. Technol.,* 14:75 (1977)
219. J. A. Copeland, *IEEE Trans. Elec. Dev.*, ED-17:404 (1970)

220. A. M. Cowley, *J. Appl. Phys.,* 37:3024 (1966)

221. J. D. Wiley and G. L. Miller, *IEEE Trans. Elec. Dev.,* ED-22:265 (1975)

222. P. K. Vasudev, B. L. Mattes, E. Petras and R. H. Bube, *Solid-St. Electron.,* 19:557 (1976)

223. J. M. Borrego, R. J. Gutmann and S. Ashok, *Solid-St. Electron.,* 20:125 (1977)

224. B. Pellegrini and G. Salardi, *Solid-St. Electron.,* 21:465 (1978)

225. T. Okumura and K. N. Tu, *J. Appl. Phys.,* 54:922 (1983)

226. O. Engstrom, H. Pettersson, and B. Sernelius, *Phys. Stat. Sol.,* (a)95:691 (1986)

227. A. Tanabe, K. Konuma, N. Teranishi, S. Tohyama, and K. Masubuchi, *J. Appl. Phys.*, 69:850 (1991)

228. N. F. Mott, *Proc. Cambridge Philos. Soc.,* 34:568 (1938)

229. M. Schlüter, *Phys. Rev.,* B17:5044 (1978)

230. W. Mönch, *Surface Sci.,* 21:443 (1970)

231. J. M. Andrews and J. C. Phillips, *Phys. Rev. Lett.,* 35:56 (1975)

232. G. Ottaviani, K. N. Tu, and J. W. Mayer, *Phys. Rev. Lett.,* 44:284 (1980)

233. J. L. Freeouf, *Solid State Commun.,* 33:1059 (1980)

234. J. Tersoff, *Phys. Rev.,* B32:6968 (1985)

235. R. T. Tung, in *Silicon Molecular Beam Epitaxy,* (J. C. Bean and E. Kasper, eds.), CRC Press, Boca Raton, FL (1988)

236. R. T. Tung, J. M. Poate, J. C. Bean, J. M. Gibson and D. C. Jacobson, *Thin Solid Films,* 93:77 (1982)

237. R. T. Tung, J. M. Gibson, and J. M. Poate, *Phys. Rev. Lett.,* 50:429 (1983)

238. E. J. van Loenen, A. E. M. Fischer, J. F. van der Veen, and F. LeGoues, *Surf. Sci.,* 154:52 (1985)

239. B. D. Hunt, L. J Schowalter, N. Lewis, E. L. Hall, R. J. Hauenstein, T. E. Schlesinger, T. C. McGill, M. Okamoto, and S. Hashimoto, *Mat. Res. Soc. Symp. Proc.,* 54:479 (1986)

240. H. von Känel, T. Graf, J. Henz, M. Ospelt, and P. Wachter, *J. Cryst. Growth,* 81:470 (1987)

241. R. T. Tung, J. M. Gibson and J. M. Poate, *Appl. Phys. Lett.,* 42:888 (1983)

242. R. T. Tung, *J. Vac. Sci. Technol.,* A5:1840 (1987)

243. R. T. Tung and F. Schrey, *Appl. Phys. Lett.*, 55:256 (1989)

244. A. Ishizaka and Y. Shiraki, *Surf. Sci.*, 174:671 (1986)

245. T. R. Harrison, A. M. Johnson, P. K. Tien, and A. H. Dayem, *Appl. Phys. Lett.*, 41:734 (1982)

246. A. Ishizaka and Y. Shiraki, *Jpn. J. Appl. Phys.*, 23:L499 (1984)

247. H. Föll, *Phys. Stat. Sol.*, (a)69:779 (1982)

248. D. Cherns, G. R. Anstis, J. L. Hutchison, and J. C. H. Spence, *Philos. Mag.*, A46:849 (1982)

249. J. M. Gibson, R. T. Tung, and J. M. Poate, *Mater. Res. Soc. Symp. Proc.*, 14:395 (1983)

250. E. Vlieg, A. E. M. J. Fischer, J. F. van der Veen, B. N. Dev, and G. Materlik, *Surf. Sci.*, 178:36 (1986)

251. J. Zegenhagen, M. A. Kayed, K.-G. Huang, W. M. Gibson, J. C. Phillips, L. J. Schowalter, and B. D. Hunt, *Appl. Phys.*, A44:365 (1987)

252. E. J. van Loenen, J. W. M. Frenken, J. F. van der Veen, and S. Valeri, *Phys. Rev. Lett.*, 54:827 (1985)

253. I. K. Robinson, R. T. Tung, and R. Feidenhans'l, *Phys. Rev.*, B38:3632 (1988)

254. P. Werner, R. Mattheis, D. Hesse, R. Hillebrand, and J. Heydenreich, *Phys. Stat. Sol.*, (a)116:81 (1989)

255. R. T. Tung, *Phys. Rev. Lett.*, 52:461 (1984)

256. R. J. Hauenstein, T. E. Schlesinger, T. C. McGill, B. D. Hunt, and L. J. Schowalter, *Appl. Phys. Lett.*, 47:853 (1985)

257. M. Ospelt, J. Henz, L. Flepp, and H. von Känel, *Appl. Phys. Lett.*, 52:227 (1988)

258. J. Vrijmoeth, J. F. van der Veen, D. R. Heslinga, and T. M. Klapwijk, *Phys. Rev.*, B42:9598 (1990)

259. Y. Shiraki, T. Ohshima, A. Ishizaka, and K. Nakagawa, *J. Cryst. Growth*, 81:476 (1987)

260. A. Kikuchi, T. Ohshima, and Y. Shiraki, *J. Appl. Phys.*, 64:4614 (1988)

261. A. Kikuchi, *Phys. Rev.*, B40:8024 (1989)

262. R. T. Tung, J. P. Sullivan, F. Schrey, and A. F. J. Levi, *Mater. Res. Soc. Symp. Proc.*, 221:71 (1991)

263. J. M. Woodall, G. D. Pettit, T. N. Jackson, C. Lanza, K. L. Kavanagh, and J. M. Mayer, *Phys. Rev. Lett.*, 51:1783 (1983)

264. R. T. Tung and F. Schrey, *Phys. Rev. Lett.*, 63:1277 (1989)

265. S. D. Kevan and R. T. Tung, unpublished.

266. A. Chantre, A. F. J. Levi, R. T. Tung, W. C. Dautremont-Smith, and M. Anzlowar, *Phys. Rev.,* B34:4415 (1986)

267. T. Flohr, M. Schulz and R. T. Tung, *Appl. Phys. Lett.,* 51:1343 (1987)

268. J. H. Werner, *Appl. Phys. Lett.,* 54:1528 (1989)

269. T.-H. Shen and C. C. Matthai, *J. Phys.:Condens. Matter,* 3:613 (1991)

270. Y. Hasegawa, Y. Kuk, R. T. Tung, P. J. Silverman, and T. Sakurai, *J. Vac. Sci. Technol.,* B9:578 (1991)

271. A. Fernandez, H. D. Hallen, T. Huang, R. A. Buhrman, and J. Silcox, *J. Vac. Sci. Technol.,* B9:590 (1991)

272. H. Lim and R. E. Allen, *J. Vac. Sci. Technol.,* A4:2328 (1986)

273. J.-J. Yeh, *Appl. Phys. Lett.,* 55:1241 (1989)

274. S. Ossicini, O. Bisi, and C. M. Bertoni, *Phys. Rev.,* B42:5735 (1990)

275. O. Bisi and S. Ossicini, *Surf. Sci.,* 189/190:285 (1987)

276. N. V. Rees and C. C. Matthai, *Semicond. Sci. Technol.,* 4:412 (1989)

277. X. Yongnian, Z. Kaiming, and X. Xide, *Phys. Rev.,* B33:8602 (1986)

278. D. E. Rodrigues and F. J. Weisz, *Phys. Rev.,* B39:1622 (1989)

279. D. R. Hamann, in *Metallization and Metal-Semiconductor Interfaces,* (I. P. Batra, ed.), Plenum, New York (1988)

280. R. T. Tung, F. Schrey, and S. M. Yalisove, *Appl. Phys. Lett.,* 55:2005 (1989)

281. J. L. Batstone, J. M. Gibson, R. T. Tung, and A. F. J. Levi, *Appl. Phys. Lett.,* 52:828 (1988)

282. D. J. Eaglesham, R. T. Tung, and J. P. Sullivan, to be published.

283. Cherns, C. J. D. Hetherington, and C. J. Humphreys, *Philos. Mag.,* A49:165 (1984)

284. H. Fujitani and S. Asano, *J. Phys. Soc. Jpn.,* 60:2526 (1991)

285. C. C. Matthai, N. V. Rees, and T. H. Shen, presented at *Third Int'l. Conf. Formation Semicond. Interfaces*, Rome, May (1991)

286. D. Loretto, J. M. Gibson, and S. M. Yalisove, *Phys. Rev. Lett.,* 63:298 (1989)

287. R. T. Tung, *Mat. Res. Soc. Symp. Proc.,* 37:345 (1985)

288. J. L. Batstone, J. M. Gibson, R. T. Tung, A. F. J. Levi, and C. A. Outten, *Mat. Res. Soc. Symp. Proc.,* 82:335 (1987)

289. R. T. Tung, S. Nakahara, and T. Boone, *Appl. Phys. Lett.,* 46:895 (1985)

290. R. T. Tung and J. M. Gibson, *J. Vac. Sci. Technol.,* A3:987 (1985)

291. E. Rosencher, S. Delage, F. A. D'Avitaya, *J. Vac. Sci. Technol.,* A3:762 (1985)

292. J. R. Jimenez, L. M. Hsiung, R. D. Thompson, S. Hashimoto, K. V. Ramanathan, R. Arndt, K. Rajan, S. S. Iyer, and L. J. Schowalter, Mater. R*es. Soc. Symp. Proc.,* 60:237 (1990)

293. R. T. Tung, A. F. J. Levi, F. Schrey, and M. Anzlowar, *NATO ASI Series B: Physics,* 203:167 (1989); J. P. Sullivan, R. T. Tung, D. J. Eaglesham, F. Schrey, and W. R. Graham, *J. Vac. Sci. Technol.* (1993)

294. D. J. Eaglesham, R. T. Tung, R. L. Headrick, I. K. Robinson, and F. Schrey, *Mater. Res. Soc. Symp. Proc.,* 159:321 (1990)

295. J. M. Gibson, J. C. Bean, J. M. Poate and R. T. Tung, *Appl. Phys. Lett.,* 41:818 (1982)

296. A. E. M. J. Fischer, E. Vlieg, J. F. van der Veen, M. Clausnitzer, and G. Materlik, *Phys. Rev.,* B36:4769 (1987)

297. C. W. T. Bulle-Lieuwma, A. F. de Jong, A. H. van Ommen, J. F. van der Veen, and J. Vrijmoeth, *Appl. Phys. Lett.,* 55:648 (1989)

298. D. R. Heslinga, H. H. Weitering, D. P. van der Werf, T. M. Klapwijk, and T. Hibma, *Phys. Rev. Lett.,* 64:1585 (1990)

299. G. Le Lay, K. Hricovini, and J. E. Bonnet, *Appl. Surf. Sci.,* 41/42:25 (1989)

300. G. Le Lay and K. Hricovini, *Phys. Rev. Lett.,* 65:807 (1990)

301. J. R. Brews, K. K. Ng, and R. K. Watts, in *Submicron Integrated Circuits*, (R. K. Watts, ed.), Wiley, New York (1989)

302. D. D. Tang, *ibid.*

303. G. Baccarani, M. R. Wordeman, and R. H. Dennard, *IEEE Trans. Elec. Dev.,* 31:452 (1984)

304. J. R. Brews, W. Fichtner, E. H. Nicollian, and S. M. Sze, *IEEE Elec. Dev. Lett.,* 1:2 (1980)

305. H. D. Barber, *Solid-St. Electron.,* 10:1039 (1967)

306. K. K. Ng and R. Liu, *IEEE Trans. Elec. Dev.,* 37:1535 (1990)

307. C. Y. Chang, Y. K. Fang, and S. M. Sze, *Solid-St. Electron.,* 14:541 (1971)

308. S. E. Swirhun and R. M. Swanson, *IEEE Elec. Dev. Lett.,* 7:155 (1986)

309. M. Finetti, S. Guerri, P. Negrini, and A. Scorzoni, *Thin Solid Films,* 130:37 (1985)

310. W. J. Boudville and T. C. McGill, *J. Vac. Sci. Technol.,* B3:1192 (1985)

311. D. Mathiot and J. C. Pfister, *J. Appl. Phys.,* 55:3518 (1984)
312. M. D. Giles, *Appl. Phys. Lett.,* 58:2399 (1991)
313. P. G. Carey, T. G. Sigmon, R. L. Press, and T. S. Fahlen, *IEEE Electron. Lett.,* 6:291 (1985)
314. H. Okabayashi, E. Nagasawa, and M. Morimoto, *IEEE IEDM Tech. Dig.,* p. 732 (1982)
315. L. S. Hung, J. W. Mayer, M. Zhang, and E. D. Wolf, *Appl. Phys. Lett.,* 43:1233 (1983)
316. D. L. Kwong and N. S. Alvi, *J. Appl. Phys.,* 60:688 (1986)
317. R. Liu, D. S. Williams, and W. T. Lynch, *IEDM Tech. Dig.,* p. 58 (1986)
318. R. Liu, D. S. Williams, and W. T. Lynch, *J. Appl. Phys.,* 63:1990 (1988)
319. L. E. Terry and R. W. Wilson, *Proc. IEEE,* 57:1580 (1969)
320. R. D. Brooks and H. G. Mathes, *Bell Syst. Tech. J.,* 50:775 (1971)
321. C.-Y. Ting and C. Y. Chen, *Solid-St. Electron.,* 14:433 (1971)
322. K. K. Ng and W. T. Lynch, *IEEE Trans. Elec. Dev.,* 34:503 (1987)
323. G. Baccarani and G. A. Sai-Halasz, *IEEE Elec. Dev. Lett.,* 4:27 (1983)
324. K. K. Ng and W. T. Lynch, *IEEE Trans. Elec. Dev.,* 33:965 (1986)
325. H. Murrmann and D. Widmann, *IEEE Trans. Electron. Dev.,* 16:1022 (1969)
326. H. H. Berger, *Solid-St. Electron.,* 15:145 (1972)
327. S. J. Proctor and L. W. Linholm, *IEEE Elec. Dev. Lett.,* 3:294 (1982)
328. R. L. Gillenwater, M. J. Hafich, and G. Y. Robinson, *IEEE Trans. Elec. Dev.,* 34:537 (1987)
329. M. Finetti, A. Scorzoni, and G. Soncini, *IEEE Elec. Dev. Lett.,* 6:184 (1985)
330. T. F. Lei, L.-Y. Leu, and C. L. Lee, *IEEE Trans. Elec. Dev.,* 34:1390 (1987)
331. W. L. Yang, T. F. Lei, and C. L. Lee, *Solid-St. Electron.,* 32:997 (1989)
332. S. S. Cohen, P. A. Piacente, G. Gildenblat, and D. M. Brown, *J. Appl. Phys.,* 53:8856 (1982)
333. L. van den Hove, R. Wolters, K. Maex, R. F. de Keersmaecker, and G. J. Declerck, *IEEE Trans. Elec. Dev.,* 34:554 (1987)
334. R. W. Bower, *Appl. Phys. Lett.,* 23:99 (1973)
335. K. Nakamura, S. S. Lau, M.-A. Nicolet, and J. W. Mayer, *Appl. Phys. Lett.*, 28:277 (1976)

336. C. Y. Ting, S. S. Iyer, C. M. Osburn, G. J. Hu, and A. M. Schweighart, *Proc. Electrochem. Soc.*, 82-2:224 (1982)

337. J. Amano, P. Merchant, T. R. Cass, J. N. Miller, and T. Koch, *J. Appl. Phys.*, 59:2689 (1986)

338. M. E. Alperin, T. C. Hollaway, R. A. Haken, C. D. Gosmeyer, R. V. Karnaugh, and W. D. Parmantie, *IEEE Trans. Elec. Dev.,* 32:141 (1985)

339. C. K. Lau, Y. C. See, D. B. Scott, J. M. Bridges, S. M. Perna, and R. D. Davies, *IEEE IEDM Tech. Dig.,* p. 714 (1982)

340. T. Okamoto, K. Tsukamoto, M. Shimizu, and T. Matsukawa, *J. Appl. Phys.,* 57:5251 (1985)

341. D. C. Chen, T. R. Cass, J. E. Turner, P. Merchant, and K. Y. Chiu, *IEEE IEDM Tech. Dig.,* p. 413 (1985)

342. H. K. Park, J. Sachitano, M. McPherson, T. Yamaguchi, and G. Lehman, *J. Vac. Sci. Techno.,* A2:264 (1984)

343. T. E. Tang, C.-C. Wei, R. A. Haken, T. C. Holloway, L. R. Hite, and T. G. W. Blake, *IEEE Trans. Elec. Dev.,* 34:682 (1987)

344. D. B. Scott, R. A. Chapman, C. Wei, S. S. Mahant-Shetti, R. A. Haken, and T. C. Hollaway, *IEEE Trans. Elec. Dev.,* 34:562 (1987)

345. F. d'Heurle, E. A. Irene, and C. Y. Ting, *Appl. Phys. Lett.*, 42:361 (1983)

346. K. Maex and R. F. De Keersmaecker, *Physica.,* B129:192 (1985)

347. M. Delfino, E. K. Broadbent, A. E. Morgan, B. J. Burrow, and M. H. Norcott, *IEEE Elec. Dev. Lett.,* 6:591 (1985)

348. E. Nagasawa, H. Okabayashi, and M. Morimoto, *IEEE Trans. Elec. Dev.,* 34:581 (1987)

349. T. Larsson, U. Wennström, H. Norström, H.-O. Blom, S. Berg, and I. Engström, *Solid-St. Electron.,* 32:385 (1989)

350. K. L. Wang, T. C. Holloway, R. F. Pinizzotto, A. P. Sobczak, W. R. Hunter, and A. F. Tasch, Jr., *IEEE Trans. Elec. Dev.,* 29:547 (1982)

351. H. Norström, K. Maex, and P. Vandenabeele, *MRS Symp. Proc.*, 181:173 (1990)

352. J. L. Regolini, D. Bensahel, G. Bomchil, and J. Mercier, *Appl. Surf. Sci.,* 38:408 (1989)

353. E. Mastromatteo, J. L. Regolini, C. d'Anterroches, D. Dutartre, D. Bensahel, J. Mercier, C. Bernard, and R. Madar, *Ext. Abs. Electrochem. Soc. Mtg.,* Seattle (1990)

354. Z. G. Xiao, G. A. Rozgonyi, C. A. Canovai, and C. M. Osburn, *MRS Symp. Proc.,* 202 (1991)

355. L. Van Den Hove, K. Maex, L. Hobbs, P. Lippens, R. De Keersmaecker, V. Probst, and H. Schaber, *Appl. Surf. Sci.,* 38:430 (1989)

356. J. R. Phillips, P. Revesz, J. O. Olawolafe, and J. W. Mayer, *MRS Symp. Proc.,* 181:159 (1990)

357. R. T. Tung, A. F. J. Levi, and J. M. Gibson, *Appl. Phys. Lett.,* 48:635 (1986)

358. S. M. Yalisove, R. T. Tung, and D. Loretto, *J. Vac. Sci. Technol.,* A7:1472 (1989)

359. S. M. Yalisove, D. J. Eaglesham, and R. T. Tung, *Appl. Phys. Lett,.* 55:2075 (1989)

360. B. D. Hunt, N. Lewis, E. L. Hall, L. G. Turner, L. J. Schowalter, M. Okamoto, and S. Hashimoto, *Mat. Res. Soc. Symp. Proc.,* 56:151 (1986)

361. J. Henz, M. Ospelt, and H. von Känel, *Solid State Commun.,* 63:445 (1987)

362. B. D. Hunt, N. Lewis, L. J. Schowalter, E. L. Hall, and L. G. Turner, *Mat. Res. Soc. Symp. Proc.,* 77:351 (1987)

363. J. Henz, M. Ospelt, and H. von Känel, *Surf. Sci.,* 211/212:716 (1989)

364. A. E. White, K. T. Short, R. C. Dynes, J. P. Garno, and J. M. Gibson, *Appl. Phys. Lett.,* 50:95 (1987), and *Mater. Res. Soc. Symp. Proc.,* 74:481 (1987)

365. J. C. Barbour, S. T. Picraux, and B. L. Doyle, *Mat. Res. Soc. Symp. Proc.,* 107:269 (1988)

366. A. H. van Ommen, J. J. M. Ottenheim, A. M. L. Theunissen, and A. G. Mouwen, *Appl. Phys. Lett.,* 53:669 (1988)

367. K. Kohlhof, S. Mantl, B. Stritzker, and W. Jäger, *Nucl. Instrum. Methods* , B39:276 (1989)

368. E. Rosencher, S. Delage, Y. Campidelli, and F. A. d'Avitaya, *Electron. Lett.,* 20:762 (1984)

369. M. L. A. Dass, D. B. Fraser, and C.-S. Wei, *Appl. Phys. Lett.,* 58:1308 (1991)

370. S. L. Hsia, T. Y. Tan, P. Smith, and G. E. McGuire, *J. Appl. Phys.,* 70:7579 (1991)

# 4

# Insulator/Semiconductor Contacts

*Harry H. Wieder*

## 1.0 INTRODUCTION

The contact between insulators and semiconductors involves a number of complex issues which concern the bulk and interfacial structure, composition, electrical, optical and electro-optic characteristics of the insulators and semiconductors. One of the aims of past and current studies concerned with these issues is the development of unambiguous, realistic models of such contacts in terms of first principles. Another objective of such research is that of obtaining better technological control of dielectric passivation of electronic and electro-optic devices and integrated circuits. Dielectric materials in the form of thin layers are required for gate and field isolation in field effect transistors, and for optical waveguides; used as substrates such materials may serve to reduce the parasitic interaction between adjacent devices, to suppress short channel effects and stray capacitances in field effect transistors and to increase their radiation resistance. Such insulators may be either crystalline or amorphous; on a macroscopic scale they are likely to be homogeneous and isotropic. For technological applications the thermal expansion coefficients of the insulators and semiconductors need to be closely matched in order to prevent the formation of lattice defects at their interfaces. Such defects may affect adversely their insulating and passivation characteristics. If the insulator and semiconductor represent a crystalline heterojunction, then their lattice constants must be closely matched across the interface for the same reason. The insulator should have a large fundamental bandgap, a low doping density, and the band offsets between the insulator and semiconductor should be large enough to prevent the flow of any conduction currents. In either case, the properties of the interfacial transition region are crucial; they have been, and remain, under intensive investigation using a variety of experimental techniques. A review and assessment of their current status is one of the principal objectives of this paper.

### 1.1 The Ideal Semiconductor/Insulator Interface

In the context of this paper an ideal insulator is either an amorphous or a crystalline material with well defined dielectric properties, no free charge carriers, no fixed charges and no ionic or electronic space charge distributions. An ideal insulator, shielded from high energy perturbations, should sustain only an electric field-induced polarization and no remanent polarization. Its dielectric constant should have no frequency dispersion, at least not from DC through the microwave range. Unlike metal-semiconductor contacts, a contact between an ideal insulator and a semiconductor should not affect or alter the band-bending of the semiconductor surface in the absence of any electrical, optical or thermal excitation.

An ideal insulator-semiconductor (i-s) interface should contain no lattice defects or localized non-stoichiometric inclusions. The interface should represent a high energy barrier for electron injection into the semiconductor for high electric fields, optical or thermal excitations. Ideal insulator-semiconductor interfaces should be mechanically stable and chemically inert.

The thermally-grown Si/$SiO_2$ interface approaches, in many respects, the properties of ideal i-s interfaces; it serves as a template for emulation by other i-s interfaces and is one of the principal reasons for the dominance of the silicon technology in present-day microelectronics. The native oxides of other elemental or compound semiconductors are far from ideal. Although great strides have been made in understanding many aspects of passivation and technological control of compound semiconductor surfaces, a fundamental understanding of their i-s interfaces is still in a rudimentary stage compared to that of the $SiO_2$ system.

## 2.0 METAL-INSULATOR-SEMICONDUCTOR (MIS) STRUCTURES

MIS capacitors consist of an insulating layer grown or deposited on a semiconducting material, a metal electrode superposed upon it called a *gate*, and an ohmic contact applied to the obverse side of the semiconductor. Its electrical properties may be defined in terms of a static capacitance, $C_{stat} = Q_T / V_g$, where $Q_T$ is the total charge density on the capacitor and $V_g$ is the potential applied to it, and a differential capacitance, $C = dQ_2 / dV_g$ which is a nonlinear function of the applied bias. In an ideal metal-insulator-semiconductor (MIS) structure the potential division is

Eq. (1) $$V_g = -V_{FB} + (Q_s / C_i) + V_s$$

where $V_{FB}$ is the flatband voltage, $C_i$ is the insulator capacitance per unit area, $Q_s$ is the induced charge per unit area consisting of both mobile charge carriers and charge in the depletion layer, and $V_s$ is the surface potential of the semiconductor. The departure from ideality of MIS structures may be used to calculate theoretically and to determine experimentally the specific energy, distribution, capture cross-sections and time constants of bulk and interfacial trapping centers. The procedures, with specific reference to Si MOS structures, are described in detail in the monograph of Nicollian and Brews.[1]

MOS structures have, at room temperature, small, and in most instances negligible, DC drift currents. There is a barrier for electronic conduction through Si/$SiO_2$ interfaces of the order 3 - 5 eV. This barrier can be overcome by high field electron injection (in the dark) and by low field photoemission in the presence of ultraviolet light. If the electric field in the oxide is large enough, then electrons can tunnel through the triangular energy barrier at the interface producing a current which varies exponentially with the applied bias voltage, $V_g$. Electrons may gain sufficient energy from absorbed ultraviolet radiation to penetrate the insulating layer where their mobility may be of the order 20 $cm^2 / V_s$.

Avalanche breakdown which involves the creation of electron-hole pairs by scattering of individual electrons having kinetic energies greater than the fundamental bandgap takes place in Si; it does not occur in $SiO_2$ in fields up to 15 MV/cm and tunneling breakdown is not likely to occur for fields < 30 MV/cm.

## 2.1 Bulk, Surface, and Interfacial States

Attempts to reconcile experimental measurements with models based on ideal MIS structures led to the necessity for postulating the presence of an interfacial charge density. This charge density depends the physical and chemical properties of the i-s interfaces which are closely related to the technological procedures used for making them. A considerable research effort has been expended world-wide during the past three decades in an attempt to reach complete understanding and control of the properties of silicon MOS structures. Experimentally observed interfacial charge distributions have been explained in terms of

1. Fast surface states located at the Si/$SiO_2$ interface limited to a frequency response < 1 MHz
2. Fixed positive surface charge density located at, or near the interface
3. Fixed and mobile ionic charges (such as Na), located within the $SiO_2$ layer, produced by contamination during fabrication or induced by incident ionizing radiation
4. Impurity redistribution and defect generation during thermal oxidation of the Si surface

A basic assumption is that charge trapping occurs during current flow because mobile electrons or holes are captured by localized states associated either with defects present in the insulator or at the i-s interface. Such trapped charges cause instabilities in Si MOS structures because of charge-induced shifts of their flatband voltages. It is not possible to eliminate all of the charges trapped at an interface or those trapped within a bulk insulator as a result of charge injection. A defect free insulator is probably unattainable. However, silicon MOS devices with fewer than $10^{10}$ electrical defects/cm$^2$ have been produced. A reduction of device dimensions in integrated circuits by scaling rules to less than 1 μm requires a commensurate decrease in the thicknesses of their $SiO_2$ layers. This places a greater emphasis on the few nm containing the transition region between amorphous $SiO_2$ and crystalline Si.

Although there is a superficial resemblance between MOS and MIS structures, analyses and modeling based on similarities between their experimentally observed properties should be approached with caution. The interfacial characteristics of semiconducting III-V compound/insulator interfaces are more complicated than those of Si because native or thermally grown oxides, as well as most synthetic insulators deposited upon them, are not suitable for MIS applications. Such layers have, as a rule, much higher conductivities than $SiO_2$. Their MIS interfaces exhibit high densities of discrete fast surface states, energy distributions of such states and fixed charges whose parameters depend on the preparation of the semiconductor surfaces prior to deposition of the insulator, on the growth of the insulator and on subsequent annealing procedures.

## 2.2 Low Frequency Differential Capacitance

Let $V = V_{DC} + dV(t)$ be the potential difference between the metal electrode and the semiconductor, where $V_{DC}$ is the DC component and $dV(t) = V_O \sin \omega t$ is a small amplitude AC potential with $\omega$ chosen so that the surface space charge layer and surface states are in equilibrium with the applied potential. With

Eq. (2) $$dV(t) = dV_i(t) + dV_s(t)$$

(here $V_i$ is the potential drop across the insulator), a charge, $-dQ_T$, is induced on the metal plate and a corresponding charge, $dQ_T$, on the semiconductor surface, such that $-dQ_T = C_i\, dV_i(t)$ and

Eq. (3) $$dQ_T = dQ_{sc} + dQ_{ss}$$

The surface charge on the semiconductor, $Q_s$, is shared between the space charge, $Q_{sc}$ and the surface state charge, $Q_{ss}$. By defining the differential capacitances $C_{sc} = -dQ_{sc}/dV_s$ and $C_{ss} = -\,dQ_{ss}/dV_s$, the relation

Eq. (4) $$C_i\, dV_i = [C_{sc} + C_{ss}]\, dV_s$$

can be represented by the equivalent circuit shown in Fig. 1. The space charge capacitance, $C_{sc}$, may be expressed in the form

Eq. (5) $$C_{sc} = (\varepsilon_s / L_D)\,|dF_s/DV_s|$$

where $L_D$ is the Debye length, $\varepsilon_s$ is the dielectric constant of the semiconductor, and $F_s$ is a function obtained from a numerical solution of Poisson's equation in accumulation, depletion and inversion which must be calculated numerically; it is presented graphically in the monograph of Many et al.[2] The surface state capacitance for a discrete trap of density, $N_t$, and energy $E_t$ is

Eq. (6) $$C_{ss} = (q^2 / kT) \bullet N_t\, f_t\, (1 - f_t)$$

where the occupation probability, $f_t$, is

Eq. (7) $$f_t = \left(1 + \exp \frac{E_t - E_f - qV_s}{kT}\right)^{-1}$$

and for states having a distribution in several discrete energy levels, $E_t^i$ (i = 2, ... , n), $C_{ss}$ is

Eq. (8)
$$C_{ss} = \frac{q^2}{kT} \sum_{i=1}^{n} N_t^i f_t^i (1 - f_t^i)$$

This equation can be extended to surface states with a continuous distribution in energy such that

Eq. (9)
$$C_{ss} = \frac{q^2}{kT} \int_{-\infty}^{\infty} N_t(E_{ts})\, f_t(E_{ts}) [1 - f_t(E_{ts})] dE_{ts}$$

where $N_t(E_{ts})\, dE_{ts}$ represents the number of states in the energy interval $E_{ts}$, $E_{ts} + dE_{ts}$. If $N_t(E_{ts})$ varies slowly with $E_{ts}$ then it can be shown that $C_{ss}$ is approximately equal to $qN_t(E_F)$, the density of states at the Fermi level $E_F$.

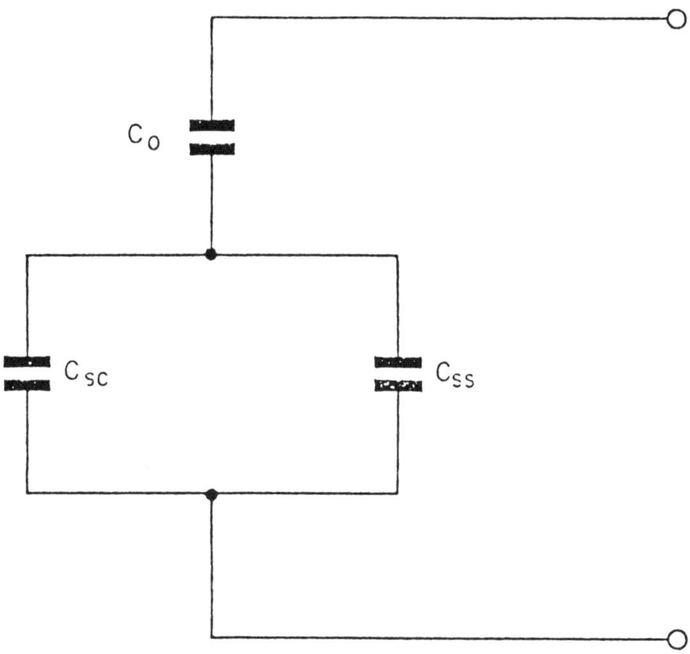


**Figure 1.** Equivalent circuit of MIS capacitor including space charge capacitance, $C_{sc}$, insulator or oxide capacitance, $C_0$, and surface state capacitance, $C_{ss}$.

## 2.3 High Frequency Differential Capacitance

In the case of accumulation or depletion, the space charge layer response to moderate or high frequencies is essentially the same as that at low frequencies. However, surface states have a finite relaxation time and their response to an AC excitation, $V_0e^{i\omega t}$ will decrease with increasing frequency if $1/\omega$ is comparable or smaller than the relaxation time, $\tau_0$, for electrons,

Eq. (10) $$\tau_0 = 1/[C_n (n_s + n_t)]$$

where $C_n = \sigma_n v_n$ is the probability per unit time for an electron to be captured by an empty surface state, with $\sigma_n$ the capture cross-section and $v_n$ the thermal velocity component of the electrons. The surface charge density, $n_s$, defined in terms of $N_c$, the density of states of the semiconductor and the thermal energy, kT, is

Eq. (11) $$n_s = N_c \exp[(E_F - E_c + qV_s) / kT]$$

$$n_t = N_c \exp[(E_t - E_c) / kT]$$

The admittance, $V_{ss}$, of the surface states as a function of $\omega$ can be expressed in the form

Eq. (12) $$Y_{ss} = [C_{ss} / (1 + \omega^2\tau_0^2)] (i\omega + \omega^2\tau_0)$$

it can be represented[1] by the lumped equivalent circuits, shown in Fig. 2a, in which

Eq. (13) $$C_p = C_{ss} / (1 + \omega^2\tau_0^2)$$

$$G_p = (\omega^2\tau_0 C_{ss}) / (1 + \omega^2\tau_0^2)$$

or the alternative circuit of Fig. 2b, in which the series resistance is

Eq. (14) $$R_s = \tau_0/C_{ss}$$

If the surface states have several energy levels, $E_t^i$ (i = 1, 2, 3, ..., n) then the total admittance can be represented by

Eq. (15) $$Y_{ss} = i\omega \sum_{i=1}^{n} C_p^i + \sum_{i=1}^{n} G_p^i$$

For a continuous distribution of surface states Lehovec has shown[3] that

Eq. (16) $$C_p = \frac{qN_t(E_F)}{\omega\tau_m}\tan^{-1}(\omega\tau_m), \quad C_p = \frac{qN_t(E_F)}{2\tau_m}\log(1+\omega+\tau_m^2)$$

where $N_t(E_F)$ is the density of surface states at the Fermi level and

Eq. (17) $$\tau_m = [C_n(E_t s)\, n_s]^{-1}$$

In inversion the high frequency space charge capacitance $C_{sc} = \varepsilon_s/w$; the depletion depth, w, increases with $V_s$ with the effective depletion depth w approaching the value $L_D(2\sqrt{U_b} + \sqrt{2})$ when $V_s$ becomes very large.

Fixed charges produce a shift in the flatband position of the capacitance and cause the parallel displacement along the voltage axes of C-V curves. Mobile ionic or electronic space charges in the insulator or charge carriers injected from the semiconductor into the insulator can produce hysteresis and drift of C-V parameters which depend on the magnitude and time scale of the applied electric fields. In some cases all of these processes may be present simultaneously creating considerable difficulties for unambiguous characterization of MIS structures.

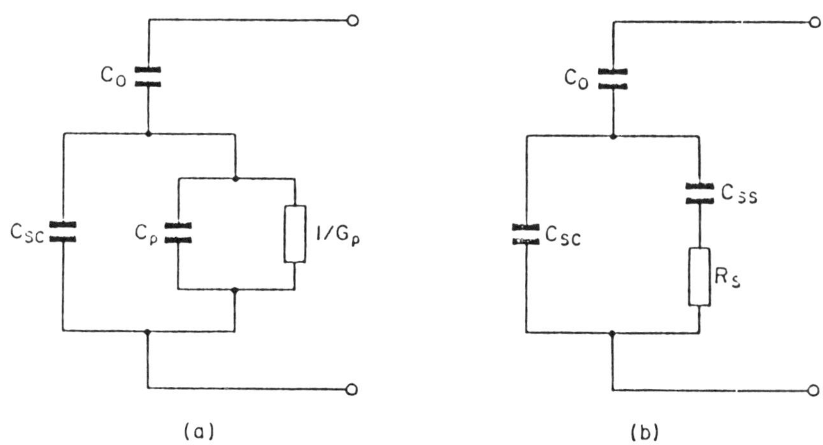


**Figure 2.** Equivalent circuit impedance of MIS capacitor in accumulation and depletion in terms of: *(a)* parallel capacitance, $C_p$, and parallel conductance, $G_p$; *(b)* in terms of a series resistance, $R_s$, and surface state capacitance, $C_{ss}$.

## 2.4 Characterization of MIS Structures

One of the most widely used methods for determining the high frequency capacitance vs. voltage (C-V) dependence of insulator-semiconductor interfaces is that developed by Terman[4] in which the capacitance is measured as a function of gate bias with a superposed small signal AC of frequency high enough so that no interface states can respond to it. However such states can follow slow changes in bias producing a distortion of experimentally measured, compared to theoretically calculated, C-V curves. By comparing a theoretically derived capacitance vs. surface potential curve with an experimentally obtained capacitance vs. gate voltage the dependence of $V_s$ on $V_g$ can be obtained and $C_{ss}$ ($V_s$) can be calculated by means of

Eq. (18) $$C_{ss}(V_s) = C_i\left[\left(\frac{dV_s}{dV_g}\right)^{-1} - 1\right] - C_{sc}(V_s)$$

One of the problems associated with such measurements is the frequency of the AC signal. While 1 MHz is adequate for silicon MOS structures, it is not suitable for measurements made on GaAs MIS capacitors; for the latter, frequencies in excess of 0.1 GHz are required to eliminate the response of any surface states. A detailed discussion of errors involved in high frequency C-V measurements appears in the monograph of Nicollian and Brews.[5]

Low frequency C-V measurements developed originally by Berglund[6] are based on the use of a low enough frequency so that all of the interface states are essentially in equilibrium with the AC excitation. Since the surface state capacitance is in shunt with the space charge capacitance, the low frequency capacitance is

Eq. (19) $$C_{LF} = [C_i\,(C_{sc} + C_{ss})] / (C_{sc} + C_{ss} + C_i)$$

and $C_{ss}$ can be determined from the measured values of $C_{LF}$ and the calculated $C_{sc}$ using Eq. (19). An alternative method used for this purpose is that described by Kuhn.[7] A linear voltage ramp is applied to the gate of an MIS structure whose back contact is connected to the input of a high impedance operational amplifier so that the output voltage of the amplifier is

Eq. (20) $$V_0\,(t) = -RC_{LF}\,(dV/dt)$$

A more accurate although a more complex method of measuring $C_{LF}$ is that developed by Ziegler and Klausmann.[8] It is based on measuring the potential V(t) developed across a bias-independent capacitor, $C_x$, in series with, and much larger than, the MIS structure. V(t) is measured as a function of a slowly rising ramp voltage so that the charge Q(t) = V(t) $C_x$. The Q-V data can then be used to determine the shape of the $V_s$ vs. $V_g$ dependence.

Commercially available admittance bridges can be used for measuring the frequency and gate bias-dependence of the differential capacitance and of the parallel conductance of MIS structures provided that they can cover an adequate frequency range and take due account of lead inductance. Such instruments should be capable of making three terminal capacitance measurements in order to eliminate parasitic stray capacitances to ground which, in some cases, may be comparable to the specimen capacitance.

Deep level transient spectroscopy (DLTS) is also a useful technique for determining[8]-[10] surface or interface states in MIS structures. It is based on filling such states in accumulation by means of an appropriate applied bias. Then the MIS capacitor is pulsed into depletion and, as the capacitance relaxes to steady state, a change in capacitance is observed because interface states are emitting their charge. The spectrum of their emission time constants can be obtained and, because the emission rate is thermally activated, the trap spectrum is obtained by scanning through a temperature range with different selected time delay windows. A problem associated with DLTS measurements is the analytical deconvolution of the time constant spectrum into an energy spectrum.

Illumination of MIS structures with bandgap radiation[11] provides additional possibilities for determining, independently, majority and minority charge carrier cross-section and photoexcitation of interface states. Infrared radiation in conjunction with electrical measurements has been used[12]-[14] to determine the cross-section of optical transitions near the Fermi level or quasi-Fermi levels. Internal photoemission can be used for determining the energy barrier of insulator-semiconductor interfaces or the energy barrier between a metal electrode and an insulator, such as the barrier determined by Williams[15] between the valence band of silicon and the effective conduction band of $SiO_2$. Fowler-Nordheim tunneling currents can provide information[16][17] on the depth distribution profile of interface states in thin MIS tunneling structures which have a high interface state density produced by carrier injection.

Other methods used principally for investigating surfaces, such as x-ray photoemission spectroscopy, (XPS)[18] have also been used principally for probing the properties of thin oxide or insulating layers on semiconductors. XPS is one of a family of spectrometric techniques which include Auger electron spectroscopy (AES), low energy electron loss spectroscopy (EELS) and ultraviolet emission spectroscopy (UPS). These methods employ a high resolution electron energy analyzer to determine the energy distribution of the electrons emitted by the high energy excitation. The application of XPS for determining the properties of i-s interfaces is complicated by the charging of the insulator, by lattice damage produced during depth profiling required in order to obtain depth-related data, and by radiation damage-induced perturbations produced by the high energy incident radiation.

## 3.0 THE Si/$SiO_2$ INTERFACE

The world-wide interest in the scientific and practical applications of the Si/$SiO_2$ technology is related to its use in electronic devices and integrated circuits. This has led to intensive and extensive research on the properties of interfaces between crystalline Si and amorphous $SiO_2$. The reduction in $N_{ss}$ which has occurred during the past three decades in this material system is related to procedural improvements in handling silicon wafers, preventing surface contamination prior to and during oxidation, and developing adequate post-oxidation annealing procedures. A comprehensive review by Deal[19] attributes fast states present at the Si/$SiO_2$ interface to dangling bonds, and trapped charge in the oxide to broken bonds in the oxide. Deal and others proposed that fixed oxide charges were due to excess Si or to an oxygen deficiency near the interface which might be circumvented by post-oxidation thermal annealing procedures. Experimental investigations using XPS revealed[20] that the Si/$SiO_2$ interface is abrupt on a scale approaching the $SiO_4$ tetrahedra, and flat to at least several hundreds of nanometers, that some of the interfacial bonds are strained, that non-stoichiometric bridges may be present within 0.6 nm of the interface, and that hydrogen impurities are incorporated as Si-H defects. However, the total concentration of hydrogen species determined chemically as Si-OH and loosely bound water is considerably greater than the concentration of electrically active defects. A comprehensive review by

Grunthaner and Grunthaner[21] places emphasis on the structure and composition of the interface based on data obtained by XPS, nuclear magnetic resonance ellipsometry and other experimental measurements and techniques. They visualize bulk $SiO_2$ to be made up of $SiO_4$ tetrahedra joined at the edges forming Si-O-Si bonds interconnected in a random network. Planar rings within the glass comprise 3 to 8 tetrahedral members. On the silicon side of the interface, there is evidence of localized strain in the vicinity of the chemical discontinuity within 15 - 30 Å of the interface. The chemical transition region is ~ 5 Å in extent and consists principally of a monolayer of oxidized Si. The extent of the interface roughness is related to the process of growing the oxide and to the net stress produced by the oxide at the interfacial boundary.

Photo-injection studies made on the Si/$SiO_2$ interface were performed by Williams,[15] Powell[22] and DiStefano.[23] The application of large electric fields across the oxide results in an interfacial potential barrier whose width and height decrease with increasing electric field, with the potential maximum in the oxide located within several angstroms of the interface.

Schwarz and Schulz have described in their review[24] the interface morphology and trap distribution of Si/$SiO_2$ interfaces. The perturbation in the electronic distribution and composition at the interface produce interfacial traps which interact with free charge carriers in the Si substrate. These traps contain majority carriers when the interface is in accumulation and the charge in these states changes as the surface is driven into depletion or inversion. The fixed and mobile charges in the bulk of the oxide do not interact with the semiconductor except after carrier injection induced by photoexcitation or high fields at elevated temperatures.

A typical high frequency C-V curve superposed on the low frequency curve of a p-type MOS structure produced by a slowly rising ramp voltage of 10 mV/s is shown in Fig. 3. A Terman-type analysis indicated that its minimum $N_{ss}$ is less than $10^{11}/cm^2$-eV while admittance and DLTS measurements show that $N_{ss}$ is of the order $10^8/cm^2$-eV. For specimens annealed in hydrogen at temperatures of the order 800 - 1000°C, the interface state density is of the order $10^{10}/cm^2$-eV in the vicinity of the Si conduction band decreasing as the Fermi level is displaced towards the valence band reaching a minimum of the order $10^8/cm^2$-eV. A typical capture cross-section for electrons near midgap is between $10^{-17}$ and $10^{-14}$ $cm^2$; it is of the order $10^{-19}$ in the vicinity of the conduction band edge.

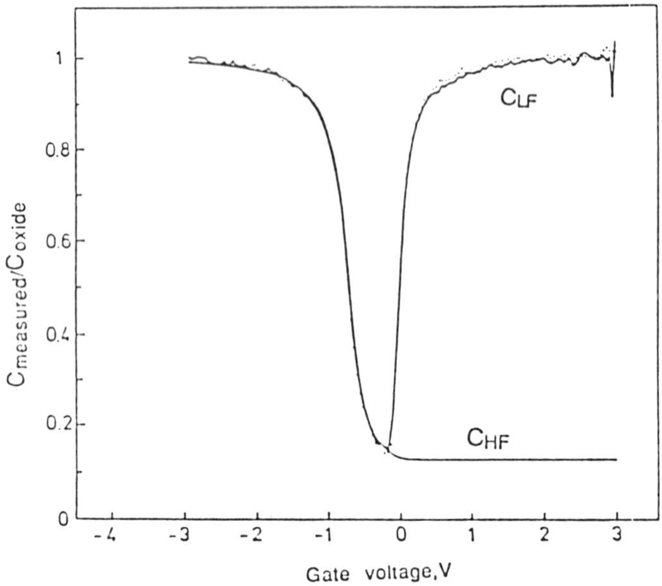


**Figure 3.** Superposed high frequency and low frequency capacitance vs. voltage curves of a p-type silicon MOS capacitor.

## 4.0 GALLIUM ARSENIDE-INSULATOR INTERFACES

The search for semiconductors with charge carrier transport parameters superior to those of silicon led to the evolution of the III-V intermetallic semiconductor technology and, as a corollary, to investigations of their s-i interfaces. GaAs is of specific interest for such applications, and homomorphic as well as heteromorphic insulators were grown or deposited on n and p-type GaAs. The interfacial properties of such MIS structures are still under active investigation world-wide. A recent review by Meiners and Wieder[25] contains a summary of past and current research and a critical evaluation of their properties.

Measurements made over an extended frequency range on GaAs MIS structures with an anodic oxide insulating layer are shown in Fig. 4. The quasi-static C-V data is very shallow and integration of the area between this curve and the horizontal line through the insulator capacitance intended to provide $V_s(V_g)$ suggests a total change in surface potential $\Delta V_s = 0.4$ V which is not consistent with modulation of the surface potential from accumulation to inversion; this requires $\Delta V_s = 1.2$ V. If a Terman analysis is made of the 1 MHz data using the electron density determined from Hall measurements then the total surface potential excursion is, again, $\Delta V_S = 0.4$ V. The data suggest that the Fermi level can be displaced from ~ 0.6 eV below the conduction band minimum (CBM) to ~ 1.0 eV below the CBM; therefore, no accumulation of electrons is feasible. Measurements made on p-type GaAs MIS structures in the frequency range from 0.0025 Hz to 150 MHz have demonstrated conclusively that $\Delta V_S$ = 0.4 V; thus neither accumulation or inversion are feasible due to pinning of the GaAs Fermi level. The properties of GaAs MIS structures cannot be described without considering the nature of Fermi level pinning, i.e., the obstacle imposed by various surface absorbates to the displacement of the surface potential within the fundamental bandgap of GaAs. These include most of the homomorphic and heteromorphic insulators grown or deposited thus far on GaAs. Figure 5 represents the characteristic dependence of fast interface states on the position of the surface potential within the bandgap of GaAs and demonstrates the nature of Fermi level pinning. Essentially similar results have been obtained for most homomorphic and heteromorphic GaAs MIS structures.

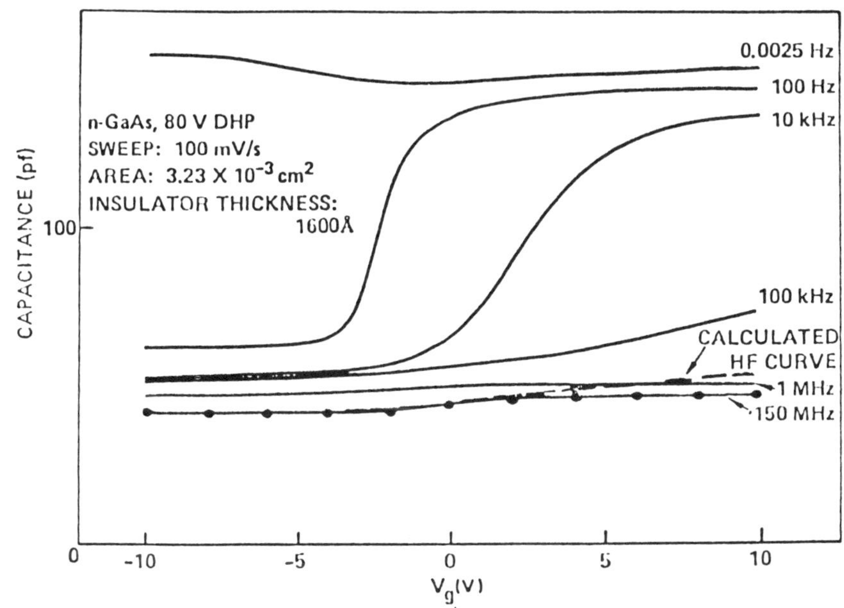


**Figure 4.** Measured capacitance vs voltage curves, with frequency as a parameter, of an n-type GaAs MIS structure with an anodized insulating layer; also shown is the calculated high frequency curve represented by the dashed line [after Meiners and Wieder, *Mater. Sci. Rep'ts,* 3, 139 (1988)].

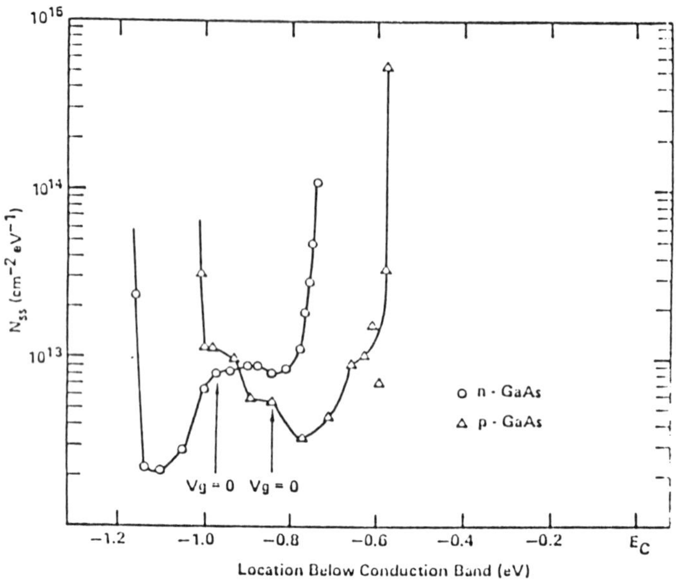


**Figure 5.** Surface state density as a function of surface potential of n-type and p-type GaAs derived from high frequency C-V measurements [after L. G. Meiners, "Dielectric-semiconductor interfaces of GaAs and InP," Ph.D Dissertation, Colorado State University, SF-19 (1979)].

The original and enhanced model of Spicer et al.[26] attributes Fermi level pinning to lattice defects induced by absorbates. The disorder induced gap states model of Hasegawa and Ohno[27] ascribes Fermi level pinning to band tailing within the forbidden gap. Tersoff's model[28] involves an intrinsic charge neutrality level whose ability to affect band-bending depends critically on the boundary conditions between the GaAs and its absorbate. The effective work function model[29] argues that Fermi level pinning is due to the presence of arsenic clusters at the native oxide interface. Anomalous frequency dispersion of the accumulation capacitance observed on GaAs/$SiO_2$ MIS structures has been attributed to the presence of an interfacial native oxide layer.[30]

A technique used by Hattangady et al.[31] has led to a sharp reduction of interface state densities with a dramatic unpinning of the Fermi level. It consists of the deposition of a thin, 10 to 20 Å thick, Si layer on a clean GaAs surface prior to the remote, plasma enhanced, low temperature, chemical vapor deposition (CVD) of the $SiO_2$ insulating layer. They argue that the Si interlayer serves the dual purpose of providing a heterojunction interface which, as in the case of $Ga_x Al_{1-x}As/GaAs$ heterojunctions, leads to low interface state densities and second, that it preserves the excellent dielectric properties of the $Si/SiO_2$ interface. It is similar to the earlier work of Tiwari et al.[32] who used molecular beam epitaxial (MBE) techniques to deposit the GaAs/Si heterostructure. They found, for p-doped GaAs, a flatband voltage of ~ 1.5 V with a hysteresis of ~ 1 V and an $N_{ss}$ of the order $10^{12}$ $cm^2$-eV; however, instabilities in the quasi-static C-V data suggested the presence of a high density of deep level traps. Hattangady et al.[31] determined, from HF and quasi-static C-V measurements, $\Delta V_s = 0.83$ V, a hysteresis of 0.77 V for a ramp rate of 10 mV/s at 10 MHz suggesting that the introduction of a Si intermediate layer eliminates Fermi level pinning producing, in consequence, a surface potential displacement from accumulation to inversion.

Is the heterojunction interface responsible for the reduction of $N_{ss}$ and unpinning of the Fermi level? The XPS and the spectroscopic ellipsometric measurements of Freeouf et al.[33] suggest that the reduced $N_{ss}$ is not necessarily due to the presence of unoxidized silicon; in any event, not in the amount necessary to form a Si/GaAs heterojunction. However, one cannot exclude the possibility that a $SiO_x$ compositionally graded interface might be responsible for the beneficial reduction in $N_{ss}$, the increase in $\Delta V_s$, and the unpinning of the Fermi level.

There are several factors which can contribute to unsatisfactory GaAs/ heteromorphic insulator interfaces; for example: etch-cleaning the semiconductor surfaces prior to deposition may generate intrinsic defects or may introduce new ones. Organic liquids and acids may inadvertently leave a residue of surface contaminants. The insulating layer may develop voids, it may be nonstroichiometric, or it may include impurities such as C or H; there also may be parasitic and subcutaneous interactions at the interface between the semiconductor and the deposited insulator such as oxidation reactions which generate a thin layer of native oxide under the heteromorphic layer. This may have a significant effect on the electrical properties of MIS structures as shown by Lucovsky et al.[34] who found that such native oxides generally degrade the electrical characteristics of MIS structures. Stresses between heteromorphic insulating layers and their GaAs substrates may also affect the electrical properties of MIS structures.

Hashimoto et al.[35] found by means of Raman spectroscopy that the stresses between GaAs encapsulated in $Si_3N_4$ layers depend on the thicknesses of these layers. They attributed the correlation between the stress and Raman spectra to the variation in depth of the space charge depletion layers modified by stress. Klose et al.[36] have investigated the properties of $Si_3N_4$ layers on GaAs grown by pyrolyzing $SiCl_4$ and $N_2H_4$ using either Ar or $N_2$ as the carrier gas. The C-V curves of their MIS structures show negligible dispersion measured at 3, 5 and 10 MHz suggesting that accumulation may have been obtained in positive gate voltages. GaAs MIS structures with amorphous $Al_2O_3$ insulating layers have been made[37]-[39] by the pyrolysis of aluminum isopropoxide in argon and by the plasma oxidation of vacuum or MBE-deposited Al layers. No evidence has been presented that such MIS structures have a lower $N_{ss}$ or a greater displacement of $V_s(V_g)$ than those of other heteromorphic MIS structures. Pincik et al.[40] have determined some of the properties of such MIS structures using DLTS. They found an effective activation energy for the dominant DLTS peak, $E_c$ - 0.75 eV, and attributed it to the out-migration of Ga atoms through the oxide layer, and to the formation of a high resistivity oxygen-compensated interfacial layer between the silicon nitride insulator and the GaAs substrate.

It is no mere coincidence that the position within the fundamental bandgap of GaAs at which the slopes of $N_{ss}(V_s)$ reach maxima in Fig. 5 are in reasonable agreement with the Fermi level pinning position determined[41] on (110)-surface of ultra-high vacuum cleaved GaAs coated with metals such as Cs, or to chemisorption of a fractional monolayer of oxygen. It suggests that, for oxides or heteromorphic insulating layers on GaAs, it is the density of surface states on the GaAs surface rather than the properties of the insulator which determine Fermi level pinning. Chemical treatment of GaAs surfaces which eliminates native oxides and residual contaminants was considered to be a requirement for reducing $N_{ss}$ and Fermi level pinning. However, there might be more than one origin for the states which affect the experimentally observed properties of MIS structures. The XPS measurements of Muench et al.[42] and Ismail et al.[43] indicate that the Fermi level position is dependent on the thickness of the oxide layer, suggesting that interface states and states within the oxide might contribute to the experimentally observed $N_{ss}$ .

Offsey et al.[44] used a photochemical method based on the dissolution of the fraction of the native oxide containing $As_2O_3$ (and of any As clusters present) in oxygenated water in the presence of light to obtain a substantial reduction in the nonradiative surface recombination velocity; the resultant quantitative reduction, at equilibrium, in the density of interface states is still in dispute. If the applicable solid state reaction is 2GaAs + $As_2O_3 \rightarrow Ga_2O_3$ + 4As, then removal of As from the interface should leave a layer of $Ga_2O_3$ and a low $N_{ss}$ should be present at the GaAs/$Ga_2O_3$ interface. A reduction in $N_{ss}$ might be associated with GaAs/$Ga_2O_3$

interfaces. Callegari et al.[45] have investigated the properties of MIS structures made by the reactive electron beam-deposition of $Ga_2O_3$ on GaAs following the plasma cleaning of its surface in $H_2$ and $N_2$. Their C-V measurements, made only at 1 MHz, show a large modulation of the differential capacitance as well as hysteresis which suggests the presence of slow states. Their claim of reaching inversion and accumulation, effectively reducing $N_{ss}$ to less than $10^{12}$ $cm^2$-eV, based only on a comparison of their measurements with a theoretical high frequency C-V curve, cannot be considered as conclusive at this stage.

More persuasive are the results obtained by Yablonovitch et al.[46][47] using optically injected charge carriers into alkaline sulfide-covered GaAs surfaces to obtain substantial modulation of the surface potential as a function of the intensity of the incident radiation. They derived band-bending from the change in conductance affected by optical modulation of the depletion depth of a GaAs layer covered by a polycrystalline layer of $Na_2S \cdot 9H_2O$. This layer is deposited on the GaAs surface from a $Na_2S$ aqueous solution which is then spun dry, at 5000 rpm. Immediately after deposition the total interfacial charge density of such a structure is estimated to be $\sim 5 \times 10^{11}/cm^2$. However, this condition is not stable with time; after 20 to 30 minutes the band-bending increases to a value comparable to that of other GaAs/insulator interfaces whose surface Fermi level is pinned.

Results comparable and superior to the above were obtained by Sandroff et al.[48] by first depositing elemental selenium from an aqueous solution on a GaAs surface followed by immersing it into a sodium sulfide solution. This dissolves the Se but, as determined from XPS measurements, leaves upon the surface a number of chemisorbed complex selenides and sulfides. Such chemically modified surfaces have their photoluminescence efficiency enhanced by a factor of four hundred compared to untreated surfaces and a band-bending, estimated from Raman scattering measurements, of ~ 0.1 eV. This suggests a considerable reduction of the interface states, at least between midgap and the CBM. This condition lasts somewhat longer that obtained with $Na_2S \cdot 9H_2O$ layers, but it is still time dependent.

## 5.0 INDIUM PHOSPHIDE-INSULATOR INTERFACE

InP-insulator interfaces and MIS structures are, in most respects, different from those of GaAs primarily because $N_{ss}$, at midgap, is more than one order of magnitude smaller in InP than in GaAs, and because chemical surface preparation procedures can be used to overcome Fermi level pinning, which is in the vicinity of the CBM rather than at midgap as in GaAs.

Slow states present at the interface within InP MIS structures produce C-V hysteresis, as shown in Fig. 6, and a slow drift of the flatband voltage. The model proposed[49] by Van Vechten and Wager attributes these effects to hopping of one of the four nearest neighbor In atoms into a P vacancy thus creating a defect complex consisting of an In vacancy and an InP antisite defect. Because such a reaction requires four electrons, it is considered to occur when the MIS structure is biased into accumulation. An alternative explanation for interface state generation has been proposed by Gergel and Solyakov[50] who suggested that trapping of electrons in the accumulation regime may, itself, be responsible for the electrostatic generation of new traps.

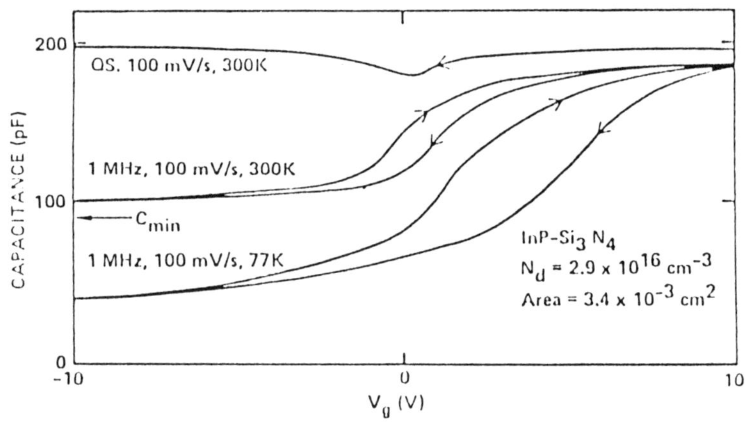


**Figure 6.** Dependence of the capacitance on gate voltage of an InP/$Si_3N_4$ MIS structure. The InP surface was etched in hydrazine prior to the deposition of the insulating layer [after Meiners and Wieder, *Mater. Sci. Rep'ts,* 3, 139 (1988)].

A model for the slow drift of the channel current of metal-insulator InP field effect transistors (MISFET) has been proposed by Goodnick et al.[51] in terms of thermally assisted electron tunneling to a conducting $In_2O_3$ layer within the native oxide interfacial layer of MIS structures rather than to discrete trap levels. Assuming plausible values for the bulk and interfacial parameters, they were able to simulate the time dependence of the drift; however, their model cannot explain the absence of such drift at low temperatures.

Van Staa et al.[52] have proposed a model for this drift based on the assumption of traps distributed in space and in energy within the insulator in the immediate vicinity of the interface. Electrons in these traps interact with the InP conduction band by tunneling during their capture and by tunneling in conjunction with a thermal activation process during electron emission. Their own constant capacitance DLTS measurements made on InP/$SiO_2$ MIS structures, as well as those of Meiners,[53] who found a linear relation between interfacial charge and the electric field induced across the insulator on a variety of homomorphic and heteromorphic structures, appear to be in good agreement with the features of this model. However, information on the specific location of these traps is lacking. Kulish and Kassing[54] arrived at the conclusion that these are located within a thin native oxide present at the interface between the heteromorphic $SiO_2$ insulator and the InP substrate. They also found that low temperature plasma enhanced chemical vapor phase deposition (PECVD) of $SiO_2$ on InP in the presence of P vapor yields slow state concentrations of the order $10^{11}$/ $cm^2$-eV.

The composition of the $SiO_2$/InP interface was investigated by Wager et al.[55] using photoemission and electron loss spectroscopy. They found an intermediate native oxide consisting of a thin, primarily $InPO_4$, layer and a thicker $In_2O_3$ layer near the InP surface. This composition is similar to that of thermally-grown oxides on InP. Liliental et al.[56] found that such oxides grown at 350°C consist of a 3 - 7 nm thick $InPO_4$ and a 6 nm thick outer layer of $In_2O_3$.

Low coverage (110)-oriented oxygen-exposed n-InP surfaces are depleted; this surface depletion is attributed to an acceptor level located at 0.45 eV relative to the CBM.[57] Increasing the oxide thickness produces Fermi level pinning at ~ 0.2 eV relative to the CBM. Experimentally obtained evidence[58][59] suggests that removal of this oxide returns the equilibrium surface Fermi level to 0.45 eV. This suggests that the shallow donors located at ~ 0.1 eV relative to the CBM which compensate the 0.45 eV acceptors might be located in the oxide rather than on the InP surface. The position of the equilibrium Fermi level is consistent with that measured on (100)-cleaved bulk InP by Spicer et al.[60] The position of the low oxide-coverage equilibrium surface Fermi level of ~ 0.5 eV below the CBM is consistent with theoretical calculations[61] of energy levels associated with donor-like phosporus vacancies, Vp, and with acceptor-like cation antisite defects, InP.

The surface potential modulation achieved with InP MIS structures using $SiO_2$ as an insulator depends[62] on the deposition process, as shown in Fig. 7; it varies from about 0.4 eV to 1.2 eV while that with $Si_3N_4$ insulating layers is ~ 0.6 eV. The typical dependence of $N_{ss}$ on the surface potential, $V_s$, is U-shaped with a steep rise in $N_{ss}$ for $E_c - E \geq 0.4$ eV, a value of the order $10^{11}$/ cm$^2$-eV at $E_c - E \geq 0.25$ eV and a well defined peak at 0.1 eV below the CBM.

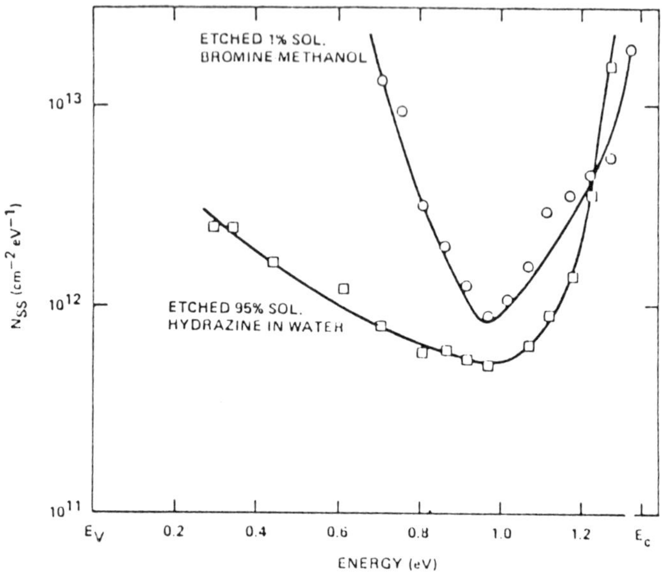


**Figure 7.** Surface state density as a function of energy within the fundamental bandgap of InP/$SiO_2$ MIS structures subjected to different surface treatments [after L. G. Meiners, *J. Vac. Sci. Technol.* 19, 373 (1981)].

An indirect plasma-enhanced low pressure chemical vapor deposition system was used by Chang et al.[63] to grow In $P_xO_y$ insulators on InP. Such layers were found to have resistivities in the range of $10^{13}$ to $10^{15}$ ohm-cm, fast interface state densities, determined from C-V measurements, with a minimum of the order $10^{10}$ /cm$^2$-eV, rising towards the valence band, and conduction band edges in excess of $10^{12}$ / cm$^2$-eV, with a hysteresis of the high frequency C-V curves of 0.2 eV for a gate voltage swing between +5 V and -5 V.

Hasegawa and Sawada[64] have proposed that the existence of a non-localized density of interface states distribution in experimentally measured MIS structures, with anodized $Al_2O_3$ insulating layer on InP, is due to the presence of a highly disordered non-stoichiometric interfacial region.

Hasegawa and Ohno have attributed[65] their experimentally observed wide distribution of interface states to interfacial disorder-induced gap states. However, Iliadis et al.,[66] using DLTS measurements made on $SiO_2$/InP MIS structures, found not a continuous energy distribution of interface traps but distinct states clustered around three discrete electron traps: a trap with an energy distribution, centered near midgap, attributed to intrinsic surface states, $N_{ss}$ = 2.2 to 3.6 x $10^{14}$/cm$^2$-eV; a second trap (observed only in MBE-grown InP) located within 0.22 to 0.24 eV of the CBM, $N_{ss}$ = 0.23 to 3.4 x $10^{14}$/cm$^2$-eV; and another trap located between 0.44 eV and 0.58 eV of the CBM, $N_{ss}$ = 2.2 to 2.4 x $10^{12}$/cm$^2$-eV. In the vicinity of the valence band, Lepley et al.,[67] using optical DLTS, found $N_{ss}$ > 5 x $10^{13}$/cm$^2$-eV at $E - E_V < 0.1$ eV which is in reasonable agreement with the data of Moisson et al.,[68] $N_{ss}$ > 3 x $10^{13}$/cm$^2$-eV for oxygen treated or ion bombarded InP surfaces.

Chemical treatment of InP surfaces before deposition or growth of a homomorphic or a heteromorphic insulating layer can have significant and beneficial effects on the properties of MIS structures. Wilmsen et al.[69] have investigated sulfurized InP surfaces by means of x-ray and Auger photoelectron spectroscopy. Their data suggests that sulfur replaces surface phosphorus or fills surface P vacancies forming an $In_2S_3$ /InP heterojunction. The MIS and MISFET properties of devices made with a sulfur pretreatment investigated by Iyer et al.[70] have $N_{ss}$ in the range $10^{10}$ to a few $10^{11}$/cm$^2$-eV with low to moderate drain current drift of < 5% over a period of 12 hrs.

Annealing of InP surfaces in an overpressure of phosphorus prior to the deposition of an insulator produces intermediate polyphosphide layers which have beneficial effects on the insulator/semiconductor interfaces by reducing $N_{ss}$ as shown, for example by Pande et al.[71] The properties of such polyphosphides, principally $In(PO_3)$, grown by wet anodization were investigated by means of x-ray photoemission and ellipsometric measurements by Robach et al.[72] They found this polyphosphate to be homogeneous, chemically stable and nonreactive; it has a high resistivity, large energy gap and a low $N_{ss}$ at InP interfaces.

Chemical modification of an InP surface prior to the deposition of an $Al_2O_3$ insulating layer by subjecting it to an As overpressure of $10^{-6}$ to $10^{-4}$ torr in a vacuum chamber held at $10^{-9}$ torr or by etching it in a $H_3AsO_4$ solution was investigated by Chave et al.[73] They found that this leads to reduction or elimination of the frequency dispersion of the capacitance in accumulation and depletion, a reduction in $N_{ss}$ near the conduction band edge and, in the upper half of the bandgap of InP, the elimination of hysteresis and a reduction in the time-dependent drift. These results are similar to those reported by Yamaguchi[74] with $PAs_xN_y$ layers as the insulator in InP MIS structures.

The dielectric and interfacial properties of plasma-grown oxides on InP were investigated by Al-Refaie[75] using conductance measurements and Cole-Cole-type plots of the frequency dependence of the complex dielectric constant of InP metal-insulator-metal (MIM) and MIS capacitors. He found a broad distribution of relaxation times in the range from a fraction of a Hz to tens of kHz. He attributed the frequency dispersion to trapping of electrons in localized energy levels of the oxide bandgap as well as at the InP interface. At low fields the electrons are injected from the Al electrode by tunneling into the oxide gap states, and at high fields they are injected directly into the oxide "conduction band" over the Al/oxide energy barrier. Acting as a stored charge, these trapped electrons would then modify the effective electric field in the oxide.

Low temperature (25 - 300°C) growth of silicon dioxide from monosilane and nitrous oxide using a mercury vapor-sensitized photochemical vapor deposition process, which produces near ideal C-V characteristics in narrow bandgap semiconductors such as InSb, is satisfactory but not particularly outstanding as an insulator when applied to InP. Su et al.[76] have shown that such MIS structures have a minimum in $N_{ss}$ of $1.2 \times 10^{11}$/cm$^2$-eV. Their high frequency C-V curves exhibit hysteresis and they have an oxide breakdown strength of $7.0 \times 10^6$ V/cm. Thus, preservation of the interfacial stoichiometry presumably maintained during low temperature insulator processing may be a necessary although not a sufficient condition for approaching ideal InP MIS and MISFET characteristics. The variety of experimental observations made on the static and dynamic characteristics of MIS structures using different chemical treatments of the InP surfaces prior to and/or during deposition of heteromorphic insulators could, therefore, be related to the evolution of the different interfacial oxide phases as well as to defect-related complexes. A few monolayers or even a fractional monolayer of such an oxide may be sufficient to induce a high density of acceptor and donor interface states which sets the equilibrium position of the surface Fermi level and determines its dynamic displacement. The motion of the Fermi level as a function of oxide thickness with increasing exposure to oxygen is complex; it moves, initially, away from and then back towards the CBM as different oxide phases are formed. The equilibrium position of the Fermi level may also be affected[77] by charges incorporated in the oxide.

## 6.0 HETEROJUNCTION QUASI-INSULATOR INTERFACES

The conditions which must be satisfied in order to provide an adequate crystalline heterojunction interface include the need for matching the lattice constants across the junction or, if that is not feasible, then providing elastically deformed (pseudomorphic) interfaces which prevent the generation of interfacial defects. Such defects may relieve some or all of the interfacial mismatch strain and produce, in consequence, interface states associated with such defects. If a heterojunction consists of a large bandgap, totally depleted, semiconductor in contact with a lower bandgap semiconductor whose surface charge can be depleted, accumulated, and inverted, and if the band edge discontinuities between them are large enough to prevent the flow of any conduction current, then, ostensibly, the large bandgap material performs the function of an insulator or that of a quasi-insulator, i.e., it performs the function of a dielectric material in contact with a semiconductor. If such a structure is provided with metal electrodes, then its C-V characteristics can be described in conventional MIS terms.

The alkaline earth fluorides, such as $CaF_2$ and $SrF_2$ satisfy some of these requirements. They do not dissociate when evaporated in high vacuum; instead, they sublimate and can be used, therefore, to grow stoichiometric layers by MBE on other semiconductors. The $CaF_2$/Si and the $SrF_2$/Si interfaces have been investigated with a variety of surface and interface sensitive techniques. High resolution soft x-ray photoemission spectroscopy shows that most of the interface bonding occurs between Ca and Si which induces large energy shifts of the interfacial Ca and Si core levels. The valence band edge discontinuities determined from photoelectron spectra[78] are in the range between 7.3 and 8.3 eV.

The situation is much more complex in the case of compound semiconductors. The $SrF_2$/GaAs lattice mismatch is 2.56%; however, the lattice constants of $SrF_2$ and $CaF_2$ bracket that of GaAs. Assuming the validity of Vegard's law, $Ca_{0.44}Sr_{0.56}F_2$ can match the lattice constant of GaAs. Siskos et al.[79] have deposited by MBE such a solid solution fluoride on (100) GaAs substrates. Electron diffraction and Raman spectroscopic measurements have confirmed the epitaxial character and the transitivity of GaAs/$(CaSr)_2F_2$ heterojunctions. However, the difference in chemical bonding, essentially covalent in GaAs and ionic in the fluorides, as well as the much greater elastic stiffness of the fluorides, may be responsible for the TEM and RHEED observations[80]-[82] that the interfaces are not coherent.

C-V measurements made on $CaF_2$ /GaAs MIS structures by Sinhroy et al.[83] indicate the presence of a high density of interface states which pin the Fermi level of GaAs. In order to overcome these problems, Waho et al.[84] have grown pseudomorphic fluoride layers of (111) GaAs at substrate temperatures between 450 and 700°C. They used these to make MIS structures and determined the crystallographic and electrical properties of the insulator and of its interfaces. They found that two-dimensional nucleation of $CaF_2$ takes place on GaAs (111B)-surfaces and that temperatures in excess of 580°C are required to promote Ca-As bonding within the first monolayer and for subsequent heteroepitaxial growth of pseudomorphic layers. In contrast with this, two-dimensional growth becomes three-dimensional after the first monolayer on (100)-oriented GaAs surfaces. C-V measurements made on such MIS structures exhibit, at room temperature, a counterclockwise hysteresis suggesting the presence of mobile ions within the insulator and a carrier injection-type clockwise hysteresis at 77 K. The accumulation capacitance remained constant from 120 MHz to 500 MHz and agrees with calculated values, and the C-V data suggest a substantial reduction in $N_{ss}$ between the CBM and midgap. It is not yet clear from their data whether a similar reduction is present between midgap and the valence band because, in high reverse bias, their measurements were affected strongly by leakage currents.

The interfaces formed between (110)-cleaved GaAs surfaces and $CaF_2$ or $SrF_2$ layers deposited in high vacuum were investigated by Mao et al.[85] They found that the deposition of the fluoride layers induced an initially large, 0.9 eV band-bending on p-type GaAs and a small initial band-bending on n-type GaAs. The final pinning positions of the Fermi level for about 1 ML of $CaF_2$ is at ~ 0.95 eV above the VBM for p-type GaAs and 0.25eV below the CBM for n-type GaAs. The band offsets are estimated to be 7.7eV and 8.0 eV for $CaF_2$ and $SrF_2$ , respectively. Unlike the $CaF_2$/Si interface they found no measurable Ca-F interfacial bonding.

A different approach was taken by Barriere et al.[86] They subjected (100)-oriented GaAs to 1 bar of fluorine at a temperature greater than 300°C to produce, on their surfaces, $GaF_3$ quasi-insulating layers. MIS structures made by such a fluoridation process exhibit a small hysteresis and well developed 1 MHz curves, however, the C-V data suggest the presence of fixed negative charges in the insulator, and the density of interface states as well as their energy distribution has, as yet, not been determined.

The 5.7% lattice mismatch between $BaF_2$ and InP and the 6.9% mismatch between $CaF_2$ and InP are not impediments in developing adequate MIS structures. Farrow et al.[87] have investigated the heteroepitaxial deposition by MBE of $BaF_2$ on InP. Sullivan et al.[88] found that $CaF_2$ layers can be grown on InP substrates heated to ~ 350°C. The MBE synthesis of $CaF_2$/InP/$CaF_2$ double heterostructures, as well as $Ba_xSr_{1-x}F_2$ / InP MIS structures, was investigated by Tu et al.[89] who found that the InP/$CaF_2$ heterojunction is not transitive. While the deposition of InP on $CaF_2$ is accommodated by pseudomorphic growth, the converse does not apply. The double heterostructures were found to have poor electrical characteristics suggesting the presence of deep level traps located either at the heterostructure interfaces or within the insulator. The synthesis and properties of $Ba_{1-x}Sr_xF_2$ on InP with fractional Ba concentrations, x = 0, 0.5 and 1 were also investigated by Paul and Bose[90] who found that such layers have a resistivity of the order $10^{12}$ ohm-cm and a breakdown strength > 5 x $10^5$ V/cm. Typical C-V characteristics showed negligible hysteresis, $\Delta V_s$ covering most of the fundamental bandgap and for thermally annealed $BaF_2$ MIS structures, $N_{ss}$ = 5 x $10^{10}$/cm$^2$-eV at $E_c$ - 0.47 eV. However, annealing produces a high density of cracks in the layers due to the difference in their thermal expansion coefficients and residual strain.

The application of smaller bandgap totally depleted quasi-insulators in MIS structures has been investigated extensively on the GaAs/$Ga_xAl_{1-x}As$ heterojunction system. Casey et al.[91] and Gossard et al.[92] have measured the I-V and the C-V characteristics of such two-terminal MIS structures. Similar measurements made as a function temperature by Solomon et al.[93] on $Ga_{0.6}Al_{0.4}As$/GaAs capacitors show that at low temperatures these behave like semiconductor-insulator-semiconductor (SIS) structures with interface barrier heights of 0.38 and 0.40 eV for the bottom and the top interfaces. The I-V curves exhibit rectification with thermionic emission in low bias values and the C-V curves exhibit depletion in reverse bias and accumulation > $10^{12}$ electrons/cm$^2$ in forward bias. Comparing SIS theory and experimental data, they suggest the presence of a negative fixed charge in the GaAlAs of ~ 4 x $10^{16}$/cm$^2$ which gives rise to a flatband voltage shift of 0.16 V. Delagebeaudeuf et al.[94] have observed Fowler-Nordheim tunneling in such SIS structures and resonant tunneling was obtained at 4.2 K by Hickmott et al.[95]

The role of interfacial defects in GaAs/GaAlAs heterojunctions was investigated theoretically by Zur and McGill.[96] They show that defect densities of ~ $10^{12}$/cm$^2$ can affect the position of the Fermi level at the interface, however, simple electrostatic considerations suggest that interface densities of the order $10^{13}$ - $10^{14}$/cm$^2$ are required to pin the Fermi level.

DLTS measurements were used by As et al.[97] to determine the presence of deep electron traps at the interface between n-isotype GaAs/

$Al_xGa_{1-x}As$ quantum wells with x in the range between 0.24 and 0.39. They found five deep electron traps, $E_c$ - $E_t$ , at 0.12, 0.22, 0.29, 0.52 and 0.63 eV, in addition to the well known DX center located on defects in the upper AlGaAs barrier layer, within a 15 nm wide region; these defects are attributed to less than optimum growth conditions for the first 10 - 20 nm $Al_xGa_{1-x}As$.

A different method of making such quasi-insulators, described by Casey et al.,[98][99] intended to avoid the formation of interface states at the insulator/semiconductor junction, is based on the use of oxygen-doped $Al_{0.5}Ga_{0.5}As$ as the insulator in GaAs MIS structures and Cr/Au metal electrodes. MBE growth conditions were so chosen that oxygen is readily incorporated into $Al_xGa_{1-x}As$ but not in GaAs. C-V measurements, such as that shown in Fig. 8, show no hysteresis. DLTS measurements made on such MIS structures indicate the presence of a relatively small interface trap concentration. Admittance measurements made as a function of frequency and temperature show a deep level at 0.64 ± 0.4 eV with a concentration of $4 \times 10^{16}/cm^3$ and the presence of space charge limited currents for either polarity bias voltage. The energy band diagram of such a structure is shown schematically in Fig. 9.

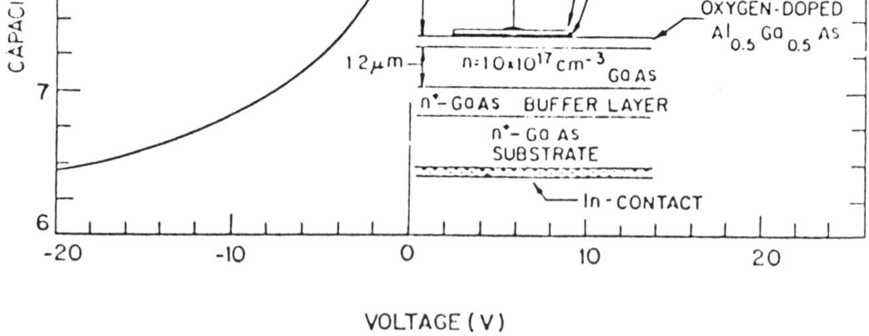


**Figure 8.** High frequency capacitance as a function of gate voltage for a GaAs MIS structure with an $Al_xGa_{1-x}As$ quasi-insulator [after H. C. Casey, et al., *J. Appl. Phys.* 50, 3484 (1979)].

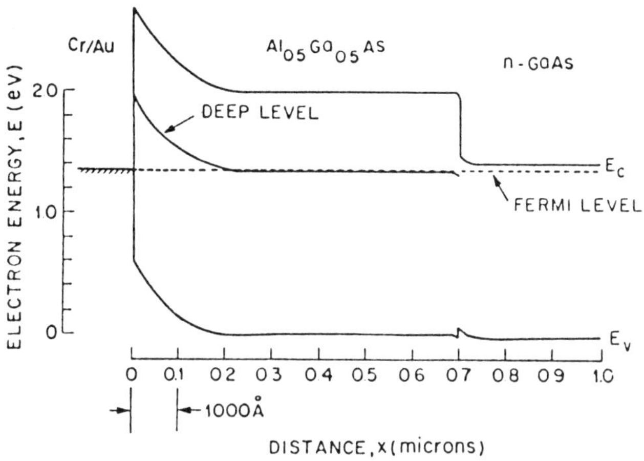


**Figure 9.** Energy-band diagram of a MIS heterostructure such as shown in the inset of Fig. 8 [after H. C. Casey, et al., *J. Appl. Phys.* 50, 3484 (1979)].

Subsequent measurements made by Casey et al.[100] have demonstrated the advantages of choosing a wide bandgap quasi-insulator. For the AlAs fraction x = 0.5 the fundamental bandgap is indirect, $E_g = 2.0$ eV , and the electron mobility of the n-type material is ~ 200 $cm^2$/V-s; the resistivity of thin layers is high enough for conduction currents to be negligible. Admittance measurements made on such structures suggest a simple equivalent circuit of the depletion capacitance in series with the quasi-insulator capacitance in low to moderate reverse bias, and the emission rate of the deep level traps in the $Al_{0.5}Ga_{0.5}As$ is too small at room temperature to contribute significantly to the MIS capacitance. Their data suggests that flatband conditions can be expected near zero bias, however no inversion layer is induced in high fields due to leakage of minority carriers through the quasi-insulator. The observed photoluminescence intensity of such structures is larger by a factor of fifty-two compared to that of bare GaAs, suggesting a reduction in the surface recombination velocity produced at the quasi-insulator/GaAs interface. Andre et al.[101] have performed further investigations on similar MIS structures, grown by MOVPE, with the $Al_{0.5}Ga_{0.5}As$ doped either with oxygen or chromium. In accumulation the measured capacitance corresponds to that of the calculated quasi-insulator thickness and dielectric constant. They confirmed the absence of inversion and find hysteresis to be present in the oxygen-doped, but not in the chromium-doped specimens. The latter were also found to have lower leakage currents and higher breakdown fields, > 7 x $10^5$ V/cm , than the O-doped MIS structures. The interface state density of heterojunction MIS structures of $In_{0.52}Al_{0.48}As/In_{0.53}Ga_{0.47}$ as grown on InP substrates by MBE were investigated by Hong et al.[102] using frequency-dependent admittance measurements. They found $N_{ss}$ to be in the range of 1 - 8 x $10^{12}$/cm$^2$-eV with emission time constants from these states of 100 to 700 ns. The interface trap density of the n-isotype $In_{0.1}Ga_{0.9}As$/GaAs heterostructures was investigated experimentally and modeled theoretically by Jeong et al.[103] by solving Poisson's equation with the interface trap density in the form of a box charge located at the heterojunction interface. They found qualitative agreement between experiment and theory if traps located at $E_c$ -0.13 eV with a concentration -8 x $10^{10}$/cm$^2$ in a 300 Å box-like distribution are introduced in the calculation. Experimentally, DLTS data suggests that the interface traps are located at $E_c$ -0.13 eV and at $E_c$ -0.17 eV with corresponding capture cross-sections of 2 x $10^{-14}$ and 1 x $10^{-15}$ cm$^2$ .

The properties of $In_xAl_{1-x}As$/InP MIS structures in which the quasi-insulator is deliberately lattice mismatched in order to gain an advantage in terms of a larger conduction band edge discontinuity with decreasing x were investigated by Hanson and Wieder[104] The equilibrium surface potential is $E_c$ -0.005 eV and the Fermi level can be displaced over most of the InP bandgap without, however, achieving inversion. No long term drift of the flatband potential was found[105] suggesting the absence of deep level interfacial traps. Figures 10 and 11 show that the density of fast interface states determined from C-V measurements is in the $10^{10}$/cm$^2$-eV at midgap, and admittance measurements made as a function of frequency suggest an energy-independent interface state density in the low $10^{11}$/cm$^2$-eV range.

Boron nitride/InP heterojunctions were made by Yamaguchi and Minakata[106] by a low temperature pyrolytic vapor deposition process using the reaction of $NH_3$, $B_2H_6$, and $PH_3$. XPS and AES measurements show the boron nitride layers to be $B_xN$ with 1.3 ≤ x ≤ 2. MIS heterostructures with boron nitride as the insulator (which has a fundamental bandgap of 4.1 eV and a dielectric constant of 3.28) were found to have a $N_{ss}$ of the order $10^{10}$/cm$^2$-eV , a breakdown strength > 3 x $10^6$ V/cm, and a resistivity > $10^{16}$ ohm-cm. Similar results with boron nitride layers deposited on InP were obtained by Paul et al.[107] using a laser assisted deposition process. On Al/BN/InP MIS structures they obtained $N_{ss}$ = 6.2 x $10^{10}$/cm$^2$-eV for $E_c$ -0.5 eV. Mizuta et al.[108][109] and Fujieda et al.[110] have grown aluminum nitride on N-type GaAs and on InP by a low pressure MOCVD process using trimethyl aluminum and hydrazine as the sources, respectively, of Al and of N.

Current voltage (I-V) characteristics of AlN/GaAs MIS structures exhibit ohmic transport for electric fields lower than $10^6$ V/cm and a resistivity > $10^{16}$ ohm-cm, and in higher fields the current is of Poole-Frenkel type. The dielectric constant of AlN is in the range between 5 and 7.5, depending on the substrate temperature used during growth, and its breakdown field is 4 x $10^6$ V/cm. Using C-V and isothermal transient capacitance spectroscopic measurements made on such MIS structures, two discrete levels were identified: one at $E_c$ -0.7 eV and another at $E_c$ -0.9 V. Between the conduction band and midgap $N_{ss}$ is of the order $10^{12}$/cm$^2$-eV and $\Delta V_s$ is nearly twice that of GaAs MIS structures using conventional homomorphic or heteromorphic insulators. On AlN/InP MIS structures preliminary C-V data suggests a minimum $N_{ss}$ of the order $10^{11}$/cm$^2$-eV. The effects of surface chemical treatment prior to the deposition of AlN on InP was investigated by Fujieda et al.[111] They found that a thermally induced surface reaction with $PH_3$ prior to the deposition of the AlN produces a large frequency dispersion in the depletion/inversion regime of the MIS structures while specimens treated with $AsH_3$ had a reduced frequency dispersion but a larger injection-type hysteresis.

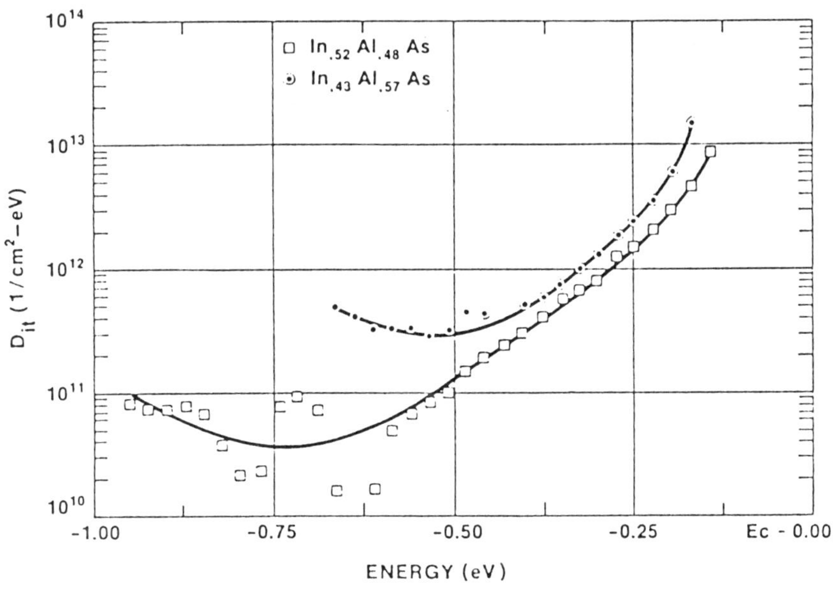


**Figure 10.** Interface state density as a function of energy for the lattice matched heterojunction InP/$In_{0.52}Al_{0.48}As$ and for a similar mismatched structure with an $In_{0.43}Al_{0.57}As$ quasi-insulator derived from C-V measurements [after C. M. Hanson and H. H. Wieder, *J. Vac. Sci. Technol.* B5, 971 (1987)].

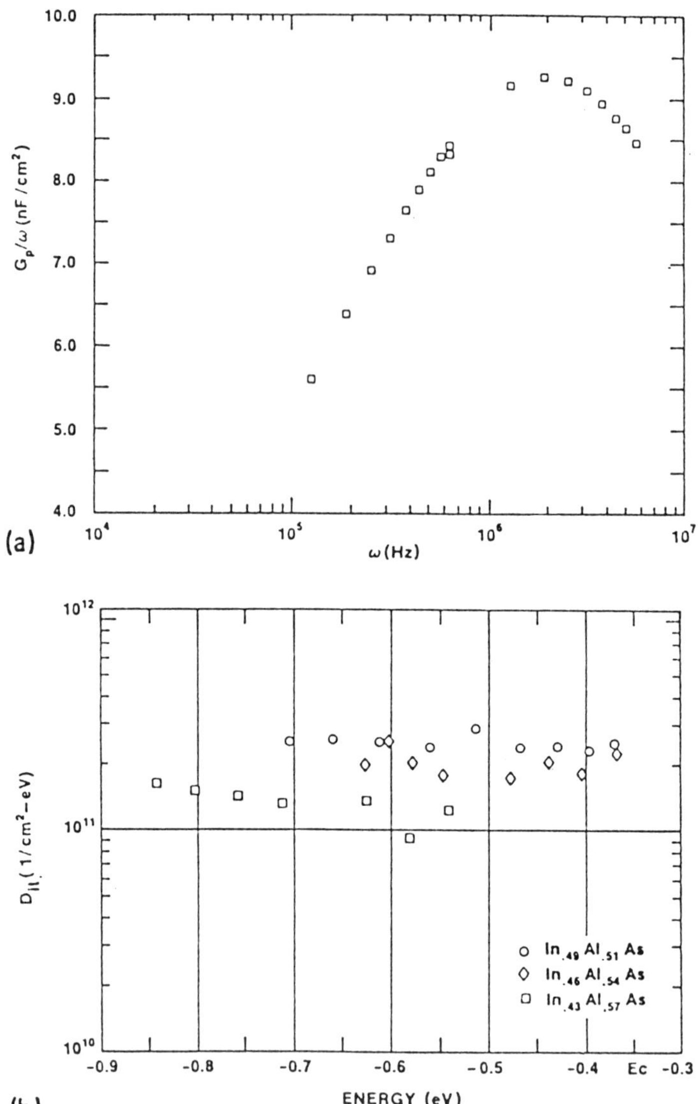


**Figure 11.** Conductance measurements on $InP/In_xAl_{1-x}As$ heterojunction MIS capacitors; *(a)* equivalent parallel conductance of an $InP/In_{0.49}Al_{0.51}As$ heterojunction capacitor as a function of frequency; *(b)* density of interface states as a function of energy for $InP/In_xAl_{1-x}As$ lattice-mismatched heterojunctions with various fractional indium concentrations and thicknesses of 0.1 µm of the quasi-insulators. The interfacial strain relaxation produced by the formation of lattice defects does not produce a substantial change in $N_{ss}$ [after C. M. Hanson and H. H. Wieder, *J. Vac. Sic. Technol.* B5, 971 (1987)].

The search for adequate gate insulators for GaAs MISFET has led Pruniaux et al.[112] and Fleming et al.[113] to investigate proton bombardment of GaAs surfaces with protons of 25 keV and fluence of $10^{14}/cm^2$ in order to produce quasi-insulating layers on GaAs surfaces. They found that the lattice disorder produced by the ion bombardment generated a low conductivity, essentially semi-insulating layer, between a metal gate and a GaAs substrate. Their C-V and I-V measurements made on such MIS structures with a resistivity of $10^8$ ohm-cm indicated the presence of a fixed positive interfacial charge ~ $5 \times 10^{11}/cm^2$. Detailed C-V measurements made by Meiners[114] as a function of frequency, up to 150 MHz, on GaAs surfaces subjected to Ar ion bombardment[115] fabricated into an MIS configuration indicate that, contrary to the results reported by Pruniaux et al., flatband is not reached, the Fermi level is pinned at $E_c$ 0.75 eV, and the surface charge density is of the order $10^{12}/cm^2$. This suggests that the lattice disorder produced by the $Ar^+$ bombardment is sufficient to generate a high enough density of interface states to pin the Fermi level.

One of the advantages of heterojunction/insulator and quasi-insulator MIS structures compared to conventional homomorphic and heteromorphic structures is their potentially unitary synthesis and growth; semiconducting, insulating and perhaps metallic layers might be deposited in some desired sequence upon each other without interruption or breaking vacuum integrity which might affect their interfacial properties. A further advantage of such structures is the absence of the long term drift and instability associated with conventional MIS structures. The energy difference between the larger bandgap insulator of quasi-insulator and the lower bandgap semiconductor as well as the band offsets between them must be large enough to prevent charge carrier conduction.

For a given doping level the quasi-insulator thickness should be smaller than its depletion depth; however, residual ionized impurities may lower the threshold for impact ionization and its breakdown strength. It is, therefore, desirable to reduce residual impurities to a minimum. Furthermore, it might be desirable to choose an indirect bandgap material for quasi-insulator purposes.

## 7.0 SOME AFTERTHOUGHTS

While a substantial amount of progress has been made during the past three decades in improving the $Si/SiO_2$ technology, primarily by empirical procedures, models of this interface on a microscopic level which correlate most of the available experimental observations are of comparatively recent origin. Neither cohesive models nor adequate technologies have been developed, as yet, for the interfaces between other insulators and semiconductors. Two different but not necessarily exclusive methods intended to provide comparable or, at least adequate, insulator/semiconductor interfaces have been investigated on GaAs and InP. One of these is based on the preservation of the interface between the native oxide or a selected phase of this oxide and the semiconductor surface, the other is based on a controlled modification of the surface chemistry of the semiconductor intended to prevent, or eliminate, pinning of the Fermi level by reducing $N_{ss}$ as well as avoiding the formation of deep level interfacial traps. Preservation of the native oxide is not a adequate method for eliminating Fermi level pinning in MIS structures. It comes as no surprise that the position of the equilibrium Fermi level in conventional homomorphic or heteromorphic MIS structures is consistent with that measured on metal/GaAs interfaces exposed to a fractional monolayer of oxygen, because most of the technological procedures employed in the synthesis of such layers involve a sufficient amount of oxygen to produce Fermi level pinning. Methods such as plasma cleaning may produce sufficient lattice damage and interface states pinning the Fermi level, in analogy to that produced by a high density of cleavage steps on the (110)-surface of GaAs.

Chemical modification of the GaAs surface by the chemisorption of S, Se, or Si prior to the deposition, or as a precursor to the deposition of heteromorphic insulating layers might overcome some of the impediments associated with their synthesis, growth and annealing procedures used thus far. It is too early for speculation on the prospects for obtaining, eventually, a stable, long term reduction of $N_{ss}$ and a $\Delta V_s$ range from accumulation through depletion, under equilibrium conditions.

The situation is much more favorable in the case of InP, in part because of the location of Fermi level; on both, n-type and p-type InP surfaces exposed to the ambient environment it is close to the CBM; and, in part due to the lower $N_{ss}$, smaller by an order of magnitude compared to GaAs. Removal of surface contaminants and special preparation procedures of InP surfaces, lowering the substrate temperature during growth of the insulating layers, the introduction of a phosphorus environment during growth and/or annealing have contributed to the elimination of Fermi level pinning, the reduction of interface state densities to low $10^{11}$/cm$^2$-eV and clear evidence that accumulation as well as inversion can be obtained with such MIS structures.

There is sufficient experimental evidence available to assign at least some of the interface states present on InP/insulator interfaces to lattice defects present at specific location within the fundamental bandgap of InP. However, they cannot be used to explain all of the experimental data, in particular that associated with the long term drift of the flatband voltage.[116]-[118] While the phosphorus vacancy hopping mechanism proposed by van Vechten and Wager[49] can explain the beneficial effects of processing MIS structures in a phosphorus environment, the activation energy of this process, estimated to be 1.2 eV, should lead to a strong temperature dependence of the drift instability, contrary to experimental evidence. The disorder induced gap states (DIGS) model of Hasegawa et al.[119] is in better accord with such observations. The main features of their model include the following considerations: *(a)* $N_{ss}(V_s)$ is U-shaped; the curvature of this relation and its minimum, $N_{ss}$(min), location within the fundamental bandgap of a specific semiconductor are variables which do not depend on singularities associated with discrete states; *(b)* the magnitude of $N_{ss}$ and the range of surface potential modulation depend on the nature of the insulator and on the insulator/semiconductor interfacial chemistry; *(c)* the distribution of time constants associated with the interface states is wide producing slow hysteresis drift as well as high frequency dispersion of the capacitance; *(d)* the location in energy of $N_{ss}$ (min) is characteristic of the specific semiconductor and it is the locus of the charge neutrality point of the interface state distribution.

Experimental data obtained by means of cross-sectional transmission electron microscopy and Rutherford backscattering measurements revealed a strong correlation between lattice disorder and $N_{SS}$. They add support to the DIGS model which proposes that random stress resulting from band mismatch modifies the interfacial bond length and the bond angle, producing gap states.

If the DIGS model is generic, then, as proposed by its authors, it ought to apply to heterojunctions as well as to MIS interfaces. However the high frequency, quasistatic, and isothermal capacitance transient spectroscopic measurements of Mochizuki et al.[120] made on AlN/GaAs MIS structures cast doubt on such an interpretation. They suggest that there might be two different types of energy distributions of the interface states near midgap in both n-type and p-type GaAs MIS structures: one is U-shaped and the other has discrete features. They find that the presence of a high density of discrete localized traps at 0.9 eV below the CBM favor a defect-related origin for the states that pin the Fermi level.

Only a limited number of investigations have been made, as yet, concerning the interfaces of crystalline insulator/semiconductor heterojunctions such as those employing lattice matched MBE-grown fluorides on GaAs or InP substrates. It is not known at this time if any long term drift and instability of the flatband voltage is also present in such structures although it is clear that heterojunctions which consist of a lower bandgap quasi-insulator/semiconductor, such as $In_xAl_{1-x}As/InP$, even if not lattice matched, have an $N_{ss}$ comparable with the best conventional MIS structures and also have a negligible drift instability. To first order, the lower resistivity of such quasi-insulators is not an impediment to their use in MIS structures, to the measurement and characterization of such structures in conventional terms, nor given appropriate band edge discontinuities, is it a hindrance in obtaining inversion or accumulation. Conventional homomorphic or heteromorphic insulators used in MIS structures require a high resistivity for the formation of an equilibrium inversion layer. It is not clear, at this time, if there is a minimum in resistivity required for inversion and the dependence of this minimum on the properties of the insulator and of the semiconductor. In heterojunctions the electrostatic forces associated with the band edge discontinuities and charge transfer across the quasi-insulator interface by modulation doping can produce an equilibrium accumulation or inversion layer on the semiconductor.

The drift and instability of the flatband voltage associated with conventional insulator-semiconductor interfaces is absent in metal-semiconductor interfaces, presumably because of the good communication between the traps and the conduction band of the semiconductor. In view of its negligible effect in some heterojunction interfaces, it is tempting to speculate on dependence of such instabilities on the dielectric relaxation time. A reduction in the dielectric relaxation time might be sufficient to modify the interfacial charge trapping and release time constants, and to reduce such drift to negligible proportions, in spite of the presence of high densities of interfacial traps.

Thus far, at least, the $Si/SiO_2$ remains unique. All other insulator/semiconductor interfaces have $N_{ss}$ greater by, at least, two to three orders of magnitude and no immediate prospects are available for reducing them to comparable values. While substantial progress has been made in determining many of the properties of such interfaces, a comprehensive model is not yet at hand.

## REFERENCES


1. E. H. Nicollian and J. R. Brews, *MOS (Metal Oxide Semiconductor) Physics and Technology,* John Wiley & Sons (1982)
2. A. Many, Y. Goldstein and N. B. Grover, *Semiconductor Surface,* pp. 138-148, North Holland Publishing Co, Amsterdam (1965)
3. K. Lehovec, *Appl. Phys. Lett.,* 8:48 (1966)
4. L. M. Terman, *Solid-State Electron.,* 5:285 (1962)
5. E. H. Nicollian and J. R. Brews, *MOS (Metal Oxide Semiconductor) Physics and Technology,* pp. 333-356, John Wiley & Sons (1982)
6. C. N. Berglund, *IEEE Trans. Electron Dev.,* ED-13:701 (1966)
7 M. Kuhn, *Solid-State Electron.,* 13:873 (1970)
8. K. Ziegler and E. Klausmann, *Appl. Phys. Lett.,* 26:400 (1975)
9. M. Schultz and E. Klausmann, *Appl. Phys.,* 18:169 (1979)
10. H. Goto, Y. Adachi, and T. Ikoma, *Jpn. J. Appl. Phys.,* 18:1979 (1979)
11. T. C. Poon and H. C. Card, *J. Appl. Phys.,* 51:588 (1980)
12. D. W. Grewe and W. E. Dahlke, *Inst. Phys. Conf. Serv.,* 50:107 (1979)
13. R. F. Pierret and B. R. Roesner, *Appl. Phys. Lett.,* 24:366 (1974)
14. J. G. Swanson and V. J. Montgomery, *Electron. Mat.,* 19:13 (1990)
15. R. Williams, *Phys. Rev.,* A140:569 (1965)
16. J. Maserjian and N. Zamani, *J. Appl. Phys.,* 53:559 (1982)
17. J. A. Shimer and W. E. Dahlke, *Appl. Phys. Lett.,* 40:734 (1982)
18. L. C. Feldman and J. W. Mayer, *Fundamentals of Surface and Thin Film Analysis,* pp. 213-230, North Holland, New York (1986)
19. B. J. Deal, *Electrochem. Soc.,* 121:198C (1974)
20. P. J. Grunthaner, R. P. Vasquez, B. F. Lewis, J. Maserjian, and A. Madhukar, *Phys. Rev. Lett.,* 43:1638 (1979)
21. F. J. Grunthaner and P. J. Grunthaner, *Mat. Sci. Rep.,* 1:69 (1986)
22. R. J. Powell, *J. Appl. Phys.,* 40:5093 (1969)
23. T. H. DiStefano, *Appl. Phys. Lett.,* 19:280 (1971)
24. S. A. Schwarz and M. J. Schulz, *VLSI Electronics,* Vol. 10, Ch. 2, (N. G. Einspruch and R. S. Bauer, ed.), pp.30-70, Academic Press (1985)
25. L. G. Meiners and H. H. Wieder, *Mat. Sci. Rep.,* 3:143 (1988)

26. W. E. Spicer, Z. Liliental-Weber, E. Weber, N. Newman, T. Kendelewicz, R. Cao, C. McCants, P. Mahowald, K. Miyano, and I. Lindau, *J. Vac. Sci. Technol.,* B6:1245 (1988)

27. H. Hasegawa and H.Ohno, *J. Vac. Sci. Technol.,* B4:1130 (1986)

28. J. Tersoff, *J. Vac. Sci. Technol.,* B3:1157 (1985)

29. J. L. Freeouf and J. Woodall, *Appl. Phys. Lett.,* 39:727 (1981)

30. N. Suzuki, T. Hariu, and Y. Shibata, *Appl. Phys. Lett.,* 33:176 (1978)

31. S. V. Hattangady, G. G. Fountain, D. J. Viktavage, R. A. Rudder, and R. J. Markunas, *J. Electrochem Soc.,* 136:2070 (1989)

32. S. Tiwari, S. L. Wright, and J. Batey, *IEEE Electron Dev. Lett.,* EDL9:488 (1988)

33. J. L. Freeouf, J. A. Silberman, S. L. Wright, S. Tiwari, and J. Batey, *J. Vac. Sci. Technol.,* B7:854 (1989)

34. G. Lucovsky, S. S. Kim, D. V. Tsu, G. G. Fountain, and R. J. Markunas, *J. Vac. Sci. Technol.,* B7:861 (1989)

35. A. Hashimoto, T. Kamijah, H. Takano, and M. Sakuta, *J. Electrochem. Soc.,* 134:153 (1987)

36. H. Klose, Y. E. Maronchuk, and O. V.Senoshenko, *Phys. Stat. Sol.,* 21a:659 (1974)

37. S. Yokoyama, K. Yukitomo, M. Hirose, Y. Osaka, A. Fisher, and K. Ploog, *Surf. Sci.,* 86:835 (1979)

38. K. Kamimura and Y. Sakai, *Thin Solid Films,* 56:215 (1979)

39. Y. Hirayama, F. Koshiga, and T. Sugano, *J. Appl. Phys.,* 52:4697 (1981)

40. E. Pincik, B. Matatko, J. Bartos, I. Thurzo, M. Grendel, V. Nadazdy, M. Zubekova, and M. Morvic, *Phys. Stat. Sol.(a),* 106:659 (1988)

41. E. Spicer, P. W. Chye, P. R. Skeath, C. Y. Su, and I. Lindau, *J. Vac. Sci. Technol.,* 16:1422 (1979)

42. W. Muench, *Appl. Surf. Sci.,* 22/23:705 (1985)

43. A. Ismail, J. M. Palau, and J. Lassabatere, *J. Appl. Phys.,* 60:1730 (1986)

44. S. D. Offsey, J. M. Woodall, A. C. Warren, P. D. Kerchner, T. I. Chapell, and G. D. Petit, *Appl. Phys. Lett.,* 48:475 (1986)

45. A. Callegari, P. D. Hoh, D. A. Buchanan, and D. Lacey *Appl. Phys. Lett.,* 54:332 (1989)

46. E. Yablonovitch, C. J. Sandroff, R. Bhat, and T. J. Gmitter, *Appl. Phys. Lett.,* 51:439 (1987)

47. E. Yablonovitch, B. J. Skromme, R. Bhat, J. P. Harbison, and T. J. Gmitter, *Appl. Phys. Lett.,* 54:555 (1989)

48. C. J. Sandroff, M. S. Hegde, L. A. Farrow, R. Bhat, J. P. Harbison, and C. C. Chang, *J. Appl. Phys.,* 67:586 (1990)

49. J. A. Van Vechten and J. F.Wager, *J. Appl. Phys.,* 57:1956 (1985)

50. V. A. Gergel and A. N. Solyakov, *Sov. Phys. Semicond.,* 17:641 (1983)

51. S. Goodnick, T. Hwang, and C. W. Wilmsen, *Appl. Phys. Lett.,* 44:453 (1984)

52. P. van Staa, H. Rombach, and R. Kassing, *J. Appl. Phys.,* 54:4014 (1983)

53. L. G. Meiners, *J. Vac, Sci. Technol.,* 19:373 (1981)

54. W. Kulish and R. Kassing, *J. Vac. Sci. Technol.,* B5:523 (1987)

55. J. F. Wager, K. M. Geib, C. W. Wilmsen, and L. Kazmerski, *J. Vac. Sci. Technol.,* B1:778 (1983)

56. Z. Liliental, O. L. Krivanek, J. F. Wager and S. M. Goodnick, *Appl. Phys. Lett.,* 46:889 (1985)

57. K. A. Bertness, T. Kendelewicz, R. S. Last, M. D. Williams, I. Lindau, and W. E. Spicer, *J. Vac. Sci. Technol.,* A4:1424 (1986)

58. L. J. Brillson, *Surf. Sci. Rept.,* 2:123 (1982)

59. W. Moench, *Surf. Sci.,* 168:577 (1985)

60. W. E. Spicer, P. W. Chye, P. R. Skeath, C. Y. Su, and I. Lindau, *J. Vac. Sci. Technol.,* 16:1422 (1979)

61. P. Boher, M. Renaud, J. M. Lopez-Villegas, J. Schneider, and J. P. Chane, *Appl. Surf. Sci.* 30 (1987)

62. L. G. Meiners, *J. Vac. Sci. Technol.,* 19:373 (1981)

63. H. L. Chang, L. G. Meiners, and C. J. Sa, *Appl. Phys. Lett.,* 48:375 (1986)

64. H. Hasegawa and T. Sawada, *J. Vac. Sci. Technol.,* 21:457 (1982)

65. H. Hasegawa and H. Ohno, *J. Vac. Sci. Technol.,* B4:1130 (1986)

66. A. Iliadis, S. C. Laih, D. E. Ioannou, and E. A. Martin, *Ins. Phys. Conf.,* Ser. No. 91:219 (1988)

67. B. Lepley, A. Bath, C. Carabatos-Nedelec, G. J. Papaioannou, P. C. Euthymiou, and S. Ravelet, *Phys. Stat. Sol.,* 108 (1988)

68. J. M. Moisson, M. Van Rompay, and M. Bensoussan, *Appl. Phys. Lett.,* 48:1362 (1986)

69. C. W. Wilmsen, K. M. Geib, J. Shin, R. Iyer, D. L. Lile, and J. J. Pouch, *J. Vac. Sci. Technol.,* B7:851 (1989)

70. R. Iyer, R. R. Chang, and D. L. Lile, *Appl. Phys. Lett.,* 53:134 (1988)

71. K. P. Pande and D. Guttierrez, *Appl. Phys. Lett.,* 46:416 (1985)

72. Y. Robach, J. Joseph, E. Bergignat, B. Commere, G. Hollinger, and P. Viktorovich, *Appl. Phys. Lett.,* 49:1281 (1986)

73. J. Chave, A. Choujaa, C. Santinelli, R. Blanchet, and P. Viktorovitch, *J. Appl. Phys.,* 61:257 (1987)

74. E. Yamaguchi, V. Hirota, and M. Minakata, *Thin Solid Films,* 103:201 (1983)

75. S. N. Al-Refaie, *Appl. Phys.,* A48:575 (1989)

76. Y. K. Su, C. R. Huang, and Y. C. Chou, *Jpn. J. Appl. Phys.,* 28:1664 (1989)

77. L. Koenders, F. Bartels, H. Ulrich, and W. Moench, *J. Vac. Sci Technol.,* B3:1107 (1985)

78. M. A. Olmstead, R. I. G. Uhrberg, R. D. Bringans, and R. Z. Bachrach, *Phys. Rev.,* B35:7526 (1987)

79. S. Siskos, C. Fontaine, and A. Munoz-Yague, *Appl. Phys. Lett.,* 44:1146 (1984)

80. J. M. Philips and J. M. Gibson, *Mat. Res. Soc. Proc.,* 25:381 (1984)

81. C. Fontaine, J. Castagne, E. Bedel, and A. Munoz-Vague, *J. Appl. Phys.,* 64:2076 (1988)

82. T. Waho and F. Yanagawa, *J. Crystal Growth,* 95:515 (1989)

83. S. Sinhroy, R. A. Hoffman, R. F. C. Farrow, J. D. Warner, and K. Basin, *Proc. Symp. Dielectric Films on Comp. Semicond.,* 86-3:313 Electrochem. Soc., Pennington, NJ (1985)

84. T. Waho, F. Yanagawa, Y. Yamada-Maruo, and H. Saeki, *Solid-State Electron.,* 33 Supl. 253 (1990)

85. D. Mao, K. Young, A. Kahn, R. Zanoni, J. McKinley, and G. Margaritondo, *Phys. Rev.,* B39:12735 (1989)
86. A. S. Barriere, G. Couturier, H. Guegan, T. Seguelong, A. Thabti, P. Alnot, and J. Chazelas, *Appl. Surf. Sci.,* 41/42:383 (1989)
87. R. F. C. Farrow, P. W. Sullivan, G. M. Williams, G. R. Jones, and D. C. Cameron, *J. Vac. Sci. Technol.,* 19:415 (1981)
88. P. W. Sullivan, R. F. C. Farrow, and G. R.Jones, *J. Crystal Growth,* 60:403 (1982)
89. C. W. Tu, S. R. Forrest, and W. D.Johnston, Jr., *Appl. Phys. Lett.,* 43:569 (1983)
90. T. K. Paul and D. N. Bose, *J. Appl. Phys.,* 67:3744 (1990)
91. H. C. Casey, A. V. Cho, D. V. Lang, E. H. Nicollian, and P. W. Foy, *J. Appl. Phys.,* 50:3484 (1979)
92. A. C. Gossard, W. Brown, C. L. Allyn, and W. Wiegmann, *J. Vac. Sci. Technol.,* 20:694 (1982)
93. P. M. Solomon, T. W. Hickmott, H. Morkoc, and R. Fisher, *Appl Phys. Lett.,* 42:82 (1983)
94. D. Delagebeaudeuf, P. Delescluse, P. Etienne, J. Massies, M. Laviron, J. Chaplart, and N. T. Linh, *Electron Lett.,* 18:85 (1982)
95. T. W. Hickmott, P. M. Solomon, R. Fisher, and H. Morkoc, *Appl. Phys. Lett.,* 44:90 (1984)
96. A. Zur and T. C. McGill, *J. Vac. Sci. Technol.,* B2:440 (1984)
97. D. J. As, P. W. Epperlein, and P. M. Mooney, *J. Appl. Phys.,* 64:2408 (1988)
98. H. C. Casey, A. Y. Cho, and E. H. Nicollian, *Appl. Phys. Lett.,* 32:679 (1978)
99. H. C. Casey, A. Y. Cho, D. V. Lang, and E. H. Nicollian, *J. Vac. Sci. Technol.,* 15:1408 (1978)
100. H. C. Casey, A. Y. Cho, D. V. Lang, E. H. Nicollian, and P. W. Foy, *J. Appl. Phys.,* 50:3484 (1979)
101. J. P. Andre, C. Schiller, A. Mitonneau, A. Briere, and J. Y. Aupied, *Inst. Phys. Conf.,* Ser #65, 117 (1982)
102. W-P. Hong, J-E. Oh, P. Bhattacharya, and T. E. Tiwald, *IEEE Trans. Electron. Dev.,* 35:1585 (1988)

103. J. Jeong, T. Schlesinger, and A. Milnes, *IEEE Trans Electron. Dev.,* 34:1911 (1987)

104. C. M. Hanson and H. H. Wieder, *J. Vac. Sci. Technol.,* B5:971 (1987)

105. C. M. Hanson and H. H. Wieder, *Thin Solid Films,* 153:497 (1987)

106 E. Yamaguchi and M. Minakata, *J. Appl. Phys.,* 55:3098 (1984)

107. T. K.Paul, P. Bhattacharya, and D. Bose, *Electron. Lett.,* 25:1602 (1989)

108. M. Mizuta, S. Fujieda, T. Jitsukawa, V. Matsumoto, and T. Kawamura, *Jpn J. Appl. Phys.,* 25:L945 (1986)

109. M. Mizuta, S. Fujieda, T. Jitsukawa, and Y. Matsumoto, *Inst. Phys. Conf.,* Ser. 83:153 (1987)

110. S. Fujieda, M. Mizuta, and Y. Matsumoto, *Jpn. J. Appl. Phys.,* 27:L296 (1988)

111. S. Fujieda, K. Akimoto, I. Hirosawa, J. Mizuki, Y. Matsumoto, and J. Matsui, *Jpn, J. Appl. Phys.,* 28:L16 (1989)

112. B. Pruniaux, J. North, and A. Payer, *IEEE Trans. Electron Dev.,* ED-19:672 (1972)

113. P. L. Fleming, A. Meulenberg, Jr., and H. E. Carlson, *IEEE Electron Dev. Lett.,* EDL-3:104 (1982)

114. L. G. Meiners, *J. Appl. Phys.,* 50:1154 (1979)

115. H. M. Macksey, D. W. Shaw, and W. R. Wisseman, *Electron. Lett.,* 12:192 (1976)

116. D. Fritzsche, *Inst. Phys. Conf.,* Ser. #50:258 (1980)

117. M. Okamura and T. Kobayashi, *J. Appl. Phys.,* 19:2143 (1980)

118. D. L. Lile and M. J. Taylor, *J. Appl. Phys.,* 54:260 (1983)

119. H. Hasegawa, H. Ohno, H. Isshii, T. Haga, Y. Abe, and H. Takahashi, *Appl. Surf. Sci.,* 4:372 (1989)

120. V. Mochizuki, M. Mizuta, S. Fujieda, and V. Matsumoto, *J. Appl. Phys.,* 67:2466 (1990)

# 5

# Interface States

*Leonard J. Brillson*

## 1.0 INTRODUCTION

This chapter aims to provide a comprehensive overview of the physical mechanisms which can produce localized charge at metal-semiconductor interfaces. As already described in preceding chapters, the electronic properties of metal-semiconductor junctions are a critical component of solid state device performance. Likewise, such properties depend upon the fundamental nature of charge transfer and localization at microscopic heterointerfaces. Such mechanisms may now be addressed at the atomic scale. Enabling this development has been the advent of ultrahigh vacuum (UHV) techniques, which permit electronic, chemical and structural properties of clean and atomically-controllable interfaces to be examined at the submonolayer scale and as the interface builds up, layer upon deposited monolayer.

Over the past decade, there has been major progress in characterizing such properties for many of the semiconductor systems described in previous chapters. Indeed these studies have uncovered a variety of novel physical phenomena which suggest new mechanisms for Schottky barrier formation. In this chapter, we will first examine the role of localized charge in Schottky barrier formation and its relation to conventional charge exchange between metal and semiconductor. The next section provides descriptions of many new techniques to characterize interface states which complement more conventional methods. These modern techniques provide detailed information on energies, densities, charge nature and even spatial position of the interface charge. They provide the most direct evidence thus far on the atomic-scale nature of interface states.

The following sections categorize and describe interface states in detail. Here we address the central importance of localized charge for semiconductor contacts and describe such interface states in their various forms. In particular, we discuss both states intrinsic to the bulk materials forming the contact as well as those which arise out of features extrinsic to either contacting material. In turn, extrinsic states are subdivided according to their macroscopic, conventional or microscopic, interface-specific origin. In each case we present evidence to support the existence of such states.

The wide variety of systematic barrier variations with atomic-scale chemical changes suggests a dominant influence of extrinsic mechanisms in most contact structures and in particular at the junctions described in earlier chapters. A picture then emerges of an extended metal-semiconductor interface with localized charge contributions from several potential mechanisms, more than one of which contribute simultaneously. Based on the existence of such chemical mechanisms, the following section presents evidence for systems with low and/or controllable densities of interface states, where it is possible to achieve a high degree of control over the Schottky barrier formation.

Building on the success achieved to date in controlling the semiconductor interface state properties, the final section presents a variety of potential opportunities to further characterize, understand, and control the electronic properties of metal-semiconductor and semiconductor-semiconductor contacts.

## 2.0 TRAPPED CHARGE AT INTERFACES

### 2.1 Ideal Schottky Barrier Formation

Interface states manifest themselves at metal-semiconductor junctions as deviations from otherwise ideal Schottky barrier formation—the charge exchange and rectification resulting only from the difference in potential between the two separate media. As such, these interface states are localized within a few atomic layers of the intimate metal-semiconductor junction with energies which fall inside the forbidden band gap. Their effect on the contact rectification diminishes the further away from the interface they lie spatially. Such states may extend away from the interface spatially or away from the forbidden gap energetically—with correspondingly reduced effect.

Schottky barrier formation depends strictly on the potential difference between the Fermi level ($E_F$) in the metal and the majority carrier band edge of the semiconductor. Figure 1 illustrates the band bending scheme for a metal at the surface of an n-type semiconductor.[1] For the case of a high work function metal shown in the upper panel, electrons flow from the semiconductor to the metal after contact, depleting a characteristic surface region in the semiconductor of electrons. With the two Fermi levels $E^M{}_F$ and $E^{SC}{}_F$ aligned, a double layer forms with a voltage drop of $qV_B$ equal to the contact potential difference between the metal and the interior of the semiconductor. The double layer consists of a surface space charge region, typically $10^{-4}$ - $10^{-6}$ cm thick, and an induced charge on the metal surface. The n-type depletion region pictured is a layer of high resistance. Thus a voltage applied to this junction will fall mostly across the surface space charge region. The barrier height $\Phi_{SB}$ depends on the difference in thermionic work function and in the simple Schottky-Mott-Davidov model[1]-[3] is expected to be

Eq. (1) $$\Phi_{SB} = \Phi_m - X_{SC}$$

where $\Phi_m$ and $X_{sc}$ are the metal work function and semiconductor electron affinity respectively, defined with respect to the vacuum level $E_{VAC}$.[4] According to this model, the potential $\Phi$ within the semiconductor satisfies Poisson's equation

Eq. (2) $$\nabla^2\Phi(x) = -4\pi p(x)/\varepsilon_s$$

where $\rho$ is the charge density in the surface space charge region of width w, x denotes the coordinate axis normal to the metal-semiconductor interface, and $\varepsilon_s$ is the static dielectric constant of the semiconductor. This relation defines a parabolic band bending region for the abrupt metal-semiconductor junction. Furthermore, by assuming $\rho$ = qN, the bulk concentration of ionized impurities within the surface space charge region, one obtains a depletion layer width of

Eq. (3) $$w = [\varepsilon_s (V - V_O) / 2\pi N]^{1/2}$$

For a very low work function metal, band bending with the opposite sign can produce majority carrier concentrations higher at the surface than in the bulk (accumulation region), resulting in no barrier to majority carrier transport across the interface. Analogous conclusions follow for p-type semiconductors.

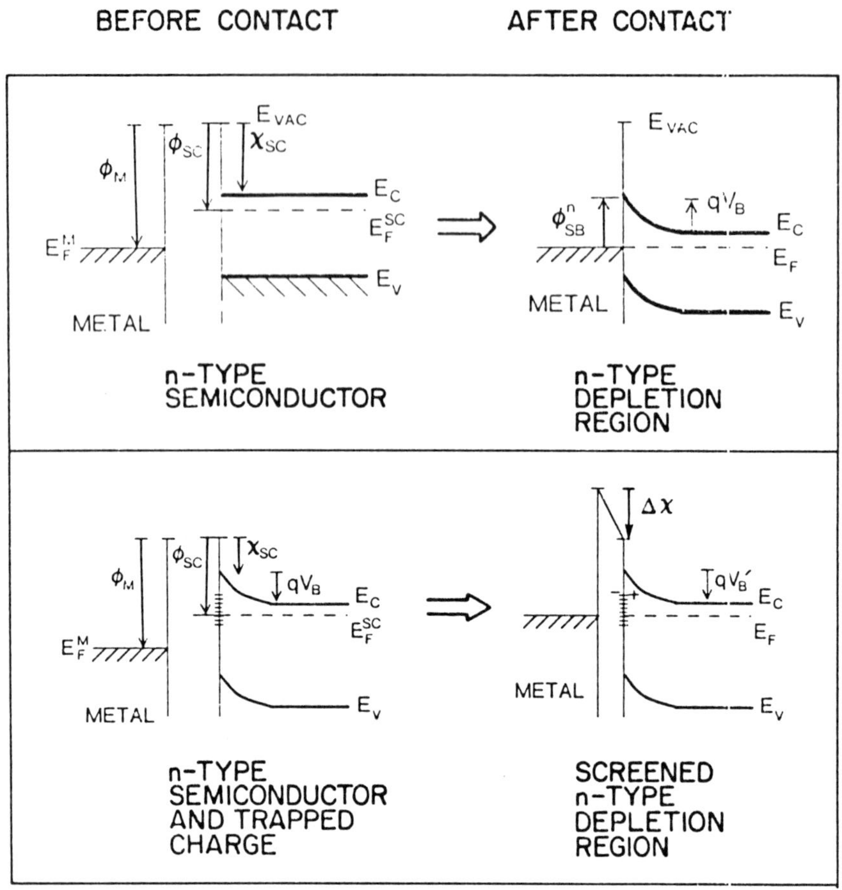


**Figure 1.** Schematic diagram of band bending before and after metal-semiconductor contact for interfaces without *(upper)* and with *(lower)* trapped charge states. Trapped charge which screens the contact potential difference between metal and semiconductor may exist on the semiconductor surface before contact or may be created upon metallization.

Conventional techniques for measuring barrier heights, and indirectly, interface states, macroscopically include: current-voltage methods for gauging charge transport across the interface, capacitance-voltage methods for determining the height and width of the barrier region, and internal photoemission spectroscopy for identifying optical transitions from the

Fermi level to the majority carrier band edge. On a microscopic scale, new techniques such as soft x-ray photoemission, surface photovoltage, cathodoluminescence, photoluminescence, Raman and scanning tunneling charge spectroscopies are now available which address band bending and localized states on an atomic scale. Descriptions of these new techniques appear in Sec. 3.

## 2.2 Evidence for Localized Charge

For actual metal-semiconductor interfaces, Eq. 1 does not accurately describe the dependence of Schottky barrier heights on metal work function. Up until the last five years, barrier height measurements have indicated only a weak dependence of semiconductor height on the particular metal contact. The earliest definitive demonstration of interface states was the field effect experiment of Schockley and Pearson, showing the insensitivity of Si surface conductivity to applied gate bias.[5] In the case of intimate metal-Si interfaces, metals with work functions ranging over 2 eV produce only 0.5 eV variation in barrier height.[6] Likewise, common metals on compound semiconductors yield only a few tenths of an eV variation in band bending. Until recently, metal-GaAs junctions exhibited $E_F$ movement restricted to a range of only 0.2 - 0.25 eV.[7] In other words, for actual metal-semiconductor interfaces, Eq. 1 does not accurately describe the dependence of band bending on metal work function.

## 2.3 Interface Dipoles

The insensitivity of semiconductor barrier heights to different metals can be explained by the presence of interface dipoles due to the additional, localized charge. The lower portion of Fig. 1 illustrates the formation of this dipole either before or after contact. If the existence of such states and dipoles reduces the movement of $E_F$ within the semiconductor band gap with charge transfer, due to work function differences or an applied electrical bias, then $E_F$ is said to be *pinned* as opposed to merely stabilized at a given energy. Such states can be due to a variety of factors, which we discuss in following sections.

In fact, the rectification process involves not just one but several dipoles. Consider the intrinsic process of interface charge transfer: the potential distributions for isolated metal and semiconductor surfaces are shown schematically in Fig. 2a. The work function of the metal $\Phi_m$ is comprised of two parts, an internal potential $S_m$ and a surface dipole $V_m$. Correspondingly, the semiconductor electron affinity $X_{sc}$ consists of the internal potential $S_{sc}$, a surface dipole $V_s$, and $\zeta_n$, the bulk electron $E_F$ relative to the bottom of the conduction band. The surface dipoles are associated with electronic tunneling into the vacuum and are required to confine electrons within the solid at the surface. Figure 2b illustrates the potential distribution for the metal and semiconductor in contact. The Fermi level is constant across the junction so that the potential drop across the interface must equal the difference in internal potentials. The surface dipoles are replaced by an interface dipole $V_i$, and the potential drop is shared between $V_i$ and the band bending $V_b$ within the surface space charge region (depletion width $\lambda$) of the semiconductor. The relative magnitudes of $V_i$ and $V_b$ depend on the amount of charge localized in the interface dipole relative to the charge transferred into or out of the surface space charge region. From the relations illustrated in Fig. 2, one obtains

Eq. (4) $$V_b = (\Phi_m - X_{SC} + \zeta_n) + (V_s - V_m - V_i)$$

and the measured (n-type) barrier height $\Phi_B$ from the Fermi level to the conduction band edge ($X_{sc} + \zeta_n$) is thus the sum of two terms, the classical expression involving the difference between the two observables $\Phi_m$ and $X_{sc}$, and a term involving the difference in dipoles.[8] (See also Ch. 8.) The p-type semiconductor expression is analogous. In practice, the dipole terms are not directly observable—one derives their magnitudes experimentally only from their effects on the observables $\Phi_m$, $X_{sc}$, and $V_B$.

Equation (4) shows that band bending and barrier height will vary as $(\Phi_m - X_{sc})$ if the charges associated with the metal and semiconductor surfaces do not change when the interface forms—that is, the dipole difference term is zero. This corresponds to the simple band bending model described by Eq. (1). However, for most interfaces studied previously,

Eq. (5) $$\Phi_B \neq (\Phi_m - X_{sc})$$

as evidenced by a weak dependence of $\Phi_B$ on $\Phi_m$ for different metals on a given semiconductor. Thus barrier variations for metals on GaP, GaAs, Si, and CdS prepared under high vacuum conditions in the 1960's showed slopes $d\Phi_B/d\Phi_m$ of only 0.27, 0.07, 0.27, and 0.38, respectively, instead of 1.0.[6] This weak dependence implies $(V_s - V_m - V_i) \neq 0$, or, in other words, a screening of the potential difference between metal and semiconductor by additional charge trapped at the junction.

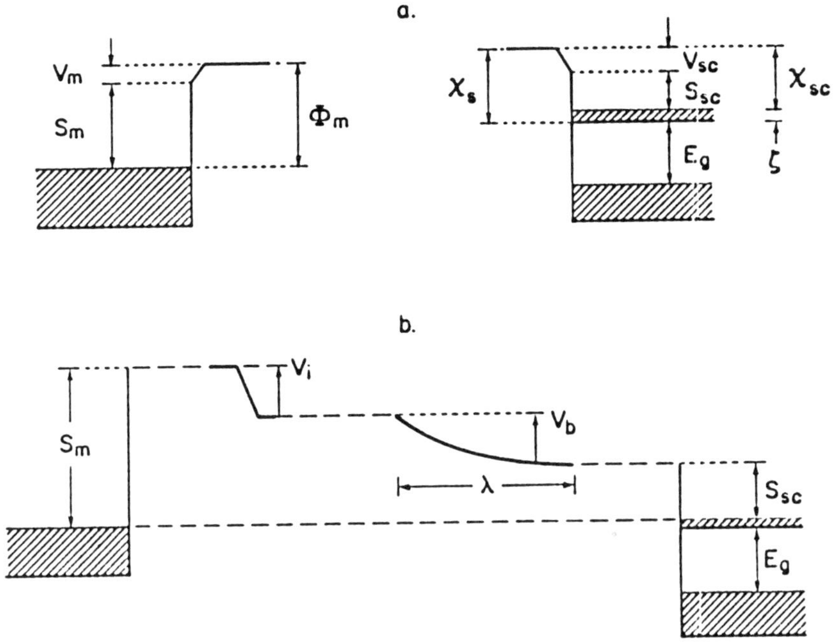


**Figure 2.** Potential distributions for *(a)* separated metal and semiconductor surfaces and *(b)* the metal-semiconductor interface. The observable work function $\Phi_m$ (electron affinity $X_{sc}$) consists of a calculated internal potential $S_m$ ($S_s$) plus a surface dipole $V_m$ ($V_s$). The local interface dipole $V_i$ plus the dipole $V_b$ of the surface space charge region account for the difference in internal potentials when metal and semiconductor are joined. See Ref. 8.

## 2.4 Interface State Classification

Bardeen first showed how such charge accumulated at the metal-semiconductor junction reduces the effective potential difference between the two materials and thereby the charge transfer into or out of the semiconductor's surface space charge region.[9] Such interface states can arise due to several factors: *(i)* intrinsic surface states of the semiconductor which arise from the discontinuity in lattice potential and bonding at the vacuum-solid interface;[9] *(ii)* localized, metal-induced gap states due to wave function tunneling from the metal into the semiconductor band gap—also intrinsic since they depend only on the bulk properties of the two constituents; *(iii)* extrinsic surface states due to contamination or lattice imperfections; *(iv)* metal-induced extrinsic features due to localized atomic bonding, interdiffusion, or chemical reaction. These different types of interface states are reviewed in Secs. 4, 5, and 6. First, we describe the advanced techniques now available to characterize the physical and chemical properties of these interface states.

## 3.0 MODERN INTERFACE STATE CHARACTERIZATION

Conventional techniques for measuring the presence of interface states are based on transport or electrostatic gauges of band bending across the semiconductor surface space charge region. Actual densities of states, energies, and donor or acceptor type can be inferred from absolute barrier values as well as the dependence of band bending upon bias, bias hysteresis, as well as the detailed barrier height dependence on metal work function. Such current-voltage and capacitance-voltage techniques[4][10] are described in previous chapters. These techniques have provided a basis for interface state characterization since the early 1960's. However, they are at best indirect measurements of interface states because they measure the effect of localized charge on extended features of the junction, namely the change in energy band position and charge density within the surface space charge region. Furthermore, they may be complicated by a number of physical processes. For current-voltage measurements, these include tunneling through the barrier,[11][12] intermediate layers with new dielectric and transport properties,[12][13] and recombination or trapping at states near the interface and within the semiconductor band gap.[13] Possible sources of error in capacitance-voltage measurements include the effects of an insulating layer between metal and semiconductor,[14] variation of the semiconductor surface charge (population of interface states) with applied voltage, series resistance of the junction, traps within the depletion region, and variations in the effective contact area with depletion layer width.[15] Perhaps more reliable than either of these two methods is the internal photoemission technique,[10] which relies on direct optical excitation of carriers from $E_F$ to the band edges and which is less susceptible to tunneling through the barrier region, deep level trapping, and recombination. All three techniques may be influenced slightly (<0.05 eV) by image-force lowering of the barrier.

Over the past two decades, a rich variety of surface sensitive techniques have been developed to probe interface states and relation to atomic-scale properties. Chief among methods to characterize surface electronic structure has been photoelectron spectroscopy. Here an incident photon beam excites core and valence electrons out of the solid through vacuum and into an energy analyzer. Depending on the excitation energy, these photoelectrons can have kinetic energies from only a few volts to a few thousand volts. Most surface-sensitive techniques such as photoemission make use of the extremely short (only a few angstroms) scattering length of electrons with kinetic energies in the 50 - 100 eV range to promote emission from only the near-surface region. Figure 3 illustrates the production of both primary and scattered secondary electrons with incident photon energy hν. Both core and valence band features are evident with kinetic energies dependent on hν. Kinetic energy $E - E_F$ depends upon the position of $E_F$ relative to the filled states. As $E_F$ shifts in the band gap with band bending (or doping), the kinetic energies shift accordingly. These rigid shifts can thereby provide a measure of band bending during interface formation. In practice, such measurements rely on core level rather than valence band features since the latter are less pronounced and can change shape substantially upon chemisorption of new species. (Nevertheless, valence band spectra can reveal new features with adsorption corresponding to localized charge in filled states outside the band gap.)

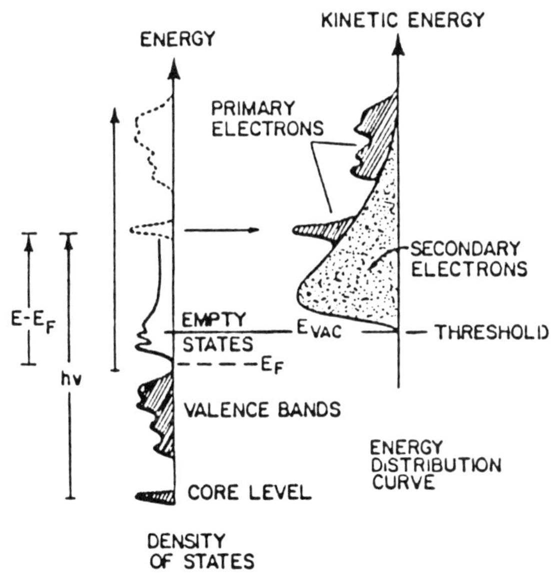


**Figure 3.** Schematic diagram of photoelectron spectroscopy. Incident photons of energy hν generate photoelectrons from occupied valence bands and core levels. "Primary" electrons above the vacuum level which escape without energy loss *(dark shading)* and "secondary" electrons which have scattered elastically *(light shading)* comprise the energy distribution curve (EDC). $E-E_F$ governs the inelastic mean free path and escape depth of electrons at a given energy within the EDC. See Ref. 10 and references therein.

Considerable research has been devoted to $E_F$ movements in the band gap as a function of metal overlayer thickness. For metals on clean semiconductor surfaces, $E_F$ moves to its equilibrium position with coverages of only a few monolayers or less. Advantages of photoemission spectroscopy include the ability to: *(a)* probe interface electronic structure during the initial stages of Schottky barrier formation, before and during the formation of the metallic layer; *(b)* monitor both electronic and chemical properties simultaneously; *(c)* correlate these electronic and chemical changes, layer upon atomic layer. Disadvantages include: *(i)* relatively poor sensitivity to interface states—typically 0.01 monolayer equivalent or ~$10^{13}$ $cm^{-2}$; *(ii)* an inability to probe the electronic and chemical properties more than a few monolayers below the surface, at the "buried" interface; *(iii)* susceptibility to light-induced changes in band bending, the surface photovoltage effect—especially for nonconducting surface layers and at low temperature, where thermionic emission Is reduced.[16][17] While there exist some photoemission measurements of intrinsic surface states within the valence band,[18] surface state features above the valence band edge have been difficult to distinguish from simple metallization of the overlayer.

Variations on the photoemission technique such as constant initial state[19] and final state[20] spectroscopies allow detection of unoccupied states in the band gap as well as in conduction band states. Likewise, Haight and coworkers have used two-photon laser-pumped photoemission spectroscopy to detect the presence of states induced by metal overlayers[21] as well as to measure semiconductor heterojunction band discontinuities.[22] While this laser-pumped technique presents considerable challenges in terms of detection limits versus maximum input power densities, it possesses advantages in terms of high optical energy resolution and of temporally resolving charge kinetics on a picosecond scale.

Electrostatic techniques can provide higher sensitivity to surface state concentrations but without the atomic-scale, chemical information of electron emission spectroscopies. Surface photovoltage spectroscopy represents perhaps the simplest of such electrostatic techniques, involving only a vibrating Kelvin probe in a feedback circuit to provide a contact potential difference as a function of incident excitation wavelength with respect to the semiconductor surface.[23][24] Because sensitivity is typically on the order of a millivolt and surface charge densities of $10^{12}$ - $10^{13}$ $cm^{-2}$ can produce barrier heights of a volt, the technique can detect charge densities of as little as $10^{9}$ - $10^{10}$ $cm^{-2}$. Figure 4 illustrates detection of metal-induced deep levels in the GaAs band gap by submonolayers of Au deposited under ultrahigh vacuum (UHV) conditions.[25] The main spectral feature is at the

GaAs band gap and corresponds to band-to-band transitions which act to flatten the bands. Below the band edge energy, there appear several new features upon metal deposition which change with increasing submonolayer coverage. At the highest nonmetallic coverage, a dominant sub-band gap feature appears at 0.9 eV, corresponding to a state 0.9 eV below $E_c$, the conduction band edge. The inset illustrates this transition schematically. The interface state position agrees with the $E_F$ position determined from photoemission spectroscopy[26] and conventional barrier height measurements[27] for Au on this cleaved GaAs (110) surface. A distinct advantage of this technique is that the slope change of the features indicates the conduction or valence band nature of the sub-band gap transitions—a feature lacking in luminescence spectroscopies. Conversely, the top curve in Fig. 4 indicates a drawback to the technique: spectral sensitivity diminishes as the semiconductor surface potential assumes a constant value with metallic coverage. Hence this surface potential technique is not useful for buried interface studies.

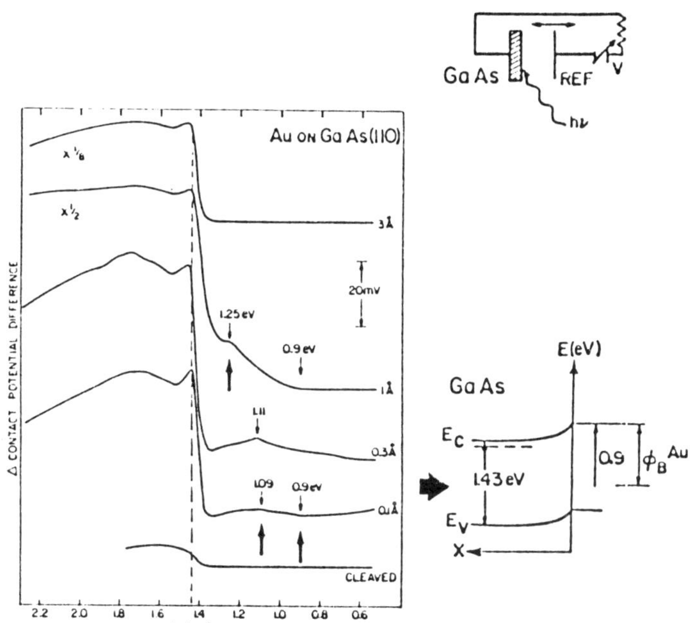


**Figure 4.** Surface photovoltage spectra of GaAs(110) cleaved in ultrahigh vacuum and with increasing thicknesses of deposited Au. The *dashed line* denotes the bulk absorption edge. *Arrows* indicate new optical transitions induced by the metal overlayer. The upper inset illustrates the vibrating capacitor circuit used for surface photovoltage spectroscopy (SPS) while the lower inset illustrates schematically the primary SPS transition within the GaAs band gap. See Ref. 25.

The scanning tunneling microscope (STM) is a significantly more sophisticated technique for measuring electronic properties of adsorbate states, while providing laterally-specific spatial information on an atomic scale.[28] The STM technique involves scanning an atomically-sharp metal tip across a surface while monitoring the tunneling current between the tip and the surface. Again using feedback techniques, one can monitor surface features with, in this case, atomic spatial resolution and, by scanning across a surface, produce images of atomic arrays and their geometric reconstructions of surface bonding.

The STM can detect the presence of filled or unfilled states by measuring conductivity versus tip-to-specimen voltage. Figure 5 illustrates such conductivity versus voltage spectra for Au clusters on cleaved GaAs (110) in UHV. Here the sample voltage corresponds to the energy of the state relative to $E_F$. The clean surface spectrum *(d)* clearly illustrates the gap region of 1.4 eV, while the Au cluster regions exhibit an additional peak at valence band edge $E_v$ + 1.06 to 1.4 eV. The energy of the unfilled state appears to depend on the spatial position either on top or on the side of the cluster. From careful scanning across such atomic sites, Feenstra[29] attributes such metal-induced states to Au-Ga bonds. Another state 0.7 eV below $E_v$ is attributed to hybrid metal-GaAs states. Both states appear to be common to several adsorbate-GaAs interfaces.[29]

Because STM senses potential variations on an atomic scale, it also provides a measure of the surface dielectric screening.[30] When coupled with surface illumination, one can measure variations in band bending near adsorbate or morphological features. These provide a measure of the screening length (~15 - 25 Å at moderately high doping levels) for surface recombination centers.[31] The ability to monitor surface recombination velocity near atomic features is likely to yield valuable information on the relationship between atomic structure and bonding to this macroscopic parameter.

Other techniques are available which require additional interpretation, including for example, near-surface capacitance-voltage,[32] or which require highly constrained surface geometries, such as surface reflectivity.[33]

## 3.1 Buried Interface Techniques

While the surface-sensitive techniques described thus far provide considerable information on chemical and electronic properties of monolayer adsorbates on semiconductor surfaces, they are by the same token less effective at probing features of interfaces more than a few monolayers below the semiconductor surface. Several new techniques—variations on more conventional, bulk measurement methods—can in fact probe such buried interfaces.

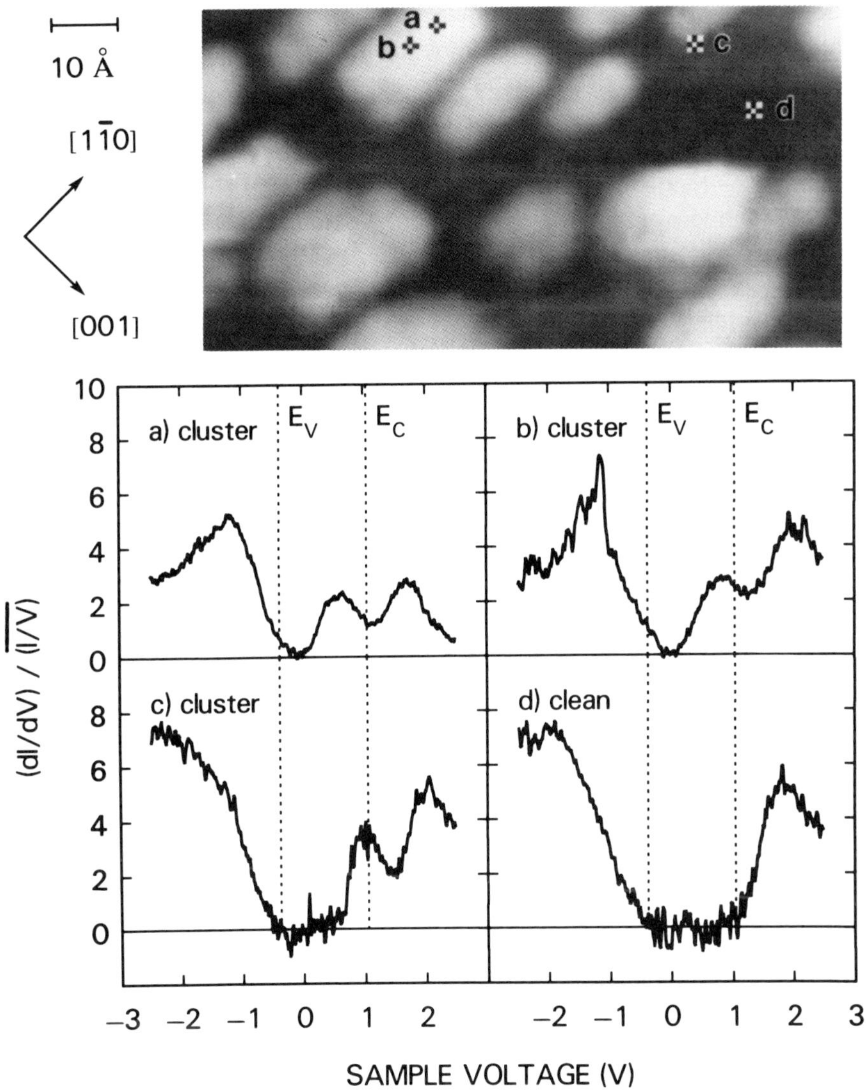


**Figure 5.** Normalized STM conductivity versus voltage, on and off Au clusters deposited on UHV-cleaved GaAs(110). The 100 x 55 Å$^2$ STM image of the surface appears in the upper part of the figure. *Checkered markers* indicate locations corresponding to the four spectra shown. See Ref. 29.

Low energy cathodoluminescence spectroscopy (CLS) has been particularly effective at detecting interface states at buried metal-semiconductor interfaces. This variation on the high energy electron microscope technique employs electrons of only a few kilovolts to penetrate metallic and/or semiconductor layers of thicknesses several nanometers or more, depending on the optical transmission of the overlayer,[34]-[36] and excite electrons and holes into various luminescence transitions. Such transitions include band-to-band recombination, deep level transitions due to impurities of native defects and even new compound formation. See Fig. 6. An additional feature of CLS is the capability of varying the incident beam energy to emphasize surface versus bulk features. Excitation depths for CLS in the low energy regime have been calculated to be on the order of several hundreds of angstroms or less.

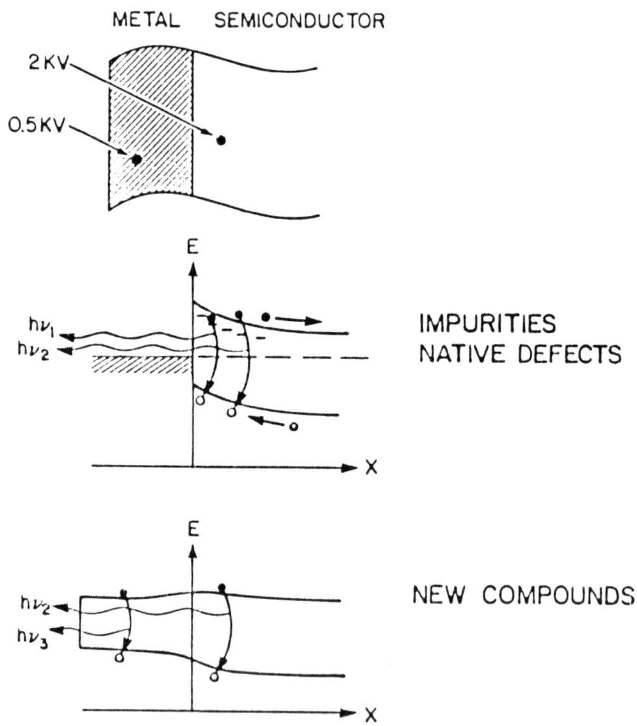


**Figure 6.** Schematic diagram of low energy cathodoluminescence spectroscopy (CLS). The finite penetration of 0.5 - 5 kV electrons through overlayers permit optical characterization of "buried" metal-semiconductor interfaces. Observed optical features correspond to band-to-band and deep level/band edge transitions. These permit identification of impurities, native defects, and new compounds, as well as changes in band bending. See Ref. 35.

Figure 6 also serves to illustrate the excitation processes for photoluminescence spectroscopy as well, where a laser or monochromatic light source of sufficiently high excitation energy penetrates the semitransparent overlayer and generates similar electron-hole pair recombination. The incident wavelength defines a fixed penetration and excitation depth. Because of the semiconductor's high optical refractive index, off-axis illumination is only marginally effective emphasizing surface versus bulk features. Instead, one can vary the incident laser wavelength since the penetration depth is defined by $\lambda/4\pi\kappa$, where $\lambda$ = incident wavelength and $\kappa$ = extinction coefficient[37] and the penetration depth is typically ~$\lambda/10$ or less.

Both cathodo- and photoluminescence spectroscopies provide a means to excite electron-hole pairs within the surface space charge region and thereby promote additional recombination involving surface optical transitions. Figure 6 illustrates the movement of majority carriers away from (minority carriers toward) the surface due to band bending. As a result, additional minority carriers are available for recombination surface or interface states. The role of band gap states at the surface as a recombination sink for charge generated within the surface space charge region accounts for the high surface sensitivity—despite the order-of-magnitude higher diffusion lengths for the excited charge.

Figure 7 illustrates the detection of interface states for metals on InP cleaved (110) surfaces.[38] The clean surface exhibits no gap features, only band-to-band recombination. With deposition of monolayer amounts of metals, one observes discrete luminescence features which evolve in energy and intensity with metal thickness. Additionally, these spectra display a reduction in band-to-band luminescence intensity as the band bending increases. Different metals produce features with different energies, different evolution with coverage, and different attenuation of the band-to-band features. The energies are consistent with Schottky barrier heights measured by electrical and photoemission techniques. Their evolution is consistent with photoelectron spectroscopy observations of $E_F$ movements (and chemical bonding changes) versus metal coverage. Likewise, overall changes in band-to-band luminescence agree with differences in absolute barrier height.

A limitation of these luminescence techniques is their semiquantitative nature. The luminescence intensities do not provide a straightforward measure of the interface state densities because the effects of $E_F$ position, surface band bending voltage and space charge width on the surface recombination must be taken into account. Furthermore, the luminescence transitions may involve either valence or conduction band. Here surface photovoltage spectroscopy has provided additional information supporting CLS assignments.[39]

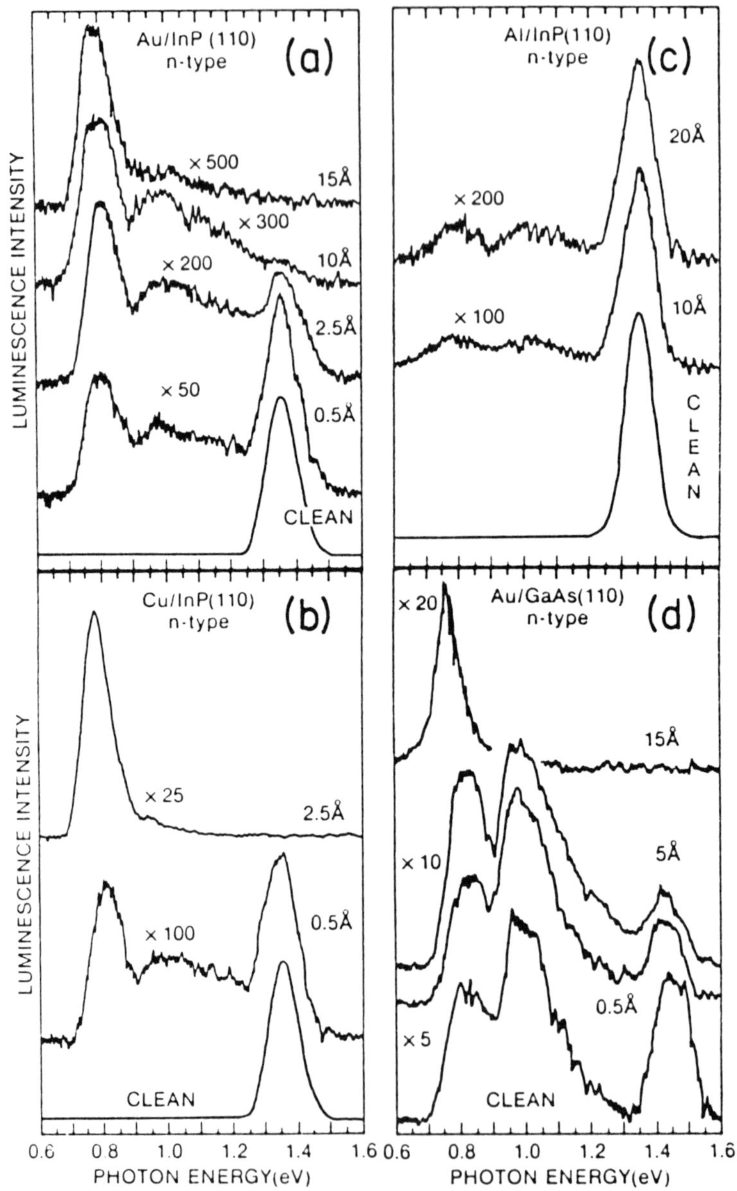


**Figure 7.** Direct observation of interface states by low energy cathodoluminescence spectroscopy. Figures 7*(a)* - 7*(d)* illustrate detection of optical transitions involving discrete states within the band gap which depend on the particular metal and which evolve with multilayer coverages. See Ref. 38.

More recent CLS work has demonstrated the growth of states with temperature[40] for Al/GaAs interfaces. Likewise, cathodo- and photoluminescence spectroscopies have highlighted the differences between the same semiconductors grown by different methods. For example, CLS has provided evidence for orders-of-magnitude higher densities of mid-gap states for bulk-grown versus MBE-grown GaAs.[41] Likewise, photoluminescence spectroscopy has highlighted the wide variability in CdTe bulk gap states from source to source and, in particular, the low gap state densities for MBE-grown crystals.[42][43]

Cathodoluminescence spectroscopy has provided evidence for discrete deep levels which can in fact account for the photoemission-measured Schottky barrier dependence on metals according to a self-consistent electrostatic calculation.[8][44] Duke and Dandrea describe such calculations in Ch. 8 of this volume. By matching the dependence of the Schottky barrier height versus metal work function to well-defined generic curves for discrete donor/acceptor levels in the band gap, it is possible to extract both donor and/or acceptor energies and densities. The results are fully consistent with the observed optical transitions.

Other optical techniques are also available for characterizing interface states. Raman scattering provides (indirect) information on interface states by virtue of its dependence on electric fields within the surface space charge region. This approach derives from the amplitude of symmetry-forbidden scattering of longitudinal-optical (LO) phonon modes, which are permitted by electric field-induced lowering of the symmetry within the surface space charge region.[45] The technique provides information for coverages up to several nanometers on the evolution and magnitude of band bending. For a recent review, see Ref. 46. The technique is also useful for studying heterojunction fields.[47] Modulation spectroscopy is based on the amplitude of space charge fields as well, yielding band bending as a function of small shifts in critical point energies within the modulation spectra.[48] Along with Raman scattering and other fully optical techniques, this spectroscopy has the advantage of not requiring high vacuum during measurement. A complication of both Raman and modulation spectroscopy is their sensitivity to the shape of the band bending region. Furthermore, the Raman technique requires crystal symmetries and experimental orientations which can probe symmetry-forbidden scattering.

An exciting new variant on the STM technique permits spatial mapping of Schottky barrier height across the initially-metallized surface.[49] This ballistic electron energy microscopy (BEEM) technique involves monitoring the onset in electron (hole) injection from the STM tip into the semiconductor conduction (valence) band(s), as shown in Fig. 8.[50] Here variations in BEEM features with morphology are illustrated and compared for Au /GaAs junctions on a near-atomic scale. The lack of significant correlation supports the electronic rather than structural nature of BEEM features. The considerable heterogeneity of the Au/GaAs surface contrasts with the homogeneity of BEEM features for other junctions such as Au on Si.[49] On the other hand, Ludeke and Prietsch and Hasagawa et al. have addressed the role of surface morphology on the BEEM features in more detail and shown that angle-dependent tunneling constraints across the interface and field pinching near domains can produce significant BEEM changes as well.[51] At Au/CdTe interfaces with relatively uniform morphology, Williams et al.[52] have demonstrated Schottky barrier heterogeneity amounting to as much as 0.4 eV from submicron location to location.

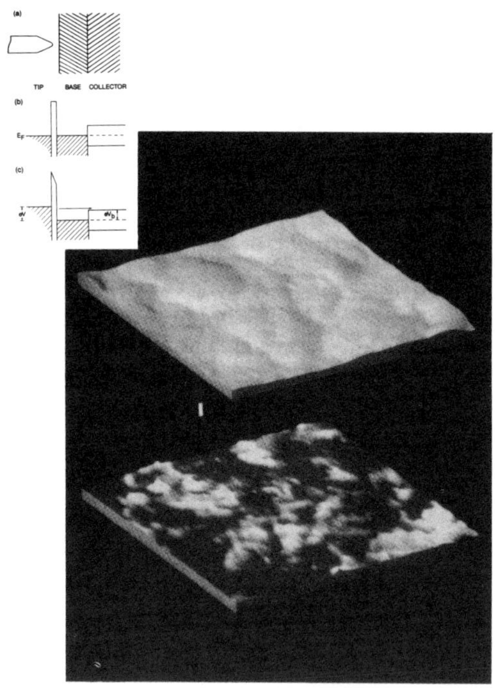


**Figure 8.** STM topographic and BEEM images of a Au-GaAs(100) interface prepared by chemical etching without air exposure before metal deposition. The STM *(upper)* and BEEM *(lower)* images were acquired simultaneously. Both images display a 510 x 390 Å$^2$ area. Minimum to maximum topographical height is 63 Å. Local current is represented by topographic altitude. Dark regions indicate zero detectable collector current. Heterogeneity in BEEM Schottky barrier height is independent of surface topographical features. Inset shows schematic BEEM geometry and transport mechanism. See Ref. 50.

The techniques described in this section yield considerable electronic information during the initial stages of Schottky barrier formation, information which is sensitive to the chemical and structural properties of the junction. Hence, there appear to be numerous, powerful methods for addressing interface states on an atomic scale.

## 4.0 INTRINSIC STATES

In this section, we examine sources of trapped charge which arise as a result of features intrinsic either to the surface structural and electronic properties of the semiconductor alone—intrinsic surface states—or to the intrinsic bulk electronic properties of the semiconductor and its metallic contact—intrinsic interface states.

### 4.1 Intrinsic Surface States

Intrinsic surface states may arise at the semiconductor surface, where the lattice potential experiences an abrupt change and where atomic bond strength and geometry may vary. Here the semiconductor surface dipole ($V_s$ in Fig. 1) will depend upon the extent to which the lattice disruption causes bond charge redistribution. Since the lattice discontinuity at covalent semiconductors affects bond charge more strongly than at ionic semiconductors, intrinsic surface states were expected to have the largest effect on barrier formation for semiconductors such as GaAs and Si. Indeed, Fig. 9 illustrates the trend of higher coefficient of interface behavior $S = d\Phi_B / d\O_M$ with increasing semiconductor ionicity.[53] Here the horizontal, ionicity axis expresses an intrinsic property of the semiconductor. The exact S values are the subject of some controversy[54] but nevertheless reflect a clear trend. Extensive research of clean semiconductor surfaces in the 1970's revealed that, with but a few exceptions, such states do not exist in the semiconductor band gap, where they are required in order to stabilize $E_F$.[10] Even in cases such as Si[55][56] and GaP[57] where intrinsic states are present, subsequent metallization on a monolayer scale removes them. The absence of states in the gap for compound semiconductors such as GaAs led to a re-examination of theoretically predicted states in the band gap. As a result of surface relaxation, semiconductor bond lengths and angles can readjust to minimize their free energy, thereby moving the so-called dangling bond states out of the band gap. Several techniques have confirmed these surface relaxations self-consistently. Hence intrinsic surface states do not play a significant role in electrical barrier formation.

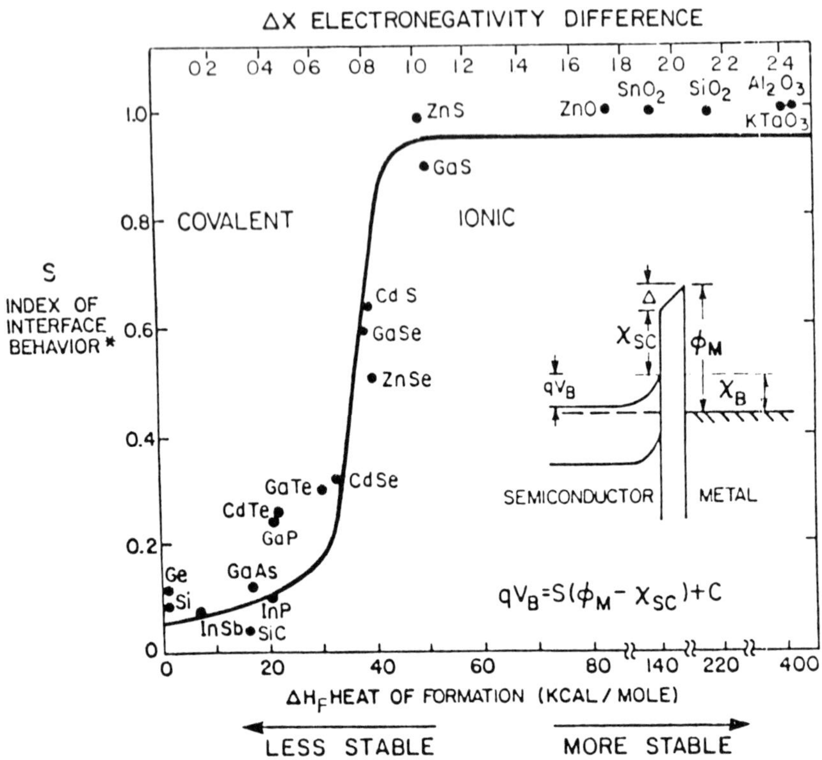


**Figure 9.** Transition in coefficient of interface behavior (see Ref. 53) between covalent and ionic semiconductors plotted versus chemical heat of formation *(lower scale)* and versus electronegativity difference *(upper scale)* of the semiconductor. See Ref. 10.

## 4.2 Intrinsic Interface States

Intrinsic interface states involve charge localization near the boundary of a metal with a semiconductor without a change in the chemical or electronic properties of either medium. Several models are available which satisfy these requirements. The basis for many of such models is the wave function tunneling from the continuum of metal traveling wave states into the semiconductor. The wave function tailing into the semiconductor band gap shown in Fig. 10 (b) constitutes an atomic-scale dipole which offsets part of the potential difference between metal and semiconductor. Heine[58] argued that such tails would be on the order of 10 Å or less for simple metals on Si. Refinement of this concept by various workers in the early 1970's involved substituting a featureless "jellium" of continuum states for the metal's electronic structure—thereby de-emphasizing localized chemical bonding and any other structural properties. See, for example, Schlüter's review, Ref. 59, and Ch. 8, this volume. By taking properties of different metals and semiconductors into account via electron density and dielectric screening, respectively, these workers could test barrier sensitivity to different metal work functions for semiconductors across a wide spectrum. Such calculations showed a weak dependence for high dielectric constants of covalent semiconductors such as GaAs and Si and much less potential screening for ionic semiconductors such as CdS and $SiO_2$. Fig. 9 supports this approach as well.

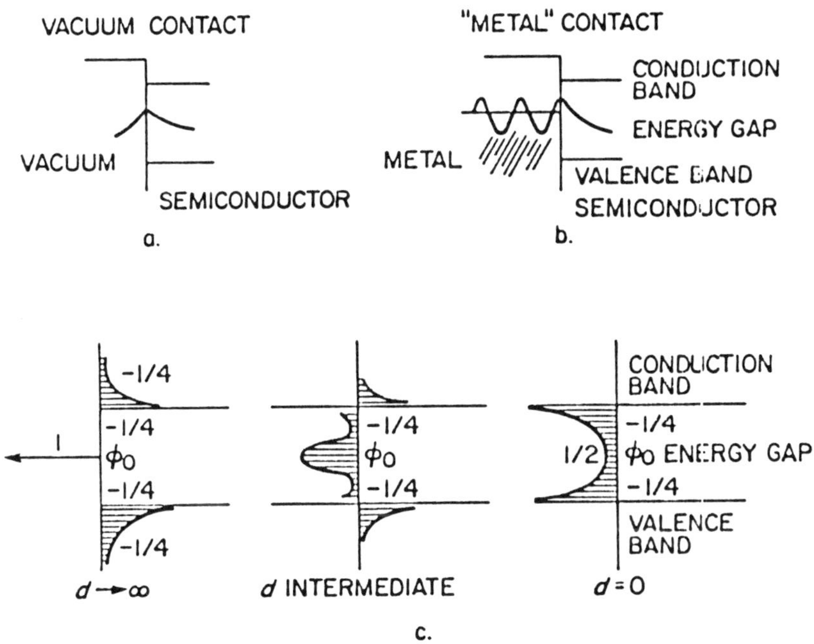


**Figure 10.** Localized wave function tunneling *(a)* into both vacuum and the semiconductor band gap at a metal-semiconductor interface. The exponentially-decaying behavior leads to a continuous density of interface states shown in *(c)* for a one-dimensional model of a covalent semiconductor-metal interface, where $\Phi_0$ defines a charge neutrality energy of the semiconductor and *d* is the metal-semiconductor separation. See Ref. 63.

Tersoff[62] and Tejedor et al.[63] provided an alternative model involving the existence of a charge neutrality level in the semiconductor band gap, minimizing the influence of metal work function on $E_F$ position. Two major points of this induced density of states model are: *(i)* wave function tunneling into the semiconductor band gap and *(ii)* a new density of states due to the metal wave function compensated by a decrease in semiconductor valence and conduction band density of states. Hence one can define a charge neutrality level such that the interface density of states below that level for the energy gap and the valence band compensate each other locally.[63] Figure 10c illustrates initial conduction and valence band densities of states without metal, the formation of discrete gap states at intermediate distances, and the formation of a continuum of states at the intimate contact, i.e., the resonance extends throughout the energy gap. Here the transfer of charge to or from the metal sets up a local dipole which tends to restore equilibrium of the metal $E_F$ and charge neutrality levels and renders barrier height insensitive to metal work function.

Eq. (6) $$D = (E_F - \Phi_o) / S$$

where D is the induced interface dipole, $\Phi_o$ is the charge neutrality level and S is a parameter which can be related to semiconductor dielectric properties.[63][64] For S ~0.1 - 0.3 and $E_F - \Phi_o$ differing by 0.1 eV, the induced dipole D can be on the order of an eV. Metal wave function tunneling is calculated to produce large local dipole charge of ~$10^{14}$ - $10^{15}$ $cm^{-2}$ $eV^{-1}$. As pointed out by Duke in Ch. 8, the calculated magnitude of this charge transfer depends sensitively on the assumed boundary conditions for charge penetration and may at best require experimental measurement of the $Ø_B$ dependence to establish unequivocally.

One can extend the charge neutrality model to heterojunctions by substituting a second charge neutrality level for $E_F$ in Eq. (6). Tersoff[60] has also related the charge neutrality level to the branch point (the gap energy between conduction- and valence-like states) of the semiconductors complex band structure.[65] Cardona and Christensen have calculated similar values.[66] An attractive feature here is that the semiconductor band structure is the only input. Masri[67] has presented the charge neutrality level in terms of metallic and metal-independent, semiconductor-like states. More recently, Flores et al.[68] have endeavored to factor in local atomic bonding; they find significant variations in charge neutrality level, depending on the metal-anion vs. metal-cation bonding. Schmid[69] and Mönch[70] have illustrated a monotonic dependence of $E_F$ "pinning" on the electronegativity of adsorbates on a semiconductor over several orders of charge magnitude. Figure 11 illustrates this dependence for GaAs with charge neutrality level (CNL) indicated.

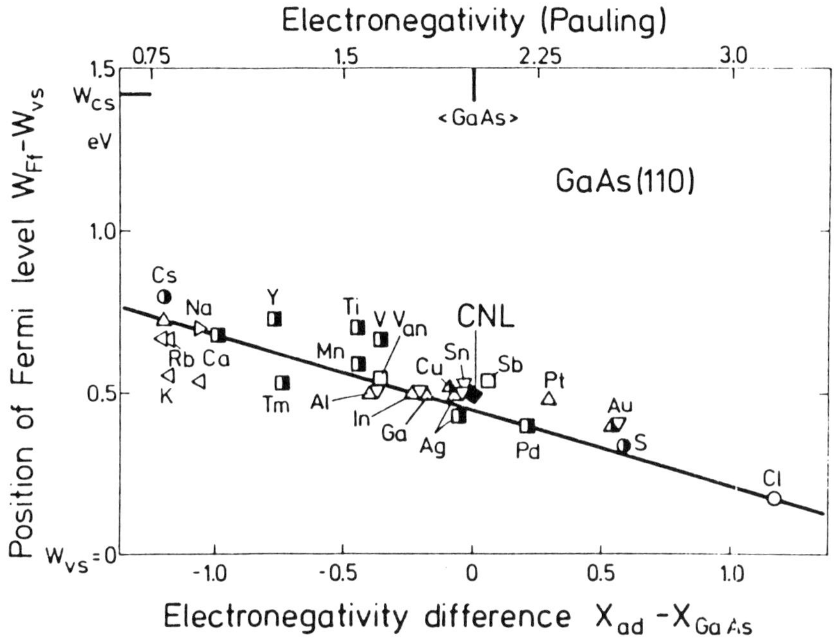


**Figure 11.** Final pinning positions of the Fermi level above the valence band edge versus electronegativity difference for adsorbates on UHV-cleaved GaAs(110) surfaces. CNL denotes the calculated charge neutrality level from Ref. 62. Data points compiled in Ref. 70 from references therein.

Potential screening by metal-induced gap states serves to reduce electrostatic effects of any surface defects. However, Zhang et al.[71] calculated that "metallic" behavior of metal-induced gap states is insufficient to screen out strong local interface potential effects. Local interface structure can produce several tenths of a volt barrier variation, due to a variation in the number of unpaired dangling bonds resulting from, for example, different substrate orientations, steps or point defects with high density, ~$10^{14}$ $cm^{-2}$ or higher. Hence the relative density of defect versus induced gap states may define a range of stabilization energies during barrier formation.[71][72]

Other theories of intrinsic $E_F$ pinning include: *(i)* Harrison's mechanism for shifting electrostatically the dangling bond hybrid orbitals in conduction and valence bands[73] to keep $E_F$ *pinned* at a constant gap energy, *(ii)* a *negative U* model of negative electron correlation between interface electrons due to atomic disorder and electron localization,[59] *(iii)* a narrowing of the semiconductor band gap at the interface, due to reduction of exchange-correlation contributions to the band gap,[59][74] and *(iv)* combinations of metal-induced gap states and defects, discussed in Sec. 6.

Tests for metal-induced gap states require atomically-abrupt metal-semiconductor interfaces, i.e., without additional chemical phases with their own dielectric properties. As will be seen, such abrupt interfaces are, in practice, difficult to obtain. Of the few such junctions, perhaps most notable are the epitaxial Si-silicide interfaces. In Ch. 3, Tung discusses the different barriers observed for different orientations of $NiSi_2$ on Si(111)[75] and Pb silicide on Si(111).[76] Such effects argue against the dominance of metal-induced gap states. Similarly, Ir silicide/Si interfaces exhibit transport properties consistent with a high density of interface states.[77] However the temperature dependence of the barrier heights differs from the predictions of a charge neutrality level model,[78] while correctly describing trends of the semiconductor's indirect band gap.

## 5.0 EXTRINSIC STATES: CONVENTIONAL

In general, it is quite difficult to obtain the ideal metal-semiconductor junctions pictured in standard solid-state physics texts. Rather, a host of phenomena over and above the bulk properties of metal and semiconductor can contribute to the electronic features of their interface. In this section, we review those phenomena related to conventional deviations from ideal chemical or structural properties of the metal and/or semiconductor constituents. In the next section, we will discuss those phenomena arising from detailed microscopic interactions between metal and semiconductor.

### 5.1 Surface Chemical Contaminants

Chemisorbed species can produce new electronic states in the semiconductor band gap, an effect noticed rather early in Schottky barrier studies.[79][80] Recent photoemission studies of $E_F$ movements at low temperatures (50 - 100 K)—low enough to retard clustering significantly and/or chemical bonding—exhibit *overshoot* phenomena believed to be due to charge exchange involving adsorbate donor or acceptor states.[64][81][82] Figure 12 illustrates the $E_F$ dependence on metal coverage and its relation to the proposed adsorbate level. For example, the effect of a positively charged donor is to raise $E_F$ closer to the donor level energy. Relating this adsorbate level energy to the properties of the isolated adatom—for example, the first ionization potential—one obtains the linear dependence shown in the lower portion of the figure. This is analogous to adsorption on ionic surfaces[83] and in accordance with Mönch's model.[84] It is in fact possible to infer bonding configurations from electronic spectra altered by chemisorption.[86][87] Strongly bonded adsorbates alter the surface electronic properties for more ionic semiconductors as well.[88][89]

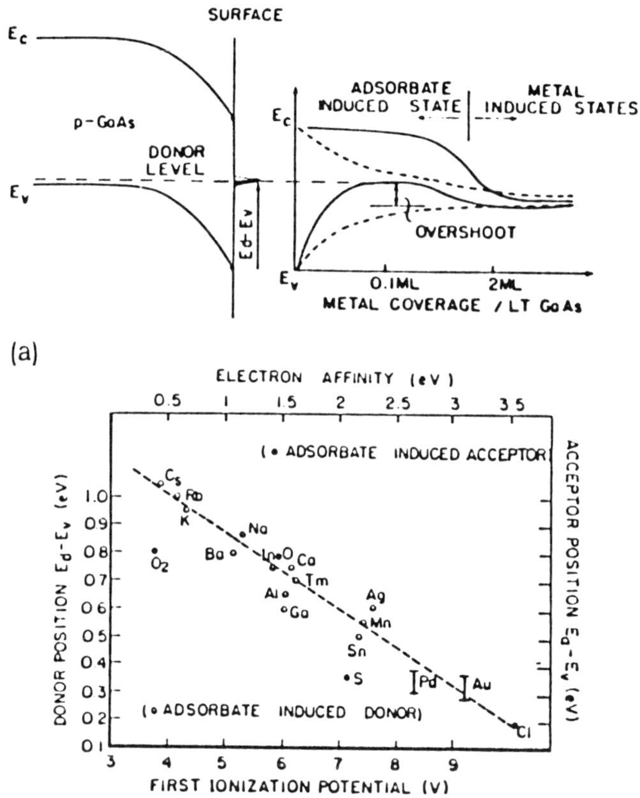


**Figure 12.** Characteristic Fermi level movement (*a*) as a function of low coverage of metal deposited on GaAs(110) at room temperature *(dashed line)* and low temperature *(full line).* The adsorbate-induced donor level shown at left corresponds with the initial maximum $E_F$ energy for p-type GaAs. At higher coverages, the $E_F$ convergence is attributed to metal-induced states. In (*b*), the adsorbate induced donor (acceptor) level position extracted from the initial $E_F$ movement on low temperature p-type (n-type) GaAs appears as a function of the adatom's first ionization energy (electron affinity). See Ref. 85 and data references therein.

Chemical contamination on a monolayer scale produces dramatic changes in interface $E_F$ movement. Exposure of clean surfaces to atmosphere can result in dramatic changes in the evolution of band bending with subsequent metallization. Figure 13 illustrates SXPS measurements of such $E_F$ movement for Au on *(a)* clean InAs and *(b)* InGaAs (100) surfaces with and without air exposure.[90] As shown, air exposure alters both the premetallized band bending and final Schottky barrier height. The air-exposed values agree rather well with diode measurements of Au on the same semiconductors prepared under low-vacuum conditions.[91] For diodes prepared under UHV conditions, Missous et al.[92] found that residual contamination had pronounced effects on the MBE growth and J-V characteristics of epitaxial AlGaAs (100) junctions. Newman et al.[93] have reported that air exposure significantly degrades the stability of metal/GaAs(110) diode barriers with annealing.

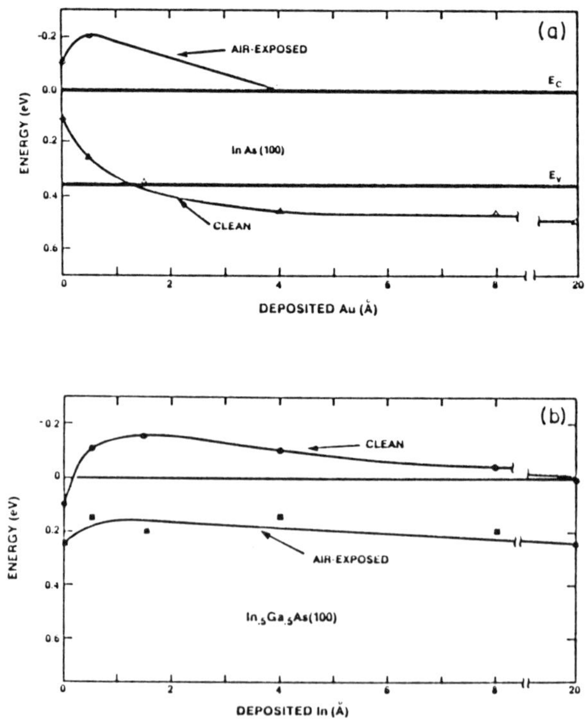


**Figure 13.** Fermi level movements as a function of metal deposition for Au on UHV-cleaved InAs(110) (*a*) and $In_{0.5}Ga_{0.5}As(100)$ (*b*). The effect of air exposure is to alter dramatically the $E_F$ movement and ultimate Schottky barrier height versus the clean surface. See Ref. 90.

## 5.2 Surface Structural Imperfections

With the introduction of UHV techniques it is now possible to detect and control surface chemical contamination on a scale of 0.01 monolayer (~$10^{13}$ atoms/cm$^2$) or less, enabling researchers to study the microscopic properties of clean surfaces and interfaces. Even for such clean semiconductor surfaces, numerous structural imperfections exist which can introduce new states into the band gap. These include exposed lattice steps due to cleavage or misorientation as well as surface point defects.

Cleavage steps are well-known sources of electronic states in the semiconductor band gap.[94] Early work by van Laar and coworkers[95][96] demonstrated that cleavage steps are responsible for $E_F$ pinning of UHV-cleaved GaAs (110) near mid-gap. Henzler[97][98] used LEED as a diagnostic tool to determine step heights and terrace widths. Rowe et al.[99] found a strong influence of surface state photoemission to such cleavage step density. Furthermore, such steps can act as catalysts for chemical processes and, thereby, additional electronic changes—for example, the dependence of oxygen chemisorption on cleaved Si step density.[100]

Steps associated with misorientation are also electrically active. As will be described in Sec. 7, vicinal GaAs (100) planes exhibit interface states deep in the semiconductor band gap whose density scales linearly with the density of exposed, active chemical sites.[101] This relationship is inclusive of both misorientation direction and angle. Misorientation angles of 2 - 4 degrees, common in MBE growth, produce acceptor level densities near mid-gap of 0.5 - 1 X $10^{14}$ cm$^{-2}$ and change Al/GaAs barrier heights by nearly half a volt.[102] Consistent with such effects, Pashley has used scanning tunneling microscopy to observe pronounced differences in exposed site densities, depending on the orientation of step edges for such vicinal GaAs(100) surfaces.[103] Significantly, such sites appear charged only after metallization,[102] suggesting that the interface state forms as a result of adsorbate-step site bonding.

Point defects are yet another source of surface and interface charge. Thus ion bombardment creates states within the band gap while otherwise cleaning the semiconductor surface. For example, low energy Ar Ion bombardment of CdS produces S interstitials and vacancies[104] and vacancy loops/dislocations for trigonal Se.[105] STM measurements by Whitman et al. have provided evidence for such Schottky point defects at GaAs surfaces.[106] Radiation effects by high energy electrons and ions are well-known to generate native defects.[107][108] Walukiewicz[108] has noted the correspondence between the energy levels of such defects and the stabilization energy range for metals on melt-grown GaAs. See also Ref. 109.

## 5.3 Bulk Impurities

Foreign atoms within the semiconductor lattice can introduce deep levels which can alter the surface space charge region. See, for example, Refs. 110 and 111. Such impurities may be associated with the semiconductor growth process (i.e., C in GaAs), nonstoichiometry of the growth phase (i.e., excess As in GaAs) or as a result of in-diffusion during metallization and processing,[112] (i.e., the electronic effects of Au diffusion into GaAs). Hall effect measurements show that the metal produces deep acceptor levels at 0.4 and 0.05 eV above $E_{VBM}$.[112] Likewise, C-V measurements provide evidence for deep donor levels 0.9 - 1.2 eV below $E_c$.[113] As in III-V compound semiconductors,[110] metal impurities and their complexes with native defects in II-VI compound semiconductors can introduce a multiplicity of deep levels which span the band gap. Halsted et al.[111] have provided deep level luminescence energies for a variety of metal impurity centers and native defects in CdS, CdSe, CdTe, ZnS, ZnSe, and ZnTe. Correspondingly, Shaw et al.[116] have used a combination of J-V, C-V, and SXPS to show that Au overlayers give rise to a compensated region near the interface which extends several thousand angstroms into the semiconductor.

In general, noble and near-noble metals such as Au, Ag, Cu, and In tend not to react strongly with semiconductors and diffuse rapidly at relatively low temperatures.[114][115] Such diffusion is particularly pronounced for the more ionic semiconductors, where relatively large lattice constants and vacancy concentrations assist impurity motion.

## 5.4 Bulk Structural Imperfections

Imperfections within the semiconductor's local or extended lattice structure are a prime source of electronic states. These include: misfit dislocations, antiphase domain boundaries, and native defects produced during growth and/or processing such as interstitials, vacancies, antisites, and other compound defects.

Misfit dislocations are a common feature of epitaxially grown semiconductor layers and have electrical activity. Woodall et al.[117] have demonstrated their influence on electrostatic barriers between GaInAs epilayers and their GaAs substrates. Figure 14 illustrates schematically the energy band bending and $E_F$ position across the surface with misfit dislocations. Each such line imperfection contains charge which gives rise to a depletion region extending radially outward. As misfit density increases, depletion regions can overlap, thereby increasing the barrier height over a greater portion of the surface area.

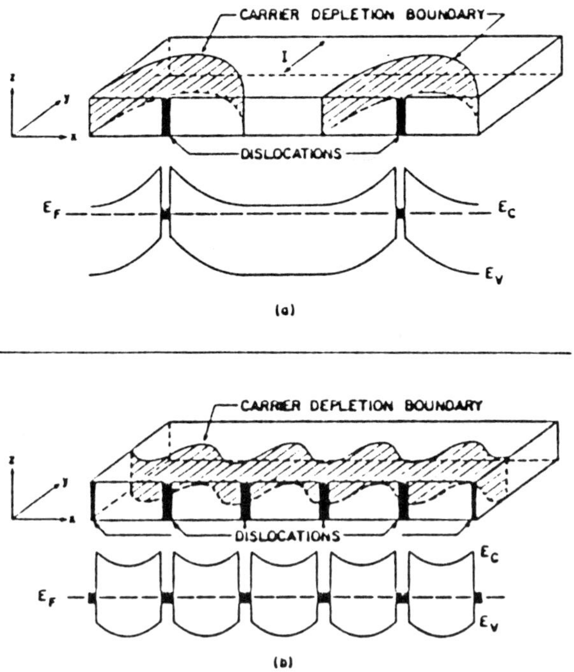


**Figure 14.** Schematic representation of the energy band bending and $E_F$ position across a surface with misfit dislocations. With increasing density of misfit dislocations, the depletion regions overlap, increasing the effective Schottky barrier height. See Ref. 117.

Metallization of semiconductor structures can also lead to deep levels as a result of interfacial stress. Ti silicide gate electrodes over $Si/SiO_2$ interfaces exert a compressive stress upon cooling from the preparation temperature.[118] As a result, deep levels form both in the $SiO_2$ and at the $Si/SiO_2$ interface. Plastic deformation is known to create bulk defects with deep levels in compound semiconductors. For GaAs, such levels include, but are not limited to, the well-known EL2 mid-gap level.[119][120] Antiphase domains are believed responsible for deep levels affecting Au-βSiC contacts.[121]

Native defects in the semiconductor lattice provide a multiplicity of deep levels in semiconductors.[122] See for example, Fig. 15. Weber and Liliental-Weber treat their influence on barrier properties at semiconductor contacts in Ch. 6. Here, we note that semiconductor growth conditions may have a pronounced effect on the energies and densities of such deep levels. Thus, for example, Fig. 16[41] illustrates the orders-of-magnitude difference in bulk deep level emission from GaAs grown from the melt (liquid encapsulated Czrochalski [LEC]) versus by MBE. The LEC spectrum exhibits two orders-of-magnitude greater emission from deep states, including a 0.9 eV emission feature not evident in the MBE spectrum. Insets illustrate schematically the difference in deep levels and recombination processes giving rise to the observed photoluminescence. Analogous differences are evident in the CLS spectra from near-surface regions of both GaAs types. The presence of the level at 0.9 eV is especially significant, given the tendency of $E_F$ to stabilize near this energy for melt-grown GaAs contacts.

Similar variations in deep level emission appear for photoluminescence studies of CdTe crystals. Here again, these deep levels can appear at energies throughout the band gap.[41] Shaw et al.[123]-[125] have provided evidence that such levels can have a strong influence on the Schottky barrier formation. Figure 17 illustrates the comparison of luminescence and interval photoemission spectra for two CdTe crystals.[41] Specimen A exhibits emission from a deep level 1 eV above $E_V$ (~0.6 eV below $E_{CBM}$). The corresponding internal photoemission thresholds exhibit dramatically different barrier heights for the same Au contact. Interestingly, the 0.91 eV barrier agrees with that expected for the 5.15 eV work function of Au vs. the 5.78 eV ionization potential[126] of CdTe. Conversely, the 0.67 eV barrier corresponds almost exactly to the deep level energy in the CdTe band gap.

Shaw et al. have demonstrated the influence of such deep levels on the $E_F$ movement at other metal / CdTe interfaces, notably In / CdTe—where incremental laser annealing of the metallized semiconductor produces staged changes in interface bonding and associated $E_F$ movements to several energy plateaus.[124][125] These plateaus correspond to the multiple deep level emission energies observed. Hence the presence of bulk deep levels appear to have an influence on the $E_F$ movement and barrier heights for metals on clean semiconductor surfaces.

A common assumption has been that such imperfections are homogeneously distributed with low enough density that they play at most a secondary role in interface band bending. However, if segregated to the surface or interface, their effect on the band bending may well increase. Surface segregation of chemical species and consequent changes in electronic states are evident in numerous studies. Bartels et al.[127] have reported surface segregation of As at cleaved surfaces of melt-grown GaAs, which typically contains an excess of As to reduce the density dislocations.[128][129] Consistent with this As movement, Makram-Ebeid et al.[130] have attributed the orders-of-magnitude decrease in EL2 concentration near the surface of GaAs annealed to 850°C to the out-diffusion of As—even with cap layers under a hydrogen flux.

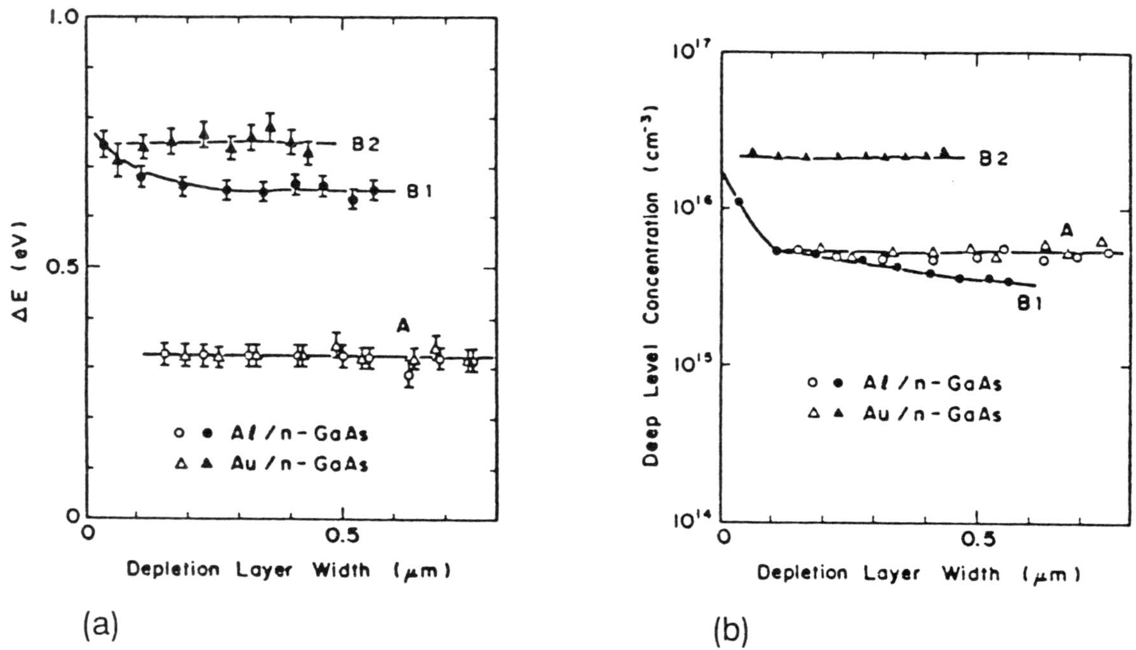


**Figure 15.** Activation energy depth profiles (*a*) and deep level concentration profiles (*b*) from DLTS spectra for deep levels A(O) and B1(●) in Al/n-GaAs and for A(Δ) and B2(▲) in Au/n-GaAs. For deep level B1 activation energy and deep level concentration increase toward the metal/semiconductor interface. See Ref. 131.

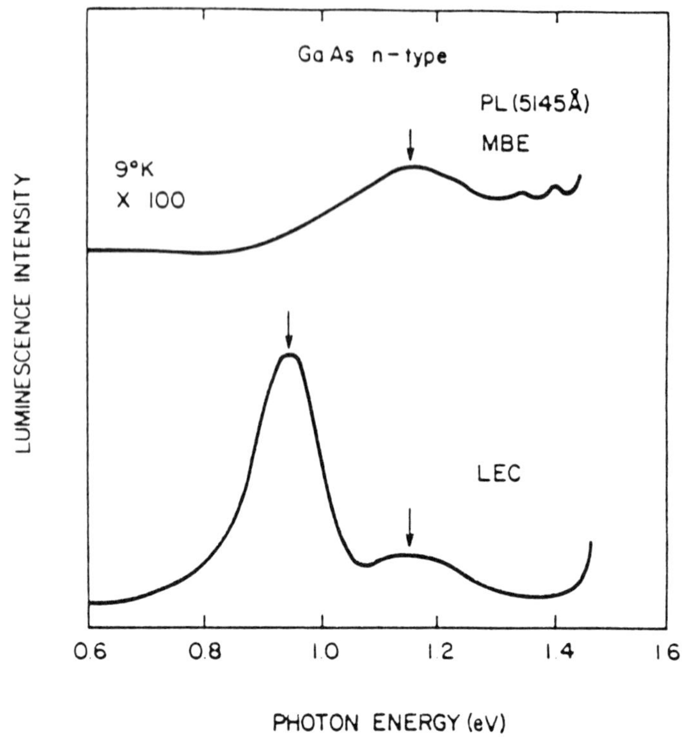


**Figure 16.** Low temperature (9°K) photoluminescence spectra of deep level bulk emission from GaAs(100) surfaces grown via liquid encapsulated Czrochalski (LEC) versus molecular beam epitaxy (MBE) using excitation from a 5145Å laser. The LEC crystal exhibits orders-of-magnitude higher deep level emission intensity near mid-gap. See Ref. 41.

The presence of metal at the semiconductor surface appears to play a role in deep level segregation. Yahata and Nakajima[131] found nearly an order-of-magnitude increase in deep level concentration at Al/GaAs contacts as measured by DLTS. As represented in Fig. 15, changes in both density and energy of only one level are evident and appear to depend on the particular metal. However, segregated concentrations in excess of $10^{18}$ $cm^{-3}$ are required to produce measurable electrostatic effects at the interface, e.g., near-surface concentrations of $\sim 10^{13}$ - $10^{14}$ $cm^{-2}$. Since bulk deep level concentrations for, e.g. EL2, are typically in the $10^{16}$ $cm^{-3}$ range, concentration increases of one to two orders of magnitude due to segregation may be sufficient to produce first-order electrostatic effects.

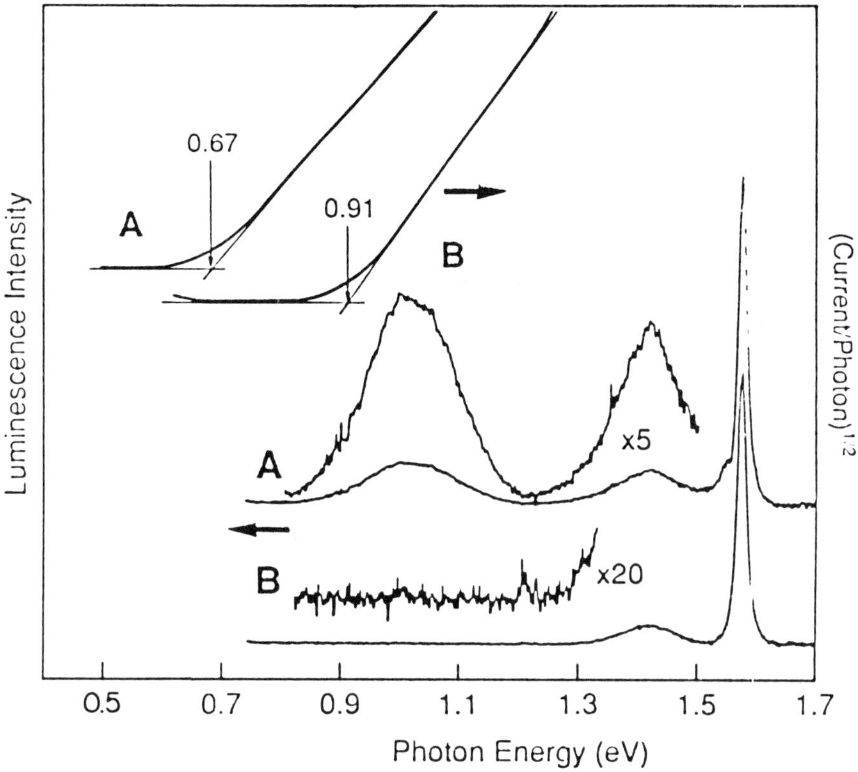


**Figure 17.** Deep level photoluminescence spectra *(left axis)* and internal photoemission barrier measurements *(right axis)* for Au deposited on two UHV-cleaved CdTe(110) surfaces. The 0.91 eV barrier in B corresponds to the Schottky limit value anticipated for Au/CdTe. The 0.67 eV barrier in A corresponds to a $E_F$ position near the additional deep level energy evident near 1 eV in the CdTe band gap. See Ref. 41.

Overall, these extrinsic features highlight the importance of the bulk and surface crystal quality in Schottky barrier formation. Deviations from ideal crystal properties on an atomic, nanometer, or micron scale contribute numerous electrically-active sites which compete with interface-specific phenomena to control the contact rectification.

## 6.0 EXTRINSIC STATES: INTERFACE SPECIFIC

In this section we present evidence for the role of interface-specific phenomena in determining the localized charge states affecting Schottky barrier formation. Such evidence derives from a variety of macroscopic and microscopic observations, portions of which are still controversial. Such controversy arises because *(a)* in some cases, more than one mechanism can be used to account for the measurements, and *(b)* the mechanisms themselves are to some degree interrelated. We examine three types of interface-specific extrinsic states: localized states due to metal-semiconductor bonding and interdiffusion, defect formation, and new phase formation. All three are sensitive to the kinetics and thermodynamics of interface chemistry, although in varying degrees.

### 6.1 Chemically-Induced States

Evidence for the primary role of chemically-induced states in Schottky barrier formation has until recently been phenomenological.[10] Andrews and Phillips's classic study[132][133] presented a linear correlation between the barrier heights of transition metal silicide/Si junctions in terms of heats of formation (enthalpy changes) of the transition metal silicide. They interpreted their barrier vs. $H_F$ ( $\Delta H = \Delta G + T\Delta S$, where $\Delta G$ and, $\Delta S$ are changes in Gibbs' free energy and entropy, respectively) in terms of local bonding at the interface. Starting with a Bardeen-type barrier,[9] they proposed shifts in barrier height proportional to the charge transfer of new chemical bonds and hybridized bonding between metal and Si atoms to be linear in $H_F$. Although the extension to noble and near-noble metal silicides is less convincing[69][134] these results are the first to emphasize the dependence of macroscopic electronic properties on microscopic chemical bonding of the interface.

Barrier heights for transition metals on Si[135] also exhibit a linear dependence on a measured thermodynamic property of the interface—the eutectic temperature for transition metal silicide/Si systems. The particular eutectic used was that closest on the metal side for silicides whose growth is dominated by metal diffusion and vice versa for interfaces with growth dominated by Si diffusion. As shown in Fig. 18, $\Phi_B$ increases with decreasing eutectic temperature, extends over nearly half the band gap in energy, and represents a correlation based on a true interfacial parameter. Besides reflecting the importance of thermodynamic bonding, this relationship suggests that band bending is related to the presence of a microscopic interfacial layer.

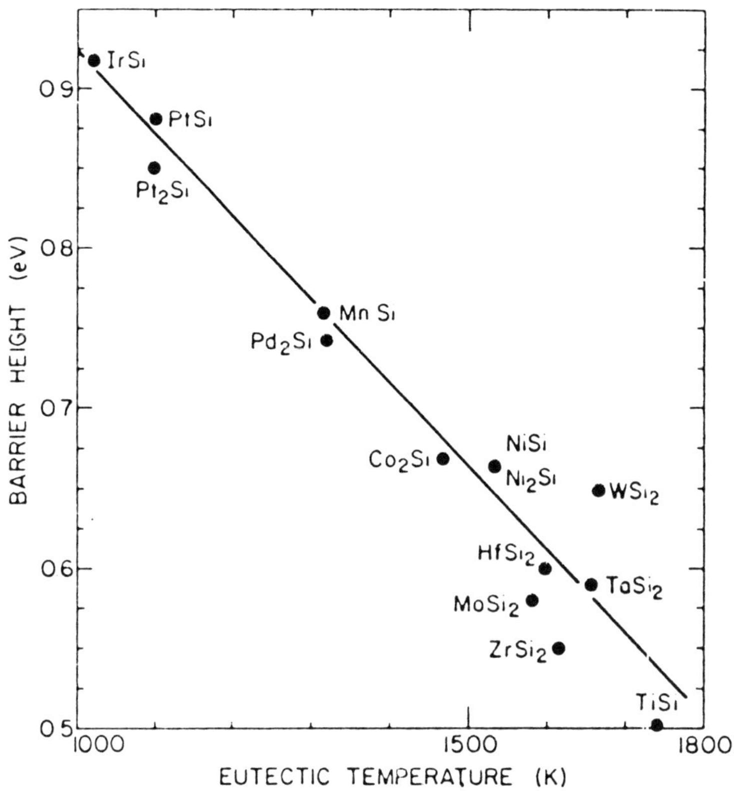


**Figure 18.** Schottky barrier height $Ø_{SB}{}^{N}$ (n-type) as a function of eutectic temperature for selected silicide-forming contacts. See Ref. 135.

In an extensive review of Si/transition metal interface compounds, Calandra[136] emphasized the importance of geometric effects—namely, the detailed positions of metal atoms at the interface—in determining the density of states in the band gap region. Barrier measurements of epitaxial $NiSi_2$-Ni interfaces by Tung[75] emphasize this geometric bonding aspect of interface charge transfer. As reviewed in Ch. 3, Tung et al. reported significant differences In $\Phi_B$ for two epitaxial interfaces of $NiSi_2$/Si: 0.65 eV for type-A and 0.78 eV for type-B epitaxy, respectively.[75] Liehr et al.[137] have challenged these measurements on the basis of possible differences in structural perfection rather than specific type of epitaxy. While misfit dislocations and residual impurities can indeed play a role in such $\Phi_B$ differences (as discussed in the previous section), these results have been confirmed experimentally[138]-[140] to some extent and theoretically justified in terms of tunneling charge transfer.[141][142] Recently, analogous $\Phi_B$

differences have been reported for Pb/Si (111) interfaces differing only in the structure of the first layer of Pb and Si atoms at the interface.[76][143]

Numerous photoelectron spectroscopy studies performed in the 1970's and 1980's underscore the role of localized bonding at the Si-metal junction. These results have emphasized the specific nature of the metal-Si bonding for metals such as V,[144] In,[145] Al,[145] Ga,[145] Ni,[75] Ir,[77] and Pd.[146][147] Besides the role of d-bonding informing barriers for many metals, Hara and Ohdomari[148] have noted a correlation between heats of silicide formation and the metal d-band occupancy as well. Figure 19 serves to underscore the relationship between local bond charge transfer and metal-Si $\Phi_B$'s. Here Schmid[69] first showed that both transition metal silicides and noninteracting metals on Si exhibit a near-linear dependence on (Miedema) electronegativity. In a comprehensive discussion of factors associated with such a relationship, Schmid advanced a model based on charge neutrality (see Sec. 4) with interfacial density of states determined by metal-Si bonding within the interfacial region. Such a charge neutrality level depends more sensitively on electronegativity differences than predicted by a model based on the semiconductor band structure alone. Figure 19 also suggests a measurable but secondary role for discrete defect states in the band gap.

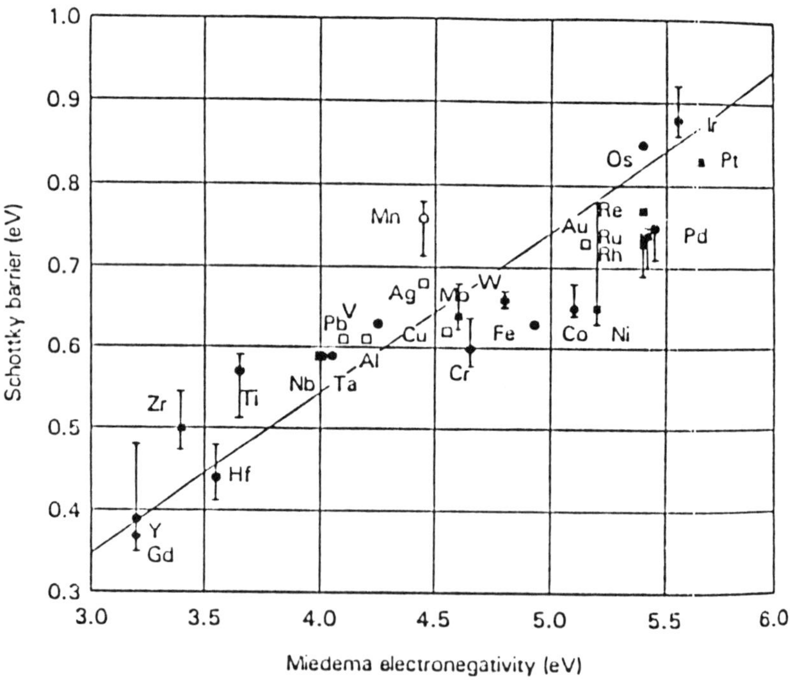


**Figure 19.** Barrier heights of transition metal silicides and nonreacting metal Schottky barriers on n-type silicon. Data points correspond to silicides (■ and ●) and nonreacting metals (□). See Ref. 69 and literature (●) references therein.

A strong correlation between $\Phi_B$'s and interface thermodynamics is evident for compound semiconductors as well. Figure 20 illustrates a pronounced transition in Schottky barrier heights for metals (M) on compound semiconductors (cation C and anion A) which depend on the heat of interface reaction, calculated per metal atom according to the reaction[149]

Eq. (7) $$M + (1/x)\ CA = (1/x)\ [M_XA + C]$$

so that

Eq. (8) $$\Delta H_R = (1/x)\ [\ H_F\ (CA) - H_F\ (M_XA)].$$

As shown, this transition in $\Phi_B$'s[150] (determined from internal photoemission) falls at the chemical boundary between reactive and unreactive interfaces, $\Phi H_R = 0$ for all four semiconductors, regardless of ionicity. Furthermore, the transition agrees with experimental observations of interface chemical bonding. Numerous studies have confirmed this $\Phi_B$ dependence on interface chemical reactivity,[151]-[155] even to an extent for GaAs.[156] This apparent relationship between barrier height and interface chemical reactivity suggests that chemical stability is a factor in the $\Phi_B$ dependence on different metals. This is borne out in Fig. 9 where, instead of electronegativity, the same index of interface behavior can be expressed in terms of semiconductor heat of formation.[10][149] Hence, semiconductors with low chemical stability exhibit the least $\Phi_B$ dependence on different metals whereas highly stable semiconductors exhibit the most.

SXPS measurements reveal that both anion and cation out-diffusion into the metal overlayer are significant for clean interfaces, even at room temperature. Indeed, the extent of this out-diffusion depends monotonically on the semiconductor stability, i.e., the heat of formation.[156a] For reactive interfaces, however, the strong metal-anion bonding at the intimate contact acts to inhibit the anion out-diffusion *(chemical trapping)*, resulting in cation-rich out-diffusion.[157] Such chemical trapping is illustrated in Fig. 21 for the metal-GaAs interface. Here, the presence of monolayers or even submonolayers of a metal which reacts strongly with the out-diffusing anion (e.g., Al with As) can alter the ratio of Ga to As diffusion into a Au overlayer by orders of magnitude. Compositional profiling of such interfaces confirms the accumulation of anion at the reacted junction. In addition, the extent of chemical trapping varies with the metal anion reactivity.[157a] Thus thermodynamics appears to have a systematic influence on the dissociation and out-diffusion of semiconductor species during contact formation. See also Ch. 7.

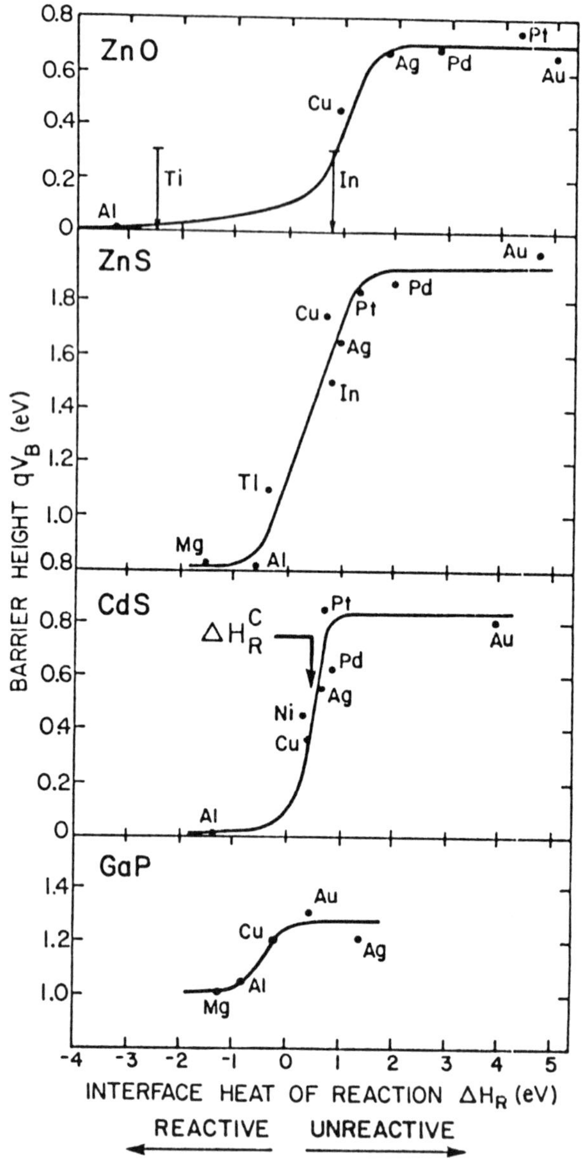


**Figure 20.** Correlation between barrier heights measured by internal photoemission (Ref. 150) as a function of interface heat of reaction (see Eqs. 7 and 8). See Ref. 149. The transition from low to high barrier height occurs at $\Delta H_R{}^C$, observed spectroscopically. The same qualitative behavior occurs for a wide variety of semiconductors, regardless of ionicity.

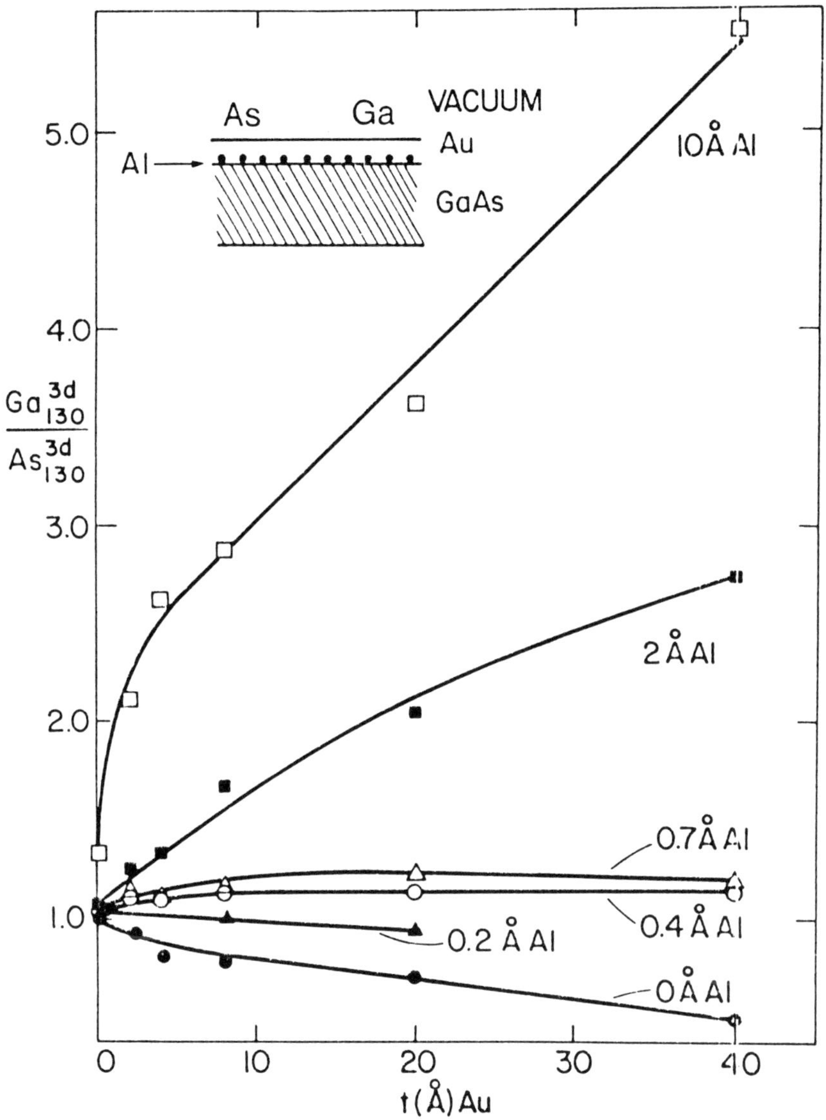


**Figure 21.** Evidence for chemical trapping of outdiffusing anions by a reactive metal interlayer. Here monolayer or submonolayer amount of Al between a Au overlayer and a clean, UHV-cleaved GaAs(110) surface trap outdiffusing As atoms and alter the resultant Ga/As ratio within the Au overlayer by nearly two orders of magnitude. See Ref. 157.

This influence of chemical reactivity on compound semiconductor out-diffusion in fact provides a chemical basis for Schottky barrier formation in these materials. Figure 22 shows SXPS measurements of anion/cation intensity ratios measured for the given thicknesses of different metal overlayers. Depending on the metal, anion-rich or cation-rich out-diffusion is possible for InP.[158][159] The inset shows the corresponding $\Phi_B$ vs. $\Delta H_R$ plot for InP according to Eqs. (7) and (8) above. Significantly, metals producing anion-rich out-diffusion lead to high $\Phi^n_B$'s whereas, without exception, metals leading chemical trapping and cation-rich out-diffusion produce low $\Phi^n_B$'s.[151] Here then, the link between out-diffusion stoichiometry and $E_F$ position suggests that the semiconductor out-diffusion can produce electrically-active defect sites with different deep level energies.[158] Thus chemical reactivity provides a basis for both localized defect and new phase formation at metal-semiconductor interface.

## 6.2 Localized Defect States

As already discussed in the previous section and by Weber and Liliental-Weber in Ch. 6, deep level defects are the subject of considerable interest due to their influence on bulk semiconductor properties. For compound semiconductors such as GaAs, there are almost three dozen possible defects involving simple vacancies, antisite defects, and interstitials, which can have energies spanning the band gap.[110][122][160][161] The association of these defects with metal impurities further extends this range of properties. Along with Wieder,[162] and following on suggestions of Mark et al.,[163] and Lindau et al.,[164] Spicer et al.,[165] and Williams[166] proposed that adsorbate chemisorption on semiconductor surfaces produced surface defects by breaking surface bonds, that these surface defects produce deep level densities sufficient to pin the $E_F$, and that the $E_F$ position is independent of chemisorbed element for a given surface. See, for example Fig. 23. A basis for proposing the surface bond breaking mechanism was the rapid stabilization of $E_F$ for GaAs with only submonolayer deposition and similar behavior for both metals and oxygen chemisorption. SXPS studies of metal deposition on semiconductors at low temperatures (see Weaver, Ch. 7) have revealed substantial effects of temperature on deposited film morphology and the tendency of metals—particularly those with no strong interface chemical reaction—to cluster at room temperature. Zunger[167] showed that such cluster formation could release sufficient energy to overcome the activation barriers for breaking surface bonds. Furthermore, such clusters could exhibit near-metallic (delocalized) electron states even for submonolayer surface-averaged depositions. Thus photoemission measurements of $E_F$ movement may, in fact, average potential variations due to clean and metallized areas, especially at room temperature.

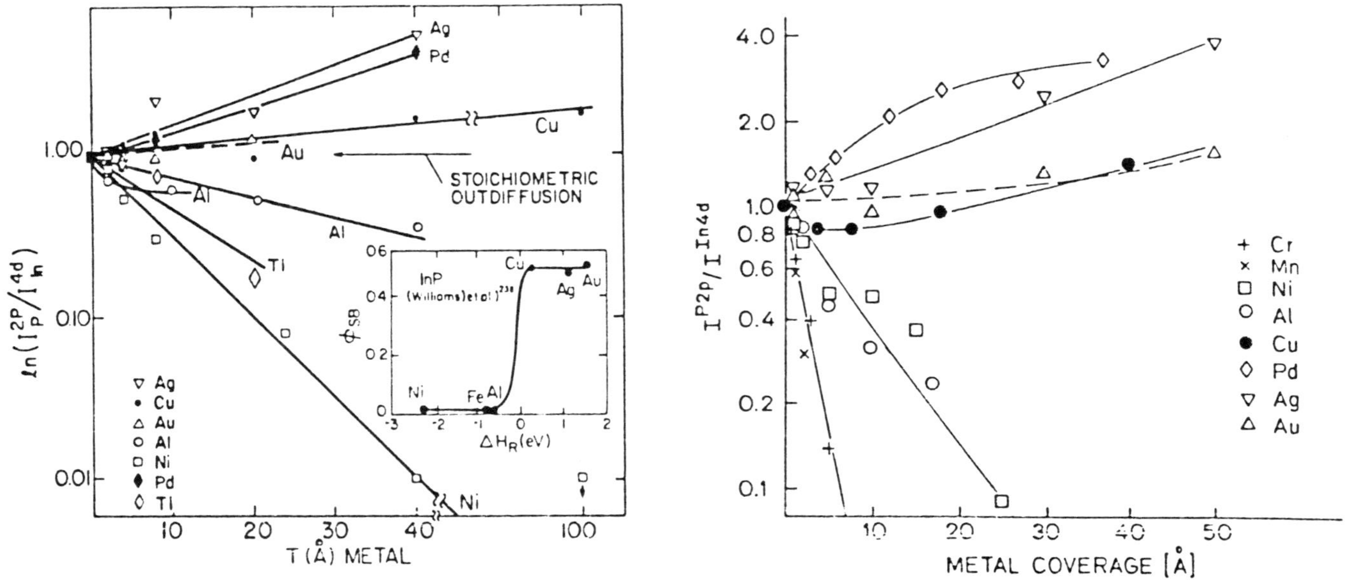


**Figure 22.** (*a*) Ratio of surface anion/cation core level intensities $I_p^{2P}/I_{in}^{4d}$ measured by SXPS versus metal coverage on InP(110) relative to the UHV-cleaved surface ratio for Ag, Pd, Cu, Au, Al, Ti, and Ni. The inset contains a barrier height $Ø_{SB}$ versus $\Delta H_R$ (See Ref. 151.) and emphasizes the correlation between $Ø_{SB}$ and the stoichiometry of out-diffusion. See Ref. 158. (*b*) Similar data for Ni, Al, Cu, Pd, Ag, and Au as well as for Cr and Mn appears in Ref. 159.

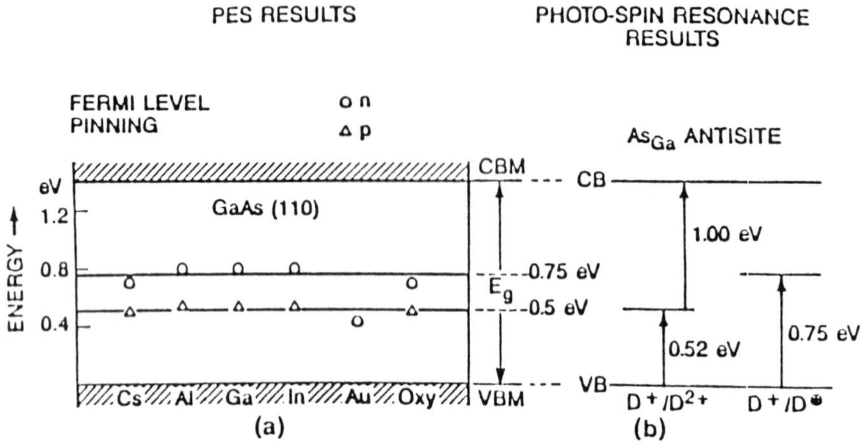


**Figure 23.** Photoemission spectroscopy (PES) (Ref. 165) and photoinduced electron spin resonance measurements (Ref. 181) of defect levels in GaAs and their relation to $E_F$ pinning. Pinning positions for n-type (O) and p-type (Δ) GaAs appear for the indicated elements in the room temperature 1.42eV GaAs band gap indicate defect energies at 0.65 and 0.5 eV, respectively. The photospin resonance results indicate nearly identical energies of the $As_{Ga}$ antisite levels above the valence band maximum at 8ºK (Eg = 1.52 eV). See Ref. 183.

Considerable theoretical activity has focused on the deep level energies of vacancies[168]-[171] and antisite[172]-[174] defects at or just below the semiconductor surface and their trends with particular metal or semiconductor composition.[91][175][176] Early interface defect models incorporated surface anion and cation vacancies.[162][165] However, VanVechten[177] has shown that antisite defects typically have lower activation energies than simple defects in III-V compounds. Given the limitations in predicting absolute energies from these theoretical models (see Duke, Ch. 8), comparison of theoretical and experimentally-observed energies is not sufficient to rule out pinning by one or the other type of defect.

Several groups have investigated the J-V and C-V properties of metal/GaAs interfaces, including GaAs (110) cleaved in UHV,[178] as well as GaAs (110) and (100) polished, etched and annealed in UHV.[166][179] These measurements all show a relatively narrow range of energies, consistent with photoelectron spectroscopy measurements. In contrast, Waldrop and Grant[179] found changes in InP $E_F$ position of up to 0.6 eV with different chemical treatments, which they attributed to multiple states of a single defect. These observations confirmed the earlier work of Williams et al.[166]

Furthermore, a 0.5 eV range is evident from more recent photoemission studies of Aldao et al.[180] Thus GaAs appears to have one of the narrowest ranges of $E_F$ stabilization energies for III-V compounds studied.

Weber et al.[181][182] have noted the correspondence between energy levels shown in Fig. 23 and those observed for the deep level EL2. They have proposed an AsGa antisite defect as the basis for $E_F$ pinning at metal/ GaAs interfaces. This most recent embodiment of the defect model for interface states, termed the *Advanced Unified Defect Model* (AUDM),[183] involves the formation of EL2 donor defects in the near surface region with the two donor levels indicated in Fig.23. Furthermore, the crystal must have additional minority acceptor levels below 0.5 eV in order to account for pinning at mid-gap in n-type GaAs. Here a $Ga_{As}$ antisite defect is proposed.[184] In this defect model, the $E_F$ position depends to some extent on interface chemistry and in particular on the interface anion/cation stoichiometry. The n- and p-type stabilization in the 0.5 - 0.75 eV energy range shown requires $AS_{GA}$ antisite concentrations close to twice those of the $Ga_{As}$ antisites. In turn, this requires the interface to be As-rich in general—consistent with the growth of the melt-grown cleavage material from the As-rich end of the GaAs bulk phase diagram. Less As-rich interfaces would then lead to pinning at the 0.5 eV level or below. Therefore the particular metal-GaAs chemistry will determine the Ga/As concentrations near the interface, the ratio of different antisite defects, and thereby the $E_F$ position in the range between the defect levels shown. Alternatively, this range had been attributed to differences in electronegativity.[179] Fermi level positions outside this narrow range of energies can be attributed to formation of additional defect complexes. Stoichiometry effects on the type and density of bulk GaAs deep levels are well-known. Studies of GaAs $E_F$ vs. surface reconstruction behavior with annealing,[185][186] Ga films on GaAs,[187] and $LaB_6$ Schottky barrier behavior with annealing[188] provide support for this variation In $E_F$ position with GaAs interface stoichiometry, consistent with the AUDM (and with other composition-based models discussed below).

On the basis of photocapacitance measurements, Nishizawa et al.[189] have also proposed stoichiometry-dependent deep levels in addition to EL2. Based on irradiation-induced defect studies on GaAs, Walukiewicz[190][191] has proposed an amphoteric defect model depending both on semiconductor doping and analogous stoichiometry-dependent behavior. Here the defects $V_{Ga}$ (acceptor) or $Ga_{As}$ + $V_{Ga}$ for n-type and $V_{As}$ or $AS_{GA}$ + $V_{As}$ (donor) for p-type are proposed and the tendency of low electronegativity metals to form Ga-rich interfaces is noted. There is a dependence on $\Phi_M$, which is, however, limited by the defect energies. Thus defect models appear capable of accounting for variations around mid-gap of metals on cleaved GaAs (110) surfaces and metals on polished and etched GaAs (100) wafers.

Defects near semiconductor interfaces may also compete with other mechanisms during the initial stages of Schottky barrier formation. For example, Cao et al.[82] have interpreted the overshoot in $E_F$ movement pictured in Fig. 12 as due to the initial influence of work function prior to formation of large densities of defects at higher metal coverages. Mönch[84] has proposed a combination of virtual gap states of the complex semiconductor band structure perturbed by electronic levels of defects created in the semiconductor close to its metal interface during its formation. The data, similar to Fig. 19, suggest a significant but secondary role for defects in Si-metal junctions, consistent with either low densities or completely filled or empty levels. For GaAs, the role of fabrication-induced defects appears to be greater. Figure 24 illustrates the variation in barrier height position vs. electronegativity for a wide variety of metals measured by the J-V technique,[179][193]-[195] shown in comparison with $E_F$ positions for Cs, S, and Cl adsorbates. The solid line corresponds to the $\Phi_B$ variation based on charge transfer due to electronegativity difference in the absence of defects. As with the metal-induced gap state model described in Sec. 4, the slope depends on the density of tunneling states and the extent of wave function penetration. The dashed line corresponds to the $E_F$ stabilization energy predicted by the largest density of defects compatible with the experimental data and assuming a single defect level at $E_{CBM}$ - 0.65 eV. Such defect densities appear to vary considerably, with the lowest densities for the rare-earth metals Ti, Pb, Ag, Cu, and Ru, and largest for the transition metals Fe, Co, and Ni.

Alternatively, Ludeke et al.[196][197] have proposed that interface impurity levels from within the semiconductor band gap become delocalized through interaction with states of the metal overlayer. Figure 25a illustrates the energy diagram for such an impurity level $E_o$ within the band gap, broadened by energy $\Gamma$ due to the presence of metal and shifted by energy $\Delta E_i$ due to screening (image force) effects. Here $\Delta\Phi$ represents the interface dipole contribution of this local charge redistribution and the dashed line signifies the potential of the metal-impurity system. Parameter *s* is the impurity-metal separation in an ideal geometry. Here

Eq. (9) $$E_F = (I_{sc} - \Phi_M) - 4\pi e^2 \lambda_{eff} N_i q_{eff}$$

where $I_{sc}$ is the ionization potential of the semiconductor, $\lambda_{eff}$ is the effective separation of $q_{eff}$ from its countercharge near the ion potential in the metal,

Ni is the density of defects, and $q_{eff}$ is the effective charge per defect level. In turn, $q_{eff}$ depends on $E_i^0$, $\Delta E_i$, and $\Gamma$, which must be solved self-consistently. $\Gamma$ depends in part on the impurity wave function and the interface dielectric constant. Figure 25b illustrates a self-consistent fit to the data of Spicer,[198] McLean,[199] Ludeke,[200] and Kaiser[49] with corresponding changes In $q_{eff}$ (typically <0.2e) as a function of work function on GaAs (110) for all but Kaiser. Here Ludeke et al. make the simplifying assumption of a unique value of $E_i$ near midgap for all levels, in accord with contributions of a broad energy range of conduction and valence band states to this localized level, as in the charge neutrality level. Although the $E_F$ position appears consistent with this self-consistent approach, photoemission evidence for such impurity-induced levels during the initial metal deposition have to date remained controversial. For GaAs (110), McLean et al.[201] report that $E_F$ does not appear to correlate with the bulk deep impurity levels of transition metals on GaAs (110) in simple fashion. Rather $E_F$ positions appear relatively constant for transition metals[202] as opposed to the linear dependence on electronegativity for simple metals.

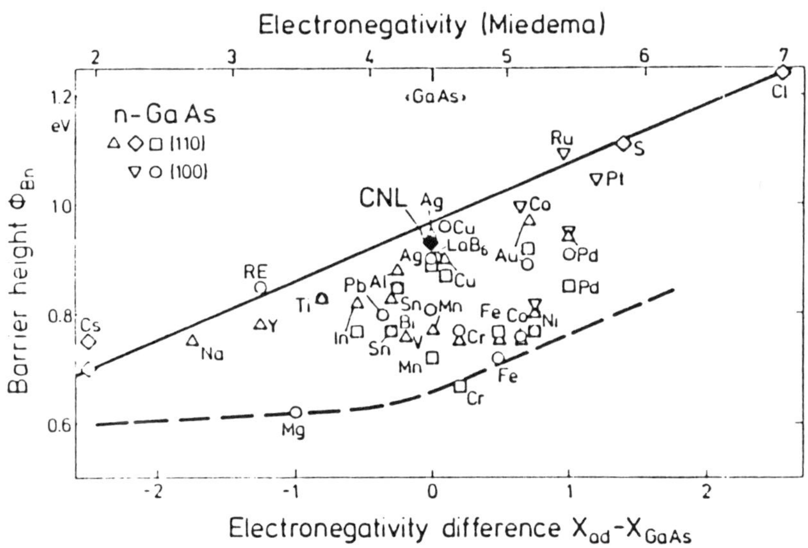


**Figure 24.** Barrier heights reported for metal-GaAs contacts and for adsorbates on GaAs(110) cleaved surfaces as a function of electronegativity difference between the adatoms and the substrate with $X_{sub}$) = 4.45. The charge neutrality level CNL is calculated in Ref. 62. The *solid line* indicates the $\Phi_B$ variation due to tunneling charge transfer as a function of electronegativity difference and no defects. The *dashed line* indicates the $E_F$ position with the largest density of defects compatible with the available data at $E_{CBM}$-0.65 eV. The deviations in $\Phi_B$ from the solid line indicate high densities of interface defects. See Ref. 192.

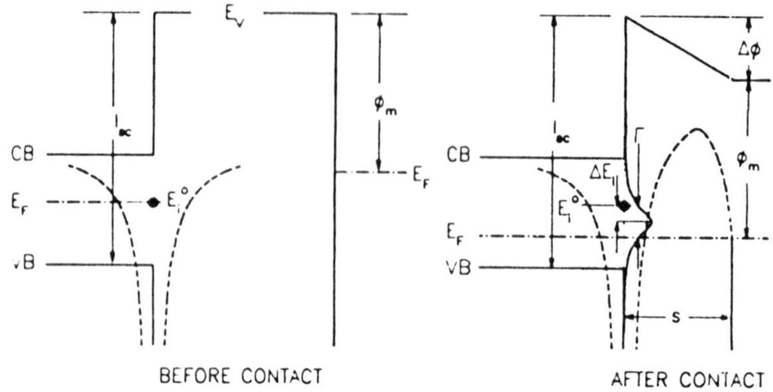


(a)

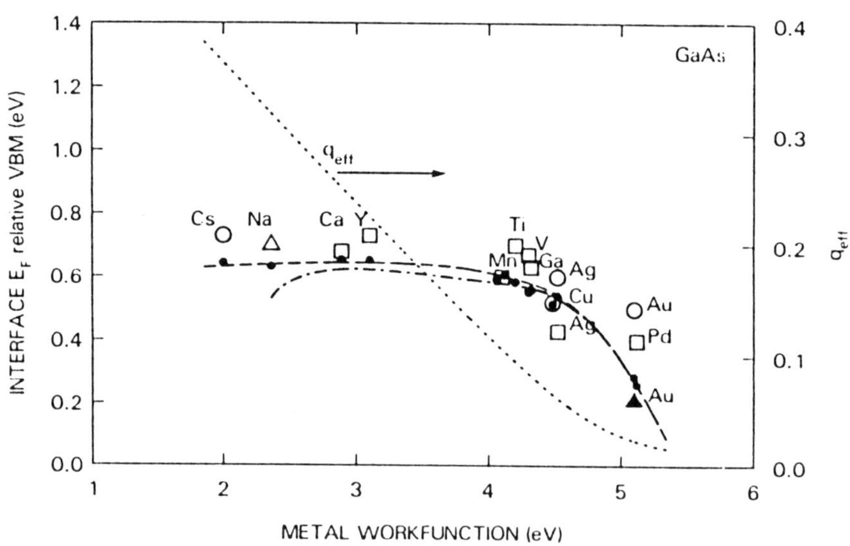


(b)

**Figure 25.** (*a*) Potential energy diagrams of a semiconductor with a surface defect level at $E_i^o$ and a metal before and after intimate contact. (*b*) Metal-GaAs interface $E_F$ position versus metal work function $\varnothing_m$. *Solid* and *dashed lines* are model predictions. A model calculation for $E_i^o$, broadened to simulate banding and disorder, *(dash-dotted curve)* provides a fit to the experimental data points, Ref. 198 (O), 199 (Δ), 196 and 200 (□), and 49 (▲). See Ref. 196.

Overall, a variety of defect and defect plus charge tunneling models appear capable of accounting for the narrow range of Schottky barrier heights reported for most melt-grown GaAs-metal interfaces, Recent defect models account for the observed metal dependence either in terms of electronegativity or stoichiometry effects on the type of defects formed.

## 6.3 Alloy Formation

An alternative chemistry-based model of Schottky barrier formation involves the formation of new alloy phases at the metal-semiconductor interface. Chemical interactions between metal and semiconductor can lead to new interfacial phases with work functions different from those of the starting materials. For metals on Si, Freeouf[203] proposed that such interfacial phases were Si-rich silicides and that the altered silicides work function, rather than that of the metal, defined the electronic charge transfer. By assuming a common (metal)$Si_4$ stoichiometry and a geometric mean $(\Phi_M \Phi_{Si}{}^4)^{1/5}$ for the interface work function, he obtained a reasonable correlation between alloy work function and $\Phi_B{}^n$. However, transmission electron microscopy studies of several silicide systems suggests abrupt bulk silicide/Si junctions and, unless changes in stoichiometry have little effect on the lattice constant, such phases must have a thickness of roughly a monolayer or less. For interfaces with III-V compound semiconductors, however, a variety of techniques[204][205] reveal the presence of excess anion species segregated to the intimate junction. Thus the work function of different metals on, for example, GaAs becomes that of As and the $E_F$ appears pinned at a common mid-gap energy defined by the classical charge transfer between the interfacial phase and the semiconductor (Eq. 1).

Figure 26 illustrates the efficacy of this effective work function model in accounting for a wide range of contact measurements on different III-V compound semiconductors.[206] Here the interface $E_F$ position for relatively unreactive (Au) contacts to various III-V compounds and pseudo-binary alloys appear on an absolute energy scale with respect to the conduction and valence band of the different semiconductors. The $E_F$ positions all lie close to the dashed line $E_{VAC} - E_F = 5.0$, corresponding to work functions 4.8 - 5.1 eV—close to the classical $\Phi_M$ of As and P. Most of the barrier heights plotted in Fig. 26 were measured for diodes prepared under low vacuum conditions, for which air-exposure acts to block As out-diffusion and further accumulate the anion at the intimate interface. See, for example, Fig. 13. Significantly, these $\Phi_B$'s are in close agreement with the $E_F$ positions determined for metals on bulk cleaved III-V crystals as well. As with the defect models, one can interpret variations In $E_F$ in relation to the changes in interface anion compositions. Interface states associated with As-rich phases may be considered chemisorption states (Sec. 5) at submonolayer coverages whose observable, mid-gap energies evolve into metallic $E_F$ edges for As clusters, precipitates, or macroscopic phases. Thus effective work function models can account for a wide range of semiconductor (and insulator) interface barrier measurements. Furthermore, it is based on microscopic but measurable chemical phases, and it provides a straightforward, Schottky model of interface charge transfer.

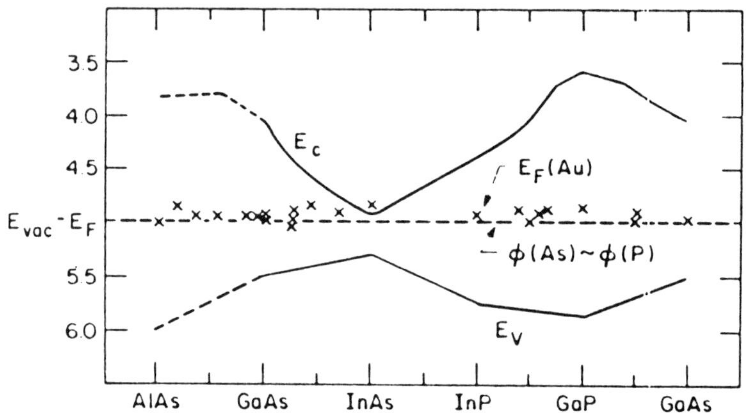


**Figure 26.** Interface Fermi level for relatively unreactive (mostly) Au contacts to various III-V compounds and alloys. On an absolute energy scale, the Fermi level pinning energies lie within the band gaps at the anion work function energy. This suggests the presence of an anion-rich interphase which dominates the Schottky-like charge transfer. See Ref. 206.

Hasagawa and Ohno[207][208] have proposed an alternative interface alloy for metal-semiconductor and insulator-semiconductor contacts which involves localized states due to lattice disorder. Such disorder arises due to fluctuations in angle and length of bonds, the presence of stress and irregularity at the boundary, and thereby a distribution of bonding and antibonding states with energies around a characteristic hybrid orbital energy. In particular, this *Unified Disorder Induced Gap State* (DIGS) model incorporates *(i)* dangling bond states with a continuum of energies in the gap, *(ii)* a minimum density near an empirical, hybrid orbital energy $E_{HO}$ of the semiconductor,[208] and *(iii)* $E_F$ pinning at this $E_{HO}$, calculated to lie in the range 5.5 to 5.75 eV below $E_{VAC}$. Hasagawa and Ohno are thereby able to obtain a plot similar to Fig. 26 (with absolute energy scale shifted to reflect the relative photo-threshold edges for different semiconductors). Disordered interfacial layers with insulating or semiconductor properties may also provide an explanation for the anomalously high $\Phi_B$ for some metal-semiconductor systems such as, for example, Ga on GaAs.[187][209] Hence DIGS incorporates elements of both defect and effective work function models but is distinct in its use of a U-shaped density of gap states and a distribution of capture cross sections for insulator-semiconductor junctions. As with the effective work function model, the basis for DIGS involves morphological features which are, at least in principle, amenable to experimental observation. However, the effective work function model is better suited than a DIGS model to account for the electronic changes observed with differences in interface chemistry and processing. Both DIGS and MIGS constructs predict a continuum of interface states with $E_F$ stabilized near a minimum density of states and with an absence of discrete gap states. The latter is in contrast with well-known discrete gap states formed by chemisorption and persisting for "buried" interfaces. See, for example, Fig. 7. Nor does DIGS provided a means to describe the interface behavior at highly-ordered, epitaxial metal-semiconductor junctions.

Overall, a rich array of chemically-driven phenomena exist which can promote electrically-active interface states. Mechanisms which incorporate such effects are capable of accounting for much of the $E_F$ pinning measurements to date—but not uniquely. In the next section, we address the reduction of such interface states densities and their implications for various interface state models.

## 7.0 INTERFACE STATE CONTROL

### 7.1 Chemical Passivation

Over the past decade, researchers have uncovered numerous metal-semiconductor systems with reduced or controllable densities of interface states. These systems fall into one of several categories: macroscopic chemical passivation, atomic-scale interlayer passivation, semiconductor crystal modification, and epitaxial interface modification.

There have been several dramatic examples of interface state reduction by macroscopic, wet-chemical treatments. Driving this work has been the need to reduce surface recombination velocity (SRV) of the bare semiconductor surface for use in, among other devices, solar cells and high speed transistors. Up until a few years ago, semiconductor passivation research was centered primarily on electrochemical solution treatments. More recently, Offsey et al.[210] found that photochemical washing of GaAs using only water with above band-gap illumination induces substantial increases in band-edge illumination, corresponding to reduced band bending in the semiconductor surface space charge region. The lower band bending provides greater spatial overlap of photo-excited electrons and holes, resulting in a stronger luminescence signal. Significantly, re-exposure of the cleaned surface to air causes gradual reduction in luminescence over the course of several hours—consistent with surface degradation by a relatively slow chemical process. Offsey et al. interpreted the SRV properties in terms of dissolution of (water-soluble) As oxides and thereby a reduction in mid-gap state density.

Yablonovitch et al.[211] and Sandroff et al.[212] have discovered a class of inorganic sulfides [$Li_2S$, $(NH4)_2S$, $Na_2S$•$9H_20$] which decrease surface recombination velocity to S = 1000 cm/sec—approaching that of the Ideal AlGaAs/GaAs interface. These and other variations in SRV are illustrated in Fig. 27. Here SRV is measured by the transient response of surface conductivity probed via RF induction.[213] Likewise, photoluminescence intensities reflect these dramatic SRV changes: here the indicated surface chemical treatments produce orders-of-magnitude changes in photoluminescence, corresponding to gap state densities as low as $10^{11}$ $cm^{-2}$.[214]

Woodall and coworkers performed an elegant experiment to demonstrate the decrease in band bending as a result of photochemical passivation. The device structure shown in Fig. 28 provides a means to monitor surface band bending via series resistance and short circuit photocurrent as a function of changes in luminescence with surface passivation.[213a] For the case of photochemical washing, series resistance decreases and short circuit photocurrent increases with increased luminescence, proving that band bending decreases with decreased surface recombination velocity. Hence, pinning by interface states is reduced as the Fermi level moves closer to flat band conditions. For the interfaces with sulfide treatments, it is believed that the sulfide solutions form stable Ga chalcogenides which are known to reduce the surface density of states, presumably by rehybridizing dangling bond states out of the semiconductor band gap.[215] Ga compounds are more likely since the sulfide solution is likely to etch away the As. Again, reoxidation due to air exposure degrades the SRV over the course of several hours. Similar interface state reductions are apparent from P surface treatments, either by $P_2S_5/(NH_4)_2S$ solution[216] or red phosphorus[217] vapor deposition on InP.

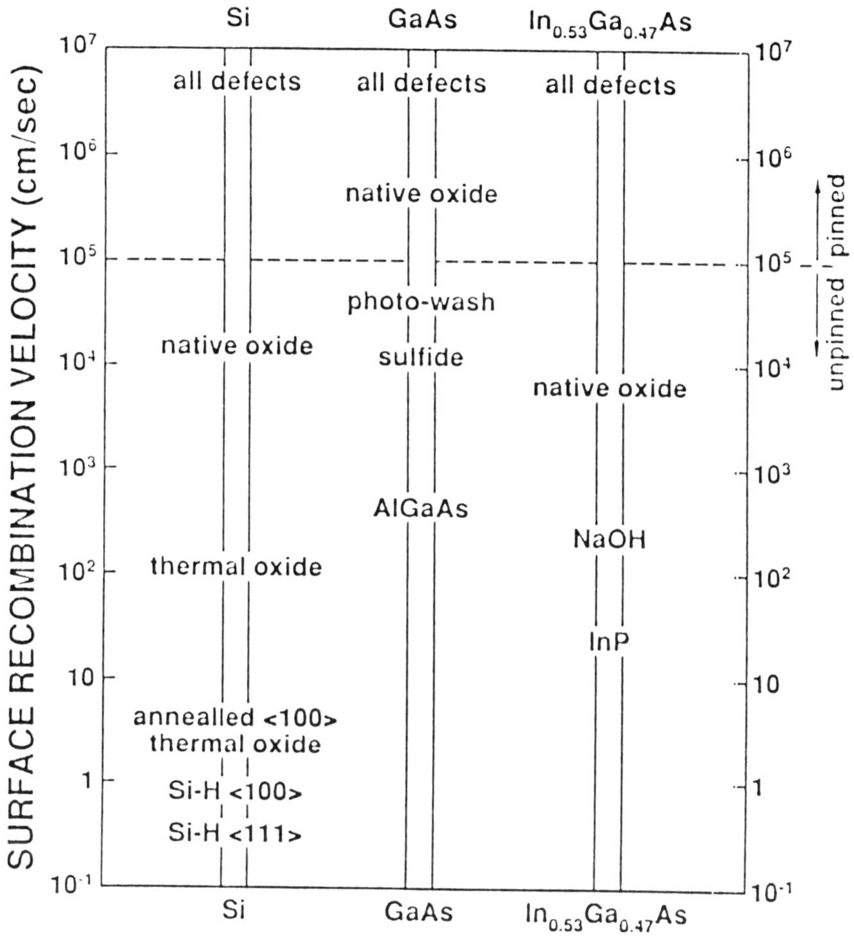


**Figure 27.** Surface recombination velocities for Si, GaAs, and $In_{0.53}$ $Ga_{0.47}$ As eptiaxial layers on GaAs(100) for various surface treatments. Low SRV values correspond to reduced interface state densities and unpinned surface levels. See Ref. 213.

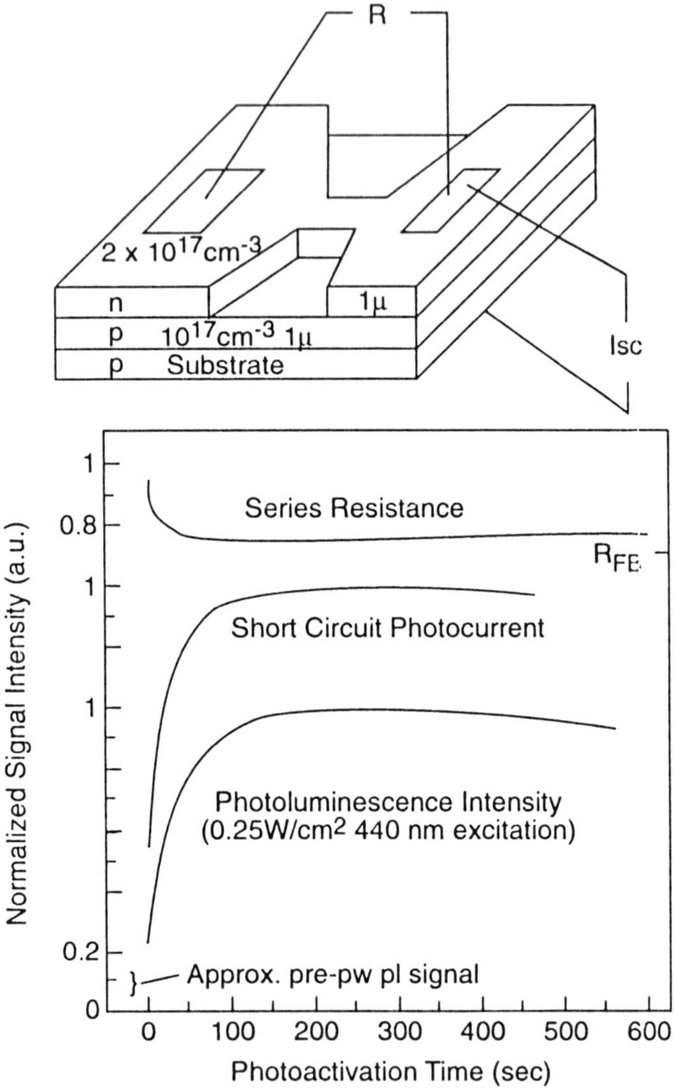


**Figure 28.** Schematic device structure (upper panel) for correlating reduction surface recombination velocity with band bending within the surface space charge region. The lower panel illustrates the reduction in series resistance and increase in short circuit photocurrent with surface passivation and increased luminescence efficiency. This result demonstrates a reduction in band bending with decreased SRV and hence a reduction in mid-gap pinning by interface states. See Ref. 213a.

Chemical treatments can produce changes in interface state energy as well as density. Shapira, Brillson, and Heller[218] showed that wet chemical treatments alter the energies and densities of InP states. Significantly, they found reduced SRV for silver cyanide treatments associated with formation of an In oxide layer, rather than with changes in the dominant deep level features. Likewise, Schmidt et al.[219] have used UV photo-oxidation to produce Bridgman-grown GaAs (100)/metal contacts with more Schottky-like $\Phi_B$ behavior. Based on a self-consistent electrostatic analysis of $\Phi_B$'s after Mailhiot and Duke[44] and Zur and McGill,[220] they argued for relatively little change in interface state density but a shift of the mid-gap level toward the band edges. Deep level transient spectroscopy (DLTS) measurements of MBE-grown GaAs interface states before and after sulfide treatment[221] indicate that both $Na_2S$ and $(NH_4)_2S$ remove a state at $E_c$ -0.58 eV, replacing it with a state at $E_c$ - 1.15 eV. Again, chemical passivation appears to improve photoluminescence intensity by moving the interface states out of the mid-gap region without decreasing band bending. Photoemission studies also derive no band bending decrease for a sulfide treatment.[222] However, such results depend sensitively on the details of surface chemical treatment. Significantly, sulfide treatments appear to have little effect on deep level features for melt-grown (LEC) GaAs, indicative of bulk crystal effects on deep level properties of the surface.

Note that band bending in the surface space charge region can be large without high densities of interface states, i.e., without $E_F$ pinning. Indeed, removal of high densities of mid-gap states provide a basis for wide variation of $E_F$ energies with different metals. Conversely, measurements of low band bending must be treated with caution because of surface photovoltage band flattening effects,[17][223] which have complicated the interpretation of numerous photoemission studies.

Besides the compound semiconductors, surface passivation produces extraordinary SRV decreases in Si via HF solutions with pH adjusted to eliminate surface roughness.[224] These ideally terminated Si (111) surfaces are microscopically smooth, involve a homogeneous monohydride phase, and display SRV values as low as <1 cm/sec. Likewise, HF treatments coupled with UV irradiation passivate the Si (100) surface against oxidation, with reactivation (and bond breaking) via a 3 kV electron beam exposure.[225] Recently, Tao et al. have found even more dramatic SRV reductions for $(NH_4)_2S$ treatment of InP(100) surfaces, which remain stable in air for up to several days.[225a] Such surfaces are believed to have a stable, ordered (1x1) atomic geometry with a monolayer of S atoms forming bridge bonds only to the In atoms. Degradation is believed to occur via P surface segregation and subsequent oxidation. Such results offer considerable hope for controlling interface electronic properties of metal-semiconductor interfaces.

## 7.2 Atomic-Scale Interlayers

A second approach to controlling interface states and semiconductor band bending is through atomic interlayers of various elements—reactive metals, unreactive adsorbates, and even other semiconductors. Reactive metal layers of only nanometer thickness produce major changes in device characteristics. As shown in Fig. 22, such reactive interlayers at the metal/III-V compound semiconductor interface act to reverse the stoichiometry of out-diffusion, leading to $\Phi_B$ changes of tenths of an eV.[151][158][226] Yb, Sm, and Sm-Al alloy interlayers at Al/GaAs (001) MBE grown surfaces also produce nearly 0.2 eV C-V and J-V $\Phi_B$ changes on n-type and p-type GaAs, along with a pronounced interlayer thickness dependence for the Sm-Al alloy.[227] Even more dramatic effects occur for Al interlayers at Au-CdS diode structures.[226] Here the reactive layers act to inhibit Cd out-diffusion, resulting in a highly n-type surface region and a $\Phi_B$ reduction due to tunneling. Similar phenomena are evident for Al at Au/CdTe contacts.[228]

Reactive interlayers which suppress anion out-diffusion at Au-CdTe (110) interfaces provide a graphic example of the effect which chemical bonding has on near-interface defects. Here, Yb interlayers only two monolayers thick effectively block Te out-diffusion and alter the production of new deep level emission.[229] Figure 29 illustrates the difference in photoluminescence spectra for clean CdTe(110) cleaved in UHV and the same surface with and without a blocking Yb interlayer at the Au/CdTe interface. As shown, Au deposition produces new deep levels at 0.8 and 1.0 eV plus strong band edge emission due to the lower band bending. SXPS core level spectra of the same surfaces reveal corresponding differences in Te bonding. Furthermore, the surface $E_F$ stabilizes at Ev + 1.1 eV for the Au / CdTe interface, consistent with an added shoulder in the spectral peak near 1.1 eV and with a defect state 1.1 eV above $E_v$ in the surface photovoltage spectra. In contrast, the altered out-diffusion at the Au/Yb/CdTe results in a 0.55 eV increase in barrier height.[229] The correlation of deep level luminescence intensity with the degree of out-diffusion provides direct evidence that diffusion-induced deep states can stabilize $E_F$ near their energy levels. Furthermore, by limiting such interdiffusion via atomic-scale chemistry, one can control Schottky barriers over a much wider energy range.

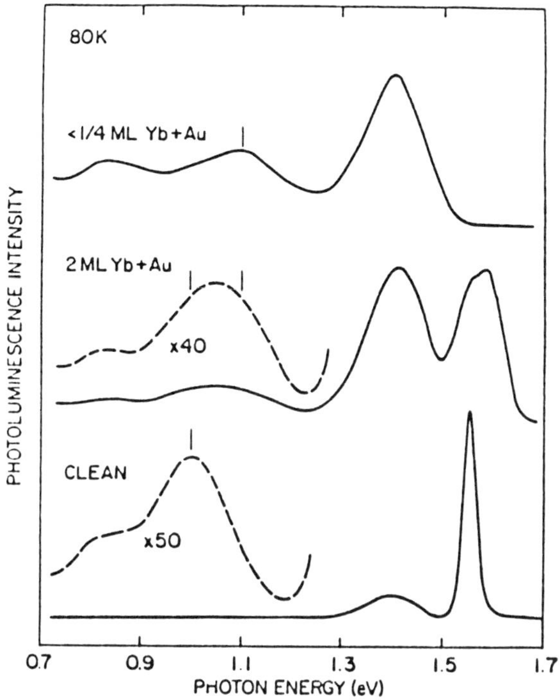


**Figure 29.** Photoluminescence spectra for clean and Au-deposited, UHV-cleaved, CdTe(110) surfaces with and without a 2-monolayer Yb interlayer. The Yb interlayer acts as a barrier against Te out diffusion resulting in reduced deep level emissions and a 0.55 eV increase in barrier height. See Ref. 229.

Unreactive adsorbates can alter contact barrier properties as well. Consider $H_2S$ on InP. At low exposures, $H_2S$ adsorbs in molecular form, as evidenced by its weak adsorption and characteristic photoelectron spectra. The effect of this adsorbate on clean, LEED-ordered InP (110) is to disorder the surface and reduce the $\Phi_B$ of the Au/InP junction from 0.5 eV to below 0.3 eV.[230] A plausible explanation appears to be surface doping by S atoms to produce shallow donor levels. The metallic overlayer may be important in initially dissociating the $H_2S$ adsorbate for such a process to occur, but further work is necessary to elucidate the exact mechanism of the barrier reduction.

Even gentler chemical treatments of the compound semiconductor interface produce significant $\Phi_B$ changes. Adsorption of Xe monolayers on

GaAs,[231] Inp,[232] and ZnSe[233] provides a means of reducing the metal-semiconductor chemistry by either physical separation and/or dissipation of the condensation/fusion energies. Following cluster formation, the Xe layer is removed by thermal desorption, placing the metal clusters in contact with the semiconductor. Such a procedure minimizes surface disruption due to the cluster formation per se. The resultant junctions present significantly different $E_F$ stabilization energies with respect to directly metallized surfaces. For example, metal-cluster formation on p-type GaAs (110) produces a metal-specific $E_F$ variation from $E_v + 0.37$ to $E_v + 0.62$ eV, in contrast to the 0.5 eV position conventionally seen (e.g., Fig. 23). Similar cluster deposition on n-type GaAs yields $E_F$ stabilization centered around $E_c$ - 0.32 ± -0.1 eV in contrast to the $E_c$ -0.7 eV typical value. Large changes in Au/InP $\Phi_B$'s result from similar changes in deposition process.

Further studies distinguish between mechanisms involving surface defect formation (via energy released by condensation and clustering) vs. interfacial chemical reaction. Ion-bombardment to introduce surface defects during metal deposition has little effect on $E_F$ movement in comparison to neutral metal deposition. Thus defects created due to substrate disruption appear to have no direct effect on the evolution of interface band bending. For ZnSe, one sees over 1 eV metal-to-metal variation with 300 - 400 eV ion acceleration having less than 0.1 eV effect on individual systems.

Semiconductors appear to be the most effective interlayers in controlling metal-semiconductor Schottky barrier heights. Introduction of the chalcogens S, Se, and Te, as well as Si or Ge interlayers, lead to large changes in $E_F$ stabilization for GaAS.[235]-[239] For nanometer thicknesses of S, Se, or Te, a representative set of metals produces a $\Phi_B$ range greater than 0.6 eV. Significantly, metals which react strongly with chalcogen layers produce the lowest $\Phi_B$'s whereas the least reactive yield the highest.[237][238] This agrees with the behavior pictured in Fig. 20 and an interpretation in terms of new defect levels heavily dependent on the nature of the interface reaction. In any event, $E_F$ is not restricted to a narrow range.

For different combinations of Au, Ge, and Ni layers on LEC GaAs (100), $\Phi_B$ ranges from ~0.25 to 0.9 eV.[238] Here different metallurgical phases form in contact with GaAs, which provide varying Ge in-diffusion to the GaAs or (not yet observed) an intermediate layer of low $\Phi_B$ Ge. In fact, a Ge interlayer doped with As appears to provide $E_F$ variations within the band gap from ~0.5 eV to 1.2 eV, depending on the substrate temperature and background deposition pressure of As or P.[238] Figure 30 illustrates $E_F$ measurements for GaAs (100) surfaces with the indicated deposition/temperature treatments.[236][237] The dashed lines indicate conduction and valence bands as measured from heterojunction band offset experiments.[240][241] In contrast, the doped Ge (As) layers move $E_F$ for various deposition temperatures close to the Ge conduction band energy. Phosphorus dopants produce similar effects. Doped Si (As) and Si (Ga) interlayers locate $E_F$ near the corresponding Si conduction and valence band edges respectively. Similar results have recently been reported by Look et al.[242] for thin (100 Å) epitaxial layers of GaAs grown at 200°C, with the As-rich surface layer producing only 0.1 - 0.2 eV band bending as measured by the Hall effect charge density. Again, the nonmetallic overlayer serves to passivate the semiconductor surface bonding without introducing significant charge densities in the overlayer. The results suggest that $E_F$ in the interlayer determines the $E_F$ of the Ge/GaAs—in agreement with Ge/GaAs studies of Chiaradia et al.[243] and in contrast to a defect picture—and that Ge diffusion into the GaAs is secondary. Furthermore, the results highlight the effect of the heterojunction offset or the work function of Ge on the Schottky barrier formation.

## 7.3 Semiconductor Crystal Modification

Semiconductor crystal quality is a key factor in determining the densities of bulk native defects. Different techniques of crystal growth can produce significant variation in stoichiometry, point defect density, dislocation density, and even elemental precipitate density. Furthermore, these differences appear to be magnified by chemical interaction at microscopic semiconductor interfaces and can give rise to significant differences in interface electronic states. In the last few years, several metal/semiconductor systems have provided examples of contact rectification exhibiting *unpinned* behavior, even for semiconductors traditionally believed to have strong $E_F$ pinning. SXPS core level measurements of band bending for metals on the $In_xGa_{1-x}As(100)$ ($0 \leq x \leq 1$) pseudo-binary alloys,[244][245] InAs (110),[246] InP (110),[181] GaP (110),[247][248] GaSe,[153] and even GaAs (100)[248]-[251] surfaces have revealed unpinned and, in some cases, near-ideal Schottky-like behavior.

GaAs/metal interfaces serves to illustrate the striking difference in interface electronic behavior with crystal growth modification. Figure 31 displays $E_F$ movement in the GaAs band gap with metal deposition for UHV-cleaved melt-grown GaAs (110)[252] (upper) vs. metals on As-decapped, MBE-grown GaAs (100) (lower) surfaces.[249][253] The melt-grown GaAs exhibits a narrow $E_F$ range, rapid $E_F$ movement to a final position with

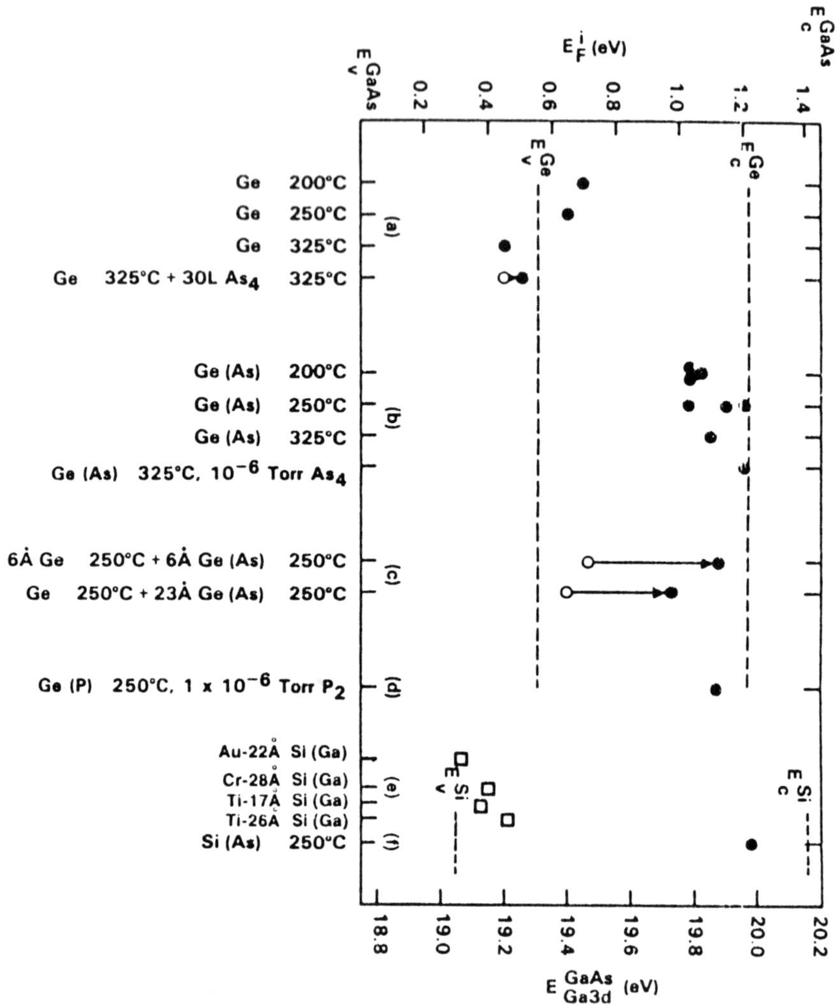


**Figure 30.** Fermi level positions within the GaAs(100) band gap for a variety of (doped) Si and Ge interlayers and metal overlayers. Unless otherwise noted, overlayer thicknesses are ~10 Å and depositions intended to incorporate As or P in the overlayer were performed in $10^{-7}$ torr background pressure of $As_4$ or $P_2$: *(a)* Ge overlayers deposited in vacuum; *(b)* Ge overlayers deposited In $As_4$; *(c)* Ge overlayers initially deposited in vacuum, but completed with deposition in $As_4$; *(d)* Ge overlayer deposited in $P_2$; *(e)* Si overlayer deposited with Ga monolayer in ~2 x $10^{-6}$ torr $H_2$ background pressure, *(f)* Si overlayer deposited in $As_4$. *Dashed lines* indicate Ge and Si band edges derived from measured band offsets. $E_F$ positions appear to range across Ge and Si band energies. See Ref. 236 and 237.

submonolayer deposition, and a 0.25 eV difference between n- and p-type crystals. In contrast, the MBE-grown GaAs exhibits a 0.7 eV range of $\Phi_B$'s at room temperature, $E_F$ movement to a final position evolving over several metal-dependent monolayers, and coincident positions for n-and p-type crystals.[249] Internal photoemission spectroscopy measurements of $\Phi_B$ for metal/MBE-grown GaAs confirm this wide spread in Schottky barrier heights.[254]

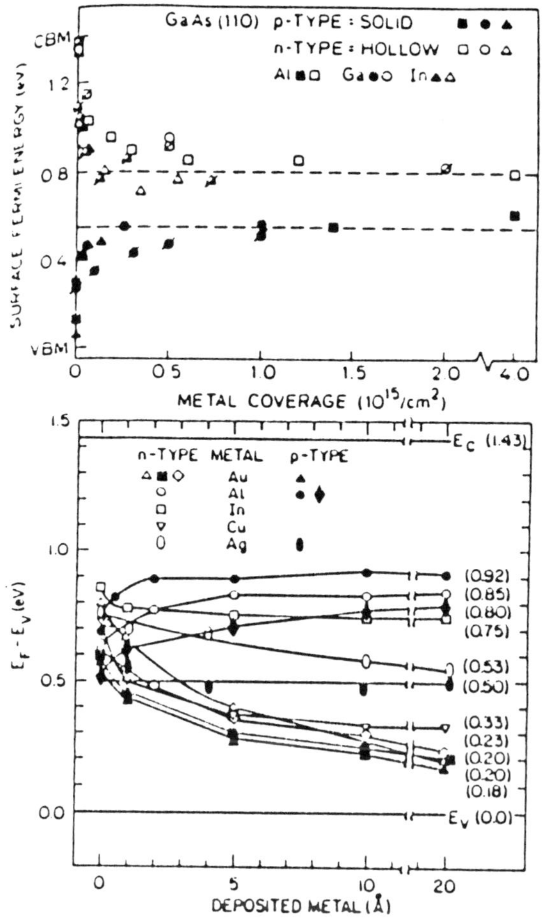


**Figure 31.** Contrast in Fermi level movement for metals on melt-grown GaAs(110) (Ref. 252) versus metals on As-decapped, MBE-grown GaAs(100) *(lower)* surfaces (Ref. 249). Metals on MBE-grown GaAs can produce a wide range of Fermi level stabilization energies, matched energies for n- and p-type GaAs, and Fermi level evolution over several metal monolayers. See Ref. 253.

Metal deposition on these two types of GaAs also produce major differences in interface electronic states. Deep level emission from melt-grown GaAs is intense and centered near mid gap; deep level emission from MBE-grown GaAs is relatively weak, and is displaced to energies away from mid-gap.[41] These differences are in close correspondence with the contrasting bulk luminescence displayed in Fig. 16. They suggest either a segregation of bulk native defects to the metal/semiconductor interface or a predisposition to form such defects near the junction which varies with the characteristic chemical composition and structure associated with a particular crystal growth method.

SXPS band bending measurement at low temperature yields an even wider range of Schottky barrier heights. Figure 32 illustrates the measured n-type $\Phi_B$'s plotted vs. metal work function[250][253] (based on most recent photoemission measurements of their clean surfaces[255][256]). The data was taken at both low temperatures (open circles) and room temperature (filled circles) and appear with respect to the Schottky limit line (Eq. 1). There is a good correspondence between the low temperature $\Phi_B$'s with the Schottky line for all but the most reactive metals (e.g., Sm and Yb). With increased temperature, several points move away from this ideal limit. SXPS core level spectra indicate significant chemical interactions for most of these metals which are otherwise inhibited at low-temperature.

Deviations from the Schottky limit line indicate the presence of significant densities of interface states or changes in semiconductor electronic structure. A self-consistent electrostatic analysis assuming discrete charge states at a depth of 10 Å inside the semiconductor determines these states to be acceptor levels at 0.2 eV with density $3 \times 10^{13}$ $cm^{-2}$ and acceptor levels at 0.8 eV with density $10^{13}$ - $10^{15}$ $cm^{-2}$, dependent upon the particular metal.[249][250][253] Both of these levels are observable via CLS, and the 0.8 eV feature intensity is indeed dependent on the particular metal.[250] Thus discrete, deep levels introduced by chemical interaction perturb the otherwise near-ideal Schottky-like behavior. For melt-grown GaAs, densities of such levels are much higher—high enough to prevent observations of such near-ideal behavior. Recent SXPS barrier measurements also show large deviations from ideal Schottky-like behavior.[250a,b] These measurements were taken on GaAs(100) surfaces which have now been shown to exhibit large variations in surface work function, electron affinity, chemical composition, and even chemical homogeneity.[250c,d,e] Also associated with such deviations are substantial variations in deep level emission.[250e] These results emphasize the process-dependent nature of interface electronic structure.

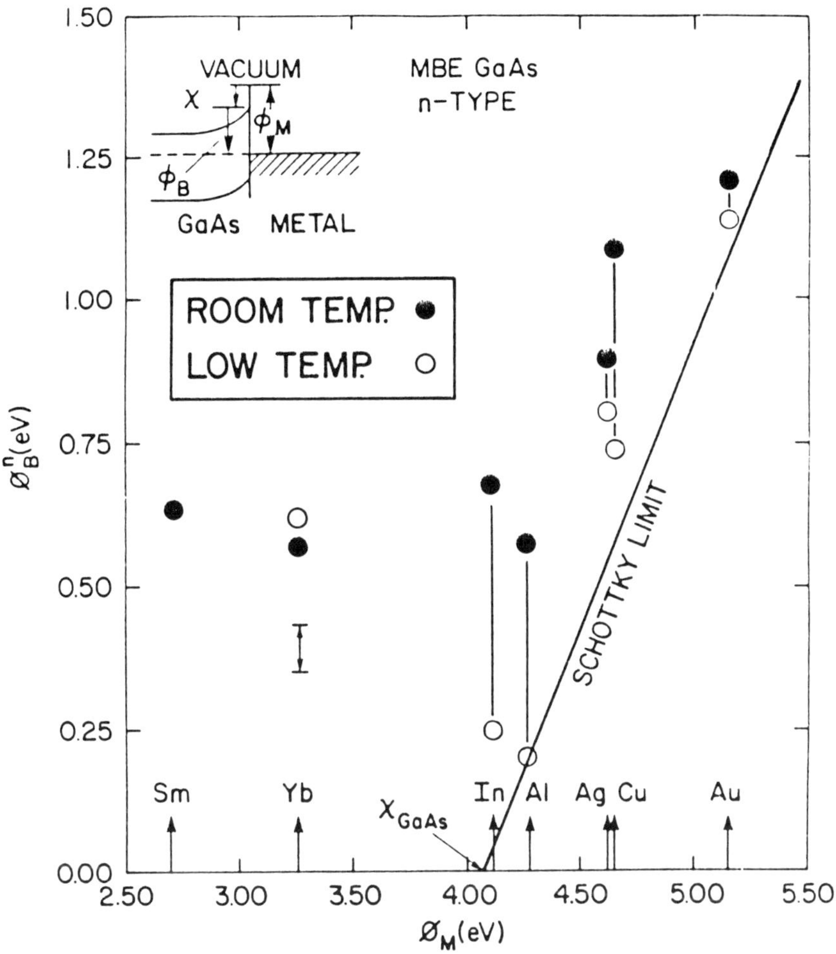


**Figure 32.** Soft x-ray photoemission-derived low temperature n-type $\Phi_B$'s for metals on As-decapped MBE-grown GaAs(100) as a function of metal work function. Equation 1 defines the Schottky limit line. Deviations form the Schottky limit line indicate the presence of significant densities of interface states, particularly at room temperature and for lowest work function (highly reactive) metals. See Ref. 250.

The absence of high densities of interface states at these metal-MBE GaAs (100) interfaces provides a test bed for gauging electronic contributions of surface structural defects. Chief among such structural defects are the steps at a vicinal surface. Such off-axis surfaces are a widely-used template for increasing the rate of epitaxial growth. The angle of misorientation defines the average site density and thereby an average density of atomic sites at the exposed step edges. Likewise, the atomic site density at step edges depends upon the direction of misorientation, e.g., [100], [110], [111]A, [111]B, etc. SXPS measurements of interface chemical reactions at Al/GaAs vicinal surfaces reveals that the extent of such reactions is proportional to the density of chemically-active edge sites—sites promoting Al-As bonding and an Al-Ga exchange reaction.[102][257] This proportionality is inclusive of step density and orientation.[258] For oriented GaAs (100) at low temperature, core level evidence for such a reaction is almost undetectable; for the [100] 4° → [111]B direction, the exchange reaction dominates the spectra. Furthermore, the reaction is localized to within 1 - 2 monolayers, as evidenced by the thickness dependence and depth sensitivity of the SXPS spectra.

The changes in local interface chemistry manifest themselves electrically as changes in Schottky barrier height. Figure 33 illustrates a ≥0.6 eV range of barriers for Al on GaAs,[258] depending upon the chemically-active site density. In contrast, no such dependence is apparent for Au / GaAs, where strong localized interface bonding is absent. Figure 33 also maps the barrier heights onto a family of curves for different interface state densities. The Fig. 33 inset represents two such states, transitions into which are visible via CLS.[259] A constant $3 \times 10^{13}$ $cm^{-2}$ density of 0.2 eV acceptor states accounts for the high work function metals in both Figs. 32 and 33. The 0.6 eV state is consistent with $E_F$ stabilization energies no lower than this energy (barrier heights no higher than 1.51 - 0.6 = 0.9 eV) and with the dominant CLS emission from states at 0.9 eV.[260] By matching the $\Phi_B$'s to the density curves, one can extract a correlation between densities of midgap states and densities of chemically-active sites. Such a relation is pictured in Fig. 34. There is a clear, linear proportionality between the chemical site and electrostatic state densities, with a slope of ≈ 2/3 charge per site and an extrapolated density of only $1 \times 10^{13}$ $cm^{-2}$ for the oriented (100) surface. (The high error bars at the highest site density are due to the nature of the calculation, for which the $\Phi_B$'s asymptote with increasing density). For the stepped surfaces, Fig. 34 indicates a large charge transfer over ≈ 10 Å due to the Al-GaAs exchange reaction. Overall, these results show the central importance of localized chemical bonding on deep level formation and subsequent Schottky barrier formation.

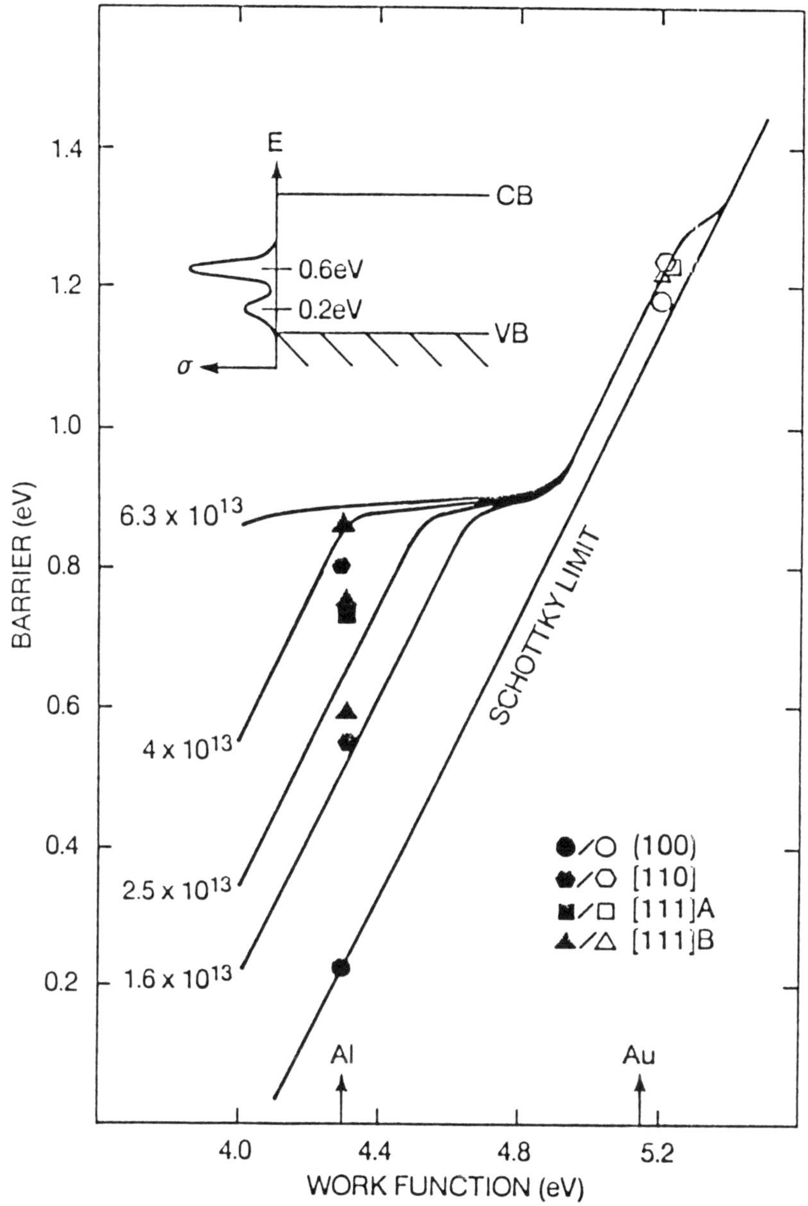


**Figure 33.** Self-consistent, electrostatic analysis of measured barrier heights at Al/GaAs(100) and vicinal surfaces. The family of curves represent a density of $0.5 \times 10^{13}$ cm$^2$ acceptors at $E_v + 0.2$ eV and various densities of acceptors at $E_v + 0.6$ eV *(inset)*. The increase in acceptor site densities causes an increase in Schottky barrier height. The densities of interface electronic states also correlate with the densities of active vicinal step sites (Fig. 34). See Ref. 257.

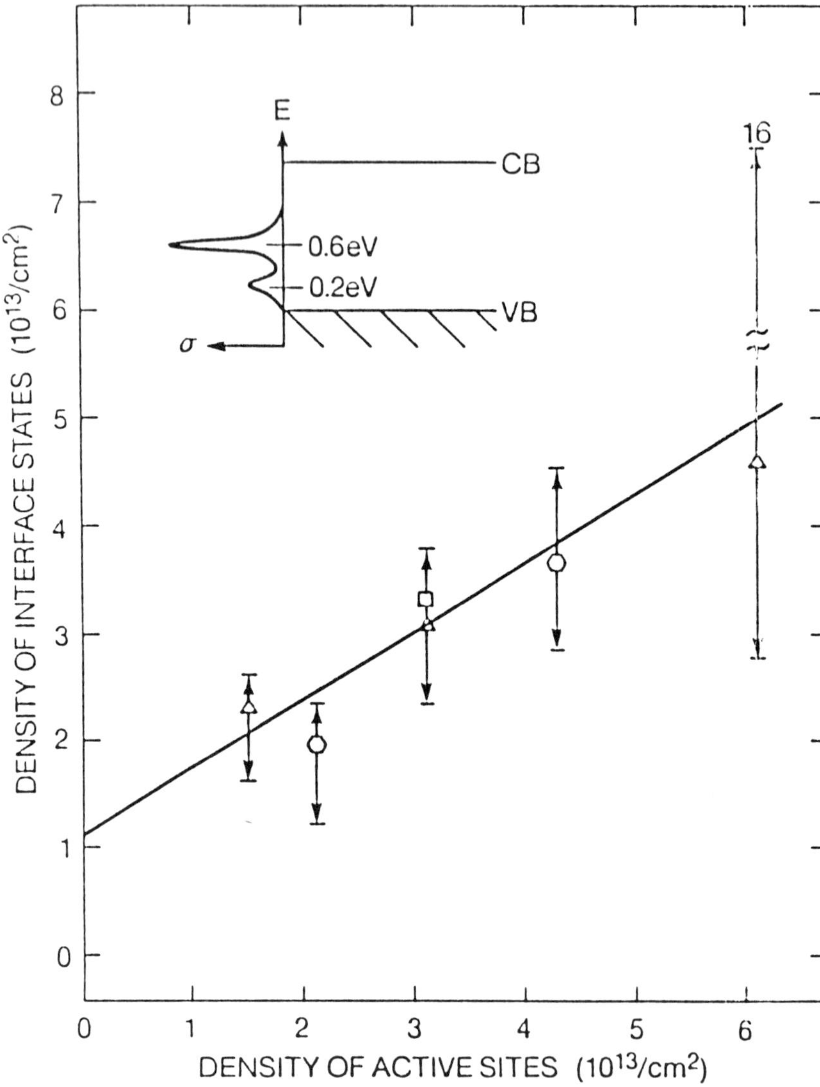


**Figure 34.** Correlation between the density of deep level acceptors at $E_v + 0.6$ eV *(inset)* determined by a self-consistent electrostatic analysis and the density of misorientation-induced active sites estimated from unreconstructed, staircase-like vicinal GaAs(100) surfaces. The nearly 1:1 correspondence emphasizes the role of step-related sites in the formation of interface states. See Ref. 257.

## 7.4 Epitaxial Interface Modification

With the improved control of microscopic lattice properties now possible via epitaxial growth techniques, additional evidence has arisen to demonstrate the role of local atomic bonding in interface state formation. Perhaps most notable have been examples of different barriers measured for different bonding arrangements at the same metal-Si interface. Tung et al.[75] provided the first evidence for this phenomenon, as described in Ch. 3. More recently, Heslinga et al.[76] found that Schottky barriers of epitaxial Pb/Si (111) contacts depend upon atomic arrangement of the first monolayer of Pb. For both Ni- and Pb-Si junctions, the abrupt metal-semiconductor interfaces are free of extended chemical reaction, diffusional mixing, and lattice disorder. Instead, variations in crystal orientation between the two lattices can produce Schottky barrier differences of ≈0.2 eV. Since the bulk lattice structures of each diode are identical, only local bonding and interface electronic structure can play a role. Indeed Lelay et al.[260] have obtained marked differences in occupied densities of states for the two Pb-Si phases via valence band photoemission. Thus the interface states depend primarily on interface atomic structure at these epitaxial interfaces.

Binary metal alloys provide additional control of interface chemical structure and have recently provided evidence for interface state controllability. As described by Sands and Palmstrøm in Ch. 2, compound metal overlayers provide additional flexibility in selecting a high melting point for reduced interface chemical activity and, simultaneously, good epitaxial lattice match. For NiAl overlayers, Chambers and Loebs[261] report 0.2 - 0.3 eV changes in GaAs $\Phi_B$ with (metallic) deposition thickness as Ni/Al surface stoichiometry and surface geometric order change. These changes appear related to the formation of an AlAs interfacial layer, Ni in-diffusion, and the properties of the GaAs grown epitaxially at low (250°C) temperatures. For $Sc_xEr_{1-x}As$ ($0 \leq x \leq 1$) metallic epitaxy on GaAs grown at higher temperatures, Palmstrøm et al.[262] have observed a strong $\Phi_B$ dependence on misorientation, similar to the result of Fig. 33. As with Al, the ScErAs epitaxy inhibits chemical reaction at intermediate metal-semiconductor interfaces. Instead, one observes >0.4 eV $\Phi B$ variation with changing orientation of the GaAs (100) growth surface.

Surface reconstruction at the semiconductor interface has a pronounced effect on interface state energies and densities at a semiconductor-semiconductor junction. Interface state densities of ZnSe-GaAs heterostructures range over at least two orders of magnitude, depending upon the

RHEED reconstruction pattern of GaAs prior to epitaxial ZnSe growth.[263] Figure 35 illustrates interface state energies and densities as extracted from a C-V analysis for several well-defined reconstructions. Densities decrease in order of decreasing surface As concentration. Energies also vary somewhat. There is an interesting correspondence with Fig. 33 with both a low energy (<0.2 eV) and 0.55 - 0.65 eV states above $E_V$. Furthermore, the reduction of states with As depletion is consistent with the thermal processing steps required to produce the GaAs barriers reported in Fig. 32. As Qiu et al.[263] suggest, however, the Ga excess may provide for the formation of a GaSe interfacial compound, which may also reduce bonding states,[248] as in the passivation treatment discussed earlier in this section. The decreased interface state density is reflected in the nearly ideal C-V characteristic and the evidence that $E_F$ is free to move across the entire GaAs band gap region. Improved electrical properties at ZnSe/GaAs heterojunctions are also confirmed by band bending measurements showing that $E_F$ is not pinned at a fixed energy, but instead follows the doping concentration-dependent $E_F$ position in ZnSe,[47] similar to earlier work of Chiaradia et al.[243] for Ge/GaAs heterojunctions.

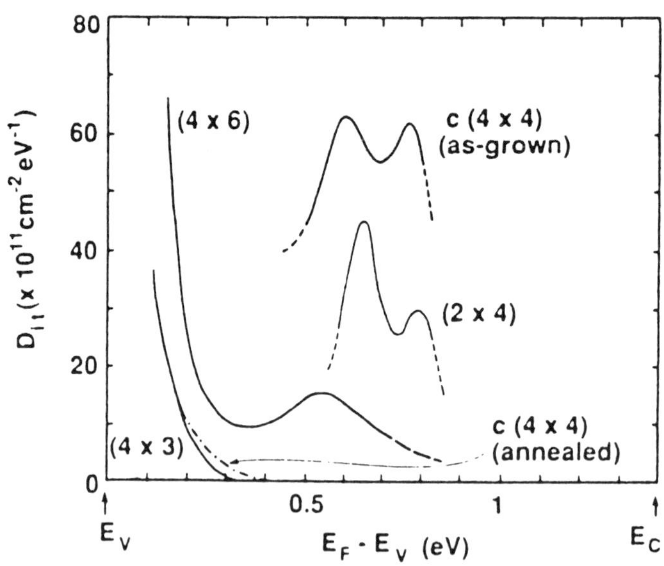


**Figure 35.** Interface state density distributions calculated from capacitance-voltage characteristics for a variety of GaAs epilayer surface reconstructions prior to ZnSe epitaxial overgrowth. Interface state densities appear to decrease as the GaAs epilayer surface becomes increasingly As-deficient. See Ref. 26.

The importance of reconstruction and localized bonding in interface electronic states highlights the electronic structure of clean semiconductor surfaces. For epitaxial growth surfaces in particular, electronic band structure depends sensitively on the detailed reconstruction. See, for example, Ref. 264 and references therein. A particular reconstruction could yield nonmetallic surface phases and/or lower interface reactivity. Thus a knowledge of surface geometric structure offers additional potential for controlling interface electronic states.

## 7.5 Comparison of Interface State Models

The electronic feature of semiconductor surfaces and interfaces presented in this section offer new evidence to distinguish between various models of interface states. There can be little dispute as to the existence of intrinsic interface states associated with wave function tunneling from metal into semiconductor. However, the magnitude of the dipole due to the charge and its penetration depends sensitively upon the details of ionic and electronic charge density assumed near the interface and, in fact, determine how dominant an effect these metal-induced gap states have on the Schottky barrier formation. Certainly the large range of barrier heights obtained for different metals on the same semiconductor over the same metal-semiconductor system with altered interface bonding is difficult to reconcile with a canonical pinning level or even one incorporating perturbation due to electronegativity differences. Similar arguments pertain to disordered interface models, especially as they relate to epitaxial junctions. Models incorporating both wave function tunneling plus discrete gap states due to defects,[72] impurities,[196] or adsorbate states[29][84][85] provide greater flexibility in accommodating the significant $E_F$ variations observed and, furthermore, they depend on the presence of discrete states, as observed experimentally.

Defect models appear to be successful in accounting for the narrow range of $E_F$ stabilization for cleaved GaAs (110), GaSb and, to a lesser extent, InP interfaces.[26] Again, discrete states are observable in the semiconductor band gap near the $E_F$ stabilization energies. However, the wide $E_F$ ranges seen for GaAs (100) and other semiconductor argue against the formation of such states at all metal-semiconductor interfaces. Indeed, the contrast between GaAs (110) and GaAs (100) grown and prepared by different methods indicate $E_F$ control via specific metal-semiconductor interactions rather than by canonical defect states.

The effective work function model and its generalization to classical Schottky barrier formation via metal-semiconductor work function differences (Eq. 1) appear to be consistent with most electronic and chemical measurements, assuming the formation of new microscopic dielectric phases. In essence, such a model derives from a low contribution of interface dipoles due to wave function tunneling to the resultant band bending in the surface space charge region. Advantages of such a model are *(a)* their conceptual simplicity, *(b)* their experimental verifiability in terms of required interface-specific changes in atomic composition and morphological structure, and *(c)* their predictive nature.

Perhaps a combination of mechanisms best accounts for the wide spectrum of interface behavior. Schottky-like charge exchange due to low interface dipole formation combined with discrete states due to *(a)* localized metal-semiconductor bonding, *(b)* adsorbate resonances, *(c)* surface segregated native defects, or other lattice imperfections offer a coherent picture of the different electronic properties for interfaces grown and processed under different conditions.

## 8.0 FUTURE DIRECTIONS

The results of the previous section offer numerous opportunities to characterize, understand, and control interface electronic states with greater refinement.

In terms of characterization, extensions of modern interface state techniques may lead to a more refined description of electronic features on an atomic scale. For CLS, these opportunities include the use of time-resolved detectors with fast gating to measure recombination cross sections of these mid-gap, "nonradiative" transitions and thereby their absolute densities.[35] Longer wavelength detectors (>2 $\mu$m) will enable observation of the lower energy transitions involving states only a few tenths of an eV from the band edges. Combining CLS with medium resolution (100 Å) electron microscopes will retain near-interface sensitivity while enabling the spatial resolution for mapping of states vs. morphology or chemical structure. For photoelectron spectroscopy, similar opportunities exist in terms of microspectroscopy of interfacial chemical bonding and electronic structure on a submicron (perhaps even 500 Å) scale.[265] Angular-dependent SXPS can provide a relatively gentle probe of geometric bonding and order of the semiconductor surface reconstructions giving rise to different interface state behavior, especially combined with new growth and preparation techniques. Furthermore, photo-electron techniques are powerful for verifying surface conditions developed for an optimized metal-semiconductor junction. For STM, a wealth of opportunities suggest themselves. BEEM techniques can be employed to map barrier inhomogeneities vs. metallization chemistry as a function of surface and interface preparation conditions, as well as for different crystal growth approaches. STM provides the most direct method of establishing the relationship of the deep levels within the band gap to specific surface atomic site or morphological features. Furthermore, the STM technique offers an opportunity to characterize the electronic properties of multiatomic adsorbate complexes and thereby monitor the transition from isolated atoms to metal.

In terms of deeper understanding of interface states, electron microscopy, BEEM, CLS, and photoluminescence of epitaxial interfaces offer the chance to relate deep level states to specific interfacial bonding and nanometer structure. Included may be the introduction of impurities or adsorbates with well-defined deep level properties which can be compared with theoretical calculations. These techniques apply equally well to semiconductor heterojunctions and metal-semiconductor junctions. In both cases, opportunities exist to study interface states as a function of growth rate, temperature, and interruption, and in addition, to correlate with roughness as exhibited by electron-beam induced luminescence.[266] Relating theoretical studies of epitaxial interfaces to such observations provides a means to test the effectiveness of such models in predicting interface state properties and their effect on interface charge transfer. Studies of interface states in reduced dimensions—ultrathin superlattice structure, for example—are possible. These measurements offer a greater understanding of electronic states in layered structures with layers of only nanometer dimensions. Furthermore, such studies have direct implications for selecting or optimizing electronic states versus growth parameters by virtue of intermediate band structures.

In terms of controlling Schottky barrier heights, the sensitivity of interface states to specific chemistry, processing, and growth, offer numerous new avenues to explore. The high temperature melting point alloys have only begun to be explored. As described in Ch. 2, such alloys offer not only a wide range of $E_F$ control but improved chemical stability at elevated temperatures. Besides other metal-metal combinations, one may envision nanometer-scale sacrificial layers which can inhibit disruption of the semiconductor lattice below while preserving desired properties of the metallic overlayer. This approach may also involve "seeding" the surface to promote specific metallic epitaxy. Alternatively, selected adsorbates may serve to rehybridize surface bonding and thereby remove electronic states from the band gap, even after metallization. New processing steps which induce self-ordering and rebonding can be monitored in situ to minimize or otherwise alter interface state densities. Thus interface state measurements within the growth and processing facilities provide a powerful approach to optimize electronically each stage of the interface formation.

## 9.0 SUMMARY

In conclusion, we have examined the diverse phenomena capable of introducing and controlling localized states at semiconductor interfaces. We have shown how such localized charge states act to modify the transfer of charge between metal and semiconductor and the associated Schottky barrier formation. We have described the role of intrinsic states due only to the properties of the noninteracting constituents, extrinsic states due to conventional features such as impurities, bulk defects, and morphological imperfections, as well as extrinsic states to interface-specific phenomena such as interface chemical reactions, altered interface bonding, interface defects, and interface chemical phases with new dielectric properties. Correspondingly, we have described the numerous approaches already taken to minimize or otherwise control interface states, such as wet chemical treatments, atomic-scale interlayers, alternative semiconductor crystal growth, and epitaxial interface modification. We have described how these observations relate to the various models proposed for interface states. Finally, we have offered a view to further advances in characterizing, understanding, and controlling interface states. Given our need for an increasingly refined knowledge of semiconductor interface state properties, this field will continue to provide rich and varied research and technology opportunities well into the future.

## REFERENCES


1. W. Schottky, *Z. Physilk,* 113:367 (1939)
2. N. F. Mott, *Proc. Roy. Soc. London,* A171:27 (1939)
3. B. Davidov, *J. Tech. Phys. USSR,* 5:87 (1938); *Soviet J. Phys.,* 1:167 (1939)
4. E. H. Rhoderick and R. H. Williams, *Metal-Semiconductor Contacts, Monographs in Electrical and Electronic Engineering,* 2nd ed., (P. Hammond and R. L. Grimsdale, eds.) Clarendon, Oxford (1988)
5. W. Schockley and G. L. Pearson, *Phys. Rev.,* 74:232 (1948)
6. S. M. Sze, *Physics of Semiconductor-Devices,* 2nd ed., Ch. 5., Wiley-Interscience, New York (1981)
7. C. J. Palmstrøm and D. V. Morgan, *Gallium Arsenide,* (M. J. Howe and D. V. Morgan, eds.), Ch. 6, John Wiley & Sons, Ltd., New York (1985)
8. C. B. Duke and C. Mailhiot, *J. Vac. Sci. Technol.,* B3:1170 (1985)
9. J. Bardeen, *Phys. Rev.,* 71:717 (1947)
10. L. J. Brillson, *Surface Sci. Rpts.,* 2:123 (1982)
11. C. B. Duke, *Tunneling in Solids,* pp. 102-110 Academic Press, New York (1969)
12. C. R. Crowell, *Solid-State Electron.,* 8:395 (1965)
13. M. P. Shaw, *Handbook on Semiconductors, Vol. 4, Device Physics,* Ch. 1, (C. Hilsum, ed.), North Holland, Amsterdam (1981)
14. A. M. Cowley, *J. Appl. Phys.,* 37:3024 (1966)
15. A. M. Goodman, *J. Appl. Phys.,* 34:329 (1963)
16. G. Margaritondo, L. J. Brillson, and N. G. Stoffel, *Solid St. Commun.,* 35:227 (1980)
17. M. H. Hecht, *Phys. Rev.,* B41:7918 (1990)
18. J. E. Rowe, *J. Vac. Sci. Technol.,* 13:798 (1976); G. Margaritondo, J. E. Rowe, and S. B. Christman, *Phys. Rev.* B14:5396 (1976)
19. D. E. Eastman and J. L. Freeouf, *Phys. Rev. Lett.,* 33:1601 (1974); W. Gudat and C. Kunz, *Phys. Rev. Lett.,* 29:169 (1972); G. J. Lapeyre and J. Anderson, *Phys Rev. Lett.,* 35:117 (1975)
20. G. J. Lapeyre, J. Anderson, P. L. Gobby, and J. A. Knapp, *Phys. Rev. Lett.,* 33:1290 (1974)
21. R. Haight and J. Bokor, *Phys. Rev. Lett.,* 56:2846 (1986)
22. R. Haight and J. A. Silberman, *Appl. Phys. Lett.,* 57:1548 (1990)

23. H. C. Gatos and J. Lagowski, *J. Vac. Sci. Technol.,* 10:130 (1973)
24. L. J. Brillson, *Surface Sci.,* 51:45 (1975)
25. L. J. Brillson, in *Proc. 14th Int'l. Conf. on the Physics of Semiconductors,* (B. L. H. Wilson, ed.), Edinburgh, (1978), 43:765, *Inst. Conf. Ser.,* Institute of Physics, Bristol, (1979); L. J. Brillson, *J. Vac. Sci. Technol.,* 16:1137 (1979)
26. W. E. Spicer, I. Lindau, P. Skeath, C. Y. Su, and P. Chye, *Phys. Rev. Lett.,* 44:420 (1980)
27. C. A. Mead, *Solid-State Electron.,* 9:1023 (1966), and references therein
28. G. Binnig, H. Rohrer, C. Gerber, and E. Weibel, *Phys. Rev. Lett.,* 49:57 (1982)
29. R. M. Feenstra, *Phys. Rev. Lett.,* 63:1412 (1989)
30. P. N. First, J. A. Stroscio, R. A. Dragoset, D. T. Pierce, and R. J. Celotta, *Phys. Rev. Lett.,* 63:1416 (1989)
31. R. J. Hamers and K. Markert, *Phys. Rev. Lett.,* 64:1051 (1990); Y. Kuk, R. S. Becker, P. J. Silverman, and G. P. Kochanski, *Phys. Rev. Lett.,* 65:456 (1990)
32. P. S. Ho, E. S. Yang, H. L. Evans, and X. Wu, *Phys. Rev. Lett.,* 56:177 (1986)
33. G. Chiarotti, S. Nannarone, R. Pastore, and P. Chiaradia, *Phys. Rev.,* B4:3398 (1971)
34. L. J. Brillson, R. E. Viturro, J. L. Shaw, and H. W. Richter, *J. Vac. Sci. Technol.,* A6:1437 (1988)
35. L. J. Brillson and R. E. Viturro, *Scanning Electron. Microsc.,* 2:789 (1988)
36. B. G. Yacobi and D. B. Holt, *Cathodoluminescence Microscopy of Inorganic Solids,* Plenum, New York (1990)
37. M. Cardona, *Modulation Spectroscopy,* Academic Press, New York (1969), and references therein
38. R. E. Viturro, M. L. Slade, and L. J. Brillson, *Phys. Rev. Lett.,* 57:487 (1986)
39. L. J. Brillson, H. W. Richter, M. L. Slade, B. A. Weinstein, and Y. Shapira, *J. Vac. Sci. Technol.,* A3:1011 (1985)
40. S. Chang, R. E. Viturro, and L. J. Brillson, *J. Vac. Sci. Technol.,* A8:3803 (1990)
41. J. L. Shaw, R. E. Viturro, L. J. Brillson, and D. LaGraffe., *J. Electron. Mat.,* 18:59 (1989)

42. J. L. Shaw, L. J. Brillson, S. Sivananthan, and J.- P. Faurie, *Appl. Phys. Lett.*, 56:1266 (1990)
43. J. L. Shaw, L. J. Brillson, S. Sivananthan, and J.- P. Faurie, *Proc. Mat. Res. Soc.*, 161, Mat. Res. Soc., Pittsburgh, PA, (1990)
44. C. Mailhiot and C. B. Duke, *Phys. Rev.*, B33:1118 (1986)
45. A. Pinczuk, A. A. Ballman, R. E. Nahory, M. A. Pollack, and J. M. Worlock, *J. Vac. Sci. Technol.*, 16:1168 (1979)
46. F. Schiffler and G. Arbstrelter, *J. Vac. Sci. Technol.*, B3:1184 (1985)
47. D. J. Olego and D. Cammack, *J. Crystal Growth*, 101:546 (1990)
48. P. Parayanthal, F. H. Pollak, and J. M. Woodall, *Appl Phys. Lett.*, 41:961 (1982)
49. W. J. Kaiser and L. D. Bell, *Phys. Rev. Lett.*, 60:1406 (1988)
50. L. D. Bell, M. H. Hecht, W. J. Kaiser, and L. C. Davis, *Phys. Rev. Lett.*, 64:2679 (1990); M. H. Hecht, L. D. Bell, and W. J. Kaiser, *Appl. Surf. Sci.*, 41/42:17 (1989)
51. R. Ludeke, *J. Vac. Sci. Technol.*, A9:885 (1991); Y. Hasaqawa, Y. Kuk, R. T. Tung, P. J. Silverman, and T. Sakurai, *J. Vac. Sci. Technol.*, B9:578 (1991)
52. A. E. Fowell, R. H. Williams, B. E. Richardson, and T.-H. Shen, *Semicond. Sci. Technol.*, 5:348 (1990)
53. S. Kurtin, T. C. McGill, and C. A. Mead, *Phys. Rev. Lett.*, 22:1433 (1970)
54. M. Schlüter, *Phys. Rev.*, B17:5044 (1978)
55. D. E. Eastman and W. D. Grobman, *Phys. Rev. Lett.*, 28:1376 (1972)
56. L. F. Wagner and W. E. Spicer, *Phys. Rev. Lett.*, 28:1381 (1972); M. Erbudak and T. E. Fischer, *Phys. Rev. Lett.*, 29:732 (1972)
57. J. van Laar, A. Huijser, and J. L. van Rosy, *J. Vac. Sci. Technol.*, 14:894 (1977)
58. V. Heine, *Phys. Rev.*, 138:A1689 (1965)
59. M. Schlüter, *Thin Solid Films*, 93:3 (1982), and references therein
60. J. Tersoff, *Phys. Rev. Lett.*, 52:465 (1984)
61. J. Tersoff, *Phys. Rev.*, B32:6968 (1985)
62. J. Tersoff, *J. Vac. Sci. Technol.*, B3:1157 (1985); *Surface Sci.*, 168:275 (1986)
63. C. Tejedor, F. Flores, and E. Louis, *J. Phys. C (Solid State Phys.)*, 10:2163 (1977); F. Flores and C. Tejedor, *J. Phys. C (Solid State Phys.)*, 20:145 (1987)

64. W. Mönch, *Festkörperprobleme-Advances in Solid State Physics,* (P. Grösse, ed.), 26:67, Vieweg, Braunschweig (1986)

65. J. J. Rehr and W. Kohn, *Phys. Rev.,* B9:1981 (1974); B10:448 (1974)

66. M. Cardona and N. E. Christensen, *Phys. Rev.,* B35:6182 (1987)

67. P. Masri, *Phys. Rev. Lett.,* 64:208 (1990)

68. F. Flores, A. Muñoz, and J. C. Duran, *Appl. Surf. Sci.,* 41/42:144 (1989)

69. P. E. Schmid, *Helvetica Physica Acta,* 58:371 (1985)

70. W. Mönch, *Appl. Surf. Sci.,* 41/42:128 (1989)

71. S. B. Zhang, M. L. Cohen, and S. G. Louie, *Phys. Rev.,* B32:3955 (1985)

72. W. Mönch, *Phys. Rev. Lett.,* 58:1260 (1987)

73. W. Harrison, *Phys. Rev. Lett.,* 37:312 (1976)

74. J. C. Inkson, *J. Vac. Sci. Technol.,* 11:943 (1974)

75. R. T. Tung, *Phys. Rev. Lett.,* 52:461 (1984)

76. D. R. Heslinga, H. H. Weltering, D. P. Van der Werf, T. M. Klapwijk, and T. Hibma, *Phys. Rev. Lett.*, 64:1589 (1990)

77. M. Wittmer, *Phys. Rev.,* B42:5249 (1990); M. Wittmer, P. Oelhafen, and K. N. Tu, *Phys. Rev.,* B35:9073 (1987)

78. M. O. Aboelfotoh, A. Cros, B. G. Svensson, and K. N. Tu, *Phys. Rev.,* B41:9819 (1990)

79. W. H. Brattain and J. Bardeen, *Bell Syst. Tech. J.,* 82:1 (1953)

80. T. A. Goodwin and P. Mark, *Progress in Surface Science,* 1:1 Pergamon, New York, (1972)

81. K. Stiles, A. Kahn, D. G. Kilday, and G. Margaritondo, *J. Vac. Sci. Technol.,* B5:987 (1987); K. Stiles, S. F. Horng, A. Kahn, J. McKinley, D. G. Kilday, and G. Margaritondo, *J. Vac. Sci. Technol.,* B6:1392 (1988); K. Stiles and A. Kahn, *Phys. Rev.Lett.,* 60:440 (1988)

82. R. Cao, K. Miyano, T. Kendelewicz, K. K. Chin, I. Lindau, and W. E. Spicer, *J. Vac. Sci. Technol.,* BS:998 (1987)

83. S. Baidyaroy, W. R. Bottoms, and P. Mark, *Surf. Sci.,* 29:165 (1971)

84. W. Mönch, *J. Vac. Sci. Technol.,* B6:1270 (1988)

85. A. Kahn, K. Stiles, D. Mao, S. F. Horng, K. Young, J. McKinley, D. G. Kilday, and G. Margaritondo, *Metallization and Metal-Semiconductor Interfaces,* (I. P. Batra, ed.), 195:163, NATO Advanced Study Institute, Series B Physics, Plenum, New York (1989)

86. G. Margaritondo, J. E. Rowe, C. M. Bertoni, C. Calandra, and F. Manghi, *Phys. Rev.,* B20:1538 (1979)

87. E. J. Mele and J. D. Joannopoulos, *J. Vac. Sci. Technol.,* 15:1287 (1987)

88. S. Baidyaroy and P. Mark, *Surface Sci.,* 30:53 (1972)

89. W. E. Spicer, I. Lindau, P. E. Gregory, C. M. Garner, P. Pianetta, and P. W. Chye, *J. Vac. Sci. Technol.,* 13:780 (1976); L. J. Brillson, *J. Vac. Sci. Technol.*, 13:325 (1976)

90. L. J. Brillson, M. L. Slade, R. E. Viturro, M. Kelly, N. Tache, G. Margaritondo, J. M. Woodall, G. D. Pettit, P. D. Kirchner, and S. L. Wright, *J. Vac. Sci. Technol.,* B4:919 (1986)

91. K. Kajiyama, Y. Mizushima, and S. Sakata, *Appl. Phys. Lett.,* 23:458 (1973)

92. M. Missous, E. H. Rhoderick, and K. E. Singer, *J. Appl. Phys.,* 60:2439 (1986)

93. N. Newman, Z. Liliental-Weber, E. R. Weber, J. Washburn, and W. E. Spicer, *Appl. Phys. Lett.,* 53:145 (1988)

94. J. van Laar and J. J. Scheer, *Surface Sci.,* 8:342 (1967)

95. J. van Laar and A. Huijser, *J. Vac. Sci. Technol.,* 13:769 (1976)

96. A. Huijser and van Laar, *Surf. Sci.,* 52:202 (1975)

97. M. Henzler, *Surface Sci.,* 19:159 (1970)

98. M. Henzler, *Surface Sci.,* 22:12 (1970)

99. J. E. Rowe, S. B. Christman, and H. Ibach, *Phys. Rev. Lett.,* 34:874 (1975)

100. H. Ibach, K. Horn, R. Dorn, and H. Lüth, *Surface Sci.,* 38:433 (1973)

101. S. Chang, L. J. Brillson, D. S. Rioux, D. Pettit, and J. M. Woodall, *J. Vac. Sci. Technol.,* B8:1008 (1990)

102. S. Chang, L. J. Brillson, Y. J. Kime, D. S. Rioux, D. Pettit, and J. M. Woodall, *Phys. Rev. Lett.,* 64:2551 (1990)

103. M. D. Pashley, K. W. Haberern, and J. M. Gaines, *J. Vac. Sci. Technol.,* B9:938 (1991)

104. L. J. Brillson, *J. Vac. Sci. Technol.,* 12:76 (1975)

105. L. J. Brillson and C. H. Griffiths, *J. Vac. Sci. Technol.,* 15:529 (1978)

106. L. J. Whitman, J. A. Stroscio, R. A. Dragoset, and R. J. Celotta, *Phys. Rev.,* B42:7228 (1990)

107. L. W. Aukermann, *Semiconductors and Semimetals,* (R. K. Willardson and A. C. Beer, eds. ) 4:343, Academic, New York, (1973)

108. W. Walukiewicz, *Phys. Rev.,* B37:4760 (1988)

109. T. M. Valahas, J. S. Sochanski, and H. C. Gatos, *Surf. Sci.,* 26:41 (1971)

110. A. G. Milnes, *Deep Impurities in Semiconductors,* Wiley-Interscience, New York (1973)

111. R. E. Haisted, M. Aven, and H. D. Coghill, *J. Electrochem. Soc.,* 112:177 (1965)

112. P. Hiesinger, *Phys. Stat. Solidi,* (a) 33:K39 (1976)

113. J. M. Borrego, R. J. Butman, and S. Ashok, *Solid-State Electron.,* 20:125 (1977)

114. B. I. Boltaks, *Diffusion in Semiconductors,* Infosearch, London (1963)

115. D. Shaw, *Atomic Diffusion in Semiconductors,* Plenum, New York (1973)

116. J. L. Shaw, R. E. Viturro, L. J. Brillson, and D. LaGraffe, *J. Vac. Sci. Technol.*, A7:489 (1989)

117. J. M. Woodall, G. D. Pettit, T. N. Jackson, C. Lanza, K. L. Kavanaugh, and J. W. Mayer, *Phys. Rev. Lett.,* 51:1783 (1983)

118. P. J. Tauters, M. Offenberg, and P. Balk, *Appl. Phys. Lett.,* 56:1903 (1990)

119. D. M. Hoffman, B. K. Meyer, J.-M. Spaeth, M. Wattenbach, J. Kroqer, C. Kisielowski-Kemmerich, and H. Alexander, *J. Appl. Phys.,* 68:3381 (1990)

120. T. Haga, M. Suezawa, and K. Sumino, *Jpn. J. Appl. Phys.,* 27:1929 (1988)

121. K. Das, H. S. Kong, J. B. Petit, J. W. Bumgarner, R. F. Davis, and L. G. Matus, *J. Electrochem. Soc.,* 137:1598 (1990)

122. A. Mirceau and D. Bois, *Gallium Arsenide and Related Compounds,* p. 82, Inst. Phys. Conf. Ser. No. 46 (1979)

123. J. L. Shaw, R. E. Viturro, L. J. Brillson, D. Kilday, M. K. Kelly, and G. Margaritondo, *J. Electron. Mat.,* 17:149 (1988)

124. J. L. Shaw, R. E. Viturro, L. J. Brillson, D. Kilday, M. K. Kelly, and G. Margaritondo, *J. Vac. Sci. Technol.,* A6:1579 (1988)

125. J. L. Shaw, R. E. Viturro, L. J. Brillson, D. Kilday, M. K. Kelly, and G. Margaritondo, *J. Vac. Sci. Technol.,* A6:2752 (1988)

126. R. K. Swank, *Phys. Rev.,* 153:844 (1967)

127. F. Barrels, H. J. Clemens, and W. Mönch, *Physica,* 117B/118B:801 (1983)

128. J. C. Brice and G. D. King, *Nature,* 209:1346 (1966)

129. D. E. Holmes, K. R. Elliott, R. J. Chen, and C. G. Kirkpatrick, *Appl. Phys. Lett.,* 40:46 (1982)

130. S. Makram-Ebeid, D. Gautard, P. Devillard, and G. M. Martin, *Appl. Phys. Lett.,* 40:161 (1982)

131. A. Yahata and M. Nakajima, *Jpn. J. Appl. Phys.,* 23:L313 (1984)

132. J. M. Andrews and J. C. Phillips, *Phys. Rev. Letters,* 35:56 (1975)

133. J. M. Andrews and J. C. Phillips, *CRC Critical Rev. Solid State Sci.,* 5:405 (1975)

134. S. P. Murarka, *Materials Lett.,* 1:26 (1982)

135. G. Ottaviani, K. N. Tu, and J. W. Mayer, *Phys. Rev. Lett.,* 44:284 (1980)

136. C. Calandra, O. Bisi, and G. Ottaviani, *Surf. Sci. Rpts.,* 4:271 (1984)

137. M. Liehr, P. E. Schmid, F. K. LeGoues, and P. S. Ho, *Phys. Rev. Lett.,* 54:2139 (1985)

138. R. T. Tung, K. K. Ng, J. M. Gibson, and F. J. Levy, *Phys. Rev.,* B33:7077 (1986)

139. R. J. Hauenstein, T. E. Schlesinger, T. C. McGill, B. D. Hunt, and L. J. Schowalter, *J. Vac. Sci. Technol.,* A4:860 (1986)

140. M. Ospelt, J. Henz, L. Flepp, and H. van Kanel, *Appl. Phys. Lett.,* 52:227 (1988)

141. H. Fujitani and S. Asano, *Phys. Rev.,* B42:1696 (1990)

142. G. P. Das, P. Blöchl, O. K. Andersen, N. E. Christensen, and O. Gunnarson, *Phys Rev. Lett.,* 63:1168 (1989)

143. D. R. Heslinga, H. H. Waitering, T. M. Klapwijk, and T. Hibma, *Proc. 20th Intern. Conf. on the Physics of Semiconductors,* Thessaloniki, (E. M. Anastassakis and J. D. Joannopoulos, eds.), p.347, World Scientific, London (1990)

144. J. G. Clabes, G. W. Rubloff, and T. Y. Tan, *Phys. Rev.,* 29:1540 (1984)

145. J. E. Rowe, G. Margaritondo and S. B. Christman, *Phys. Rev.,* B15:2195 (1977)

146. G. W. Rubloff, *Surface Sci.,* 132:268 (1983)

147. P. S. Ho and G. W. Rubloff, *Thin Solid Films,* 89:433 (1982)

148. S. Hara and I. Ohdomari, *Phys. Rev.,* B38:7554 (1988)

149. L. J. Brillson, *Phys. Rev. Lett.,* 40:260 (1978)

150. C. A. Mead, *Solid State Electron.,* 9:1023 (1966)

151. R. H. Williams, V. Montgomery, and R. R. Varma, *J. Phys. C (Solid State Phys.),* 11:L735 (1978)

152. E. Hökelek and G. Robinson, *J. Appl. Phys.,* 54:5199 (1983)

153. G. J. Hughes, A. McKinley, R. H. Williams, and 1. T. McGovern, J. Phys. C. *(Solid State Phys.),* 15:L159 (1982)

154. N. C. Wyeth and A. Catalano, *J. Appl. Phys.,* 51:2286 (1980)

155. J. Bears, D. Bassett, and M. Schulz, *Phys. Status Solidi,* (a)49:483 (1978)

156. J. F. McGilp and A. B. McLean, *J. Phys.,* C:21:807 (1988)

156a. L. J. Brillson, *Thin Solid Films,* 89:461 (1982)

157. L. J. Brillson, G. Margaritondo, and N. G. Stoffel, *Phys. Rev. Lett.,* 44:667 (1980)

157a. Y. Shapira, L. J. Brillson, A. D. Katnani, and G. Margaritondo, *Phys. Rev.,* B30:4586 (1984)

158. L. J. Brillson, C. F. Brucker, A. D. Katnani, N. G. Stoffel, and G. Margaritondo, *Appl. Phys. Lett.,* 38:784 (1981)

159. T. Kendelewicz, N. Newman, R. S. List, I. Lindau and W. E. Spicer, *J. Vac. Sci. Technol.,* B3:1206 (1985)

160. D. V. Lang, *Inst. Phys. Conf. Ser. No. 131,* 1:70 (1977)

161. P. Bhattacharya and S. Dhar, *Semiconductors and Semimetals,* (R. K. Willardson and A. C. Beer, ed.), 26:143, Academic, New York (1988)

162. H. H. Wieder, *J. Vac. Sci. Technol.,* 15:1498 (1978)

163. P. Mark, S. C. Chang, W. F. Creighton, and B. W. Lee, *CRC Critical Rev. Solid State Sci.,* 5:189 (1975); P. Mark, E. So, and M. Bonn, *J. Vac. Sci. Technol.,* 14:865 (1977)

164. I. Lindau, P. W. Chye, C. M. Garner, P. Planetta, C. Y. Su, and W. E. Spicer, *J. Vac. Sci. Technol.,* 15:1337 (1978)

165. W. E. Spicer, I. Lindau, P. Skeath, and C. Y. Su, *Phys. Rev. Lett.,* 44:420 (1980)

166. R. H. Williams, R. R. Varma, and V. Montgomery, *J. Vac. Sci. Technol.,* 16:1418 (1979)

167. A. Zunger, *Phys. Rev.,* B24:4372 (1981)

168. M. D. Daw and D. L. Smith, *Phys. Rev.,* B20:5150 (1979)

169. M. D. Daw and D. L. Smith, *Appl. Phys. Lett.,* 8:690 (1980)

170. M. D. Daw and D. L. Smith, *J. Vac. Sci. Technol.,* 17:1028 (1980)

171. G. P. Srivastava, *Phys. Status Solidi,* (b) 93:761 (1979)

172. R. E. Allen and J. D. Dow, *J. Vac. Sci. Technol.,* 19:383 (1981)

173. R. E. Allen and J. D. Dow, *Phys. Rev.,* B25:1423 (1982)

174. J. D. Dow and R. E. Allen, *J. Vac. Sci. Technol.,* 20:659 (1982)

175. J. S. Best, *Appl. Phys. Lett.,* 34:522 (1979)

176. H. W. Wieder, *Appl. Phys. Lett.,* 38:170 (1981)

177. J. Van Vechten, *J. Electrochem. Soc.,* 122:423 (1975)

178. N. Newman, W. E. Spicer, T. Kendelewicz, and I. Lindau, *J. Vac. Sci. Technol.*, B4:931 (1986)

179. J. R. Waldrop, *Appl. Phys. Lett.*, 44:1002 (1984); *ibid.*, 46:864 (1985)

180. C. M. Aldao, I. M. Vitomirov, F. Xu, and J. H. Weaver, *Phys. Rev.*, B37:6019 (1988)

181. E. R. Weber, H. Ennen, V. Kaufmann, J. Windscheif, J. Schneider, and T. Wasinski, *J. Appl. Phys.*, 53:6140 (1982)

182. E. R. Weber and J. Schneider, *Physica*, B116:398 (1983)

183. W. E. Spicer, Z. Liliental-Weber, E. Weber, N. Newman, T. Kendelewicz, R. Cao, C. McCants, P. Mahowald, K. Miyano, and I. Lindau, *J. Vac. Sci. Technol.*, B6:245(1988)

184. R. Fornari, E. Gombia, and R. Mosca, *J. Electron. Mat.*, 18:151 (1989)

185. S. P. Swenson, J. Kanski, T. G. Anderson, and P. O. Nilsson, *J. Vac. Sci. Technol.*, B2:235 (1984)

186. H. Cheng, X.-J. Zhang, and A. G. Milnes, *Solid-State Electron.*, 27:1117 (1984)

187. A. B. McLean and R. H. Williams, *Semicond. Sci. Technol.*, 21:654 (1987)

188. T. Yokotsuka, T. Narusawa, Y. Uchida, and H. Nakashima, *Appl. Phys. Lett.*, 50:591 (1987)

189. J. Nishizawa, Y. Oyama, and K. Dezaki, *J. Appl. Phys.*, 67:1884 (1990)

190. W. Walukiewicz, J. *Vac. Sci. Technol.*, B5:1062 (1987)

191. W. Walukiewicz, *J. Vac. Sci. Technol.*, B6:1257 (1988)

192. W. Mönch, *Phys. Rev.*, B37:7129 (1988); *Appl. Surf. Sci.*, 41/42:128 (1989)

193. J. R. Waldrop, *J. Vac. Sci. Technol.*, B2:445 (1984)

194. J. R. Waldrop, *Appl. Phys. Lett.*, 47:1301 (1985)

195. M. Ludwig and G. Heymann, *J. Vac. Sci. Technol.*, B4:485 (1986)

196. R. Ludeke, A. Taleb-Ibrahimi, and G. Jezequel, *Appl. Surf. Sci.*, 41/42:151 (1989), and references therein

197. R. Ludeke, A. Taleb-Ibrahimi, and G. Jezequel, *J. Vac. Sci. Technol.*, B6:1277 (1988)

198. W. E. Spicer, P. E. Gregory, P. W. Chye, I. A. Babaloia, and T. Lukegawa, *Appl. Phys. Lett.*, 27:617 (1975); K. K. Chin, S. H. Pan, D. Mo, P. Mahowald, N. Newman, I. Lindau, and W. E. Spicer, *Phys. Rev.*, B32:918 (1985)

199. A. B. McLean, D. A. Evans, and R. H. Williams, *Semicond. Sci. Technol.*, 2:547 (1987)

200. R. Ludeke, T.-C. Chiang, and T. Miller, *J. Vac. Sci. Technol.*, B1:581 (1983)

201. A. B. McLean, R. H. Williams, and J. F. McGilp, *J. Vac. Sci. Technol.*, B6:1252 (1988)

202. W. Tersoff and W. A. Harrison, *Phys. Rev. Lett.*, 58:2367 (1987) and references there in

203. J. L. Freeouf, *Solid State Commun.*, 33:1059 (1980)

204. L. J. Brillson, C. F. Brucker, A. D. Katnani, N. G. Stoffel, and G. Margaritondo, *Phys. Rev. Lett.*, 46:838 (1981)

205. Z. Liliental-Weber, R. Gronsky, J. Washburn, N. Newman, W. Spicer and E. R. Weber, *J. Vac. Sci. Technol.*, B4:912 (1986)

206. J. L. Freeouf and J. M. Woodall, *Appl. Phys. Lett.*, 39:727 (1981); J. M. Woodall and J. L. Freeouf, *J. Vac. Sci. Technol.*, 19:794 (1981)

207. H. Hasegawa and H. Ohno, *J. Vac. Sci. Technol.*, B4:1130 (1986)

208. H. Hasagawa, L. He, H. Ohno, T. Sawada, T. Haga, Y. Abe, and H. Takahashi, *J. Vac. Sci. Technol.*, B5:1097 (1987)

209. W. A. Harrison, *J. Vac. Sci. Technol.*, B3:1231 (1985)

210. S. D. Offsey, J. M. Woodall, A. C. Warren, P. D. Kirchner, T. I. Chappell, and G. D. Pettit, *Appl. Phys. Lett.*, 48:475 (1986)

211. E. Yablonovitch, C. J. Sandroff, R. Bhat, and T. Gmitter, *Appl. Phys. Lett.*, 51:439 (1987)

212. C. J. Sandroff, R. N. Nottenburg, J.-C. Bischoff, and R. Bhat, *Appl. Phys. Lett.*, 51:33 (1987)

213. E. Yablonovitch and T. Gmitter, *IEEE Electron Device Lett.*, EDL-6:597 (1985)

214. E. Yablonovitch and T. J. Gmitter, *Proc. Electrochem. Soc.*, 88-20:207 (1988)

215. R. J. Nelson, J. S. Williams, H. J. Leamy, B. J. Miller, H. C. Casey, Jr., B. A. Parkinson, and A. Heller, *Appl. Phys. Lett.*, 38:76 (1980)

216. K. C. Hwang and S. S. Li, *J. Appl. Phys.*, 67:2162 (1990)

217. R. Schachter, D. J. Olego, J. A. Baumann, L. A. Bunz, P. M. Raccah, and W. E. Spicer, *Appl. Phys. Lett.*, 47:272 (1985)

218. L. J. Brillson, Y. Shapira, and A. Heller, *Appl. Phys. Lett.*, 43:174 (1983)

219. M. T. Schmidt, Q. Y. Ma, D. V. Podlesnik, R. M. Osgood, and E. S. Yang, *J. Vac. Sci. Technol.*, B7:980 (1989)

220. A. Zur, T. C. McGill, and D. L. Smith, *Phys. Rev.*, B28:2060 (1983)

221. D. Liu, T. Zhang, R. A. LaRue, J. S. Harris, Jr., and T. W. Sigmon, *Appl. Phys. Lett.,* 53:1059 (1988)

222. C. J. Spindt, R. S. Besser, R. Cao, K. Miyano, C. R. Helms, and W. E. Spicer, *Appl. Phys. Lett.,* 54:1148 (1989)

223. S. Chang, I. Vitomirov, L. J. Brillson, D. F. Rioux, P. D. Kirchner, G. D. Pettit, and J. M. Woodall, *Phys. Rev. B, Rapid Commun.,* 41:12299 (1990)

224. G. S. Higashi, Y. J. Chabal, G. W. Trucks, and K. Raghavachari, *Appl. Phys. Lett.,* 56:656(1990)

225. T. Takahagi, A. Ishitani, H. Kuroda, Y. Nagasawa, H. Ito, and S. Wakao, *J. Appl. Phys.,* 68:2187 (1990)

225a. Y. Tao, A. Yelon, E. Sacher, Z. H. Lu, and M. J. Graham, *Appl. Phys. Lett.,* in press; Z. H. Lu, M. J. Graham, X. H. Feng, and B. X. Yang, *Appl. Phys. Lett.*, in press

226. C. F. Brucker and L. J. Brillson, *Appl. Phys. Lett.,* 39:67 (1981); L. J. Brillson, *J. Vac. Sci. Technol.,* 20:652 (1982)

227. K. Hirose, H. Tsuda, and T. Mizutani, *Appl. Surf. Sci.,* 41/42:174 (1989)

228. R. H. Williams and M. H. Patterson, *Appl. Phys. Lett.,* 40:484 (1982)

229. J. L. Shaw, R. E. Viturro, L. J. Brillson, and D. LaGraffe, *Appl. Phys. Lett.,* 53:1723 (1988)

230. V. Montgomery, R. H. Williams, and G. P. Srivastava, *J. Phys. C (Solid State Phys.),* 14:L191 (1981)

231. G. D. Waddill, I. M. Vitomirov, C. M. Aldao, and J. H. Weaver, *Phys. Rev. Lett.,* 62:1568 (1989)

232. I. M. Vitomirov, C. M. Aldao, G. D. Waddill, C. Capasso, and J. H. Weaver, *Phys. Rev.,* B41:8465 (1990)

233. M. Vos, C. M. Aldao, D. J. W. Aastuen, and J. H. Weaver, *Phys. Rev.,* B41:991 (1990)

234. J. R. Waldrop, *J. Vac Sci. Technol.,* B3:1197 (1985)

235. J. R. Waldrop, *Appl. Phys. Lett.,* 47:1301 (1985)

236. R. W. Grant and J. R. Waldrop, *J. Vac. Sci. Technol.,* B5:1015 (1987)

237. J. R. Waldrop and R. W. Grant, *Appl. Phys. Lett.,* 52:1794 (1988); J. R. Waldrop, *Appl. Phys. Lett.,* 53:1518 (1988)

238. J. R. Waldrop and R. W. Grant, *Appl. Phys. Lett.,* 50:250 (1987)

239. J. C. Costa, F. Williamson, T. J. Miller, K. Beyzavi, M. I. Nathan, D. S. L. Mui, S. Strite, and H. Morkoc, *Appl. Phys. Lett.,* 59:382 (1991)

240. J. R. Waldrop, E. A. Kraut, S. P. Kowalczyk, and R. W. Grant, *Surf. Sci.,* 132:513 (1983)

241. R. S. List, P. H. Mahowald, J. Woicik, and W. E. Spicer, *J. Vac. Sci. Technol.,* A4:1391 (1986)

242. D. C. Look, C. E. Stutz, and K. R. Evans, *Appl. Phys. Lett.,* 57:2570 (1990)

243. P. Chiaradia, A. D. Katnani, H. W. Sang, Jr., and R. S. Bauer, *Phys. Rev. Lett.,* 52:1246 (1984)

244. L. J. Brillson, M. L. Slade, R. E. Viturro, M. Kelly, N. Tache, G. Margaritondo, J. M. Woodall, G. D. Pettit, P. D. Kirchner, and S. L. Wright, *Appl. Phys. Lett.,* 48:1458 (1986)

245. L. J. Brillson, M. L. Slade, R. E. Viturro, M. Kelley, N. Tache, G. Margaritondo, J. M. Woodall, G. D. Pettit, P. D. Kirchner, and S. L. Wright, *J. Vac. Sci. Technol.,* B4:919 (1986)

246. L. J. Brillson, R. E. Viturro, M. L. Slade, P. Chiaradia, D. Kilday, M. Kelly, and G. Margaritondo, *Appl. Phys. Lett.,* 50:1379 (1987)

247. P. Chiaradia, R. E. Viturro, M. L. Slade, L. J. Brillson, D. Kilday, M. Kelly, N. Tache, and G. Margaritondo, *J. Vac. Sci. Technol.,* B5:1075 (1987)

248. R. E. Viturro, J. L. Shaw, and L. J. Brillson, *J. Vac. Sci. Technol.,* B6:1397 (1988)

249. L. J. Brillson, R. E. Viturro, J. L. Shaw, C. Mailhiot, N. Tache, J. McKinley, G. Margaritondo, J. M. Woodall, P. D. Kirchner, G. D. Pettit, and S. L. Wright, *J. Vac. Sci. Technol.,* B6:1263 (1988)

250. R. E. Viturro, S. Chang, J. L. Shaw, C. Mailhiot, L. J. Brillson, R. Zanoni, Y. Hwa, G. Margaritondo, P. D. Kirchner, and J. M. Woodall, *J. Vac. Sci. Technol.,* B7:1007 (1989)

250a. D. Mao, M. Santos, M. Shayegan, A. Kahn, G. LeLay, Y. Hwu, G. Margaritondo, L. T. Florez, and J. P. Harbison, *Phys. Rev.,* B45:1273 (1992)

250b. C. J. Spindt, M. Yamada, P. L. Meissner, K. E. Miyano, A. Herrera, W. E. Spicer, and A. J. Arko, *J. Vac. Sci. Technol.,* B9:2090 (1991)

250c. W. Chen, M. Dumas, D. Mao, and A. Kahn, *J. Vac. Sci. Technol.,* B10:1886 (1992)

250d. R. Duszak, C. J. Palmstrøm, C. J. Sandroff, Y.-N. Yang, and J. H. Weaver, *J. Vac, Sci. Technol.,* B10:1891 (1992)

250e. I. M. Vitomirov, A. D. Ralsanen, A. C. Finnefrock, R. E. Viturro, L. J. Brillson, P. D. Kirchner, G. D. Pettit, and J. M. Woodall, *J. Vac. Sci. Technol.,* A10:1898 (1992)

251. R. E. Viturro, J. L. Shaw, C. Mailhiot, L. J. Brillson, N. Tache, J. McKinley, G. Margaritondo, J. M. Woodall, P. D. Kirchner, G. D. Pettit, and S. L. Wright, *Appl. Phys. Lett.,* 52:5052 (1988)

252. W. E. Spicer, I. Lindau, P. Skeath, and C. Y. Su, *J. Vac. Sci. Technol.,* 17:1019 (1980)

253. L. J. Brillson, *Comments Cond. Mat. Phys.,* 14:311 (1989)

254. S. Chang, R. E. Viturro, and L. J. Brillson, *J. Vac. Sci. Technol.,* A8:3803 (1990); S. Chang. A. Raisanen, L. J. Brillson, P. D. Kirchner, G. D. Pettit, and J. M. Woodall, *J. Vac. Sci. Technol.,* B10:1932 (1992)

255. H. B. Michaelson, *J. Appl. Phys.,* 11:4729 (1977)

256. D. E. Eastman, *Techniques of Metals Research,* VI(1):441., Interscience, New York (1972)

257. S. Chang, L. J. Brillson, D. S. Rioux, S. Kirchner, D. Pettit, and J. M. Woodall, *J. Vac. Sci. Technol.* B8:1008 (1990).

258. S. Chang, L. J. Brillson, D. S. Rioux, S. Kirchner, D. Pettit, and J. M. Woodall, *Phys. Rev.,* B44:1391 (1991)

259. S. Chang, L. J. Brillson, Y. J. Kime, D. S. Rioux, S. Kirchner, D. Pettit, and J. M. Woodall, *Phys. Rev.,* B, in press

260. G. LeLay, K. Hricovini, and J. E. Bonnet, *Appl. Surf. Sci.,* 41/42:25 (1989)

261. S. A. Chambers and V. A. Loebs, *J. Vac. Sci. Technol.,* B8:724 (1990)

262. C. J. Palmstrøm, T. L. Cheeks, H. L. Gilchrist, J. G. Zhu, C. B. Carter, and R. E. Nahory, *Mat. Res. Soc. Extended Abstract,* EA-21:63 (1990)

263. J. Qiu, Q.-D. Qian, R. L. Gunshor, M. Kobayashi, D. R. Menke, D. Li, and N. Otsuka, *Appl. Phys. Lett.,* 56:1272 (1990)

264. G.-X. Qian, R. M. Martin, and D. J. Chadi, *Phys. Rev.,* B38:7649 (1988)

265. F. Cerrina, G. Margaritondo, J. H. Underwood, M. Hettrick, M. Green, L. J. Brillson, A. Franciosi, H. Hoechst, P. M. DeLuca, Jr., and M. Gould, *J. Nucl. Instr. & Methods*, A266:303 (1988)

266. J. Christen, M. Grundmann, and D. Bimberg, *Appl. Surf. Sci.,* 41/42:329 (1989)

# 6

# Atomic Structure of Metal/GaAs Interfaces: The Role of Defects, Epitaxy, and Morphology

*Zuzanna Liliental-Weber, Eicke R. Weber, and Nathan Newman*

## 1.0 INTRODUCTION

One of the long-standing problems in electronic materials is the fundamental understanding of Schottky barrier formation. This problem has caused an extensive controversy in the literature for many years,[1] fueled by the fundamental interest in this question and the technological requirement of reliable and reproducible metal-semiconductor contacts as a prerequisite for the fabrication of electronic devices.

In the case of GaAs and other III-V compounds, two types of contacts are required: ohmic contacts, which connect the outside world to the active portion of the device; and rectifying contacts, which utilize the properties of the depletion region to achieve the device function, e.g., as gate contacts in field effect transistors, or for charge collection in detectors and solar cells. For some device fabrication sequences, the contacts must survive 800°C or higher temperatures during the anneals needed to activate implanted dopants.

Despite the widespread use of rectifying contacts to GaAs, two important issues remain to be resolved: the basic mechanism responsible for the observed Schottky barrier heights, and the reproducibility and stability of electrical performance during annealing and aging. An unstable interface invariably results in unpredictable contact resistance for ohmic contacts and barrier height changes and increased leakage currents for Schottky contacts. Although all these effects are known to be detrimental to device performance, the changes in barrier height that cause threshold voltage shifts in MESFET devices have been particularly troublesome.

Most experimental data show that the barrier heights for metals deposited by evaporation or sputtering on GaAs fall within a few tenths of an eV in the midgap region, indicating a strong Fermi level pinning mechanism at the metal/GaAs interfaces. The measurements of barrier heights for many metals deposited in situ on ultrahigh-vacuum-cleaved GaAs (110), as determined, e.g., by Newman,[2]-[4] using I-V and C-V characteristics, seem to be very consistent. They show the same ideality factor n = 1.05 independent of the reactivity of the particular metal. The lowest barrier height found on n-GaAs was for Cr ($\Phi_b$ = 0.67 eV), and the highest was for Au ($\Phi_b$ = 0.92 eV). Generally, the sum of the barrier heights for the same metal deposited on n- and p-type substrates comes close to the GaAs bandgap, indicating that both n-type and p-type diodes have a common pinning position within the GaAs bandgap. Similar results have also been reported by McLean[5] and Waldrop.[6] To explain these results, several models have been proposed.[7]-[16] These models can be divided into two categories: those that include only the inherent properties of ideal metal/GaAs interfaces (frequently referred to as metal-induced gap states models[8]), and models that include the presence of lattice defects near the heterointerface. Currently popular defect models[1] propose either that native deep-level defects whose energy levels dominate Fermi level pinning are formed upon metal deposition,[11]-[16] e.g., due to the energy released during metal solidification, or that the effective work function differences between GaAs and microscopic near-interfacial anion-rich metallic inclusions determine the Schottky barrier height (the *Effective Work Function* model[9][10]).

In reality, most metal/semiconductor interfaces formed at room temperature are found to be non-ideal. In general, the near-surface region of the semiconductor is disrupted by the deposition of the metal. Annealing of metal films on compound semiconductor substrates at elevated temperatures can result in the dissolution of the compound semiconductor and the subsequent precipitation or regrowth of heavily doped or alloyed epitaxial layers. In each case (metal deposition or annealing), the metal/semiconductor interface is most affected by formation of defects such as protrusions, planar defects (stacking faults), line defects (dislocations), and point defects.

This chapter we will discuss the current knowledge of the structure of metal/GaAs interfaces and the influence of this parameter on technologically relevant issues such as the chemical stability and electrical behavior of the contacts. The metal/III-V contacts are considered to be model systems because the intrinsic surface states in these systems do not fall within the bandgap and are therefore not responsible for the pinning the Fermi-level in Schottky contacts. While this review covers the work on GaAs, other metal/III-V systems are found to have similar trends in properties to that on GaAs, although some quantitative differences are found. For example, Schottky barriers on InP are found to fall in a similarly narrow range of Fermi-level pinning position, as is found for GaAs, but with a position significantly higher in the bandgap. Similar reactions are found between a chosen metal and the cation and anion, although InP is in general more reactive than GaAs due its smaller heat of formation. In addition, the role of non-stoichiometric defects is believed to play an important role for all systems studied to date. Because of the similar nature of many of the metal/III-V systems, this review should be useful to investigators studying or using GaAs, as well as other semiconductor contacts.

We describe possible defects formed at the metal/GaAs interface, giving several examples of non-reactive and reactive metals used for Schottky or ohmic contacts. The influence of particular defects on the Schottky barrier height and electronic properties of those contacts is also discussed in this chapter. The chapter is organized in the following way: first, detailed characteristics of the structure of several diodes are described, followed by stoichiometry and stability studies. Finally, we show that the presence of antisite defects can be related to stoichiometry changes near the interface. These near-interfacial antisite defects are found to play an important role in Fermi-level pinning of the Schottky barriers.[14][17]

This review is mostly based on our own work in cooperation with the Stanford group. Structural studies were done using transmission electron microscopy, including high-resolution and analytical electron microscopy for studying interface abruptness, orientation relationship, new phases, and stoichiometry changes in the interfacial regions of the contacts. The Schottky barrier heights were measured using I-V and C-V characteristics. The aging of the diodes was obtained by applying current and voltage stress. The same diodes for which electrical parameters were measured were used for structural studies; therefore, it was possible to directly correlate the structural differences with the electrical properties.

## 2.0 INTERFACE FORMATION AND RESULTING MICROSTRUCTURE

Metal contacts on semiconductor surfaces generally show a flat interface after deposition. However, one technological requirement for a good metal contact is the preservation of interface flatness upon subsequent annealing during device processing, as otherwise the degradation of interface flatness can result in inhomogeneous electrical properties, leakage currents, and generally unpredictable electrical performance. As device dimensions decrease, deviations from flatness are particularly troublesome because protrusions can extend through the active portion of the device. Interface stability can be achieved either by uniform reaction at the interface or complete absence of reaction.

In this paragraph we discuss several examples of metal contacts to GaAs in order to understand the fundamental mechanisms responsible for the formation of the Schottky barrier and the factors that may result in interface degradation. An effort was made to minimize contamination at the interface in order to eliminate any unnecessary variables that could influence the contact properties.[2] Almost all of the Schottky contacts in this study were prepared on GaAs (110) cleavage surfaces.[2][18][19] Cleavage in ultrahigh vacuum (UHV) with in-situ metal deposition virtually guarantees the complete absence of impurities. (Henceforth, structures fabricated in this way are referred to as *atomically clean* interfaces.) Comparison of such contacts with those deposited after cleavage in air allow for the direct investigation of the effects of interface contamination. (We refer to these structures as *contaminated* interfaces.) It is shown that the orientation relationship between the metal grains and the semiconductor found after annealing is influenced by impurities present at the semiconductor surface prior to metal deposition, as well as by growth conditions such as deposition rate and substrate temperature. A specific orientation relationship between the metal and semiconductor is found for the atomically clean interfaces; this orientation relationship can be vastly different from metals deposited on contaminated semiconductor surfaces. In some cases, we also studied the same metal (Au,[18][19] Au-Ni-Ge,[20] or Al-Ni-Ge[21]) deposited on chemically (industrially) prepared GaAs surfaces in order to observe similarities and differences in interface reactions.

Several classes of metals are described. We summarize this section starting with the least reactive overlayers and ending with the most reactive overlayers.

The first class of metal described is represented by Ag.[3][4][22]-[24] This metal does not form compounds with GaAs even at elevated temperatures. Only for in-situ Ag depositions on UHV-cleaved GaAs does the interface and the Schottky barrier height remain stable and uniform at elevated temperature.[3] When contamination is present at the interface, large changes in the Schottky barrier height are found, and extended protrusions are formed upon annealing.[22] As can be seen, the presence of impurities at the interface influences the electrical and chemical stability of these contacts. TiN[25] and ZrN[26] are also found to be thermally stable contacts. Because these overlayers can react with and/or dissolve native GaAs oxides, contacts formed on air-exposed or chemically prepared surfaces are expected to be less susceptible to the effects of contamination. Contacts fabricated on chemically prepared surfaces are found to have flat interfaces even up to anneals at 500°C for TiN and 700°C for ZrN, with a stable Schottky barrier height to these temperatures. At 700°C and above, specific "pockets" are formed at the interface, suggesting out-diffusion of some substrate elements leading to large leakage currents in the Schottky contacts.

For slightly more reactive interfaces, such as Au on GaAs,[3][18][19][22] similar trends in the annealing-induced changes in the electrical properties are found for both atomically clean interfaces and contaminated interfaces. In both cases, if the effects of peripheral leakage currents are removed,[3] the contacts are found to have near-ideal, strongly rectifying behavior after annealing, with a barrier height reduction of ~0.15 eV.[3][22] However, morphologically large differences are found. After anneals in $N_2$ at 405°C, the atomically-clean interface is found to be flat. High-resolution electron microscopy did not detect a new crystallographic phase at the interface. However, energy dispersive x-ray (EDX) spectroscopy[18] and surface analysis by laser ionization (SALI)[30][31] detected a few percent (not more than 5%) of Ga in the Au layer and accumulation of As close to the interface. This indicates uniform out-diffusion of Ga from the GaAs interface, leaving As in the semiconductor beneath the metal. For the contaminated interface, the formation of protrusions resulted due to the selective reaction at voids and pinholes in the contamination layer.

While Au is known to react preferentially with Ga, Al is known to react with As. Al/GaAs systems, upon prolonged annealing times or extremely high annealing temperatures, form ternary (AlGa)As compounds. However, short annealing times (10 min) at lower temperatures (e.g., 400°C) lead only to reactions within a few monolayers of the interface; reactions that are not always possible to detect by transmission electron microscopy. Nevertheless, under these annealing conditions a substantial increase in the barrier height is still found.[3] Under most circumstances, the Al/GaAs interfaces are found to be flat and are not greatly influenced by the presence of impurities.[19][24][32] Al is expected to reduce any native GaAs oxide and form oxides of Al (e.g., $Al_2O_3$).

Al, Ag, and Au have a cubic structure with similar lattice parameters (0.405 nm, 0.409 nm, and 0.408 nm, respectively). All three metals are found to form the same specific orientation relationship between the metal and GaAs in annealed contacts fabricated with atomically clean interfaces. The orientation relationships are found to be greatly different when the metals are deposited on air-exposed substrates.[19] In these cases the orientation relationship follows the expected epitaxy with $\gamma$-$Ga_2O_3$, which immediately forms upon air exposure.[33]

For ohmic contacts, this protrusion formation can be prevented even when impurities are present if Al is used with Ge and Ni, instead of the conventional use of Au. It will be shown in the case of ohmic contacts (Al-Ni-Ge) that even when the flatness of the interface is reached, the lack of appropriate dopant (Ge) at the interface does not lead to a low resistivity ohmic contact.[22] Only specific layer deposition sequences that allow diffusion of selected dopant elements to the interface can fulfill the ohmic device requirements: interface flatness and doping of the semiconductor beneath the metal.

Cr is a metal that is known to react with both Ga and As, as well as with the native GaAs oxides. Differences in the interface properties due to the presence of impurities are observed for Cr.[22][34][35] For atomically clean interfaces, the Schottky barrier height is found to be stable upon annealing.[22] For contaminated interfaces, the Schottky barrier height is found to increase substantially upon annealing.[22] For the atomically clean and air-exposed contact, the interface remains flat up to anneals at 370°C, with no new large-scale reaction products formed. Annealing at higher temperatures leads to substantial reactions at the interface, and the resulting diodes contain large amounts of leakage currents.

Very reactive metals, such as Ti[36]-[38] and Pd,[39]-[41] comprise the next class of metal/GaAs interfaces. These metals are known to react at room temperature forming ternary compounds. Flat interfaces are formed with Ti, and undulating interfaces are formed with Pd case. New reaction products are formed upon annealing at elevated temperatures. Annealing also leads to substantial changes in barrier heights for both the air-exposed and clean interfaces.

## 2.1 Ag/GaAs Morphology [4][23][24]

Ag is an example of a metal that is nonreactive with GaAs. Both thermodynamic bulk data and data from surface science studies for submonolayer to several-monolayer coverages of metal show no evidence for Ag-GaAs reaction products. Upon deposition, Ag forms grains with twins, but the interface is flat for both the samples deposited in-situ in UHV on cleaved (110) GaAs surfaces and the samples deposited on air-exposed cleaved (110) GaAs. The grain size was observed to be larger and less defective for atomically clean interfaces. This is shown in Figs. 1 a, b. The oxide layer was easily detected for the contaminated interfaces, while in the atomically clean interfaces, the Ag was in intimate contact with the substrate (Figs. 2 a, b).

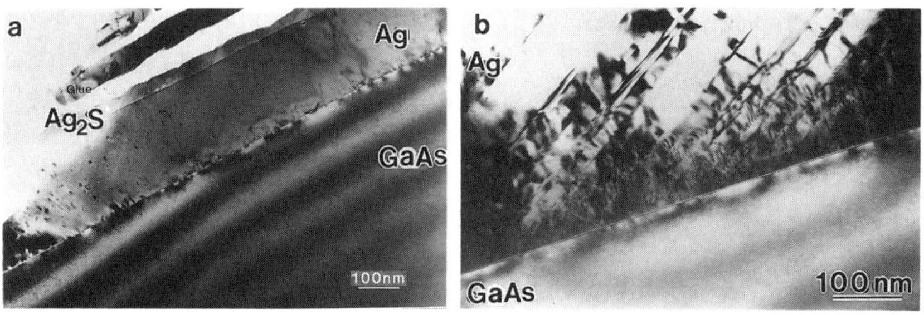


**Figure 1.** Cross-section micrographs of Ag/GaAs interfaces. *(a)* Ag deposited in-situ on UHV-cleaved (110)GaAs. Note the very large Ag grain size; *(b)* Ag deposited on air-exposed cleaved (110)GaAs, showing a high density of twins.

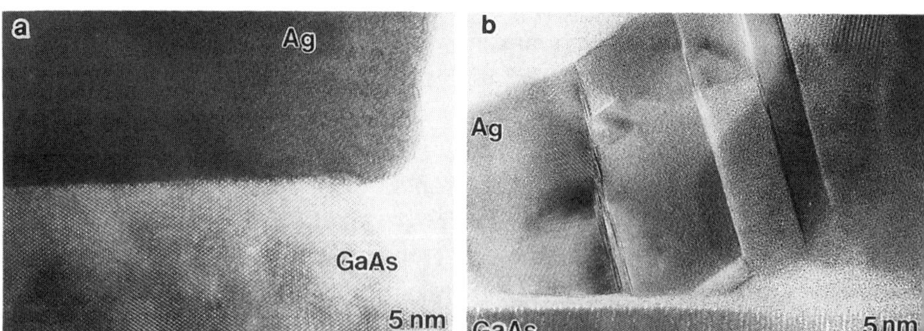


**Figure 2.** High resolution images of Ag samples deposited on *(a)* UHV-cleaved GaAs surface and on *(b)* air exposed surface. Note thick layer of oxide present at the contaminated interface.

A clear difference in the interface morphology for Ag samples deposited on air-exposed and UHV-prepared substrates was observed after annealing. The interface remained flat for the atomically-clean interface, and high-resolution electron microscopy showed that {111}Ag planes were almost parallel to the {200}GaAs planes (Fig. 3a). Dislocations were formed at the interface. For the annealed contaminated interfaces, faceted Ag protrusions were found that extended into the GaAs (Fig. 3b). The protrusions were elongated along [011]GaAs. Note that this is dramatically different from the annealed samples prepared in UHV, which were described above.

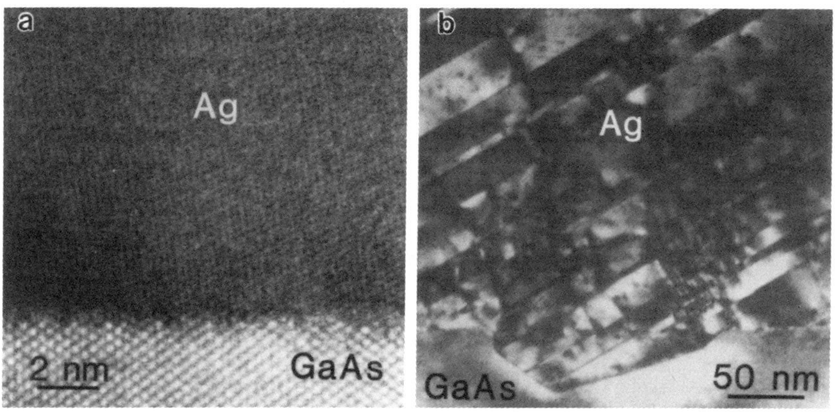


**Figure 3.** *(a)* High resolution image of the Ag deposited under the same conditions in Fig. 2a after annealing in $N_2$ at 405°C for 10 min; *(b)* protrusion formation at the Ag/GaAs air-exposed interface after the same annealing conditions as *(a)*.

Formation of protrusions in annealed air-exposed samples indicates that non-reactivity alone is an insufficient criterion for an ideal flat-metal/GaAs interface. Most probably, the impurities at the interfaces influence the protrusion formation, due either to a specific reaction with the metal or to non-uniform out-diffusion of the substrate elements.

## 2.2 Au/GaAs Morphology[18][19][22][27][29]

Au/GaAs is an example of a reactive interface that is flat after annealing when Au deposition has been done on atomically clean interfaces but that forms protrusions after annealing when the deposition was done on contaminated interfaces. The as-deposited Au layer was found to be polycrystalline, with grain diameters in the 10 - 50 nm range. The largest grain size was found in UHV-deposited Au samples on cleaved in situ (110) GaAs surfaces. Such unannealed Au layers observed in cross section show atomically flat interfaces with GaAs. Some of these grains, particularly in UHV-cleaved samples, were epitaxial, with their (211) or (011) orientation parallel to the (011)GaAs substrate orientation; but generally the grains were randomly oriented, resulting in diffraction patterns with textured rings.

Significant differences between these air-exposed samples and UHV-deposited samples occur after annealing in $N_2$ at 405°C for 10 min (Figs. 4, 5). For the UHV-cleaved samples, the interface remains flat and abrupt despite the annealing process (Fig. 4a). The entire Au layer is almost monocrystalline, with the smallest grain size ~500 nm. Both a specific semiconductor surface reconstruction beneath the metal and the formation of misfit dislocations were found at this interface (Fig. 5), similar to those observed for the annealed Ag layers.

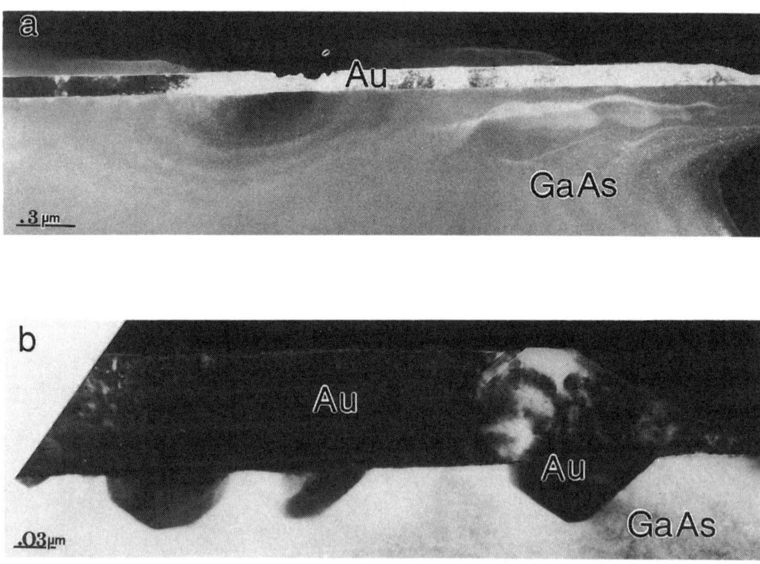


**Figure 4.** The Au/GaAs interface abruptness after annealing for 10 min at 405°C; *(a)* Au deposited in situ on a UHV cleaved GaAs surface; *(b)* Au deposited on the GaAs surface cleaved in air. Note protrusion formation at the contaminated interface.

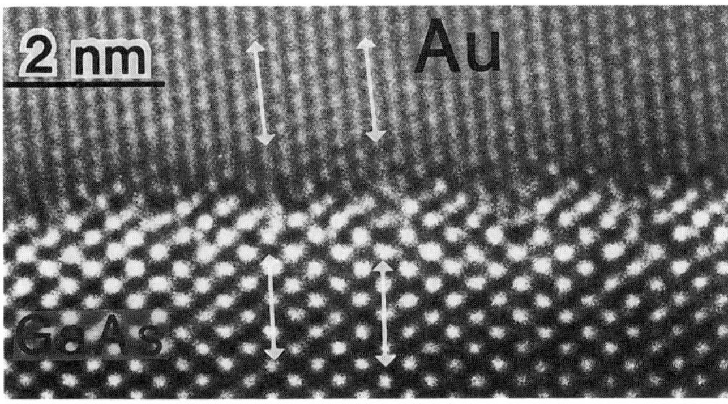


**Figure 5.** High resolution micrograph of the Au/GaAs (011) interface annealed at 405°C for 10 min in $N_2$ atmosphere. Note the lattice distortion in the interfacial area and the twisting of all Au planes toward the GaAs planes. A 10° angle was measured between the $(11\bar{1})$ Au (shown on the micrograph with weak contrast) and $(11\bar{1})$ GaAs planes. Note that six $(\bar{1}1\bar{1})$ Au planes coincide with five $(200)_{GaAs}$ planes.

The same annealing treatment for Au samples deposited on GaAs cleaved in air resulted in the formation of voids and metallic protrusions at the interface (Fig. 4b). Many small grains, highly twinned and dislocated and with irregular shapes, were observed in a plan view of these annealed contaminated interfaces. The oxide layer was still detected at the interface even after annealing. The Au layer above the oxide layer has many defects, and its grain size is much smaller than found for annealed atomically clean Au interfaces. The orientation relationship for grains at the contaminated interface was (011)Au || (011)GaAs. Au diffusion through pinholes in the oxide layer lead to protrusion formation at the interface. These protrusions are single grains of Au. For the contaminated interfaces, the volume of the remaining overlayer plus that of the protrusions is found to be equal to the volume of the Au overlayer in the as-deposited samples.

In cross section, two different shapes of protrusions extending into the GaAs were found (Fig. 4b): *(i)* triangular protrusions, whose sides are delineated by GaAs {111} planes, and *(ii)* multifaceted protrusions delineated by GaAs {111}, {110}, and {100} planes. Similar protrusions were observed in annealed Au/GaAs samples formed on chemically prepared GaAs(100) surfaces (Fig. 6a). The Au layer was separated from the GaAs substrate by a thin oxide band (Fig. 6b). The presence of oxygen at the interface was confirmed by energy-dispersive x-ray spectroscopy.

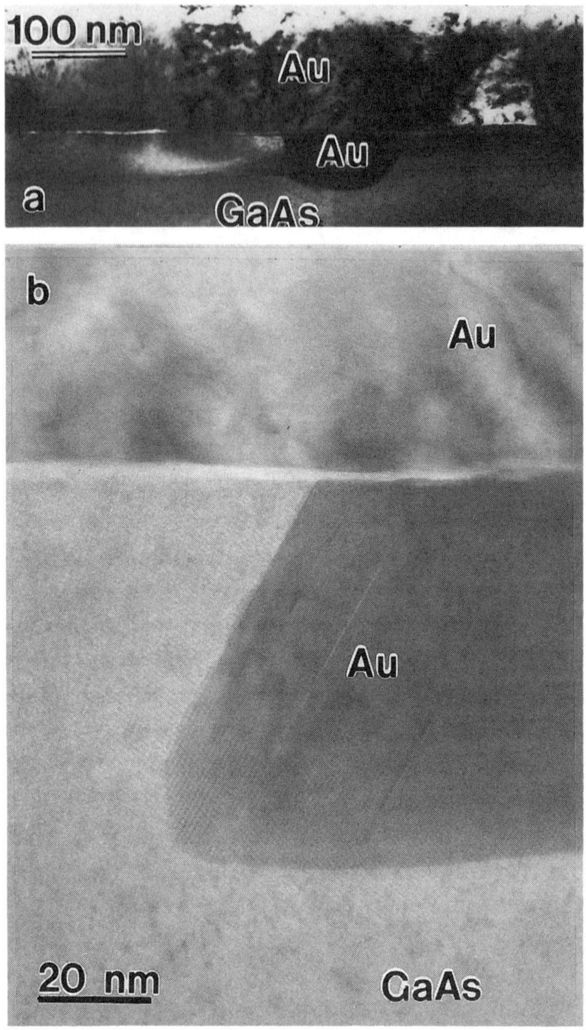


**Figure 6.** *(a)* Au deposited on a chemically cleaned (100)GaAs surface after annealing; note the Au island beneath the oxide layer *(white band marked by arrow)*. *(b)* High resolution image of the same interface showing layer of oxide and the lattice image of Au protrusion formed below the oxide layer.

The formation of protrusions at Au/GaAs and Au-based ohmic contacts on GaAs interfaces has been observed by several independent researchers, and it was concluded that elevated temperatures are a sufficient condition for their formation. Our own study[18] shows that the morphology of the interface is strongly influenced by the surface preparation prior to Au deposition. Impurities at the interface can promote different reactions at the interface, causing non-uniform out-diffusion of substrate elements and subsequent protrusion formation at the interface. This result demonstrates that the formation of protrusions is not the result of annealing at elevated temperatures alone but is clearly affected by the semiconductor surface preparation prior to metal deposition.

Generally, if a reactive elemental metal on GaAs undergoes prolonged annealing, different phases can be formed at the interface. For Au, according to the phase diagram, Au + GaAs reactions should not take place if annealing is performed in a closed system.[42] New phases were not found after annealing atomically clean interfaces for 10 min in $N_2$ at 400°C. Ga was detected in the Au grains by x-ray dispersive spectroscopy and SALI, with a Ga concentration smaller than 5%. This concentration is not sufficient to form of a new crystallographic phase, although it does indicate the formation of a weak alloy with Ga and the subsequent release of As near the interface. In an open system, a Au-Ga phase should be formed, leaving excess As behind. The formation of the $AuGa_2$ phase was indeed observed when annealing was performed in vacuum for 10 min in $N_2$ at 400°C.[19] The resulting system is found to follow the expected epitaxy of the near perfect lattice match between GaAs and $AuGa_2$ and between $AuGa_2$ and Au.

In order to avoid any semiconductor surface disruption at the interface during metal solidification, a novel technique has been developed that makes it possible to bring preformed metal clusters into contact with clean semiconductor surfaces.[43]-[45] Protective layers of ~3 nm Xe were first condensed on GaAs (110) cleaved in UHV at 60 K. A metal was then deposited by evaporation onto the solid xenon layer. As a result of this deposition, metal clusters formed in and on the xenon layer. The clusters were buffered from the semiconductor surface. The Xe buffer layers were sublimed upon warming to room temperature so that the metal clusters were brought into contact with the undisrupted GaAs surface (Fig. 7). In this way, atom deposition, cluster nucleation, and growth occurred in the overlayer before any direct substrate contact, eliminating complications from the release of energy by metal solidification. Au clusters deposited in this way are found to have an abrupt unreacted interface. High-resolution transmission electron microscopy and analytical electron microscopy show direct evidence of a defect-free interface produced by in situ cluster deposition. In contrast to interfaces produced by atom-by-atom deposition, no specific interface reconstruction or orientation relationship was observed near the interface.

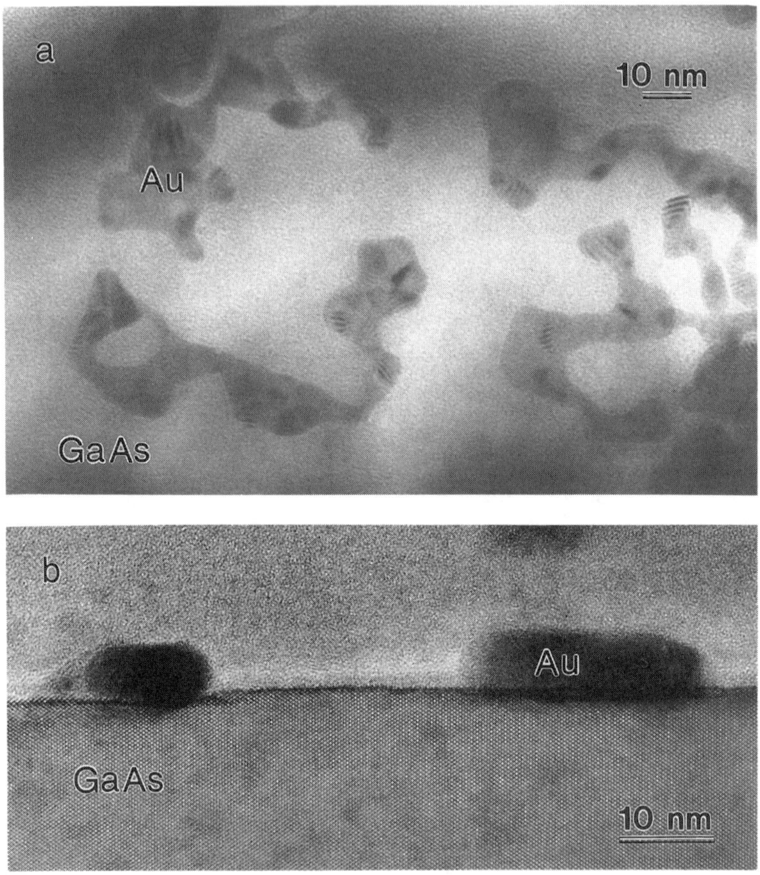


**Figure 7.** *(a)* Plan-view micrograph of Au clusters deposited on the UHV-cleaved (110)GaAs surface; *(b)* cross-section of the interface between Au clusters and the GaAs substrate.

## 2.3 Al/GaAs Morphology[24][32]

For Al grown on UHV cleaved (110) GaAs surfaces, a typical grain size of 100 - 300 nm was observed (Fig. 8a). The interface with GaAs remains flat and Al(111) planes form a 10° angle with (111)GaAs planes (Fig. 8b). This angle remains constant even for grains with different orientations. When Al react with GaAs, it is known to preferentially form Al-As bonds and release free Ga. Upon annealing at 375°C in $N_2$ for 10 min, the interface remained flat and the grain size did not increase. In some areas a very thin layer of AlGaAs was formed. The formation of AlGaAs did not occur uniformly. There were large areas where this phase was not detected.

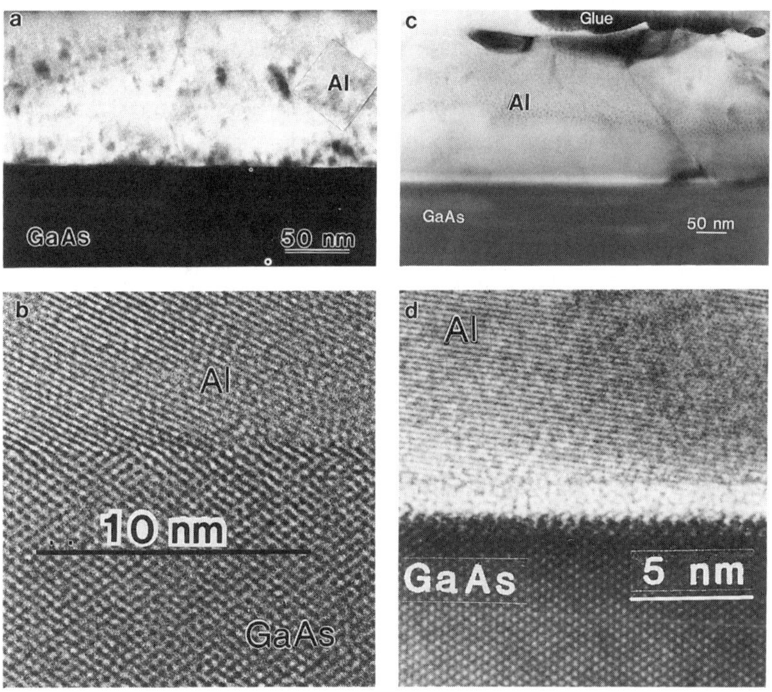


**Figure 8.** TEM micrograph of cross sections of Al/GaAs interfaces *(a)* from the sample prepared on UHV-cleaved substrate; *(b)* high resolution image of the same sample annealed at 405°C for 10 min in $N_2$ ; *(c)* from the Al sample prepared on air-exposed GaAs; high resolution image of the air-exposed sample annealed under the same conditions as in *(b)*. Note amorphous layer at the interface.

For the samples cleaved in air, the interface remained flat before and after annealing, but a significant smaller Al grain size was observed in these samples, compared to samples deposited in UHV (Fig. 8c). Only in some areas of the annealed air-exposed samples was an amorphous oxide layer detected at the interface (Fig. 8d). Above these oxide layers, the orientation of the Al was different from that found in the areas where the oxide was not detected. For Al metallization, in contrast to the other metals (Au and Ag), protrusions at the interface were not observed, even for annealed air-exposed samples. This probably can be explained by the possibility of thin oxide dispersion by Al or formation of an AlGaAs phase in intimate contact with GaAs and no As out-diffusion from the system. Void formation was not observed in as-deposited samples or annealed samples.

## 2.4. Ag, Au, and Al/GaAs Epitaxial Relationships[19]

As mentioned before, the metal/GaAs orientation relationship can be influenced by the impurities present at the substrate surface. The influence of these impurities is most visible in annealed samples. Upon deposition, the Au layer was found to be polycrystalline, independent of the surface preparation of GaAs. Large differences between these samples were observed after annealing. For atomically clean Au interfaces which annealed for 10 min at 405°C in a $N_2$ atmosphere, most of the grains were elongated along the [011]GaAs. Their orientation relationship towards the substrate was (522)Au parallel to (011)GaAs. A 10° angle was measured between the (111)Au and (111)GaAs planes. Near perfect coincidence was observed at each fifth {200}GaAs plane and each sixth {111}Au plane (Fig. 5). The existence of this orientation relationship for Au deposited in situ on UHV-cleaved GaAs was confirmed by scanning tunneling microscopy.[46] A similar orientation relationship and correlation between the {111} planes of the metal and the substrate was observed for Ag and Al.[19]

Two completely different types of orientation relationships for large grains were observed for the annealed (110) air-exposed samples and for the (100) chemically prepared samples, after subsequent annealing in the same conditions as the UHV-prepared samples: (011)Au || (011)GaAs (type I) and ($\bar{4}$11) Au || (011)GaAs (type II). The type II orientation relationship is related to the type I relationship by twinning along the (111) planes. This twinning provides a better lattice match to the substrate and releases the existing stress.

The type I orientation relationship was observed for all diodes deposited on air-exposed surfaces [Au and Ag on GaAs(110) and Au on GaAs

(100)] after annealing. This type of orientation relation was explained for Au by Yoshiie and Bauer[47] as the epitaxial relationship to the newly formed Au-Ga phase, e.g., $(0\bar{1}1)$GaAs || $(1\bar{1}0)$AuGa || $(01\bar{1})$Au with $[01\bar{1}]$GaAs || [001]AuGa || [011]Au. However, the formation of a Au-Ga phase is not necessary to achieve this orientation relationship. This orientation relationship exists in annealed Au even when a Au-Ga phase is not formed and exists for other metals, such as Ag, as well. The mechanism for this epitaxy is obviously more general than for just the case when $AuGa_2$ is formed at the interface.

A possible explanation for the type I orientation relationship is that $\gamma$-$Ga_2O_3$ grows epitaxially on GaAs in a type I orientation relationship: (011) $\gamma$-$Ga_2O_3$ || (011)GaAs with [100] $\gamma$-$Ga_2O_3$ || [100]GaAs.[19][33] This oxide provides an excellent lattice match to Au: $d_{400}(\gamma\text{-}Ga_2O_3)$ = 0.205 nm, as compared to $d_{200}$ (Au) = 0.203 nm (and very similar spacings for Ag), and $d_{044}$ ($\gamma$-$Ga_2O_3$) = 0.145 nm, compared to $d_{022}$ (Au) = 0.149 nm. This observation suggests that GaAs surfaces exposed to air form epitaxial $\gamma$-$Ga_2O_3$, and the deposited metal relates epitaxially to the oxide already existing on the surface. The $\gamma$-$Ga_2O_3$ oxide is not a continuous layer. In pinholes, twinning gives a better match at the interface, which can explain the observed type II orientation relationship.

The clear difference in the orientation relationship between air-exposed and UHV-prepared samples of the same metal (e.g., Au) provides an additional tool to distinguish between metal layers deposited on clean and on contaminated semiconductor surfaces. Surface contamination before metal deposition influences not only the orientation relationship but also the stability of metal contacts, as discussed in detail in Sec. 3 of this contribution.

For Al deposited at low substrate temperatures (<0°C) in situ in an MBE chamber on GaAs, the Al layer is epitaxially regrown on the (001)GaAs surface, with Al(001) || GaAs(001), Al[100] || GaAs[110], and [010]Al || $[1\bar{1}0]$GaAs. This is the most stable and expected Al arrangement on (001)GaAs, showing fourfold symmetry because of a good lattice match between the $GaAs_{220}$ (0.1991 nm) and $Al_{200}$ (0.2034 nm) on two perpendicular axes.[32][48][49]

When the substrate temperature was raised to 25°C during metal deposition, some of the grains showed the (110) orientation, with [002]Al || $[2\bar{2}0]$GaAs. The lattice image showed monoatomic steps (2 - 4 Å high) along the interface (Fig. 9). No exchange reaction was discovered by dark-field electron microscopy imaging. For the As-stabilized surface, c(2x8), at a substrate temperature of 150°C, triangular faceted islands with (110)Al (three-dimensional growth) orientation were observed along the interface with GaAs, with [002]Al || [$\overline{2}\overline{2}0$]GaAs (Fig. 10). Those islands were imbedded in strips of Al parallel to the interface where the [$\overline{2}\overline{2}0$]Al axis was inclined to the same axis in the triangular islands. A layer of AlGaAs was found at the GaAs interface. This AlGaAs layer was significantly larger in thickness for the samples fabricated with a substrate temperature of 400°C during Al deposition. In most cases, the Al layer adjacent to AlGaAs had a (110) orientation parallel to the (001)GaAs. Formation of triangular islands was characteristic of higher growth temperatures.

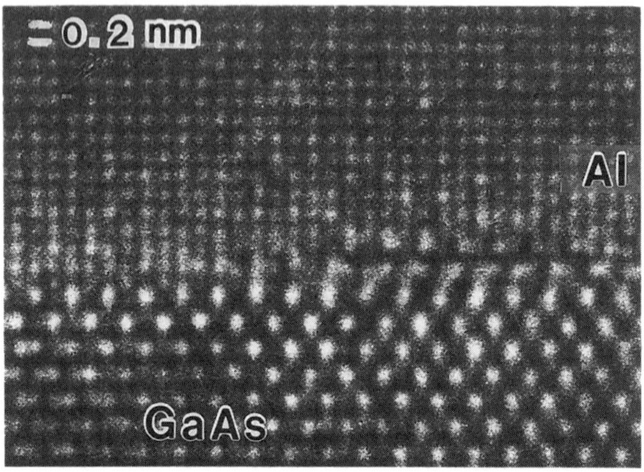


**Figure 9.** High resolution image of the Al/GaAs (100) interface grown in an MBE chamber with 25°C substrate temperature.

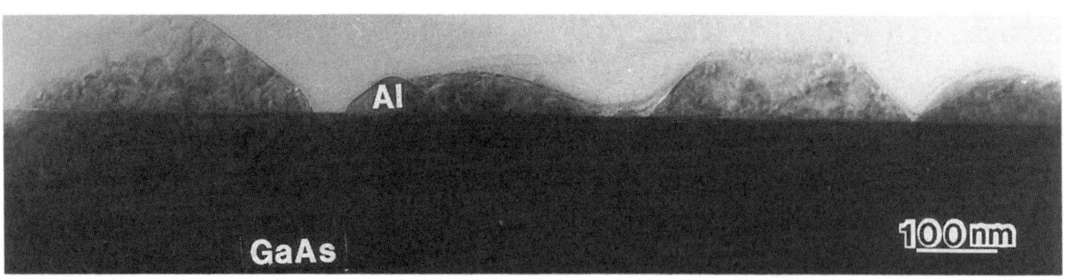


**Figure 10.** Al islands formed on the interface with GaAs for deposition in MBE chamber at the substrate temperature of 150°C. The thin layer of GaAlAs was observed by dark field imaging in these samples.

## 2.5 AuNiGe/GaAs Morphology[20]

Protrusion formation at the interface is detrimental to both ohmic and Schottky contacts. For Au-Ni-Ge contacts, formation of protrusions of different composition was observed at the interface. For ohmic contacts, these protrusions determine the electrical properties of the contact. Braslau[50] proposed that Ge accumulation surrounds such protrusions and forms an $n^+$ layer beneath the metal. It has been predicted that the tunneling current flow is controlled by the field enhancement at these penetrating points. The resulting current was predicted to be dominated by the spreading resistance, which is known to be proportional to the doping concentration. Such protrusions contain Au-Ga grains of varying compositions that can produce strain at the interface (Fig. 11a). This strain leads to formation of various defects near the interface, such as dislocations and stacking faults (Fig. 11b). All dislocations were found on the apexes of the triangular protrusions. It appears that short circuit "pipe" diffusion of the alloying elements takes place along the dislocation lines. The metal diffusion coefficient is about two orders of magnitude higher than for the fastest bulk impurity diffusion coefficients reported for GaAs.[20][51] Stacking faults were found in the corners between the triangular protrusions and the flat interface (Fig. 11c). Formation of such defects can lead to device degradation, especially when these defects propagate through the active area of the devices.

## 2.6 AlNiGe/GaAs Morphology[21]

When Al was substituted for Au in ohmic contacts (e.g., Al-Ni-Ge contacts), annealing of these contacts did not introduce any protrusions at the interface. However, the metal layer sequence during deposition was found to be a critical factor in determining the electrical contact properties. This property was associated with the dispersal of an oxide layer on the semiconductor surface after chemical cleaning. Deposition of Ni as the first layer (Ni/Al-Ge/Ni/Al) instead of Ge (Ge/Ni/Al) did not disperse the impurity layer present at the semiconductor surface even after annealing (Fig. 12a), and layers with such metal layer sequences did not produce ohmic contacts. In this case a Ge layer is formed that is separated from GaAs by an amorphous impurity layer (Fig. 12 b, c). Since low contact resistance was ascribed to Ge dopants in GaAs forming an $n^+$ layer beneath the metal, it is possible that an amorphous impurity layer at the interface did not allow Ge to penetrate. When the Ge layer is deposited directly on the semiconductor surface, then annealing leads to the formation of two phases: $Ge_3Ni$ (Fig.

12c) and $Al_3Ni$. The interface remained flat (Fig. 13b) despite annealing and formation of different phases. In addition Ge was found by SIMS and EDX in the semiconductor beneath the newly formed phases. The contact resistance after annealing at 500°C for 1 min was $1.4 \times 10^{-6}$ A $cm^{-2}$ and it was expected that a Ge doped $n^+$ layer was formed in the semiconductor and was responsible for low contact resistance.

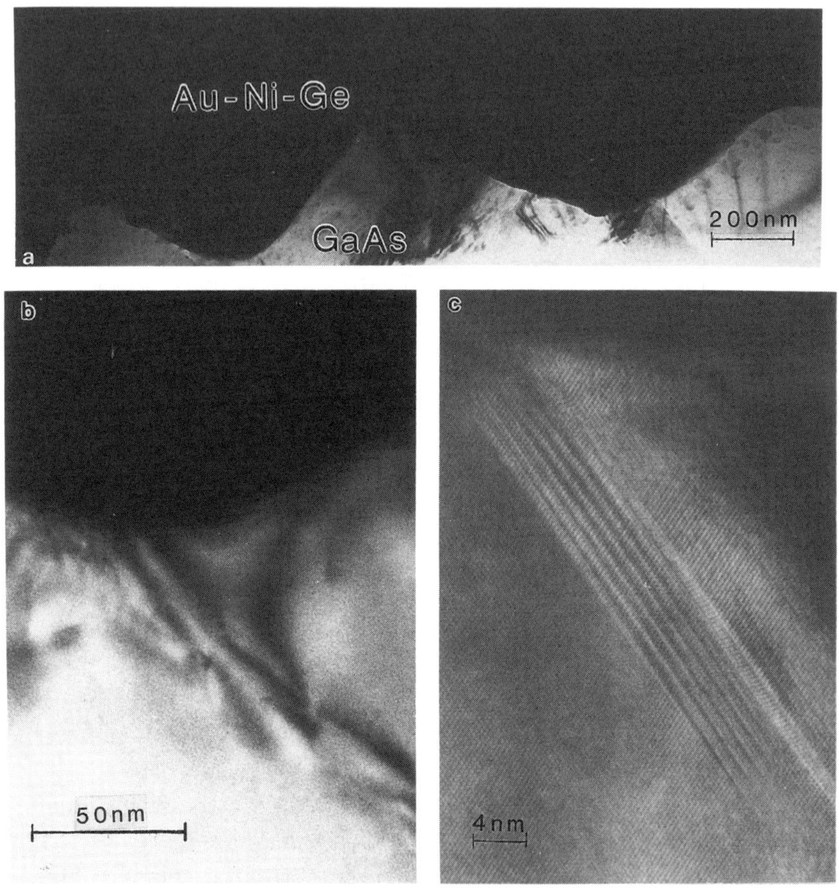


**Figure 11.** Defects formed after annealing at 450°C at the Au-Ni-Ge/GaAs interface; *(a)* protrusion formation; *(b)* a dislocation at the apex of triangular protrusion; *(c)* microtwins at the intersection of the flat interface and the protrusion.

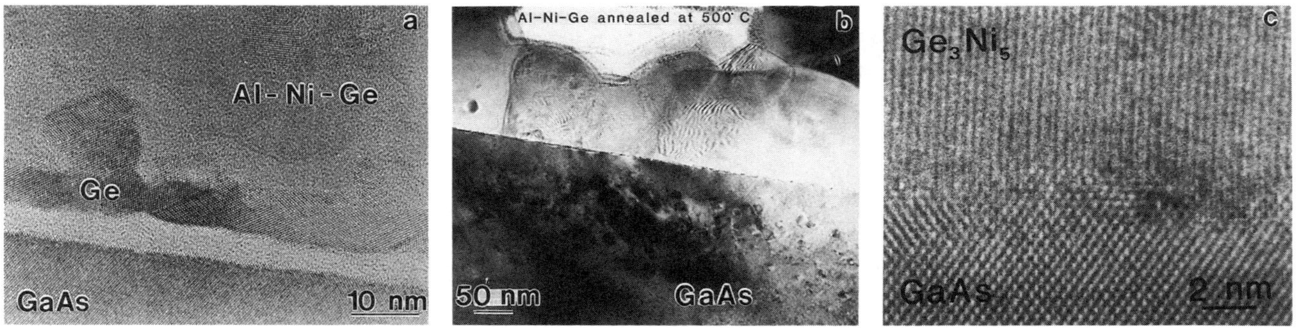


**Figure 12.** TEM micrographs of cross-section of Al-Ni-Ge contacts after annealing at 500°C for 1 min; *(a)* for the contact with Ni layer deposited directly on the GaAs. Note the amorphous oxide layer on the interface with GaAs and the Ge layer formed above it; *(b)* from the contact with Ge layer deposited directly on the GaAs showing flat interface; *(c)* high-resolution micrograph of the same interface, showing formation of $Ge_3Ni_5$ compound.

**Figure 13.** TEM micrographs from Cr/GaAs interface. *(a)* As-deposited UHV sample showing the columnar structure of Cr with voids between columns. Note that columns are almost parallel to each other and inclined ~80° toward the interface with GaAs. *(b)* Air-exposed sample showing a columnar structure of Cr with columns inclined in different directions to the substrate; *(c)* high-resolution micrograph of annealed samples deposited in UHV; *(d)* High-resolution image of annealed air-exposed samples. Note thick layer of oxide at the interface and increased buckling of lattice planes toward the top of the layer.

## 2.7 Cr/GaAs Morphology and Epitaxial Relationships[34][35][52]

Deposition of Cr on either UHV or air-exposed GaAs surfaces leads to the formation of flat interfaces in both cases (Fig. 13 a, b, c, d). Columnar grains of Cr are formed that are separated from each other by voids. However, for air-exposed samples the Cr columns are randomly oriented, and their size was 2 - 4 times smaller than the samples deposited in-situ in UHV. Diffraction patterns show ring patterns typical of polycrystalline Cr material with body-centered cubic (*bcc*) structure and no texture. The Cr layers deposited in UHV on clean-cleaved surfaces were primarily *bcc*, however, in some areas face-centered cubic (*fcc*) Cr with a lattice constant of a = 0.35 nm, was detected. The orientation relationship between the GaAs substrate and the Cr layer was {100}Cr || {100}GaAs with [100]Cr || [022]GaAs. Cr matches almost perfectly to GaAs, because the Cr lattice parameter (a = 0.288 nm) is almost exactly half of that of the GaAs lattice parameter (a = 0.565 nm).

Annealing of these samples for 10 min at 370°C in $N_2$ did not cause the formation of a new phase in the atomically clean and contaminated interfaces. This experiment shows that the structure and interface abruptness remained stable after annealing at the described conditions. High resolution TEM shows a specific contrast, independent of the defocus value, suggesting presence of specific atoms at the interface. Annealing at higher temperatures (630°C for 10 min) leads to the formation of rough interfaces, with grains protruding up to 100 nm into the GaAs (Fig. 14a). Two kinds of grains were found at this interface: $CrAs_2$ and $Ga_4Cr_3$.[53] Only where the interface is flat, as is found for $CrAs_2$ grains (Fig. 14b), is a consistent epitaxial relationship observed [$(32\bar{2})$ $CrAs_2$ || (200) GaAs and 4 $d_{32\bar{2}}$ = $2d_{200}$)]. In the areas where the interface is undulated, no specific epitaxial relationship is consistently found.

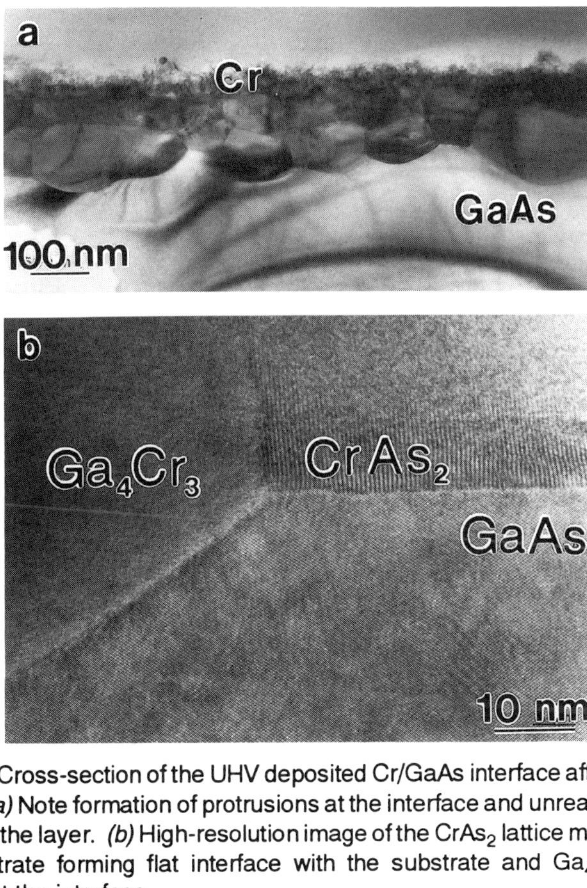


**Figure 14.** Cross-section of the UHV deposited Cr/GaAs interface after annealing at 630°C. *(a)* Note formation of protrusions at the interface and unreacted Cr layer at the top of the layer. *(b)* High-resolution image of the $CrAs_2$ lattice matched to the GaAs substrate forming flat interface with the substrate and $Ga_4Cr_3$ forming protrusion at the interface.

## 2.8 Ti/GaAs Morphology and Epitaxial Relationships[38][45]

An example of a metal that reacts with GaAs at room temperature is Ti. Even when GaAs was protected against disruption by solid Xe using cluster deposition on GaAs cleaved in UHV (see Sec. 2.2 for a description of the deposition method), Ti is found to react with GaAs, forming a ~2.5 nm-thick amorphous layer as soon as the Xe layer disperses during warming to room temperature (Fig. 15a).[45] Amorphous layers of approximately the same thickness were also observed when Ti was directly deposited on UHV-cleaved or air-exposed GaAs.[38] The interface with this amorphous layer is atomically flat, indicating that the room-temperature reaction between Ti and GaAs is very uniform. The composition of this layer is not known, since its thickness is not sufficient for EDX studies. XPS studies show the formation of Ti-As bonds,[45] which may be evidence for the initial stages of TiAs phase formation. Unreacted Ti with $(10\bar{1}0)$ planes ($d_{Ti(10\bar{1}0)} = 0.255$ nm) parallel to the (110) GaAs planes were observed above this amorphous layer. For 100 nm of Ti deposited on UHV-cleaved GaAs and annealed for 20 min at 450°C, the interface is found to be rough. The resulting structure was found to consist of three layers with different compositions (Fig. 15 b, c). The first layer adjacent to the GaAs is the TiAs layer, followed by a $Ga_3Ti_2$ layer, and then an unreacted Ti layer on the top. The TiAs layer consists of large grains (~ 500 nm) that are twinned to each other with $(\bar{1}012)$ twinning planes (Fig. 15d). Formation of particular twins allows for the best lattice match to the substrate. The main orientation relationship can be described as $[1\bar{2}10]$ TiAs || [110]GaAs with $(\bar{1}010)$ TiAs || $(\bar{1}10)$ GaAs.

## 2.9 Pd/GaAs Morphology and Epitaxial Relationships[39]-[41]

Another metal that is highly reactive with GaAs is Pd. Palladium is found to diffuse through the oxide layer, forming a hexagonal ternary phase with the grains connected to each other by low-angle boundaries (Fig. 16). Their major orientation relationship is <0001>||<011>GaAs and $\{\bar{2}110\}$|| {100}GaAs; however, the grains were only parallel to one set of (011) planes. This phase is found to increase in thickness to ~20 nm for samples annealed at 200 - 250°C. Many voids are found in the Pd film since Pd is a diffusing species. When samples are annealed at more than 250°C, another hexagonal ternary PdGaAs phase nucleates at grain boundaries of the first phase. At longer annealing times, this phase penetrates deeply into the substrate, forming many protrusions at the interface. At annealing temperatures higher than 350°C, the first phase disperses, leaving only the second phase in the layer and possibly some Pd-Ga phases on the top. The first phase, formed at room temperature, is slightly more As-rich than the phase formed at higher annealing temperatures. As described by Sands,[39][40] formation of the second phase and its composition is related to the amount of Pd unconsumed after the first phase is formed. The dependence of the formation of the different ternary phases on annealing temperature and the amount of Pd deposited on the GaAs surface makes this metal/substrate interface highly unstable. However, the fact that Pd can disperse the existing native oxide at the interface is considered a very desirable property of an ohmic contact.[53]

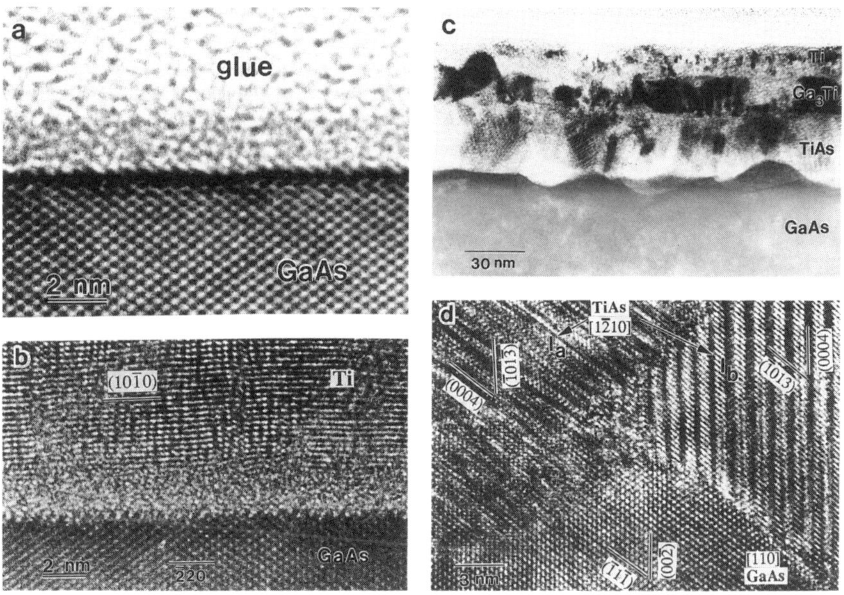


**Figure 15.** Cross-section of Ti/GaAs interface: *(a)* an amorphous layer formed for cluster Ti deposition on cleaved (110) GaAs buffered by Xe layer; *(b)* formation of the same amorphous layer for Ti deposited directly on the UHV-cleaved (110) GaAs substrate; *(c)* the layers of TiAs, $Ga_3Ti_2$, and Ti formed after annealing for 20 min at 450°C. Note undulated interface with TiAs; *(d)* high-resolution image of the interface shown in *(c)*. Note twinning of TiAs at the interface with GaAs.

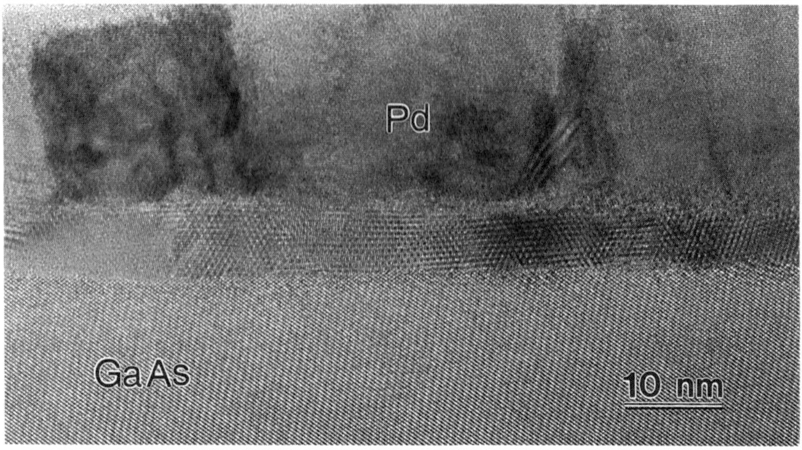


**Figure 16.** Formation of ternary Pd-Ga-As phase after Pd deposition on the UHV-cleaved (110)GaAs substrate.

## 2.10 TiN/GaAs Morphology[25]

The last type of contact we discuss is formed using metals with high temperature melting points. Thermally stable Schottky contacts are important in the self-aligned gate MESFET technology because both the gate material and ion-implanted GaAs substrate are subjected to high-temperature annealing (~800°C). Metal-nitrides are considered to be a suitable material for such contacts. The highest barrier height has been measured for ZrN[26] and TiN[25] contacts. For ZrN contacts, the interface remains flat and no degradation in the electrical properties are found up to anneals at 700°C. At higher temperatures amorphous pockets are found at the interface (Fig. 17). For TiN, amorphous pockets have been observed at temperatures as low as 500°C. The average depth of these pockets into the GaAs substrate was about 4 nm. Annealing at higher temperatures increased slightly the depth of penetration of those pockets and decreased the average spacing between pockets. The mechanism of pocket formation is not clear. It is believed that no chemical reaction should occur at the TiN/GaAs interface, but the pocket formation might be related to the coalescence of residual impurities such as oxides at the interface.

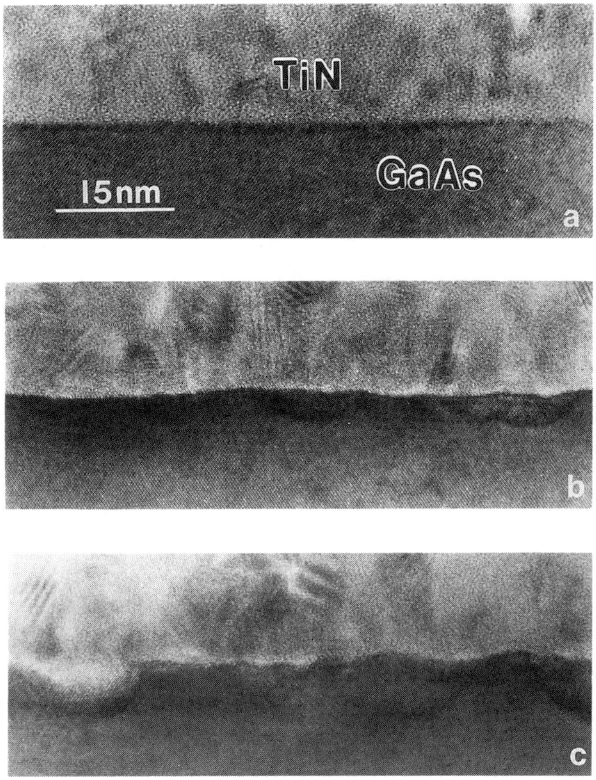


**Figure 17.** High-resolution cross-section TEM images of the TiN/GaAs *(a)* as-deposited sample, and *(b)* annealed at 500°C, and *(c)* sample annealed at 800°C. Note the "pocket" formation at the GaAs interface after annealing at the highest temperature.

It can be expected that, even though TiN thin films have been used widely as diffusion barriers in Si technology,[54][55] volatile arsenic atoms from the GaAs substrate may be able to escape through weak points and pinholes in the thin intervening layer at the interface. These atoms would then diffuse out through the columnar boundary structure of the thin film and the TiN grain boundaries during anneals at high temperatures. This extensive out-diffusion of arsenic atoms would leave excess gallium atoms at the interface near these weak points, which might form GaN within the pocket. However, this crystalline phase was not detected in our study.[25]

An alternative explanation for this pocket formation is that the fast diffusion paths provided by the columnar boundary structure of the thin film (relative to lattice diffusion) may permit a certain degree of out-diffusion of both Ga and As atoms from the substrate, resulting in pocket formation below the interface. These void-like defects would be expected to be detrimental to contact properties. It has recently been found that the number of these pockets at the interface can be related to the substrate preparation before nitride deposition.[56] These findings are consistent with the fact that impurities at the interface can influence interfacial morphology, and the defects present (protrusions, pockets, dislocations and stacking faults) can determine contact properties. Both the defects and the contamination at the interface are expected to greatly influence the kinetics of the resulting interfacial chemical reactions.

## 3.0 CONTACT STABILITY[3][22]-[24][35]

The stability of Schottky contacts deposited on UHV-cleaved and air-exposed GaAs (110) was studied by thermal annealing and electrical aging under reverse bias conditions.[22][23] The results are summarized in Table 1 and Figs.18 and 19.

**Table 1.** Results of I-V and C-V Electrical Measurements

| | Diodes formed on Air-Exposed GaAs(110) | | | Diodes formed on Clean UHV-cleaved GaAs(110) | | |
|---|---|---|---|---|---|---|
| | $\phi_{bo}$ I-V (eV) ±0.02 | n | $\phi_b$ C-V (eV) ±0.05 | $\phi_{bo}$ I-V (eV) ±0.02 | n | $\phi_b$ C-V (eV) ±0.05 |
| Al | | | | | | |
| Unannealed | 0.76 | 1.07 | 0.84 | 0.83 | 1.06 | 0.87 |
| 370°C Anneal | 0.83 | 1.07 | 0.90 | 0.90 | 1.05 | 0.92 |
| Ag | | | | | | |
| Unannealed | 0.95 | 1.07 | 1.06 | 0.89 | 1.06 | 0.94-0.97 |
| 370°C Anneal | 0.79 | 1.06 | 0.85 | 0.91 | 1.07 | 0.92-0.99 |
| Au | | | | | | |
| Unannealed | 0.83 | 1.08 | 0.94 | 0.92 | 1.05 | 1.00 |
| 370°C Anneal | 0.65 | 1.06 | 0.72 | 0.80[a] | 1.06[a] | 0.88[a] |
| Cr | | | | | | |
| Unannealed | 0.68 | 1.06 | 0.79 | 0.66 | 1.06 | 0.74 |
| 370°C Anneal | 0.76 | 1.08 | 0.88 | 0.67 | 1.06 | 0.74 |

[a] The data in this table for Au diodes formed on the clean UHV-cleaved GaAs is from a 430°C anneal.

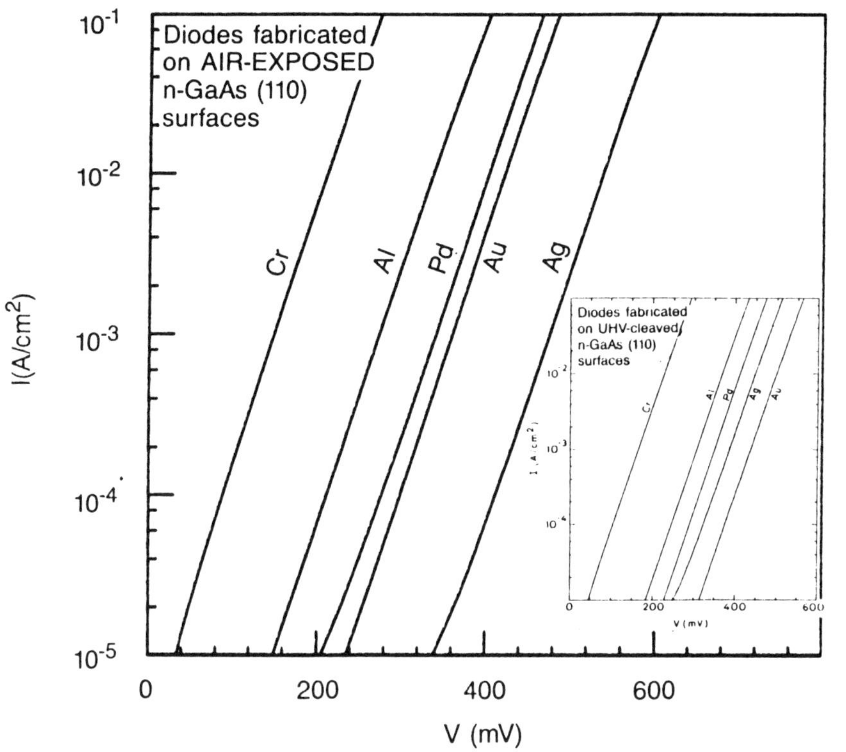


**Figure 18.** A comparison of typical forward-bias I-V data from a number of metal/ n-type GaAs diodes which were fabricated on air-exposed and UHV-cleaved *(insert)* diodes. The exponential dependence with a near unity ideality factor can be seen. For current densities less than 5 x $10^{-15}$ $A/cm^2$, non-linearities due to the effects of leakage currents can be seen in the electrical measurements of the Ag (UHV-cleaved and air-exposed) and Pd (air-exposed) diodes.

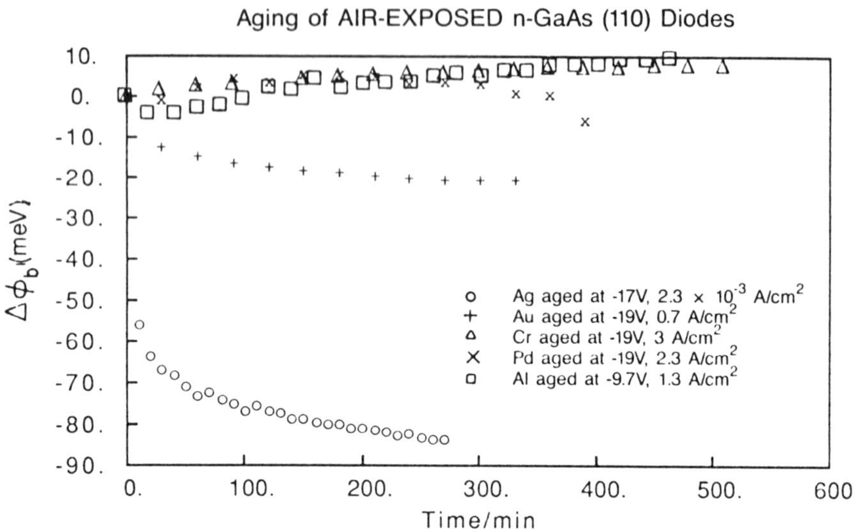


**Figure 19.** This figure illustrates typical results of electrical aging for diodes formed on cleaved air-exposed GaAs (110) surfaces. The change in barrier height is plotted as a function of the length of time which the diodes were exposed to electrical aging. The electrical aging conditions for each diode are given in the figure's key.

The most important observation is the unexpected stability of contacts deposited in situ on cleaved GaAs (110) under UHV conditions. In contrast, contacts deposited on air-exposed cleaved GaAs (110) show clearly inferior stability upon electrical and thermal stressing. This is especially evident for Ag/GaAs contacts, as shown in Fig. 20 and Fig. 21. The barrier height of these unannealed Ag diodes (obtained from I-V measurements) formed on air-exposed substrates is 0.96 eV. A significant decrease of more than 80 meV was found after 100 min of electrical aging using $2.3 \times 10^{-3}$ A/cm$^2$ current density at -17 V reverse bias voltage (Fig. 21). In contrast, the barrier height of Ag diodes formed on UHV-cleaved substrates was 0.89 eV and was stable even upon stressing with 1.4 A/cm$^2$ for 400 min. Similarly, upon annealing for 10 min at 370°C, the barrier heights of air-exposed diodes decreased by 160 meV, whereas the barrier height of the UHV-deposited Ag diode remained constant to the accuracy of the measurement.

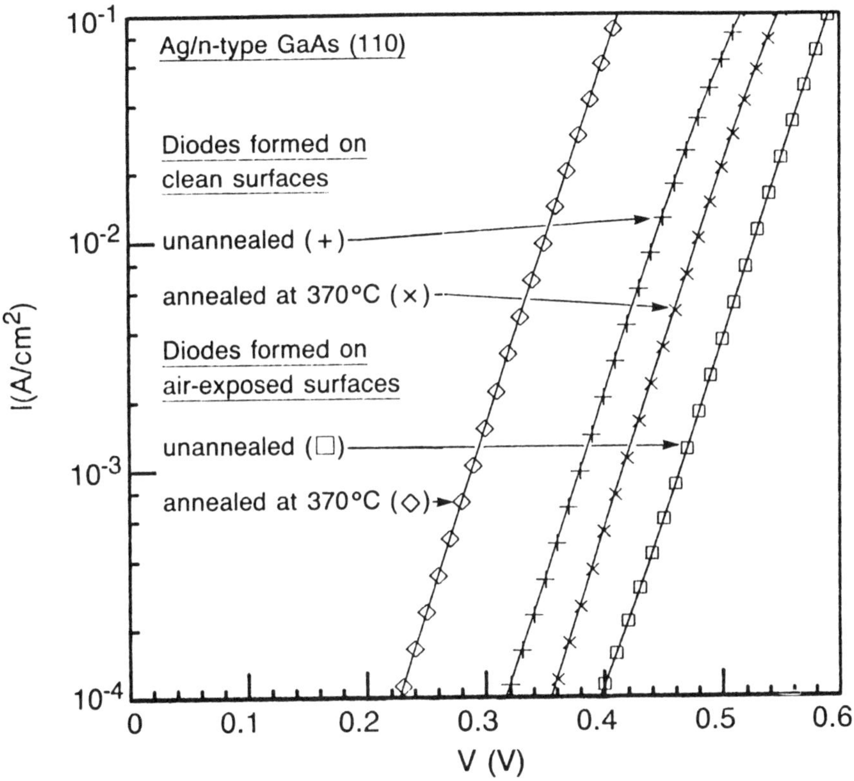


**Figure 20.** Typical current-voltage (I-V) measurements for diodes formed on clean n-type GaAs (110) surfaces prepared by cleavage in ultra-high vacuum (UHV) and on air-exposed surfaces prepared by cleavage and exposure to the atmosphere for ~1-2 hours.

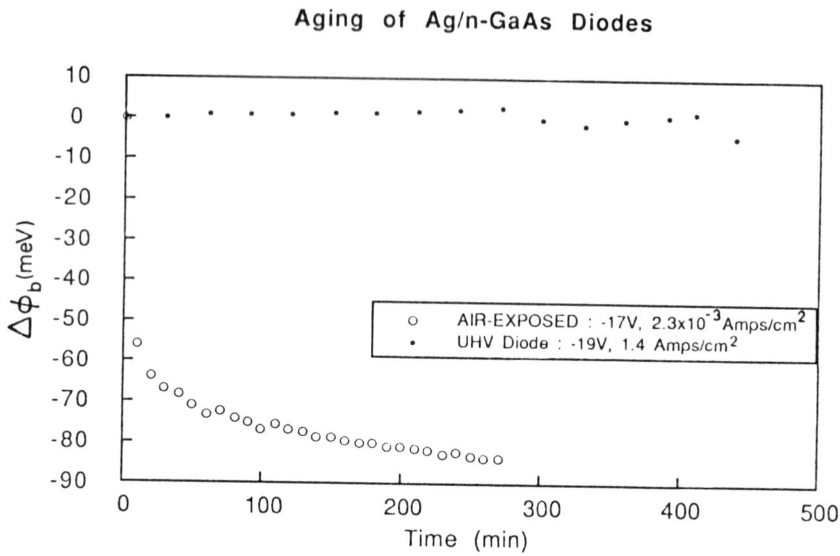


**Figure 21.** This figure illustrates typical results of electrical aging for Ag/n-type GaAs (110) diodes formed on air-exposed and UHV-cleaved GaAs (110) surfaces. The change in barrier height is plotted as a function of the length of time which the diodes were exposed to electrical aging. The electrical aging conditions for each diode are given in the figure's key.

In Sec. 2.1 significant differences in the structure of the two kinds of Ag contacts were briefly discussed. An oxide layer of 4 nm thickness was present at the interface of as-deposited air-exposed samples, but it was not present in UHV-prepared samples. The air-exposed diodes contained a higher density of twins and much smaller Ag grains compared to samples deposited in situ under UHV conditions. Current stressing did not result in any structural changes of UHV-cleaved diodes (Fig. 22a). However, structural degradation of the air-exposed samples was found with current stressing (Fig. 22b). After current stressing, a significant decrease in Ag grain size, voids separating those grains, and enhanced electromigration of Ag were observed. Thermal annealing of UHV-deposited samples did not degrade the interface flatness; however, annealed air-exposed diodes showed a high density of protrusions similar to those shown in Fig. 3b.

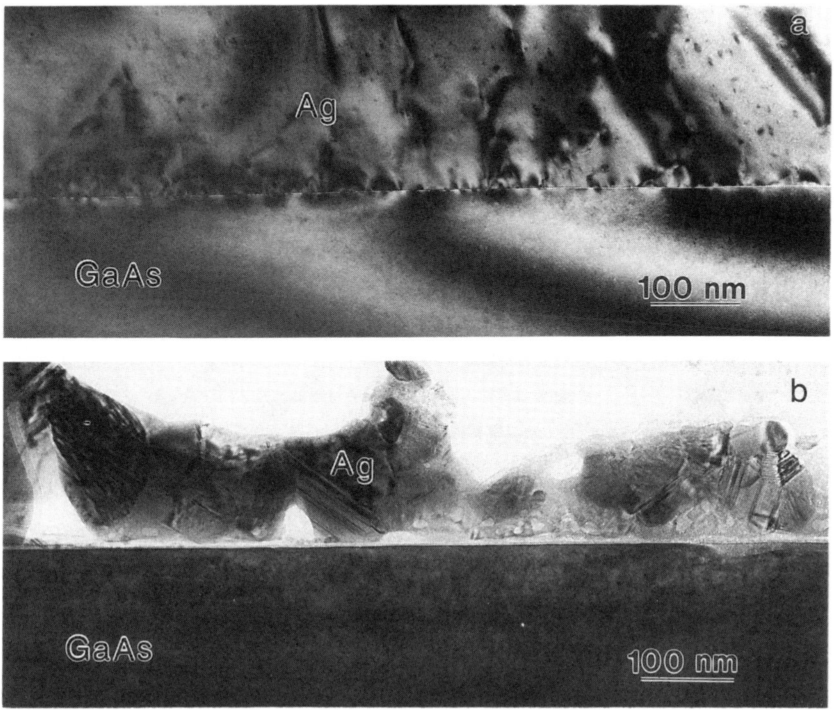


**Figure 22.** TEM micrographs of Ag/GaAs interfaces after current and voltage aging; (*a*) Ag deposited on UHV-cleaved GaAs aged for 7 hrs, -19 V, 1.4 A/cm2. (*b*) Ag deposited on air-exposed GaAs aged for 50 min, -14 V, $4.3 \times 10^{-3}$ A/cm$^2$. Note smaller grain size than before aging and void formation.

For Al/GaAs contacts, a very small increase of 9 meV in the barrier height was observed for air-exposed diodes upon electrical stressing for more than 400 min at -9.7 V with a current density of 1.3 A/cm$^2$. Thermal annealing of UHV-deposited Al contacts resulted in an increase in the Schottky barrier height from 0.83 eV to 0.90 eV. A similar increase of 70 meV was found for the samples deposited on air-exposed substrates, where the barrier height increased from 0.76 to 0.83 eV (Table 1). Although it is tempting to ascribe this increase in barrier height to the formation of near-interfacial AlGaAs with a larger bandgap than GaAs, a careful comparison of the annealing induced changes in the electrical properties of UHV-deposited Al/GaAs and Al/InP revealed that the barrier height of the contacts on p-type semiconductors had an equal and opposite change to that observed for n-type, i.e., the barrier heights before and after annealing added up to the GaAs bandgap.[3] This indicates that the changes in the barrier height upon annealing of the Al diodes can be attributed to shifts in the interface Fermi-level position within the GaAs bandgap, rather than to the formation of a larger-bandgap semiconductor (i.e., AlGaAs, AlInP).

The influence of interfacial impurities has been observed in admittance measurements of Al contacts deposited by MBE on n-GaAs (100) buffer layers.[57] For Al deposited by molecular beam epitaxy on n+ GaAs wafers, ideal current-voltage characteristics were measured. When Al was deposited using inferior vacuum conditions, nonidealities in the current-voltage characteristics and excess capacitances were seen. The onset of excess capacitances was attributed to the change in surface properties.

The I-V characteristics of Au/n-GaAs (110) diodes showed a considerable difference in barrier height for the Au diodes deposited in situ on UHV-cleaved GaAs (0.92 eV) and deposited on the samples cleaved in air (0.83 eV). After annealing at 370°C for 10 min, the barrier height decreases by 120 meV for the UHV-deposited samples and by 180 meV for air-exposed diodes (Fig. 18). It should be noted that the data for annealed Au/GaAs diodes could only be established after removing the peripheral leakage current.[3] This can be accomplished by the use of a chemical "mesa" etch.[3] The excess current is due to thin Au "fingers" formed at the thin boundary region of the Au contact. Surface recombination at these peripheral structures was shown to cause the Au/GaAs diode degradation upon annealing in the 400°C range.[58]

On air-exposed Au/GaAs (110) samples, electrical aging was performed using -19 V reverse bias at a current density of 0.7 A/cm$^2$ for 200 min. A decrease in the barrier height of over 20 meV was found (Fig. 23). With atomically clean Au interfaces, similar instabilities were observed upon electrical aging at -19 V reverse bias at a current density of 4.2 A/cm$^2$ for 200 min. A decrease in the barrier height of over 15 meV was found.

The microstructure of the Au/GaAs (110) diodes is described in detail in Sec. 2.2. It showed similarities to Ag/GaAs: annealing of UHV-deposited samples revealed preservation of a flat interface accompanied by an increase in grain size, whereas for air-exposed diodes, the formation of faceted protrusions at the interface was observed upon annealing. No significant change of the microstructure was found after electrical stressing.[59]

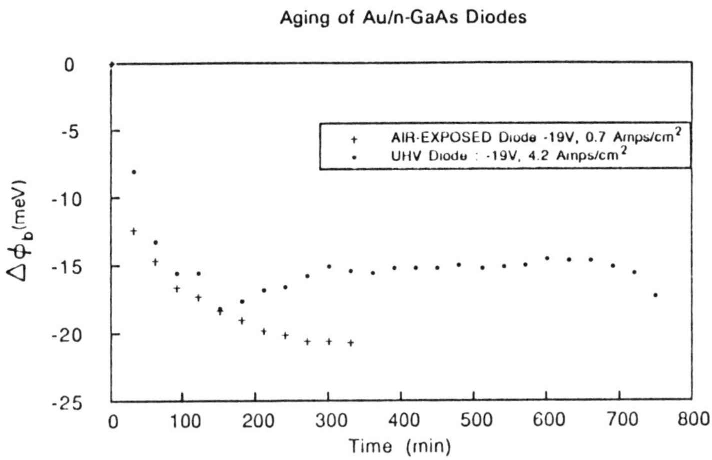


**Figure 23.** This figure illustrates typical results of electrical aging for Au/n-type GaAs (110) diodes formed on air-exposed and UHV-cleaved GaAs (110) surfaces. The change in barrier height is plotted as a function of the length of time which the diodes were exposed to electrical aging.

Electrical aging of Cr/GaAs (110) diodes at -19 V for more than 6 hr with a reverse current flow of 3 $A/cm^2$ did not change the barrier height by more than 6 meV (within experimental error) for contaminated interfaces. However, annealing of contaminated Cr interfaces up to 370°C induced an 80 meV increase in barrier height (from 0.68 to 0.76 eV), while the same thermal treatment did not lead to any significant changes in the barrier height of atomically clean interfaces (Fig. 24). The change in the barrier height for the contaminated interfaces is correlated with the smaller Cr column size found in these samples, as compared with atomically clean interfaces (Fig. 13). The high number of voids between small, randomly oriented columns may allow As to escape during annealing. Because the column size in the annealed UHV samples is larger, the number of voids is fewer, and the out-diffusion of As might be suppressed in these samples.

These observations show that surface preparation of a semiconductor before metal deposition, and impurities present during deposition, can be detrimental to metal contact performance in that they can influence the stability of the electrical properties. Moreover, the observed differences in changes of barrier height upon annealing revealed some interesting correlations with the near-interfacial stoichiometry. This topic is discussed in the next section.

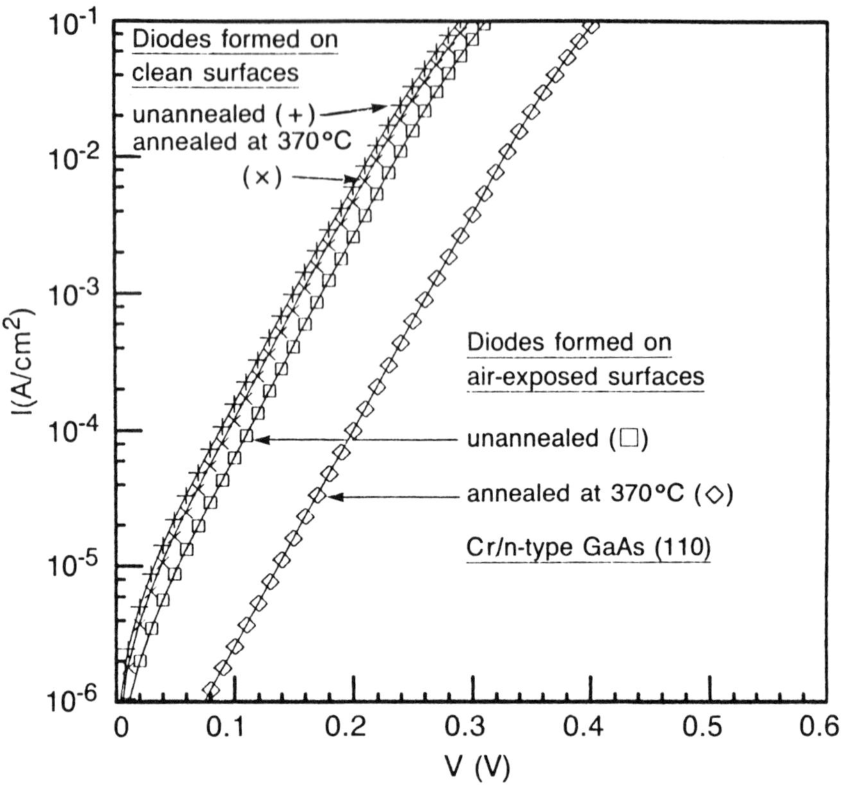


**Figure 24**. Typical current-voltage (I-V) measurements for Cr diodes formed on clean n-type (110)GaAs surfaces prepared by cleavage in UHV and on air-exposed surfaces prepared by cleavage and exposure to the atmosphere for ~1-2 hours. Note a change in the barrier height after annealing for the contaminated interfaces.

## 4.0 NEAR-INTERFACIAL STOICHIOMETRY

The metal/n-GaAs (110) Schottky barrier of in situ UHV-prepared samples was found to exhibit near-ideal electrical characteristics. The lowest barrier height was found for Cr ($\Phi_b$ = 0.67 eV) and the highest for Au ($\Phi_b$ = 0.92 eV).[2] Cross-sectional samples of these contacts were studied by conventional and analytical transmission electron microscopy. The characteristic x-ray emission (EDX) spectra consistently show a deviation in stoichiometry for the region within ~10 nm of the GaAs beneath the metal contact, as compared with the GaAs far from the interface (Fig. 25). The deviation is always As-rich when compared quantitatively to bulk GaAs. These measurements of the crystal stoichiometry are very difficult, as the excess As is volatile under the electron beam. Even when an electron beam is directed onto an As precipitate, the stronger As-$K_\alpha$ peak relative to the Ga-$K_\alpha$ peak can be observed only for a short time. Therefore, to obtain statistically reliable results at a given distance from the interface, the electron beam was scanned to several equivalent positions along the interface so that reliable data could be obtained.

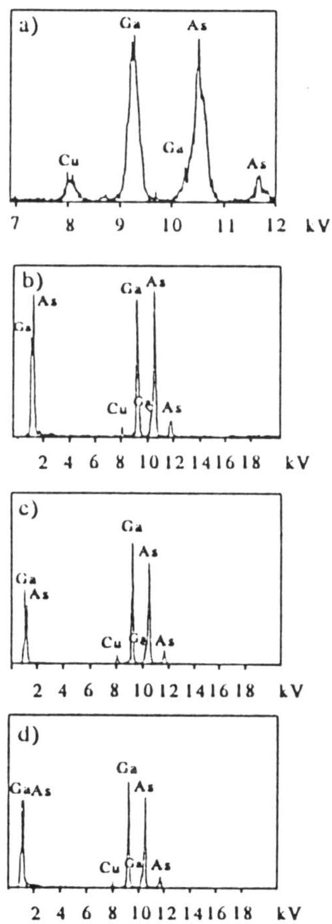


**Figure 25.** EDX spectra taken in the GaAs substrate; *(a)* below the annealed Au layer deposited on the UHV-cleaved (110)GaAs surface; *(b)* below the Cr layer deposited in the same way; *(c)* below Au cluster deposited on the UHV-cleaved (110)GaAs surface buffered by Xe; *(d)* far from the interface. The As/Ga ratio in *(c)* and *(d)* is typical for bulk GaAs. In *(a)* and *(b)* case As/Ga ratio is much higher indicating As rich interfaces.

This As instability under the electron beam was further confirmed by studying As-rich GaAs grown by MBE at low temperature (~200°C). These layers are known from conventional x-ray and PIXE analysis to contain up to 1.5% of excess As.[60][61] Studying cross sectional samples of low-temperature (LT)-GaAs by EDX showed a very similar time dependence: the x-ray signal from excess As can be seen only for a certain time. Comparison of the near-interfacial regions of metal/GaAs contacts for interfaces that are known to be As-rich (such as Cr/GaAs) to that of As-rich LT GaAs leads to the conclusion that the (average) deviation from stoichiometry in a 10 nm-deep region from the metal/GaAs interface must be on the order of 1%. The increase in As concentration (Fig. 26) near the metal interface was confirmed independently by SALI.[30][31]

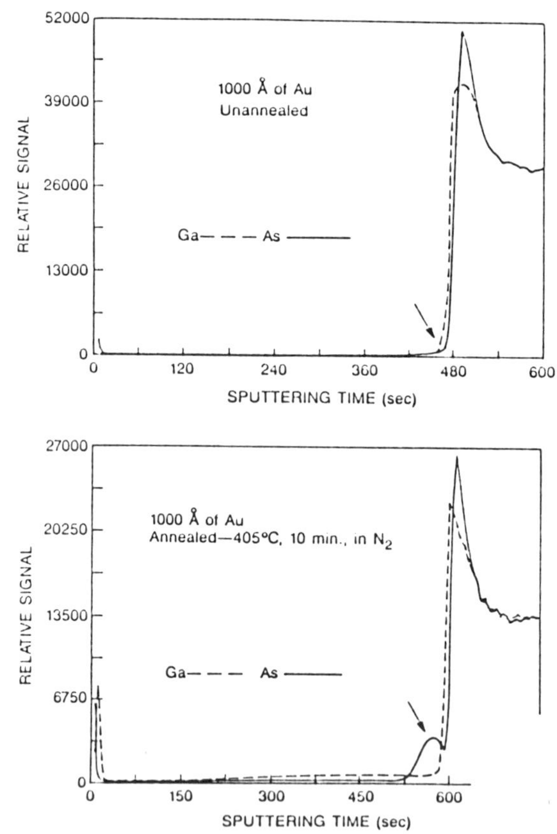


**Figure 26.** SALI depth profiles of UHV prepared Au (100 nm thick) on GaAs. A 5 keV Ar+ beam was used with ~ 30 mm diameter and a DC current of ~ 100 mm.

Comparison of the EDX spectra of different metal/GaAs contacts gives evidence that the amount of additional As found in this near-interfacial region depends on the metal used and the thermal history of the sample. For as-deposited metals, the largest deviation from stoichiometry was found for Cr, and the smallest was found for Au. After annealing above 370°C, the Au barrier height *decreased* by 120 and 180 meV for UHV-deposited and air-exposed samples, respectively.[3][22] In contrast, the Cr barrier height was stable upon thermal annealing for UHV-deposited samples but *increased* by 80 meV to 0.76 eV for air-exposed samples.[22] Air-exposed annealed Cr/GaAs samples showed less excess As than the as-deposited structures; the excess concentration was comparable to Au/GaAs in magnitude. Indeed, the barrier height of annealed Au/GaAs samples (0.80 eV) is also comparable to annealed air-exposed Cr/GaAs samples (0.78 eV).[3][22] The excess As found near the Al/GaAs interfaces was less than that found for Cr/GaAs, and showed a similar decrease upon annealing.

Nevertheless, all the Schottky barrier heights found in the course of this work[2][3][22][36] correspond to Fermi-level pinning positions between 0.75 eV and 0.5 eV above the valence band maximum. In addition, changes in barrier height upon annealing were found to correlate with changes in the near-interfacial stoichiometry. An increase in the barrier height of n-GaAs such as the one found for Cr on air-exposed GaAs (110) and for Au on UHV and air-exposed GaAs (110), is correlated with As-rich interfaces in both cases. However, the amount of excess As near the interface in the first case is much larger than in the second case. The changes in As concentration at the interface after annealing for Al, Ag, Au, and Cr contacts are generally found to follow this relationship: an increase (decrease) in As concentration results in a decrease (increase) in barrier height. This relation and the change in As concentration is especially noticeably for Au and Al. In the first case, higher As concentration is observed after annealing compared to as-deposited samples. In contrast, for Al, the As concentration is lower in annealed samples compared to as-deposited samples.

The observed changes in stoichiometry and near-interfacial reconstruction are typical for atom-by-atom deposition methods such as evaporation or sputtering in vacuum onto substrates kept near room temperature. During such a deposition, the metal atoms condense from the gas phase to the solid phase, which releases the heat of sublimation and may result in disruption of the semiconductor surface. In addition, a significant amount of energy can also be released when metal atoms combine on the surface to form clusters. This is particularly important for the less-reactive metals such as Ag. This surface disruption might be avoided by cluster deposition, described in Sec. 2.2. The energetics of such a deposition are dramatically different from atom-by-atom deposition, making it possible to minimize surface disruption. Analytical electron microscopy performed in the semiconductor below the clusters (Fig. 25) did not show any deviation from stoichiometry of the sort observed for cases such as the Au atom-by-atom deposition onto the GaAs surface.[45] This shows that metal/semiconductor interfaces free of bulk defects can be formed. Photoemission measurements gave evidence for Fermi-level pinning of cluster-deposited Schottky contacts outside the usually observed range of energies.[44][45]

## 5.0 NEAR-INTERFACIAL ELECTRICALLY ACTIVE DEFECTS

In Sec. 2 - 4 of this chapter, a detailed structural analysis of metal/GaAs contacts is summarized. In all cases investigated, the interfaces were far from ideal: for in-situ UHV deposition, a distinct deviation from perfect stoichiometry was found; for diodes deposited onto air-exposed GaAs (110) surfaces, an interfacial oxide layer can also be detected that influences both the chemistry and orientation relationship of the metal grains to the substrate. Moreover, evidence for distinct changes of the near-interfacial stoichiometry upon annealing of the diodes was found.

It is difficult to reconcile these experimental results with models that ascribe the narrow range of Fermi-level pinning positions to the extra states in the semiconductor band gap induced by the metal contact.[8] The dependence of the pinning position on the near-interfacial stoichiometry strongly suggests the influence of near-interfacial electrically active defects.

Two defect-related models currently receive the greatest level of acceptance: *(i)* the effective work-function model that ascribes the Fermi level pinning to the work function of As-rich metal compounds that are hypothesized to be formed at the metal/semiconductor interface,[9][10] and *(ii)* the antisite defect model,[14][17][63] which was based originally only on the experimental determination of the energy levels of $As_{Ga}$ antisite defects at $E_c$ -0.75 eV and $E_v$ +0.52 eV.

The structural studies summarized in this chapter do not support the effective work function model.[9][10] For the model to be applicable, the presence of anion-rich clusters or anion-rich near-interfacial phases is required for all the Schottky contacts studied here. The limit of detection is the resolution limit of high-resolution electron microscopy, which can easily detect clusters above 2 nm diameter or thin-film phases above one monolayer in thickness. It is worth noting that the situation is completely different for low-temperature grown GaAs, in which *after annealing above 500°C*, a high density of As precipitates is found.[64][65] It has been suggested that these As precipitates are responsible for the Fermi-level pinning in these semi-insulating layers.[66]

It is worth considering in detail whether the antisite defect model finds support in this and other recent studies. This model assumes that the surface disruption upon metal sublimation from the gas phase onto the GaAs surface releases enough energy to displace atoms in the first few monolayers of the crystal, so that a certain fraction of these distorted atoms comes to rest in the wrong sites, effectively forming a high concentration of antisite defects of both kinds, $As_{Ga}$ and $Ga_{As}$. If the near-interfacial region is anion-rich, as experimentally observed, it is expected that $As_{Ga}$ will dominate, partly compensated by $Ga_{As}$. This situation of a high $As_{Ga}$ concentration partly compensated by other native defects is found as well after heavy n-irradiation of bulk GaAs[17] or in as-grown MBE layers of GaAs grown at low temperatures (LT-GaAs).[59][60]

The range of Fermi-level pinning positions determined electrically with well-defined contacts on GaAs correspond very well to the two donor levels of the anion antisite defect at $E_v + 0.52$ eV and $E_c - 0.7$eV.[17]

Changes of near-interfacial stoichiometry are predicted in this model to result in characteristic changes in the Fermi level pinning position. More As-rich interfaces with little compensation should be dominated more by the upper, near midgap donor level of $As_{Ga}$, corresponding to lower barrier heights for n-GaAs; while less excess of As at the interface should result in stronger compensation and thus stronger domination of the second donor level at $E_v + 0.5$ eV, i.e., in larger barrier heights on n-GaAs. This prediction of the antisite defect model is without exception supported by the experimental observations of correlated changes of barrier height and near-interfacial stoichiometry described in the previous section. Additional examples of correlations between interfacial stoichiometries and barrier heights are discussed in detail by Spicer.[63]

Direct evidence for the creation of deep-level defects upon metal deposition on GaAs was found by cathodoluminescence experiments of Chang et al.[67][68] These authors detected the evolution of an emission peaking near 0.85 eV upon Au, Cu, and Al deposition on (100)GaAs grown by MBE, together with other emissions that were more typical of the respective metals. It is worth noting that this 0.85 eV emission corresponds quite well to near-midgap electronic energies, although a direct correlation with an emission known from bulk GaAs defects is not yet possible.

If the Fermi level pinning position were indeed dominated by deep-level defects, then application of hydrostatic pressure should result in similar changes in the barrier height and the defect ionization energy. Shan et al.,[69] studied the pressure dependence of the Pt/n-GaAs(100) Schottky barrier height and found a parabolic behavior according to:

$$e\Phi_b(P) \;=\; 0.703\,[\mathrm{eV}] + 11\,[\mathrm{meV/GPa}]\,P - 2.6\,[\mathrm{meV/GPa^2}]\,P^2$$

For the midgap level of EL2 (which is generally accepted to be essentially an $As_{Ga}$ antisite defect) a distinctly different pressure dependence was found by DLTS:[70]

$$(\mathrm{EL2})^{0/+} \;=\; 0.703\,[\mathrm{eV}] + 44\,[\mathrm{meV/GPa}]\,P - 1.1\,[\mathrm{meV/GPa^2}]\,P^2$$

However, DLTS measures the electron emission barrier, which is the sum of the true thermal ionization energy (E) plus the barrier against free carrier capture $E^{cap}$.[71] Therefore, the capture barrier has to be taken into account for a correct evaluation of the EL2 pressure dependence. Baj and Dreszer[72] recently measured the pressure dependence of the capture barrier as:

$$\partial E^{cap}/\partial P \;=\; (-49 \pm 5)\ \mathrm{meV/GPa}$$

Therefore, the pressure dependence of the EL2 ionization energy is:

$$E = 0.703\,[\mathrm{eV}] + 93\,[\mathrm{meV/GPa}]\,P - 1.1\,[\mathrm{meV/GPa^2}]\,P^2$$

Figure 27 presents experimentally determined values of the pressure dependence of the Schottky barrier height[69] and pressure-induced energy level changes of the $EL2^{0/+}$, $\Delta E$,[70] corrected by the pressure dependence of the capture barrier.[72] It is clear that *the pressure dependence of* $\Delta E(P)$ *cannot be distinguished within the accuracy of the experimental data from the pressure dependence of the Schottky barrier* $\Delta\Phi_b(P)$ *of Pt/n-GaAs.* This conclusion is in direct contradiction to the original conclusion of Ref. 69, whose analysis neglected the pressure dependence of the capture barrier. Although this extraordinary agreement is not unique, as some other deep-level defects have similar pressure coefficients, it can be regarded as a very strong argument in favor of the antisite defect model.

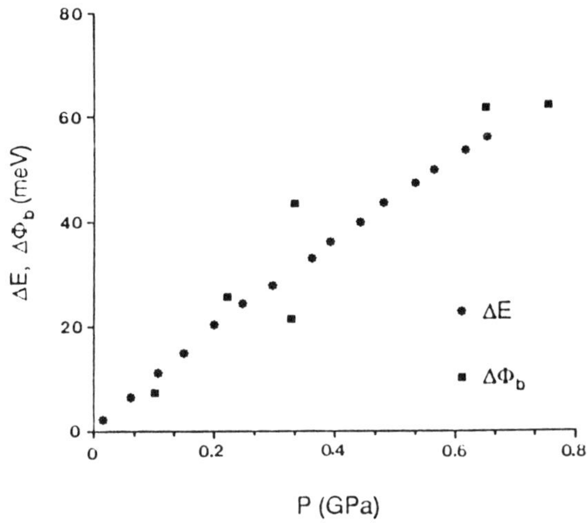


**Figure 27.** Pressure-induced change of the Schottky barrier height $\Delta\Phi_b(P)$ according to Shan et al[69] and experimental data of thermal ionization energy of the EL2$^{0/+}$ level, $\Delta E(P)$ obtained as a sum of an activation energy of the emission rate after Dobaczewski and Sienkiewicz[70] and the pressure derivative of capture barrier energy.[68] Note the excellent agreement between these two sets of data.

Moreover, the dominance of antisite defects in the Fermi level pinning process could be avoided if defect formation could be suppressed altogether, or influenced by modification of the interfacial stoichiometry. For MBE-grown metal/GaAs (100) interfaces, conflicting results were published: Viturro et al.[73] reported photoemission results indicating a very strong dependence of the barrier heights on work function for low-temperature deposited contacts of Ag, Al, Au, Cu, In, and Yb, whereas Wilks et al.[74] reported I-V measurements of similar diodes (In, Al, and Au) that revealed only the well-known "canonical" pinning positions near and slightly below midgap. This difference, e.g., in the case of Al between a barrier height near 0.2 eV[73] and 0.79 eV[74] might be due to either the methods used (photoemission vs. electrical measurements) or to the different approach: whereas Wilks et al.[74] deposited the metallization in situ on UHV-clean, freshly grown GaAs (100) surfaces, Vitturo et al.[73] utilized GaAs wafers grown ex situ that were capped with As that had to be thermally desorbed before the metal deposition. Although it is possible that this discrepancy can be due to artifacts such as surface photovoltaic effects,[75] the different surface stoichiometry due to the As-capping might play the decisive role.

Similarly, distinct changes in the barrier height (determined by photoemission) were found for off-axis substrates with a high step density,[67] which might again be related to different interfacial stoichiometry. However, a detailed study of the effective stoichiometry present at the MBE-grown interfaces has not yet been performed.

A different approach which avoid the formation of interfacial defects, has been described in Sec. 2.3. Photoemission measurements of diodes prepared by cluster deposition rather than by the disruptive atom-by-atom deposition indicate Fermi level pinning positions outside the commonly observed range.[43] In this case, analytical electron microscopy of cross-sectional samples did not find evidence for the As-rich near interfacial layer typical of contacts prepared by the more disruptive atom-by-atom deposition.

Finally, recent self-consistent local density calculations of several metal/GaAs structures have revealed that the Fermi-level pinning positions for "ideal" unreconstructed interfaces in the absence of defects should range throughout the band gap, depending on the metal and the interfacial geometry, effectively refuting the notion that metal-induced gap states should result in a universal pinning position near the experimentally observed values.[76] Moreover, it was directly demonstrated that defects present near the interface can indeed result in pinning at the bulk-like levels of the defects.[77] This study thus provides a "missing link" between the extensive experimental evidence for the role of deep-level defects in Fermi-level pinning and state-of-the-art theory of the electronic properties of heterostructures.

## 6.0 CONCLUSIONS

This study allows us to conclude that interface morphology, grain size and orientation of metal layers on GaAs depend strongly on the surface preparation of the substrate before metal deposition. The metals investigated (Au and Ag, with similar lattice parameters), deposited in situ on a UHV-cleaved GaAs surface, show very similar orientation relationships with GaAs upon annealing. This relationship changes when GaAs is exposed to air before metal deposition. For Cr, an almost perfect lattice match to GaAs was observed for UHV-deposited samples, but random orientations were observed for air-exposed samples. A pronounced correlation between changes in barrier height and near-interfacial stoichiometry was found. This correlation has important implications for the fundamental understanding of the mechanism of Fermi-level pinning. High-resolution electron microscopy showed evidence of lattice reconstruction and defect formation in the interfacial area, supporting defect models of Schottky barrier formation.

Among the defect models for Fermi-level pinning, the antisite defect model seems to be especially attractive, as it directly relates to the experimentally determined barrier heights and the observed deviation from stoichiometry. Besides the agreement with a large number of well-known experimental results,[62] this model received additional support through the new observation that the pressure dependence of a typical Schottky barrier (Pt with $\Phi_b$ = 0.70 eV) corresponds within the measurement error to the pressure dependence of EL2, which is composed of an As antisite defect. On the other hand, a defect model requires the possibility of avoiding or modifying the defect formation, and this is indeed the case, as the examples of MBE-deposition and cluster-deposited contacts seem to indicate.

Therefore, a realistic model of the metal/semiconductor interface has to take into account intrinsic and extrinsic interfacial processes: intrinsic effects such as metal-induced gap states and interfacial dipoles can be dominating if defect formation is effectively suppressed, but we conclude that extrinsic effects take over in most metal/GaAs rectifying contacts of practical relevance that show the well-known small range of Fermi-level pinning positions.

Despite the large body of circumstantial evidence in favor of the defect model that was discussed in detail in Sec. 5, further experiments are required to finally directly *identify* the defects that are involved in the Fermi-level pinning process. As we are reaching the limits even of high-resolution electron microscopy, this final clarification might come from scanning tunneling microscopy, a technique that should allow us to directly image individual point defects.

## ACKNOWLEDGMENT

This review was based on the work supported by the U.S. Department of Energy under contract DE-ACO3-76SF00098 for studying the structure of Au-, Ag, and Al-GaAs Schottky contacts and by the SDIO/IST program administrated by the Office of Naval Research under the contract N00014-86-K-0668 for studying the reliability of metal contacts to GaAs and the structure of refractory metal contacts.

## REFERENCES


1. See, e.g., *J. Vac. Sci. Technol.,* B4,903:1085 (1986); K. Stiles, A. Kahn, and D. G. Kildey, *ibid.* 5:987 (1987); K. Winer and L. Ley, *ibid.* 6:1165 (1988); D. Heskett, T. M. Wong and J. A. Smith, *ibid.* 7:915 (1989); X. Yin, H.-M. Chen, F. H. Pollak, Y. Cao, P. A. Montano, P. D. Kirchner, G. D. Pettit, and J. M. Woodall, *ibid.* 9:2114 (1991)
2. N. Newman, M. van Schilfgaarde, T. Kendelewicz, M. D. Williams and W. E. Spicer, *Phys. Rev.,* B33:1146 (1986)
3. N. Newman, W. E. Spicer, and E. R. Weber, *J. Vac. Sci. Technol.,* B5:1020 (1987); N. Newman, W. G. Petro, T. Kendelewicz, S. H. Pan, S. J. Eglash, and W. E. Spicer, *J. Vac. Sci. Technol.,* 57:1247 (1985); N. Newman, K. K. Chin, W. G. Petro, T. Kendelewicz, M. D. Williams, C. E. McCants, and W. E. Spicer, *J. Vac. Sci. and Technol.,* A3:996 (1985)
4. Z. Liliental-Weber, N. Newman, C. Jou, W. E. Spicer, J. Washburn, and E. R. Weber, *Mat. Res. Soc. Symp. Proc.*, 102:241 (1988)
5. A. B. McLean and R. H. Williams, *J. Phys.,* C21:783 (1988)
6. J. R. Waldrop, *J. Vac. Sci. Technol.,* B24:445 (1984)
7. V. Heine, *Phys. Rev.,* A138:1689 (1965)
8. J. Tersoff, *Phys. Rev. Lett.,* 52:465 (1984); C. Tejedov, F. Flores, and E. Louis, *J. Phys.,* C10:2163 (1977); S. G. Louie and M. L. Cohen, *Phys. Rev.,* B13:4172 (1976)
9. J. L. Freeouf and J. M. Woodall, *Appl. Phys. Lett.,* 39:727 (1981)
10. J. L. Freeouf, *Appl. Phys. Lett.,* 41:285 (1982); J. L. Freeouf and J. M. Woodall, *Surf. Sci.,* 168:518 (1986); 168:522 (1986)
11. W. E. Spicer, P. W. Chye, P. R. Skeath, C. Y. Su, and I. Lindau, J. *Vac. Sci. Technol.*, 16:1427 (1979); 17:1019 (1980); *Phys. Rev. Lett.*, 44:420 (1980)
12. W. Walukiewicz, *J. Vac. Sci. Technol.*, B6:1257 (1988); *Mat. Res. Soc. Symp. Proc.*, 148:137 (1989)
13. W. Mönch, *Chemistry and Physics of Solid Surfaces V*, (R. Vanselov and R. Howe, ed.) p. 529, Springer, Berlin (1984)
14 W. E. Spicer, Z. Liliental-Weber, E. Weber and N. Newman, *J. Vac. Sci. Technol.,* B6:1245 (1988)
15. W. E. Spicer, T. Kendelewicz, N. Newman, R. Cao, C. McCants, K. Miyano, I. Lindau, and E. R. Weber, *Appl. Surf. Sci.,* 33/34:1009 (1988)

16. Z. Liliental-Weber, E. Weber, N. Newman, W. E. Spicer, R. Gronsky, and J. Washburn, Proc. *14th Int'l Conf. on Defects in Semiconductors 1986,* (J. Bourgoin, M. Lannoo, and H. J. von Bardeleben, eds.), p. 1223, Trans Tech Publications (1986)
17. E. R. Weber, H. Ennen, V. Kaufmann, J. Windscheif, J. Schneider, and T. Wosinski, *J. Appl. Phys.,* 53:6140 (1982); E. R. Weber and J. Schneider, *Physica,* B116:398 (1983)
18. Z. Liliental-Weber, R. Gronsky, J. Washburn, N. Newman, W. E. Spicer, and E. R. Weber, *Appl. Phys. Lett.,* 49:1514 (1986)
19. Z. Liliental-Weber, *J. Vac. Sci. Technol.,* B 5:1007 (1987)
20. Z. Liliental-Weber, R. W. Carpenter, and J. Escher, *Ultramicroscopy,* 14:135 (1984)
21. Z. Liliental-Weber, J. Washburn, C. Musgrave, E. R. Weber, R. Zuleeg, W. V. Lampert, and T. W. Haas, *Mat. Res. Soc. Symp. Proc.,* 126:295 (1988)
22. N. Newman, Z. Liliental-Weber, E. R. Weber, and J. Washburn, *Appl. Phys. Lett.,* 53:145 (1988)
23. A. Miret-Goutier, N. Newman, E. R. Weber, Z. Liliental-Weber, and J. Washburn, *J. Appl. Phys.,* 63:2006 (1988)
24. Z. Liliental-Weber, *Point, Extended and Surface Defects in Semiconductors,* (C. Benedek, A. Cavallini, and W. Schrotered, ed..), p. 165, Plenum Publishing Corporation (1989)
25. J. Ding, Z. Liliental-Weber, E. R. Weber, J. Washburn, R. M. Fourkas, and N. W. Cheung, *Appl. Phys. Lett.,* 52:2160 (1988)
26. J. Ding, B. Lee, D. Chin, R. Gronsky, T. Van Duzer, and J. Washburn, *Appl. Phys. Lett.,* 52:135 (1988)
27. Z. Liliental-Weber, R. Gronsky, J. Washburn, N. Newman, W. E. Spicer, and E. R. Weber, *J. Vac. Sci. Technol.,* B4:912 (1986)
28. Z. Liliental-Weber, N. Newman, W. E. Spicer, R. Gronsky, J. Washburn, and E. R. Weber, *Mat. Res. Soc. Symp. Proc.,* 54: 415 (1986)
29. D. Coulman, N. Newman, G. Reid, Z. Liliental-Weber, E. R. Weber, and W. E. Spicer, *J. Vac. Sci. Technol.,* A5:1521 (1987)
30. J. B. Pallix, C. H. Becker, and N. Newman, *MRS Bulletin,* 12:52 (1987)
31. J. B. Pallix, C. H. Becker, and N. Newman, *J. Vac. Sci. Technol.,* A6:1049 (1988)
32. Z. Liliental-Weber, C. Nelson, R. Gronsky, J. Washburn, and R. Ludeke, *Mat. Res. Soc.Symp. Proc.,* 77:229 (1987)
33. T. Sands, J. Washburn, and R. Gronsky, *Mater. Lett.,* 3:247 (1985)

34. Z. Liliental-Weber and M. A. O'Keefe, *J. Vac. Sci. Technol.,* 7:1022 (1989)

35. Z. Liliental-Weber, N. Newman, J. Washburn, E. R. Weber, and W. E. Spicer, *Appl. Phys. Lett.,* 54:356 (1989)

36. C. E. McCants, T. Kendelewicz, P. H. Mahowald, K. A. Bertness, M. D. Williams, N. Newman, I. Lindau, and W. E. Spicer, *J. Vac. Sci. Technol.,* A6:1466 (1988)

37. S. Ogawa, T. Kouzaki, T. Yoshida, and R. Sinclair, *Mat. Res. Soc. Symp. Proc.*, 181:139 (1990)

38. X. Lin, Z. Liliental-Weber, W. Swider, T. McCants, N. Newman, W. E. Spicer, J. Washburn, and E. R. Weber, Mat. Res. Soc. Symp. Proc., 281 (in press 1993)

39. T. Sands, V. G. Keramidas, K. M. Yu, J. Washburn, and K. M. Krishnan, *J. Appl. Phys.,* 62:2070 (1987)

40. T. Sands, V. G. Keramidas, A. J. Yu, K. M. Yu, R. Gronsky, and J. Washburn, *Mat. Res. Soc. Symp. Proc.,* 54:367 (1986)

41. Z. Liliental-Weber, N. Newman, J. Washburn, E. R. Weber, and W. E. Spicer, to be published

42. R. S. Williams, J. R. Lince, T. C. Tsai, and J. H. Pugh, *Mat. Res. Soc. Proc.,* (R. J. Nemanich, P. S. Ho, and S. S. Lau, ed.), 54:335 (1986)

43. G. D. Waddill, I. M. Vitomirov, C. M. Aldao, and J. Weaver, *Phys. Rev. Lett.,* 62:1568 (1989)

44. G. D. Waddill, I. M. Vitomirov, C. M. Aldao, S. G. Anderson, C. Capasso, J. H. Weaver, and Z. Liliental-Weber, *Phys. Rev.,* B41:5293 (1990)

45. Z. Liliental-Weber, E. R. Weber, J. Washburn, and J. H. Weaver, *Appl. Phys. Lett.,* 56:2507 (1990)

46. R. M. Feenstra, *J. Vac. Sci. Technol.,* B7:925 (1989)

47. T. Yoshiie and C. L. Bauer, *J. Vac. Sci. Technol.,* A1:554 (1983)

48. R. Ludeke and G. Landgren, *J. Vac. Sci. Technol.,* 19:667 (1981)

49. G. Landgren, R. Ludeke, and C. Serrano, *J. Cryst. Growth,* 60:391 (1982)

50. N. Braslau, *J. Vac. Sci. Technol.,* 19:803 (1981)

51. L. Holland, J. P. Hallais, and J. C. Brice, *Current Topics in Materials Sciences,* (E. Kaldis, ed.), 5:1, North Holland, Amsterdam, (1980)

52. Z. Liliental-Weber, N. Newman, X. Lin, J. Washburn, E. R. Weber, and W. E. Spicer, to be published

53. E. D. Marshall, W. X. Chen, C. S. Wu, S. S. Lau, and T. F. Kuech, *Appl. Phys. Lett.,* 47:298 (1985)
54. M. A. Nicolet, *Thin Solid Films,* 52:415 (1978)
55. M. Wittmer and H. Melchior, *Thin Solid Films,* 93:397 (1982)
56. P. Phatak, Z. Liliental-Weber, and E. R. Weber, *Mat. Res. Soc. Symp. Proc.,* 240:455 (1992)
57. P. Muret, D. Elguennouni, M. Missous, and E. H. Rhoderick, *Appl. Phys. Lett.,* 58:155 (1991)
58. R. P. Leon, N. Newman, Z. Liliental-Weber, E. R. Weber, J. Washburn, and W. E. Spicer, *J. Appl. Phys.,* 66:711 (1989)
59. Z. Liliental-Weber - unpublished
60. M. Kaminska, Z. Liliental-Weber, E. R. Weber, T. George, J. B. Kortright, F. W. Smith, and A. R. Calawa, *Appl. Phys. Lett.,* 54:1881 (1989)
61. M. Kaminska, E. R. Weber, Z. Liliental-Weber, R. Leon, and Z. Rek, *J. Vac. Sci. Technol.,* B7:710 (1989)
62. Z. Liliental-Weber, W. Swider, K. M. Yu, J. B. Kortright, F. W. Smith, and A. R. Calawa, *Appl. Phys. Lett.,* 19:2153 (1991)
63. W. E. Spicer, *Semiconductors and Semimetals,* (E. R. Weber, ed), to be published
64. Z. Liliental-Weber, *Mat. Res. Soc. Symp. Proc.*, 188:371 (1990)
65. Z. Liliental-Weber, A. Claverie, J. Washburn, F. Smith, and A. R. Calawa, *Appl. Phys.,* A53:141 (1991)
66. A. C. Warren, J. M. Woodall, J. L. Freeouf, D. Grischkowsky, D. T. McInturff, M. R. Melloch, and N. Otsuka, *Appl. Phys. Lett.,* 57:1331 (1990)
67. S. Chang, L. J. Brillson, Y. J. Kime, D. F. Rioux, P. D. Kirschner, G. D. Pettit, and J. M. Woodall, *Phys. Rev. Lett.,* 64:2551 (1990)
68. S. Chang, I. M. Vitomirov, L. J. Brillson, D. F. Rioux, P. D. Kirchner, G. D. Pettit, and J. M. Woodall, *J. Vac. Sci. Technol.*, B9:2129 (1991)
69. W. Shan, M.F. Li, P.Y. Yu, W.L. Hansen, and W. Walukiewicz, *App. Phys. Lett.,* 53:974 (1988)
70. L. Dobaczewski and A. Sienkiewicz, *Acta. Phys. Pol.,* A71:341 (1987)
71. D. V. Lang, *Thermally Stimulated Relaxation in Solids,* (P. Braunlich, ed.), p.93, Springer, New York (1979)
72. P. Dreszer and M. Baj, *Acta. Phys. Pol.,* A73:219 (1988)

73. R. E. Viturro, S. Chang, J. L. Shaw, C. Mailhiot, L. J. Brillson, A. Terrasi, Y. Hwo, G. Margaritondo, P. D. Kirchner, and J. M. Woodall, *J. Vac. Sci. Technol.,* B7:1007 (1989)

74. S. P. Wilks, J. I. Mirros, D. A. Woolf, and R. H. Williams, *J. Vac. Sci. Technol.*, B9:2118 (1991)

75. M. H. Hecht, *Phys. Rev.,* B41:7918 (1990)

76. N. Newman, M. van Schilfgaarde, and W. E. Spicer, *Phys. Rev.,* B35:6298 (1987)

77. M. van Schilfgaarde and N. Newman, *J. Vac. Sci. Technol.,* B9:2140 (1991)

# 7

# Atomic-Scale Chemistry of Metal-Semiconductor Interfaces

*Celso M. Aldao and John H. Weaver**

## 1.0 INTRODUCTION

Interface research focuses on the unique properties of the transition region between two different materials. It seeks to understand how the properties of boundary regions differ from those of the bulk solids on either side and how these regions influence the behavior of the composite. Phenomena occurring at interfacial zones are the main issue in areas such as catalysis, corrosion, and microstructure electronics. In the last, miniaturization is becoming increasingly critical as the technology pushes toward the physical limitations of circuit density. Moreover, the understanding of interfaces has determined the development of new devices and the control of their electrical properties.

The interface zone can be atomic in scale for abrupt interfaces or it can extend for tens or hundreds of angstroms for interfaces where the two materials react chemically. The properties of the interface region are influenced by, for example, dimensional constraints, disorder, defects, the formation of compounds, heterogeneity, and kinetics. Boundaries are therefore fascinating regions, rich in both scientific challenge and technological importance. Interface research is inherently interdisciplinary, blending experimental and theoretical work in science and engineering.

---

*This work was supported by the Office of Naval Research, the Army Research Office, and the National Science Foundation.

Here, we will focus on a subclass of interface phenomena, those at metal-semiconductor interfaces, which play a central role in microelectronic devices. When designing and optimizing a multilayer device, the challenge is to determine which elements, compounds, or alloys are most suitable in a particular application, to investigate the properties of those materials when they form interfaces, and to assess the stability of the interface during processing and operation. The structure of a metal-semiconductor interface, made of only two components, is simple at first sight, but in fact many parameters are responsible for its final properties. Considerable problems arise for the theorist in modeling junctions and for the experimenter in creating well-characterized interfaces. However, the reward is very attractive. Just as we are now able to design bulk semiconductors, the study of semiconductor interfaces will allow us to construct novel devices having useful electronic and optical characteristics.

The application of ultrahigh vacuum surface science techniques has modified drastically the physical picture of metal-semiconductor interfaces and this brought significant progress in understanding their microscopic morphology. Atomic-scale studies have shown that metal adatoms can interact strongly with clean semiconductor surfaces and that junction electronic structures cannot be predicted simply from bulk properties. Instead, simultaneous and diverse processes can be present during interface formation, including adatom clustering, substrate disruption, the release and out-diffusion of substrate atoms, and formation of new compounds. Regarding the interface chemistry, the scientific goal is to identify the composition, electronic structure, and microstructure of the interfacial region as well as the parameters that control processes occurring during contact formation.

Among many fascinating properties of metal-semiconductor interfaces, Schottky barrier formation takes a particularly prominent place from both the scientific and technological point of view. Many decades ago, the rectifying behavior associated with metal-semiconductor contacts was related to the formation of a depletion region on the semiconductor side of the interface. The potential barrier related to that layer, the Schottky barrier, is the most important parameter controlling the transport properties of the junction. Many studies have focused on understanding the microscopic mechanisms that control the Schottky barrier and, in particular, on the relation between interfacial chemistry and barrier height. As yet, the resolution of issues related to Schottky barrier formation is elusive, although recent detailed studies have helped to orient and simplify the work of the theoretician who endeavors to model the junction.

This chapter is based on photoemission investigations of metal-semiconductor interfaces conducted in recent years. Those studies have been interesting and important with respect to interfacial phenomena including adatom-induced substrate disruption, chemical reaction, out-diffusion of one or both substrate species, and varying degrees of surface segregation. In addition, different overlayers followed different growth mechanisms, ranging from ordered epitaxial growth (e.g., Co/GaAs) to overlayer clustering with virtually no substrate disruption (e.g., Ag/GaAs). Temperature variations are known to have a profound impact on atom distributions at metal-semiconductor contacts, and such effects will be explored in detail.

In order to minimize the amount of substrate disturbance caused by exposure of semiconductor surface to metal flux during conventional vacuum depositions, we have pioneered a novel cluster deposition technique. This technique involves cooling of the semiconductor surface, condensation of a noble gas buffer layer on top of it, and then metal deposition atop this structure. Adatoms agglomerate on and in the noble gas overlayer, and the clusters thus formed are subsequently brought into contact with the semiconductor by warming the system and desorbing the inert gas interlayer. Such depositions result in abrupt interfacial structures with unique electrical characteristics.

This chapter emphasizes the formation and morphology of metal-semiconductor interfaces. After an overview in interface morphology, we will consider the experimental probes used to study them with emphasis on photoelectron spectroscopy. We will then discuss data analysis techniques for core level spectrum. We will stress experimental work performed on Si and especially III-V semiconductors, particularly GaAs and InP. The final section will review recent results and will summarize future research topics.

## 2.0 INTERFACE MORPHOLOGIES

Figure 1 defines an interface and some phenomena that are observed during its evolution. Although the drawings are highly simplified compared with the situation in an actual device, they allow us to identify many issues of concern in interface research.[1] A perfect array of atoms represents the clean surface on which an overlayer will be created. When foreign atoms are deposited onto this substrate, they can be adsorbed. This is the case at the top drawing in the figure where isolated adatoms with different coordination are sketched. Surface science has long been interested in chemisorption and physisorption, and a great deal has been learned about surface bonding, surface reconstruction, and the formation of ordered and disordered monolayers. Such information is very important for modeling overlayer growth.

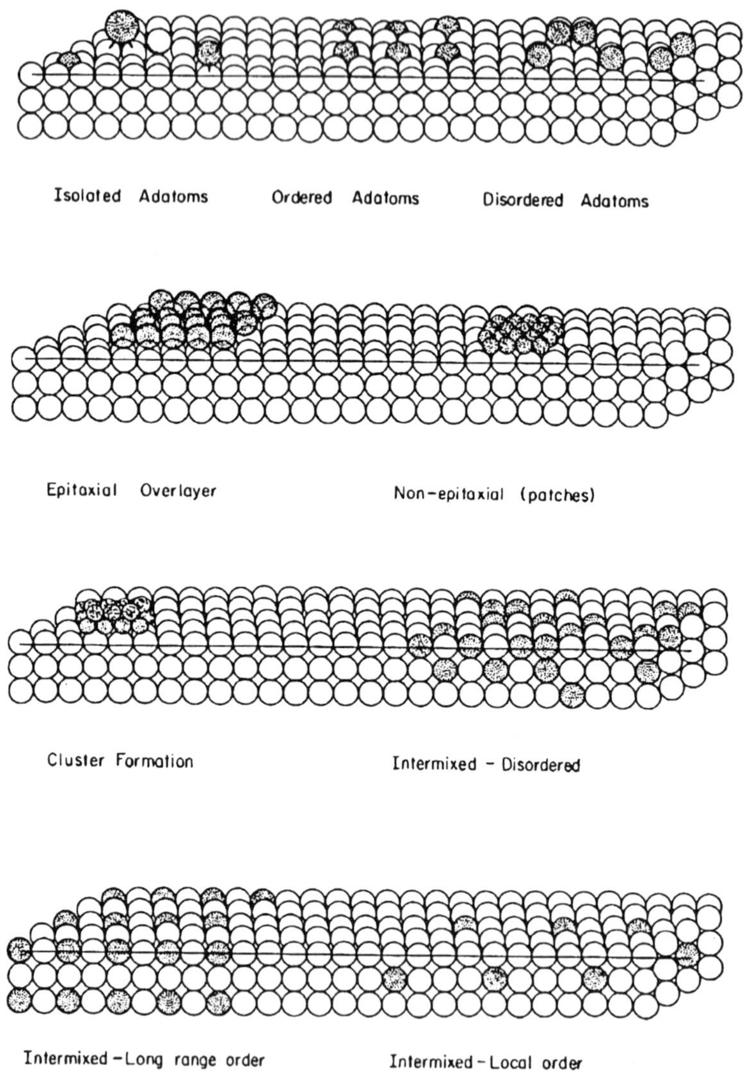


**Figure 1.** Evolution of an interface (from Ref. 1). This pictorial description of interface development begins with isolated adatoms on a surface and proceeds through an intermixed boundary layer. This chapter focuses on the situation in the bottom two panels, which depict cluster formation and intermixing for compounds and solutions.

With increasing coverage, adatom-adatom interactions are possible, and the growth of an array begins. The second drawing in Fig. 1 shows the formation of a monolayer of adatoms. Epitaxial growth of perfect overlayers in crystallographic registry with the substrate is most likely when the parameters of the overlayer closely match those of the substrate and when the tendency for intermixing is not strong. Nonepitaxial growth is more likely when there is a large lattice mismatch. Layer-by-layer growth represents only one of the common growth modes. Other modes include the formation of three-dimensional clusters over one or more monolayers or directly over the substrate. Hence, three basic growth modes have been discussed in the literature of thin films: the layer-by-layer or Frank-van der Merwe mode, the island or Volmer-Weber mode, and the layer-plus-island or Stranski-Krastanov mode.[2]

In the Volmer-Weber mode, small clusters nucleate on the substrate surface and grow into islands of the condensed phase. This happens when the atoms of the deposit are more strongly bound to each other than to the substrate. In the Frank-van der Merwe mode, atoms are more strongly bound to the substrate than to each other. Therefore, the first atoms condense to form a monolayer on the surface. If a nearly perfect match of the lattice constants leads to a low strain energy, the layer growth mode is obtained. The Stranski-Krastanov growth mode is an intermediate case. After forming the first monolayer or few monolayers, layer growth is unfavorable and islands form on the top of this intermediate stage. The possible reasons for this growth mode are the mismatch of lattice parameters and structural differences which result in a high free energy for subsequent layers.

During deposition, atoms arrive at a particular rate, R, and the population of single atoms builds up. These single atoms can then diffuse over the surface until they are lost by one of several processes.[3] When intermixing is not present these processes include re-evaporation, nucleation of clusters, capture by existing clusters, and capture at defect sites. Each of these processes can be phenomenologically described by characteristic times. If such processes are thermally activated, these times will be controlled by activation energies and frequency factors in the form $\tau = \upsilon \exp(-E/kT)$. For diffusion, as an example, $\tau$ represents the mean time spent by the adatom on a site before hopping to the next one.

Crystal growth is then a non-equilibrium kinetic event, and the final macroscopic state of the system depends on the above processes and their competition. Certain parts of the forming interface may be in local thermodynamic equilibrium while others will be kinetically limited. Therefore the state obtained is not necessarily the most stable. For example, at very low temperatures or high arrival rates, the atoms stay where they arrive and an irregular bed of atoms is built up. On the other hand, when R and T permit the rearrangement of atoms, the activation energies determine the regime of nucleation and growth.

The growth patterns presented above describe abrupt interfaces formed between the overlayer and the substrate. These growth modes are applicable when there is minimal atomic intermixing and when convergence to bulk properties is rapid on both sides of the sharp boundary. Our focus here, however, is on intermixing, depicted by the bottom two pictures in Fig. 1. For a single-phase bulk system in equilibrium, we expect intermixing to produce either a solid solution or a compound. When the interactions of an adatom with an atom of the substrate or with another adatom are chemically equivalent, a solid solution is expected where the driving force is entropic. In the other case, a compound forms with a well- defined chemical environment as shown in the bottom diagram. This description of the interface in terms of compounds assumes that the system is in equilibrium and that bulk thermodynamics applies to ultrathin layers. Interface research generally deals with metastable or nonequilibrium systems with interfacial or diffusion constraints. Disruptive interfaces can be complex and unstable and, in general, they are not completely understood. The final morphology of the interface depends on the atomistic processes present for abrupt interfaces and also on the chemical reactions between adatoms and substrate atoms. Both kind of processes which take place while the interface is formed can depend on preparation conditions.

Even though metal-semiconductor interfaces have been studied with many techniques, photoelectron spectroscopy has been one of the more important because of its surface sensitivity and its ability to examine chemical processes.[4][5] This technique is described before returning to the discussion of metal-semiconductor interface formation.

## 3.0 EXPERIMENTAL PROBES

The goal is to determine the properties and spatial extent of the species that form during intermixing. The challenge consists in characterizing a thin, inherently-multiphase region that may be subject to change with operating conditions. The experimental techniques used to examine evolving interfaces include those developed for surface studies as well as those that can probe deeper into the solid. Figure 2 depicts several techniques and the means by which they probe an interface, which is shown evolving as an ordered overlayer *(left)*, as an unreacted cluster *(center)*, and as an intermixed system *(right)*.

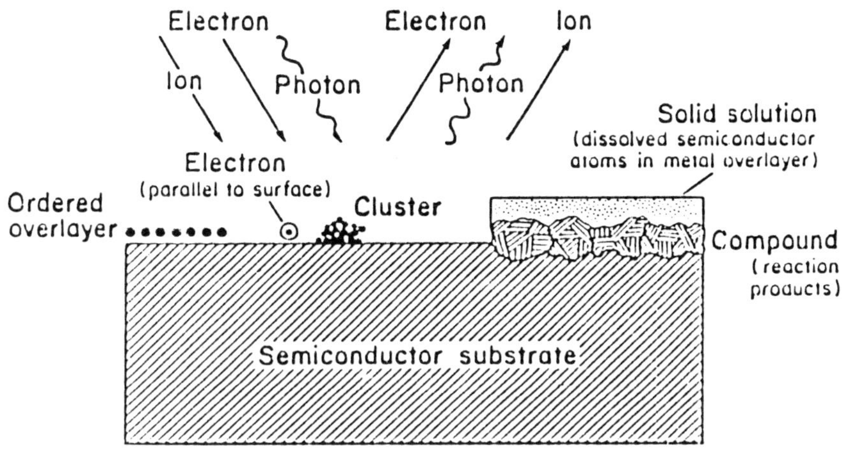


**Figure 2.** Techniques used to examine evolving interfaces (from Ref. 1). Indicated here are photoemission (photons in, electrons out), Auger spectroscopy (electrons in, electrons out), inverse photoemission spectroscopy (electrons in, photons out), low-energy electron diffraction (electrons in, diffracted electrons out), ion scattering (ions in, ions out), scanning electron microscopy (electrons in, electrons out), and transmission electron microscopy (electrons in parallel to surface, electrons out for lattice imaging). Scanning tunneling microscopy should also be included because of the ability to sense changes in the charge distribution on the atomic scale.

Photoemission spectroscopies are sensitive to the chemical state of atoms in the overlayer as well as to their distribution. Low-energy electron diffraction reveals whether there is ordering on the surface and gives structural information. Auger and angle-resolved Auger spectroscopies use electrons to identify the atomic species present and to determine their spatial distribution; these spectroscopies also reveal the structure of the overlayer through diffraction modulation. Inverse photoemission, in which electrons stimulate the emission of photons, gives detailed information about the empty electronic states. Surface EXAFS (extended x-ray absorption fine structure), in which photons stimulate the emission of electrons, gives information about bond lengths and surface geometries. The scattering of medium- and high-energy ions (energies in the keV and

MeV ranges) is useful in studies of thin overlayers and thick films, respectively. Ion milling, in which ions remove target atoms, is used extensively to obtain profiles of the atomic composition of thick reacted overlayers. Transmission, scanning electron, and scanning tunneling microscopies give images that show the morphology of the interface.

### 3.1 Photoelectron Spectroscopy

In a photoemission experiment, electromagnetic radiation impinges on the surface under investigation and, if the energy of the photon is sufficiently high, some electrons will be emitted.[6] Outside the solid, electrons can be analyzed according to their kinetic energy regardless of emission angle, i.e., angle-integrated photoemission, or one may collect electrons emitted into a small solid angle, i.e., angle-resolved photoemission.

The transport of a photoelectron toward the sample surface is governed by inelastic events involving a single electron or collective excitations. Photoelectrons in the solid can reach the surface without being scattered but many of them suffer inelastic scattering. Substantial loss of energy is involved in the creation of electron-hole pairs or by the generation of plasmons. If inelastically scattered electrons have sufficient energy, they reach the surface and can escape from the solid. These scattered electrons contribute a generally-structureless background to the energy distribution curves.[7]

One of the advantages of using photoemission as a tool to investigate solid surfaces is the fact that the inelastic scattering length of electrons in solids is short, less than 20 Å for electrons in the energy range from 10 eV to several keV. Figure 3 shows inelastic scattering lengths for excited electrons in solids as a function of electron energy above the Fermi level.[8] The scattering length exhibits a minimum for energies around 40 eV and, due to the Pauli principle, increases rapidly for lower electron energies. Therefore, since we are interested in studying the spatial distribution of atomic species during the formation of the interface, the ability to vary the photon energy is very important.

The fundamental ideas which describe the photoemission process were introduced by Einstein as early as 1905. Einstein's fundamental hypothesis was that photoelectrons escape after absorbing the energy of a photon $h\upsilon$. If the photon energy is sufficiently high, electrons can be emitted. In angle-integrated photoemission, the technique used in the present studies, electrons are analyzed according to their kinetic energy, regardless of emission angle. Information about chemical bonding, interface morphology, and Schottky barrier formation can be obtained from the energy distribution curves.[9][10]

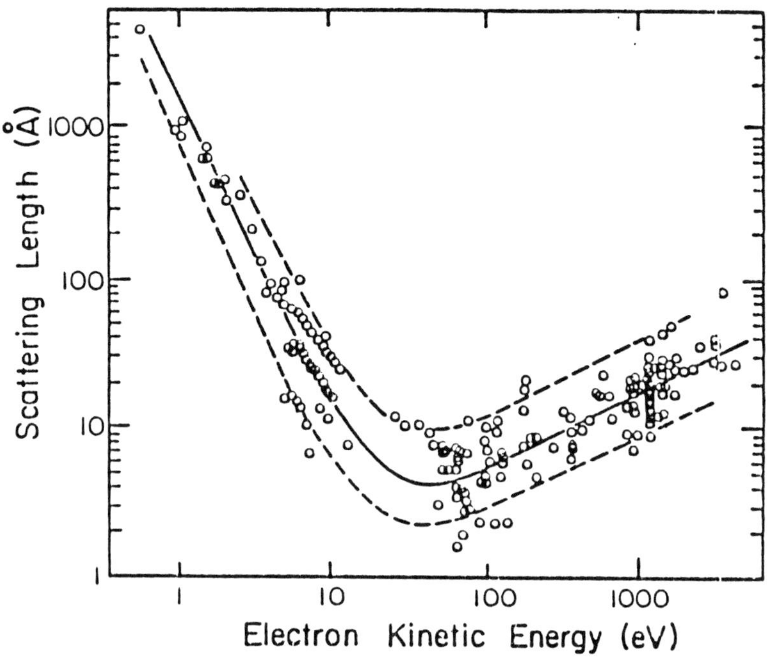


**Figure 3.** Scattering length vs. electron energy, compiled by M. P. Seah and W. A. Dench (Ref. 8). The scattering length is given in monolayers as a function of energy above the Fermi level.

The relevant energy conservation equation in the photoemission process is

Eq. (1) $$h\upsilon + E_i(N) = E_k{'}(l) + E_f(N-1)$$

where $E_i(N)$ is the total energy of the initial ground state, $E_f(N-1)$ is the total energy of the final state after the ejection of one electron from the level l, and $E_k{'}(l)$ is the kinetic energy of the photoelectron.[11] The electron binding energy relative to the sample vacuum is defined as

Eq. (2) $$E_{bv}(l) = E_f(N-1) - E_i(N)$$

Substituting Eq. (2) into Eq. (1) yields

Eq. (3) $h\upsilon = E_k'(l) + E_{bv}(l)$

Since sample and analyzer are in thermodynamic equilibrium, their Fermi levels are equalized. Binding energies measured with respect to the Fermi level can be determined if the analyzer work function, $\phi_a$, and the electron kinetic energy in the analyzer, $E_k(l)$, are known. The relationship shown in Eq. (3) becomes

Eq. (4) $h\upsilon = E_k(l) + E_b(l) + \phi_a$

where $E_b(l)$ is the electron binding energy with respect to the Fermi level. The region probed by photoemission is much smaller than the thickness of the depletion layer over which band bending occurs in a semiconductor. Thus, any change in band bending can be measured by a change in the kinetic energy of electrons excited from substrate atoms.

### 3.2 Photon Sources and Electron Energy Analyzers

X-rays are usually produced in the lab by bombarding a suitable target with electrons which create vacancies in the desired inner shell. An x-ray results from the filling of the holes with electrons by means of radiative transitions. The two most important x-ray sources used in photoelectron spectroscopy are the $K_\alpha$ lines of magnesium and aluminum at 1253.6 and 1486.6 eV, respectively. These yield the narrowest lines that can be obtained, 0.9 eV for Al and 0.8 eV for Mg. The intensity of x-ray production just above threshold is very low but increases rapidly with the energy of the bombarding electron. For an efficient production of x-rays with Al and Mg, it is necessary to use exciting electron energies about an order of magnitude higher than the threshold energy. The intensity of x-rays produced depends directly on the current of the electron beam striking the anode.

The most important sources of radiation for photoelectron spectroscopy of the outer shells are rare gas discharge lamps. The spectra created in these lamps contain lines which are intense and have a small natural width of only a few millivolts. The most widely used wavelengths in the UV region are the He I line (21.22 eV) and He II line (40.82 eV). Since their widths are only a few meV, monochromatization is not necessary but a low resolution monochromator is needed to select one of the lines present.

The photon sources discussed up to now have available only a few lines at fixed photon energies. Ideally one would like to have a continuously tunable source at all the desired energies. It is well known that when a charged body is accelerated it emits radiation. This effect is used in a synchrotron where high energy electrons are confined by magnetic fields to move in a nearly circular orbit.[12] The emitted radiation approximates a continuous spectrum whose intensity maximum is proportional to the radius of curvature and inversely proportional to the cube of the electron energy. When coupled with a suitable monochromator, synchrotron radiation provides a tunable light source covering all energies of interest for photoelectron spectroscopy.

The device that measures the energies of the emitted electrons is the heart of any spectrometer. During the development of photoemission and Auger spectroscopies, various types of analyzers were employed. Because magnetic fields are difficult to produce and handle, most analyzers are of the electrostatic type. The optics of LEED led to the first attempt to implement an electron energy analyzer of this kind, the retarding field analyzer (RFA). A negative voltage applied to one of the concentric grids sets up a potential barrier that the photoelectrons overcome before they reach the collector. A plot of the collector current vs. the barrier potential represents the integral energy spectrum. The energy distribution of the electrons is obtained through differentiation by superimposing a small modulating voltage on the retarding potential. The modulated collector current is synchronically detected and the signal at the fundamental frequency gives the energy distribution. In spite of its simplicity, the RFA is not suitable for many applications since all electrons with energies greater than the retarding potential are allowed to reach the collector. Shot noise is generated by many electrons and the signal to noise ratio is poor. Differential analyzers solved the two main problems of the RFA. They do not need the modulation technique and, since they are dispersive, only electrons of energies to be analyzed reach the collector. The electrons are deflected between two capacitor plates which are at different potentials. The relationship between the applied voltage and the energy of transmitted electrons is linear, hence the energy distribution is obtained by sweeping the analyzer voltage. To optimize the focusing properties of the analyzer, the condenser plates can be shaped cylindrically, as in the cylindrical mirror analyzer (CMA), or spherically, as in the concentric hemispherical analyzer (CHA). CMA and CHA are analyzers available commercially.

### 3.3 Core Levels

The core level structure is characteristic for a given element. Therefore, it can be utilized for an elemental analysis of the studied sample. Figure

4 shows an x-ray photoelectron spectrum of silver excited by Mg $K_\alpha$ radiation.[13] We can see that this radiation ($h\upsilon$ = 1253.6 eV) is energetic enough to probe the core levels up to the 3s shell and that core levels are doublets except for $l = 0$ states. These doublets are a consequence of spin-orbit coupling. Two states characterized by the quantum number j ($j = l + s$) are possible when $l \neq 0$, $|l + 1/2|$ and $|l - 1/2|$. The magnitude of the energy separation increases with the atomic number Z for a given subshell and decreases with the quantum number l for a given quantum number n. The relative areas of the doublet peaks are related to the ratio of the respective degeneracies, $(2j + 1)$. Therefore, these ratios are 1:2 for a p shell, 2:3 for a d shell, and 3:4 for an f shell.

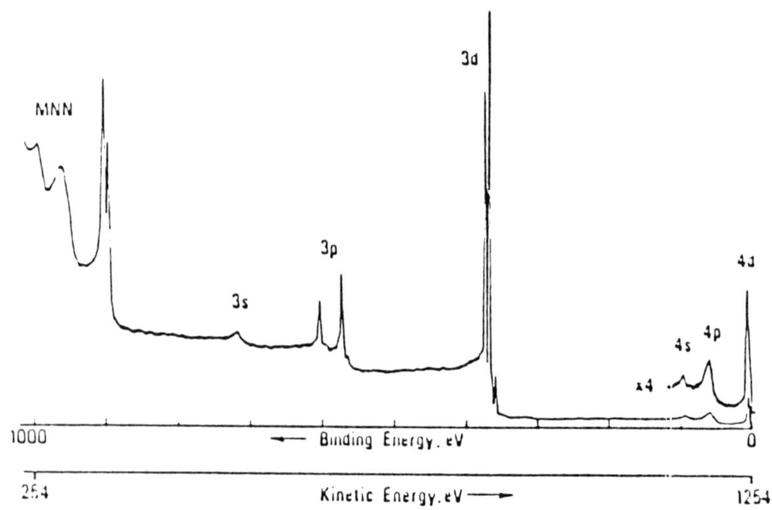


**Figure 4.** X-ray photoelectron spectrum of silver excited by Mg Kα relocation. (From Ref. 9.)

The peak width ΔE measured for a core level emission spectrum is a convolution of several contributions, namely

Eq. (5) $$\Delta E = [(\Delta E_n)^2 + (\Delta E_p)^2 + (\Delta E_a)^2]^{1/2}$$

where $\Delta E_n$ is the natural or inherent width of the core level, $\Delta E_p$ is the width of the photon source, and $\Delta E_a$ the analyzer resolution. This relation is a good approximation but it is only valid if all components have a Gaussian line shape. What can be called the *natural width* of a line is determined by the lifetime $\tau$ of the hole left behind in the photoemission process. This width is related to $\Delta E$ through the Heisenberg uncertainty relation.

The exact value of the binding energy for a given element depends on the chemical environment of the atoms of that element. The non-equivalence of atoms can arise from differences in oxidation state, molecular environment, or lattice site. These shifts may, at first, appear confusing since the inner electrons do not take part in bonding. However, a simple analysis illustrates the origin of the chemical shifts. A change in the chemical environment of a particular atom involves a spatial rearrangement of its valence charges so that a different effective potential is observed by the inner shell electrons. This picture gives a relationship connecting the binding energies for an atom in two different compounds,

Eq. (6) $$\Delta E_i(A,B) = K\,(q_i^A - q_i^B) + (V_i^A - V_i^B)$$

where the first term describes the difference in the electron-electron interaction between the core orbital and the valence charge and the second term describes the potential difference due to the surrounding atoms. If charges q are considered as screening ones, then the potential at the nucleus due to $q_i$ is equal to $q_i/r_v$, where $r_v$ is the average valence orbital radius. A change in the valence electron charge of $\Delta q_i$ changes the potential by $\Delta q_i/r_v$. Thus the binding energy of all core levels will change by that amount. The constant K takes into account the overlap between core and valence electrons. For cases where core-valence interaction is small, the observable core levels from a given atom experience a similar shift. The second term considers the influence of the atoms bonded to atom i in molecular solids, but in ionic solids the whole lattice contributes. In this case, it has the character of a Madelung potential and adopts the form

Eq. (7) $$V_i = \sum_{i \neq j} \frac{q_j}{R_{ij}}$$

which expresses the sum over potentials arising from ionic charges $q_j$ centered at positions $R_{ij}$ relative to the atom i.

Equation 6, the so-called *charge potential model,* has been very successful for small molecules[14] but it fails for many compounds. A major simplification of the charge potential model is that energies involved in the drastic rearrangement of the system upon ionization of the atom are not included. No account is taken of the polarizing effect of the core hole on the surrounding electrons, both intra-atomic and extra-atomic. This involves a flow of negative charge towards the hole created in the photoemission process in order to screen the sudden appearance of a positive charge. This screening lowers the energy of the hole state and then lowers the measured binding energy. This phenomenon is referred to as the relaxation energy $E_R$. The magnitude of $E_R$ is expected to differ for the same atom in different environments and a term $(E_R{}^A - E_R{}^B)$ has to be added in Eq. (6).

## 4.0 DATA ANALYSIS AND EXPERIMENTAL DETAILS

Not many years ago most experiments using photoemission involved valence band spectra or low resolution core level studies. Nowadays, the application of high resolution core level photoelectron spectroscopy has substantially increased our understanding of metal-semiconductor interfaces. As a consequence of adatom deposition, emission intensities and lineshapes of the core levels suffer changes due to the appearance of new species. Curve fitting is needed to find the relative concentration of the components related to the different species at the forming interface.

Intrinsic parameters of these core levels include the binding energy, spin orbit splitting, branching ratio, and the width of the core level. Extrinsic effects that affect the electron curve densities (ECDs) can be a consequence of the photoemission process itself, such as the background and mean free path, and those due to the equipment or measurement techniques. The latter include the instrument response involving the monochromator and the analyzer characteristics.

### 4.1 Curve Fitting

Curve fitting is the method most widely used to separate overlapping peaks in composite profiles. A spectrum can be synthesized by summing a series of functions representing individual peaks to produce a final function that closely represents the experimental spectrum. The model function must contain the physics of the photoemission process: the lifetime of the core hole, phonon broadening, photon source and analyzer resolutions, and, for metals, the final state screening asymmetry. We have then to make an initial guess on the number of peaks, position, intensity, and width.

The natural peak shape due to a finite core level hole lifetime is a Lorentzian function and the effect of phonon broadening is essentially Gaussian. The correct way of combining two broadening processes is through convolution. Thus the core lineshape can be calculated as

Eq. (8)
$$f(E) = \int_{-\infty}^{\infty} g(E')\,h(E-E')\,dE$$

where g(E) is a Gaussian and h(E) is a Lorentzian function. The result of this convolution is known as the Voigt function. In the case of metals, the effect of the conduction electrons was calculated by Doniach and Sunjic,[15]

Eq. (9)
$$f(E) = \frac{\Gamma(1-\alpha)\cos\left[\dfrac{\pi\alpha}{2+(1-\alpha)\arctan(E/\gamma)}\right]}{\left(E^2+\gamma^2\right)^{\frac{(1-\alpha)}{2}}}$$

where $\gamma$ is the half-width at half maximum of the Lorentzian and $\alpha$ the singularity index. Thus the convolution of the Doniach-Sunjic lineshape with a Gaussian provides an appropriate function to represent peaks in a core electron photoemission spectrum.

The standard method to evaluate the quality of the curve fit is the least squares criterion. Let the experimental data be $Y_i$ and the fit function $F(x, p_1, ..., p_m)$, where $p_1, ..., p_m$ are the unknown parameters. The best fit is then obtained when the parameters are chosen in order to minimize the function,

Eq. (10)
$$S(p_1, ..., p_m) = \sum_{i-1}^{n} \left(F_i - Y_i\right)^2$$

In general F is not a linear function of the m parameters and the optimum values of $p_1, ..., p_m$ can not be calculated readily from Eq. (10). However, if approximate values $p_{1'}, ..., p_{m'}$ of the parameters can be guessed, $F_i$ can be taken as a linear function of the adjustments $\Delta p_j = p_j - p_{j'}$ required to optimize the parameters by using Taylor's expansion. That gives

Eq. (11)
$$F_i = F_{i'} + \sum_{j=1}^{m} \frac{\partial F_{i'}}{\partial p_{j'}} \Delta p_j$$

The condition that S in Eq. (10) is a minimum requires $\delta S = 0$ and, together with Eq. (11), a set of m simultaneous equations is obtained, namely

Eq. (12)
$$\sum_{i=1}^{n} \left(y_i - F_{i'}\right) \frac{\partial F_{i'}}{\partial P_k}$$

where k = 1, ..., m. These equations can be solved by an iterative process and, since F is not a linear function of the parameters, the corrections calculated from Eq. (12) will not completely optimize the parameters and the process must be repeated. If the initial choice of parameters is close to optimum, only a few cycles of iteration are needed, but if the initial choice is poor, the process may not converge and the method fails since there may be local minima in the parameter space.[16][17] Valuable information about the quality of the fit can be obtained from the residuals, defined as the point by point difference spectrum. For an ideal fit, the residuals must contain only the random statistical fluctuations of the data. The presence of any systematic oscillation is evidence of the model function inadequacy.[18]

## 4.2 Core Level Spectra Interpretation

The photoelectron intensity $I_i(\theta)$ corresponding to the ith species that originate from substrate atoms with atomic density $C_i$ after the deposition of an overlayer coverage can be written as

Eq. (13)
$$I_i(\theta) = \sigma_i(E)\, C_i\, T(KE_i)\, I(E) \int \exp[(-\mu_i x) - (x/\lambda_i)]dx$$

where $\sigma_i$ is the photo-absorption cross-section, $T(KE_i)$ considers the transmission properties of the analyzer at a kinetic energy $KE_i$, $I(E)$ is the photon beam intensity, $\mu_i(E)$ is the photon absorption coefficient, and $\lambda_i(E)$ is the photoelectron escape depth.

The photon penetration depth is several orders of magnitude larger than the photoelectron mean free path. Thus, we can write $1/\lambda_i + \mu_i \approx 1/\lambda_i$. When the ith phase is built with a regular thickness over the substrate, Eq. (13) becomes

Eq. (14) $$I_i(\theta) = \sigma_i(E)\, C_i\, T(KE_i)\, \lambda_i\, I(E)\, [(1 - \exp(-\theta/\lambda_i)]$$

For the substrate in the case $\theta = 0$, we can write

Eq. (15) $$I_s(0) = \sigma_0(E)\, C_o\, T(KE_0)\, \lambda_0\, I(E)$$

The ratio of Eqs. (14) and (15) is given by

Eq. (16) $$\frac{I_i(\theta)}{I_s(0)} = \frac{\sigma_i(E)\, C_i\, \lambda_i(E)\, T(KE_i)\, [(1 - \exp(-\theta/\lambda_i)]}{\sigma_0(E)\, \lambda_0(E)\, T(KE_0)}$$

where $C_i$ is now the ratio of the atomic densities $C_i/C_0$.

The transmission function of the analyzer is a slowly varying function of the kinetic energy. Since we are working with a range of photoelectron energies that are only a few eV, we let $T(KE_i) = T(KE_0)$. Moreover, the ratio $\sigma_i/\sigma_0$ can be approximated as unity, and this should be good far from the threshold. With these approximations, Eq. (16) becomes

Eq. (17) $$\frac{I_i(\theta)}{I_s(0)} = C_i\left(\frac{\lambda_i}{\lambda_0}\right)\left[1 - \exp\left(-\frac{\theta}{\lambda}\right)\right]$$

This is a fundamental equation for interpreting the results of attenuation curves and for quantitative evaluation of atomic concentrations.[19]

The starting point for modeling the evolution of the semiconductor core levels during interface formation is the fitting of the clean semiconductor surface core level EDCs. Due to differences in bond length, coordination number, and nearest neighbor positions, atoms at the surface generally present core levels with binding energies different than atoms at the bulk. Because of the high surface sensitivity and resolution of synchrotron radiation photoemission, this phenomenon can be readily detected in many cases.[20]-[24]

Figure 5 shows the As 3d core level lineshape for the cleaved GaAs(110) surface held at 60 K. The photon energies are 90 eV and 58 eV and cubic polynomial backgrounds have been substracted from the spectra. The experimental data are shown as individual points, and it is obvious that one doublet is not enough to fit this EDC. The presented fitting is the output from a computer program using fifteen adjustable parameters and a nonlinear least squares minimization routine performed with an IBM RT computer to achieve the best fit to the original data.[25] Bulk component and surface components are shown with dashed lines and the sum as full line. Different ratios of the surface to bulk intensities are a consequence of the dependence of the electron mean free path on the photoelectron kinetic energy. The total resolution combining the effects of the monochromator and the electron analyzer is calculated to be 90 meV.

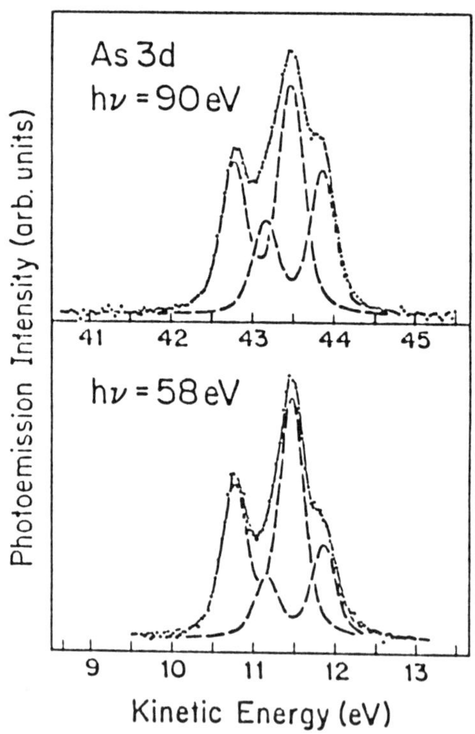


**Figure 5.** Nonlinear least squares computer fit of the As 3d core level for cleaved GaAs(110) taken with hv = 90 eV and 58 eV. The EDCs are background substracted and normalized to the same height. The fits are composed of a bulk component (lower kinetic energy) and a surface component (higher kinetic energy).

The adjustable parameters used as inputs to the fitting program include four background parameters, the spin-orbit splitting, the branching ratio, the Lorentzian width, and the intensity, energy position and Gaussian width for each component. The values of the obtained fitting parameters for the clean surface are the starting point to fit spectra during interface formation. The Lorentzian width, spin-orbit splitting, and branching ratio are then kept fixed for subsequent spectra. The variables used in our fits are the position and intensity of the spin-orbit doublets. The Gaussian broadening function contains contributions from instrumental resolution, inhomogeneous band bending, phonon broadening, and inequivalent sites in the interface region. Hence, Gaussian widths are allowed to vary for all components. As mentioned above, a Doniach-Sunjic lineshape is used to replaced the Lorentzian for reacted species in metallic environments. This method of analysis is essential to reveal the details of interface evolution and to achieve an accurate determination of the energy position for the different components and then for the Fermi level position in the gap. Results inferred by-eye for photoemission spectra cannot be expected to yield the precision needed for a detailed analysis. Indeed, much of the confusion found in the literature reflects this by-eye approach to analysis of complex spectra. (Of course, instances can be found where computer-aided lineshape decompositions produced non-physical results, but such errors can generally be minimized.)

## 4.3 Experimental Details

Most of the interfaces included in this study were investigated using synchrotron radiation photoemission spectroscopy, and the experiments were performed at the Wisconsin Synchrotron Radiation Center using the Aladdin storage ring that operated at 800 MeV. The beamlines used included the grasshopper monochromators Mark II and Mark V and the Minnesota-Argonne extended range grasshopper (ERG) monochromator.

To make quantitative measurements of core level intensities during the development of the interface, the photon flux needs to be accurately measured. This was accomplished by placing a tungsten mesh in the optical path and measuring the current due to photoelectrons leaving the mesh. Thus, the photon flux can be monitored continuously while collecting data, and a proper normalization takes into account photon flux variations during the experiments.

The measurement chamber is an ultrahigh vacuum stainless steel system. It has two 220 liter/sec ion pumps and a titanium sublimation pump to trap reactive background gases. The entire system is routinely baked to

150°C and has a routine operating pressure of ~5 x $10^{-11}$ torr. The photoemission section of the chamber contains the analyzer, electron gun for Auger analysis, sputter gun, leak valve, quadrupole mass spectrometer, sample manipulator, crystal thickness monitor, evaporator, sample bank, and cleaving tool.

Studies in the range $20 \leq T \leq 300$ K were made experimentally feasible by soldering the sample holders to a Cu tank that was rigidly attached to the cold head. A heater adjacent to the sample holder made it possible to raise the temperature to melt the Ga solder while continuing to operate the cold head. This had the advantage of rapid sample interchange, thermal stability, and minimal change in pressure.

A Physical Electronics model 255 double pass cylindrical mirror analyzer (CMA) was used to measure electron kinetic energies. The CMA can be operated in two modes, constant retarding ratio and constant pass energy. The latter is used in photoemission studies because the resolution is constant for any kinetic energy. A control unit was built to provide the voltage supplies needed to use the CMA for high resolution core level photoemission. This control unit can operate in constant pass energy mode for photoemission and constant retarding ratio for Auger spectroscopy. The pass energy is continuously variable from 1.5 to 45 eV. This control unit contains meters for the pass energy as well as the kinetic energy of the photoelectrons. The offset voltage from which to start a scan and the width of the scan can be independently selected from the CMA control unit or from an interfaced IBM AT computer. The data acquisition program runs in a multiscan mode allowing the operator to control the acquisition time. Data files record the number of counts and mesh current at each data point. They can be later transferred to an IBM RT for analysis.

The starting point for the studies presented here is the preparation of a clean semiconductor surface. Most of them were prepared by cleaving under UHV conditions at the start of each experiment to ensure a substrate free of contaminants. The quality of the surface is then assessed by visual inspection to verify that a smooth mirror-like surface was obtained. Next, a valence band is taken to check for its characteristic features. The final and most stringent test of cleave quality for GaAs(110) and InP(110) consists of checking the kinetic energy position and lineshape of the substrate core levels. The energy position of the core level is a direct measurement for the flat band condition, and the lineshape indicates the possible presence of inequivalent sites related to defects. Only those cleaves presenting the Fermi level within 60 meV of the conduction band minimum, CBM, or valence band maximum, VBM, at 300 K should be used.

Overlayers are usually deposited by evaporation from hot sources, and the material to be evaporated is held in tungsten or tantalum boats or baskets which are heated resistively. Extensive degassing of the evaporation sources after the chamber bakeout and water cooling of the evaporator makes evaporation possible at pressures of 0.7 - 4 x $10^{-10}$ torr. The overlayer thickness is measured by a quartz crystal oscillator monitor. Stable evaporation rates of up to 1 Å/min are established before the sample is exposed to the vapor flux.

## 5.0 ROOM TEMPERATURE METAL-SEMICONDUCTOR INTERFACE FORMATION

In the past few years much experimental and theoretical work has focused on the properties of evolving metal-semiconductor interfaces.[26][27] It is now generally accepted that *many* deposited metal atoms induce substrate disruption and release the semiconductor species into the overlayer. The resulting reaction products can be quite complicated and more than a single phase can be formed. First works on metal-semiconductor interfaces were done on metal-Si contacts stimulated by their central role in Si-device or integrated circuit technology and by the need to gain fundamental understanding from a simple elemental semiconductor. Of the compound semiconductors, GaAs and InP have received the most attention, and the general behavior of such interfaces at room temperature is reasonably well delineated. In these studies, surface-sensitive photoemission with synchrotron radiation has made it possible to follow interface formation at very low metal coverage and to discern the presence of substrate atoms in the overlayer at higher coverage.[28]-[44] It is not our intention to review the vast body of research in the field but to focus on general trends and some particularly exciting results.

### 5.1 Metal-Si Contacts

A considerable amount of experimental work indicates that metal Si interfaces are, in general, non-abrupt and provides evidence for chemical reactions and diffusion at metal-Si interfaces. In general, a sequence of phases can evolve which depend on the specific type and amount of metal and the temperature at which the interface is formed. On the other hand, migration of Si atoms into the growing metal overlayer can proceed without the formation of any alloy or compound phase.

With the continuous shrinkage of Si device dimensions, the phenomena occurring at interfaces can no longer be ignored. Due to its low electrical resistivity, high corrosion resistance and adhesion to $SiO_2$, Al was used as a contact metal for devices. However, Al reacts with Si at low temperatures thereby inducing an alloyed interface originated from intermixing or mass transport across the Al/Si contact. The phenomenon of alloy penetration seriously damages the device function. Experiments have also shown that Si atoms can be easily supplied from a Au/Si interface and that they migrate through a Au film. Hence, Al and Au are metals which do not form stable compounds (or silicides) with Si and that may be responsible for the alloy penetration which is characterized by the non-uniformity of the interface structure and composition.[45]

Ag/Si is an exceptional case of an unreactive interface. Direct observation by scanning electron microscopy (SEM) reveals that silver grows in a Stranski-Krastanov growth mode, with a very strong dependence of the island density on the deposition temperature.[46] The resulting islands are easily visualized by SEM above 200°C. The islands approximate truncated cones with a (111) top face, parallel to the (111) Si substrate, and with faceted sides, which have the tendency to be (111) faces. The density of islands which forms is a kinetic problem, though of course strongly linked to free energy of an atom in an island relative to in the adsorbed layer. This energy difference dominates the observed temperature dependence.

A very different behavior is observed for Si-transition metal contacts where reactions lead to the formation of silicides that have a much higher thermal stability with Si than does Al.[47][48] Most of these compounds are metals, so the reaction replaces the transition metal/Si interface with a more stable silicide/Si interface. They are, therefore, very attractive candidates for metal contacts with improved interface chemical behavior. With photoemission spectroscopy it is possible to see how silicide formation reaction begins and how the final metal/semiconductor interface atomic structure appears.

As an example of a very reactive interface we will describe the Ti/Si formation. Due to the relatively large core-level shifts it is possible to identify and follow the evolution of three reacted components for this interface.[49] The first Ti-induced species is associated with a bonding configuration of empirical formula $TiSi_4$ and forms only during the deposition of the first angstrom of Ti. The second component is associated with TiSi, and this phase forms in the coverage range $0.5 \leq \theta \leq 8$ Å. The final Si configuration reflects a phase of variable stoichiometry and binding energy, corresponding to a solution of Si in Ti. Titanium metal accumulates at coverages above ~14 Å. The profile and morphology of this heterogeneous interface is controlled by Ti and Si diffusion through the reaction products and their grain boundaries. This graded profile seems to be a common feature for most refractory-metal Si interfaces where differences are related to the energetics, character, and morphology of the reaction products and the kinetics of atomic motion through the evolving overlayer.

With reference to Fig. 6, the formation of the Ti/Si(111) interface can be described as follows.

1. Coverage range 0 - 0.5 Å. When Ti atoms are first deposited onto the semiconductor surface, they react to produce a bonding configuration that is rich in Si, namely a $TiSi_4$-like configuration. This species probably forms as patches as Ti atoms disrupt the Si substrate. Little is known about the diffusion of Ti on Si at 300 K and the atomistic behavior in this regime. Scanning tunneling microscopy studies of such interfaces would be of great value, as has been done for Ti/GaAs. With increasing coverage, this disruption spreads across the surface but appears to be complete by about 1 Å.
2. Coverage range 0.5 - 1 Å. This region corresponds to the onset of silicide formation. In particular, processes occur that reflect the completion of the $TiSi_4$ transition region and the development of the silicide. The morphology of the interfacial zone in this coverage regime is heterogeneous with islands of the second reacted phase growing over the first. At the same time, the first phase (or transition region) grows laterally until the surface is covered at 1 Å. It does not form thereafter, but phase two does.
3. Coverage range 1 - 6 Å. In this coverage range, only the TiSi phase forms. As sketched in Fig. 6, islands of TiSi that nucleated at lower coverage coalesce to form a polycrystalline layer. The result is that emission from Si atoms below the TiSi layer is attenuated exponentially.
4. Coverage range 6 - 14 Å. This is a region in which the TiSi reaction product is completed and the solid solution of Si in Ti forms. The solution phase grows because the TiSi layer acts as a barrier, preventing the in-diffusion of Ti atoms that would continue the disruption of the substrate and release Si. The Si 2p core level line shape changes with coverage in this solution phase, indicating that the concentration of Si decreases with distance from the interface.
5. Coverages exceeding ~14 Å. For Ti depositions exceeding this coverage, a Ti layer forms over the reacted regions with Si content that is below the level of detectability. This completes the room-temperature evolution of the Ti/Si interface. Thereafter, Ti metal grows.

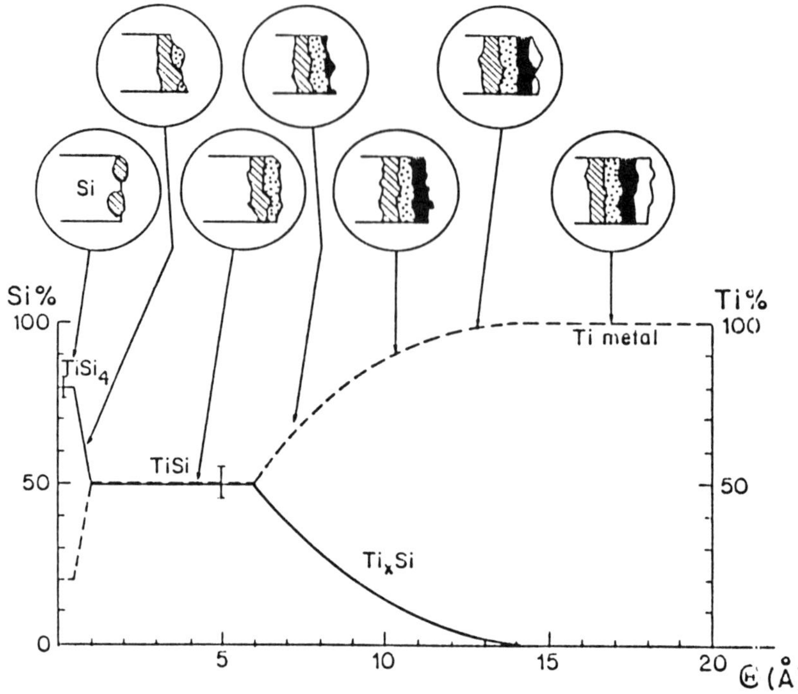


**Figure 6.** Schematic of Ti/Si(111) interface evolution. The left axis corresponds to the Si content in atomic percent and the right axis is the Ti content. The circled insets show the morphology at different stages of film growth. (From Ref. 6.)

These results indicate that a complex, heterogeneous interface is formed at room temperature due to the competition between reaction and diffusion processes at the interface. One must expect that an increase of the substrate temperature would allow more diffusion across the TiSi barrier so that the concentration profile of Fig. 6 would be stretched to higher values of θ. This diffusion process would reflect the activation energies for grain boundary and bulk diffusion.

As already mentioned, the Au/Si interface has attracted a great deal of attention because of the use of Au metallization in integrated circuit technology. Early work envisioned Au as a prototype metal overlayer because Au is a noble metal, but recent studies have shown that Au contacts are remarkably complex. There is consensus that Au deposition onto Si induces disruption of the Si surface, although the extent of the intermixing has not been clear. Studies that emphasize the surface region of a Au/Si junction show a region rich in Si, and this has been variously termed an alloy or a $Au_3Si$-like compound.[50]-[57]

As an example of interface engineering, it can be shown that Si substrate disruption and Au-Si intermixing can be controlled by the deposition of small amounts of Co onto Si. Co was chosen as an interlayer material because the epitaxial growth of $CoSi_2$ on Si has been well documented, even at room temperature,[58]-[62] and because $CoSi_2$ offers the opportunity of forming a diffusion barrier free of grain boundaries. By examining the details of Au/Si interface formation with interlayers of Co grown by depositing 0.5 to 12 Å, the coverage at which the Si surface is passivated against Au-induced disruption has been determined. This passivation makes it possible to form a Si-free Au overlayer on a stable $CoSi_2$ interlayer.

In order to characterize reaction and intermixing at Au/$CoSi_2$/Si interfaces, the Si 2p core level emission was examined as a function of Au and Co coverage using surface and bulk sensitive photoemission.[63] Figure 7 shows Si 2p core spectra acquired at $h\upsilon$ = 135 eV for representative Au coverages with Co interlayer thicknesses of zero (left panel), 3 Å (middle panel), and 4.5 Å (right panel). All EDCs have been background substracted and normalized to highlight line shape changes. Band bending effects have been removed to facilitate assessments of chemical changes. The bottom-most curve on the left for the cleaved Si(111)-2x1 surface shows the bulk and surface-shifted $2p_{1/2,3/2}$ components. The surface components are rapidly lost when metal adatoms are deposited.

Room-temperature Au-induced disruption of the Si substrate can be easily seen through the appearance of a shoulder on the high binding energy side in the Si 2p EDC, even for $\theta$ = 1 Å (left panel of Fig. 7). In the low coverage range ($\leq$ 5 Å), this feature grows in relative intensity, shifts from the substrate peak to greater binding energy, and broadens. This suggests that a heterogeneous, intermixed region forms with a variety of Au-Si bonding configurations. With increasing Au coverage, however, the Si 2p EDCs sharpen while continuing to shift, slowly, to higher binding energy. This indicates that the distribution of Si atoms in the near surface region becomes more uniform (dilute) and that Si atoms are dissolved in the Au matrix or are dispersed on the surface.

Additional insight into the distribution of Si in the overlayer can be gained by plotting the Si 2p core level intensity as a function of coverage. In the upper panel of Fig. 8 we show these attenuation curves for two photon energies where the energies were chosen to vary the probe depths, $3\lambda$, from ~10 Å for 135 eV to ~60 Å for 109 eV. The attenuation is rapid at low coverage but it is much more gradual after ~10 Å deposition. It is observed that the greater the surface sensitivity the greater the relative emission at high coverage. These results provide clear evidence that there is Si enrichment of the surface and near-surface region relative to regions deeper in the film.

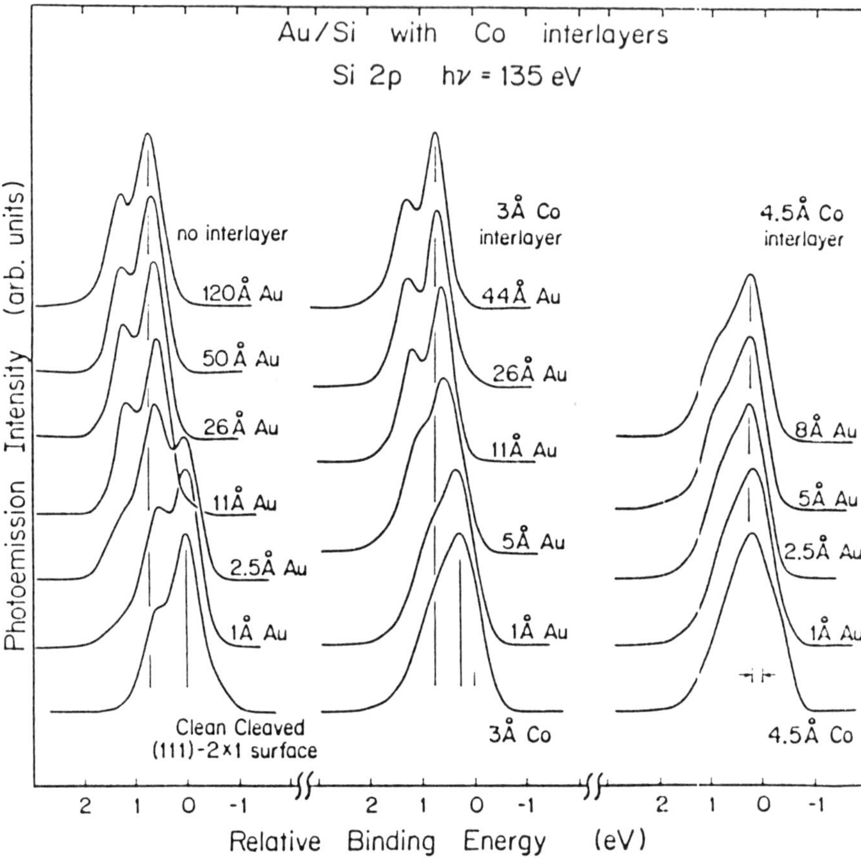


**Figure 7.** Si 2p core level energy distribution curves for representative Au coverages following the deposition of 0 Å *(left panel)*, 3 Å *(center panel)*, and 4.5 Å *(right panel)* of Co taken with a photon energy of 135 eV to maximum surface sensitive. Band bending has been removed and the spectra have been normalized to emphasize line changes. The persistence of Si emission to very high coverage is characteristic of the binary Au/Si system, but intermixing is very sensitive to the morphology of a thin $CoSi_2$ interlayer. (From Ref. 6.)

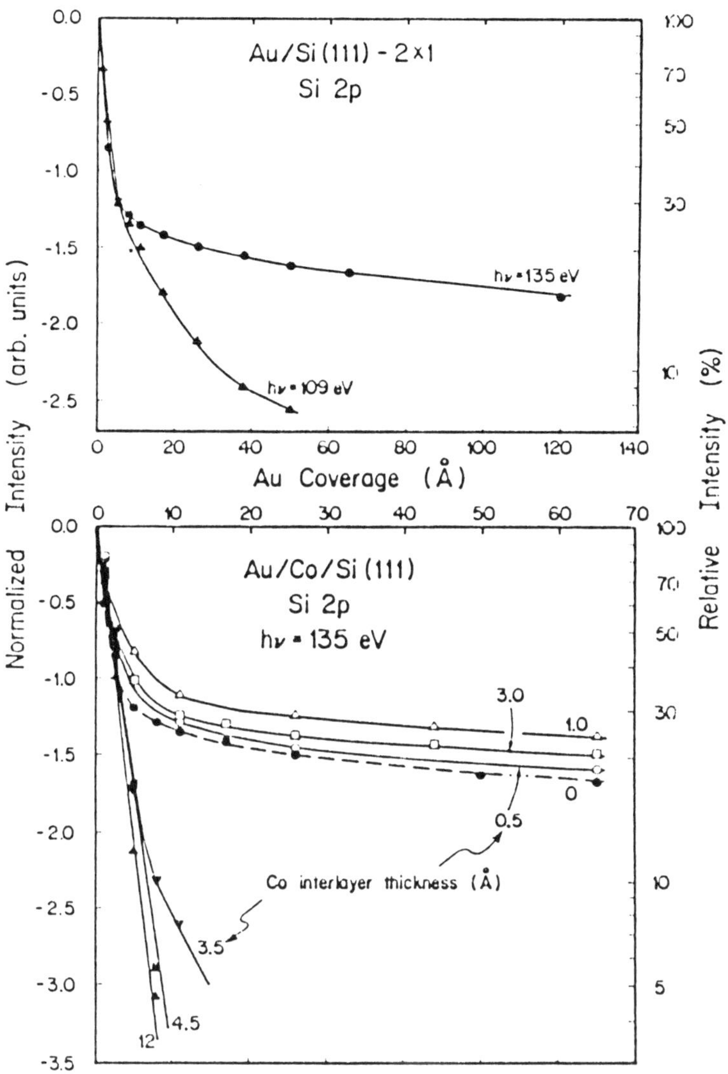


**Figure 8.** *Upper panel.* Si 2p core level attenuation curves taken at 135 eV and 109 eV for Au/Si showing that the greater the surface sensitivity the greater the Si signal. These results indicate that intermixing is not uniform and that Si segregates to the near surface region of the Au layer. (From Ref. 63.) *Bottom panel.* Surface sensitive Si 2p attenuation as a function of Au coverage for representative Co interlayers. Normalization has been done to the value of each starting Co/Si configuration, regardless of the amount of Co deposited. The exponential Si 2p emission decay for nominal Co coverages ≥3.5 Å indicates that the epitaxial $CoSi_2$ layer is an excellent barrier against Au-Si intermixing once it has sealed the Si surface.

Information about growth morphologies can also been obtained by following changes in the valence- or d-band spectra.[51] Our results show that, with increasing Au deposition, the binding energy positions of the Au 5d-derived states continuously shift toward the Fermi level and the valence-band splitting increases from 1.75 eV at 1 Å to a final value of 2.2 eV at ~11 Å. Again, it is concluded that Au adatoms are intermixed with released substrate Si on the surface during the initial stages of interface formation. Even at very high coverages of Au (~120 Å), the probed region contains Si. This does not imply that the Au film is homogeneously composed of Si.

The purpose of growing a thin reactive metal interlayer before Au is deposited onto Si is to produce a barrier that can alter the Si profile summarized by Figs. 7 and 8. Co reacts with Si to produce a thin $CoSi_2$ layer. Indeed, Boscherini et al.[58] and others proposed that $CoSi_2$ nucleates at low coverage and grows laterally to cover the surface. When the Co deposition exceeds 3.5 Å, these $CoSi_2$ clusters have coalesced to seal the surface against further in-diffusion of Co and a metallic Co layer grows on the silicide. To test the effectiveness of $CoSi_2$ as a barrier against Au-Si intermixing and to test the cluster model of $CoSi_2$ growth, Co in amounts of 0.5, 1, 3, 3.5, 4.5, 6, and 12 Å were deposited onto Si(111) prior to Au overlayer formation.

The results summarized in Figs. 7 and 8 show that Au-Si interaction and the extent of intermixing do depend critically on the amount of Co at the interface, i.e., on the coverage-dependent morphology of the Co/Si interface. In the center panel of Fig. 7 representative Si 2p EDCs for different Au coverages with 3 Å Co interlayer are presented. The energy zero corresponds to that of clean Si and the 0.2 eV energy shift to higher binding energy reflects $CoSi_2$ formation.[58] The deposition of even a small amount of Au changes this Si 2p line shape, as reflected by the extra shift to higher binding energy. At high Au coverages, the Si 2p spectra show the same characteristics of Au-Si intermixing and Si segregation as those for the binary Au/Si junction (left panel). Analogous results for lesser amounts of Co show the same trend. Indeed, the attenuation curves of Fig. 8 demonstrate that the amount of Si in the surface region of the evolving Au layer continues to be very high and has the same dependence on Au coverages as it does in the binary Au/Si system (dashed line). (The attenuation curves have been normalized to the emission of the starting surface, regardless of the Co thickness. An alternative form of presentation would have been to offset each of the curves by the attenuation induced by the Co deposition, thereby simultaneously showing the attenuation induced by Co and Au depositions. These would have shown that the amount of Si in the Au layer decreases with nominal Co interlayer thickness, as expected.)

Dramatic changes in the Si 2p core level line shape evolution occur when the Co deposition reaches 3.5 Å. From Fig. 8, we can see that Si 2p emission is rapidly attenuated for all higher Co coverages. These results provide strong experimental support for the coalescing cluster model. Indeed, by 3.5 Å of Co the Au adatoms deposited onto the silicide layer are blocked from reaction with the Si substrate. The absence of significant Au-Si intermixing is seen from the EDCs since the Si 2p line shape does not change. For higher Co coverages, the amount of excess Si is very small and the effect of Au is to attenuate the Si signal with a 1/e length of ~3 Å. The Au overlayer then grows in a layer-by-layer fashion on the silicide.

To further verify the effects of the interlayer in controlling Au-Si intermixing, the valence band spectra after 8 Å of Au deposition were measured as a function of Co deposition. For Co depositions below 3 Å, the results were similar to those for the binary system, indicating that the surface region contained Si and was not pure Au. A dramatic change was found when the Co deposition exceeded 3.5 Å as the Au-derived 5d valence-band features shifted toward $E_F$ and the splitting between the two structures increased to 2.5 eV. Comparison with the valence-band spectra for pure Au indicates that intermixing is no longer detectable and that the Au overlayer is elemental.

The results presented above are summarized pictorially in Fig. 9 where the Au/Si interface is shown with no Co, with <3.5 Å of Co, and with ≥3.5 Å of Co. Au-Si intermixing occurs for the binary system and segregation of Si in indicated. Intermixing and Si surface segregation continue to occur for Co depositions <3.5 Å, but to a lesser extent than for Au/Si. This can be understood in terms of incomplete coverage of the Si surface by $CoSi_2$ clusters and a diffusion barrier that is only partially effective. The deposition of ≥3.5 Å of Co seals the surface with a $CoSi_2$ layer, and substrate disruption by Au is sharply curtailed with no Si atoms in the growing Au overlayer. Clearly, the details of the chemical reactions and the morphologies of the reaction products at metal/semiconductor interfaces play a crucial role in affecting interdiffusion processes and determining the sharpness and stability of a junction.

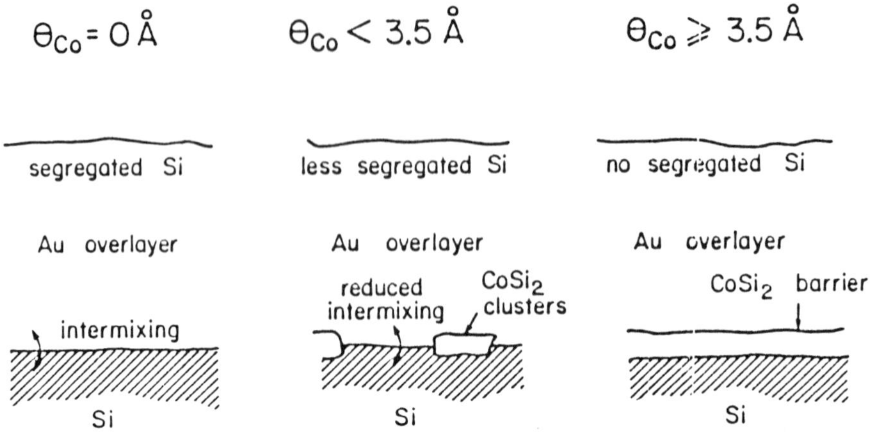


**Figure 9.** Schematic of interface formation for Au/$CoSi_2$/Si. Intermixing is strong and Si surface segregation occurs for the binary system. Once Co-Si reaction is complete and $CoSi_2$ seals the surface, the amount of intermixing is dramatically reduced and a Au layer free of Si can form. (From Ref. 63.)

## 5.2 Metals on III-V Semiconductors

The properties of interfaces formed by vapor deposition of metal adatoms at room temperature onto III-V compound-semiconductor surfaces have been studied extensively. A wide range of surface-science techniques have been used to study microscopic interactions and Schottky barrier formation for metals deposited on clean and oxide-covered III-V semiconductors. We stress here the early formation that corresponds to ultra-thin metal overlayers deposited on clean semiconductor interfaces. Contacts free of contaminants are referred to as intimate contacts and will be the subject of our studies.

It is now generally accepted that deposited metal atoms often induce substrate disruption and release one or both of the semiconductor species into the overlayer. For the reactive metal deposition, there are reactions that involve the metal and the anion. The reaction products can be complicated and more than a single phase can be formed. Likewise, the cations released by substrate disruption can mix in the growing metal overlayer to form solid solutions. These arguments are supported by thermodynamic considerations. Indeed, inspection of the thermodynamic heats of formation shows that metal-anion compound formation will usually be energetically favorable.[64] In contrast, metal-cation reactions for many systems are not favorable and are not generally observed. (A notable exception to this trend corresponds to Pd/III-V semiconductors systems which show preferential anion segregation.[65])

The typical boundary regions formed between transition-metal overlayers and III-V semiconductor substrates are characterized by disruption of the semiconductor accompanied by atomic interdiffusion and compound formation. The spatial extent of this boundary region and its heterogeneity depend on the details of the system. Although local chemical reactions can be rationalized from thermodynamic data associated with bulk compounds, there is no definitive way to predict the morphology, species profile, and stability of this region.

For a wide variety of metal III-V compound-semiconductor interfaces, it is found that cation and/or anion species segregate to the vacuum surface of the metal overlayer. The presence of these atoms accounts for the slow attenuation of the observed core emission intensities. Surface chemical reactions induce morphology changes and intermixing in the low-metal-coverage range at room temperature, but reaction becomes diffusion limited and can be neglected as the amount of deposited metal increases. If the intermixed region is covered in a layer-by-layer fashion by the metal overlayer, then the attenuation of the substrate emission should have a characteristic length scale that is the mean free path of the photoelectrons. If the experimental results show slower attenuation at high coverages, there is reason to consider metal-cluster formation, continued mixing of semiconductor atoms with the overlayer, or semiconductor-atom segregation to the free surface.

With the exception of a few metals, such as Al, Ag, and In deposited on III-V semiconductor substrates, there is no strong tendency for metal-cluster formation at high coverages. This is especially true for the reactive metals where the photoemission intensity of the substrate signal is rapidly attenuated, even if the total is not. From experimental results for GaAs and InP substrates, we can see that the attenuation curves for In have lower slopes than those of Ga, implying that In has a greater tendency for surface segregation than Ga. Comparison of metal-InP, metal-GaAs, and metal-InSb systems for the same metal overlayer show that Sb has the strongest surface segregation, As has medium surface segregation, and P has very weak or negligible surface segregation. Weaver and co-workers were able to model this tendency to surface segregate.[66] Although the model is simple, it works for many systems. This success implies that the cohesive energy and the strain energy are key factors that control surface segregation.

As an example of a well-characterized transition metal on a III-V semiconductor interface, we will present detailed studies of the room temperature formation of Co overlayers on cleaved GaAs(110) and on sputter-annealed GaAs(100)-c(8x2) substrates. We will show that Co induces limited disruption of the substrate, and thereby promotes Ga and As out-diffusion, but that a template suitable for epitaxy is left intact. The result is the formation of weakly ordered bcc Co. There is segregated As on the vacuum surface and the amount decreases with Co thickness as As atoms are dissolved in the increasingly-thick Co layer.

In Fig. 10 we show representative Ga and As 3d core-level energy distribution curves for Co/GaAs(110).[30] These synchrotron radiation photoemission spectra have been background substracted and normalized to highlight changes in line shape induced by the deposition of Co. Chosen photon energies give maximum surface sensitivity ($\lambda \sim 4$ Å). Band-bending-induced shifts of both the Ga and the As core levels have been removed by aligning the substrate core-level components. The bottommost EDCs of Fig. 10 are for the clean surface where emission from surface atoms (labeled s) is shifted 0.28 eV to higher binding energy for Ga and 0.38 eV to lower binding energy for As.

Room temperature intermixing of Ga and As with the deposited Co is evident from Fig. 10 through examination of the evolving 3d core-level emission. The chemically-shifted feature on the low binding energy side of the spectra is induced by Co deposition. For coverages as low as 2 Å this feature is dominant, indicating that disruption of the substrate has produced significant amounts of intermixed Ga in the overlayer. There is no evidence for Ga surface segregation since it is no longer possible to detect Ga within the probe depth (outermost ~12 Å) by the time that 20 Å of Co have been deposited onto GaAs(110).

Typical core level line shape decompositions are shown as dashed lines in Fig. 10. The Ga behavior is analogous to that observed for other transition-metal-GaAs systems. Comparison of these results with those for Fe/GaAs[67] shows that the substrate component attenuates more rapidly for Co overlayers, consistent with thermodynamic predictions of greater Co-Ga interaction compared to Fe-Ga (heat of formation for FeGa is -18 kJ/mol compared to -31 kJ/mol for CoGa). Examination of the Ga 3d core-level behavior for many other metal-GaAs systems shows that the total Ga core-level binding energy shifts vary directly with the Pauling electronegativity difference between the overlayer and Ga.[68]

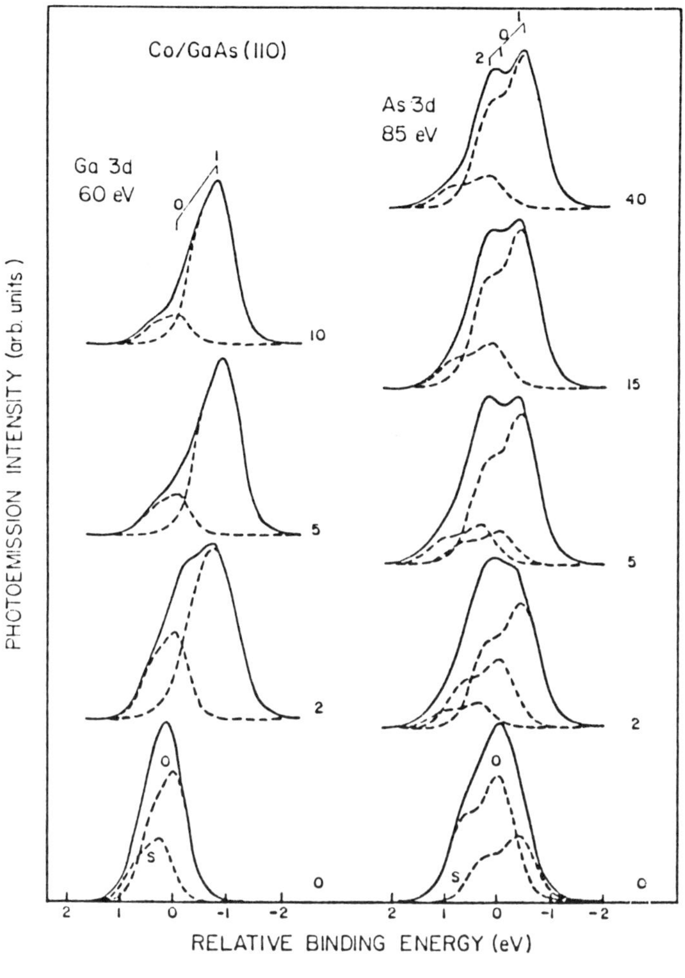


**Figure 10.** Representative Ga and As 3d core-level energy distribution curves showing lineshape evolution with Co deposition on GaAs(110). These photon energies result in very high surface sensitivity (~4 Å photoelectron mean free path). The bottom pair of curves are for the clean surface showing both substrate (0) and surface shifted (s) emission. For Ga, a chemically shifted feature, labeled *1*, appears at lower binding energy and shifts away from the substrate line with increasing coverage (total shift -0.9 eV). For As, the Co-induced components, labeled *1* and *2*, have distinct energy shifts of -0.45 and 0.25 eV relative to the substrate. (From Ref. 30.)

Figure 10 shows representative As 3d core-level EDCs and their decompositions. Lineshape analysis indicates that adequate fits could not be obtained with only two components. Inspection of the raw data at high coverage shows that this might have been expected because of the anomalous As 3d branching ratio. When a third component is introduced, labeled *2* in Fig. 10, the agreement is excellent and internally consistent. The component at lower binding energy can be associated with As atoms on the Co surface while the component shifted to higher binding energy corresponds to As in solution within the Co matrix.

Figure 11 presents the attenuation curves for Ga and As 3d emission intensities as a function of cobalt thickness, defined as $\ln[I(\theta)/I(0)]$, where $\theta$ is the coverage. The total intensities are shown as solid lines, and the dashed curves for the different components were obtained from the core-level decompositions. The total rate of attenuation is much faster for Ga than for As and the emission persists to high coverage. Component-specific profiles show that the rates of attenuation of substrate Ga and As are equal with 1/e values of ~1 Å at the earliest stages of formation because substrate disruption has the effect of increasing the distance from the undisturbed substrate and the surface. By ~3 Å, the decay is slower (1/e value of ~4 Å), consistent with photoelectrons propagating through a layer-by-layer growing overlayer. Figure 11 shows that for coverages exceeding 5 Å, the emission from the intermixed Ga atoms also diminishes exponentially with the same 1/e length as the substrate. These results indicate that Ga is unable to diffuse away from the interface and is covered up by further metal deposition.

Photoemission results reveal that the As behavior is very different from that of Ga because emission from the intermixed and surface segregated components persists with little attenuation as the Co overlayer grows. Since there is no evidence of extended disruption of the substrate after a coverage of ~5 Å, these As atoms are those disrupted during the early stages of deposition that are only partially trapped at the interface. These atoms form weak bonds with Co and can be displaced by Co atoms in favor of the formation of epitaxial bcc Co. They could also be incorporated into the Co matrix, although the solubility of As in Co is low (but not zero). Most are therefore displaced to the surface where they are partially coordinated with Co. The concentration profile near the surface is not a step function because the amount of As decays with distance from the surface.

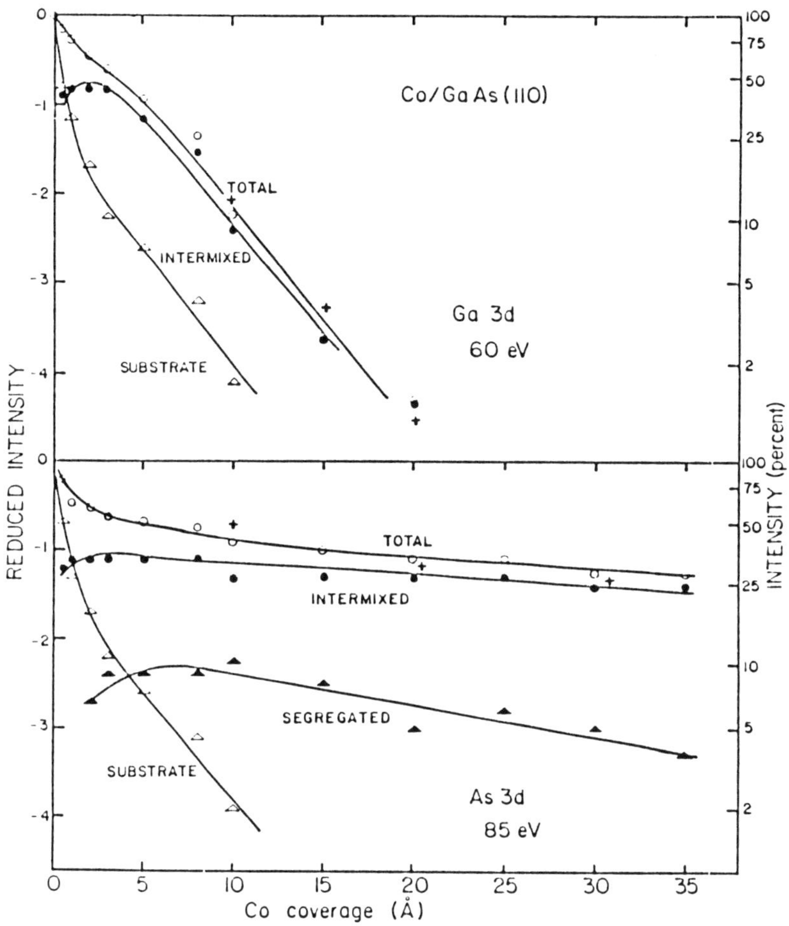


**Figure 11.** Normalized attenuation curves using the results of Fig. 10 to show the rate of change of the As and Ga content of the probed region as a function of Co coverage. For Ga, the substrate emission diminishes rapidly due to disruption at low coverage and covering up at higher coverage *(open triangles).* The amount of Ga dissolved in the Co matrix likewise diminishes exponentially after ~5 Å coverage, indicating very limited out-diffusion. For As, the substrate intensity is the same as that for Ga. The *solid circles* correspond to As on the free surface and the *solid triangles* correspond to As coordinated with Co in the intermixed region close to the free surface. (From Ref. 30.)

Angle-dependent x-ray photoelectron spectroscopy measurements can further quantify the spatial distribution of Ga and As atoms normal to the interface. By changing the polar angle of emission from grazing to normal, it is possible to vary the region that is sampled ~3λsinθ. Figure 12 presents Ga and As 3d XPS spectra at two different polar angles for a 30 Å Co coverage under conditions for which λ = 17 Å. The dashed curves show decompositions of the total emission based on the line shape of the clean surface. For these decompositions, the overall resolution is about 0.9 eV and it is not possible to distinguish the subtle chemical shifts found with synchrotron radiation. Instead, each spectrum has been fit to a pair of doublets corresponding to the substrate and intermixed species. These results, and those taken at other coverages (15 - 150 Å) and other angles (10 - 90°), form the basis for assessing the distribution of Ga and As atoms in the Co overlayer.

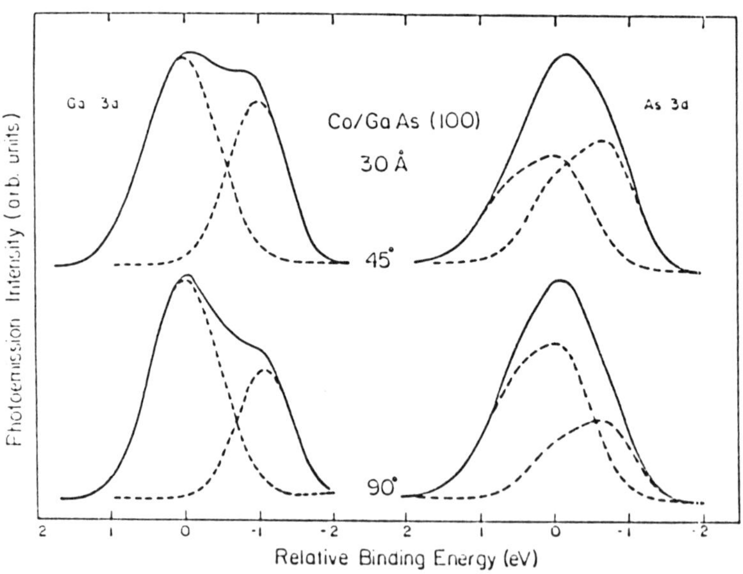


**Figure 12.** Polar-angle-dependent x-ray photoemission spectra for the Ga and As 3d core levels after the deposition of 30 Å of Co onto GaAs(110). The changes in line shape reflect variations in the probe depth, given by λ sinθ, for angles of 45° and 90°. For these spectra, the mean free path is far greater than that of Fig. 10 (17 Å vs. 4 Å). (From Ref. 30.)

Figure 13 presents reduced total Ga and As 3d emission intensities that are analogous to the surface sensitive results of Fig. 11. These results show the difference of attenuation rate due to the sampling depth. For Ga the intensity detected at any angle decreases rapidly with coverage. For As the rate decreases only slightly with increasing coverage, indicating the persistence of segregated As. (The drawn *solid lines* are predicted attenuation curves as discussed below.)

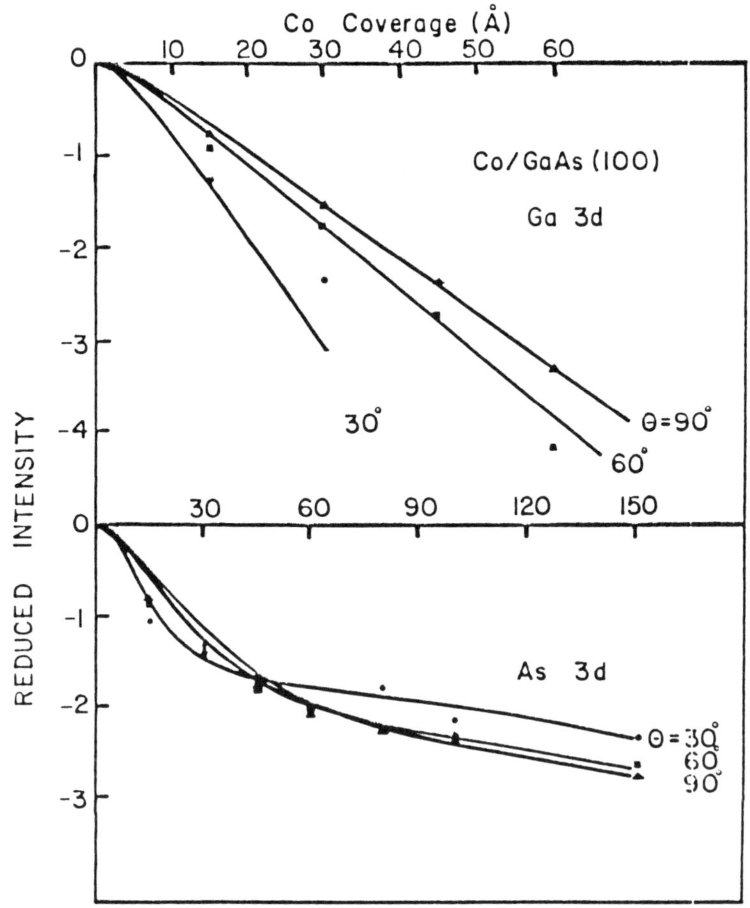


**Figure 13.** Ga and As 3d emission intensities obtained from the polar-angle dependent XPS spectra showing the variation in atom concentration sensed as a function of probe depth. The angle of emission of each family of points is indicated. The *solid lines* were calculated from the distribution function of Fig. 14. For As, the results show surface segregation; for Ga, they show rapid decay with distance from the surface. (From Ref. 30.)

Figure 14 shows the normalized Ga 3d and As 3d emission relative to the total detected intensity (the sum of the As 3d, the Ga 3d, and the Co 3p emission) as a function of polar angle for several coverages. Each intensity is an integration of the appropriate core-level photoelectron peak after background substraction. Normalization of the emission intensities has been done by determining the Co 3p emission for a thick Co film and the Ga and As 3d emission for the GaAs(100) surface under identical measurement conditions (spot size, x-ray intensity, pass energy of the analyzer, detection geometry). This takes into account the photo-ionization cross sections and the analyzer response function; it eliminates the reliance on calculated cross sections or results from other experimental systems. As shown in Fig. 14, the relative amount of Ga in the near-surface region is much less than that closer to the interface, as evidenced by the decrease in the ratio at shallow angles. These results confirm that Ga atoms are concentrated close to the interface and that the diffusion length of Ga into the overlayer is short.

In Fig. 14 we also show the polar-angle-resolved total intensity of As 3d emission relative to the total Ga, As, and Co intensities at high coverages where the Ga content is negligible and there is no contribution from the substrate. A strong As 3d signal is easily detected at coverages as high as 150 Å. The percentage of As increases as $\theta$ decreases, indicative of segregation of As in the surface and near-surface region, in contrast to the Ga case.

A simple continuum model was developed to fit the attenuation curves of Fig. 13 and the relative intensities of Fig. 14. Density distribution functions of the elements contained in the overlayer, $\rho(z)$, can be extracted by considering the overlayer as a uniform slab that attenuates the photoelectron intensity from atoms contained therein. The density distributions that lead to the most satisfactory fitting of the Ga and As profiles of Fig. 13 and Fig. 14 are shown in bimodal Fig. 15. In summary, Ga atoms are concentrated near the surface, and As atoms exhibit a decay profile away from the buried interface and the free surface. Only a small quantity of As remains dissolved in the Co matrix. These results are similar to those for Fe/GaAs but different from most other transition-metal-GaAs systems. While the behavior observed for Ga is very similar to that observed for other GaAs interfaces, there is evidence of reaction-induced disruption and the formation of As-based compounds or interface phases for the others. In these cases, the energetics of reaction favor the formation of As-based phases at the interface—not simply the dissolving of released As atoms in the thickening overlayer matrix.

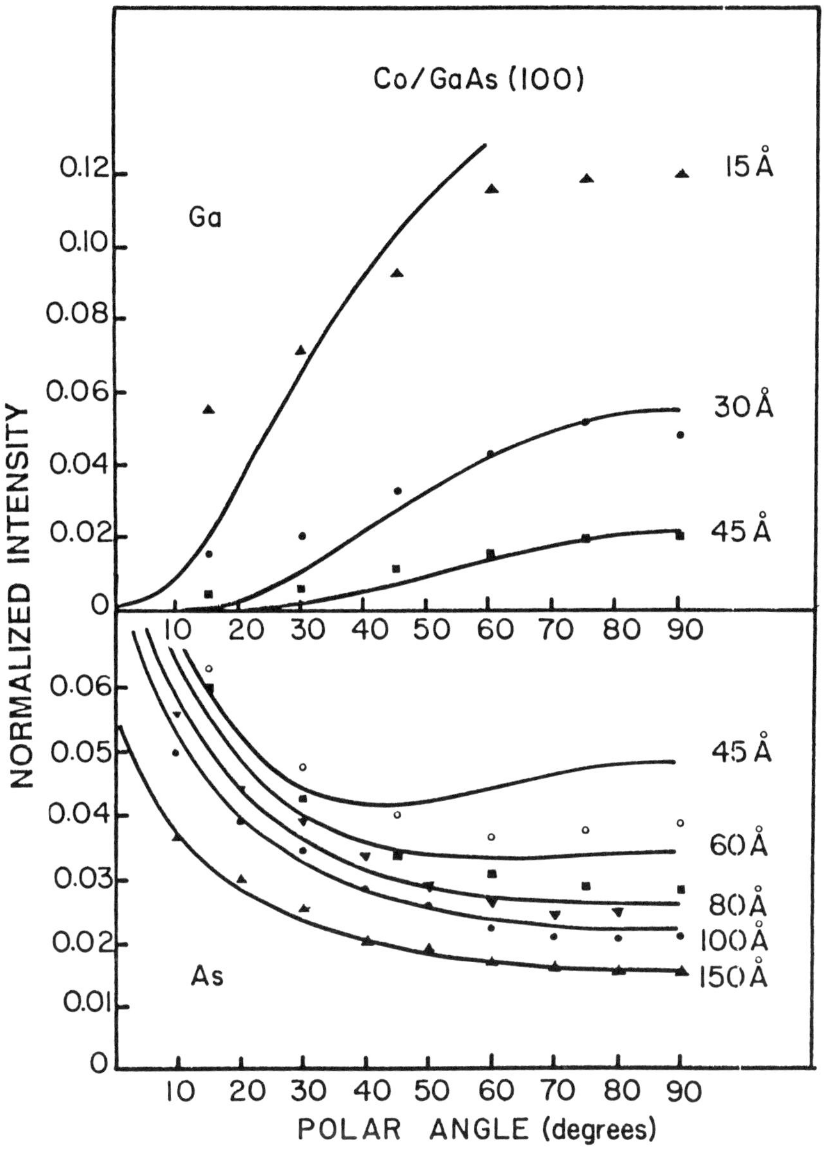


**Figure 14.** Normalized Ga 3d intensity, $I_{Ga}/(I_{Ga}+I_{As}+I_{Co})$ as a function of detection angle θ for Co coverages of 15, 30, and 45 Å. Corresponding results for As are shown at coverages of 80, 100, and 150 Å. The *solid lines* are the fits based on the model of Ref. 30.

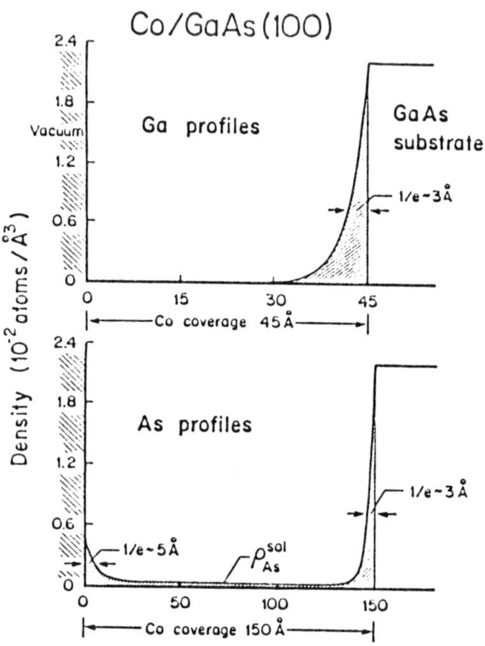


**Figure 15.** Distribution functions for Ga and As at coverages of 45 and 150 Å. For Ga, the assumed form is given by an exponential decay into the Co overlayer from the GaAs substrate (1/e length corresponding to $b_{Ga}$ = 3 Å). For As, the results indicated the need for surface segregation. The concentration on the free surface decreased with coverage. At any coverage, its decay into the Co layer from the free surface could be characterized by another exponential function. Far from either boundary, the As solubility is 0.26 at.%. (From Ref. 30.)

By combining high resolution x-ray photoemission with $Ar^+$ ion bombardment, Weaver and co-workers[69] demonstrated that the concentration profiles were determined in large measure by the detailed chemical reactions between the deposited metal adatoms and the released anions. When strong metal-anion reaction occurs, there is expulsion of cations from the region where compounds form so that a region forms that is deficient in cations near the interface. In these experiments the sample was alternately sputtered with 3.5-keV $Ar^+$ ions and measured with XPS for the metal and semiconductor atoms. The sputter profiles were obtained by determining the integrated core emission of the various elements as a function of sputter time.

Inspection of the results presented in Fig. 16 shows that the cation intensities exceed those of the anions after prolonged sputtering as the substrate is reached. This preferential anion sputtering and cation enrichment is a well-known and unavoidable phenomenon for III-V semiconductors.[70][71] Moreover, extended sputtering can increase surface roughness and alter the erosion rates. These effects make quantitative analysis impossible, and the intensity profiles should not be seen as the actual concentrations profiles. However, they provide qualitative information about the species distribution normal to the surfaces and they are particularly useful for interfaces with reactive metals where systematic behaviors are sought.

The concentrations curves of Fig. 16 present a very interesting common feature, namely, that the anion distributions in the overlayer are smooth and almost structureless while those of the cations exhibit a distinct valley-like structure or a plateau. This valley reflects cation depletion near the buried interface and a piling-up in front of the deficient region. Although the valley-like structure is common for reactive metal overlayers, it is not always as pronounced as for Ti/GaAs or Cr/InP. Indeed, the sputter profile for 100 Å Co/InP shows it to be reduced to a plateau. This similarity in cation depletion, but the variability in the amount of cation accumulation, suggests that these distributions reflect a common expulsion mechanism with differences related to the amount of surface segregation and interface solubilities.

In order to determine whether these anion and cation profiles were metastable with respect to temperature, a 100 Å Cr/GaAs interface was formed at room temperature and the profiles were compared to those obtained after annealing at 360°C for 75 min.[69] It was found that annealing increases the Ga and As content of the surface and the near-surface region, and the Ga and As profiles at the buried interface become sharper. It can be concluded that substantial amounts of dissociated Ga and As atoms weakly bond to Cr exist in the intermixed region near the buried interface as a consequence of room-temperature substrate disruption. Heating increases the diffusion of these impurity-like atoms and promotes their segregation to the surface. The result is that the Cr layer becomes more perfect and the free energy of the system is lowered.

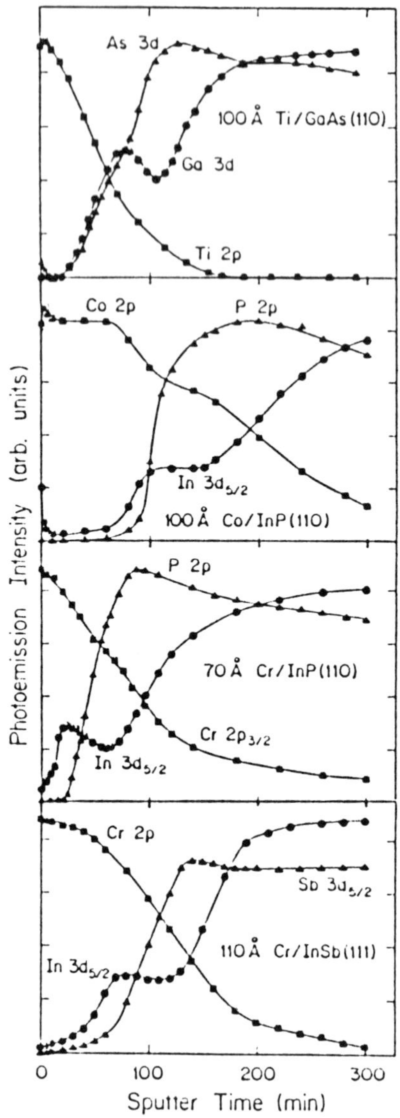


**Figure 16.** Sputter depth profiles for 100 Å Ti/GaAs(110), 100 Å Co/InP(110), 70 Å Cr/InP(110), and 110 Å Cr/InSb(111). The vertical axes correspond to the integrated core intensities measured with XPS at an emission angle of 60° with respect to the surface normal. The As and Ga intensities are corrected by their cross section ratios as measured for the clean surface. The valley-like structures in the cation concentration curves are a result of cation expulsion from the region where metal-anion compound formation occurs. (From Ref. 69.)

An abrupt change in the sputter-depth profile could be observed for the highly reactive Ti/GaAs interface after annealing at relatively low temperature.[72] A 100 Å Ti/GaAs system was formed at room temperature and subsequently annealed at 205, 285, and 365°C for 2.5 h. These annealing assignments were done on freshly prepared samples without a previous heating history. The sputter profiles showed that there are no qualitative changes in the interface morphology when the system is heated to 205 and 285°C for 2.5 h. Detailed comparison of these curves indicates, however, that heating promotes Ti-As reaction, substrate retreat, and the broadening of the reacted Ti-As region. When the 100 Å Ti/GaAs interface is annealed at 365°C, the depth profile changed dramatically. At this temperature, the As content of the surface region was still very low. Ti can diffuse readily to the GaAs substrate where Ti-As reaction is enhanced and, hence, a substantial thickening of the Ti-As region is observed. This wide Ti-As layer appears to expel the already existing Ga atoms, thereby reducing the Ga concentration to only a few percent. This supports the early conclusion that it is the growth of extended amounts of Ti-As which is responsible for the Ga expulsion and the resulting Ga valley structure in the room temperature sputter profiles.

## 6.0 LOW TEMPERATURE METAL-SEMICONDUCTOR INTERFACE FORMATION

Recently, motivated by Schottky barrier formation studies, interest has centered on the effects of cooling the semiconductor during metal atom deposition.[73]-[79] These low temperature studies were performed with the original purpose of simplifying the development of the interface because the fundamental mechanism(s) involved in Schottky barrier formation remain controversial.[73]-[85] Results indicate that clustering and atomic interdiffusion are strongly inhibited at low temperature but the amount of substrate disruption is not significantly affected.[86]-[89] Indeed, synchrotron radiation photoemission studies show that strong surface layer disruption/reaction is not reduced even at 60 K but that kinetic processes involving substrate atom out-diffusion and surface segregation are greatly reduced.

### 6.1 Ti/GaAs(110)

Previous studies of Ti/GaAs(110) interface development at 300 K showed that reaction is present from very low coverages.[35][90][91] Strong chemical reactions were clearly evident from the Ga and As 3d core level evolution, and the core level intensity variations as a function of metal coverage indicated intermixing over a scale of several monolayers. As the amount of Ti increased and the film thickened, there was also expulsion of the substrate atoms with As out-diffusion and dynamic changes in atomic concentrations.[67][90] In the following, we compare these results with those observed for low temperature in order to correlate the dependence of the fundamental properties of the interface with the temperature.

The left panel of Fig. 17 shows the Ga 3d core level emission for representative Ti coverages for deposition at 60 K on GaAs(110). The spectrum at the bottom corresponds to the clean cleaved surface, with decompositions that show the contributions from bulk and surface atoms, components *1* and *2* respectively. Disruptive interaction between the Ti adatoms and the substrate is observed from the lowest coverage and is evident as a rather broad Ti-induced feature. The Ga 3d spectrum corresponding to 0.2 Å shows clearly the development of a new feature, component *3,* at 0.4 eV lower binding energy than the bulk peak. Simultaneously, emission from the surface-shifted component was rapidly diminished and was not needed to fit the core level lineshape by 1 Å Ti deposition.

Component *3* shows increased broadening with deposition and it is featureless by 2 Å because of a wide range of inequivalent configurations for Ga atoms dissociated from the substrate. Component *3* sharpens at higher coverages (e.g., 16 Å of Fig. 17), indicating that the environment of the released Ga atoms becomes better defined within the probed region. After ~12 Å, the Ga 3d emission is shifted 1.67 eV, in agreement with the final shift observed at 300 K.[90] The increasing shift of component *3* to lower binding energy makes it possible to determine with confidence the intensity and binding energy of the attenuated substrate doublet. Analysis of the As 3d spectra is more complex than for Ga but, significantly, comparison with the 300 K results reveals the absence of a doublet corresponding to segregated As atoms at 60 K, which indicates that As out-diffusion is kinetically limited.

Figure 18 shows the total and component-specific Ga 3d attenuation curves, defined as $\ln[I(\theta)/I(0)]$ where $I(\theta)$ is the intensity at deposition $\theta$ and $I(0)$ is the total intensity for the clean surface. The total attenuation curve for interface development at 60 K has a slope with 1/e decay length of ~3.8 Å. However, the very rapid decay of the substrate component indicates disruption of the semiconductor surface as liberated substrate atoms form an intermixed region with the Ti atoms. For Ga, component *3* corresponds to Ga atoms in this region. Indeed, the distinguishability of the substrate and

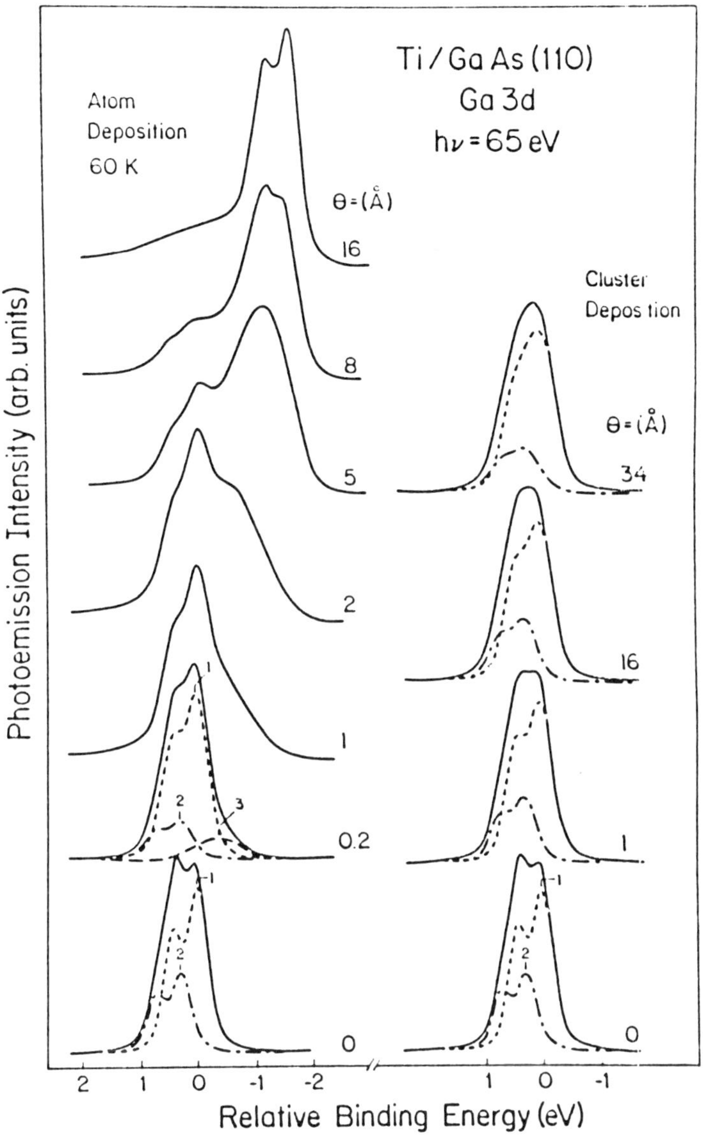


**Figure 17.** Representative Ga 3d core-level EDCs and lineshape decompositions for Ti-atom deposition at 60 K and for Ti-cluster deposition (discussed in Sec. 7). Component *3* results from emission of intermixed Ga atoms produced by atom deposition. This component is broad for intermediate coverages because a variety of inequivalent Ga-bonding configurations are present at the interface, but it sharpens at higher coverage as Ga bonding becomes more homogeneous. (From Ref. 104.)

intermixed components makes it possible to easily follow the growth of component *3,* as given in Fig. 18. For coverages above ~5 Å, the rates of attenuation of the substrate and component *3* are quite similar, indicating uniform coverage of substrate and disrupted regions. This is quite different from the behavior at 300 K where Ga atoms intermix with the thickening Ti film over a more extended scale. Low temperature thereby provides kinetic trapping near the buried interface of atoms disrupted from the substrate.

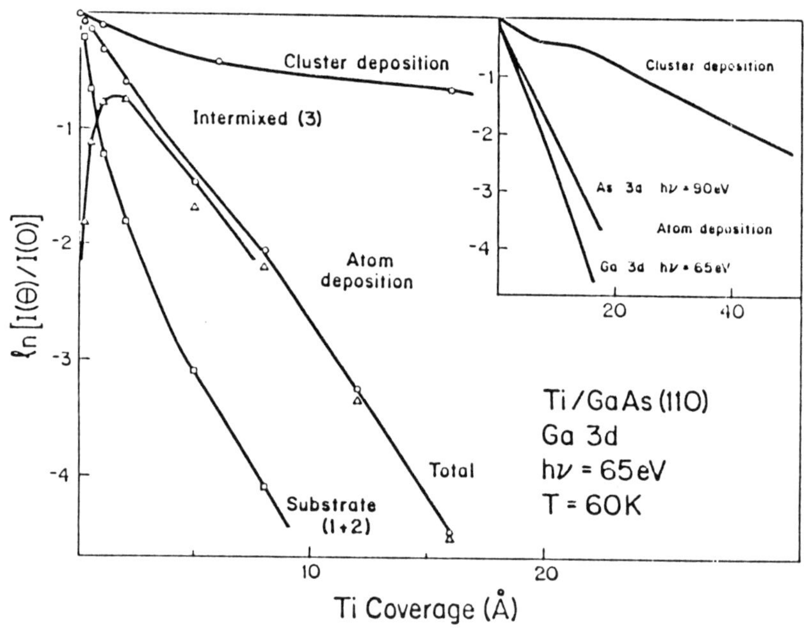


**Figure 18.** Component-specific attenuation curves after Ti deposition at 60 K showing substrate attenuation and the growth and attenuation of the emission from intermixed Ga. The retention of Ga near the buried interface is associated with kinetic trapping and contrasts to more extended intermixing at 300 K. The total Ga 3d core level intensity for cluster deposition shows poor coverage of the surface. The inset compares total attenuation results for Ga and As for atom and cluster deposition. (From Ref. 89.)

From the attenuation curves, we can estimate the amount of disruption. For both 300 K and 60 K atom deposition, the substrate disruption or retreat is equivalent to ~3 ML. There is no evidence for any meaningful difference that might reflect thermal quenching of chemical reactions that lead to the early stages of substrate disruption. The same final binding energy for component *3* at 60 and 300 K and its well-defined lineshape suggest that Ga atoms released from the substrate are present in an equivalent dilute form in the intermixed region.

The inset in Fig. 18 shows total attenuation curves for Ga and As for atom deposition at 60 K. While both substrate species are rapidly attenuated, the As intensity is higher than that of Ga at all coverages, even though the photoelectron mean free paths are the same. This shows that the As atoms have a greater tendency to out-diffuse when not bound in Ti-As bonds. This tendency for As expulsion is more clearly manifest at 300 K and is reflected by the presence of As on the surface of relatively thick films.[90] Such atom distributions are metastable, and it is likely that the amount of As in the overlayer is greater than would be predicted from an equilibrium phase diagram, even at 300 K. Low temperature formation creates an even less stable buried interface region because atom movement is kinetically limited. (The stability against annealing of interfaces formed at 300 K was discussed in the previous section.)[69]

### 6.2 Al/GaAs(110)

Of the many metal-GaAs interfaces examined, Al/GaAs has probably received the most attention, in part because Al is a relatively simple metal from a theoretical standpoint (no d electrons),[92] because Al-Ga exchange reaction occurs at the surface, and because of the importance of $Al_xGa_{1-x}As$/GaAs heterojunctions. Figure 19 shows Ga 3d EDCs for representative depositions at 300 K *(left)* and 60 K *(right)*. Lineshape decomposition shows the familiar bulk and surface-shifted components for the clean surface (labeled *1* and *2*, respectively). The clean surface EDC measured at 60 K exhibit reduced (Gaussian) broadening compared to that at 300 K, as expected.

As is well known, Al atom mobility on GaAs(110) is relatively high at 300 K so that widely separated clusters form and substantial amounts of the surface remain exposed.[93] Hence, the Ga 3d EDCs of Fig. 19 show the persistence of the surface-shifted component until ~16 Å. Where Al-GaAs reaction occurs, this surface component disappears and a third component appears at lower binding energy. The large width of this new structure indicates that Ga atoms released from the substrate exist in a variety of chemical environments in the developing overlayer.

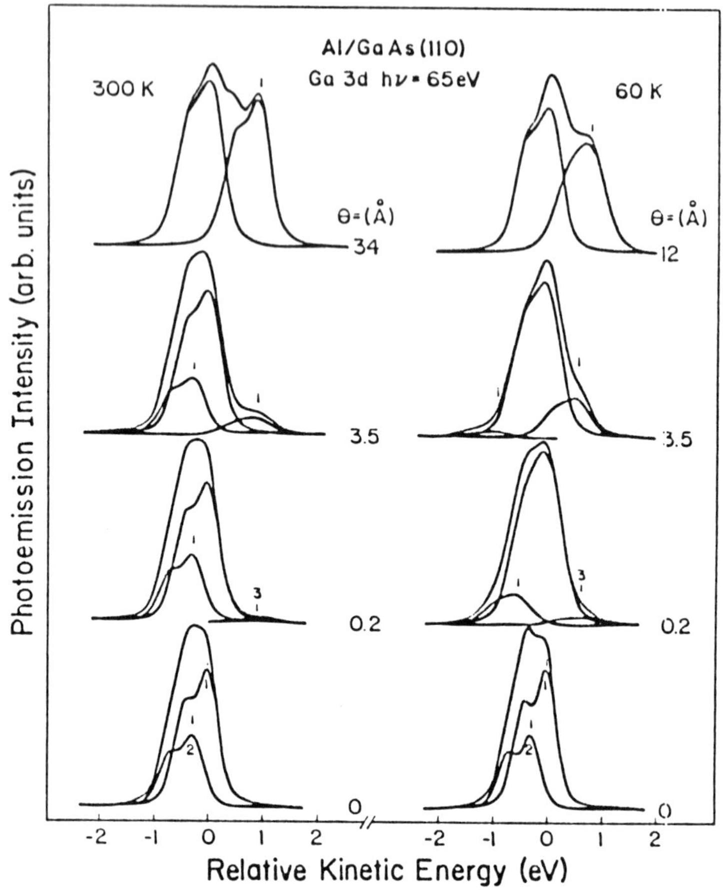


**Figure 19.** Ga 3d core-level EDCs for representative Al depositions at 60 K and 300 K showing the formation of new Ga features *(labeled 3)*. The energy position of component *3* shifts gradually to lower binding energy at 60 K but is invariant at 300 K. This reflects differences in Ga coordination in the evolving overlayer. Although different morphologies are observed at 300 and 60 K, there is approximately the same amount of substrate reaction. (From Ref. 97.)

The overlayer formed at 60 K is much more uniform than at 300 K because Al surface mobility is orders of magnitude smaller. As a result, the rate of attenuation of the surface component is faster at low temperature (compare Ga EDCs at 3.5 Å). Detailed analysis of the 60 K results shows that the surface-shifted component moves to higher binding energy as it decays. This finding was suggested to reflect surface unrelaxation and surface bonding of Al to Ga.[94] The observed increase in surface-to-bulk peak spacing supports the calculations of Priester et al.[95] that predicted Ga surface core-level shifts to be 0.58 eV for the unrelaxed surface and 0.38 eV for the relaxed surface.

Analysis of the relative intensities for Ga component *3* (Fig. 19) suggests that the amount of released Ga is greater for 60 K deposition than for 300 K at low coverages (compare EDCs at 3.5 Å) because differences in overlayer morphology exist. Deposition onto cold substrates produces a more uniform film than for 300 K deposition. Consequently, a greater proportion of the surface is perturbed by adatoms. When the entire surface has been covered, the total amount of substrate disruption is expected to be approximately the same. It can be noted that the position of component *3* relative to the bulk feature is not the same at 60 and 300 K because released Ga atoms are coordinated with Al differently in the two cases. In particular, Ga atoms released by reaction at low temperature are kinetically trapped in the thin Al layer near the interface. In contrast, deposition of Al at 300 K results in Ga atoms residing in, and on, the growing Al clusters. The different bonding configurations inferred from the Ga 3d spectra are supported by analysis of the Al 2p core levels which are narrower and more bulk-like at 300 K than at 60 K for equivalent depositions.

Figure 20 provides quantitative information about the attenuation of each of the Ga 3d spectral features. Comparison of the total Ga 3d emission shows much faster reduction at 60 K than at 300 K because cluster formation at 300 K does not effectively cover the surface. For 300 K deposition, emission from the surface component is still evident at 10 - 15 Å at the ~5% level, indicating that the clusters have covered most of the surface, but that some areas remain exposed. In contrast, the surface-shifted component is quenched by ~3.5 Å at 60 K with a 1/e length of ~1.4 Å. The signal from the disrupted feature increases until 3.5 Å with deposition at 60 K and is then attenuated relatively fast (1/e length of 4.5 Å). In contrast, the corresponding reaction-induced feature at 300 K exhibits a plateau at high coverage. These results suggest that Ga atoms released from the substrate are kinetically trapped near the interface at 60 K but segregate to the surface at 300 K. While differences in morphology for 300 and 60 K deposition make qualitative comparisons difficult, the maximum emission from the released component is approximately the same for both temperatures, suggesting that the total amount of substrate disruption is similar.

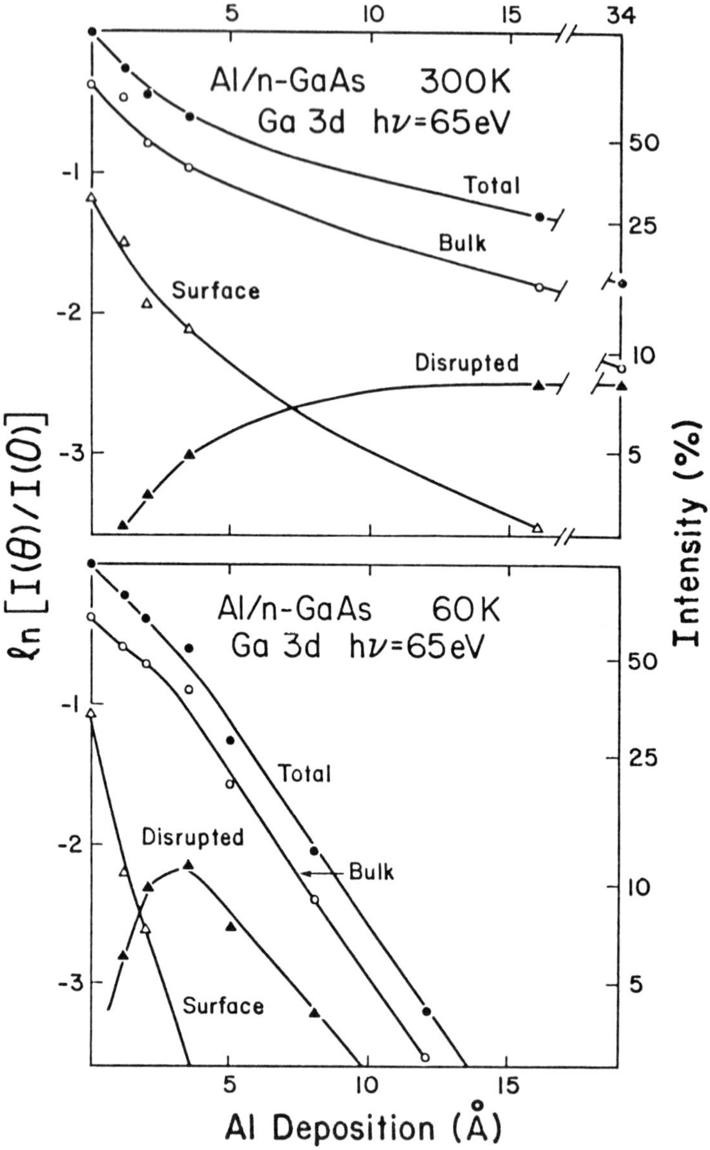


**Figure 20.** Attenuation curves for the Ga 3d core-level intensities as a function of Al deposition at 300 K and 60 K. Reduced adatom surface mobility at low temperature inhibits three-dimensional cluster formation, thereby leading to more uniform coverage and faster substrate attenuation. Ga atoms released by Al-substrate interaction are kinetically trapped near the interface at 60 K but they can segregate at 300 K. (From Ref. 97.)

Several research groups have assumed that low temperature Al/GaAs interface formation is accompanied by the inhibition of reaction and defect formation compared to 300 K deposition.[73]-[79] The results of Fig. 19 indicate that disruption is not quenched and that the amount of reaction is not affected significantly when depositions are made at 60 K.[96] We conclude that thermal effects do not eliminate chemical processes and the release of substrate atoms. At the same time, however, temperature does modify the redistribution of released Ga and As atoms. Because diffusion is a thermally activated process, kinetic trapping of the disrupted substrate atoms near the interface is expected to occur at low temperature.

### 6.3 Metal/InP(110)

This section extends previous discussion of GaAs to InP for two quite different interfaces, namely nonreactive Ag/InP(110) and highly disruptive Ti/InP(110). Figure 21 summarizes the In 4d core level evolution for Ag/InP interfaces formed at 300 K and 60 K by atom deposition. Except for phonon broadening, the clean surface spectra show no change between 300 K *(left)* and 60 K *(right)*. They are decomposed into the substrate bulk emission *(1)* and the surface-shifted component *(2)*. The latter corresponds to semiconductor atoms at the relaxed free surface. For Ag deposition at 300 K, the surface-shifted component remains for the full range of investigated coverages ($0.01 \leq \theta \leq 15$ ML) at the same relative binding energy ($0.30 \pm 0.03$ eV) as measured at the clean surface. The surface-to-bulk emission intensity ratio (S/B) is initially reduced by Ag deposition, but it remains unchanged at coverages beyond ~3 ML. These results indicate that the Ag overlayer formed at 300 K clusters spontaneously and that there is incomplete substrate covering. Numerous studies of Ag overlayers on other semiconductors have indeed revealed a strong tendency for clustering at 300 K.[97] At the same time, Ag atom deposition gives rise to a new component *3* due to substrate disruption which is not observed for other systems and will be discussed below.

Clear evidence for overlayer clustering is provided by the attenuation plots of Fig. 22. In the upper panel we show changes in the component-specific and total emission intensities for In 4d as a function of Ag coverage at 300 K. The results reveal nearly identical attenuation rates for the substrate and surface-shifted components. (Bulk and surface-shifted components attenuate with 1/e lengths of ~13 Å for $\theta \geq 4$ ML.) This slow decay indicates formation of a clustered overlayer since a uniform overlayer would reduce substrate emission with a 1/e decay length equal to the photoelectron mean free path of ~3.5 Å. The third component in the attenuation plots (discussed below) corresponds to the In atoms released from the substrate.

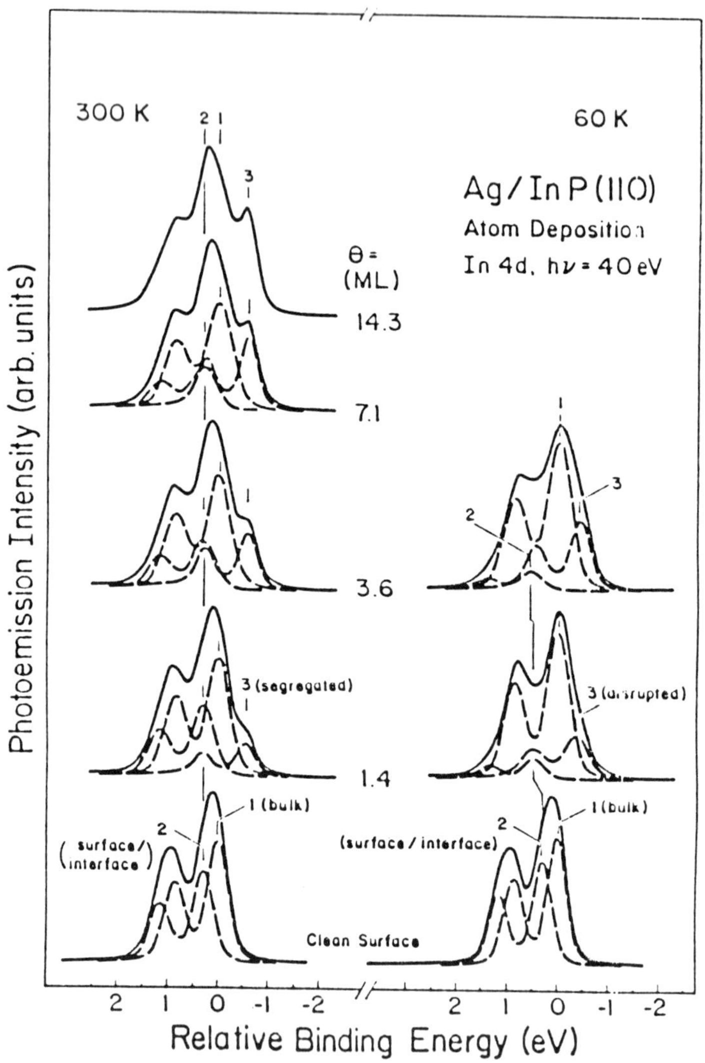


**Figure 21.** Representative In 4d core level EDCs and lineshape decompositions for 300 K and 60 K atom deposition of Ag on InP(110). Components *1* and *2* correspond to emission from bulk and surface atoms, respectively. Component *3* corresponds to In atoms released from the substrate and indicates equal amounts of substrate disruption at 300 K and 60 K. (From Ref. 105.)

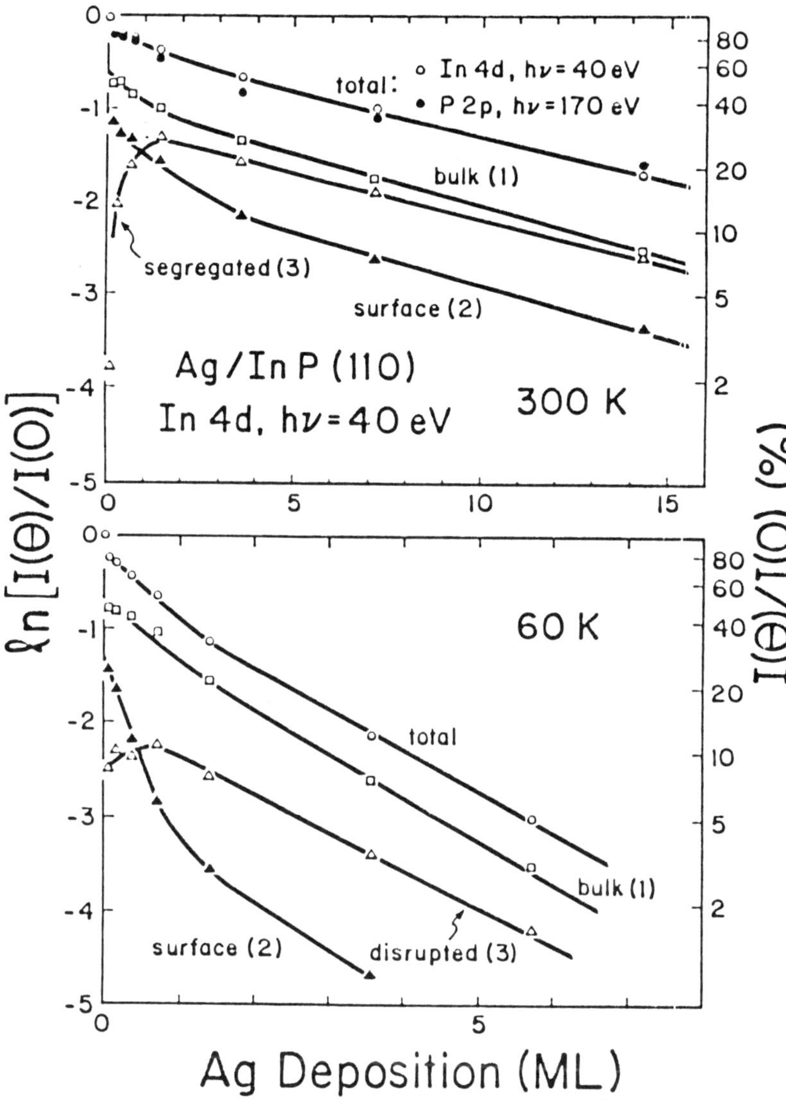


**Figure 22.** Logarithmic core level emission attenuations for 300 K *(top)* and 60 K *(bottom)* atom deposition of Ag. Note the factor of two difference in the horizontal scales for the 300 K and 60 K results. Gradual attenuation of the substrate signal at 300 K indicates overlayer clustering, whereas at 60 K the overlayer grows uniformly. To estimate the scale of In segregation for clustered overlayers at 300 K, the intensity of component *3* has been renormalized to the covered portion of the surface. Such renormalization reveals pronounced segregation of In atoms at 300 K and their kinetic trapping at 60 K. (From Ref. 105.)

A very different overlayer morphology is observed for interfaces formed by Ag atom deposition at 60 K. As shown in Fig. 21, the In 4d S/B ratio is much more rapidly reduced at 60 K than at 300 K and the surface component exhibits a definite shift to higher binding energy. This shift, emphasized by the vertical guides-to-the-eye in Fig. 21 *(right),* saturates at -0.56 eV for 3.6 ML and yields a slight shoulder on the high binding energy side of the 60 K spectra. Since this new component is attenuated at the same rate as the substrate above ~1.4 ML (see *bottom panel* of Fig. 22), we associate it with In atoms in the transition layer between the substrate and the Ag overlayer. It is analogous to the surface-shifted component but reflects a change in local bonding due to the immediate proximity of Ag atoms. The equivalent transition from the surface to the interface component is not evident in the 300 K spectra because the clusters are relatively large and they attenuate the emission from beneath them (a simple hemispherical cluster model[88] gives an estimated cluster radius of ~ 10 Å at 4 ML with ~95% of surface covered). Observation of this interfacial component at 300 K is further frustrated because of the near-degeneracy with the much stronger surface-shifted component. Based on these lineshape considerations and the substrate emission attenuation shown in the lower panel of Fig. 22. (1/e length of ~2.4 ML or 3.5 Å), we conclude that the interface formed at 60 K is uniformly covered after ~1.4 ML. This reflects the much reduced Ag surface mobility at 60 K. This interface morphology is metastable with respect to changes in temperature, and results of interface annealing to 300 K (not shown) reveal spontaneous clustering of the Ag overlayer (see Ref. 98 for results for Ag/GaAs).

The lineshape decompositions shown in Fig. 21 also provide valuable information regarding the temperature dependence of adatom-induced substrate disruption. It was recently suggested that Ag/InP(110) presents a rather unique system in which interface reactivity and defect formation are severely reduced at low temperature.[79][99] Our results unequivocally show that disruption does occur even at 60 K. In particular, disruption can be monitored through component *3* in Fig. 21 which corresponds to In atoms released from the substrate into the overlayer. At 300 K, component *3* is clearly resolved as a pronounced shoulder on the low binding energy side of the substrate emission. It is not as obvious to the eye in the 60 K spectra because of its smaller binding energy shift of -0.39 eV compared to -0.54 eV for 300 K. We emphasize, however, that the fitting of the In 4d emission for 60 K deposition requires this third spin-orbit doublet, and fittings with deliberate omission of it resulted in unphysical broadening of the substrate Gaussian width. The observed difference in the relative binding energy is interesting and can be understood in terms of the different local environments of liberated In atoms. At 60 K, In out-diffusion is kinetically limited, and the In concentration trapped in the Ag overlayer is much higher than at 300 K where the kinetic constraints are removed and the In atoms are expelled at the free surface. Indeed, many previous studies have established that dissolution of segregated cations in metallic overlayers leads to the gradual decrease in the binding energy of their core electrons, and the increased shift of the disrupted component for Ag/InP(110) at 300 K agrees with this trend.[100]

In order to quantitatively estimate the amount of substrate disruption and cation release, it is instructive to follow the emission intensity of component *3*. For the 300 K results, the signal intensity of the released In atoms has been normalized to the portion of the surface covered by Ag clusters. This procedure is motivated by the fact that In atoms segregate only from beneath the clusters and the remainder of the surface is still pristine InP. It thereby provides a more realistic indication of atomic distribution at heterogeneous interfaces and enables their comparison to systems presenting a layered growth mode. As shown in Fig. 22, the signal intensity of component *3* rises until ~1.5 ML and is then gradually attenuated (1/e length ~12 ML). This decay length is significantly shorter than the photoelectron mean free path in silver (~2.4 ML, as determined from the substrate attenuation at 60 K), and it reflects pronounced segregation to the surface of the Ag cluster of In atoms from the substrate. At 60 K, however, component *3* emission reaches its maximum at ~1 ML. Thereafter, it decays at approximately the same rate as the substrate. This behavior reflects kinetic trapping of released In atoms at the interface at 60 K.

In summary, Ag/InP interfaces formed by atom deposition exhibit limited substrate disruption at 300 K and 60 K. At 300 K, spontaneous clustering of Ag adatoms results in a heterogeneous interface with significant numbers of liberated cations. Based on the cation attenuation, we conclude that these released In atoms are expelled to the free surface of the Ag clusters. Calculations based on bulk thermodynamics yield a small negative heat of solution of In in Ag (-6.7 kJ/mole) indicative of a very weak tendency for intermixing.[101] These calculations, however, are of limited value in estimating actual preferential location of the released In atoms in a heterogeneous Ag overlayer having a constantly evolving distribution of cluster sizes. Following more elaborate calculations of cohesive and atomic energies, Weaver et al.[64][102] predicted In segregation at this interface. Intriguingly, P 2p EDCs do not show any disruption-induced feature at either temperature. The P 2p emission is attenuated in a like manner to the In 4d emission and this indicates that disrupted P atoms from the substrate are also segregated to the free surface of Ag clusters at 300 K, in agreement with the segregation trends predicted in Ref. 102. The lack of any change in the P 2p core level lineshape is then attributed to the accidental degeneracy in the P 2p binding energy between the substrate atoms and atoms released in the overlayer. At 60 K, liberated substrate atoms are kinetically trapped at the interface, and no evidence for segregation is found. Reduced temperature also quenches adatom surface mobility and results in nearly-uniform overlayer growth.

Figure 23 shows representative In 4d EDCs for Ti/InP(110) interfaces formed at 300 K *(left)* and 60 K *(right)* to emphasize temperature-induced changes in interface morphology. Lineshape decompositions for 0.1 ML show a shoulder labeled *3* on the low binding energy side of the bulk substrate component at both temperatures. This shoulder is associated with In atoms released from the disrupted semiconductor surface. Interfacial reaction that leads to substrate disruption and cation release replaces the surface layers of InP(110) with Ti-P bonding configurations and the surface-shifted component *2* disappears completely by 1 ML. For coverages below 0.3 ML there is little difference in the 300 K and 60 K spectral lineshapes, and this again demonstrates that reaction is independent of temperature.

Ti-P bond formation at the interface releases In atoms from the substrate, and these cations are expelled toward the free surface as regions of the reacted phase grow and coalesce.[100] As shown in Fig. 23, the released-In emission intensity grows rapidly and exceeds that of the bulk substrate after 0.3 ML. Beyond 0.3 ML, the FWHM of component *3* decreases as the (heterogeneous) local environment of liberated In atoms converges to a more uniform segregated phase at 300 K. At 60 K, reduced cation mobility again kinetically traps these expelled atoms in the thickening Ti film. Kinetic trapping of these released In atoms at 60 K in a heterogeneous environment causes the Gaussian FWHM to remain broad (~700 meV) for $0.3 \leq \theta \leq 3$ ML. At higher coverage the width decreases rapidly; it is reduced to 440 meV by 10 ML, equaling that of the 300 K spectra.

Additional information regarding the local environment of the liberated cations is provided by their binding energy (for metal/III-V interfaces in general, the relative binding energy of this component decreases with coverage[100]). For Ti/InP(110), the shift is less pronounced at 300 K than at 60 K, amounting to -0.37 eV and -0.47 eV at 0.3 ML and -0.70 eV and -0.96 eV at 3.5 ML for 300 K and 60 K, respectively. While the reduction in FWHM can be attributed to the increasingly homogeneous environment of the segregated cations at 300 K, the shift to lower binding energy indicates

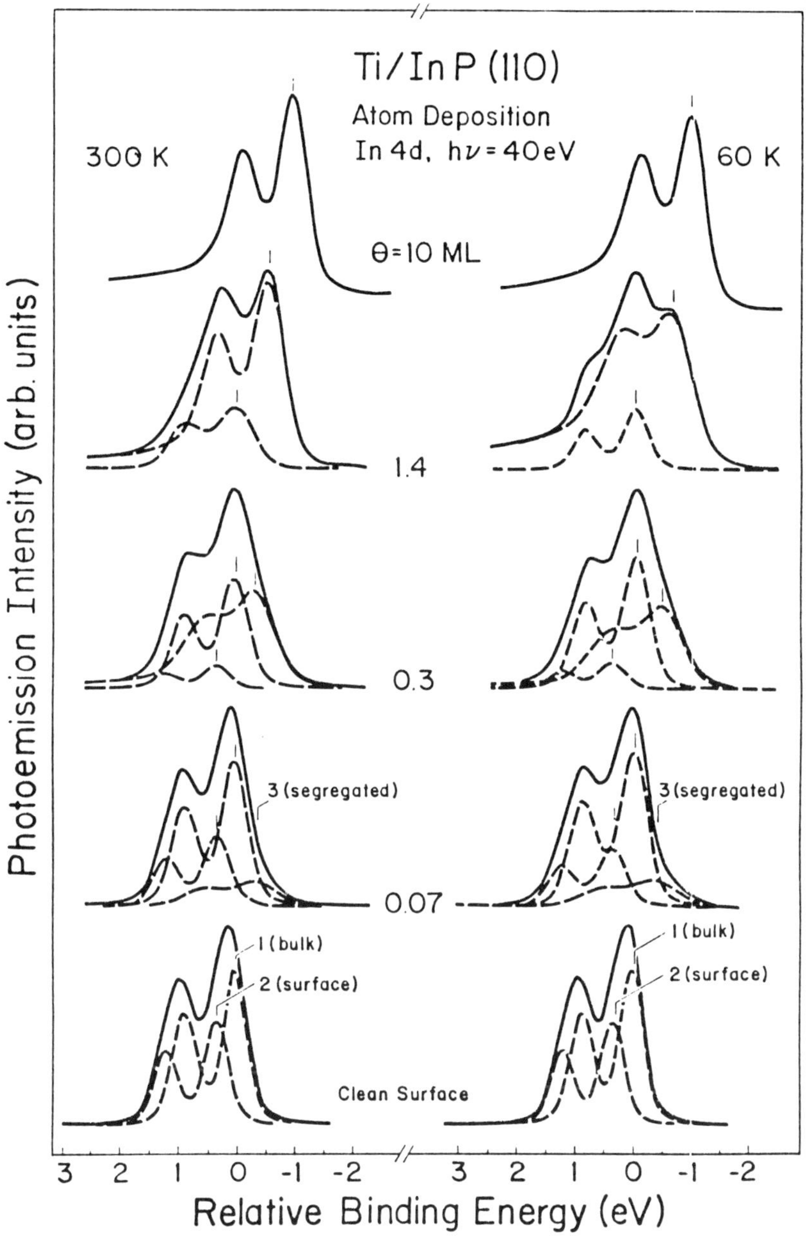


**Figure 23.** Representative In 4d core level EDCs and lineshape decompositions for Ti atom deposition at 300 K and 60 K. Strong interfacial reaction releases In atoms whose emission results in the rapid development of component *3*. Ti deposition disrupts ~4.4 ML of the substrate at both 300 K and 60 K. (From Ref. 105.)

that they become progressively dispersed at the Ti surface. Segregation is very limited at 60 K as the cations become kinetically trapped. This metastable trapping is reflected in the gradual shift to lower binding energy of the emission from these dissociated In atoms. This spectral feature is asymmetric at both 300 K and 60 K and this reveals a metallic environment of released In atoms.[15] The asymmetry parameter required for successful fits at high coverages was ~0.12 for 300 K and ~0.20 for 60 K interfaces. This difference, noticeable in Fig. 23, possibly reflects the more Ti-rich environment at 60 K.

In Fig. 24 we show logarithmic attenuation curves for Ti/InP(110) formed at 300 K *(top)* and 60 K *(bottom)*. Ti-P chemical reaction leads to substrate attenuation below ~1.5 ML. At higher coverages, the rate of attenuation is smaller [1/e lengths of ~2 ML (2.9 Å) at 300 K and ~2.8 ML (4.1 Å) at 60 K]. The slightly higher substrate attenuation rate at 300 K possibly indicates that reaction proceeds more slowly after ~1.5 ML at 300 K but is kinetically inhibited at 60 K. A simple calculation based on substrate attenuation profiles establishes that 1.5 ML Ti deposition leads to disruption of ~4.4 ML of substrate at both 300 K and 60 K.

Chemical reaction and cation release induce an abrupt increase in the segregated In component *3* for $\theta \leq 1.5$ ML regardless of temperature. At 60 K, the segregated component intensity then remains unchanged at ~40% of the clean surface emission until ~3.6 ML and attenuates at the same rate as the substrate signal at higher coverages. The constant intensity between 1.5 and 3.6 ML and attenuation above 3.6 ML correlate well with the previously discussed variations in the Gaussian width of component *3* (constant large Gaussian width below ~3 ML with sharpening thereafter). From this and the changes in the substrate In 4d emission, we can discern three distinct steps in the Ti/InP interface formation at 60 K. There are *(i)* initial disruption and cation release, *(ii)* trapping of the released cations in the Ti matrix, and *(iii)* the final covering of the interface with very limited segregation of In atoms. For 300 K, however, thermodynamically preferred cation segregation results in substantially higher emission from the segregated In phase. Its intensity saturates between 6 ML and 10 ML at ~80% of the clean surface signal and attenuates very slowly for coverages above ~10 ML. Indeed, the emission from segregated In atoms equals ~30% of the clean surface In 4d signal even for 55 ML Ti coverage.[100]

Figure 25 shows the P 2p core level lineshape evolution for the Ti/InP interface formed at 60 K. The component-specific attenuation profiles for this interface are shown in the inset. Fitting of the clean surface spectrum required two components, namely substrate emission and a surface-shifted

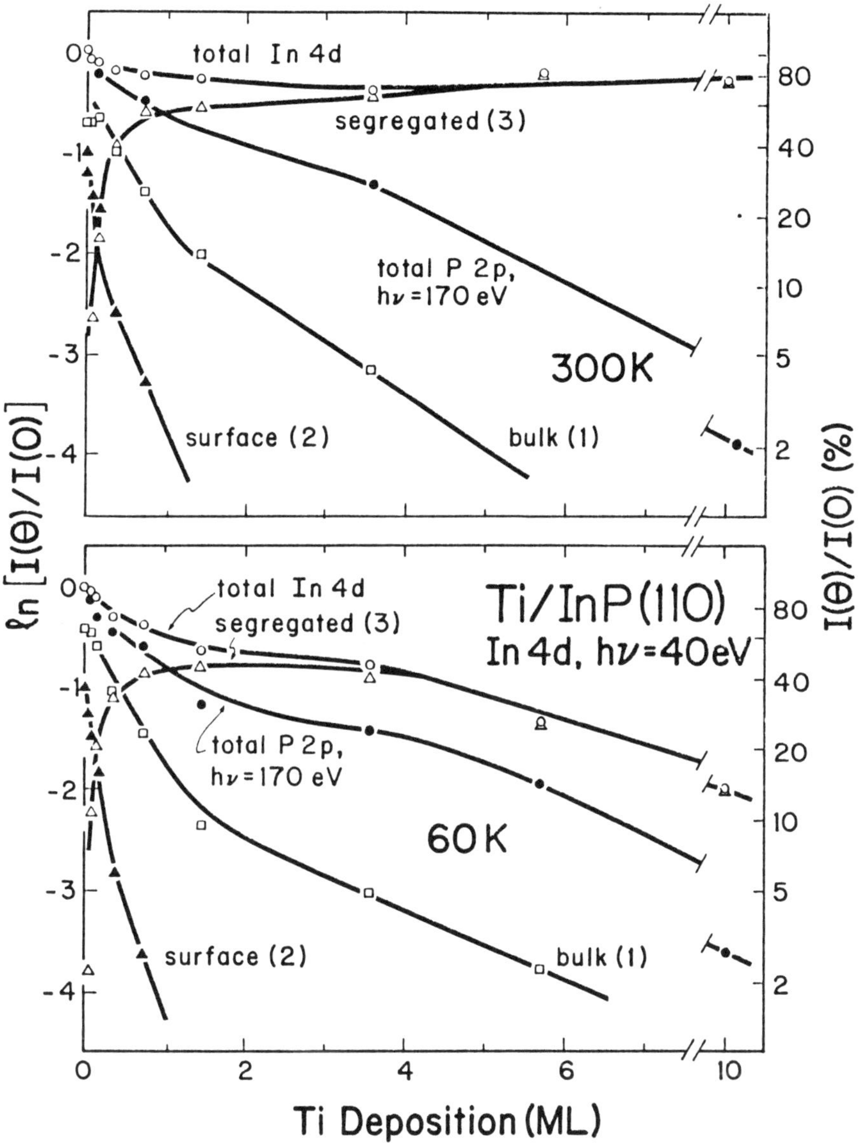


**Figure 24.** Attenuation curves for the Ti/InP(110) interface at 300 K *(top)* and 60 K *(bottom)*. At 300 K, strong In segregation results in continuous increase of the component *3* emission for $\theta_{Ti} \leq$ 10 ML. The very slow subsequent decay of this component is due to the metastable trapping of these In atoms in Ti. At 60 K, In segregation is very limited above ~4 ML of Ti. Total P 2p attenuations differ very little between 300 K and 60 K and reveal pronounced chemical trapping of released P atoms at both temperatures. (From Ref. 105.)

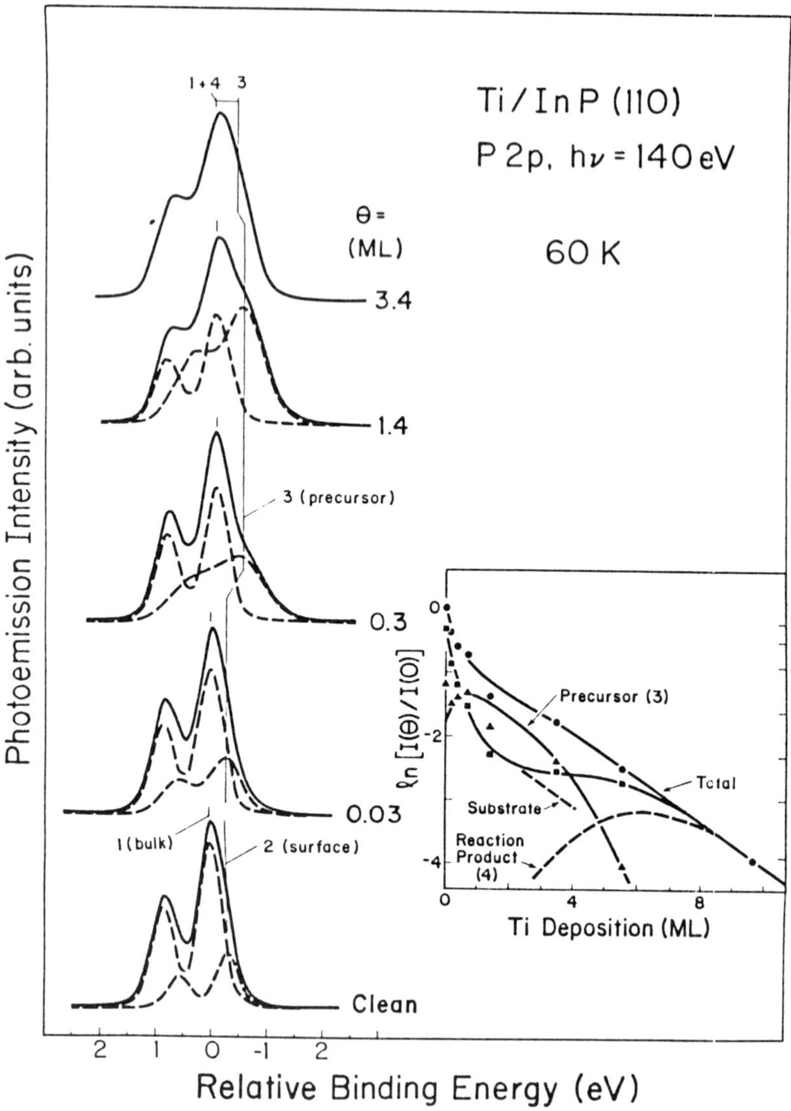


**Figure 25.** Representative P 2p core level EDCs and lineshape decompositions for Ti/InP(110) formed at 60 K. Components *1* and *2* correspond to the P atoms in the semiconductor bulk and at the surface, respectively. Component *3* reflects the metastable (precursor) Ti-P bonding configuration that is replaced by fully developed Ti-rich phosphide at higher coverage, evidenced in component *4*. Components *1* and *4* are degenerate in energy and their component specific attenuations shown in the inset were obtained using the In 4d substrate attenuation. (From Ref. 105.)

feature at -0.30 ± 0.03 eV. The S/B ratio for $h\nu$ = 140 eV is ~33%, corresponding to the photoelectron mean free path of ~8 Å. Submonolayer Ti deposition resulted in enhanced emission at lower binding energy that is associated with surface disruption and Ti-P bond formation. This new feature (labeled *3*) is degenerate with the surface-shifted component below ~0.1 ML. At higher coverage it grows and shifts to lower binding energy so that a distinct shoulder in the P 2p emission is observed at 0.3 ML. Above 1.4 ML, the intensity of this component decreases with a rate exceeding that of the total P 2p emission. Likewise, the binding energy of this component shifts back toward that of the substrate. This indicates that component *3* is a precursor to the more stable Ti-P phase in the sense that its bonding configuration is not yet completely defined. The P 2p binding energy of the fully-developed Ti-P configuration is degenerate with that of bulk InP, but its existence is clear from the attenuation results shown in the inset. In particular, the extrapolated substrate signal attenuation *(dashed line)* is used to remove the substrate contribution from the completely-reacted component *(4)*. This procedure shows that the component *(4)* is the only remaining feature in the spectrum for $\theta \geq 7$ ML, and its emission is attenuated exponentially at higher coverage. The total P 2p emission attenuation at 300 K is identical to that at 60 K (Fig. 24) and this reflects temperature-independent interfacial chemical reaction and anion retention in the reacted region. Finally, the P 2p lineshape evolution at 300 K is the same as that at 60 K, again emphasizing that adatom-substrate reaction is not altered and that the interface region is dominated by Ti-P bonding at both 60 K and 300 K.

## 7.0 ABRUPT INTERFACES FORMED BY CLUSTER DEPOSITION

Studies of Schottky-barrier formation have been complicated by the difficulty of producing an abrupt, "ideal" interface. As shown in detail in previous sections, commonly-used methods of producing a metal-semiconductor interfaces result in substrate disruption, atomic interdiffusion, alloy or compound formation, and structural changes of the substrate surface. The nature of such interfaces makes it difficult, from a fundamental point of view, to identify the mechanism(s) that dominate Schottky-barrier formation for the various stages of its development.

A number of studies have been performed at low temperature (60 - 200 K) in an attempt to inhibit certain interfacial interactions and thereby exclude their effect on Schottky-barrier formation. However, we showed in the previous section that the primary difference between interfaces formed at 60 K and 300 K involves the distribution of liberated substrate atoms in the evolving interface.

To create more "ideal" boundary regions, we have developed a method of joining preformed metal clusters with atomically clean semiconductor surfaces.[87][88][103] Our approach has been to isolate the substrate from the impinging metal-atom flux and thereby avoid the complexity at the surface due to adatom impact and bonding. In this way it is possible to prevent adatoms from interacting with the surface until they have agglomerated into clusters. To achieve this, we first condense a thin layer of solid Xe on the clean substrate at 60 K and then deposit metal atoms onto this buffer layer. Adatom mobility is sufficient to assure the formation of metallic clusters on and within the Xe. These clusters come into contact with the clean surface when the Xe buffer is desorbed.

## 7.1 Metal-cluster Morphologies on Xe and on GaAs(110)

In Fig. 26 we show the rate of attenuation of the Xe 4d emission from a 30-Å-thick buffer layer as a function of nominal coverage for Ag, Al, Au, Co, Ga, and Ti. The emission from the Xe layer was normalized to the value for zero metal coverage, taking the initial total Xe 4d emission as unity. As can be seen, there is attenuation for all metals at low coverage as the Xe emission is reduced to ~70% of its starting value for coverages below 1 Å. This indicates that the surface is covered by small clusters. For metal coverages exceeding ~5Å, attenuation is virtually constant. The absence of significant deviations in the attenuation for all six metals indicates that the average nucleation and growth of clusters on the interlayer is largely metal independent.

Clusters are brought into contact with the GaAs(110) surface when the Xe buffer layer is desorbed. In Fig. 26 we compare the attenuation results for clusters on Xe to those after the clusters reached the GaAs surface. In the latter case, the Ga and the As 3d emission from the clean cleaved surface is taken as unity. If Xe desorption does not produce any morphology change, then the attenuation results should be indistinguishable. Indeed, the two curves are indistinguishable below ~10 Å for the six different metals. We conclude that the GaAs(110) surface is covered by small metal clusters below 1 Å and by larger clusters between 1 and 10 Å, as for the Xe interlayer, leaving large areas of the substrate exposed. Above ~10 Å the metal clusters attenuate emission from GaAs much more rapidly than emission from the Xe interlayer. This suggests that there are structural changes when the clusters reach the semiconductor.

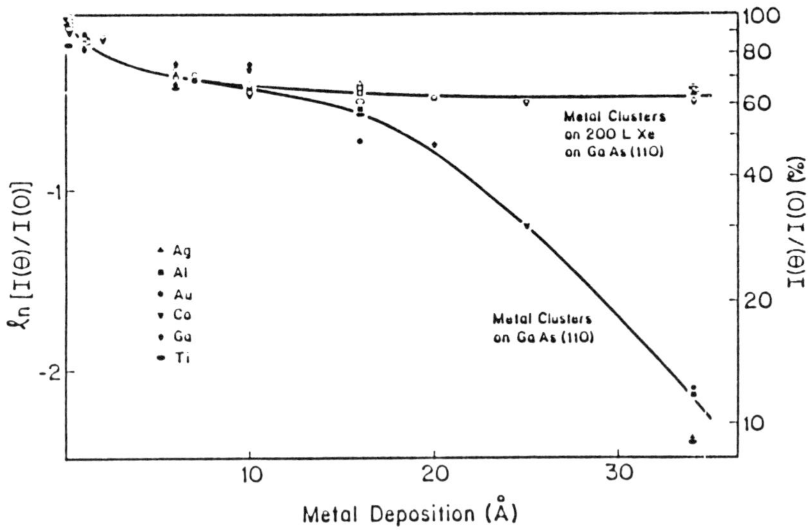


**Figure 26.** *Upper curve* with *open symbols* shows the rate of attenuation of emission from Xe 4d core levels caused by spontaneous metal-cluster growthon a Xe buffer layer on GaAs(110). Cluster growth produces equivalent attenuation for all these metals, indicating analogous surface morphologies. The *lower curve* with *solid symbols* shows the GaAs substrate attenuation observed when the Xe buffer layer is desorbed and the clusters cover the substrate. (From Ref. 104.)

An estimate of the size of the clusters can be obtained by assuming them to be uniform hemispheres distributed across the surface. In this case, the normalized substrate emission (either Xe or GaAs) can be written

$$\frac{I(\theta)}{I(0)} = 1-\left(\frac{3\theta}{2R}\right)\left\{1+\frac{2}{R^2}\left[\lambda(R+\lambda)\exp\left(-\frac{R}{\lambda}\right)-\lambda^2\right]\right\}$$

where $\lambda$ is the photoelectron mean free path, R is the cluster radius, and $\theta$ the amount of metal deposited in Å. From the data of Fig. 26, we find R ~ 10 Å for $\theta$ = 1 Å, corresponding to ~100 atoms in the cluster. For depositions between 5 and 15 Å on Xe, R varies between 30 and 40 Å (3000 - 6000 Å atoms per hemisphere). This large estimated cluster size is consistent with the high mobility of metal atoms on Xe surfaces where they are only weak metal-Xe interactions. The fact that the measured attenuation rate between 5 and 15 Å is slow suggests that the number of clusters is essentially the same as at lower coverages and that most of the deposited atoms become incorporated in those large clusters.

We speculate that the nearly constant Xe attenuation for coverages higher than 15 Å implies that some of the metal clusters are embedded in the Xe layer or that Xe covers the clusters. Xe displacement would not be surprising, given the relative heats of formation for Xe and any of these metals. Likewise, the presence of Xe on the clusters is consistent with the very broad Xe 4d emission observed upon cluster formation which reflects spatial inhomogeneities. It is also possible that some cluster sintering occurs within the Xe layer, thereby providing another explanation for the increased GaAs attenuation relative to that for Xe at higher metal coverages. Although the detailed morphology of the cluster/Xe interface is not known completely, the conclusions for the cluster/GaAs interface do not depend on detailed knowledge of those interfaces. It suffices to note that the buffer layer prevents isolated metal atoms from inducing GaAs surface disruption.

To gain further information regarding the structure and morphology of the metal-cluster/GaAs(110) interface, scanning- and transmission-electron-microscopy studies were undertaken.[104] In Fig. 27 we show results for 7 Å Au(cluster)/GaAs(110). As can be seen, by 7 Å there has been considerable sintering and network formation, but large portions of the surface are exposed. Direct evidence for sintering of several clusters can be seen in Fig. 27a in a plan-view TEM micrograph. The size of the large central aggregate is approximately 1000 Å. In Fig. 27b we show a transmission electron micrograph of a cross section through the Au cluster/GaAs interface. For this section, the lateral spacing between clusters along the [110] direction is 200 - 250 Å, and there is intimate contact between the clusters and the substrate. Figure 27c shows a high-resolution lattice image of one of the clusters. Inspection of the left portion of this cluster shows lattice images of (111) Au planes separated by 2.5 Å. The region at the right also appears well defined but the orientation of Au(111) planes is different. This indicates that one Au cluster consists of several grains of Au. From this image we can see an abrupt interface without preferential lattice orientation between the metal cluster and the semiconductor surface. The thickness of the cluster is ~60 Å. There is no evidence for reaction between Au and the GaAs substrate. The reason for the slightly fuzzy image in the central part of the metal cluster close to the substrate is probably related to differently oriented small grains or distortion at the grain boundary where lattice spacing between planes is smaller than the resolution of the microscope, i.e., below 1.4 Å.

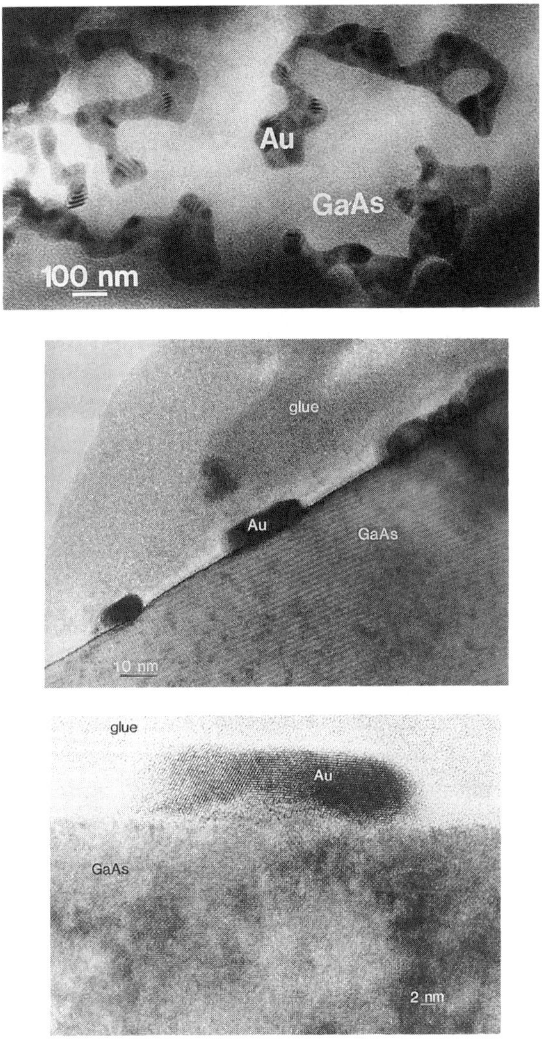


**Figure 27.** *Upper panel* shows a plan view of the 7 Å Au(cluster)/GaAs interface with interconnected metallic clusters. *Center panel* shows several clusters in cross section. *Bottom panel* shows high-resolution images that reveal good contact to the GaAs surface and evidence for sintering of two or more microcrystallites of Au. (From Ref. 104.)

From the above it is clear that metal aggregates formed by cluster deposition onto GaAs(110) are rather extended, even for 7 Å nominal coverage. Indeed, the onset of metallicity has been detected from the valence band evolution.[89] Cluster deposition gives rise to a metallic Fermi-level cutoff from very low coverages. This is expected because the attenuation measurements indicate that the clusters are 10 Å in radius (~100 atoms) for $\theta$ = 1 Å and such clusters should be metallic.

## 7.2 Core-level Lineshape Studies

In order to determine the morphology of the contacts we will compare the Ga and As 3d core-level emission as a function of coverage after cluster deposition and atom deposition at 300 and 60 K. Figure 28 compares As 3d EDCs taken at h$\nu$ = 90 eV for cluster and atom deposition of Co at 60 K. The clean surface spectra were collected at 60 K but spectra following cluster deposition and Xe desorption were collected between 100 and 300 K. While this made it possible to investigate the temperature stability of the interface, it introduced changes in the core-level line shapes associated with thermal broadening. These changes are reflected in the increase in Gaussian width for the core-level spectra following cluster deposition shown in Fig. 28. These spectra reveal a decrease in the emission intensity from surface atoms (feature *2)* relative to that from bulk atoms (feature *1)* as the nominal coverage is increased. Most important, no new adsorbate-induced features are required to fit the core-level spectra. The stability of these interfaces and the equilibrium character was demonstrated by the absence of any band bending or line-shape changes while warming to 300 K over a period of ~3 h.

In contrast to the cluster-deposition results, examination of the As core-level spectra for atom deposition at 60 K shows two Co-induced features that dominate after ~2 Å, namely components *3* and *4* that correspond to emission from As atoms released from the substrate and bound in different chemical environments at the interface. Similar line-shape changes were found for atom deposition at 300 K as shown in Sec. 5.

In Fig. 17 we show Ga 3d core-level EDCs following Ti-cluster and atom deposition. For atom deposition at 60 K, substantial changes in the Ga 3d line shape occur as the Ti coverage increases. Substrate disruption is observed from the lowest coverage and is evident as a rather broad Ti-induced feature (labeled *3)* for $\theta$ = 0.2 Å. As feature *3* grows, the intensity of the surface component (labeled *2)* decreases rapidly, indicating release of the Ga atoms from surface-bonding configurations into the evolving interface. The large width of feature *3* for $\theta \leq$ 8 Å indicates a variety of inequivalent chemical environments for the released Ga atoms. It sharpens for θ ≥ 10 Å as the environment of the Ga atoms in the Ti overlayer becomes more uniform. Ti-atom deposition disrupts ~3 ML of the substrate at both 60 and 300 K and there is kinetic trapping of the released substrate atoms near the interface at low temperature.

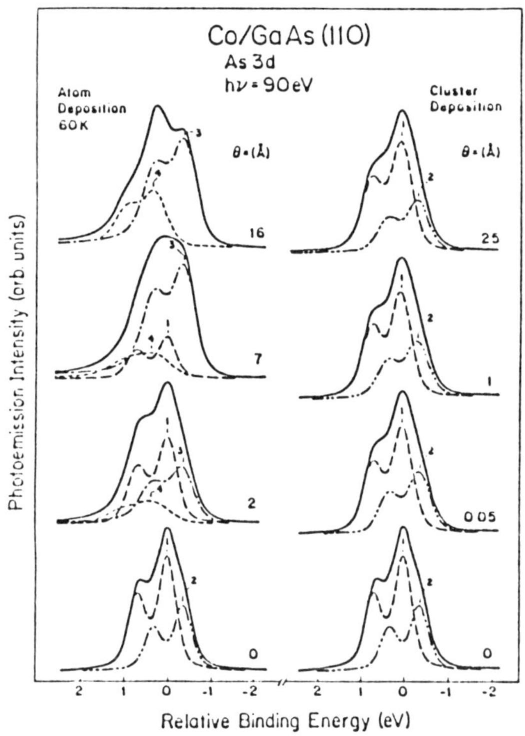


**Figure 28.** Representative As 3d core-level EDCs and lineshape decompositions for Co-cluster deposition and for Co-atom deposition at 60 K. Components *1* and *2* correspond to emission from bulk and surface As atoms. Components *3* and *4* are induced by atom deposition at 60 K and correspond to As atoms released from the substrate. Cluster deposition alters the relative intensity of the surface and bulk components but introduces no new spectral features. (From Ref. 104.)

Inspection of the results of Fig. 17 for Ti cluster deposition shows that there is no evidence for Ti-cluster-induced features in the Ga 3d EDCs. Indeed, the only changes following cluster deposition reflect an increase in phonon broadening and a decrease in the relative intensity of emission from surface-shifted Ga atoms compared to bulk Ga atoms. Due to the large size of the clusters compared to the photoemission probe depth (~80 Å clusters thickness versus ~10 - 20 Å probe depth), we cannot exclude the possibility of metal-substrate interactions beneath the clusters. Indeed, reaction does occur beneath the Ti clusters when they are warmed to 300 K, a conclusion reached based on the movement of the Fermi level.

From the Co- and Ti-overlayer results in Figs. 28 and 17, we can conclude that there are significant differences in the core-level line shapes for interfaces formed by atom and cluster deposition. The most prominent difference is the absence of any reaction-induced feature in EDCs following cluster deposition. Additional studies with Al, Ag, Au, and Ga clusters show that the only line-shape changes observed in both the As 3d and Ga 3d EDCs involve phonon broadening and the decrease in the surface to bulk emission intensity ratio as the nominal cluster coverage increased. The absence of cluster-induced features suggests that this new deposition process produces abrupt junctions. In contrast, there are radical line-shape changes following atom deposition of Al,[17] Au,[18] Co,[19] and Ti[16] on GaAs(110), and various amounts of substrate disruption, atomic intermixing, reaction, and clustering are revealed by careful line-shape analysis. Similar morphological differences between atom and cluster depositions were found for InP.[88][105]

## 8.0 INTERFACES FORMED BY ION DEPOSITION

Several recent studies have focused on changing the morphology of the interface so as to isolate different processes that occur as contacts are formed and to distinguish the roles of each in their morphology and electrical properties. In Sec. 6 we presented studies of the effects associated with cooling the semiconductor during metal atom deposition. They showed that processes related to nucleation and growth are temperature dependent, and detailed studies of chemical changes have shown that the amount of substrate disruption and related defect creation is not dependent on temperature. Other experiments that altered the parameters of interface formation brought preformed metal clusters into contact with clean GaAs(110) and InP(110) surfaces, as presented in Sec. 7. This made it possible to change the energy-releasing processes associated with adatoms impinging on the semiconductor, adatom-adatom bonding at the semiconductor surface, and (isolated) adatom-semiconductor bond formation.

In this section we examine results from an investigation of metal-semiconductor interface formation in which neutral and ionized atoms were deposited onto GaAs(110), InP(110), and ZnSe(100) surfaces at 300 K.[106][107] The goal of these studies was to *enhance* adatom-substrate interactions so as to again alter the parameters of interface formation. Synchrotron radiation photoemission results showed that the acceleration of the ions to 200 eV makes it possible to vary the amount of surface disruption. Parallel studies of Fermi level movement induced by neutrals and ions were particularly significant in the context of Schottky barrier formation. It is observed that $E_F$ gradually moves toward the middle of the gap but this movement does not substantially depend on the deposition technique.

Ion beams used in thin film deposition technologies can be divided into two basic configurations.[108] In one, the desired material is ionized and is deposited at low energy (~100 eV) directly onto a substrate primary ion beam deposition. In the other, the desired material is sputtered from a target by an energetic beam of inert or reactive gases. The deposition of ions from a charged particle beam to form a contact presents attractive features compared to neutral deposition. In particular, it affords control over deposition energy and gives the ability to direct the beam and mass-analyze it to yield a pure deposited species. The reason for using ions for overlayer growth is to impart kinetic energy and thus affect film kinetics and growth processes.[109][110] Ion-surface interactions can include enhancement of adatom migration, intermixing, defect creation, sputtering, and implantation. Thus, it is of interest to investigate the possibility of using ions to change film growth characteristics and to correlate surface effects with the electronic properties and the Schottky barrier.

## 8.1 The Ion Source

For our studies, we constructed a compact single-grid, ultrahigh-vacuum-compatible, low-energy ion gun. The ion source is bakeable, capable of operating at high temperature to produce ions of high-boiling-point solids, able to yield ion fluxes of >20 nA/cm$^2$ and an efficiency greater than 1% at deposition rates of ~1 ML/min. Uniform ion-beam intensities were obtained over a 10-cm diameter at a distance of ~40 cm from the source. The design allows the temperature of the entire ion source to be maintained high enough to avoid condensation of the ions or evaporated atoms within the source.

The ionization mechanism used involves electron-impact in a low voltage gas discharge. The neutral-atom density can be several times the ion density, but the ionization efficiency can be high enough to produce rates of 4% for Ag, 7% for In, and 10% for Bi since the ions are extracted more efficiently by an acceleration potential. The fraction of ionized metal atoms is determined by comparing the amount of metal deposited (measured with a crystal thickness monitor) with the ion current (measured with a Faraday cup).

Figure 29 shows a schematic of the cylindrically symmetric ion source. The evaporants are heated in an alumina crucible and the evaporated atoms are injected into the discharge chamber or anode. The discharge is sustained by thermionic emission of electrons from a spiral filament. In operation, the anode is maintained at a positive potential and the target is grounded. The ion density in the discharge chamber thus depends on the electron mean free path (via the pressure and the ionization cross section), the electron emission current, and the total electron path length. A fraction of the ions accelerated by the extraction grid voltage pass through the grid apertures and form the beam. Grid voltages of 200 V extracted ion-beam currents up to 100 nA/cm$^2$ for the metals Bi, In, and Ag.

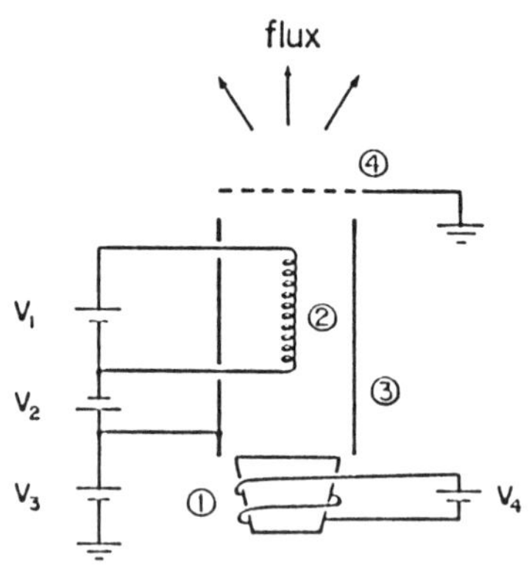


**Figure 29.** Schematic of the ion source. The metal to be evaporated is heated by a tungsten coil wrapped around the alumina crucible *(1)*. A tungsten filament *(2)* provides energetic electrons to ionize the metal atoms within the anode *(3)*. A grounded grid *(4)* extracts the ions from the plasma established in the source. $V_1$ is the filament power supply, $V_2$ is the discharge power supply, $V_3$ controls the ion energy, and $V_4$ is the power source used to evaporate the metals from the crucible. (From Ref. 107.)

## 8.2 In/GaAs(110)

Early studies of interface development for In/GaAs(110) at 300 K have shown overlayer growth that was characterized by In clusters and Fermi level movement in the gap that was slow compared to other metals.[111]-[113] No evidence was found for an exchange reaction of In with Ga at 300 K. Figure 30 shows a series of Ga 3d and In 4d core level spectra taken after the deposition of neutral In atoms *(on the left)* and In ions *(on the right)*. The spectra at the bottom, corresponding to Ga 3d for clean cleaved surfaces of GaAs(110), present contributions from bulk (low binding energy doublet) and surface atoms (high binding energy doublet). As In is deposited, new structures appear that correspond to emission from the spin-orbit-split In 4d core level. Detailed line-shape analysis shows that the introduction of one new doublet is sufficient to achieve a good fit for the Ga 3d and In 4d emission at all coverages (two components for Ga to represent bulk and surface atoms and one for In to represent In clusters). No changes were observed in the Ga emission, except for the loss of intensity related to overlayer coverage. This loss was slow because the adatoms formed clusters. These findings reflect the development of an abrupt, non-reactive interface from which neither Ga nor As atoms were released by adatom-induced disruption, in agreement with previous reports (Ref. 113 and citations therein).

The *right* position of Fig. 30 summarizes results obtained following the deposition of a partially ionized beam of In atoms on GaAs(110). Inspection quickly shows that the In 4d core level lineshape differs from that obtained for neutral atom deposition. Attempts to fit these results in a manner similar to that for atom deposition failed. To obtain acceptable fittings for the ion-deposition results, two new components were needed, one corresponding to a new In doublet, labeled *5*, and one to a new Ga doublet, labeled *3*, as shown in Fig. 31. Indium feature *5* is shifted 0.95 eV relative to feature *4*, i.e., relative to metallic In. The magnitude of this shift agrees with results observed when comparing metallic and covalently bonded In [Ref. 114 discussed studies of In-overlayer growth on InP(110)]. The Ga 3d core level shows a new doublet at lower binding energy that corresponds to atoms released from the substrate and intermixed in the forming overlayer. These new results can be understood in terms of an exchange reaction involving In and Ga that is stimulated by ion deposition but not by neutral-atom deposition. It should be noted that a similar assignment to component *5* was proposed by Chin *et al.*[115] when they induced chemical reaction by heating 10 ML In/GaAs(110) above the melting point of bulk In.

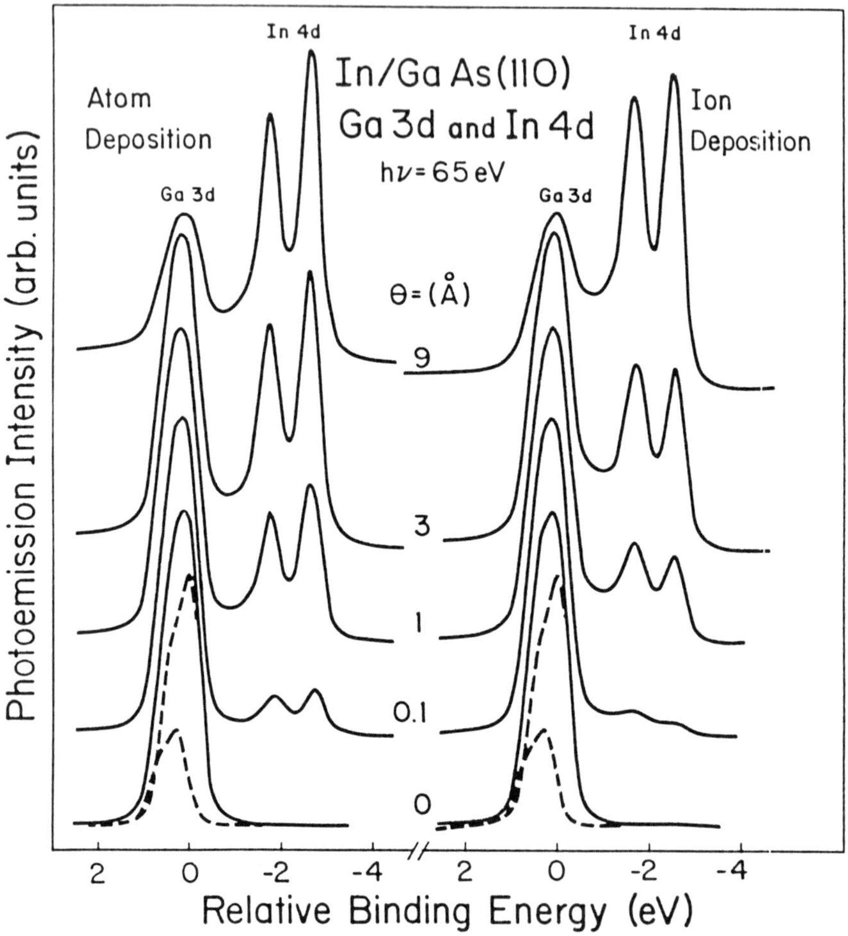


**Figure 30.** Representative Ga 3d and In 4d EDCs for In neutral-atom and ion depositions on GaAs(110) at 300K. The EDCs are normalized to the same height and shifted in energy to account for band bending. The two components for the clean cleaves correspond to emission from the bulk and surface atoms. The spectra can be fitted by adding a single doublet for metallic In, but additional components are needed to fit the results obtained after ion deposition, as shown in Fig. 31. (From Ref. 107.)

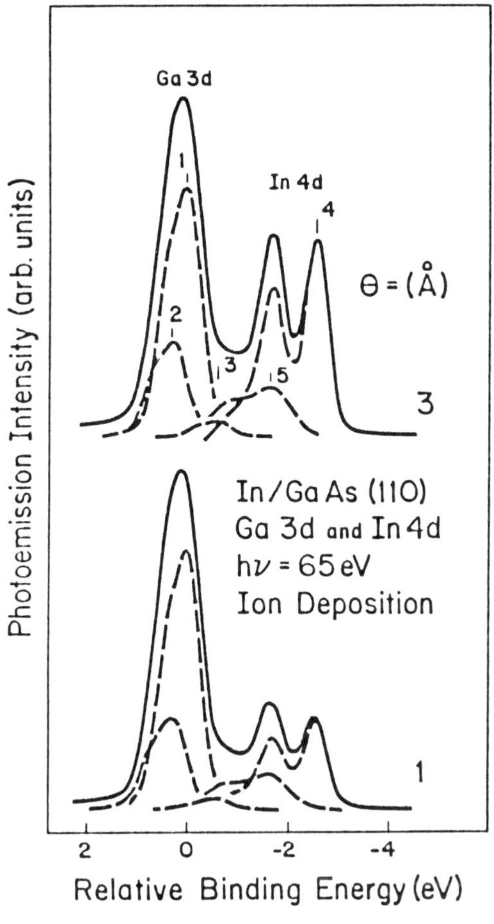


**Figure 31.** Line-shape decomposition for In-ion deposition. Features resulting from bulk and surface Ga atoms at the relaxed GaAs(110) surface are labeled *1* and *2*. Feature *4* corresponds to metallic In. These three features are enough to fit spectra obtained after atom deposition (Fig. 30), but two additional features, labeled *3* and *5*, are needed for ion deposition. Feature *3* corresponds to Ga released from the substrate and feature *5* to covalently bonded In. (From Ref. 107.)

Studies of the As 3d core level emission showed equivalent behavior for deposition of neutral In and ionized In on GaAs(110). For both deposition methods the As 3d core level presents two doublets which correspond to emission of atoms from the bulk and the surface of the semiconductor because large areas were not covered and remain unaffected. As a consequence of disruption some As atoms must be in new environments such as In-As configurations and, probably, intermixed in the overlayer, but they were not spectroscopically identifiable. These core-level results and the attenuation profiles based on them show that In-ion deposition leads to overlayer clustering that is similar to that for atom deposition, but that the formation of these clusters is accompanied by substrate disruption. The amount of disruption can be estimated from the intensity of the ion-induced In component to be approximately 0.3 ML after 15 Å deposition.

## 8.3 Bi/GaAs(110)

Bi/GaAs(110) interface formation is characterized by overlayer growth that is free of substrate disruption. Several previous studies have shown[116]-[122] that Bi deposition at 300 K produces ordered zig-zag chains on the surface along the [110] direction for the first monolayer with an array of missing rows to accommodate strain, and that Bi forms clusters over this monolayer, i.e., Stranski-Krastanov growth. Studies of the Ga 3d core-level emission at relatively high Bi coverages show contributions from substrate emission (Ga bonded to As) and boundary-layer emission (Ga bonded to Bi).[120] As an example, we show in Fig. 32 the Ga 3d emission measured after 25 Å neutral-atom deposition. The total Ga 3d emission intensity at this coverage is only 10% of the clean-surface intensity, consistent with the growth of a nondisruptive but irregular overlayer of Bi. Low-energy electron diffraction (LEED) studies show the persistence of a 1x1 pattern for neutral-atom deposition.[116]-[119]

The formation of the Bi/GaAs(110) interface is quite different when the Bi flux is partially ionized and accelerated to 200 eV. First, a diffuse background appears with growing coverage so that the LEED spots are no longer visible at 20 Å. This indicates that the partially ionized beam alters the long-range order of the Bi overlayer at 300 K and prevents spontaneous ordering. Further evidence for disorder is found by examining the Ga 3d emission of Fig. 32, where the Gaussian widths of the core-level emission increase ~30% compared to atom deposition and the Ga-bonded-to-Bi component is substantially larger. Hence, ion deposition modifies the GaAs surface and boundary layer, as well as the growing Bi overlayer, and this accounts for a wider variety of bonding configurations. It is also likely that Ga atoms are released from the substrate. From the attenuation curves (not shown), there is evidence that ion deposition produces clusters prior to completion of the first layer and that the clusters attenuate the substrate more slowly than neutral-atom deposition, indicating that they are larger.

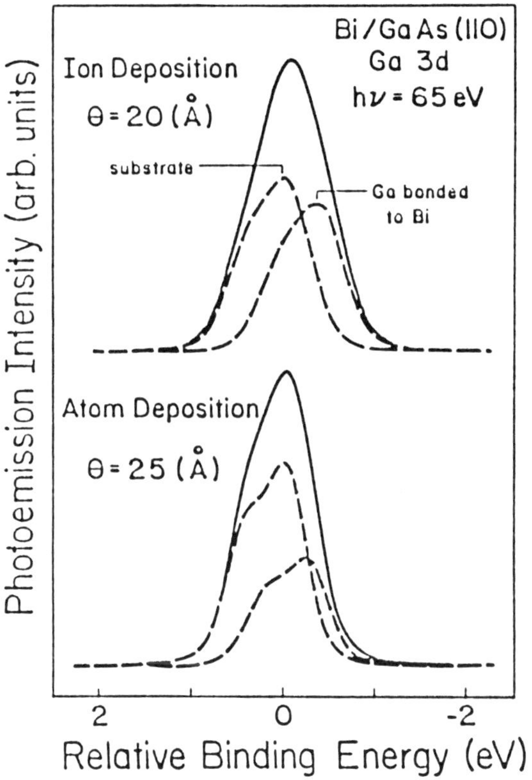


**Figure 32.** Ga 3d core-level EDCs for Bi-ion deposition and neutral-atom deposition. The enhanced Bi-bonded component and the much greater width of the spectral features for ion deposition indicate that the ions cause substrate and overlayer disorder. No disruption is observed for neutral atom deposition. (From Ref. 107.)

Ion bombardment has been observed to favor the more rapid growth of larger islands while inhibiting the rate of secondary nucleation and providing a more uniform island size distribution.[109] This can be explained as being due to a combination of enhanced adatom surface diffusion and the dissociation of small island into small clusters and adatoms with higher surface mobility. Heating of the substrate results in increased adatom migration during standard vapor deposition and similar effects can be expected by using accelerated ion beams without actually heating the substrate.

We have presented results of a new manner of forming a metal-semiconductor interfaces in an UHV system with the purpose, in contrast to previous intents, of controlling the interface morphology to enhance the adatom-substrate interactions. Indeed, that goal was achieved and enabled us to evaluate the process of Schottky barrier formation while altering the chemistry and morphology of the interface. Results have shown that the Fermi level evolution is quite insensitive to those changes indicating that the specific adatom-semiconductor interaction plays the central role in the Schottky barrier formation process.

## 9.0 RECENT RESULTS AND FINAL COMMENTS

Scanning tunneling microscopy (STM) has been used with great success to determine electronic density contours and surface reconstruction for surfaces of many semiconductors and metals.[122] Extensions of STM techniques to the study of adsorbed atoms and interfaces have given new insight into low-coverage bonding configurations and overlayer morphologies for adsorbed gases and nondisruptive metal overlayers.[123]-[131] Imaging reactive overlayers with STM poses additional challenges but gives insight into chemical reaction during the earliest stages of interface formation.[132][133]

### 9.1 Photoemission Spectroscopy and Scanning Tunneling Microscopy Studies of Sm/GaAs(110)

Some interfaces are believed to evolve through distinct steps with thresholds that reflect changes in bonding and structure. Indeed, recent studies for the Sm/GaAs(110) system have shown that the low coverage regime is characterized by weak interaction while interface growth is characterized by disruption and chemical intermixing after a specific coverage.[134]-[136] Photoemission results accompanied with STM studies were recently used to investigate the growth structures of Sm on GaAs(110) and the stability of growth structures at 300 K.

Figure 33 shows STM images for GaAs(110) with Sm coverages of 0.01 and 0.05 ML, along with a geometric model depicting the bonding sites for the observed growth structures.[137] These images were stable, and the surface could be scanned repeatedly. For 0.01 ML deposition, most of the GaAs surface is exposed, as shown in Fig. 33a. Significantly, the adatom-induced bright spots appear to be grouped rather than randomly dispersed. Analysis for coverages below ~0.01 ML indicates that the Sm atoms, highlighted by black dots in Fig. 33, bond to surface Ga atoms. Comparison of the number of defects observed for the clean surface to that for the 0.01 ML Sm coverage (dark regions) gives no indication of disruption or defect formation induced by Sm condensation.

Figure 33b shows a 55 x 55 Å$^2$ region of the surface following 0.05 ML deposition. The zig-zag chain adatoms define a low density structure formed by Sm atoms in Ga-bonded sites, as in Fig. 33a, but linked along [$1\bar{1}0$] across adjoining substrate rows. At 0.01 ML coverage, zig-zag and linear chain growth structures are observed in a ratio of about 10 to 1. A count of the Sm atoms in zig-zag chain structure gives a typical number of 5 - 9 for coverages of 0.05 to 0.66 ML. In some cases, the zig-zag chains bridge three rows (as in Fig. 33b), but none form long structures along [$1\bar{1}0$] of the sort reported for Cs/GaAs(110) despite the fact that the Cs atoms occupy the same sites as Sm.[138] These ordered low coverage structures on the GaAs(110) template are preferred over randomly dispersed adatoms. The formation of such ordered structures at very low coverage demonstrates that Sm atoms are mobile on GaAs(110) at 300 K. The Sm-Sm distances for the nondisruptive zig-zag formation are 6.9 Å and 7.9 Å for the first and second nearest neighbors, dimensions that are much larger than in bulk rhombohedral Sm[139] where the nearest neighbor distance is 3.588 Å for atoms in the close packed (111) plane. This demonstrates the critical role of the template in establishing the overlayer structure since direct Sm-Sm bonding must be small.

In Fig. 34a we show a drift-corrected 160 x 160 Å$^2$ image of 0.075 ML Sm/GaAs(110). Inspection shows two distinct structures corresponding to ordered zig-zag chains (boxes labeled *A*) and linear chains (boxes labeled *B*). In the linear chain structure, the Sm atoms bridge along [$1\bar{1}0$] and bond near surface As atoms, Fig. 33c. In this structure, the Sm-Sm separation of

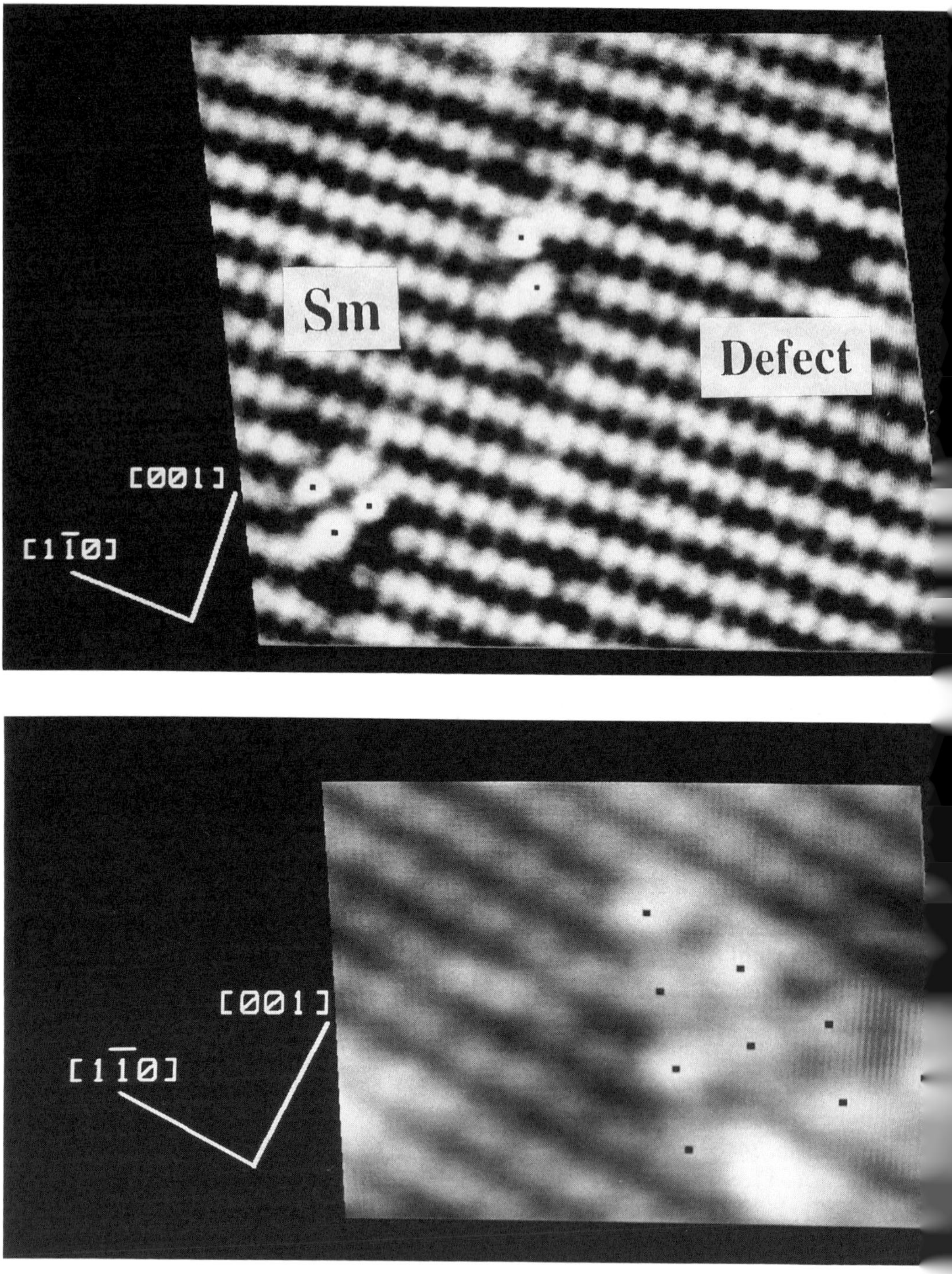


**Figure 33.** (Continued next page)

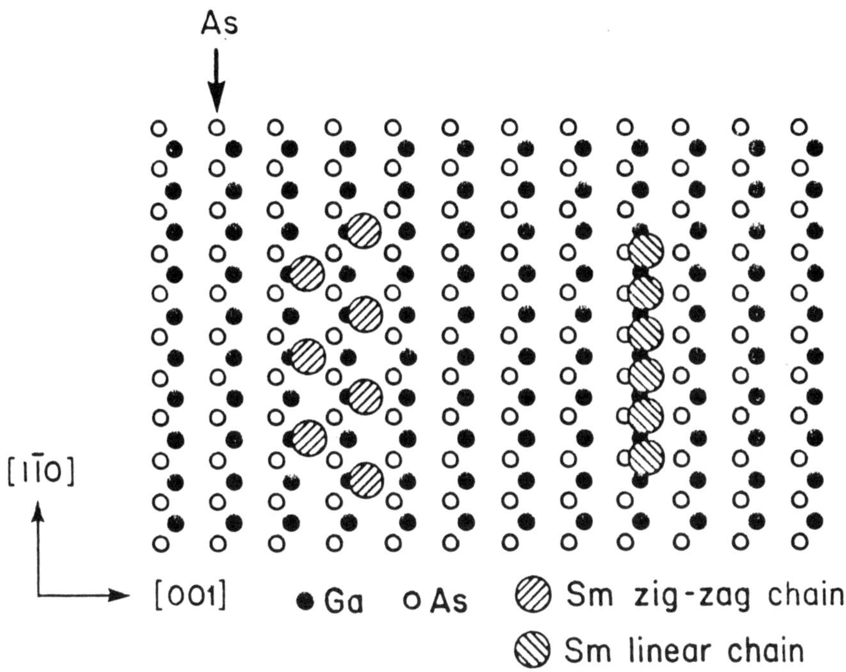


**Figure 33 (Cont'd).** STM images for *(a)* 0.01 and *(b)* 0.05 ML of Sm on GaAs(110) for areas of 80 x 80 Å$^2$ and 55 x 55 Å$^2$ acquired with a negative sample bias of 2.5 V and a tunneling current of 0.1 nA. *(c)* is a schematic of the GaAs surface with Sm placed in sites corresponding to the zig-zag and linear chains. (From Ref. 137.)

4 Å is closer to that of bulk Sm. Counting indicates that ~60% of the adatoms are bonded in zig-zag chains sites and 40% are in linear chain sites at 0.075 ML. The linear chains become more numerous at the expense of the lower density zig-zag chains, as shown in Fig. 34b for 0.33 ML of Sm. At this coverage, there are trace amounts of zig-zag chains and substantial contributions from linear chains. Significantly, another structure is evident as a cluster where individual atoms cannot be imaged (boxes labeled *C*). These clusters typically cover the equivalent of 3 - 5 rows of the surface and appear to be disrupted regions of the surface. With increased deposition, the amount of surface conversion from chains to clusters increases, and growth to 1 - 2 ML produces an overlayer where atom resolution is not possible.

Comparison of STM and photoemission results makes it possible to correlate changes in surface geometry with Sm bonding and valence states. The STM results show a transition from Sm zig-zag chains to linear chains and a second transition from linear chains to a surface dominated by clusters. Lineshape changes in valence bands reflect variations that can be associated with the zig-zag and linear chain structures reported from STM results. The proportion of $Sm^{3+}$ emission to the total Sm 4f emission (~25% at 0.33 ML) is comparable to the proportion of Sm atoms in clusters observed with the STM (~20% at 0.33 ML by calculating the area of surface affected by clusters).

The evolving structure and chemical bonding for Sm/GaAs(110) at 300 and 20 K can be summarized by the sketches of Fig. 35. For growth at 300 K, ordered zig-zag chains derived from <10 atoms appear from very low coverage. Sm atoms in these chains are ~7 Å apart and are bonded to surface Ga atoms in positions equidistant from 4 As atoms. (This would correspond to a c(2x2) structure if the surface were uniform covered.) With increased deposition to 0.1 ML, linear chains of Sm appear along $[1\bar{1}0]$ with shorter Sm-Sm distances of ~4 Å. The Sm atoms in these chain configurations are divalent and those in the zig-zags present higher binding energies than those in the linear chains. Further deposition induces conversion from ordered chains to unresolved structures extending across several rows of the substrate (reacted clusters sketched in Fig. 35b). Photoemission shows a fully-shifted As component for these reacted clusters which indicates Sm-As bonding without long range three-dimensional order. Continued growth at 300 K converts the surface into a reacted interfacial layer (Fig. 35c). Ultimately, Sm film growth occurs and by ~7 ML the surface of the film is indistinguishable from that of Sm, although As atoms decorate it.

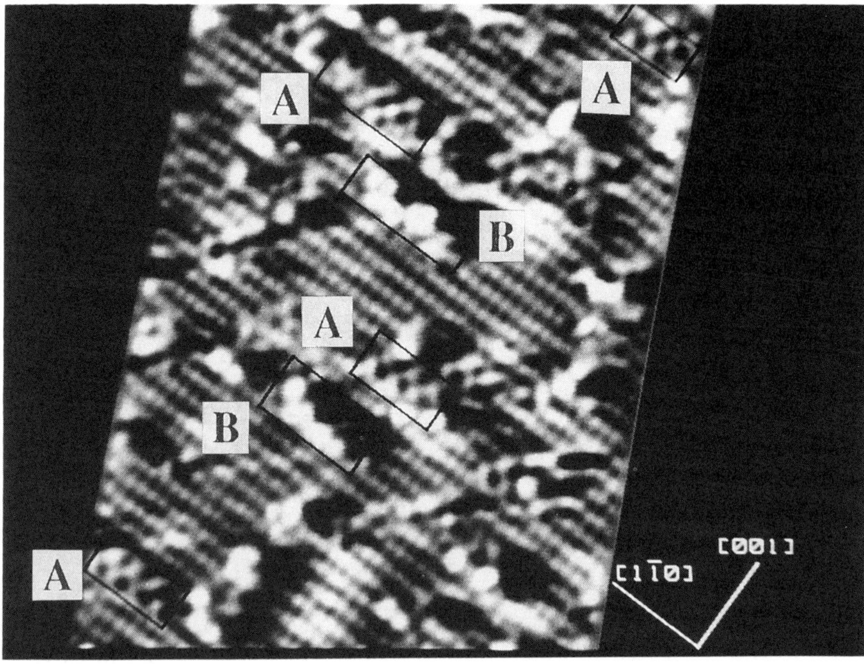


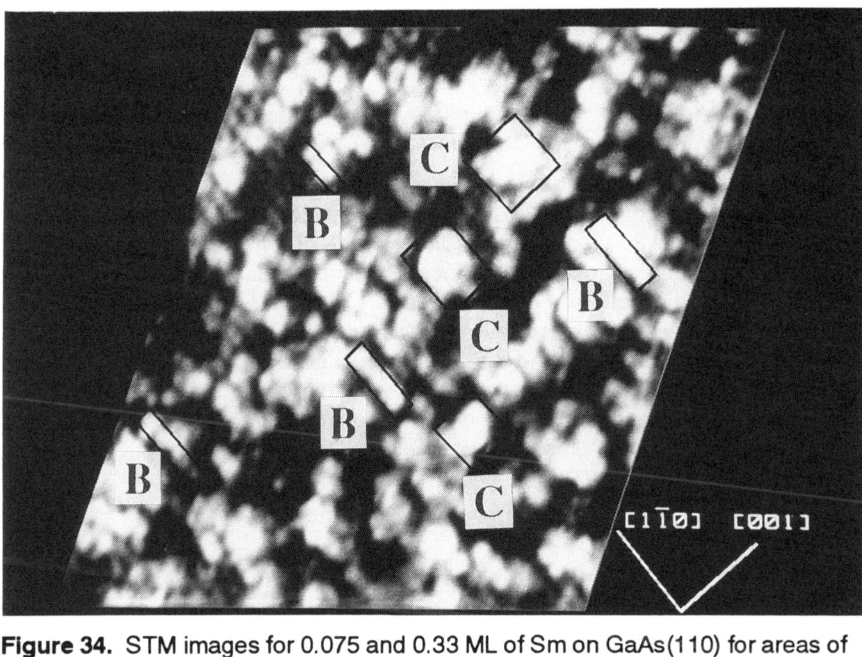


**Figure 34.** STM images for 0.075 and 0.33 ML of Sm on GaAs(110) for areas of 160 x 160 Å$^2$ and 195 x 195 Å$^2$ under the same conditions as for Fig. 33. The boxes highlight zig-zag chains *(A)*, linear chains *(B)* and clusters *(C)*. It is likely that the linear chain shown in box *B* in the upper center of 34a is not well ordered. (From Ref. 137.)

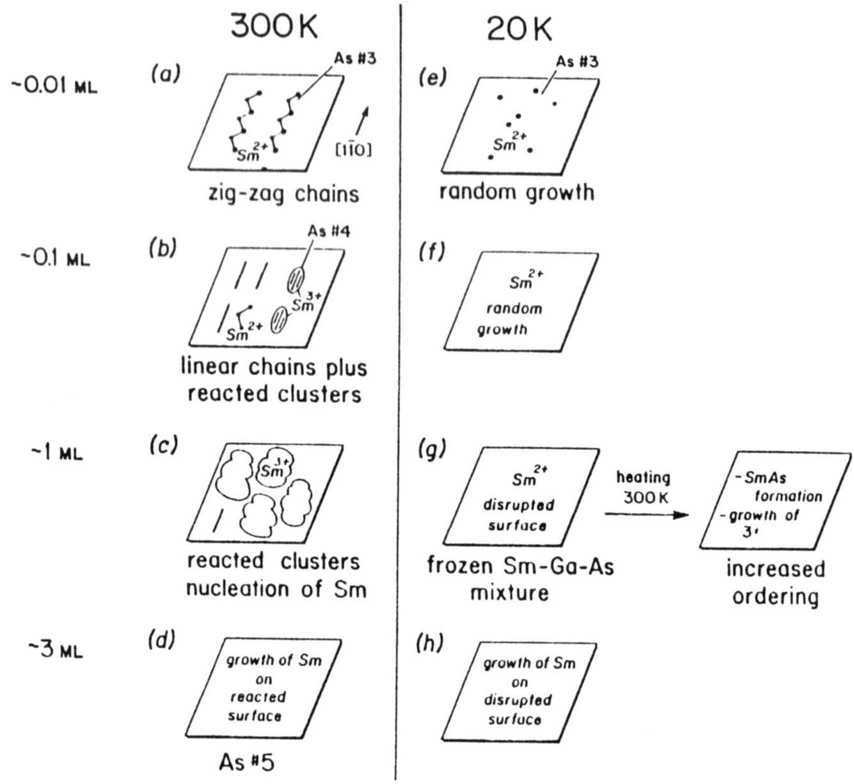


**Figure 35.** Schematic of the evolution of the Sm/GaAs(110) interface. For low coverage growth at 300 K, zig-zag and linear chains of $Sm^{2+}$ produce a new feature in the As 3d core level, called component *3* (components *1* and *2* correspond to emission from the bulk and surface atoms). At higher Sm coverage, surface disruption produces $Sm^{3+}$ and a new As component, *#5*. For 20 K growth, disruption of the surface occurs when the Sm local density exceeds a critical size, but the resulting mixture is kinetically constrained in an amorphous structure. Heating to 300 K releases these constraints and allows atom redistribution for very thin films. Growth at both temperatures produces Sm films at higher coverage. (From Ref. 136.)

Low temperature Sm growth on GaAs(110) differs from that at 300 K in several important ways. At 20 K, the reduced Sm mobility frustrates the formation of ordered chains, and we envision a surface randomly decorated with $Sm^{2+}$ atoms that form groups upon further deposition. The number of atoms in each group is determined by the statistical nature of the deposition process (Figs. 35e and f). From the Ga and As core levels, intermixed regions of Sm-Ga-As form but Sm atoms in the frozen mixture remain $2^+$ and SmAs-like bonding is not observed. Continued deposition again leads to Sm nucleation in the intermixed region and kinetics limits further mixing.

From the above, it is clear that kinetics play a very important role in overlayer growth, limiting disruption to a monolayer or two at 300 K and keeping the Sm-Ga-As mixture frozen at 20 K. To determine whether the frozen mixture formed by 1 ML deposition at 20 K would remain unchanged during thermal activation, the film was progressively annealed. The results showed the enhancement of $Sm^{3+}$ emission and the appearance of a fully-reacted As component by ~50 K, with continued conversion thereafter. Comparison to results obtained after 1 ML deposition at 300 K showed only minor quantitative differences.

Additional insight into the stability of a high-density Sm layer in contact with GaAs(110) can be gained by examining cluster-assembled interfaces. Here, we were interested in forming Sm clusters so that the stability of the cluster-GaAs interface could be assessed. Small amounts of Sm (~0.1 Å corresponding to ~$2.5 \times 10^{13}$ atoms/cm$^2$) were deposited onto Xe at 20 K, the system was warmed to ~100 K to desorb the Xe, and the clusters were brought into contact with the substrate. The widely separated clusters formed on Xe were very small because the $Sm^{2+}$ emission was much greater than the $Sm^{3+}$ (estimated separation ≥50Å). When the clusters were brought into contact with GaAs(110) at ~100 K, the ratio of $Sm^{3+}$ to $Sm^{2+}$ emission increased to nearly unity. We conclude that the small Sm clusters produced reacted regions upon contact at ~100 K. Upon warming to ~200 K, the relative amount of $Sm^{3+}$ increased and the As 3d core levels broadened. We conclude that high density Sm layers are unstable when in contact with GaAs(110), even when formed by cluster assembly.

## 9.2 Final Comments

This chapter has focused on issues related to the atomic-scale chemistry of metal-semiconductor interfaces, using examples to elaborate on those issues. Unique aspects have dealt with the variation of the chemical parameters made possible by cluster deposition, ion deposition, and low temperature growth. We anticipate that further control of the many parameters related to interface and thin film growth will offer additional insight in the near future, and enhanced use of STM will make it possible to correlate chemical and physical properties.

## ACKNOWLEDGMENTS

It is a pleasure to draw attention to the excellent work by G. D. Waddill, I. M. Vitomirov, M. Vos, F. Xu, S. A. Chambers, J. J. Joyce, M. del Giudice, and M. W. Ruckman as reviewed in this chapter. Interactions with them, with many other members of the Electronic Materials Group of the University of Minnesota, and with the community at large have been stimulating and rewarding.

This work has been supported by the Office of Naval Research, the National Science Foundation, and the Army Research Office. The synchrotron radiation photoemission experiments were done at Aladdin, a user facility supported by the National Science Foundation and operated by the University of Wisconsin. The assistance of the staff of that laboratory is gratefully acknowledged. Data acquisition and data analysis has been enhanced immeasurably by access to IBM minicomputers obtained through a Materials Science and Processing Grant from IBM.

## REFERENCES

1. J. H. Weaver, *Physics Today,* 39:24 (1986)
2. E. Bauer and H. Poppa, *Thin Solid Films,* 12:167 (1972)
3. J. A. Venables, G. D. T. Spiller and M. Harbücken, *Rep. Prg. Phys.,* 47:339 (1984)
4. J. H. Weaver, *Analysis and Characterization of Thin Films*, (K. N. Tu and R. Rosenberg, eds.), Ch. 2, Academic Press, New York (1987)
5. G. Margaritondo and J. H. Weaver, *Methods of Experimental Physics: Surfaces*, Ch. 4, Vol. 22 (R. L. Park and M. G. Lagally, eds.) Academic Press, Orlando (1985)
6. *Appl. Phys.,* Vols. 26, 27 (M. Cardona and L. Ley, eds.) Springer-Verlag, New York (1978)
7. B. Feuerbacher and R. F. Willis, *J. Phys.,* C9:169 (1976)
8. M. P. Seah and W. A. Dench, *Surf. and Int. Analysis,* 1:2 (1979)
9. D. Briggs and J. C. Riviere, *Practical Surface Analysis by Auger and X-ray PhotoelectronSpectroscopy*, (D. Briggs and M. P. Seah, eds.) Wiley, New York (1984)
10. J. H. Weaver, Z. Lin, and F. Xu, *Surface Segregation and Related Phenomena*, (P. A. Dowben and A. Miller, eds.), Ch. 10, CRC Press, Boca Raton (1990)

11. R. H. Williams, G. P. Srivastava and I. T. McGovern, *Rep. Prog. Phys.,* 43:87 (1980)
12. C. Kunz, in *Appl. Phys.,* Vol. 27, Ch. 6, (M. Cardona and L. Ley, ed.) Springer-Verlag, New York (1978)
13. D. Briggs and M. P. Seah, *Practical Surface Analysis by Auger and X-ray Photoelectron Spectroscopy,* Wiley, New York (1984)
14. K. Siegbahn, *ESCA Applied to Free Molecules,* North Holland, London (1969)
15. S. Doniach and M. Sunjic, *J. Phys.,* C3:285 (1970)
16. K. Levenberg, *Quart. Appl. Math.,* 2:164 (1944)
17. D. W. Marquardt, *J. Soc. Indust. Appl. Math.*, 11:431 (1963)
18. G. K. Wertheim and S. B. Dicenzo, *J. Electron Spec. and Rel. Phen.,* 37:57 (1985)
19. R. A. Butera, M. del Giudice, and J. H. Weaver, *Phys. Rev.,* B33:5435 (1986)
20. M. del Giudice, J. J. Joyce, M. W. Ruckman, and J. H. Weaver, *Phys. Rev.,* B35:6213 (1987)
21. S. Brennan, J. Stöhr, R. Jaeger, and J. E. Rowe, *Phys. Rev. Lett.,* 45:1414 (1980)
22. F. J. Himpsel, P. Heimann, T.-C. Chiang, and D. E. Eastman, *Phys. Rev. Lett.,* 45:1112 (1980)
23. T. Miller, T. C. Hsieh, and T.-C. Chiang, *Phys. Rev.,* B33:6983 (1986)
24. D. E. Eastman, T.-C. Chiang, P. Heimann, and F. J. Himpsel, *Phys. Rev. Lett.,* 45:656 (1980)
25. J. J. Joyce, M. del Guidice, and J. H. Weaver, *J. Elect. Spect. and Rel. Phenom.,* 49:31 (1989)
26. For work until 1982, L. J. Brillson, *Surf. Sci. Rep.* 2:123 (1982)
27. For more recent work, see E. H. Rhoderick and R. H. Williams, *Metal-Semiconductor Contacts*, 2nd ed. Clarendon Press, Oxford (1988)
28. J. J. Joyce, M. Grioni, M. del Giudice, M. W. Ruckman, F. Boscherini, and J. H. Weaver, *J. Vac. Sci. Technol.,* A5:2019 (1987)
29. M. Grioni, J. J. Joyce, and J. H. Weaver, *J. Vac. Sci. Technol.,* A3:918 (1985)
30. F. Xu, J. J. Joyce, M. W. Ruckman, H.-W. Chen, F. Boscherini, D. M. Hill, S. A. Chambers, and J. H. Weaver, *Phys. Rev.,* B35:2375 (1987)
31. P. Skeath, I. Lindau, C. Y. Su, and W. E. Spicer, *Phys. Rev.,* B28:7051 (1983)

32. W. G. Petro, T. Kendelewicz, I. A. Babalola, I. Lindau, and W. E. Spicer, *J. Vac. Sci. Technol.,* A2:835 (1984)

33. T. Kendelewicz, R. S. List, M. D. Williams, K. A. Bertness, I. Lindau, and W. E. Spicer, *Phys. Rev.,* B34:558 (1986)

34. G. Hughes, R. Ludeke, F. Schäffler, and D. Rieger, *J. Vac. Sci. Technol.,* B4:924 (1986)

35. R. Ludeke and G. Landgren, *Phys. Rev.,* B33:5526 (1986)

36. R. H. Williams, A. McKinlay, C. J. Hughes, T. P. Humphreys, and C. Maani, *J. Vac. Sci. Technol.,* B2:561 (1984)

37. R. H. Williams, A. B. McGean, D. A. Evans, and W. G. Herrenden-Harker, *J. Vac. Sci. Technol.,* B4:966 (1986)

38. J. R. Waldrop, S. P. Kowalczyk, and R. W. Grant, *J. Vac. Sci. Technol.,* 21:607 (1982)

39. N. G. Stoffel, M. Turowski, and G. Margaritondo, *Phys. Rev.,* B30:3294 (1984)

40. T. Yoshiie and C. L. Bauer, *J. Vac. Sci. Technol.,* A1:554 (1983)

41. T. Yoshiie, C. L. Bauer, and A. G. Milnes, *Thin Solid Films,* 111:149 (1984)

42. T. G. Anderson, J. Kanski, G. Lelay, and S. P. Svensson, *Surf.Sci.,* 168:301 (1986)

43. J. Massies, P. Delescluse, P. Etienne, and N. T. Linh, *Thin Solid Films,* 90:113 (1982)

44. J. Massies and N. T. Linh, *J. Cryst.Growth,* 56:25 (1982)

45. A. Hiraki, *Surf. Sci. Rep.,* 3:357 (1984)

46. J. A. Venables, J. Derrien, and A. P. Janssen, *Surf.Sci.,* 95:411 (1980)

47. G. W. Rubloff, *Surf.Sci.,* 132:268 (1983)

48. C. Calandra, O. Bisi, and G. Ottaviani, *Surf. Sci. Rep.,* 4:271 (1985)

49. M. del Giudice, J. J. Joyce, M. W. Ruckman, and J. H. Weaver, *Phys. Rev.,* B35:6213 (1987); STM studies of Ti/GaAs have been reported by Y. N. Yang, B. M. Trafas, R. L. Siefert, and J. H. Weaver, *Phys. Rev.,* B44:5720 (1991)

50. G. Le Lay, *Surf.Sci.,* 132:169 (1983)

51. L. Braicovich, C. M. Garner, P. R. Skeath, C. Y. Su, P. W. Chye, I. Lindau, and W. E. Spicer, *Phys. Rev.,* B20:5131 (1979)

52. I. Abbati, L. Braicovich, and A. Franciosi, *Solid State Commun.,* 33:881 (1980)

53. L. J. Brillson, A. D. Katnani, M. Kelly, and G. Margaritondo, *J. Vac. Sci. Technol.,* A2:551 (1984)

54. M. Hanbucken, Z. Imam, J. J. Metois, and G. Le Lay, *Surf. Sci.,* 162:628 (1985)

55. A. Franciosi, D. G. O'Neill, and J. H. Weaver, *J. Vac. Sci. Technol.,* B1:524 (1983)

57. A. Franciosi, J. H. Weaver, and D. G. O'Neill, *Phys. Rev.,* B28:4889 (1983)

58. F. Boscherini, J. J. Joyce, M. W. Ruckman, and J. H. Weaver, *Phys. Rev.,* B35:4216 (1987) and references therein.

59. S. A. Chambers, S. B. Anderson, H. W. Chen, and J. H. Weaver, *Phys. Rev.,* B34:913 (1986)

60. C. Pirri, J. C. Peruchetti, G. Gewinner, and J. Derrien, *Phys. Rev.,* B29:3391 (1984)

61. J. Derrien, *Surf. Sci.,* 168:171 (1986)

62. C.-D. Lien, M.-A. Nicolet, C. S. Pai, and S. S. Lau, *Appl. Phys.,* A36:153 (1985)

63. F. Xu, C. M. Aldao, I. M. Vitomirov, and J. H. Weaver, *Appl. Phys. Lett.,* 51:1946 (1987)

64. *Handbook of Chemistry and Physics*, 66th ed., (R. C. Weast, ed.) CRC, Boca Raton, Florida (1985)

65. I. M. Vitomirov, C. M. Aldao, Z. Lin, Y. Gao, B. M. Trafas, and J. H. Weaver, *Phys. Rev.,* B38:10776 (1988),

66. Z. Lin, F. Xu, and J. H. Weaver, *Phys. Rev.,* B36:5777 (1987)

67. M. W. Ruckman, J. J. Joyce, and J. H. Weaver, *Phys. Rev.,* B33:7029 (1986)

68. M. Grioni, J. J. Joyce, and J. H. Weaver, *J. Vac. Sci. Technol.,* A4:965 (1986)

69. D. M. Hill, F. Xu, Z. Lin, and J. H. Weaver, *Phys. Rev.,* B38:1893 (1988)

70. I. L. Singer, J. S. Murday, and L. K. Cooper, *Surf. Sci.,* 108:7 (1981)

71. Y.-X. Wang and P. H. Holloway, *J. Vac. Sci. Technol.,* B2:613 (1984)

72. F. Xu, D. M. Hill, Z. Lin, S. G. Anderson, Y. Shapira, and J. H. Weaver, *Phys. Rev.,* B37:1893 (1988)

73. C. R. Bonapace, K. Li, and A. Kahn, *J. Phys. (Paris) Colloq.,* 45:C5-409 (1984)

74. K. Stiles, A. Kahn, D. G. Kilday, and G. Margaritondo, *J. Vac. Sci. Technol.,* B5:987 (1987)

75. K. Stiles, S. F. Horn, A. Kahn, J. McKinlay, D. G. Kilday, and G. Margaritondo, *J. Vac. Sci. Technol.,* B6:1392 (1988)

76. K. Stiles and A. Kahn, *Phys. Rev. Lett.,* 60:440 (1988)

77. K. K. Chin, T. Kendelewicz, C. McCants, R. Cao, K. Miyano, I. Lindau, and W. E. Spicer, *J. Vac. Sci. Technol.,* A4:969 (1986)

78. R. Cao, K. Miyano, T. Kendelewicz, K. K. Chin, I. Lindau, and W. E. Spicer, *J. Vac. Sci. Technol.,* B5:998 (1987)

79. W. E. Spicer, R. Cao, K. Miyano, C. McCants, T. T. Chiang, C. J. Spindt, N. Newman, T. Kendelewicz, I. Lindau, E. Weber, and Z. Liliental-Weber, *Metallization and Metal-Semiconductor Interfaces,* (I. P. Batra, ed.), Vol. 195, Series B, Physics, NATO Advanced Study Institute, Plenum, New York (1991)

80. A. B. McLean and R. H. Williams, *J. Phys.,* C21:783 (1988)

81. J. L. Freeouf, *Surf. Sci.,* 132:233 (1983)

82. R. Ludeke, *Surf. Sci.,* 168:290 (1986)

83. W. E. Spicer, Z. Liliental-Weber, E. Weber, N. Newman, T. Kendelewicz, R. Cao, C. McCants, P. Mahowald, K. Miyano, and I. Lindau, *J. Vac. Sci. Technol.,* B6:1245 (1988)

84. W. Mönch, *Phys. Rev. Lett.,* 58:1260 (1987)

85. J. Tersoff, *J. Vac. Sci. Technol.,* B3:1157 (1985)

86. G. D. Waddill, C. M. Aldao, I. M. Vitomirov, Y. Gao, and J. H. Weaver, *J. Vac. Sci. Technol.,* A7:865 (1989)

87. G. D. Waddill, I. M. Vitomirov, C. M. Aldao, and J. H. Weaver, *Phys. Rev. Lett.,* 62:1568 (1989)

88. C. M. Aldao, G. D. Waddill, I. M. Vitomirov, and J. H. Weaver, *Appl. Phys. Lett.,* 53:2647 (1988)

89. C. M. Aldao, G. D. Waddill, S. G. Anderson, and J. H. Weaver, *Phys. Rev.,* B40:2932 (1989)

90. M. W. Ruckman, M. del Giudice, J. J. Joyce, and J. H. Weaver, *Phys. Rev.,* B33:2191 (1986)

91. R. Ludeke, D. Straub, F. J. Himpsel, and G. Landgren, *J. Vac. Sci. Technol.,* A4:874 (1986)

92. S. B. Zhang, M. L. Cohen, and S. G. Louie, *Phys. Rev.,* B34:768 (1986). The reverences cited in this paper provide an overview of theoretical progress made in understanding Al/GaAs interfaces.

93. A. Kahn, *J. Vac. Sci. Technol.,* A1:684 (1983)

94. J. R. Chelikowski, D. J. Chadi, and M. Cohen, *Phys. Rev.,* B23:4013 (1981)

95. C. Priester, G. Allan, and M. Lannoo, *Phys. Rev. Lett.,* 58:1989 (1987)

96. S. G. Anderson, C. M. Aldao, G. D. Waddill, I. M. Vitomirov, S. J. Severtson, and J. H. Weaver, *Phys. Rev.,* B40:8305 (1989)

97. G. D. Waddill, C. M. Aldao, I. M. Vitomirov, S. G. Anderson, and J. H. Weaver, *J. Vac. Sci. Technol.,* B7:950 (1989),

98. I. M. Vitomirov, G. D. Waddill, C. M. Aldao, S. G. Anderson, C. Capasso, and J. H. Weaver, *Phys. Rev.,* B40:3483 (1989)

99. R. Cao, K. Miyano, T. Kendelewicz, I. Lindau, and W. E. Spicer, *Phys. Rev.,* B39:11146 (1989)

100. C. M. Aldao, I. M. Vitomirov, F. Xu, and J. H. Weaver, *Phys. Rev.,* B37:6019 (1988)

101. A. K. Niessen, F. R. de Boer, R. Boom, P. F. de Chatel, W. C. M. Mattens, and A. R. Miedema, *CALPHAD,* 7:51 (1983)

102. J. H. Weaver, Z. Lin, and F. Xu, *Surface Segregation and Related Phenomena*, (P. A. Dowben and A. Miller, eds.), Ch. 10, CRC Press, Boca Raton (1990)

103. C. M. Aldao, G. D. Waddill, I. M. Vitomirov, and J. H. Weaver, *J. Vac. Sci. Technol.,* A7:817 (1989)

104. G. D. Waddill, I. M. Vitomirov, C. M. Aldao, S. G. Anderson, C. Capasso, J. H. Weaver, and Z. Liliental-Weber, *Phys. Rev.,* B41:5293 (1990)

105. I. M. Vitomirov, C. M. Aldao, G. D. Waddill, C. Capasso, and J. H. Weaver, *Phys. Rev.,* B41:8465 (1990)

106. M. Vos, C. M. Aldao, D. J. W. Aastuen, and J. H. Weaver, *Phys. Rev.,* B41:991 (1990)

107. C. M. Aldao, D. J. W. Aastuen, M. Vos, I. M. Vitomirov, G. D. Waddill, P. J. Benning, and J. H. Weaver, *Phys. Rev.,* B42:2878 (1990)

108. J. M. E. Harper *Thin Film Processes,* (J. L. Vossen and W. Kern, eds.), Academic Press, New York (1978)

109. M.-A. Hasan, S. A. Barnett, J.-E. Sundgren, and J. E. Greene, *J. Vac. Sci. Technol.,* A5:1883 (1987)

110. S.-N. Mei, T.-M. Lu, and S. Robert, *IEEE Elec. Dev. Lett.,* 8:503 (1987)

111. R. R. Daniels, T.-X. Zhao, and G. Margaritondo, *J. Vac. Sci. Technol.,* A2:831 (1984)

112. K. Stiles, A. Kahn, D. G. Kilday, and G. Margaritondo, *J. Vac. Sci. Technol.,* B5:987 (1987)

113. K. K. Chin, T. Kendelewicz, R. Cao, K. Miyano, I. Lindau, and W. E. Spicer, *J. Vac. Sci. Technol.,* A4:969 (1986)

114. F. Xu, C. M. Aldao, I. M. Vitomirov, Z. Lin, and J. H. Weaver, *Phys. Rev.,* B36:3495 (1986)

115. K. K. Chin, T. Kendelewicz, N. Newman, I. Lindau, and W. E. Spicer, *J. Vac. Sci. Technol.,* B4:955 (1986)

116. R. Ludeke, A. Taleb-Ibrahimi, R. M. Feenstra, and A. B. McLean, *J. Vac. Sci. Technol.,* A7:936 (1989)

117. J. J. Joyce, J. Anderson, M. M. Nelson, C. Yu, and G. J. Lapeyre, *J. Vac. Sci. Technol.,* A7:859 (1989)

118. J. J. Joyce, J. Anderson, M. M. Nelson, C. Yu, and G. J. Lapeyre, *Phys. Rev.,* B40:10412 (1989)

119. Y. Hu, T. J. Wagener, M. B. Jost, and J. H. Weaver, *Phys. Rev.,* B40:1146 (1989)

120. G. D. Waddill, C. M. Aldao, C. Capasso, P. J. Benning, Y. Hu, T. J. Wagener, M. B. Jost, and J. H. Weaver, *Phys. Rev.,* B41:5960 (1990)

121. Y. Hu, T. J. Wagener, M. B. Jost, and J. H. Weaver, *Phys. Rev.,* B41:5817 (1990)

122. Y. Kuk and P.J. Silverman, *Rev. Sci. Instrum.,* 60:165 (1989)

123. R. M. Feenstra, J. A. Stroscio, J. Tersoff, and A. P. Fein, *Phys. Rev. Lett.,* 58:1192 (1987)

124. J. A. Stroscio, R. M. Feenstra, and A. P. Fein, *Phys. Rev. Lett.,* 58:1668 (1987)

125. J. A. Stroscio and R. M. Feenstra, *J. Vac. Sci. Technol.,* B6:1472 (1988)

126. R. J. Wilson and S. Chiang, *Phys. Rev. Lett.,* 58:369 (1987)

127. R. J. Wilson and S. Chiang, *Phys. Rev. Lett.* 58:2575 (1987)

128. J. Nogami, S. Park, and C. F. Quate, *Appl. Phys. Lett.,* 53:2086 (1988)

129. R. M. Feenstra, *Phys. Rev. Lett.,* 63:1412 (1989)

130. Y. -M. Mo, B. S. Schwartzentruber, R. Kariotis, M. B. Webb, and M. G. Lagally, *Phys. Rev. Lett.,* 63:2393 (1989)

131. Y.-M. Mo, D. E. Savage, B. S. Schwartzentruber, and M. G. Lagally, *Phys. Rev. Lett.,* 65:1020 (1989)

132. P. Avouris and R. Wolkow, *Phys. Rev.,* B39:5091 (1989)

133. J. S. Villarrubia and J. J. Boland, *Phys. Rev. Lett.,* 63:306 (1989)

134. A. Fujimori, M. Grioni, and J. H. Weaver, *Phys. Rev.,* B33:726 (1986)

135. M. Grioni, J. J. Joyce, and J. H. Weaver, *Phys. Rev.,* B32:962 (1985)

136. T. Komeda, S. G. Anderson, J. M. Seo, M. Schabel, and J. H. Weaver, *J. Vac. Sci. Technol.,* A9:1964 (1991)

137. B. M. Trafas, D. M. Hill, R. L. Siefert, and J. H. Weaver, *Phys. Rev.,* B42:3231 (1990)

138. P. N. First, R. A. Dragoset, J. A. Stroscio, R. J. Celotta, and R. M. Feenstra, *J. Vac. Sci. Technol.,* A7:2868 (1989)

139. B. D. Cullity, *Elements of X-ray Diffraction,* p. 507, Addison-Wesley, Reading, MA (1978)

# 8

# Survey of Recent Developments in the Theoretical Description of the Properties of Semiconductor Interfaces

*Charles B. Duke and Robert G. Dandrea*

## 1.0 INTRODUCTION

The purpose of this chapter is to provide a broad yet concise overview of the development of theoretical models and concepts in semiconductor interface science. Its intended audience consists of individuals who want a general understanding of the topics of current interest in this area and/or an entrée into the literature on these topics. Both the ground-state properties (e.g., atomic geometries) and excitation spectra (e.g., surface states in the vicinity of semiconductor-vacuum surfaces) are considered. Emphasis is placed on the development of theoretical concepts which enable the synthesis of the extensive experimental results on semiconductor interfaces and on specific calculations and data analyses which have appeared since 1987. Work prior to 1988 has been reviewed in two collections which appeared that year edited by King and Woodruff[1] and by de Wette.[2] These collections serve as the starting point for our consideration of the subsequent literature.

## 2.0 MODEL CONCEPTS, METHODS, AND ACCOMPLISHMENTS

### 2.1 Property Taxonomy

A review of the classification scheme for the properties of semiconductor interfaces is useful because theoretical models often are developed to describe only a limited subset of these properties. Confusion can arise if these models subsequently are applied to properties for which they are inappropriate.

Ground state properties of a semiconductor interface include its structure (i.e., atomic positions) and electronic charge density. An internally consistent theory of the electronic properties of a semiconductor interface (e.g., heterojunction or metal-semiconductor contact) must predict the ground state properties *a priori* and use these as the basis for the discussion of excited state properties like the flow of current through the junction upon the imposition of an external voltage. The use of models which are incompatible with the structure of semiconductor interfaces to describe their transport properties has led to considerable misunderstanding in the literature.

Even if the electrons of a semiconductor remain in their ground state, the nuclei can move generating both interface and bulk vibrational excitations as well as diffusion. Such atomistic motions often are described using models in which the nuclear species are regarded as interacting through effective two (or many) body interactions.[3]-[5] The utility of such models at semiconductor surfaces and interfaces is limited because of the new forms of chemical bonding (often accompanied by reconstructed atomic geometries) which occur at these interfaces.

The primary topic of interest for many semiconductor interfaces (e.g., in heterostructures or contacts) is the transport of electronic excitations in the vicinity of the interface.[6][7] In general, however, models which yield adequate descriptions of ground state properties require extension to achieve comparable successes for the electronic excitation spectra.[8]-[10] Still further extensions are required to describe the transport properties of these excitations.[11]-[12] Moreover, their transport properties often are dominated by structural and chemical defects both in the bulk materials and near the interface[6][7][13] which are notoriously difficult to characterize accurately. Thus, a grasp of the difficulty of a given problem is required in order to appreciate recent developments in its theoretical description.

## 2.2 Model Taxonomy

**Phenomenological Models:** Inspection of almost any text on junction transport properties reveals that such properties usually are described by lumped-parameter models in which the parameters are evaluated phenomenologically from the transport experiments themselves.[6][7][11]-[15] Thus, transport coefficients (e.g., mobilities), electrostatic and magnetic fields, and junction potential parameters (e.g., Schottky barrier heights) are embedded in these models *a priori* and often extracted from the measurements which they purport to interpret. Even if the transport coefficients are calculated rather than regarded as lumped parameters, phenomenological quantities like trap densities, trapping cross sections, scattering-rates or electron-phonon coupling constants enter the analysis.[11]-[14] It is possible to try to correlate structural information (e.g., dislocation density, interface perfection) with the values of these phenomenological parameters, but such correlations are established only *ad hoc* by virtue of their nature.

The overwhelming value of these models lies in electronic device design and the quantitative measurement of the effects of process control on device fabrication.[16] Essentially every manufacturer of Si and GaAs chips has its own internal set of design and process models which are used in the fabrication of these devices. Numerous versions of these models are available as commercial software packages for popular personal computers and workstations.[17]

***Ab Initio* Models:** As their name implies, *ab initio* models purport to provide descriptions of ground-state properties as well as elementary excitation spectra in terms of fundamental physical parameters (charges and masses of the electrons and nuclei) and interactions (coulomb interactions). At present, they are useful for the calculation of the structure, the atomic geometry, vibrational excitation spectra, and electronic excitation spectra of surfaces and interfaces.[18]-[22] Basically, the model calculations consist of four steps: *(i)* specify the Hamiltonian of the system; *(ii)* select a functional form for a trial ground-state wavefunction or charge density; *(iii)* minimize the ground-state energy by varying the parameters in the trial functional form and the positions of the nuclei; *(iv)* extend the analysis to encompass electronic excited states. Such calculations result in the ground-state energy and charge density and in the associated eigenvalue spectra of electronic excited states. Vibrational excitations are described by use of calculated atom-atom interactions in a dynamical description of the vibrations of a suitably chosen slab.[22] The atom-atom interactions are evaluated using the electronic density response function calculated for the ground state.

Perhaps the best known of this class of models are the self-consistent-field, linear combination of atomic orbitals (SCF-LCAO) methods utilized in quantum chemical calculations for molecules and clusters.[23] If the electronic wave function is taken to be a single one-electron determinant, one obtains the *Hartree-Fock* limit; corrections to the associated energy are referred to as the *correlation energy*. For molecules, these corrections are typically evaluated by using more complicated wave functions in a systematic way called *configuration interaction* calculations. These also are used to evaluate electronic excitation spectra. Early versions of many-electron perturbation theory for solids were built on the Hartree-Fock limit as the starting wave function.[8] In calculations for surfaces and interfaces, however, the SCF-LCAO methods are rarely used because they have been supplanted by methods based on local density functional theory.[24][25]

Essentially all modern *ab initio* calculations of interface structures and electronic excitation spectra are based on density functional models as embedded in various computer programs.[9][21][26]-[32] These models result from obtaining approximate representations for the electron density which minimizes an expression for the ground-state energy which is a functional of this density. A comprehensive review of the history, substance, and applications of local density functional theory and methods has been given recently by Dreizler and Gross.[24] In their original form, the methods were suitable only for the description of ground-state properties, so that considerable effort recently has been invested in extending them to encompass electronic excitation spectra, especially those of semiconductors.[9][10][21][27]-[29] Confusion still arises in the literature, however, because the need for these extensions to extract the parameters associated with semiconductor heterojunctions and metal-semiconductor contacts is not universally recognized.[28] Models based on these methods are capable of predicting both the atomic geometries and the interfacial electronic excitation spectra of semiconductor surfaces and interfaces provided, of course, that the methods are exercised in a fashion suitable for the calculation to which they are applied.

**Hybrid Models:** Over the years in quantum chemistry, approximate versions of SCF-LCAO methods have been formulated to attack problems too complicated (e.g., large molecules, chemical reactions) for the economical application of the full *ab initio* methodology. A substantial literature has evolved around these models[23][33] which have independent etymologies in quantum chemistry and solid-state physics. At the most elementary level, one finds models in which no self-consistency is attempted and certain electronic interactions are approximated by adjustable parameters, i.e., the Hückel model in quantum chemistry and its equivalent tight-binding model in solid-state physics. Beyond that, additional coulomb interactions can be treated approximately, yielding an alphabet soup of models (e.g., Complete Neglect of Differential Overlap or CNDO, Incomplete Neglect of Differential Overlap or INDO) which require a dictionary[23][33] just to understand the labels. We refer to these as hybrid models because they utilize the mathematical format of *ab initio* models to introduce a phenomenological parameterization of certain integrals. They are most useful when the parameters can be determined on "simple" systems (e.g., small molecules) and then used to determine the properties of "complex" systems (e.g., large molecules with the same constituents and bonding types as the small ones used to determine the parameters). A description of a successful application of this philosophy to large aromatic organic molecules may be found in an overview of the CNDO/S3 model.[34]

These models are important for semiconductor interfaces because extensions of the tight-binding model to give total energies, developed by Chadi[35] and Pandey[36][37] have proven remarkably effective at predicting surface and interface properties based on parameters determined from bulk properties alone. While their utility has been recognized for some time,[1][2][18][19] extensions to Si(111)-(7x7),[38] II-VI cleavage faces,[39] oxide cleavage faces,[40] and III-V(110)-p(1x1)-Sb[41] have been reported recently which continue to yield new insights about the structure and bonding of these surfaces. Such models also are widely used in the study of semiconductor heterostructures,[7] in which case we extend the nomenclature "hybrid" to encompass models like the effective mass approximation and k•p perturbation theory.

## 2.3 Highlights of Developments During 1988 - 1990

Perhaps the most pervasive development since 1987 in the theory of semiconductor interfaces is the increasingly widespread utilization of both *ab initio* and hybrid models to predict, either quantitatively or semiquantitatively, the atomic geometries of a wide range of clean surfaces. In the case of Si(100), the predictions of an elegant analysis[42] of the effect of external stress on the relative sizes of inequivalent (2x1) and (1x2) stress domains were confirmed experimentally.[43] Interest in the energetics of the (2x1) dimer structure continues unabated, with indications emerging from *ab initio* calculations that the low-energy dimers may be untilted.[44][45] The detailed energetics of the dimer-adatom-stacking fault (DAS or Takayanagi) model[18] for Si(111)-(7x7) and Ge(111)-(7x7), however, remain elusive.[38][44][46] Analyses of the surface structures of the cleavage surfaces of tetrahedrally coordinated compound semiconductors were extended to encompass the prediction of the zincblende(110) surface structures of ZnS,[47] ZnSe,[48] and CdTe,[49] as well as of $(10\bar{1}0)$ and $(11\bar{2}0)$ surfaces of wurtzite structure ZnS,[47] ZnSe,[48] ZnO,[50] CdS,[51] and

CdSe.[51] The zincblende(110) and ZnO($10\bar{1}0$) calculations yielded previously determined structures.[19] The CdSe predictions were confirmed *a posteriori* by low-energy electron (LEED) and positron (LEPD) diffraction experiments.[52] A series of predictions for cubic, β-cristobalite and β-tridymite oxide surfaces has been reported[40] but not yet tested experimentally. The atomic geometries of ordered adsorbate structures also have been predicted, especially for Sb on III-V cleavage faces[41] and alkali metals on Si(100),[53] although this topic lies beyond the scope of our review.

An important use of surface structural calculations is the identification of general insights concerning surface chemistry and relaxation. One such insight has emerged from examination of studies of Si(100)-(2x1),[45] zincblende, and wurtzite cleavage faces,[39] and charge-neutral oxide surfaces,[40] i.e., the critical role of surface topology in determining surface reconstruction. If the topology permits (e.g., dimerization of atomically flat cleaved Si(100) or bond-length-conserving relaxations for zincblende, wurtzite, β-cristobolite, and β-tridymite charge neutral surfaces) then activationless relaxation or reconstruction occurs leading to spontaneously deformed surfaces. If activation barriers exist (e.g., the DAS Si(111)-(7x7) structure), then the surface geometry depends on the processing conditions and/or defect structure of the initial surface. While this concept is hardly new [recall, e.g., the temperature and sample dependence of the structures formed when Al is deposited on GaAs(110)],[19] its quantitative elucidation via total energy calculations permits the *a priori* prediction of the atomic geometries of activationless systems, e.g., the case of the cleavage surfaces of CdSe.[39][52] Another such insight is the recognition that a common mechanism of activationless surface relaxations can yield "universal" surface geometries in suitably scaled atomic units for stoichiometric non-polar cleavage surfaces of compound semiconductors.[20][54] This analysis establishes the dominance of the effects of surface topology and chemical rehybridization over the specific details of the materials involved for these relatively simple surfaces. Finally, analogous general principles of surface reconstruction have been sought in the more complex cases of elemental semiconductors and polar surfaces of compound semiconductors for which the surface stoichiometry is not established *a priori*. For example, in the latter case, a particular set of such hypothetical principles has been applied to speculations about the mechanisms of both III-V[55] and II-VI[56] molecular beam epitaxy. Despite considerable effort, however, an accepted, internally consistent set of structure-determination principles involving saturation of surface valences, metallic versus insulating surface electronic excitation spectra, and the role of strain has not yet been generated either for elemental semiconductors[18][44][46] or the polar surfaces of compound semiconductors.[57][58]

Turning from solid-vacuum surfaces to solid-solid interfaces, three important new complications appear. Foremost among these is the failure of many of the interfaces of interest (e.g., those in Si and GaAs microelectronics technology) to be at equilibrium. Thus, one is confronted with the dual problems of constructing models of the "actual" interface and of performing studies of simpler "model" interfaces in order to generate the insight required to build models of the technologically important ones.[59] Second is the paucity of detailed structural information about even the most perfect model interfaces due in part to the experimental challenges of quantitative structure analyses, and in part to the difficulty of preparing defect-free, epitaxial interfaces. Third is the traditional focus on the electrical or opto-electronic characteristics of devices built embodying the interfaces of interest. These characteristics tend to be dominated by chemical and structural defects which are difficult to characterize either experimentally or theoretically.

The most fully characterized solid-solid interfaces are nearly-lattice-matched semiconductor heterostructures.[1][7][60] In the absence of interdiffusion, the interface geometries can be determined experimentally in certain cases[61] and calculated using modern *ab initio* methods.[30][60] Except for thin (e.g., up to ten monolayers) superlattice structures, the properties of interfaces per se are incorporated into models of the electronic structure and transport properties as boundary conditions,[7][13][60] specifically as energy shifts (band offsets) and continuity conditions on the excitation wave functions. Recent theoretical developments in this area include the use of several approaches to describe the electronic excitation spectra of heterostructures subject to such boundary conditions[7][60] and *ab initio* calculations of the structure and electronic excitation spectra of both thin superlattices and heterojunctions.[29][30] These are reviewed more extensively in Sec. 4.

Finally, metal-semiconductor contacts continue to attract considerable interest. *Ab initio* calculations of the effective potential governing the dynamics of electronic excitations in the vicinity of such contacts (Schottky barrier potentials) are beginning to appear.[28][31][32] These interfaces are characterized by all of the complications noted above, however, in addition to which a clear distinction between calculations suitable for ground and excited state properties is not always made.[28] Thus, model calculations of the features of these Schottky barrier potentials still fail to give a fully satisfactory account of available experimental results[62] at least in part due to the important role of defects in determining these potentials. These calculations are examined more thoroughly in Sec. 5.

## 3.0 SEMICONDUCTOR VACUUM INTERFACES

### 3.1 Relaxation and Reconstruction

Unlike most metals, for which atoms near close-packed faces experience small relaxations relative to the bulk under the influence of unbalanced electrostatic forces near the surface,[63] most semiconductors exhibit large ($\Delta a \sim 1$Å) changes in atomic positions for surface species for all surfaces. If the symmetry of the surface unit cell remains identical to that of the bulk, such changes are called *relaxations.* If the symmetry is lowered relative to the bulk, they are referred to as *reconstructions.* Both are associated with major electronic charge rearrangements ("rehybridization" of the surface chemical bonds) caused by breaking of the directional covalent bonds in the bulk solid upon formation of a surface.[18]-[20] The precise nature of the relaxed or reconstructed surface structure is determined by numerous variables: surface stoichiometry, atomic topology, energies of the various possible structures, and processing conditions. Our purpose in this section is to provide a brief survey of a few special cases which are indicative of the factors which influence these structures.

The most important of these factors seem to be surface stoichiometry and topology. If the stoichiometry is fixed, e.g., on the charge neutral cleavage faces of compound semiconductors, then surface topology governs the allowed modes of activationless surface relaxation and leads to simple, perhaps even universal,[54] surface atomic geometries.[40] In the cases of elemental semiconductors and polar faces of compound semiconductors, the surface stoichiometry is typically not fixed *a priori* and hence is determined by the processing conditions. Si(100)-(2x1) exhibits a single family of stable dimer structures in analogy with compound semiconductor cleavage faces.[18][45] Si(111) surfaces are prepared both by cleavage and by ion or laser bombardment and anneal cycles leading to a variety of structures determined by a combination of the relaxation kinetics (i.e., processing conditions) and energetics of the possible structures. This is characteristic of surfaces for which an activationless relaxation can occur at the cleavage stoichiometry [i.e., the Pandey $\pi$-bonded Si(111)-(2x1) chain][18][37] but for which different effective surface stoichiometries lead to lower-energy structures that require, however, activation to be formed [i.e., the DAS Si(111)-(7x7) structure].[18] Polar surfaces of compound semiconductors afford a yet more complex case in which the surface stoichiometry is not self-selected by kinetics and energetics but is controlled externally, typically by dosing or fluxes in a molecular beam epitaxy (MBE) machine.[64] In the remainder of this section, we examine selected recent model calculations that illuminate the factors involved in the formation of this rich variety of surface structures.

### 3.2 Zincblende Cleavage Faces

The (110) surfaces of zincblende structure compound semiconductors are the charge neutral cleavage faces consisting of zig-zag chains containing equal numbers of anions and cations as shown in Fig. 1.[19][20] They may be prepared either by vacuum cleavage or by suitable ion bombard and anneal cycles.[65] For III-V semiconductors several theoretical models of their relaxations, via an approximately bond-length-conserving rotation by $\omega = 29° \pm 3°$ from the surface plane, have been available since the early 1980s,[19] but these analyses had not been extended to II-VI compounds which exhibit the same surface structures[66] but significantly different small-molecule coordination chemistries. Recent extensions of tight-binding total energy models to predict the (110) surface atomic geometries of ZnS,[47] ZnSe,[48] and CdTe[49] yielded quantitative predictions of the experimentally determined structures[19] for these II-VI compounds as well, suggesting a topology-dominated mechanism for the surface relaxations. This mechanism was proposed[48] to be the lowering of the anion *dangling bond* surface-state energies upon bond-length-conserving rotations of the top-layer chains accompanied by rehybridization of the dangling bond charge into back and surface bonds. This proposal was confirmed by comparison of the model predictions with angle-resolved photoemission spectra for ZnSe(110)[48] and CdTe(110).[49] These results gave rise to the proposition that the surface topology of these precisely charge neutral, stoichiometric zincblende (110) surfaces, which permits bond-length-conserving rotations of the top layer chains, was the dominant ingredient in achieving activationless relaxations to approximately the same universal surface structure (with distances measured in units of the bulk lattice constant) for the surfaces of all III-V and II-VI compound semiconductors.[39] It was not known at the time this article was written (1991) whether this proposition is extensible to materials with different coordination chemistries which can be grown in the zincblende structure, e.g., CuCl.

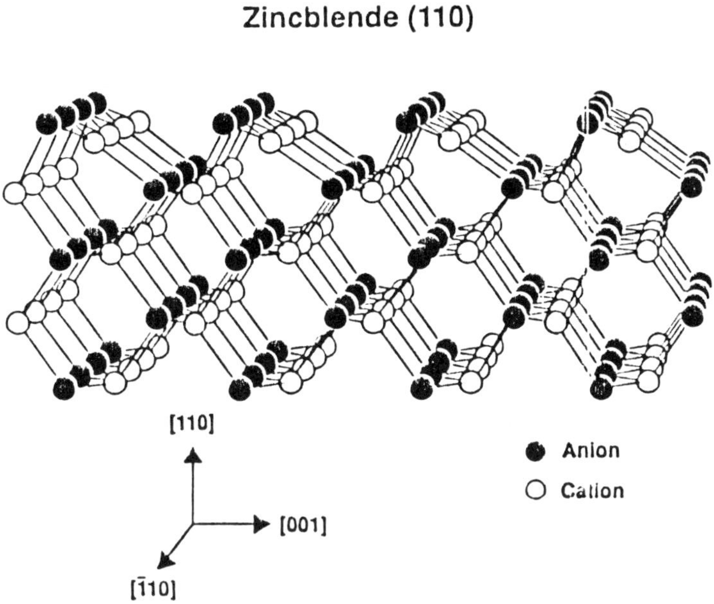


**Figure 1.** Illustration of the relaxed zincblende(110) surface. The relaxations of the surface atoms from their truncated-bulk positions lead to a raising of the top-layer anions and lowering of the top-layer cations, so that the surface cations form a nearly planar $sp^2$ arrangement with their three nearest neighbors. (After Duke.[19])

### 3.3 Wurtzite Cleavage Faces

Wurtzite structure compound semiconductors, mostly II-VI compounds, exhibit two cleavage faces. Both cleavage faces are charge neutral, i.e., they contain equal numbers of anions and cations. The $(10\bar{1}0)$ surfaces consist of anion-cation pairs, each anion and cation being connected to each other via a single surface bond and to the substrate by two back bonds. The $(11\bar{2}0)$ surfaces consist of anion-cation chains, analogous of zincblende(110), in which each anion is bonded to two surface cations and one substrate cation. Prior to 1988 no quantitative structural analyses had been performed for these surfaces except for an old study of ZnO$(10\bar{1}0)$.[67]

Tight-binding total energy calculations for both surfaces of ZnS,[47] ZnSe,[48] ZnO,[50] CdS,[51] and CdSe[51] revealed that all of these surfaces were relaxed; that the predicted relaxations were universal in the same sense (i.e., linear scaling of the surface structural parameters with the bulk lattice constant) as the zincblende(110) surfaces; and that the same topology-dominated surface-state-lowering mechanism was responsible for the activationless surface relaxations.[54] These predictions subsequently have been confirmed via a series of surface-structure determinations[52][68][69] of CdSe($10\bar{1}0$) and CdSe($10\bar{2}0$), photoemission measurements of the corresponding surface state eigenvalue spectra,[70] and a structure analysis of CdS($11\bar{2}0$).[71]

These results confirm the notions of universality developed to describe zincblende(110) surfaces. The topologies of the wurtzite ($10\bar{1}0$) and ($11\bar{2}0$) surfaces both permit bond-length-conserving relaxations which convert high-energy anion dangling bond surface states on the truncated bulk surface into lower energy surface- bonding and back-bonding states on the relaxed surfaces. These relaxations are activationless as in the case of the zincblende(110) surfaces. The resulting relaxed structure parameters scale linearly with the bulk lattice constant. Thus, this recently developed body of calculations and structure determinations of the cleavage surfaces of III-V and II-VI semiconductors supports quantitatively the concepts that the relaxed structures of these surfaces are approximately universal (i.e., that the dependence on the specific materials is described by the linear scaling of the surface structured parameters with the bulk lattice constant); that the relaxations are activationless; that both of these results are consequences of the surface topology admitting bond-length-conserving relaxations; and that the detailed mechanism of surface relaxation is the lowering in energy of surface states as they change from primarily anion dangling bond states on the truncated bulk surface to surface-bonding and back-bonding states on the relaxed surface.

Although the surface atoms at the zincblende and wurtzite cleavage surfaces experience large relaxations from their bulk positions, the symmetry of the surface unit cell remains identical to that of the truncated bulk lattice. Therefore, the surface is said to be *relaxed*, i.e., to exhibit a (1x1) surface structure. Polar surfaces of zincblende and wurtzite crystals, as well as all surfaces of elemental semiconductors, exhibit surfaces structures with lower symmetries parallel to the surface than the truncated bulk crystal. Such surfaces are said to be *reconstructed.*

### 3.4 Si(100)

Although the (100) surface of Si is not a cleavage face, it is the surface of an elemental semiconductor most analogous to the charge neutral surfaces of compound semiconductors in that its surface stoichiometry seems rather well fixed. As shown in Fig. 2, the surface species form dimers, roughly analogous to the wurtzite($10\bar{1}0$) surface, and hence are threefold coordinated with one surface and two back bonds.[18] This produces a (2x1) reconstruction of the truncated bulk structure. Since the dimers can form along either of two inequivalent directions, both (1x2) and (2x1) domains can form. It was recently shown that these are associated with a mosaic of elastic stress domains separated by steps and giving equal areas of (1x2) and (2x1) domains in the absence of external stress.[42][72]

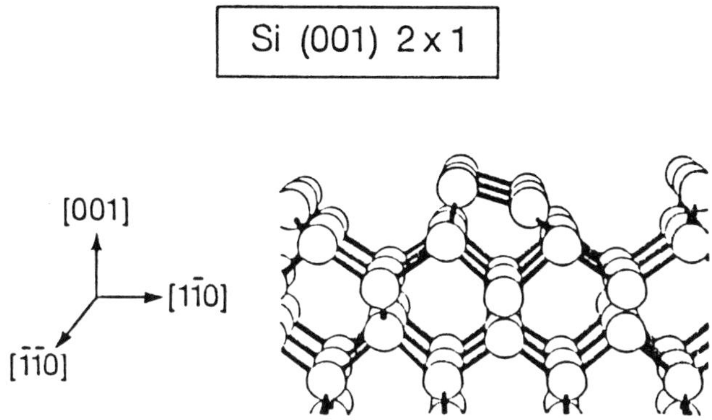


**Figure 2.** Illustration of the (001) silicon 2x1 buckled dimer reconstruction. The dimers are drawn tilted, i.e., with one dimer atom higher than the other. In the symmetric dimer structure, the dimers are not tilted. (After MacLaren et al.[171])

Another significant recent development is a new *ab initio* local density functional calculation of the energies of various possible dimer structures,[45] motivated by an analysis of the strong coulomb interaction limit within the symmetric dimer model.[73] These calculations suggest that a symmetric dimer structure exhibits the lowest energy ($\Delta\varepsilon$ = -1.17 eV per atom relative to the truncated bulk structure) but is nearly equal in energy with a buckled dimer structure. Reconstruction of the truncated bulk structure to the symmetric dimer minimum in the potential energy surface (PES) is activationless. An asymmetric (i.e., buckled) dimer minimum in the PES also occurs, however, and is nearly degenerate ($\Delta\varepsilon \sim$ 0.01 eV/atom) with the symmetric dimer minimum. It is separated from the symmetric dimer minimum by an activation barrier, also of the order of 0.01 eV/atom.

The picture of the Si(100)-(2x1) structure which emerges from these calculations embodies a symmetric dimer as the minimum energy structure which exhibits surface states with a gap associated with the occurrence of antiferromagnetic order between the spins of the electrons on the two atoms of the surface dimers.[44][45][73][74] This state is nearly degenerate with a buckled dimer minimum in the PES but is separated from this minimum by a low activation barrier. Since the symmetric structure is reached without activation from the truncated bulk structure, it is presumably the actual ground state of the (2x1) structure, although room temperature processing could lead to appreciable regions of the surface being in the asymmetric configuration. This picture of nearly degenerate symmetric and buckled dimers, both of which exhibit semiconducting one-electron surface-state excitation spectra, is qualitatively consistent with the large body of experimental data which has been acquired on the Si(100)-(2x1) surface.[18][73][75] This is a significant improvement over the situation in 1987,[18] although a quantitative analysis of these data has not yet been achieved.[45][75] Since an alternative model of the surface is that of asymmetric (i.e., buckled) dimers which can rapidly interconvert due to a low energy barrier for motion between their two degenerate minimum-energy positions, a quantitative analysis of the available data is required to determine which model actually describes the atomic geometry of Si(100)-(2x1).

### 3.5 GaAs(100)

Relative to Si(100), predictions of the surface structures of GaAs(100) must contend with two additional complications. First, since Ga and As are electronically inequivalent, the difference in their electronic structure (and hence chemical bonding) must be incorporated into the analysis. Second, in general the surface stoichiometry is not known so that predictions of the structure must be performed as a function of

Eq.(1) $$\Delta n = n_a - n_c$$

in which $n_a$ and $n_c$ are the surface concentrations of the anion (As) and cation (Ga) respectively.

The first factor is traditionally incorporated into the analysis by a two-step procedure.[76] Since the anion dangling bond states lie lower in energy than the corresponding cation states, a semiconducting surface is formed by choosing a structure in which the anion-derived states are full while the cation-derived ones are empty. Second, the resulting structure is allowed to relax, which reduces the energies of the (occupied) anion derived states while increasing those of the (empty) cation derived states. In this fashion possible reconstructed structures can be postulated so that their total (relaxed) energies can be calculated and compared.[57][76]

The second factor is included by minimizing the Gibb's free energy

Eq. (2) $$G = E - (\mu_a + \mu_c)n_t/2 + (\mu_c - \mu_a)\Delta n/2$$

Eq. (3) $$n_t = n_c + n_a$$

in which E is the total energy and $\mu_a$ ($\mu_c$) is the anion (cation) chemical potential. The $n_t$ term is presumed to be constant ($\mu_a + \mu_c$ being taken to be the bulk chemical potential of GaAs) so that the minimum of G is a function of $\Delta\mu = (\mu_c - \mu_a)$ which can be varied experimentally via, e.g., adjusting the fluxes in an MBE machine.[64] This procedure produces minimum-energy structures which are explicit functions of surface composition, $\Delta n$.[57][77]

A major study of GaAs(001) using both *ab initio* and hybrid models has been reported recently by Qian et al.[57] This calculation illustrates clearly the nature of these analyses. Its predictions for the missing-row dimer model of the GaAs(100)-2x4 structure, and by implication the c(2x8) structure shown in Fig. 3, have been confirmed experimentally.[78][79] Results for GaAs($\bar{1}\bar{1}\bar{1}$) suggest, however, that not all observed structures exhibit semiconducting electronic surface state behavior,[80] so that more general classes of structures need to be considered in the theoretical analyses.

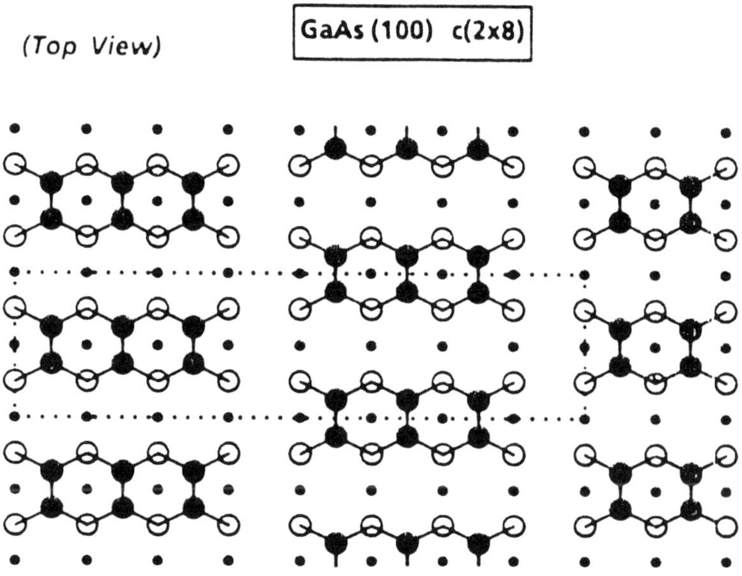


**Figure 3.** Top view of the c(2x8) GaAs(100) surface structure. The top-most dimer As atoms are depicted with large solid circles, while open circles are the second-layer Ga atoms. The dotted rectangle outlines the c(2x8) unit cell. One out of every four As dimers is missing, and the dimer row flips position at the position of the missing dimer. (After Biegelsen et al.[79])

The picture of the atomic geometry of GaAs(100) which emerges from these studies is similar to that of Si. For As stabilized surfaces, the As species form dimers containing one As-As surface bond with each As backbonded to two subsurface species. These dimers are stabilized by large contractions of the back bonds. More complex structures with differing As-Ga ratios are obtained by arranging these dimers on the surface in various configurations to obtain (2xn) structures. For enough excess As, two layers of As are believed to occur on top of the first layer of Ga so that As dimers form on an As substrate to give the GaAs(100)-c(4x4) structure. All of these structures exhibit semiconductor surface-state electronic excitation spectra in which the As-derived surface states are occupied and the Ga-derived states are empty. The Ga rich end of the surface composition spectrum yields a c(8x2) structure which is less thoroughly studied but is believed to be formed by pairs of Ga dimers on stepped As sublayer. A recent synopsis of the features of the various surface structures has been given by Biegelsen et al.[79]

### 3.6 Band Bending and Surface Dipoles

At the surface of a solid, electrons tunnel out into the vacuum leading to a dipole layer which contributes to the *work function* (or electron affinity and ionization potential). The associated charge densities and potentials are shown in Fig. 4 for the (111) surface of Al. In addition, charge can be trapped in the vicinity of the surface (e.g., by chemisorption-induced surface states) leading to band bending near the surface. Both phenomena are of interest in studies of metal overlayers on semiconductors as precursors to metal-semiconductor contacts,[62][81] as well as in models of metal-insulator-semiconductor junctions.[82] A unified treatment of both the vacuum surface and junction cases has been given by Zur et al.[81] An important feature of the vacuum surface is that only trapped charge at the surface leads to band-bending associated with the compensation of this surface charge by bulk space charge. Such surface charge can be associated with defects (e.g., steps) or chemisorbed species at the surface. The microscopic origin of this surface/interface charge becomes an important issue in the deposition of metals to form metal-semiconductor contacts in which the image charge associated with the space charge in the semiconductor is usually taken to be the interfacial charge.[81]

The presence of light alters surface space charge effects by virtue of modifying the occupancy of surface traps and generating carriers in the space charge region. This effect, while known for years, was re-recognized to be of importance in low temperature photoemission studies of surface band bending in 1990.[83] The correction of misunderstandings in the literature concerning the determination of surface and interfacial band bending via photoemission[83] constituted another significant development during the 1988-90 period covered by this review.

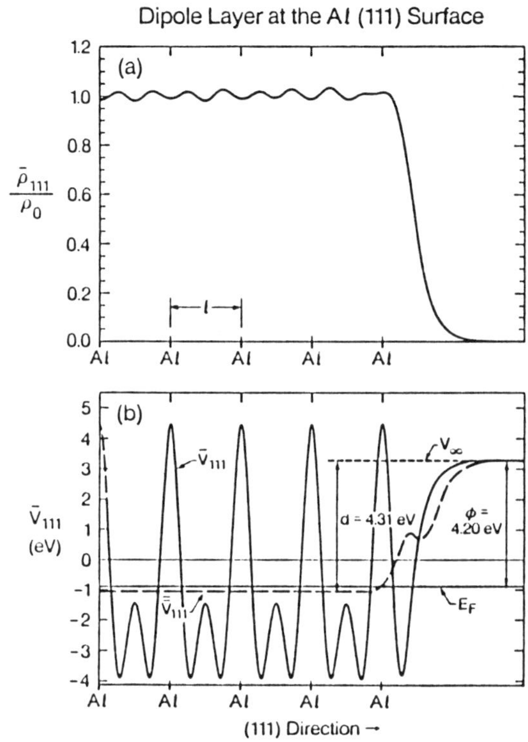


**Figure 4.** Results of a local density functional calculation of the dipole layer at the Al(111) surface. Five Al atomic planes are shown, with the vacuum region being on the far right. Figure (a) shows $\bar{\rho}_{111}$, the Al valence charge density averaged in planes orthogonal to (111), in units of its mean bulk value $\rho_0$. The tunneling of charge into the vacuum region is apparent. Figure (b) shows the resulting dipole and work function. $\bar{V}_{111}$ is the planar-averaged electrostatic potential, with $V_\infty$ being its value far in the vacuum region. $\bar{\bar{V}}_{111}$ is the result of averaging $\bar{V}_{111}$ over a length scale equal to the Al-Al interplanar distance. The calculated work function is $\phi = V_\infty - E_f = 4.20$ eV. The experimental value is $\phi$[Al(111)] = 4.18 eV. In terms of the surface dipole $d = V_\infty - \bar{\bar{V}}_{111}(-\infty) = 4.31$ eV, the work function can be written $\phi = d - \mu$, where $\mu = E_F - \bar{\bar{V}}_{111}(-\infty) = 0.11$ eV. [After Dandrea and Duke (unpublished).]

# 4.0 SEMICONDUCTOR HETEROJUNCTIONS

## 4.1 Introduction: Band Structure Engineering

Research into semiconductor heterostructures such as quantum wells and superlattices is spurred largely by the important role they play in opto-electronic devices.[84] The goal of this research is largely that of band structure engineering:[85] learning how to control semiconductor properties such as optical gaps and carrier mobilities that directly affect device characteristics. This band structure engineering proceeds by control of any of the several available parameters:

(1) Controlling the width of the quantum wells determines the degree of confinement near the band edges in thick layer structures[7] and strongly affects the intervalley mixings and splittings in ultrathin superlattices.[86][87]

(2) Similarly, control of the growth direction affects the confinement in thick quantum wells by determining the pertinent effective mass in the case of an ellipsoidal band edge valley. In thin-layer superlattices, the energies of the band edges depend critically on the growth direction, because the growth direction determines the pertinent band folding relationships.[87]

(3) In the case of a heterojunction with one constituent being an alloy, alloy composition is frequently used as a tuning parameter. For example, the mole fraction x in $Al_xGa_{1-x}As/GaAs$ structures directly determines the barrier heights for both electrons and holes.[88]

(4) Another degree of freedom is afforded by the recent development of strained layer epitaxy of lattice mismatched materials, whereby a thin epilayer can be grown coherently on a substrate with a different lattice spacing. Controlling the resulting strain configuration of the epilayer (for example, by controlling the substrate lattice constant) allows tunability of the band edge states through the resulting deformation potential shifts and strain splittings of the pertinent energy levels.[89]

Although in this review we discuss primarily these aspects of band engineering applied to a two-dimensional planar heterojunction geometry, both one-dimensional (quantum wire) and zero-dimensional (quantum dot) geometries also have recently been explored.[90]

Another major aspect of heterojunctions discussed below is the growth process itself. Large scale simulations using classical atomic potentials have been performed for Si and Ge, giving insight into how surface reconstructions affect the growth process. Growth of heterovalent semiconductor interfaces (such as GaAs/Si) has been a long-standing goal, due to the promise of combining the optical activity of III-V materials with the ubiquitous Si technology. The high-quality interfaces required for such devices have not, however, yet been realized experimentally. We discuss below the large theoretical effort invested in understanding both the heterovalent growth process and the electronic structure of the resulting heterostructures.

## 4.2 Studies of Semiconductor Growth

In this section we review recent developments in the theoretical modeling of epitaxial growth. Developments prior to 1984 are reviewed by Venables et al.[91] These developments are based on the use of phenomenological two- or multiple-body atom-atom interactions to describe the (classical) dynamics of the growth process and the use of *ab initio* methods to discuss the energetics of various steps in this process.

There has been a substantial effort in the past decade to improve the range of applicability of empirical interatomic potentials, largely because of their use in simulations of large systems. Potentials now exist for C, Si, and Ge that reproduce at least some of the experimentally determined surface reconstructions.[4] We indicate below some of the important information regarding semiconductor growth that has been obtained through atomistic simulations using such potentials.

Early simulations of thin film formation on a substrate used spherically symmetric Lennard-Jones potentials. For example, Grabow and Gilmore[92] have performed molecular dynamics simulations to explore the dependence of growth mode on film-substrate lattice mismatch and on the relative strength of film-substrate and film-film interaction energies. They find that for large mismatch and/or small film-substrate interactions the equilibrium growth mode is that of clusters on a bare substrate (Volmer-Weber), whereas in the opposite regime they find clusters growing on a few layers of uniform film (Stranski-Krastanov). They also calculate the dependence of coherent versus incoherent growth on both mismatch and film-substrate interaction strength, and demonstrate that in some cases strain can be relieved more favorably with partial, rather than full, misfit dislocations. Such results obtained with spherically symmetric potentials are not, however, directly applicable to semiconductor systems with directional covalent bonding. Schneider et al.,[93] for example, note that spherically symmetric pair potentials lead to homoepitaxial growth at all growth temperatures, whereas elements that exhibit directional bonding frequently grow in an amorphous rather than epitaxial manner below a certain growth temperature. Thus, simulations recently have been done using 3-body atomic potentials that reflect the directional bonding pertinent to semiconductor systems.

Dodson,[94] for example, has performed Monte Carlo calculations of the equilibrium critical thickness (the thickness beyond which an incoherent film is favored) for a GeSi alloy on a Si substrate. He finds good agreement with elasticity theory at small lattice mismatch, whereas for larger mismatches the strain energy associated with misfit dislocations becomes greater than the continuum value, resulting in a smaller critical layer thickness. Srivastava et al.[95] have reported molecular dynamics simulations of the initial stages of growth (<6 film monolayers) of Si and Ge on the (2x1) reconstructed Si(001) surface. They identify several mechanisms by which surface dimer bonds are broken during epitaxy, including direct insertion, concerted hops, concerted exchanges for opening a single dimer, and also a correlated adatom motion perpendicular to dimer rows (i.e., in the slow diffusion direction) that leads to multiple dimer openings. Khor and Das Sarma[96] have similarly used an empirical atomic potential to study the growth direction dependence of SiGe on Si. They find that twinning does not relieve misfit strain and that, in agreement with experiment, the (001) direction is best for film growth. The (111) direction is worse due to the high cost in energy for (001) and low cost for (111) for introducing twins.

One conspicuous absence from classical dynamics approaches to growth modeling is a description of growth involving polar semiconductors (III-V's or II-VI's). Although atomic potentials that adequately describe some of the bulk elastic and cohesive properties exist for these materials,[97] no potential currently exists that is capable of accurately describing the wide-ranging chemistry (including varying degrees of charge transfer and rehybridization) that would occur during their growth. Development of such a potential is an important goal of future research.

The study of the electronic structure and energetics of the heterovalent (e.g., Si/GaAs) interface has largely been done with *ab-initio* techniques. The general features of charge-neutral nonpolar interfaces [such as (110) and (211)] and charged polar interfaces [such as (001) and (111)] were first summarized by Harrison et al.[98] They concluded that the relative independence (seen experimentally) of band offsets on interface polarity implies that the polar interfaces must reconstruct. Dandrea et al.[30] demonstrated through local density functional calculations that the microscopic mechanism that leads to the macroscopic charge described by Harrison was that of a self-compensation charge transfer from donor states localized at the Si-As bonds to acceptor states at the Si-Ga bonds. This stabilizing self-compensation occurs optimally at an interface with a non-polar III-V surface, because in this case both donor and acceptor bonds are present and nearby each other. Such theoretical concepts were validated by Wright et al.,[99] who experimentally demonstrated that the (211) direction was optimal for growth of GaP on Si because the (211) GaP surface is nonpolar and the (211) Si surface has inequivalent fcc sublattices (thus minimizing antiphase disorder).

Northrup[100] has studied the initial stages of GaAs epitaxy on Si(111). He concludes that the interface is more likely to have the intermixed structure ...SiSiGaAsSiAs because local density functional calculations show the intermixed interface is 0.57 eV per surface atom lower in energy and because the calculated energies of its lone pair As electrons agree with the results of angle resolved photoemission experiments.

The initial stages of growth of GaAs on Si(100) recently have been studied extensively. Northrup[101] has explained the experimental preference for three-dimensional island growth over two-dimensional layer-by-layer growth by performing first-principles calculations of the free energy of differently-terminated GaAs/Si(100) structures as a function of Ga and As chemical potentials. Due to both GaAs strain energy and to unfavorable heterovalent chemistry (i.e., non-octet Si-Ga and Si-As bonds), the system's free energy can be lowered by moving a GaAs molecule from a strained epilayer to a bulk-like unstrained GaAs island. Kaxiras et al.[102] have extended this work to explain the experimentally observed island growth at stationary steps on vicinal Si(100). They point out that layer growth on the terraces is inhibited by an energetic preference for a mixed GaAs overlayer and thus consider Si double layer steps as nucleation sites for GaAs island formation. The model proposed for step-edge growth leads to growth along a [211] direction.

A final important topic is spontaneous ordering in semiconductor alloys. LeGoues et al.[103] have offered a microscopic mechanism to explain the observed (111) ordering of SiGe grown on the Si(001) 2x1 surface. Assuming that growth occurs by the motion of double steps, they note that the observed ordering results from atomic segregation in the top four layers induced by atomic-scale surface stress (due to the dimer reconstruction and different sizes of Si and Ge). Khor and Das Sarma[96] have recently suggested this mechanism to be incomplete and have conjectured that growth at a moving kink at a double step corner would be needed to ensure long range order. Froyen and Zunger[104] have used local density functional techniques to explain the (111) ordering of GaInP alloys grown on GaAs(001) without involving surface steps. By comparing the energies of several cation-terminated dimer reconstructions, they conclude that the (buckled and tilted) reconstruction which is the most stable has a surface ordering of cations that would lead to the observed three-dimensional structure. The common theme in all these calculations of spontaneous ordering is that surface reconstructions may stabilize certain two-dimensional structures that are then found in three-dimensional bulk form because they are frozen in kinetically during growth.

## 4.3 Electronic Structure of Semiconductor Heterojunctions

Having examined the recent progress in understanding growth of heterostructures, we now consider recent theoretical work describing the atomic and electronic structure of the heterojunctions themselves.

Several reviews recently have been published summarizing the numerous theoretical methods used to study the electronic properties of semiconductor heterojunctions.[7][60][105][106] We relate the material covered in these reviews to the model taxonomy introduced in Sec. 2.2 by providing a qualitative comparison between *ab initio* and hybrid methods. The methods which we classify as *hybrid* are labeled as *empirical* in the review of Smith and Mailhiot.[60] Hybrid methods such as envelope function,[7] tight-binding,[105] and k•p[60] combine fitted descriptions of the band structures of the individual constituents with a band offset treated as an input parameter to provide heterojunction electronic structure as output. Their advantage lies in their simplicity of implementation and in their ability to characterize arbitrarily long quantum well structures. Their disadvantage lies in the fact that the band offset and the constituent band structures are needed as input so that their applicability to novel uncharacterized systems is limited. Furthermore, these methods generally neglect any details of the interface and are thus not well-suited for describing short-period superlattices (where the interface region is a substantial portion of the total volume) or for studying the variation of electronic structure with atomic structure at the interface. These shortcomings of the hybrid methods are largely corrected by *ab initio* local density functional schemes, which self-consistently allow for a microscopic description of the actual interfacial charge density. Although numerically quite demanding, these methods yield reasonably accurate calculations of the band offsets[107] and of short-period superlattice electronic structure.[87] Recent work has taken great strides in correcting the two most obvious shortcomings of *ab initio* methods: First, Hybertson and Louie[9] have shown that the local density functional underestimation of band gaps is corrected through solution of a self-energy Dyson equation. Second, in calculating the transmission coefficient through a Si(111) twist boundary, Stiles and Hamann[108] have demonstrated that the density functional method can be implemented in a complex wave vector scheme, making it applicable to arbitrarily long heterostructures. Therefore, both classes of models play important and complementary roles in the computation of heterojunction properties.

Before calculating the electronic structure of a heterostructure, one must determine the atomic structure at the interface. Dandrea et al.[109] have studied the strain-induced interfacial relaxations at coherent interfaces of lattice mismatched semiconductors. They find that certain growth directions [e.g., (110) and (201)] lead to particularly stable interfaces because such directions allow for interfacial relaxations that relieve much of the interface strain. Although most theoretical studies assume an ideal atomically abrupt interface, Kim et al.[110] have used transmission electron microscopy imaging to show that the composition actually changes over a 4 - 10 monolayer wide interfacial region and that the width of this region depends sensitively on the temperature history of the sample. More work on the effects of interfacial interdiffusion on the heterojunction electronic structure is needed.

**Band Offset Calculations:** We now review recent work on the calculation of band offsets at semiconductor interfaces. The band line-up at the interface is perhaps the single most important quantity determining the electronic structure of the heterojunction. As such, it is fundamental to band structure engineering. It is perhaps surprising that our understanding of band offsets is still a debated issue, with several disparate models being offered in the literature. Since an extensive survey of these models was given by Tersoff[111] in 1987, we confine our discussion to a summary of their current status.

The offset between a band level on either side of an A/B semiconductor interface can be written

Eq. (4) $$\Delta E = (E_A - V_A^{bulk}) - (E_B - V_B^{bulk}) + \Delta V_{A\text{-}B}^{IF}$$

as indicated in Fig. 5. The first two terms are the band energies in the respective bulk materials relative to the average electrostatic potential, while the third term is the interface dipole, i.e., the difference in the average electrostatic potential across the A/B interface. Early theories tended to describe the individual constituents A and B as semi-infinite bulk slabs with a surface and then to use the potential at infinity, $V_\infty$, as a common reference. The significance of this approach can be seen by rewriting Eq. (4) to read

Eq. (5) $$\Delta E = (E_A - V_\infty) - (E_B - V_\infty) + (\Delta V_{A\text{-}B}^{IF} - \Delta V_{A\text{-}B}^{SURF})$$

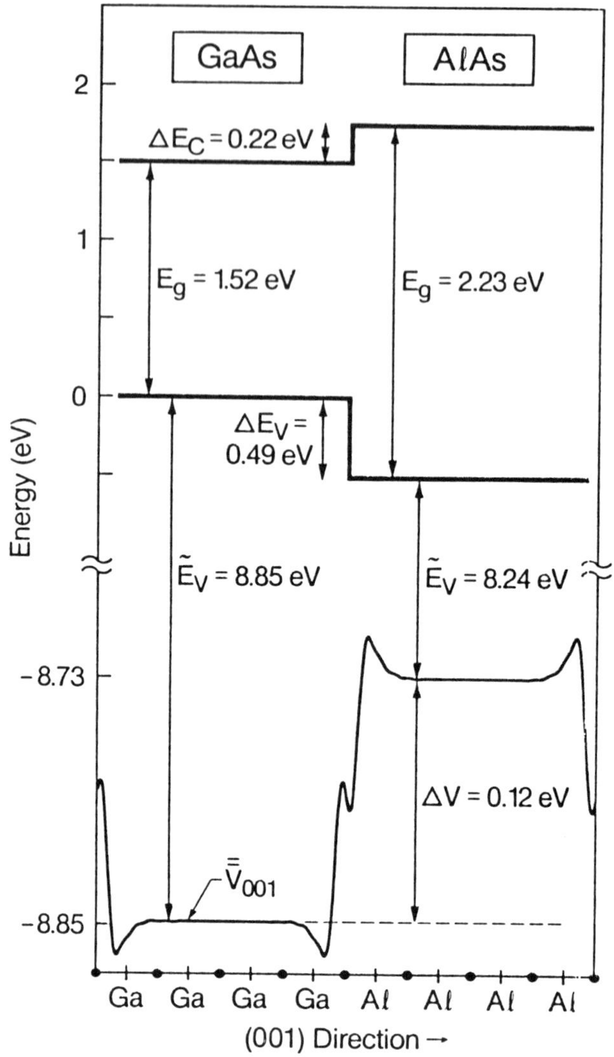


**Figure 5.** Results of a density functional calculation of the band offset at an (001) GaAs/AlAs interface. The valence offset is $\Delta E_V = \tilde{E}_v(\text{GaAs}) - \tilde{E}_v(\text{AlAs}) + \Delta V^{IF}$, where $\tilde{E}_v = E_v - \langle V_{bulk} \rangle$ is the position of the valence edge with respect to the average bulk potential, and $\Delta V^{IF}$ is the change in the average bulk potentials upon crossing the interface. $\overline{\overline{V}}_{001}$ is a suitable averaged electrostatic potential normal to the (001) surface. Its calculation is described in the caption to Fig. 4. The density functional calculation of $\Delta V$ shown is for a $(\text{GaAs})_4(\text{AlAs})_4$ (001) superlattice. [After Dandrea (unpublished).]

where $\Delta V_{A-B}^{SURF} = (V_A^{bulk} - V_\infty) - (V_B^{bulk} - V_\infty)$ is the difference in surface dipoles. If the band level in question is the conduction edge, the first two terms in Eq. (5) are simply the difference in electron affinities. Thus, as shown by Mailhiot and Duke,[112] the electron affinity rule (which asserts the equivalence of the conduction band offset and the difference in electron affinities) is valid only when the interface dipole is identical to the difference in surface dipoles. Tersoff[111] has summarized the problems with this assertion and its quantitative inaccuracy. Incomplete treatment of the last term in Eq. (5) also has limited the success of Harrison's tight-binding model of band offsets.[113] Another early model which did not survive the test of time is the common anion rule which states that the valence band offset should vanish across an interface with a common anion. The large valence offset of approximately 0.5 eV now measured for the GaAs/AlAs interface (111) demonstrates the invalidity of this assertion.

The most accurate band offset calculations to date have been performed using local density functional methods to calculate the interface dipole self-consistently. Van de Walle and Martin[107] presented such calculations for a large number of systems, both lattice matched and strained layer. They demonstrate the transitivity and orientation independence of the offsets. Although these local density calculations give an accurate description of the final (dipole) term in Eq. (4), they suffer from some inaccuracy in the first two terms because the density functional band energies are not interpretable as physical quasiparticle energies. This source of error, however, can be corrected by performing full Dyson equation calculations on the separate constituents.[9] A recent compilation of these calculations for a large number of systems by Zhu and Louie[10] shows the correction to average 0.12 eV for the eight interfaces studied.

One important recent output of these band offset calculations has been the prediction of the possibility of controllably tuning heterojunction band offsets. In particular, Munoz et al.[114] have demonstrated through density functional calculations that a bilayer of epitaxial Ge grown at a polar GaAs/AlAs interface will change the band offset by about 1 eV. The reason for this is simply that the bilayer of Ge at a polar III-V interface means that the usual III-V-III-V ionic sequence is replaced by the sequence III-IV-IV-V, which creates an extra interfacial dipole. This prediction is characteristic of polar surfaces and does not hold for non-polar surfaces. It was born out by the experiments of Sorba et al.,[115] who used MBE shuttering techniques to grow a Si interlayer at a (001) AlAs/GaAs interface with the particular As-Si-cation atomic sequence. For bilayer Si coverages $x < 0.25$, they find a band offset change of approximately 2x eV, in good agreement with calculations of Peressi et al.[116] For larger Si coverages, $0.25<x<1$, the measured change in offset falls to 0.2 eV, indicating formation of antiphase domains.[116]

Although quite accurate, these density functional calculations offer limited insight into the fundamental nature of band offsets. For example, the transitivity and orientation independence mentioned above suggest the existence of reference levels in separate semiconductors that line up with each other at an interface. Further evidence is given by Christensen and Brey,[117] who showed the A/B offset to be fairly independent of the insertion of a few monolayers of a third material C at a nonpolar A/B interface. Two such reference level theories have recently been suggested. Van de Walle and Martin[107] have noted that approximating the A/B heterojunction charge density as the sum of spherically symmetric atomic charges allows the interface dipole terms $\Delta V_{A\text{-}B}{}^{IF}$ in Eq. (4) to separate into the difference of two completely independent terms. This separation leads to a band offset which is the difference of two independent reference levels. By performing density functional calculations for a number of systems, they found these reference level offsets to differ by less than 0.2 eV from the more exact values obtained using the fully self-consistent interface, thereby demonstrating the approximate validity of their assumptions. A second reference level theory has been proposed by Tersoff.[111] He suggests that a charge neutrality level in the semiconductor gap lines up across an interface due to a self-consistent screening mechanism. Using an approximate calculation scheme for obtaining the neutrality reference level, he finds offsets that agree with experiment to within about 0.2 eV. Cardona and Christensen[118] have suggested an alternative method for calculating the neutrality level, based on a dielectric midgap concept.

Although these reference level models were proposed as being admittedly approximate, one elegant theoretical effort rigorously explains the linearity of offsets for lattice matched systems. Baroni et al.[119] have shown that the offsets at both isovalent and heterovalent lattice matched interfaces can be calculated accurately within linear response theory and that this implies that such offsets are both transitive and (in the isovalent case) independent of the particular atomic arrangement at the interface. The only shortcoming of this theory is that it requires all atoms to sit on a single underlying zincblende mesh and thus can not deal with cases where lattice mismatch or heterovalent chemistry cause relaxations away from such a common cubic parent lattice.

In summary, band offsets can now be calculated using *ab initio* methods with reasonable accuracy. Although several interpretations have been suggested to explain the linearity of band offsets, none is universally valid or accepted at this time.

**Heterojunction Band Structure Calculations:** We conclude this section on semiconductor interfaces with a brief survey of recent work on band structure calculations at a heterointerface. Given the high level of interest in the AlAs/GaAs system, we begin by mentioning the multitude of theoretical work on $(Al_xGa_{1-x}As)_n$ $(Al_yGa_{1-y}As)_m$ superlattices grown along the (001) direction. Lu and Sham[86] have used the tight-binding method to study the nature of the subbands in $(GaAs)_n$ $(AlAs)_m$ superlattices. As indicated in Fig. 6, for n = m < 12 they find the fundamental gap to be Type

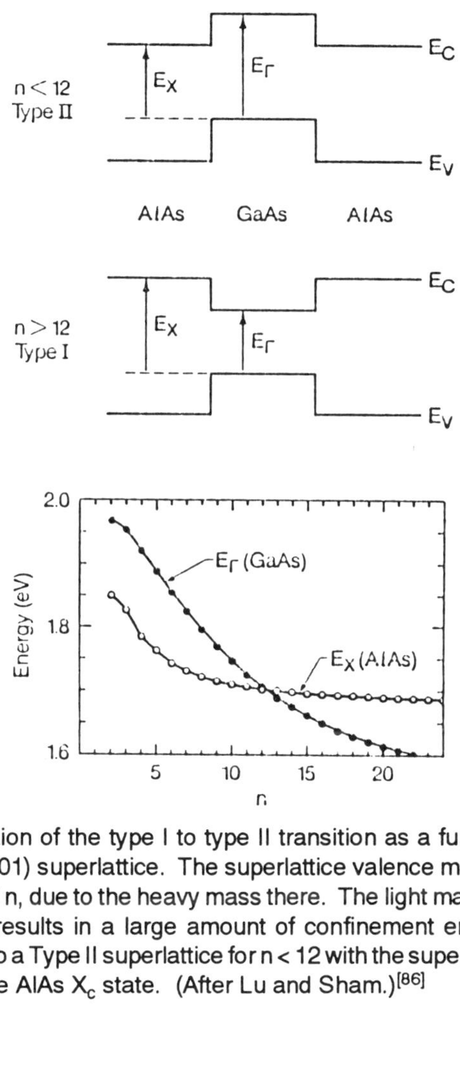


**Figure 6.** Illustration of the type I to type II transition as a function of n for the $(GaAs)_n(AlAs)n$ (001) superlattice. The superlattice valence maximum lies in the GaAs region for all n, due to the heavy mass there. The light mass of the GaAs $\Gamma_C$ conduction edge results in a large amount of confinement energy for small n, however, leading to a Type II superlattice for n < 12 with the superlattice conduction minimum being the AlAs $X_c$ state. (After Lu and Sham.)[86]

II (i.e., the valence band maximum and conduction band minimum lie in separate components of the superlattice, with the conduction band minimum being largely the AlAs $X_{1c}$ state), while for n = m > 12 the gap is Type I (i.e., the band gap of (GaAs)n lies entirely within that of $(AlAs)_m$ with the conduction band minimum being the GaAs $\Gamma_{1c}$ state). Similarly, Gell et al.[88] have used an empirical pseudopotential method to study the effects on the superlattice band edges due to *(i)* alloying (i.e., x, y variation), *(ii)* superlattice period (n, m) variation, and *(iii)* hydrostatic pressure. Ting and Chang[120] performed similar calculations, studying in detail the amount of Γ-X mixing in the superlattice conduction minimum, since that determines the system's optical activity. The effect of applied electric and magnetic fields have been studied via envelope function calculations on a $(GaAs)_9$ $(AlAs)_3$ superlattice[121] and on a single GaAs quantum well.[122] Finally, Bastard et al.[123] have shown that quasi-two-dimensional electrons in GaAs/AlAs heterostructures have excitonic binding energies that are enhanced over their three-dimensional values.

Although the above studies all involve lattice matched systems, there has also been considerable work on strained layer heterojunctions. Osbourn[89] has demonstrated how such systems allow independent variability of structural properties (the lattice constant), optical properties (the band gap), and transport properties (carrier mobilities). Smith and Mailhiot[60] have noted that the piezoelectric activity of polar semiconductors means that (111) oriented strained layer superlattices made of such materials will exhibit a spontaneous polarization. They have used the k•p method to study the effect of the induced electric fields on the superlattice band structure.

Besides the hybrid model calculations typified by the above citations, extensive *ab initio* studies of heterojunctions also have been reported. This work has largely been limited to supercell calculations of short-period structures, due to the computational complexities involved. These short-period structures provide an added dimension to band structure engineering, since the band edges in such structures are not determined by the qualitative concepts embedded in envelope function models which apply to longer period structures. Rather, as has been recently summarized by Dandrea and Zunger,[87] the band edges in these systems are determined by band foldings and intervalley mixings which can lead to level splittings and repulsions as large as 1 eV.

Figure 7 demonstrates this for the case of a $(GaAs)_1(GaP)_1$ (001) superlattice. Figure 7a shows the energy bands of a zincblende material Ga⟨AsP⟩ whose anion is the average of As and P, folded into the Brillouin zone of the n = 1 (001) superlattice. Figure 7b shows the resulting splittings in these bands after the virtual anion is changed into As and P in alternating (001) layers. The band edges in such short-period superlattices can thus be viewed as arising from those of a random alloy material, with some splittings caused by band foldings into the smaller superlattice Brillouin zone.

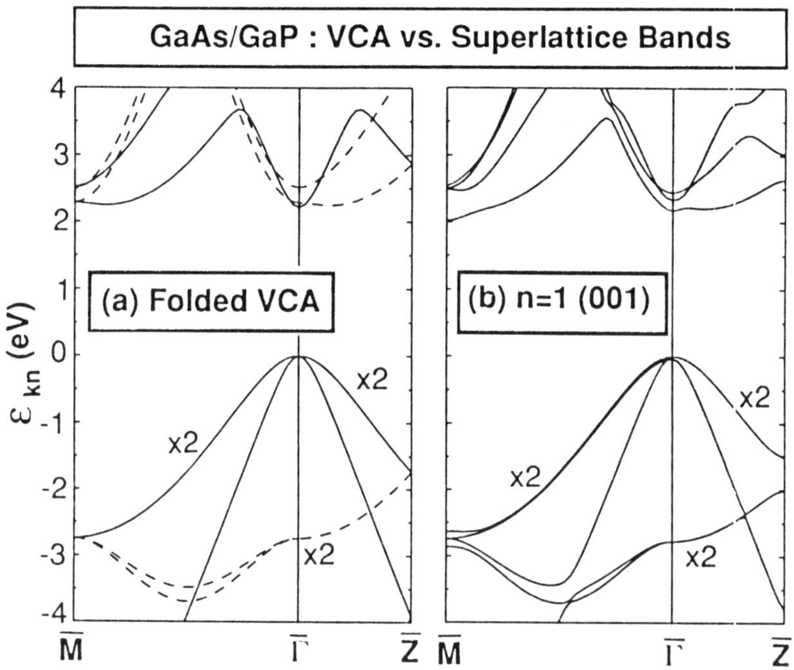


**Figure 7.** Comparison of the band structures along the ΓM and ΓZ directions in the superlattice Brillouin zone of *(a)* a zincblende material Ga⟨AsP⟩ whose anion is the average of As and P [the virtual crystal approximation (VCA)] and whose bands have been folded into the n = 1 (001) superlattice Brillouin zone and *(b)* a $(GaAs)_1(GaP)_1$ (001) superlattice. Note that the bands of an ultrathin superlattice are nearly identical to those of the alloy-like zincblende parent material, except for some large splittings at certain places in the Brillouin zone where the folded alloy bands interact strongly. In the superlattice zone Γ = (0,0,1), Z=(2π/a) (0,0,1/2) and M=(2π/a) (1,0,0). (After Dandrea and Zunger[87])

One important example of band structure engineering involving such thin-layered structures has been the study of Si/Ge strained layer superlattices. Several groups[124]-[126] have reported *ab initio* calculations which show that although both Si and Ge are separately indirect gap materials, short period Si/Ge superlattices grown on a Ge substrate should exhibit a direct band gap with nonvanishing oscillator strength. The reason for this is generic and also holds for short period GaAs/GaP superlattices. Grown with an in-plane lattice constant equal to $a_{GaP}$ or $a_{avg}$, these superlattices are indirect, whereas they are direct gap materials when grown pseudomorphically on GaAs. Figure 8 illustrates the physics behind this phenomenon. For an (001) superlattice, the anisotropic strain in the smaller Si or GaP grown pseudomorphically on the larger Ge or GaAs results in downward deformation potential shifts of the $\Gamma$ and $X_z$ conduction states which fold to the superlattice zone center, while the $X_{xy}$ conduction states are raised in energy (see Fig. 8c). Furthermore, the $\Gamma$-$X_z$ intervalley mixing results in a direct superlattice band gap with nonvanishing oscillator strength (labled f in Fig. 8a). Figures 8b and 8d show that the same physics describes the case of (111) oriented structures.

Another important example of *ab initio* band structure engineering is the work of Wei and Zunger,[127] who have reviewed previous methods for engineering infrared band gaps and suggested a new methodthat of a short period (111) superlattice ordering. Finally, we mention the *ab initio* studies of $(GaAs)_n(AlAs)_n$ (001) superlattice system. For $n \geq 5$ the empirical methods are applicable, and the band structure studies using these methods were summarized above. For $n \leq 4$ (approximately), the first-principle methods are needed to supply accurate interfacial charge densities. Zhang et al.[29] have performed density functional and Dyson equation calculations on the $n = 1$ and $n = 2$ superlattices. In both cases they find the fundamental gap to be indirect.

These examples of heterojunction band calculations are representative of the current state-of-the-art. They demonstrate the complimentary character of the hybrid and *ab initio* methods: the *ab initio* methods are calculationally viable for those systems (e.g., short period heterostructures) where the hybrid methods are least applicable.

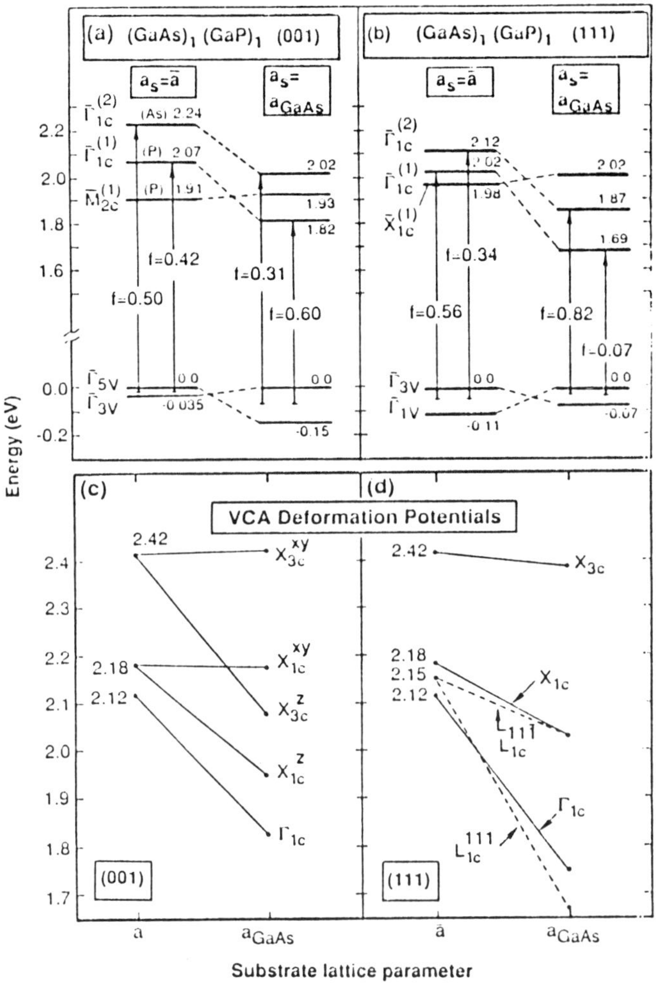


**Figure 8.** Comparison of calculated $(GaAs)_1(GaP)_1$ superlattice energy bands for growth with an in-plane lattice constant $\bar{a} = a_{avg}$ or $a_{GaAs}$. Panels *(a)* and *(c)* are for an (001) oriented superlattice; *(b)* and *(d)* are for the (111) case. Panels *(c)* and *(d)* show the shifts of the conduction band edges of a virtual crystal approximation (VCA) Ga⟨AsP⟩ zincblende material. In both the (001) and (111) cases, the zincblende states which fold to the superlattice zone center have strongly negative deformation potentials. Panels *(a)* and *(b)* show the manner in which these deformation potentials manifest themselves in the actual superlattice energy gaps: the gaps decrease and go from indirect to direct when the in-plane lattice constant is changed from $\bar{a}$ to aGaAs. (After Dandrea and Zunger.)[170]

## 5.0 METAL-SEMICONDUCTOR CONTACTS

### 5.1 Composition and Structure

Unlike semiconductor heterojunctions, the vast majority of metal-semiconductor contacts are not epitaxial, and most exhibit considerable diffusion or chemical reactions at the interface.[62][128]-[131] Contacts often are prepared by metal deposition onto a doped semiconductor substrate. Therefore, between diffusion, chemical reactions and inhomogeneous doping, even the chemical composition of the interfacial regions are unknown in most cases. Their atomic geometries are almost never known except in a few cases of epitaxial growth without mixing, for example, p(1x1) overlayers of Sb and Bi on GaAs(110)[19][61][132] and related III-V substrates.[133]

Another complication occurring at metal-semiconductor contacts is that the electrostatic potential in the vicinity of the contact may be dominated by charged defects.[81][134][135] Since these defects can rarely be detected using modern interfacial structural analysis methods, the possibility exists that the electrical properties of the contacts are governed by phenomena associated with such defects rather than the "ideal" atomic geometry of an epitaxial layer.[62][136] Even in the case of nonreactive epitaxial systems, this phenomenon occurs, as is well documented in the case of GaAs(110)-p(1x1)-Sb overlayers.[137]-[140]

A final important complication associated with metal-semiconductor contacts is the non-equilibrium character of the structure of the underlying interface(s). Specifically, the structural and electrical properties of these contacts depend upon how they are formed.[130][131][137][138][140] This fact leads to considerable difficulty in constructing models of these contacts because the appropriate structural model cannot be ascertained with clarity.

Taken in concert, the above three complications reveal that the description of the electrical properties of typical metal-semiconductor contacts is essentially more difficult than that of clean semiconductor surfaces or epitaxial semiconductor heterojunctions. During the remainder of this section we review the dimensions of this problem in which significant progress has been made during 1988 - 1990.

### 5.2 Schottky Barriers

When a metal-semiconductor contact is formed, a majority carrier depletion region typically occurs at the interface as shown in Fig. 9 with a height, at the interface relative to the Fermi energy, of $\phi_B$, and width $d_s$ which depends on the doping and thermal carrier density in a well-known fashion.[6][128] An extensive phenomenology of electronic transport across these depletion regions, commonly called *Schottky barriers,* has been developed, as described, e.g., by Sze[6] or Tyagi.[141] The interface science of Schottky barriers is primarily associated with the dependence of $\phi_B$ on the components, structure, and preparation conditions of the contact.[62][128]-[131]

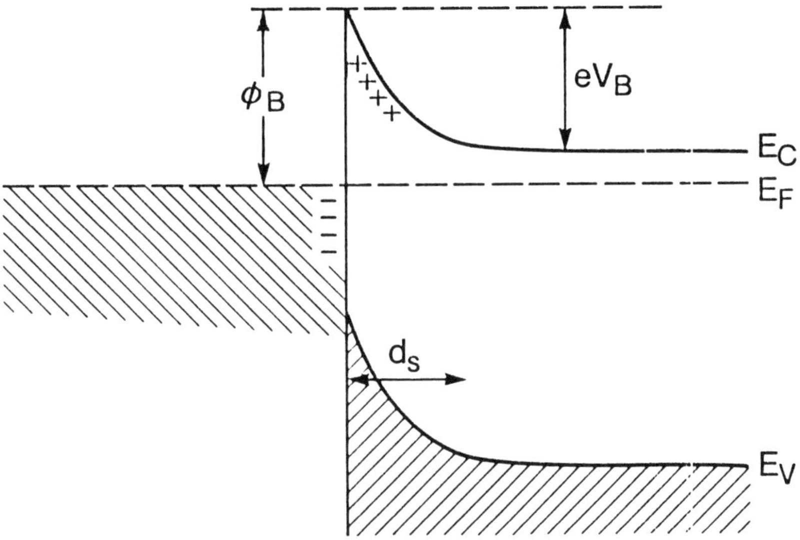


**Figure 9.** Energy level diagram for a metal-semiconductor junction. In this case, the semiconductor is n-type.

A wide variety of phenomenological models of $\phi_B$ have been proposed. One class seeks to relate $\phi_B$ to differences in reactivity or ionicity of various constituents of the contact[62][128] in order to identify chemical trends. A similar class attributes $\phi_B$ to chemical-reaction-induced morphological changes in structure at the interface.[142] A third class, called reference-level models, determines $\phi_B$ as the difference of two reference levels in the bulk metal and semiconductor. A popular model in this class is the local charge neutrality model developed by Tersoff.[143][144] Finally, there is a large class of models which attribute $\phi_B$ to charged defects in the vicinity of the interface screened by their associated "image" charges in the metal.[134]-[136] Various authors apply these models either singly or in combinations to analyze experimental measurements of $\phi_B$.

In order to discriminate between models, a popular approach to studying Schottky barriers is the building up of metal-semiconductor contacts atom-by-atom by depositing metals on semiconductor surfaces in an ultrahigh vacuum system.[62][128]-[131] Called the *surface science approach* by Mönch,[62] this methodology permits a dissection of the phenomena which occur at various stages of Schottky barrier formation and illuminates the role of the kinetics of metal film growth (and hence morphology) on the value of $\phi_B$. Currently three separate stages of Schottky barrier growth are envisaged:[62][130][131][137]-[139] isolated ad-atoms, island or cluster formation, and coalescence of the clusters to form a metallic film. Weaver[131] has developed a technique to deposit the clusters directly and finds rather different dependencies of $\phi_B$ on metal species and coverage: a direct manifestation of the non-equilibrium character of the metal-semiconductor structures formed by the various techniques.[130]

Two major developments occurred during 1988 - 1990 in the application of phenomenological models to describe Schottky barrier formation. First, the self-consistent charged defect model of Zur, McGill, Smith, Mailhiot, and Duke[134][135] has been shown to be capable of describing quantitatively the systematics of Schottky barrier formation for several diverse systems and processing conditions.[145]-[149] While yielding no insight *a priori* into the origin of such charged defects, the model permits a quantitative recasting of Schottky barrier variations with metal, semiconductor, and process changes in terms of energy levels and positions of charged defects which could cause these variations. Hence, it permits systematic searches for such defects using spectroscopies (e.g., cathodoluminescence) which are sensitive enough to detect them.[146] The reference barrier height which would occur in the absence of defects is not specified intrinsic to the model, but is usually taken to be the phenomenological Schottky barrier result on the basis of a many-electron analysis by Duke and Mailhiot[112][135] which suggests that the interface dipole is almost exactly canceled by the surface dipoles of the two components of the contact.

In recent years the charged-defect model has been extended in two important respects. One type of extension consists of building microscopic (hybrid) models of these defects for the low-metal-coverage limit based on the chemisorption bond between the adatom and the constituents of the semiconductor substrate.[150-152] These models permit interpretation of trends in the low-overage Fermi energy position with metal species[62][153] as well as of the spectroscopically determined details of the adsorbate induced electronic excitation spectra.[152] The second type of extension consists of recognizing the resonant character of the defect states near a metal overlayer and the consequences of coloumb interactions among these states.[154]-[156] As in the case of the original charged defect model,[134][135] this extension does not offer explicit insight into the microscopic origin of charged defects near metal-semiconductor interfaces, but it does account for the fact that the electronic excitation spectra of such defects consist of resonances due to their coupling to the continuum of delocalized one-electron states in the metal. This phenomena seems to be relevant in discussing the variation of Schottky barrier height with the work function of the metal.[156]

The second major development in Schottky-barrier phenomenology since 1987 is the re-recognition that surface photovoltage effects can exert a major influence on surface Fermi energy positions measured by photoemission spectroscopy, especially for low metal coverages at low temperatures on lightly-doped semiconductor substrates.[83][157]-[167] Two theoretical models, one with[159] and one without[83][157] tunnel injection at the surface have been developed which give satisfactory semiquantitative accounts of available experimental results.[83][157]-[161] An important outcome of these studies is the elucidation of the role of conductivity within the metal overlayer in shunting the surface photovoltaic effect in examinations of surface Fermi level positions as a function of metal coverage.[162]

### 5.3 *Ab Initio* Analyses

Given the intrinsic complexity of the problem, it is perhaps not surprising that *ab initio* approaches to calculating Schottky barrier potentials have met with limited success. Three issues confront these approaches: structural, conceptual, and technical. We discuss the recent literature within the context of these issues.

As noted above, the structures of metal-semiconductor contacts are not in equilibrium and, in general, are not accessible to direct experimental determination. Thus, the question arises of how to construct structural models of these contacts for calculations of their electronic properties. To date, models have been based on idealized model structures for both silicide/silicon contacts[31] and metal compound-semiconductor contacts,[32] although a few additional studies of defects near these interfaces have been reported.[168][169] These structures almost certainly are not minimum-energy structures such as those used in heterojunction analyses (see Sec. 4.3), nor are they experimentally-determined interface atomic geometries.

Therefore, the issue of the pertinence of such calculations to experimentally measured Schottky barrier properties arises from the outset.

A second fundamental issue was raised by Mailhiot and Duke,[112][135] in their many-electron theory of semiconductor heterojunctions and contacts based on a jellium model for both semiconductors and metals, namely, What are the appropriate thermodynamic boundary conditions for the calculation of interface electronic structure? Theirs is the only *ab initio* analysis to date which explicitly incorporates both the mechanical and electron-transfer boundary conditions associated with thermodynamic equilibrium and the presence of dopants (i.e., a space charge region) in the semiconductor.

The third issue concerns the technical suitability of the application of a ground-state local density functional formalism to describe Schottky barrier heights which are properties characteristic of the electronic excitation spectrum. This concern has been raised explicitly in the literature for the silicide/silicon calculation[28][31] but applies to later work[32][169] as well.

Taken together, these three issues suggest that while local-density-functional methods potentially are suited to the study of metal-semiconductor contacts, refinements relative to existing calculations are needed.

## 6.0 SYNOPSIS

During the past five years, significant advances have been made in the theoretical prediction of the atomic geometries of semiconductor surfaces and heterostructures. Not only the minimum-energy geometries but also the kinetics of processes required to achieve these geometries have been elucidated for a variety of semiconductor surfaces and heterojunctions. General principles and scaling laws have been proposed to describe the geometries of various classes of surfaces, most particularly the cleavage surfaces of tetrahedrally-coordinated compound semiconductors. Novel heterojunction ordering phenomena have been predicted successfully. Predicted surface and heterostructure electronic excitation spectra have been validated quantitatively. Descriptions of metal-semiconductor contacts have lagged those of semiconductor surfaces and heterojunctions because of the greater intrinsic complexity of the issues involved and the inherently non-equilibrium character of the structures generated by commonly used processing techniques.

## REFERENCES


1. "Surface Properties of Electronic Materials," *The Chemical Physics of Solid Surfaces and Heterogeneous Catalysis,* (D. A. King and D. P. Woodruff, eds.), Vol. 5, Elsevier Science Publishers, Amsterdam (1988)
2. "Solvay Conference on Surface Science," (F. W. de Wette, ed.), *Springer Series in Surface Sciences,* (G. Ertl, ed.), Vol. 14, Springer-Verlag, Berlin (1988)
3. O. Madelung, *Springer Series in Solid State Sciences,* (M. Cardona, P. Fulde, and H. J. Quiesser, eds.) 2:129-155, Springer-Verlag, Berlin (1978)
4. B. C. Bolding and H. C. Anderson, *Phys. Rev.,* B41:10568-10585 (1990)
5. A. E. Carlson, *Solid State Phys.,* (H. Ehreneich and D. Turnbull, eds.) 43:1-91, Academic Press, New York (1990)
6. S. M. Sze, *Physics of Semiconductor Devices,* pp. 363-504, John Wiley and Sons, New York (1969)
7. G. Bastard, J. A. Brum, and R. Ferreira, *Solid State Phys.,* (H. Ehrenreich and D. Turnbull, eds.) 44:229-415, Academic Press, New York (1991)
8. L. Hedin and S. Lundquist, *Solid State Phys.,* (F. Seitz, D. Turnbull, and H. Ehrenreich, eds.) 23:2-181, Academic Press, New York (1969)
9. M. S. Hybertson and S. G. Louie, *Phys. Rev.,* B34:5390-5413 (1986)
10. X. Zhu and S. G. Louie, *Phys. Rev.,* B43:14142-14156 (1991)
11. O. Madelung, *Springer Series in Solid State Sciences* (M. Cardona, P. Fulde, and H. J. Quiesser, eds.) 2:175-227, Springer-Verlag, Berlin (1978)
12. C. Hamann, H. Burghardt, and T. Frauenheim, *Electrical Conduction Mechanisms in Solids,* VEB Deutscher Verlag der Wissenschaften, Berlin (1988)
13. C. W. J. Beenakker and H. van Houten, *Solid State Phys.,* (H. Ehrenreich and D. Turnbull, eds.) 44:1-228, Academic Press, New York (1991)
14. M. A. Lampert and P. Mark, *Current Injection in Solids,* Academic Press, New York (1970)
15. C. B. Duke, *Solid State Phys.,* (F. Seitz, D. Turnball, and H. Ehrenreich, eds.) Suppl. 10, Academic Press, New York (1969)

16. W. R. Runyan and K. E. Bean, *Semiconductor Integrated Circuit Process Technology,* Addison-Wesley Publishing Company, Reading (1990)

17. G. Kaplan, *IEEE Spectrum,* 47(11): 47-102 (1990)

18. M. Schlüter, *The Chemical Physics of Solid Surfaces and Heterogeneous Catalysis,* (D. A. King and D. P. Woodruff, eds.), 5:37-68, Elsevier Science Publishers, Amsterdam (1988)

19. C. B. Duke, *The Chemical Physics of Solid Surfaces and Heterogeneous Catalysis,* (D. A. King and D. P. Woodruff, eds.) 5:69-118, Elsevier Science Publishers, Amsterdam (1988)

20. C. B. Duke, *Springer Series in Solid State Science,* (J. Chelikowski and A. Franciosi, eds.), 95:113-133, Springer-Verlag, Berlin (1991)

21. X. Zhu, S. B. Zhang, S. G. Louie, and M. L. Cohen, *Phys. Rev. Lett.,* 63:2112-2115 (1989)

22. A. G. Eguiluz, A. A. Maradudin, and R. F. Wallis, *Springer Series in Surface Sciences,* 14:291-296 (1988)

23. J. P. Lowe, *Quantum Chemistry,* pp. 309-346, Academic Press, New York (1978)

24. R. M. Dreizler and E. K. U. Gross, *Density Functional Theory: An Approach to the Quantum Many-Body Problem,* Springer-Verlag, Berlin (1990)

25. N. D. Lang, *Theory of the Inhomogeneous Electron Gas,* (S. Lundquist and N. H. March, eds.), pp. 309-389, Plenum Press, New York (1983)

26. G.-X. Qian, R. M. Martin, and D. J. Chadi, *Phys. Rev.,* B37:1303-1307 (1988)

27. R. W. Godby, M. Schlüter, and L. J. Sham, *Phys. Rev.,* B37:10159-10175 (1988)

28. R. W. Godby, L. J. Sham, and M. Schlüter, *Phys. Rev. Lett.,* 65:2083 (1990)

29. S. B. Zhang, M. S. Hybertson, M. L. Cohen, S. G. Louie, and D. Tomanek, *Phys. Rev. Lett.,* 63:1495-1498 (1989)

30. R. G. Dandrea, S. Froyen, and A. Zunger, *Phys. Rev.,* B42:3213-3220 (1990)

31. G. P. Das, P. Blöchl, O. K. Andersen, N. E. Christensen, and O. Gunnarsson, *Phys. Rev. Lett.,* 63:1168-1171 (1989)

32. M. Van Schilfgaarde and N. Newman, *Phys. Rev. Lett.,* 65:2728-2731 (1990)

33. J. A. Pople and D. L. Beveridge, *Approximate Molecular Orbital Theory,* McGraw Hill Book Co., New York (1970)
34. C. B. Duke, *Int. J. Quant, Chem.: Quant. Chem. Symp.,* 13:267-281 (1979)
35. D. J. Chadi, *Phys. Rev.,* B19:2074-2082 (1979)
36. K. C. Pandey and J. C. Phillips, *Phys. Rev.,* B13:750-760 (1976)
37. K. C. Pandey, *Phys. Rev. Lett.,* 47:1913-1917 (1981)
38. G.-X. Qian and D. J. Chadi, *Phys. Rev.,* B35:1288-1293 (1987)
39. C. B. Duke and Y. R. Wang, *J. Vac. Sci. Technol.,* A7:2035-2038 (1989)
40. J. P. LaFemina and C. B. Duke, *J. Vac. Sci. Technol.,* A9:1847-1855 (1991)
41. J. P. LaFemina, C. B. Duke, and C. Mailhiot, *J. Vac. Sci. Technol.,* B8:888-895 (1990)
42. O. L. Alerhand, D. Vanderbilt, R. D. Meade, and J. D. Joannopoulos, *Phys. Rev. Lett.,* 61:1973-1976 (1988)
43. M. B. Webb, F. K. Men, B. S. Swartzentruber, and M. G. Lagally, *J. Vac. Sci. Technol.,* A8:2658-2661 (1990)
44. M. C. Payne, N. Roberts, R. J. Needs, M. Needels, J. D. Joannopoulos, *Surf. Sci.,* 211/212:1-20 (1989)
45. I. P. Batra, *Phys. Rev.,* B41:5048-5054 (1990)
46. D. Vanderbilt, *Phys. Rev.,* B36:6209-6212 (1987)
47. Y. R. Wang and C. B. Duke, *Phys. Rev.,* B36:2763-2769 (1987)
48. Y. R. Wang, C. B. Duke, and C. Mailhiot, *Surf. Sci.,* L708-L716 (1987)
49. Y. R. Wang, C. B. Duke, K. O. Magnusson, and S. A. Flodström, *Surf. Sci.,* 205:L760-L770 (1988)
50. Y. R. Wang and C. B. Duke, *Surf. Sci.,* 192:309-322 (1987)
51. Y. R. Wang and C. B. Duke, *Phys. Rev.,* B37:6417-6424 (1988)
52. T. N. Horsky, G. R. Brandes, K. F. Canter, C. B. Duke, S. F. Horng, A. Kahn, D. L. Lessor, A. P. Mills, Jr., A. Paton, K. Stevens, and K. Stiles, *Phys. Rev. Lett.,* 62:1876-1879 (1989)
53. I. P. Batra, *Phys. Rev.,* B43:12322-12334 (1991)
54. C. B. Duke and Y. R. Wang, *J. Vac. Sci. Technol.,* A6:692-695 (1988)
55. H. H. Farrell, J. P. Harbison, and L. D. Peterson, *J. Vac. Sci. Technol.,* B5:1482-1489 (1987)

56. H. H. Farrell, M. C. Tamargo, and J. L. de Mignel, *J. Vac. Sci. Technol.*, B6:767-768 (1988)

57. G.-X. Qian, R. M. Martin, and D. J. Chadi, *Phys. Rev.*, B38:7649-7663 (1988)

58. D. K. Biegelsen, R. D. Bringans, J. E. Northrup, and L. E. Swartz, *Phys. Rev. Lett.*, 65:452-455 (1990)

59. C. B. Duke, *Appl. Surf. Sci.*, 11/12:1-19 (1982)

60. D. L. Smith and C. Mailhiot, *Rev. Mod. Phys.*, 62:173-234 (1990)

61. C. B. Duke, *J. Vac. Sci. Technol.*, A6:1957-1962 (1988)

62. W. Mönch, *Rep. Prog. Phys.*, 53:221-278 (1990)

63. J. E. Inglesfield, *Prog. Surf. Sci.*, 20:105-164 (1985)

64. B. A. Joyce, P. J. Dobson, and P. K. Larsen, *The Chemical Physics of Solid Surfaces and Heterogeneous Catalysis,* (D. A. King and D. P. Woodruff, eds.), 5:271-307, Elsevier Science Publishers, Amsterdam (1988)

65. A. Kahn, *Surf. Sci. Reps.*, 3:193-300 (1983)

66. C. B. Duke, *J. Vac. Sci. Technol.*, B1:732-735 (1983)

67. C. B. Duke, R. J. Meyer, A. Paton, and P. Mark, *Phys. Rev.*, B18:4225-4240 (1978)

68. C. B. Duke, A. Paton, Y. R. Wang, K. Stiles, and A. Kahn, *Surf. Sci.*, 197:11-23 (1988); (E) 214:334-335 (1989); (E) 221:619 (1989)

69. C. B. Duke, D. L. Lessor, T. N. Horsky, G. Brandes, K. F. Canter, H. Lippel, A. P. Mills, Jr., A. Paton, and Y. R. Wang, *J. Vac. Sci. Technol.*, A7:2030-2034 (1989)

70. Y. R. Wang, C. B. Duke, K. Stevens, A. Kahn, K. O. Magnusson, and S. A. Flodstrom, *Surf. Sci.*, 206:L817-L823 (1988)

71. A. Kahn, C. B. Duke, and Y. R. Wang, *Phys. Rev.*, B44:5606-5615 (1991)

72. D. Vanderbilt, O. L. Alerhand, R. D. Meade, and J. D. Joannopoulos, *J. Vac. Sci. Technol.*, B7:1013-1016 (1989)

73. E. Artacho and F. Yndurain, *Phys. Rev. Lett.*, 62:2491-2494 (1989)

74. B. I. Craig and P. V. Smith, *Surf. Sci.*, 218:569-579 (1989)

75. R. I. G. Uhrberg and G. V. Hansson, *Crit. Rev. Solid State Mater. Sci.*, 17:133-185 (1991)

76. D. J. Chadi, *J. Vac. Sci. Technol.* A5:834-838 (1987)

77. E. Kaxiras, K. C. Pandy, Y. Bar-Yam, and J. D. Joannopoulos, *Phys. Rev. Lett.*, 56:2819-2822 (1986)

78. M. D. Pashley, H. W. Haberern, W. Friday, J. M. Woodall, and P. D. Kirchner, *Phys. Rev. Lett.,* 60:2176-2179 (1988)

79. D. K. Bieglesen, R. D. Bringans, J. E. Northrup, and L. E. Swartz, *Phys. Rev.,* B41:5701-5706 (1990)

80. D. K. Biegelsen, R. D. Bringans, J. E. Northrup, and L. E. Swartz, *Phys. Rev. Lett.,* 65:452-455 (1990)

81. A. Zur, T. C. McGill, and D. L. Smith, *Phys. Rev.,* B28:2060-2067 (1983)

82. S. M. Sze, *Physics of Semiconductor Devices,* pp. 425-504, John Wiley and Sons, New York (1969)

83. M. H. Hecht, *J. Vac. Sci. Technol.,* B8:1018-1024 (1990)

84. L. Esaki, *Molecular Beam Epitaxy and Heterostructures,* (L. L. Chang and K. Ploog, eds.) pp. 1-36, Martinue Nijhoff, Dordrect (1985)

85. *Band Structure Engineering in Semiconductor Microstructures,* (R. A. Abram and M. Jaros, eds.), Plenum Press, New York (1989)

86. Y.-T. Lu, and L. J. Sham, *Phys. Rev.,* B40:5567-5582 (1989)

87. R. G. Dandrea and A. Zunger, *Phys. Rev.,* B43:8962-8989 (1991)

88. M. A. Gell, D. Ninno, M. Jaros, D. J. Wolford, T. F. Keuch, and J. A. Bradley, *Phys. Rev.,* B35:1196-1222 (1987)

89. G. C. Osbourn, *J. Vac. Sci. Technol.,* B1:379-382 (1983)

90. D. S. Citrin and Y.-C. Chang, *J. Appl. Phys.,* 68:161-168 (1990)

91. J. A. Venables, G. D. T. Spiller, and M. Hanbucken, *Rep. Prog. Phys.,* 47:399-459 (1984)

92. M. H. Grabow and G. Gilmore, *Semiconductor Based Heterostructures: Interfacial Structure and Stability,* (M. L. Green, ed.), pp. 3-19, Metallurgical Soc. Inc. (1987)

93. M. Schneider, A. Rahman, and I. K. Schuller, *Phys. Rev. Lett.,* 55:604-606 (1985)

94. B. W. Dodson, *Semiconductor Based Heterostructures: Interfacial Structure and Stability,* (M. L. Green, ed.) pp. 21-27, Metallurgical Soc. Inc. (1987)

95. D. Srivastava, B. J. Garrison, and D. W. Brenner, *Langmuir,* 7:683-692 (1991)

96. K. E. Khor and S. Das Sarma, *Phys. Rev.,* B43:9992-9995 (1991)

97. T. Ito, K. E. Khor, and S. Das Sarma, *Phys. Rev.,* B41:3893-3896 (1990)

98. W. A. Harrison, E. A. Kraut, J. R. Waldrop, and R. W. Grant, *Phys. Rev.,* B18:4402-4410 (1978)

99. S. L. Wright, H. Kroemer, and M. Inada, *J. Appl. Phys.,* 55:2916-2927 (1984)

100. J. E. Northrup, *Phys. Rev.,* B37:8513-8515 (1988)

101. J. E. Northrup, *Phys. Rev. Lett.,* 62:2487-2490 (1989)

102. E. Kaxiras, O. L. Alerhand, J. D. Joannopoulos, and G. W. Turner, *Phys. Rev. Lett.,* 62:2484-2486 (1989)

103. F. K. LeGoues, V. P. Kesan, S. S. Iyer, J. Tersoff, and R. Tromp, *Phys. Rev. Lett.,* 64:2038-2041 (1990)

104. S. Froyen and A. Zunger, *Phys. Rev. Lett.,* 66:2132-2135 (1991)

105. J. N. Schulman and T. C. McGill, *Synthetic Modulated Structures,* (L. L. Chang and B. C. Giessen, eds.), pp. 77-112, Academic Press, Orlando (1985)

106. L. J. Sham, *Superlatt. Microstruc.,* 5:335-340 (1989)

107. C. G. Van de Walle and R. M. Martin, *J. Vac. Sci. Technol.,* B4:1055-1059 (1986)

108. M. D. Stiles and D. R. Hamann, *Phys. Rev.,* B38:2021-2037 (1988)

109. R. G. Dandrea, J. E. Bernard, S.-H. Wei, and A. Zunger, *Phys. Rev. Lett.,* 64:36-39, (1990)

110. Y. Kim, A. Ourmazd, M. Bode, and R. D. Feldman, *Phys. Rev. Lett.,* 63:636-639 (1989)

111. J. Tersoff, *HeterojunctionBand Discontinuities: Physics and Device Applications,* (F. Capasso and G. Margaritondo, eds.), pp. 3-57, North Holland, Amsterdam (1987)

112. C. Mailhiot and C. B. Duke, *Phys. Rev.,* B33:1118-1133 (1986)

113. W. A. Harrison and J. Tersoff, *J. Vac. Sci. Technol.,* B4:1068 (1986)

114. A. Munoz, N. Chetty, and R. M. Martin, *Phys. Rev.,* B41:2976-2981 (1991)

115. L. Sorba, G. Bratina, G. Ceccone, A. Antonini, J. F. Walker, M. Micovic, and A. Franciosi, *Phys. Rev.,* B43:2450-2453 (1991)

116. M. Peressi, S. Baroni, R. Resta, and A. Baldereschi, *Phys. Rev.,* B43:7347-7350 (1991)

117. N. E. Christensen and L. Brey, *Phys. Rev.,* B38:8185-8191 (1988)

118. M. Cardona and N. E. Christensen, *Phys. Rev.,* B35:6182-6194 (1987)

119. S. Baroni, R. Resta, A. Baldereschi, and M. Peressi, *Spectroscopy of Semiconductor Microstructures,* (G. Fasol, A. Fasolino, and P. Lugli, eds.), NATO ASI Series B, Vol. 206, Plenum, New York (1989)

120. D. Z.-Y Ting and Y.-C. Chang, *Phys. Rev.,* B36:4359-4374 (1987)

121. N. J. Pulsford, R. J. Nicholas, P. Dawson, K. J. Moore, G. Duggan, and C. T. B. Foxon, *Phys. Rev. Lett.,* 63:2284-2287 (1989)

122. J. A. Brum, L. L. Chang, and L. Esaki, *Phys. Rev.,* B38:12977-12983 (1988)

123. G. Bastard, E. E. Mendez, L. L. Chang, and L. Esaki, *Phys. Rev.,* B26:1974-1979 (1982)

124. M. S. Hybertson and M. Schlüter, *Phys. Rev.,* B36:9683-9693 (1987)

125. S. Froyen, D. M. Wood, and A. Zunger, *Phys. Rev.,* B37:6893-6907 (1988)

126. S. Satpathy, R. M. Martin, and C. G. Van de Walle, *Phys. Rev.,* B38:13237-13245 (1988)

127. S.-H. Wei and A. Zunger, *Appl. Phys. Lett.,* 58:2684-2686 (1991)

128. L. J. Brillson, *Surf. Sci. Rpts.,* 2:123-326 (1982)

129. L. J. Brillson and G. Margaritondo, *The Chemical Physics of Solid Surfaces and Heterogeneous Catalysis,* (D. A. King and D. P. Woodruff, eds.), 5:119-181, Elsevier Science Publishers, Amsterdam (1988)

130. I. Braicovich, *The Chemical Physics of Solid Surfaces and Heterogeneous Catalysis,* (D. A. King and D. P. Woodruff, eds.), 5:235-269, Elsevier Science Publishers, Amsterdam (1988)

131. J. H. Weaver, in *Springer Series in Solid State Science* (J. R. Chelikowsky and A. Franciosi, eds.) 95:135-214, Springer-Verlag, Berlin (1991)

132. W. K. Ford, T. Guo, D. L. Lessor, and C. B. Duke, *Phys. Rev.,* B42:8952-8965 (1990)

133. W. K. Ford, T. Guo, S. L. Lantz, K. Wan, S.-L Chang, C. B. Duke, and D. L. Lessor, *J. Vac. Sci. Technol.,* B8:940-947 (1990)

134. A. Zur, T. C. McGill, and D. L. Smith, *Surf. Sci.,* 132:456-464 (1983)

135. C. B. Duke and C. Mailhiot, *J. Vac. Sci. Technol.,* B3:1170-1177 (1985)

136. W. E. Spicer, Z. Liliental-Weber, E. Weber, N. Newman, T. Kendelwicz, R. Cao, C. MaCants, P. Mahowald, K. Miyano, and I. Lindau, *J. Vac. Sci. Technol.,* B6:1245-1251 (1988)

137. F. Schäffler, R. Ludeke, A. Taleb-Ibrahmi, G. Hughes, and D. Rieger, *J. Vac. Sci. Technol.,* B5:1048-1053 (1987)

138. F. Schäffler, R. Ludeke, A. Taleb-Ibrahmi, G. Hughes, and D. Rieger, *Phys. Rev.,* B36:1328-1331 (1987)

139. R. M. Feenstra and P. Mårtensson, *Phys. Rev. Lett.,* 61:447-450 (1988)

140. R. Cao, K. Miyano, T. Kendelewicz, I. Lindau, and W. E. Spicer, *Surf. Sci.,* 206:413-425 (1988)

141. M. S. Tyagi, *Metal Semiconductor Barrier Junctions and Their Applications,* (B. L. Sharma, ed.), pp. 1-60, Plenum Press, New York (1984)

142. J. L. Freeouf and J. M. Woodall, *J. Appl. Phys.,* 39:727-729 (1981)

143. J. Tersoff, *Phys. Rev. Lett.,* 52:465-468 (1984)

144. J. Tersoff, *Phys. Rev.,* B32:6968-6971 (1985)

145. L. J. Brillson, R. E. Viturro, C. Mailhiot, J. L. Shaw, N. Tache, J. McKinley, G. Margaritondo, J. M. Woodall, P. D. Kirchner, G. D. Pettit, and S. L. Wright, *J. Vac. Sci. Technol.,* B6:1263-1269 (1988)

146. R. E. Viturro, C. Mailhiot, J. L. Shaw, L. J. Brillson, D. LaGraffe, G. Margaritondo, G. D. Pettit, and J. M. Woodall, *J. Vac. Sci. Technol.,* A7:855-860 (1989)

147. R. E. Viturro, S. Chang, J. L. Shaw, C. Mailhiot, L. J. Brillson, A. Terrasi, Y. Hwu, G. Margaritondo, P. D. Kirchner, and J. M. Woodall, *J. Vac. Sci. Technol.,* B7:1007-1012 (1989)

148. M. T. Schmidt, Q. Y. Ma, D. V. Podlesnik, R. M. Osgood, and E. S. Yang, *J. Vac. Sci. Technol.,* B7:980-985 (1989)

149. S. Chang, L. J. Brillson, D. F. Rioux, Y. L. Kime, P. D. Kirchner, G. D. Pettit, and J. M. Woodall, *J. Vac. Sci. Technol.,* B8:1008-1013 (1990)

150. J. E. Klepeis and W. A. Harrison, *J. Vac. Sci. Technol.,* B7:964-970 (1989)

151. I. Lefebvre, M. Lannoo, and G. Allan, *Europhys. Lett.,* 10:359-363 (1989)

152. G. Allan and M. Lannoo, *J. Vac. Sci. Technol.,* B9:2135-2139 (1991)

153. W. Mönch, *J. Vac. Sci. Technol.,* B6:1270-1276 (1988)

154. R. Ludeke, G. Jezequel, and A. Taleb-Ibrahimi, *Phys. Rev. Lett.,* 61:601-604 (1988)

155. R. Ludeke, G. Jezequel, and A. Taleb-Ibrahimi, *J. Vac. Sci. Technol.,* B6:1277-1284 (1988)

156. R. Ludeke, *Phys. Rev.,* B40:1947-1950 (1989)

157. M. H. Hecht, *Phys. Rev.,* B41:7918-7921 (1990)

158. S. Chang, I. M. Vitomirov, L. J. Brillson, D. F. Rioux, P. D. Kirchner, G. D. Pettit, J. M. Woodall, and M. H. Hecht, *Phys. Rev.,* B41:12299-12302 (1990)

159. A. Bauer, M. Prietsch, S. Molodtsov, C. Laubschat, and G. Kaindl, *J. Vac. Sci. Technol.,* B9:2108-2113 (1991)

160. M. Alonso, R. Cimino, and K. Horn, *Phys. Rev. Lett.,* 64:1947-1950 (1990)

161. M. Alonso, R. Cimino, C. Maierhofer, T. Chassé, W. Braun, and K. Horn, *J. Vac. Sci. Technol.,* B8:955-963 (1990)

162. G. D. Waddill, T. Komeda, Y.-N. Yang, and J. H. Weaver, *Phys. Rev.,* B41:10283-10286 (1990)

163. D. Mao, A. Kahn, M. Marsi, and G. Margaritondo, *Phys. Rev.,* B42:3228-3230 (1990)

164. D. Mao, A. Kahn, G. LeMay, M. Marsi, Y. Hwu, G. Margaritondo, M. Santos, M. Shayegan, L. T. Florez, and J. P. Harbison, *J. Vac. Sci. Technol.,* B9:2083-2089 (1991)

165. K. E. Miyano, R. Cao, C. J. Spindt, T. Kendelwicz, I. Lindau, and W. E. Spicer, *J. Vac. Sci. Technol.,* B9:2100-2107 (1991)

166. C. J. Spindt, M. Yamada, P. L. Meissner, K. E. Miyano, A. Herrera, and W. E. Spicer, *J. Vac. Sci. Technol.,* B9:2090-2094 (1991)

167. T. U. Kampen, D. Troost, X. Y. Hou, L. Koenders, and W. Mönch, *J. Vac. Sci. Technol.,* B9:2095-2099 (1991)

168. M. Van Schilfgaarde, *J. Vac. Sci. Technol.,* B8:990-994 (1990)

169. M. Van Schlifgaarde and N. Newman, *J. Vac. Sci. Technol.,* B9:2140-2145 (1991)

170. R. G. Dandrea and A. Zunger, *Appl. Phys. Lett.,* 57:1031-1033 (1990)

171 J. M. MacLaren, J. B. Pendry, P. J. Rous, D. K. Saldin, G. A. Somorjai, M. A. van Hove, and D. D. Viedeusky, *Surface Crystallographic Information Service: A Handbook of Surface Structures,* D. Reidel, Dordrecht (1987)

# 9

# Atomic-Scale Control of Heterojunction Band Lineups

*James McKinley and Giorgio Margaritondo*

## 1.0 INTRODUCTION

The interface between two different semiconductors is the building block of a variety of heterojunction devices, whose importance is continuously growing in modern microelectronics.[1][2] The properties of interfaces of this kind are dominated by the presence of discontinuities in the valence-band minimum and in the conduction-band maximum,[1] that accommodate the difference between the forbidden gaps of the two materials (see Fig. 1). The lineup of the two band structures that determines the relative magnitudes of the conduction and valence band discontinuities is a fundamental open problem in today's solid-state science.[1]

The basic and applied interest of this problem has stimulated much work, both experimentally and theoretically, in recent years. The status of this field has been extensively described by recent reviews[1][2] that illustrate both the progress made in understanding band lineups and its somewhat frustrating slow pace.

We dedicate this presentation to a novel aspect of this research: the possibility of manipulating band lineups and discontinuities through controlled, atomic-level contamination of the interface. This possibility was raised a few years ago, and has been positively verified in a number of cases. The result is certainly very interesting for applications since it may open up the possibility of tuning band discontinuities to specific applications of specific devicesan important aspect of the so-called *bandgap engineering.* At the same time, there is a fundamental interest, since learning to manipulate band lineups also increases our knowledge of their nature.

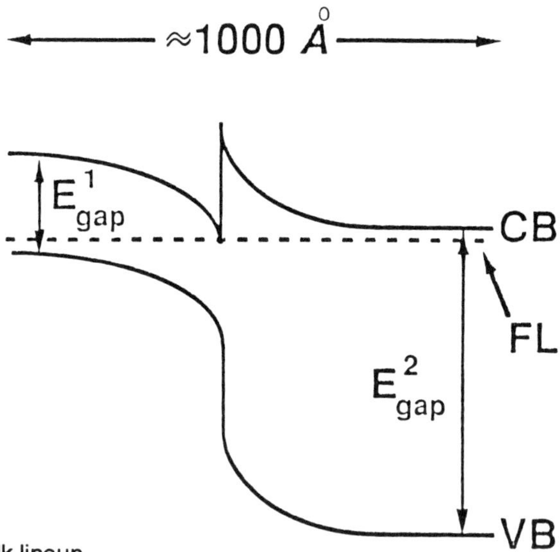


**Figure 1.** Bulk lineup.

## 2.0 CONTROL OF BAND LINEUPS: GENERAL STATUS

The general philosophy of the largely empirical attempts to manipulate heterojunction band lineups is the following. Many experiments have indicated that the lineup is significantly influenced by the local interface structure, i.e., by its components, by the electronic charge distribution and by the interface's local morphology. One can predict, then, that changes in any one of these factors could lead to changes in the lineup. The changes could be induced by inserting a thin intralayer between the two sides of the interface; such an assumption was the starting point for the empirical attack of the problem.

There have been two kinds of successful approaches in this empirical attack. First, the insertion of a thin (from submonolayer to a few monolayers) intralayer between the two sides of the heterojunction, the intralayer consisting of only one element. In several cases, the intralayer has been found to change the valence band discontinuity.[3]

In the second approach, a similarly thin double intralayer is inserted between the two sides of a homojunction, thereby creating a valence band discontinuity at the interface. For example, a discontinuity between Ge and Ge by means of a double-layer Ga-As.

These successes, however limited, provided the evidence that manipulating band lineups is possible, and opened up a very exciting field of research. In the field of theory, the first successes in modifying band lineups by intralayers stimulated many efforts to clarify the corresponding mechanisms, which in turn led to refinements of established concepts such as the charge neutrality level, and in general, led to a better understanding of the factors influencing band lineups. As a byproduct, these studies also inspired the theoretical suggestion to insert double layers to create artificial homojunction band discontinuities, later successfully tested in practice.

There have not yet been attempts to transfer these results into actual devices, primarily because a much better understanding of the overall effects of intralayers on the transport properties of the junction is a prerequisite for the transfer. One can envision, however, an extensive technological impact. Diverse devices such as memory elements, particle detectors, photon emitters, etc., could greatly profit from the enhanced flexibility created by the band lineup control. The possible applications of this technology are virtually endless—but a word of caution is required: unforeseen problems may exist in the path of the transfer to technology of the phenomena treated here.

The different parts of this chapter will discuss in depth different aspects of the band lineup manipulation, first for intralayer-induced modifications of heterojunction band lineups and then for the creation of homojunction band discontinuities. In each case, theoretical considerations will be accompanied by specific examples. Because of the complexity of the phenomena, we will present an in-depth discussion of the corresponding experimental observations.

## 3.0 MODIFICATIONS OF HETEROJUNCTION BAND LINEUPS BY INTRALAYERS

As we have seen, this is the first type of manipulation of heterojunction band lineups that has been successfully tested. Such tests have been extended to a variety of interfaces involving group-IV, III-V and II-VI materials. For the purpose of illustrating the corresponding phenomena, we will discuss here, in depth, the continuous adjustment of the valence band discontinuities $\Delta E_v$'s of heterojunctions like CdS-Ge, ZnSe-Ge, and GaP-Si, achieved by varying the thickness of ultrathin metal intralayers.[1][3][4]

For CdS-Ge, both Au and Al intralayers produce increases in $\Delta E_v$ of about 0.2 eV. Aluminum intralayers induce as much as a $\approx$ 0.3 eV increase in $\Delta E_v$ for ZnSe-Ge, whereas no change is observed for GaP-Si.[3] By comparison, Au intralayers produce strong changes in the $\Delta E_V$ of both systems. ZnSe-Ge and GaP-Si are particularly important technologically because they are well lattice-matched.

Following the historical development of this field, in which empirical attempts preceded the theoretical analysis of the phenomena, we present the experimental observations first, then discuss them again in a semi-empirical way, and finally briefly comment on the status of theories concerning this kind of band lineup manipulation.

### 3.1 General Issues

Before initiating the detailed discussion of theories and experiments, it is necessary to develop some additional background to appreciate the importance of the results. As mentioned, the importance goes well beyond the possible technological applications of the control of band lineups: it impacts our fundamental understanding of the factors that *determine* the lineup.

In essence, one tries to manipulate band lineups by hypothesizing about which factors are important ones, and by modifying them. The success or failure of the attempt reveals how important the factor really is, and in which way it influences the band lineup. Thus, these experiments may shed light upon a large variety of questions concerning the band lineup mechanism.

Interface abruptness is one of these questions, and the experiments with Au intralayers are particularly important for its clarification when compared to previous work done on much more reactive Al intralayers. In general, reactive interfaces are atomically abrupt, as is the case with Al intralayers. The Al reacts strongly with the semiconductor components and remains fairly localized to the interface. By comparison, Au intralayers are much less reactive, and strong Au out-diffusion occurs on both sides of the interface. Furthermore, while Al forms smooth flat layers,[5] Au is expected to cluster so that the interface will be segregated into regions with and without Au.

Chemically reactive and unreactive intralayers also induce the semiconductor components to interdiffuse differently. In particular, thin reactive intralayers are found to increase the cation/anion out-diffusion ratio from the binary substrate into the elemental semiconductor overlayer, whereas thin Au intralayers do not cause any cation out-diffusion. The cluster energy associated with thick Au intralayers apparently leads to some disruption of the binary substrate; 10 - 20 Å Au intralayers produce substantial cation out-diffusion, though not as much as occurred for thin Al intralayers.

Since the valence band top is dominated by anion-derived states and similarly, the conduction band minimum is dominated by cation-derived states, it is reasonable to assume that any interaction that affects anions and cations differently will affect the relative fraction of the band gap difference taken up by the valence and conduction band discontinuities. Some evidence for the validity of this reasoning has already been provided by the correlation of cation/anion out-diffusion ratios with $\Delta E_v$. In addition, the formation of bonds between Al intralayer atoms and the binary semiconductor anions correlates with modifications in $\Delta E_v$.[3][4] For example, Niles found definitive evidence of $Al_2Se_3$ formation when Al intralayers were introduced at ZnSe-Ge interfaces. Niles then correlated this to the saturation of the intralayer-induced effects in that system.[3][4] Exchange reactions at metal-semiconductor interfaces involving the semiconductor cation and the metal also correlate to changes in Schottky barrier heights.[6][7]

Electronegativity arguments can be used to ascertain which exchange reactions are probable. Sanderson's electronegativities[8] and the heats of formation of the binary semiconductors are given in Table 1. In relation to metal-semiconductor interfaces, Phillips[6] suggested that exchange reactions should occur if the semiconductor cation is less electropositive than the metal. He then showed that electronegativity differences of this kind are related to anomalous behavior of the CdS Schottky barrier as a function of metal electronegativity. Similar reasoning leads us to expect strong exchange reactions for the case of Al intralayers at GaP-Si, ZnSe-Ge, and CdS-Ge interfaces, and experiment indeed confirms this prediction.

**Table 1.** Electronegativities and Heats of Formation

| Material | Heat of Formation[a] (kcal per mole) | Sanderson Electronegativity cation (anion) |
|---|---|---|
| Ge | 69.5 | 3.59 |
| Si | 77.5 | 2.62 |
| GaP | 152 | 3.23 (3.34) |
| ZnSe | 101 | 2.84 (4.25) |
| CdS | 96 | 2.59 (4.11) |
| Au (unreactive) | | 2.88 |
| Cu (intermediate) | | 2.43 |
| Al (reactive) | | 1.94 |

(a) Ref.25; (b) Ref. 8.

By comparison, exchange reactions involving an Au intralayer are indicated only in the GaP-Si case. Even this reaction will be resisted because of the unusually high heat of formation for GaP (presumably materials with larger heats of formation require larger incentives to break the original cation bonds). In the case of Au with ZnSe, Zn and Au have almost identical electronegativities so that there is little reason for the exchange. In the case of Au with CdS, the exchange is strongly prohibited. For column IV materials, Au is known to form a silicide layer at Au-Si interfaces.[2]

An alternative method of determining the probability of chemical reactions at the interface uses heats of reaction ($\Delta H_R$).[7] The $\Delta H_R$'s of AuS and AuSe (anion from CdS or CdSe) are strongly positive (+3.94 and +1.36 eV/atom) so that these chemical reactions are unlikely.[7] However, a positive $\Delta H_R$ does not entirely preclude the reaction because the "interface" molecules are not entirely dissociated from the bulk materials. Reactions with negative $\Delta H_R$'s of somewhat smaller magnitudes are known to occur strongly. For example, the respective $\Delta H_R$'s of $Al_2S_3$ and $Al_2Se_3$ are -1.43 and -0.69 eV/atom, and these reactions are observed experimentally.[3][4]

We have already mentioned some similarities in the intralayer-induced behaviors of metal-semiconductor and semiconductor-semiconductor lineups. This is viewed as particularly significant because of the relationship between heterojunction $\Delta E_v$'s and Schottky barriers.[3] Specifically, linear theories predict:

$$\Delta E_v(A,B) = \phi^M_{SB}(A) - \phi^M_{SB}(B)$$

where $\phi^M_{SB}(A)$ is the p-type Schottky barrier for semiconductor A with metal M. Because of this relationship, we will briefly review some of the properties of Schottky barriers. Reactive Al intralayers at Au-semiconductor Schottky barriers increase the cation/anion out-diffusion ratio, and this is turn, correlates to systematic changes in the Schottky barrier heights.[7] As we mentioned, reactive Al intralayers at semiconductor-semiconductor interfaces strongly increase the cation/anion out-diffusion ratio while Au intralayers do not, and as expected, there is evidence that cation exchange reactions correlate to the saturation of the Al intralayer-induced effects.

Schottky barrier heights also correlate to the $\Delta H_R$'s of *interface molecules;* larger $\Delta H_R$'s are associated with larger p-type Schottky barriers.[7] Similar effects should occur at semiconductor-semiconductor interfaces. For the case of thick metal intralayers at semiconductor-semiconductor interfaces, this corresponds to an increase in $\Delta E_v$ correlated with the difference in the $\Delta R_v$'s of the interface molecules formed between the metal and the respective semiconductors. For thinner intralayers, there might be intermixing of the interface molecules so that the effect on $\Delta E_v$ is less clear. These effects are doubtlessly important in the Al intralayer case; in the Au intralayer case it is less clear that interface molecules form at all.

### 3.2 Specific Examples

**CdS($10\overline{1}\overline{1}$)-Au-Ge.** We are now ready to begin the in-depth discussion of a few specific examples, the first being CdS-Ge. This interface differs from the other two systems to be discussed in that it is not lattice-matched. Chemically, it should be similar to the ZnSe-Ge case, and exchange reactions between the Au intralayer and the Cd are strongly prohibited on the basis of electronegativity arguments. Figure 2 shows the values of $\Delta E_v$ for various Au intralayer thicknesses, as determined by the direct method. CdS-Au-Ge seems to rapidly reach a saturation value, and the effect looks very similar to the CdS-Al-Ge case (compare Figs. 2 and 3). Like the Al intralayer case, the discontinuity is increased by $\approx$ 0.2 eV for thin Au intralayers ($\approx$ 0.5 Å). Thicker intralayers up to 4 Å cause a slow decrease from the initial increase so that with 4 Å intralayers the total induced effect is $\approx$ 0.15 eV.

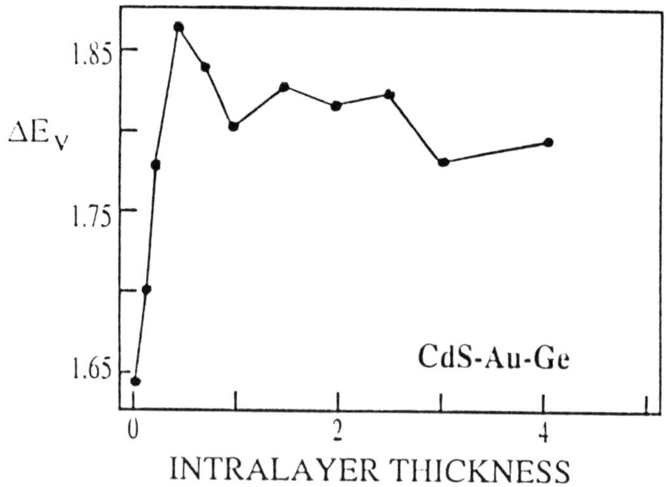


**Figure 2.** CdS-Au-Ge: $\Delta E_v$ vs. Au thickness.

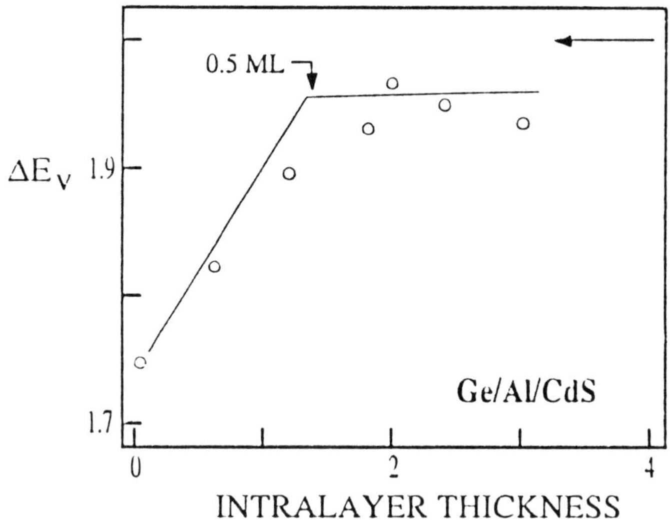


**Figure 3.** CdS-Al-Ge: $\Delta E_v$ vs. Al thickness. These data were taken from Ref. 3. The left-arrow gives the back-to-back Al Schottky barrier value.

Thick (≈ 20 Å) Au intralayers apparently reach the back-to-back Schottky barrier limit, i.e., the difference between the p-type CdS-Au and Ge-Au Schottky barrier heights. In turn, each p-type Schottky barrier height can be either directly measured or derived by subtracting the corresponding n-type barrier from the forbidden gap. These parameters are given in Table 2. Thus, the back-to-back value is (2.42 - 0.78) - (0.67 - 0.45) = 1.42 eV. This is in contrast to the result for Al intralayers which reached the back-to-back limit with intralayer thicknesses of 0.5 - 1 Å. The slower attainment of the asymptotic limit, is probably due to a highly disrupted and thus diffuse interface in the Au intralayer case. The direct method could not be applied to the thick Au intralayer measurements, but since strong chemical reactions are unlikely for this system, the indirect method should be reliable. Since the Cd4d emission is strongly diminished by this large Au coverage, we grew the Au overlayer incrementally in thicknesses of 0.06, 0.5, 8, and 20 Å. We could accurately determine the Cd4d position up through the 8 Å Au thickness. At 8 Å, the lower kinetic energy peak of the Au5d doublet was sharply defined, so that we measured further band-bending for the 20 Å total Au coverage case and all subsequent Ge coverages with the Au5d. The Au5d is probably an excellent choice for tracking band-bending because Au is unlikely to react chemically.

**Table 2.** CdS-Ge Parameters

| | | |
|---|---|---|
| Forbidden Gap (eV) | CdS:<br>Ge: | 2.42[a]<br>0.67[a] |
| Schottky Barrier Heights (eV) | n-CdS/Au<br>n-CdS/Al<br>n-Ge/Au<br>p-Ge/Au<br>n-Ge/Al | 0.78[b], 0.80[e]<br>ohmic[b], 0.2[f]<br>0.45[c]<br>0.26[e]<br>0.48[d] |
| No-intralayer $\Delta E_V$ (eV) | | 1.65[e], 1.73[c] |

(a) Ref. 1; ;(b) Refs. 7,30; (c) Ref. 15; (d) Ref. 12; (e) This work; (f) Inferred from Ref. 3.

For thick Ge overlayers, this method yielded a $\Delta E_V$ of 1.40 eV, in agreement with the expected value of 1.42 eV. We confirmed this result in another way, also. The Fermi level position with respect to the CdS valence band maximum was measured after the 20 Å Au coverage and found to be 1.63 eV. Similarly, the Fermi level to Ge valence band maximum distance for 10 Å of Au on a thick amorphous Ge film was measured to be 0.26 eV. The difference is 1.37 eV. Finally, we used the indirect method to measure a discontinuity of 1.7 eV for a 5 Å Au intralayer, so that we know that onset of the back-to-back barrier value must occur between the 5 and 20 Å Au intralayer cases.

There are two components of the Cd4d. We do not believe that this is due to a chemical reaction with the Au but rather to the fact that the Au clusters disrupt the surface and release metallic Cd into the overlayer Ge. After the Au deposition, the Cd4d moves to lower binding energy by 0.2 - 0.5 eV, with the larger number corresponding to the larger Au thicknesses (4 - 20 Å). Except for immediately after the Au deposition, the Cd4d peak appears to remain at fairly constant energy. A metallic Cd4d peak forms 0.6 - 0.7 eV to the lower binding energy side of the substrate Cd4d and is observable for certain cases. Specifically, in the 0.1 - 1 Å Au intralayer cases, this metallic Cd peak does not occur; it does appear for 2 - 3 Å Au intralayers but only after fairly thick Ge coverages (10 - 20 Å). This indicates that Au cluster energies, which would be larger for these Au thicknesses, increasingly disrupt the surface and release Cd. The metallic Cd out-diffuses into the Ge overlayer and for Ge coverages thick enough to substantially diminish the substrate Cd4d emission, the metallic Cd4d component begins to dominate. For the 4 - 5 Å Au intralayer thicknesses, the metallic component is evident even immediately after the Au deposition, and the metallic component is a substantial fraction of the total Cd4d emission by the 5 - 10 Å Ge coverage regime. For the thick 20 Å Au intralayer case the metallic peak is 50% of the total Cd4d emission immediately after the Au deposition.

Only the Cd4d and valence band spectra were taken for this system, so that no comparison of anion and cation diffusion is possible. We can say, however, that even small amounts of Au produce an Au5d emission that remains visible for 100 - 200 Å thick Ge overlayerslong after all substrate features are obscured. This behavior was observed for both Ge and Si overlayers and for all the substrate materials used in these experiments. It indicates that the Au strongly out-diffuses into the Ge overlayer, and confirms that the Au does not form any strong chemical bonds to the Ge. This contrasts sharply with the behavior of Al intralayers which react strongly with the Ge and the substrate materials and consequently remain localized to the interface.

**ZnSe(110)-Au-Ge**. Figure 4 shows the Au intralayer-induced changes at the ZnSe-Ge interface. The figure shows data derived from the direct "double-edge" method, as well as data obtained from the indirect method utilizing the Zn3d core level to track band bending.[9][10] The indirect-method data points were consistently shifted by 0.1 - 0.2 eV with respect to the direct-method ones, probably reflecting changes in the chemical shift of the Zn. Since the direct-method points are far more reliable, we corrected the indirect-method points to eliminate this shift, obtaining the open circles in Fig. 4.

ZnSe-Au-Ge is unique among the systems for which metal intralayers have been investigated, in that the smallest intralayer thicknesses induce a *decrease* rather than an increase in $\Delta E_v$. Specifically, Au thicknesses up to ≈ 0.1 Å cause a decrease in $\Delta E_v$ (see Fig. 4); the magnitude of the decrease reaches maximum of 0.2 - 0.3 eV near ≈ 1/16 Å Au intralayer thickness, and afterwards $\Delta E_v$ rapidly increases to, and then roughly stabilizes at, a value similar to that for the no-intralayer case.

Unlike Al, Au has a large density of states near the valence band maximum; this has the unfortunate effect of preventing us from using the direct method for thick Au intralayers. For the largest intralayer thicknesses for which the direct method can be used (≈ 1 Å), the discontinuity was near the intralayer-free value. The back-to-back Schottky barriers $\Delta E_v$ derived from Table 3 is ≈ 1.0 eV, and this would correspond to a 0.4 - 0.6 eV decrease with respect to the intralayer-free case. By comparison, Al intralayers of ≈ 1.4 Å were sufficient to produce the back-to-back Schottky barrier value, and after reaching this value, $\Delta E_v$ remained at the back-to-back Schottky barrier value for all larger Al intralayer thicknesses (see Fig. 5).

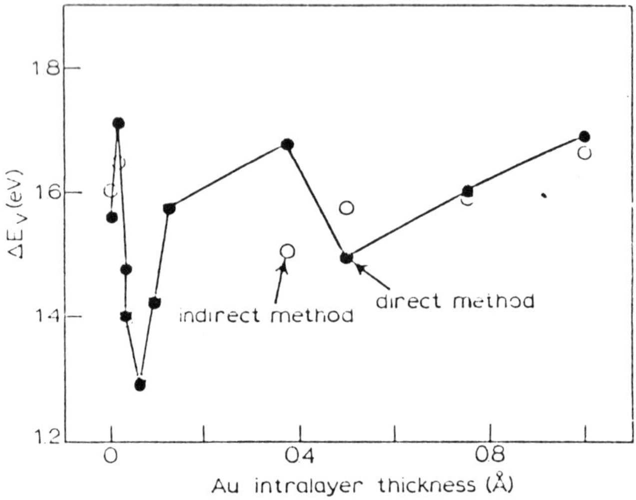


**Figure 4.** ZnSe-Au-Ge: $\Delta E_v$ vs. Au thickness. The solid dots were derived with the direct method, and the open circles with the indirect method.[1] (These latter data points were corrected for a shift, as discussed in the text.)

**Table 3.** ZnSe-Ge Parameters

| | | |
|---|---|---|
| Forbidden Gap (eV) | CdS:<br>Ge: | 2.58[a]<br>0.67[a] |
| Schottky Barrier Heights (eV) | n-ZnSe/Au<br>n-ZnSe/Al<br>n-Ge/Au<br>n-Ge/Al | 1.36[b], 1.34[c]<br>0.63[b], 0.64[c]<br>0.45[b]<br>0.48[d] |
| No-intralayer $\Delta E_V$ (eV) | | 1.40[e], 1.44[f]<br>1.56[g] |

(a) Ref. 1; (b) Ref. 31; (c) Ref. 14;
(d) Ref. 12; (e) Ref. 10; (f) Ref. 3;
(g) This work..

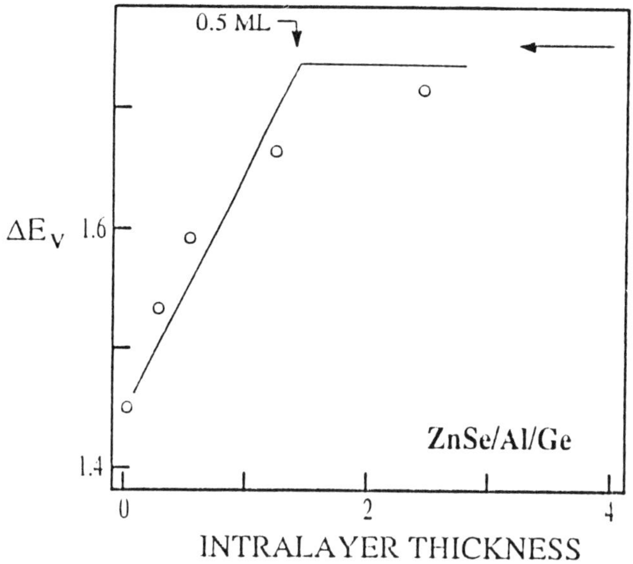


**Figure 5.** ZnSe-Al-Ge: $\Delta E_V$ vs. Al thickness. Only the direct method[1] is displayed. This plot was taken from Ref. 12. Notice that the increase is monotonic with Al thickness until it saturates near the back-to-back Schottky barrier value given by the arrow.

As in the CdS-Au-Ge case, we determined the discontinuity for 2 - 20 Å Au intralayers by the indirect method. Changes in band bending due to the Au deposition are tracked with the Zn3d, and afterwards the Au5d is used to track band bending. Because the Au out-diffuses strongly and probably does not become involved in any strong chemical reactions, the Au5d is ideal for tracking band bending. The separation of the lower kinetic energy peak of the Au5d and the Zn3d was established to be 4.05 eV with Au intralayers of 1 - 6 Å thickness with Ge coverages of 0 - 4 Å thickness. This value increased to 4.2 eV for Ge coverages greater than 10 Å thickness, which is attributed to a change in the chemical shift of the Zn (see below). For this reason, we used the first number in the following calculations. This procedure yielded values of 1.55, 1.53, 1.67, 1.68, and 1.73 eV for the 1, 2, 4, 6, and 20 Å Au intralayer cases, respectively. If we had used the larger Au5d to Zn3d splitting, the corresponding $\Delta E_V$'s would all be larger by $\approx$ 0.15 eV. The indirectly determined $\Delta E_V$'s join smoothly with the discontinuities determined by the direct method for the 0 - 1 Å Au intralayers, and apparently indicate that the thicker intralayers are still leaving $\Delta E_V$ near its intralayer-free value.

This result clearly does not agree with the back-to-back Schottky barrier value of ≈ 1.0 eV, and the trend is actually in the wrong direction. Furthermore, the disagreement is far beyond the experimental inaccuracy. Since the back-to-back value must occur for sufficiently thick intralayers, the slow attainment of the expected asymptotic value is assumed to be due to the existence of a highly disrupted and diffuse interface. Some evidence of a diffuse interface was found in the Zn3d core level spectra.

The larger Au thicknesses also evidenced a small amount of cation out-diffusion not present for the 0 - 1 Å Au cases. We doubt that a chemical reaction involving the Au is responsible. Rather, the larger Au thicknesses probably create Au clusters with sufficient cluster energy to increasingly damage the ZnSe substrate. This hypothesis is supported by the existence in the thick Au intralayer cases of a metallic Zn3d component (0.4 - 0.6 eV less bound than the substrate Zn3d) that appears only after large amounts of Ge have been deposited. For a 1 Å Au intralayer, the metallic component increases to 30% of the total Zn3d emission for the 40 Å Ge overlayer coverage. In the 6 Å Au intralayer case, the metallic component reaches 30% of the total emission by only 20 Å Ge coverage. Apparently this trend indicates that there is a small amount of metallic Zn that out-diffuses into the Ge overlayer; when the Ge overlayer is thick enough, the buried substrate Zn3d signal weakens but the out-diffusing metallic Zn signal remains strong. The fact that larger Au thicknesses produce an observable effect for smaller Ge coverages is consistent with the fact that larger Ge clusters are more disruptive to the substrate.

Interestingly, in the 20 Å Au case, an additional feature 2.3 eV less bound than the substrate Zn3d appears. This feature increases in intensity as the Ge coverage increases from 1 to 8 Å and afterward begins to diminish; the larger Ge coverages also increase its binding energy by ≈ 0.6 eV. It is not clear what this feature is, but it is consistent with the increasing disruption brought on by larger and larger cluster sizes.

We mentioned previously that we suspect a change in the chemical shift of the Zn3d. Without an intralayer, Ge coverage induces a 0.1 - 0.2 eV increase in the binding energy of the Zn3d (referred to the ZnSe valence band edge). The same conclusion is reached from the fact that the direct and indirect values of $\Delta E_v$ differ by 0.1 - 0.2 eV, and by the increase in the Au5d to Zn3d separation that occurs between the 4 Å and 12 Å Ge coverages for thick Au intralayers. Also, Kowalczyk et al.[11] observe an increase in the Zn3d-Ge3d energy separation of ≈ 0.3 eV after annealing ZnSe-Ge. The Zn3d binding energy increase observed in our experiments may be explained by an increased anion to cation ratio near the interface that occurs after some of the Zn diffuses into the overlayer; and alternate explanation is the formation of Zn-Ge bonds.

Interestingly, the Zn3d binding energy increase is not observed for 1/32 - 3/32 Å Au intralayers, the exact range for which the 0.6 eV decrease in $\Delta E_v$ occurs. We can speculate that the no-intralayer shift is due to the formation of Zn-Ge bonds inside the Ge overlayer. A thin Au intralayer prevents the out-diffusion of Zn. For thicker intralayers, the formation of clusters and the corresponding energy released can disrupt the interface, thereby making it again easy for Zn to out-diffuse. One method of testing this hypothesis is by studying the intensity decay of the Zn3d with respect to the Ge overlayer thickness. The decay was exponential for the no-intralayer case and for the case of thin Au intralayers. The 1/32 - 3/32 Å Au intralayer cases had a Zn3d decay parameter of 6 Å while the no-intralayer case and the 0.5 - 1 Å Au cases consistently had a 7 Å decay parameter. Though this is within the experimental error, it is consistent with the hypothesis that lessened Zn out-diffusion correlates with the induced decrease in $\Delta E_v$ for the 1/32 - 3/32 Å Au intralayers.

In conclusion, Niles had previously found that a strong chemical reaction takes place at the ZnSe-Al-Ge interface in which $Al_2Se_3$ is formed and large amounts of Zn are released into the Ge.[12] Furthermore, the Al intralayer itself remains localized to the interface.[12] By comparison, thin Au intralayers at a ZnSe-Ge interface do not produce strong cation out-diffusion while thicker Au intralayers do induce some cation out-diffusion. In all cases, the spectral contributions of Au demonstrate that it out-diffuses into the Ge strongly. Since the ZnSe-Au interface is diffusive,[7] we conclude that Au diffuses into both sides of the interface. There is also considerable evidence that Au clustering is responsible for disrupting and possibly segregating the interface. This may explain the noisier appearance of Fig. 4 as compared to Fig. 5.

**GaP(110)-Au-Si**. As with the other two systems, thin Al intralayers produce the back-to-back Schottky barrier value of $\Delta E_v$ at a GaP-Si interface. However, for this system, the back-to-back value is almost identical to the intralayer-free $\Delta E_v$, and consequently, no dependence on the thickness of the Al intralayer is observed. By comparison, Au intralayers at the GaP-Si interface produce some of the largest intralayer-induced effects that have yet been observed for any system. From Fig. 6 we see that a 1 Å -thick Au intralayer produces a $\approx$ 0.6 eV increase in $\Delta E_v$ with respect to the no-intralayer value of $\approx$ 1.0 eV. For comparison, a plot of $\Delta E_v$ vs. Al intralayer thickness is also provided (see Fig. 7).

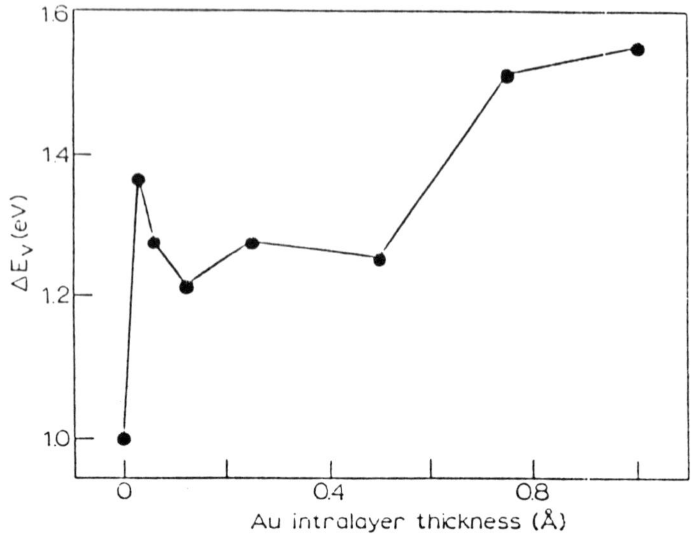


**Figure 6.** GaP-Au-Si: $\Delta E_v$ vs. Au thickness. The $\Delta E_v$ of GaP-Au-Si can be modified by varying the intralayer thickness. $\Delta E_v$ was derived by the indirect method (1), using the P2p level to estimate overlayer-induced changes in the substrate band bending.

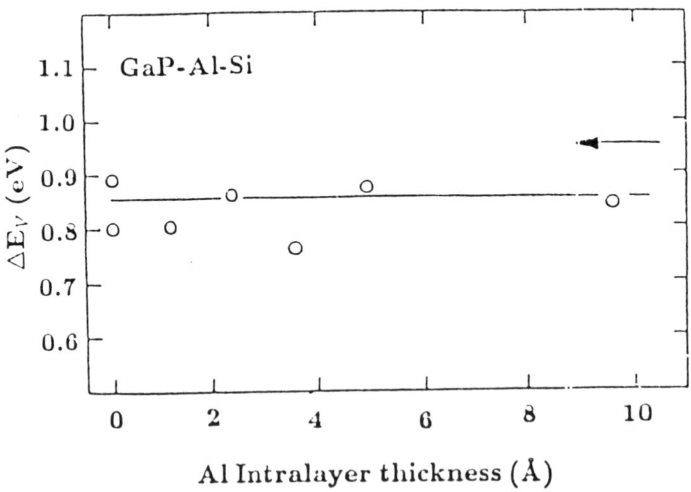


**Figure 7.** GaP-Al-Si: $\Delta E_v$ vs. Al thickness. This plot was taken from Ref. 4. It shows the dependence of $\Delta E_v$ on Al intralayer thickness. The arrow shows the predicted back-to-back Schottky barrier value.

Another difference between Au and Al intralayers is in the specific dependence of $\Delta E_v$ on the intralayer thickness. For all semiconductor pairs with Al intralayers, we found a linear dependence until saturation was reached. For Au, the data of Fig. 6 suggest instead a non-linear dependence, with a very sharp increase for the smallest thicknesses.

Note that for larger Au intralayer thicknesses—not explored by the present experiment—we would expect a decrease rather than an increase in $\Delta E_v$. In fact, in the limiting case of very thick intralayers, we expect $\Delta E_v$ to be given by the back-to-back Schottky barrier value. From the data of Table IV, we obtain a limiting $\Delta E_v$ value of (2.24 - 1.39) - 0.34 ≈ 0.5 eV.

**Table 4.** GaP-Si Parameters

| | | |
|---|---|---|
| Forbidden Gap (eV) | GaP: | 2.24[a] |
| | Si: | 1.107[a] |
| Schottky Barrier Heights (eV) | n-GaP/Au | 1.39[b] |
| | n-GaP/Al | 0.53[b] |
| | n-Si/Au | 0.8[c] |
| | p-Si/Au | 0.34[c] |
| | n-Si/Al | 0.50[c] |
| | p-Si/Al | 0.58[c] |
| No-intralayer $\Delta E_v$ (eV) | | 0.8[d], 0.89[e], 1.0[f] |

(a) Ref. 1; ;(b) Ref. 13; (c) Ref. 32; (d) Ref. 9;
(e) Ref. 12; (f) This work.

Figure 8 shows how the energy levels of the various core levels and the valence-band-maximum change as Si is deposited. We observe parallel shifts of the Ga3d and P2p levels immediately after the Au intralayer is deposited. The observed shifts agree with those derived from Ref. 13, although the initial positions of the peaks vary from cleave to cleave (as also reported in Ref. 13), apparently due to differences in cleave quality. After Si is deposited on the Au intralayer, the distance in energy between Ga3d and P2p increases. This indicates that at least one of these levels is affected by changes in the chemical shift as well as changes in the substrate band bending. Although there are no dramatically obvious new components in the substrate core level spectra, as was the case with Al intralayers, we know that chemical reactions have occurred: the P2p width remains constant, but the Ga3d broadens from ≈ 1.4 to ≈ 2.0 eV. This suggests a change in the chemical status of the Ga atoms. Thus, P2p appears to be the best choice for tracking changes in band bending.

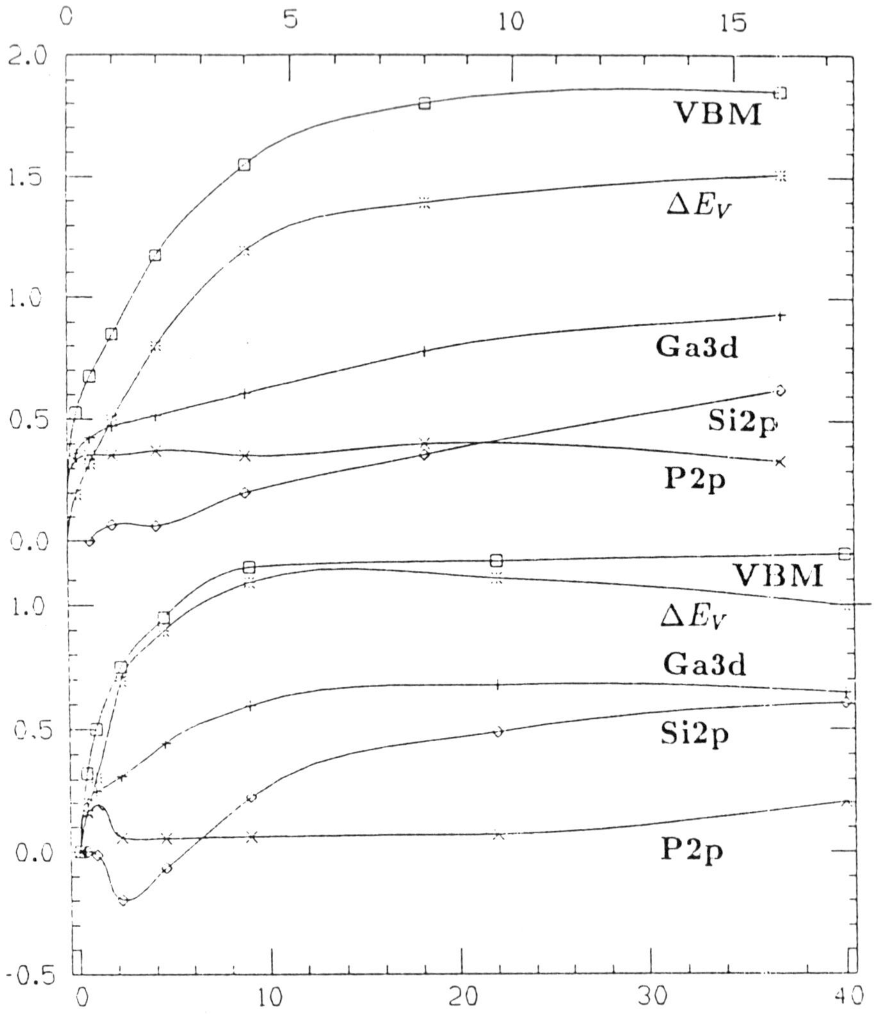


**Figure 8.** GaP-Au-Si: This plot shows how the valence band maximum (VBM), the substrate core levels (Ga3d and P2p), and the overlayer core level (Si2p) change with respect to their initial positions as the Si overlayer thickness increases. Except for the Si2p level, the initial position is taken from the cleaved GaP surface. The bottom figure shows the no-intralayer case, and the top figure shows the results when a 0.75 Å Au intralayer has been introduced.

We can also make inferences about the interface morphology by examining the dependence of the peak intensities on overlayer coverage thickness. If the substrate was remaining intact and the overlayer was covering it in a layer-by-layer fashion, we would expect to see substrate components decay with an exponential law, $\propto \exp(-l/\lambda)$, where the exponential parameter $\lambda$ is 5 - 10 Å. Similarly, overlayer components would grow $\propto (1 - \exp(-l/\lambda)$. Deviations from this dependence can indicate a variety of phenomena including clustering and interdiffusion.

With the notable exception of the first few monolayers of Si, the P2p/Ga3d intensity ratio remains approximately constant when Si is deposited on GaP, with or without a gold intralayer (see Fig. 9). Without the intralayer, we observe the Ga3d intensity to drop much faster than the P2p. Apparently, some kind of reacted layer involving Ge and P is forming at the interface and covering up the Ga from the top few monolayers of GaP. With the intralayer, this reaction is inhibited.

For the larger Si overlayer thicknesses, both substrate peak intensities decrease at the same rate and nearly exponentially with Si overlayer thickness with or without the Au intralayer; the exponential parameter is in the range 5 - 11 Å, close to the escape depth of photoelectrons from Si at these energies. We therefore conclude that the substrate Si coverage is close to being layer-by-layer, and that if there is any substrate out-diffusion, the ratio of cation to anion out-diffusion is approximately unity.

This is not at all the situation with Al intralayers. Niles reported[12] a much slower Ga3d intensity decrease with the Si thickness when a 3.6 Å Al intralayer is present, indicating that Ga is "floating" to the Si-vacuum interface. No evidence was found for P out-diffusion, in agreement with the conclusions of Ref. 7 concerning the affects of reactive metal intralayers on cation/anion out-diffusion ratios. In summary, there is negligible substrate anion out-diffusion for both Au and Al intralayers, whereas Al produces cation out-diffusion and Au does not.

In the case of Au intralayers, the fact that Au is strongly out-diffusing is indisputable. After 40 Å of Si coverage, the Au4d structure produced by a $\approx$0.1 Å thick Au intralayer is still visible. For 0.25 Å of Au, the Au4d doublet remains visible with 40 Å of Si, although the P2p has been reduced by a factor of $\approx$ 34. The Au4d doublet can be seen in the valence band spectra shown in Fig. 10. Chiaradia et al.[13] indicate that the GaP-Au interface is also interdiffused. Thus, for the GaP-Au-Si system, the Au probably diffuses into both the Si and the GaP.

Again, the case of Al intralayers is very different.[12] The Al2p core level decays exponentially with increasing Si coverage; the decay is slightly slower than that of the P2p, possibly indicating a slight Al out-diffusion. The Ga3d intensity decays more slowly than either the Al2p or P2p signals, and the decay is not even accurately modeled by an exponential. Also, the extreme reactivity of Al with Si would seem to rule out significant Al out-diffusion.

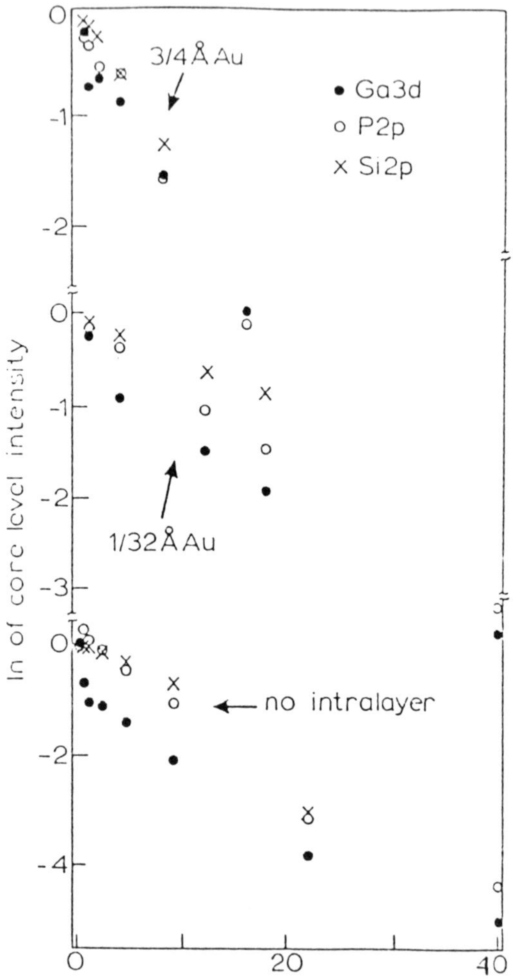


**Figure 9.** GaP-Au-Si: 2p peak intensities vs. Si coverage. A log plot of the Ga, P, and Si 2p peak intensities for the GaP-Si and GaP-Au-Si systems shows that these systems obey the ideal layer-by-layer growth model. Substrate intensities are plotted as ln $(I/I_0)$, where $I_0$ is the intensity for the cleaved surface; Si overlayer intensities are plotted as ln $(1 - I/I_\infty)$, where $I_\infty$ is the asymptotic value of the Si2p intensity for large Si thickness.

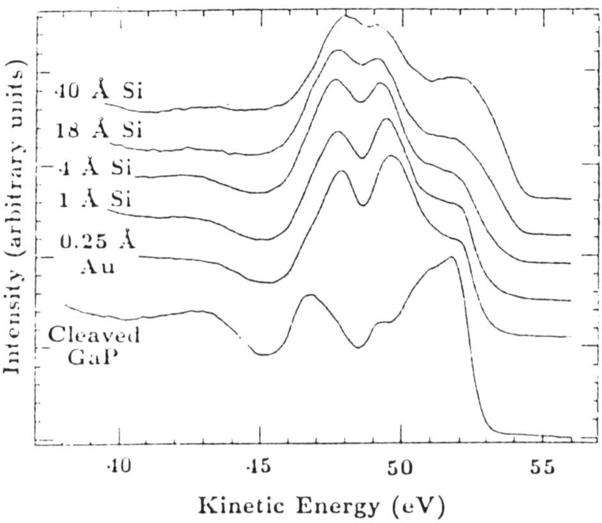


**Figure 10.** Valence Band Spectra for GaP-0.25 Å Au-Si. These spectra were taken at a photon energy of 60 eV. Notice that the Au4d doublet remains visible even when the Si valence band shape totally dominates the substrate GaP valence band maxima. There is a Si feature near the Au4d energy, but without the Au intralayer, this feature has less intensity than the Si valence band maxima for 40 Å of Si on GaP.

### 3.3 Qualitative Discussion of the Effects

Unfortunately, very little theory exists about intralayer-induced effects. The reason for the relatively small amount of theory is probably the intractable nature of the many competing and complex processes that occur in the experimental systems. The reader will shortly become convinced of this fact.

Before discussing the perturbations due to the intralayer, we briefly mention the theories governing the native (no-intralayer) interfaces. Theories of band lineups divide into Bardeen-like[1][2][14] and Schottky-like[1][15][16] classes, with the former asserting that interfacial states control Fermi pinning and the latter using the difference in work functions to determine the lineup. The Bardeen models (and in particular the Midgap Energy Rule[2][14]) are in best agreement with the no-intralayer cases of ZnSe-Ge,

GaP-Si, and CdS-Ge. This is not surprising based on the Schottky barrier behavior of these materials; Ge and Si are particularly susceptible to forming chemically reacted states near the interface. When an intralayer is introduced, a large number of extrinsic interfacial states are added to the problem.

Certain qualitative effects of varying the composition and thickness of an intralayer can be inferred from the simple *linear models*. If we use the Midgap Energy Rule[14] the interface dipole should be affected because the *induced* gap states depend upon tunneling across the interface. If the intralayer is abrupt and not segregated into clusters, the probability of semiconductor-to-semiconductor tunneling should decay exponentially with intralayer thickness and with a decay parameter of about one de Broglie wavelength. Induced gap states originating from the intralayer can also be expected if the intralayer becomes sufficiently thick.

Defect theories[17] depend on such things as vacancies and antisite defects that are produced as the interface forms; the addition of an intralayer in the growth sequence may change the kind of defects that occur. If the dominant defect energy with the intralayer differs from the dominant defect energy without the intralayer, the lineup will be altered, and in fact, there is evidence that this is the case. Niles[12] found that although Al intralayers form abrupt and ordered interfaces perpendicular to the interface plane, extreme disorder exists parallel to the interface plane. The semiconductor anions near the interface form $Al_2A_3$ with the intralayer atoms. By comparison, Au is known to have deep gap states in both Ge and Si,[18] and as Au diffuses into Si and Ge, it produces vacancy defects.[18]

Durán et al.[19] studied idealized semiconductor heterojunction interfaces using a tight-binding framework and concluded that simple alignment of bulk-derived neutrality levels is only a zero-order approximation appropriate for low-ionicity semiconductors. For this reason, he suggests that the largest interface dipoles should occur when the materials forming the interface are of greatly differing ionicity (e.g., ZnSe-Ge should have a larger interface dipole than GaP-Si). From an experimental point of view, the situation is not quite so clear. There is considerable evidence that high ionicity materials are more resistant to chemical reactions and intermixing. Since intermixing also creates interface dipoles, we must distinguish between the intrinsic dipoles of the idealized interface and the intermixing-induced dipoles of the real systems, and these two effects follow opposite trends with semiconductor ionicity.

Intralayers also affect microdiffusion processes near the interface. Experiments with Al and Au Schottky barriers on many II-VI and III-V compounds found that the more reactive the interface is, the more abrupt.[7] Al intralayers aid in the formation of abrupt interfaces for both metal-semiconductor and semiconductor-semiconductor interfaces,[7][12] while Au intralayers have the opposite effect. Thin Al intralayers at the interface of a Au Schottky barrier were found to change the out-diffusion from slightly anion-rich to strongly cation-rich[7] though more ionic semiconductors were more resistant to this effect. Furthermore, the changes in cation/anion ratios correlate to systematic changes in the Schottky barrier heights, and a similar result may occur for semiconductor-semiconductor valence band discontinuities. Since Au and Al intralayers both out-diffuse differently themselves and cause the semiconductor components to interdiffuse differently, the compositional profile of the interface may be changed on both macroscopic and microscopic scales.

On the microscopic scale, interface geometries other than the ideal one often result in smaller interface dipoles.[20] Intralayer atoms could act as both catalysts for, and participants in, the new interface geometries. Because of the large number of possibilities, the problem of alternate interface geometries is proving almost insurmountable theoretically. One method of eliminating certain interface configurations is to consider the energies of atomic substitution or exchange that must be surmounted to obtain the smaller interface dipoles.[21] Also, as already discussed, the probability of exchange reactions may be gauged on the basis of the relative electropositivity of the intralayer and semiconductor cation. Examination of photoemission core level lineshapes is one of the most powerful experimental tools for determining the kinds of atomic configurations that exist at the interface. In the CdS-Al-Ge case,[12] the semiconductor cation lineshape is affected dramatically, suggesting cation exchange reactions. By looking at the splitting of the different components and comparing them to the energies of different reacted phases, Niles was able to suggest which compounds were forming.

### 3.4 Quantitative Models

By now, the reader should be sufficiently convinced as to the complexity of the intralayer-induced dipole problem. This paves the way to a brief presentation of models that have been developed to treat this problem. Several simple "first steps" have been taken, and these fall into three categories: *(a)* tight-binding models of semiconductor heterojunctions that include an idealized intralayer, *(b)* Schottky-like corrections to the Midgap Energy Rule, and *(c)* simple capacitor models using electronegativity or bond polarity arguments to estimate charge transfer between unlike atoms.

Using a tight-binding approach, Durán et al. considered the case of a monolayer-Al intralayer at a ZnSe-Ge interface and predicted a ≈ 0.35 eV increase in $\Delta E_v$.[19] This is in excellent agreement with the experimental value of 0.2 - 0.3 eV.[3] The success in this instance is probably due to the abrupt nature of the ZnSe-Al-Ge interface which allowed Durán to assume an idealized interface geometry. Au intralayers, on the other hand, seem to develop complex interface morphologies, and for this reason, no calculations have been attempted.

Niles[3][22][23] developed a method of using Schottky-like corrections to the Midgap Energy Rule to estimate the effects of the intralayer. The band lineup determined by induced gap states and/or defects can be corrected by a linear dependence on the work functions of materials present at the interface, including intralayers. This linear dependence corresponds to the Schottky model (and to its heterojunction counterpart, the electron affinity rule) in the limiting case of no induced gap states and no defects. For the no-intralayer case,[3][24] we have:

$$\Delta E_v(0) = E_M{}^1 - E_M{}^2 + \overline{S}[(\chi^1 E_g{}^1 - \chi^2 - E_g{}^1) - (E_M{}^1 - E_M{}^2)]$$

$$\overline{S} = (S^1 + S^2)/2$$

$$S^i = aE_M^{\ i} + b$$

where the $E_M^{\ i}$'s are *midgap energies*, the $E_g^{\ i}$'s are gaps, the $\chi^i$'s are electron affinities, the $S^i$'s are pinning strength parameters, a and b are constant, and the index i refers to one of the two semiconductors that form the junction. Here a and b are 0.068 and 0.061, respectively.[3][12] We tested this theory against our experimental no-intralayer values and found that there is marginal improvement (see Table 5). The Midgap Energy Rule can also estimate the back-to-back Schottky barrier:

$$\Delta E_v(\infty) = \phi_{SB}^{p-type}\ (1,M) - \phi_{SB}^{p-type}\ (2,M)$$

$$\phi_{SB}^{p-type}\ (i,M) = E_M{}^i + S^i(\chi^i + E_g{}^i - E_M{}^i - \Phi_M)$$

where M refers to the metal and i to the semiconductor, the $\phi_{SB}^{p-type}$'s are p-type Schottky barriers, and $\Phi_M$ is the metal's work function.

With Al intralayers, the intralayer-induced change in the discontinuity, $\delta\Delta E_v$, saturates near the back-to-back value for intralayers thicknesses as thin as ½ monolayer. Thus, Niles modeled the saturated effect as:

$$\delta\Delta E_v = \Delta E_v(\infty) - \Delta E_v(0) \approx \Delta E_v(0) \times K$$

$$K = (\alpha/2)\,[(\chi^1 + E_g^{\,1} - E_M^{\,1} - \Phi_M) + (\chi^2 + E_g^{\,2} - E_M^{\,2} - \Phi_M)]$$
$$= 0.034(A - 2\Phi_M)$$

Using the midgap energies, electron affinities, and forbidden energy gaps reported in Ref. 3, we found that K was only weakly dependent on the semiconductor's components while it depended strongly on the intralayer work function: A was 10.30, 10.20, and 10.10 for CdS-Ge, ZnSe-Ge, and GaP-Si, respectively. The Al and Au work functions quoted in the literature vary significantly; we chose the photoemission work functions from Ref. 25 which are 4.08 and 4.82 for Al and Au, respectively. Table 6 shows the predictions of this theory; the agreement is reasonable for Al intralayers, and even shows the correct trend during the transition from II-VI to III-V semiconductors. The agreement for Au intralayers is very poor even for the thick intralayers for which the theory is designed.

**Table 5.** Schottky-like corrections, no intralayers: *(a)* mid-gap energies, electron affinities, and forbidden energy gaps from Ref. 3; *(b)* average value from Refs. 3, 10, 33; *(c)* average value from Refs. 3, 9, 33.

| Interface | Midgap Energy Rule[a] in eV | With Schottky-like correction in eV (pinning: $\overline{S}$) | Average experimental value in eV |
|---|---|---|---|
| CdS-Ge | 1.7 | 1.78 (0.119) | 1.72[b] |
| ZnSe-Ge | 1.4 | 1.47 (0.109) | 1.47[b] |
| GaP-Si | 0.5 | 0.53 (0.092) | 0.90 |

**Table 6.** Schottky-like corrections with intralayers: *(a)* the experimental value was used for $\Delta E_v$ and all midgap energies, electron affinities, and forbidden energy gaps were from Ref. 3 and the Al and Au work functions are from Ref. 25; *(b)* Ref. 3; *(c)* present work.

| Interface | Al-induced $\delta\Delta E_v$, theory (experiment[b]) in eV | Au-induced $\delta\Delta E_v$ theory (experiment[b]) in eV |
|---|---|---|
| CdS-Ge | 0.13 (0.2) | 0.04 (thin: 0.2, thick: -0.3) |
| ZnSe-Ge | 0.10 (0.3) | 0.03 (thin: -0.25, thick: 0.08-0.26) |
| GaP-Si | 0.06 (0.0) | 0.01 (thin: 0.6) |

The third major class of models uses electronegativity[3][22][26] or bond polarity[27][28] arguments to determine the charge transfer between unlike atoms. The average charge on each plane of atoms at the interface can be computed if the plane's areal density and atomic constituency are known. The separation of the planes can be taken to be the sum of covalent radii (disordered picture) or taken from the ideal crystal lattice (ordered picture). Poisson's equation trivially gives the dipole. We must calculate the dipole for the intralayer and no-intralayer cases and take the difference to find the intralayer-induced change in $\Delta E_v$.

Simple capacitor models have had amazing success in cases where the interface is atomically abrupt so that an idealized interface geometry can be used. An electronegativity-based scheme has been successful in predicting the Cs and H intralayer-induced changes in $\Delta E_v$ at $SiO_2$-Si interfaces.[29] Niles used the same scheme with fair success with the Al intralayer systems.[12] In the last chapter of this thesis, a bond polarity-based scheme is applied successfully to a Ga-As double layer at a Ge homojunction. However, in the case of Au intralayers, the probable non-ideality of the interface (i.e., clustering of the Au) makes it hard to justify a simple capacitor approach, and it was not attempted.

## 3.5 Intralayer Modifications of Band Lineups: Some General Conclusions

The few examples discussed in depth in this chapter make it possible to reach some preliminary but important general conclusions. The first and perhaps most important: it is, indeed, possible to modify band lineups by means of intralayers. Specifically, this has been tested with highly reactive intralayers as well as with intralayers of moderate reactivity.

The significant differences between the effects of Al and Au intralayers are particularly important in shedding light upon the factors affecting band lineups and their modifications. Aluminum intralayers favor the out-diffusion of cations from the substrate (the binary semiconductor), whereas there is no evidence of substrate anion out-diffusion. Thin Au intralayers do not induce strong out-diffusion of either substrate component, though thick Au intralayers apparently do induce some cation out-diffusion. What cation out-diffusion does occur is probably due to Au cluster energies disrupting the substrate rather than chemical reactions. Also, the Al intralayer atoms stay localized, whereas Au strongly out-diffuses—probably on both sides.

There are significant differences in the intralayer-induced changes in $\Delta E_v$. The most salient difference is that the onset of the expected back-to-back Schottky barrier value for thick intralayers occurs for 0.5 - 1 Å Al thicknesses but apparently requires much larger Au intralayer thicknesses. The back-to-back value did finally occur for 10 - 20 Å thick Au intralayers in the CdS-Au-Ge case, but it was not observed at the ZnSe-Ge interface though we used Au intralayers as thick as 20 Å.

The effects of submonolayer to several monolayer Au and Al intralayers on $\Delta E_V$ were also sharply different except in the CdS-Ge case in which both types of intralayer induce a 0.2 eV increase in $\Delta E_V$. For ZnSe-Al-Ge, there is a monotonic increase with the intralayer thickness until the saturation back-to-back Schottky barrier value is reached at ≈0.5 monolayers of Al. By comparison, ZnSe-Au-Ge exhibits a decrease for the thinnest intralayers, and this is followed by an increase which returns $\Delta E_V$ to near its intralayer-free value where it remains for all larger Au intralayer thicknesses used in these experiments (20 Å). Al causes no change in the GaP-Si system, while Au produces a substantial increase.

The complexity of the intralayer-induced effects were discussed, and several simple models were applied. The simple models had modest to good success with the Al intralayer case, but did not explain the Au intralayer-induced effects. The principal reason for this failure is probably the complex interface morphology present with Au intralayers, while by comparison, Al forms smooth flat intralayers and is consequently easier to model.

## 4.0 CREATION OF HOMOJUNCTION BAND OFFSETS VIA DIPOLE INTRALAYERS

The successes in modifying heterojunction band lineups and the subsequent theoretical work on the phenomena naturally led to the next step in band lineup modifications: the creation of artificial band discontinuities. Needless to say, such a possibility is extremely exciting for technological applications: one should consider, for example, the fabrication of nanostructures like quantum wells based on only one material, with properties that are largely complementary to those of conventional heterojunction quantum wells.

### 4.1 Examples of Successful Tests

This kind of band lineup manipulation has been successfully tested for the case of germanium homojunctions. It has been possible to fabricate such homojunctions with both positive and negative valence band offsets by introducing Ga-As dipole intralayers. The band offset was found to be 0.35 - 0.45 eV with the Ge valence band edge on the As side of the junction at lower energy (i.e., more bound). Dipole intralayers are a natural extension of previous research using monatomic intralayers to modify the band lineups at heterojunctions.[1] Intralayers modify an interface dipole in a variety of ways: they interfere with native interfacial states, may change dopant levels near the interface, and can affect tunneling across the interface. This last effect is especially important in relation to induced-gap state models.[2][14] Unlike heterojunctions, homojunctions do not have intrinsic interface dipoles that can simply be altered—rather, the intralayer itself must be dipolar.

The Ge homojunction with a Ga-As dipole intralayer was selected as the first system in our experiments because it offered several clear advantages. Foremost among them is the small lattice mismatch of only 0.08% between bulk GaAs and bulk Ge. Also, Ge(111) can be produced by cleavage, and Ge overlayers can be deposited with fairly simple techniques. In comparison, binary zincblende semiconductors cleave on nonpolar (110) planes, so that polar planes must be produced through molecular beam epitaxy or other sophisticated techniques.

## 4.2 Theoretical Predictions

The existence, sign, and magnitude of the intralayer-induced dipole are explained by the Harrison "Theoretical Alchemy" model.[1] Theoretical Alchemy has been put on a firm theoretical basis by Linear Response Theory (LRT).[34][35] A simple example of theoretical alchemy would be to transform a pair of Ge atoms into a Ga-As pair by transferring a proton between the nuclei. This imaginary charge transfer sets up a dipole. When the electronic distribution is allowed to relax, the newly created As atom (which has the extra proton) partially pulls one of the Ga valence electrons to itself. Theoretical alchemy is a clever expedient for describing what is normally explained with electronegativity arguments. These same arguments can be applied to solid interfaces.

Since the problem of a Ga-As double layer in Ge is lattice-matched, we will restrict the discussion to lattice-matched systems. The main points of LRT can be illustrated by considering an ordinary semiconductor heterojunction, A-B. The starting point is to form a virtual crystal, AB, which is an average of materials A and B. The actual interface is formed by perturbing AB so that one semi-infinite region is transformed into material

A and the other into material B. Only the difference between the charge densities of the physical system, A-B, and the virtual crystal, AB, contribute to the potential lineup. In all cases where such a difference can be dealt with by self-consistent first-order perturbation theory (i.e., LRT), it is possible to show that the potential lineup of isovalent interfaces (e.g., GaAs-AlAs) is a *bulk property* of the virtual crystal. Thus, the band offset is independent of any structural detail of the interface such as crystallographic orientation or even abruptness.

The situation is more complicated for heterovalent heterojunctions (e.g., Ge-GaAs). LRT predicts that the lineup is the sum of two contributions: an isovalent-like term independent of the microscopic details of the interface and a second term which does depend on such details. The second term coincides with the lineup generated by an assembly of classical point charges $\Delta Z_i/\varepsilon_\infty$ located at the crystal lattice sites $\mathbf{R}_i$ where $\Delta Z_i$ is the difference between the valence of the i-th virtual ion and the corresponding physical one; $\varepsilon_\infty$ is the electronic static dielectric constant of the virtual crystal. The accuracy of LRT has been tested against very accurate density-functional calculations for several systems, including Ge-GaAs,[35] and was typically within ≈ 10 meV.

Similar considerations apply to the present case of a Ga-As double layer oriented normal the [111] direction in an otherwise perfect Ge crystal. The obvious choice for the virtual crystal is a perfect Ge crystal. Since the system considered here is a homojunction, the isovalent-like term vanishes, and only the second term contributes. According to LRT, this can be straightforwardly evaluated from knowledge of the microscopic arrangement of the interface. For the ideal geometry, this amounts to a (111) plane with -e charge per lattice site (the Ga plane) and a consecutive (111) plane with +e charge per lattice site (the As plane). If we average the charge density over each plane and apply Gauss' law, the Ga side of the junction is at higher potential by $\Delta V = 4\pi d\sigma e^2/\varepsilon_\infty$ where d is the double layer planar spacing, $\sigma = 4/(\sqrt{3}a_0^2)$ is the (111) atomic surface density, and $a_0$ and $\varepsilon_\infty$ are the Ge lattice constant and dielectric constant, respectively.

Because the planar spacing in diamond structure along the [111] direction alternates between $\sqrt{3}a_0/4$ and $(\sqrt{3}a_0/4)/3$, two configurations of the double layer are possible. In Ref. 36, they are referred to as the (111)-near and (111)-far interfaces, dependent on whether the double layer planar spacing is the smaller or larger value (see Fig. 11). The (111)-near configuration has the smaller dipole and thus is energetically favorable. As it turns out, this configuration is also the most probable to be produced by the techniques used in our experiment. This is because the Ge(111) cleavage would most likely occur across the widely spaced planes, which have only a third as many bonds as the closely spaced ones; the experimentally grown Ga-As double layer would then have to be closely spaced to continue the substrate lattice structure.

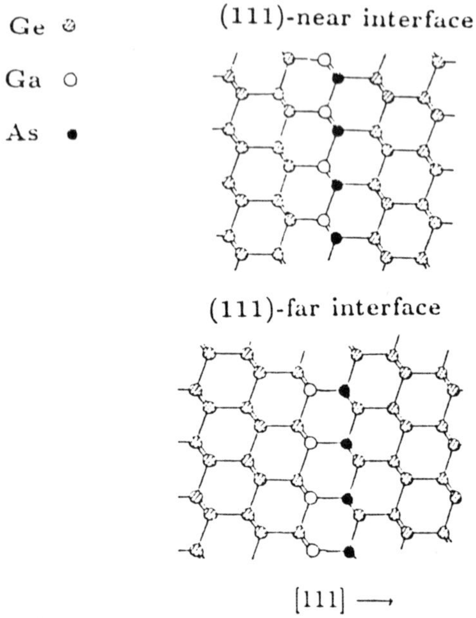


**Figure 11.** Ga-As in Ge systems. There are two possible configurations for the idealized double layer system: the (111)-near and (111)-far interfaces. As discussed in the text, the (111)-near interface is believed to possess the smaller dipole.

Experimental values for the dielectric constant of Ge range from 15.6 for crystalline Ge to 18.5 for amorphous Ge;[37] the lattice constant is 5.65 Å. Using these values, LRT predicts the dipole to be 0.57 eV and 0.69 eV for amorphous and crystalline Ge, respectively. The experimental result of 0.35 - 0.45 eV used crystalline Ge substrates and amorphous Ge overlayers. The sign and order of magnitude of the dipole are as predicted by LRT.

The difference between the experimental and theoretical values could come from several sources. The leading source is probably variation of the experimental system from the ideal geometry assumed here. Intermixing could produce alternate interface geometries and thus lead to a different prediction by LRT. Strain near the interface cannot explain the error, since this would require that the GaAs bond length vary by at least (0.57 - 0.4)/0.57 ≈ 30% which is not physically realistic.

The accuracy of LRT relative to full self-consistent dipole (SCD) calculations done for the same idealized system is probably in the vicinity of 0.1 eV. We arrived at this conclusion by considering a related system for which SCD calculations have been performed. SCD predicts a dipole of 0.74 eV for the (111)-near Ge-Ge double layer in GaAs.[36] By comparison, LRT predicts 0.98 eV. The discrepancy is believed to arise from two main sources: *(i)* there is a difference between $\varepsilon_{GaAs}$ = 10.9 used here by LRT and the dielectric constant as determined by the local density approximation used by Ref. 36; *(ii)* higher order perturbation terms neglected by LRT are larger in the present case than for Ge/GaAs for which LRT's accuracy is ≈ 0.01 eV.

## 4.3 Ga-As Dipole Intralayers Grown on Ge(111)

**Evidence of Band Offsets in the Ge3d Lineshape.** The establishment of a band offset between the two sides of the Ge homojunction should manifest itself as a splitting of the Ge3d core level into two components, one due to the Ge substrate and the other to the Ge overlayer. Such a splitting is indeed evident in Fig. 12, that corresponds to the *As-first* deposition sequence: note that the second Ge3d peak grows monotonically with the overlayer thickness. This second Ge3d peak could be caused, in principle, by a chemical shift in an intermixed phase. Evidence against this interpretation—and in favor of a band offset—is provided by several facts which are discussed below. In particular, the intensity ratios of the intralayer and substrate core levels stay constant, in conflict with strong interdiffusion across the interface.

The two Ge3d components were derived from lineshape analysis using the cleaved and 100 Å thick Ge film Ge3d lineshapes as fitting functions. Fits using only the thick film shape worked slightly better for coverages above 3 Å. When the less bound component first appears, the splitting is ≈ 0.6 eV, but in the 8 - 15 Å Ge coverage range, it decreases to 0.38 - 0.41 eV.

This decrease was repeatable. Although it could be dismissed as a surface effect, it is worth speculating that the build up of the dielectric constant around the dipole intralayer is involved. The decrease is consistent with the 1/e dependence of the dipole: when the overlayer is thin, surface effects are important and tend to reduce the effective screening. (Note: If we knew how the interface dielectric constant varied as the Ge overlayer builds up, we could estimate the change. It is interesting, though perhaps coincidental—since we do not have bulk quantities of GaAs—that:

$$(\varepsilon_{GaAs}/\varepsilon_{Ge}) \times 0.60 = 0.41 \text{ eV}$$

There are also experimental methods for measuring the interface dielectric constant. For example, Brillson used interface specific plasmon modes in an Electron Energy Loss study of the CdS-Al interface to derive its interface dielectric constant.[5] We have not attempted such an experiment.)

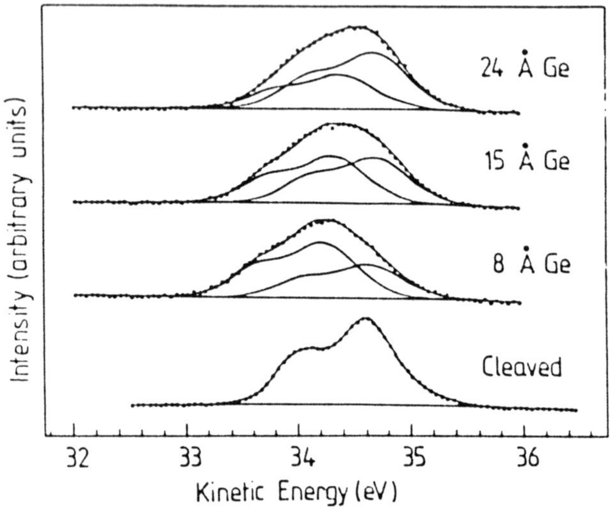


**Figure 12.** "As-first" Ge3d peaks. The splitting is ≈ 0.4 eV for all overlayer Ge thicknesses shown. These data are for a cleaved Ge(111) substrate covered (in sequence) by As and Ga monolayers and an amorphous Ge overlayer. The lineshape of the components is that of a 100 Å thick Ge film on Ge(111).

The case of the *Ga-first* deposition sequence, corresponding to Fig. 13, is more complex because of microdiffusion and reaction processes that produce a complex interface morphology. Nevertheless, we can still identify two Ge3d components that correspond to a band offset of equal magnitude and opposite sign with respect to the As-first sequence. The increased complication of the Ga-first interface is expected, because the corresponding Ga-Ge bonds are weaker than As-Ge bonds. The two Ge3d components corresponding to the Ge substrate and to the Ge overlayer displaced in energy by a Ga-As dipole correspond to the solid lines in Fig. 13. Note that the substrate Ge3d stays at 34.37 ± 0.08 eV kinetic energy. The magnitude of the splitting between substrate and overlayer peaks is initially ≈ 0.6 eV, and then it decreases to 0.33 - 0.46 eV as it does for the other deposition sequence.

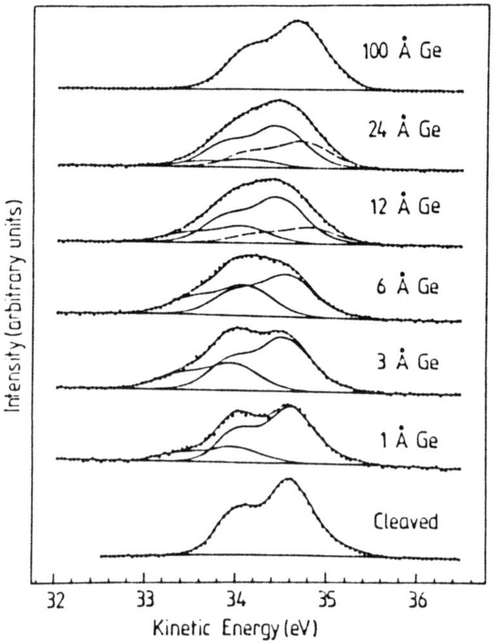


**Figure 13.** "Ga-first" Ge3d peaks. This stack plot shows the Ge3d lineshapes for a cleaved Ge(111) substrate covered (in sequence) by Ga and As monolayers and an amorphous Ge overlayer. The two solid-line components are the symmetric equivalent of those shown in Fig. 12. Above ≈ 10 Å Ge coverage, a third component appears (dashed line), whose nature is discussed in the text.

**Interface Morphology.** Evidence for a complex interface morphology in the Ga-first case is provided by a third Ge3d peak, the dashed lines in Fig. 13, appearing at Ge overlayer thicknesses above 6 Å. We explain this peak as due to regions in which an As-Ga dipole exists in spite of the Ga-first deposition sequence. Note that the splitting with respect to the substrate peak is 0.34 - 0.42 eV, consistent with this hypothesis.

This explanation appears more plausible than other hypotheses. We can rule out, for example, isolated inverted Ga-As pairs as the cause of the third Ge3d component, since the effects of such inverted pairs would be averaged out—thinking in terms of a simple capacitor model, a region of similarly oriented pairs forms two sheets of opposite charge, while isolated inverted pairs simply lower the average areal charge density. Thus, even though inverted Ga-As pairs are probable, because exchanges reactions could be driven by the huge electric fields near the interface (≈ 0.4 eV in 2.45 Å), they do not explain the third component.

The formation of regions with the inverted As-Ga dipoles can be explained with the following interface formation model. The interface morphology is influenced by the relative bond strengths of Ge-As, Ga-As, and Ga-Ge. The evidence points to Ge-As as the strongest bond. For example, experimental studies of epitaxially grown GaAs on Ge(111) or Si(111) and Ge on GaAs(100) have concluded that the final interfaces are totally dominated by Ge-As or Si-As bonds even when Ga is deposited first.[38][39] Additional discussion is given in the section on monolayer As and Ga coverages. We expect, therefore, that for the Ga-first sequence, the weakness of the Ga-Ge bond allows Ga cluster formation. In contrast, Ge(111):As forms an ideal 1x1 surface even when deposited on the cleaved 2x1 reconstruction. The Ge-As bond lengths are nearly the same as bulk Ge bond lengths.[40]

This creates regions utterly barren of Ga and others with Ga clusters. The As deposition then will cover both the Ga-free Ge regions and the Ga clusters. After enough Ge has been deposited, the excess Ga in the clusters laterally out-diffuses, finds, and bonds to As atoms in cluster-free regions by displacing Ge adatoms. This produces inverted regions with As-Ga dipoles, similar to those produced by the As-first sequence. Such regions co-exist with the normal regions with Ga-As dipoles, producing the two overlayer Ge3d peaks of Fig. 13. (Note: The hypothesis of Ga cluster diffusion was tested by using a single monolayer Ga intralayer without a corresponding As layer. In this case, we found that Ga diffusion was strongly enhanced. Previously, 20 Å of Ge completely obscured the Ga3d emission; without the As layer, the Ga3d signal was still visible for Ge overlayer thicknesses of 100 Å. Apparently, the clusters are not stable in the presence of the Ge overlayer.)

This scenario explains several experimental observations:

1. The dashed-line peak does not appear until a relatively thick Ge overlayer has been deposited.
2. The dashed-line peak becomes dominant at large overlayer thicknesses, consistent with the fact that Ga clusters cover only a small fraction of the Ge substrate; thus, most of the interface area produced by the Ga-first deposition has the same dipole sign as that of the As-first deposition.
3. Our morphological picture explains the stability of the As-first sequence. The strength of the As-Ge bond causes nearly-epitaxial growths of As on Ge and subsequently of Ga on As.

Our data interpretation is corroborated by the analysis of the Ga and As 3d peaks. Strong evidence of a charge transfer between the Ga and As monolayers is found in the fact that the As3d centroid position moves to 0.4 - 0.5 eV lower binding energy immediately after 1 monolayer of Ga is deposited on the monolayer of As (see Fig. 14); this is consistent with the As gaining electrons from the Ga.[41] Similarly, when 1 monolayer of As is deposited on a monolayer of Ga, the Ga3d centroid position moves to ≈ 0.5 eV higher binding energy, consistent with the Ga losing electrons to the As (see Fig. 15). After the dipole intralayer has been deposited, the As and Ga 3d centroids of both growth sequences remain fairly constant in energy for all subsequent Ge overlayer depositions. The possible exception is the As3d in the Ga-first case which appears to move towards slightly lower binding energy (by ≈ 0.15 eV) for Ge thicknesses above 6 Å; this may reflect the higher disorder believed to exist in the Ga-first case.

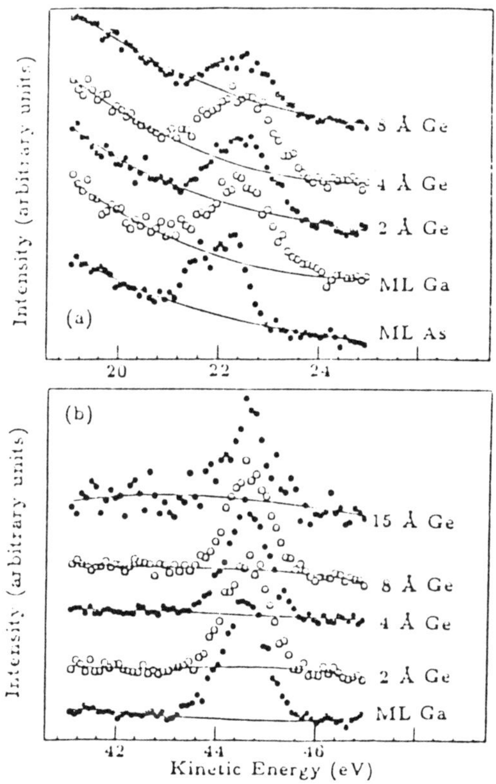


**Figure 14.** As-first As and Ga 3d peaks. These are stack plots of As3d peak *(a)* and Ga3d peak *(b)* taken at 70 eV photon energy for the As-first case on a Ge(111) substrate. They give strong evidence for a transfer of electrons from the Ga monolayer to the As monolayer.

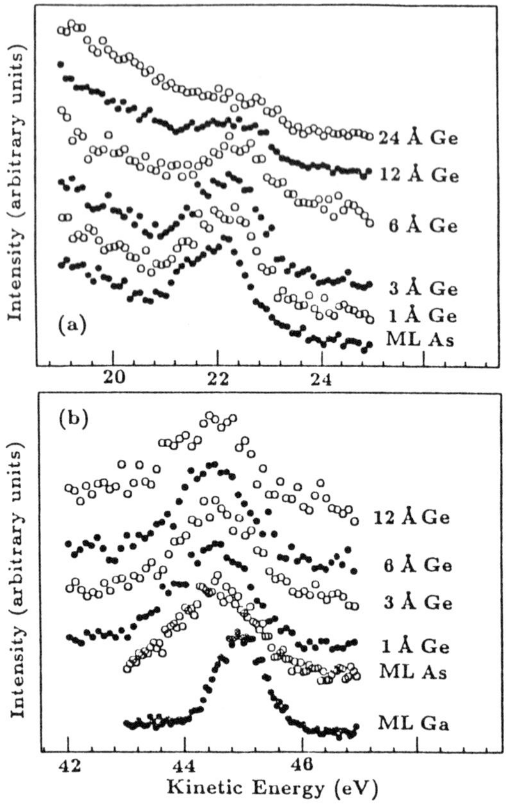


**Figure 15.** Ga-first As and Ga 3d peaks. These are stack plots of As3d peak *(a)* and Ga3d peak *(b)* taken at 70 eV photon energy for the Ga-first case on a Ge(111) substrate. They give strong evidence for a transfer of electrons from the Ga monolayer to the As monolayer.

There is also evidence of charge transfer between the dipole intralayer components and the Ge. The Ge3d substrate component increases in binding energy (relative to the cleaved surface) by ≈ 0.34 eV immediately after a monolayer As coverage, and remains constant thereafter. Again, this indicates that the substrate Ge donates electrons to the As adatoms. When a monolayer of Ga is deposited on a cleaved Ge surface, the Ge3d binding energy moves towards lower binding energy by ≈ 0.15 eV, consistent with the Ge gaining electrons from the Ga. The charge transfers would be expected on the basis of bond polarity arguments because the Ge in the layers immediately adjoining the dipole intralayers is in a nonbulk-like environment. Interestingly, LRT does *not* predict any charge centered on the adjoining Ge layers. The fact that the As-induced and Ga-induced movements have opposite signs is consistent with the bond polarity arguments, but these same arguments indicate that epitaxial Ga and epitaxial As monolayers should induce very similar amounts of charge transfer. The difference in the magnitudes of the As-induced and Ga-induced shifts is consistent with Ga clustering: since a smaller fraction of the surface is covered with Ga, the centroid of the Ge3d (which reflects both clean Ge and Ga covered Ge) shifts less in the Ga-first case.

We have mentioned that the intensity ratios of the intralayer components to the substrate are consistent with our model interpretation. For the As-first sequence, the intensity ratios rule out strong microdiffusion (see Fig. 16). In particular while the ratio of the substrate Ge3d to the shifted "overlayer" Ge3d drops by nearly a factor of 4, the ratios of the Ga3d and As3d with respect to the substrate Ge3d signal remain almost constant (22% drop). This indicates that the Ga and As are remaining at the interface. We also tried to fit the substrate and overlayer Ge3d peak intensities to simple exponentials (see Fig. 17). When we tried to fit the intensity decrease of the substrate Ge3d peak to an exponential law, $\propto \exp(-l/\lambda)$, we are forced to use an exponential parameter of $\approx$ 30 Å, much larger than the expected 5 - 10 Å photoelectron escape depth at our kinetic energies.[1] The intensity of the overlayer Ge3d peak increases with the layer-by-layer law $\propto (1 - \exp(-l/\lambda))$, where the exponential parameter $\lambda$ is now $\approx$ 8 Å, consistent with the photoelectron escape depth. This indicates that while the overlayer growth is layer-by-layer, small patches of the surface do not have the dipole so that part of the overlayer is at the same energy as the substrate, leading to the slow "substrate" Ge3d intensity decrease.

For the Ga-first sequence, substantial out-diffusion of both Ga and As is revealed by the analysis of the core level intensities. The Ga3d binding energy increases by $\approx$ 0.5 eV after the As deposition indicating loss of electrons as expected. The As3d binding energy shows a slow decrease for increasing Ge overlayer thickness that saturates at $\approx$ 0.3 eV after about 3 Å Ge, and this indicates a gain of electrons as expected. The substrate Ge3d binding energy decreases by 0.12 eV after the monolayer Ga coverage, indicating a gain of electrons from the Ga. The substrate Ge3d peak intensity exhibits a weak, slower-than-exponential decrease with the Ge overlayer. Since this decay is far slower than in the As-first case, we conclude that a larger fraction of the surface has no dipole.

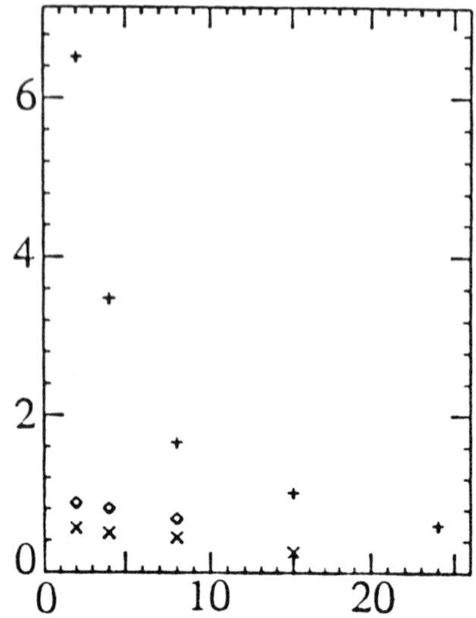


**Figure 16.** As-first core level ratios vs. Ge coverage. The ratio of the substrate Ge3d to the shifted Ge3d component (+), the As3d to substrate Ge3d ratio (◊), and the Ga3d to substrate Ge3d ratio (x) are given as a function of Ge overlayer coverage in Å.

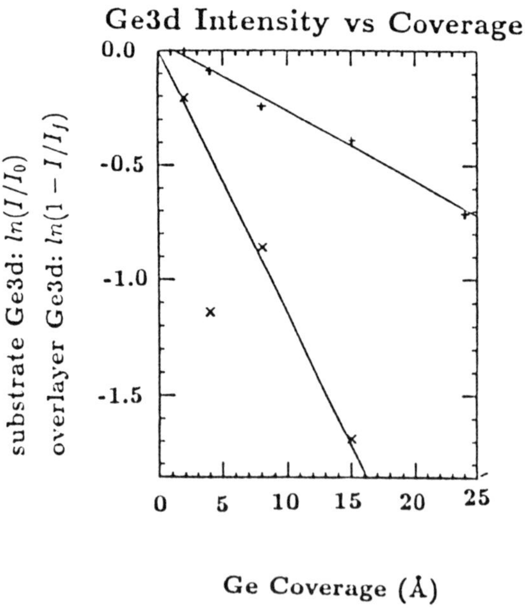


**Figure 17.** As-first Ge3d intensities vs. Ge coverage. Comparison of the decay and growth rates of the substrate and shifted "overlayer" Ge3d components gives evidence of the fraction of the surface covered by the dipole layer. The overlayer Ge3d (+) intensity increases at the expected rate for layer-by-layer coverage, but the substrate Ge3d (x) decays at a slower rate.

Our experimental findings were successfully tested for reproducibility, twice for the As-first case with a Ge(111) substrate and three times for the Ga-first case with a Ge(111) substrate. Quite interesting was the comparison of the Ga-first sequence data on crystalline Ge substrates to preliminary results obtained for the same sequence on a ≈ 50 Å thick Ge film on Ge(111). In the latter case, only two Ge3d components appeared, with a splitting of magnitude (0.37 - 0.41 eV) equal to that of the As-first sequence on Ge(111), and opposite in sign. This is precisely the result expected for the reversal of the interlayer dipole, suggesting that the factors leading to a complex interface morphology on Ge(111) are inhibited in this case. Further experiments were conducted to clarify this point and are discussed below.

**Valence Band Maxima Analysis.** Less trustworthy evidence of the dipole is provided by the valence band maxima. These were determined by extrapolation of the leading edge to the baseline, a technique with an accuracy of ± 0.1 eV.[1] In the As-first case, we observed a gradual shift to lower binding energy by ≈ 0.25 eV as Ge coverage is increased from 1 to 40 Å. The Ga-first case shows no shift.

These results are consistent with those of the core line analysis if we assume that some fraction of the surface does not develop a dipole. In this event, we may expect that the lateral average seen by the CMA (cylindrical-mirror electron energy analyzer) will have two components due to the overlayer (one coincident with that of the substrate). In the As-first case, this would broaden the edge and distort the extrapolation to the baseline, but since the leading edge for the overlayer is due to the component with the dipole, it shifts upward in energy. The fact that we see a shift but that it is less than 0.4 eV indicates that neither component overwhelmingly dominates the surface.

In the Ga-first case, the unshifted component forms the leading edge. Furthermore, it is probable that the fraction of the surface without a dipole is larger in the Ga-first case because of Ga clustering during growth. For both of these reasons, the component due to the Ga-As dipole is obscured. Furthermore, the breadth of the valence band edge makes it inherently difficult to resolve two broad edges separated by only ≈ 0.4 eV. A lineshape analysis might separate the two components, but the low cross section of Ge results in fairly poor statistics for the valence band. We did not attempt such an analysis.

**Additional Remarks.** Both overlayer components were observed to have the 0.35 - 0.45eV splitting undiminished by the other's presence. This is explained by setting a lower limit to the lateral dimensions of the oppositely oriented *patches*. Far more atoms must be well within a patch than are at its boundaries. Otherwise, antiposed capacitors would tend to severely weaken each other by lowering the average areal charge density in the boundary regions. By modeling the patch as a square of N atoms on a side, the condition that must be met is $N^2 - 4N \gg 4N$. That is, individual patches must have at least 100 Ga-As pairs.

In the As-first sequence, the majority of the surface probably retains the dipole, because the final Ge3d position seems to smoothly join to the overlayer component (see Fig. 12). This is not true in the Ga-first case which leads to the conclusion that only a small fraction of the surface develops the dipole. In fact, the overlayer component due to the dipole appears to be buried with sufficiently thick Ge overlayers. This may be due to the fact that only a small fraction of the surface actually has the dipole in the Ga-first case: as the Ge overlayer thickness increases to a substantial fraction of a Debye length, we expect that the overlayer Ge will see the average potential due to the regions with and without the dipole. Obviously, if only a small fraction of the surface has the dipole, the unshifted component will dominate in the end.

## 4.4 Ga-As Dipole Intralayer Grown on a-Ge

The initial experiments used cleaved Ge(111) substrates with the hope that the single As and single Ga monolayers would continue the substrate lattice structure and form a dipole layer. After this had been demonstrated, we decided to deal with the possible objection that our results are not due to the dipole intralayer but simply to the difference between the properties of the crystalline Ge substrate (c-Ge) and the amorphous Ge overlayer (a-Ge). The most direct way to remove this objection is to use a-Ge for both the substrate and overlayer. We used 50 - 100 Å thick Ge grown at room temperature (RT) on c-Ge(111) as substrates. Again, the principal evidence for the dipole comes from the Ge3d analysis. The Ge3d lineshapes for the As-first and Ga-first cases are shown in Fig. 18 and Fig. 19. The results of these experiments support the previous conclusions although there are a few differences. The magnitude of the dipole is $\approx$ 0.3 eV which is 25% smaller than in the c-Ge substrate case $\approx$ 0.4 eV).

We believe that this slight reduction is explained by two things. First, the dielectric constant increases from 15.6 for c-Ge to 18.5 for a-Ge.[37] Roughly, we can take the average value, 17.05, as descriptive of the c-Ge substrate with a-Ge overlayer case. The 1/e dependence of the dipole predicts that the dipole sandwiched in a-Ge will be (17.05/18.5)0.4 $\approx$ 0.37 eV. Secondly, a c-Ge substrate is expected to be more resistant to certain microdiffusion processes than a-Ge (reasons for this are discussed below). As mentioned before, smaller dipoles are expected if alternate interface geometries involving exchange reactions between the dipole intralayer and the adjoining Ge layers occur.

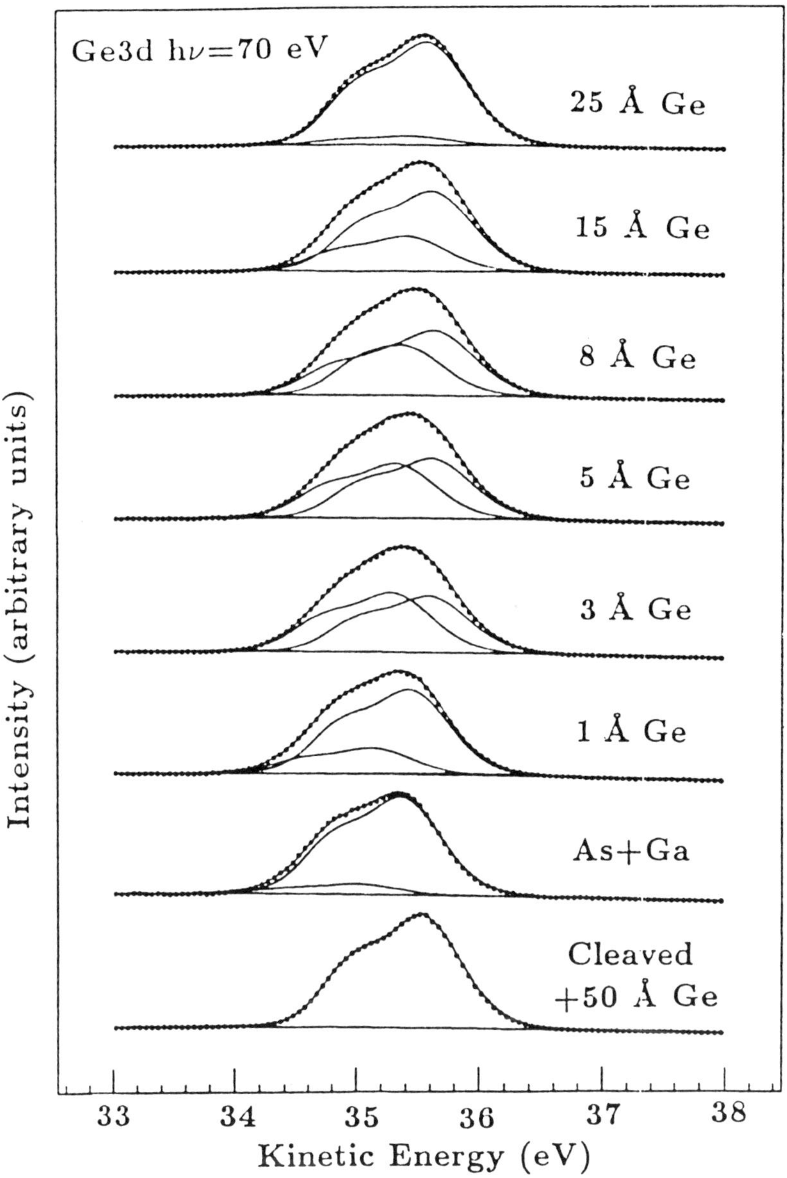


**Figure 18.** As-first Ge3d peaks. This stack plot shows the Ge3d lineshapes for an a-Ge substrate covered (in sequence) by As and Ga monolayers and an a-Ge overlayer.

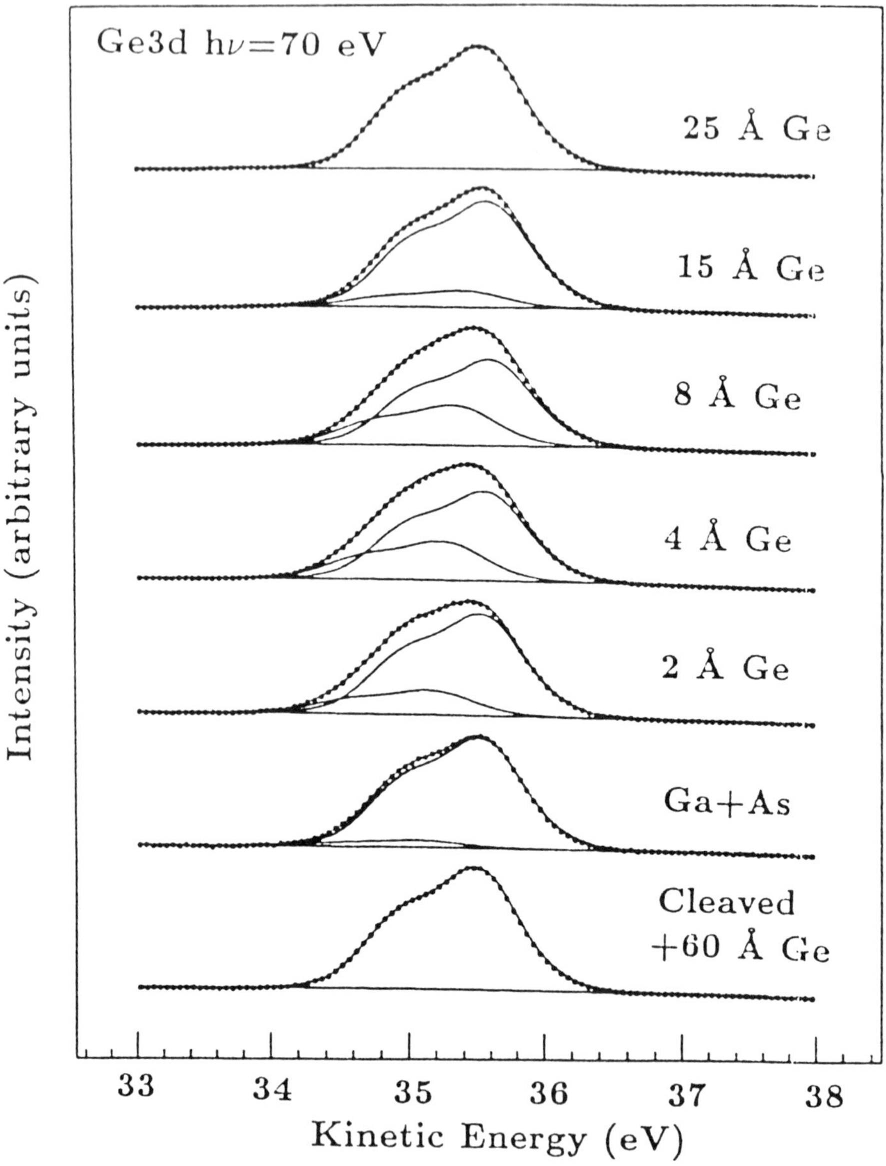


**Figure 19.** Ga-first Ge3d peaks. This stack plot shows the Ge3d lineshapes for an a-Ge substrate covered (in sequence) by Ga and As monolayers and an a-Ge overlayer.

Another very significant result of the present experiment is that the additional overlayer Ge3d component that occurred with the c-Ge substrates in the Ga-first case does not occur with a-Ge substrates in the Ga-first case. This may be explained by a lower lateral mobility of Ga deposited on an a-Ge surface; the relatively higher density of surface defects and unsaturated Ge bonds at an a-Ge surface increases the number of potential cluster nucleation sites; this, in turn, reduces the average cluster size. A large number of small clusters would cover much more of the Ge substrate than a few large clusters—more nearly approximating a two dimensional Ga layer. Thus, the morphological arguments used in the c-Ge substrate case do not apply and a third component does not develop.

This scenario is also consistent with the heightened ability of a-Ge to produce alternate interface geometries. Though unsaturated bonds reduce lateral surface mobility, the corresponding vacancy defects are a leading mechanism behind bulk diffusion. Also, the increased reactivity of the a-Ge surface could drive microdiffusion processes near the interface. Alternate interface geometries involving "interface" molecules precipitated from the bulk materials joined at the interface are known to effect both Schottky barrier heights and heterojunction valence band discontinuities: for example, Niles found definitive evidence of $Al_2Se_3$ formation when Al intralayers were introduced at ZnSe-Ge interfaces. Niles then correlated this to the saturation of the intralayer-induced effects on the valence band discontinuity in that system.[3][4]

In Fig. 20, we show the result of growing a thick 50 Å a-Ge film on c-Ge(111). Both the c-Ge and the a-Ge overlayer were p-type. Notice that the valence band maximum remains almost constant, although some rounding of the p-derived features nearest the maximum does occur. The Ge3d shape has broadened and the spin-orbit valley is more washed out; its centroid position has moved to lower binding energy by ≈0.1 eV, a slightly smaller shift than the 0.24 eV reported by Ref. 37. However, the latter number was for 200 Å thick Ge films, and the shift may be due to a dopant dipole between the two p-type materials that has not saturated for 50 Å thickness (Debye lengths are typically 100 - 1000 Å). From these results, it is clear that the effects that we have attributed to the dipole intralayer cannot be alternatively attributed to the use of c-Ge substrates and a-Ge overlayers.

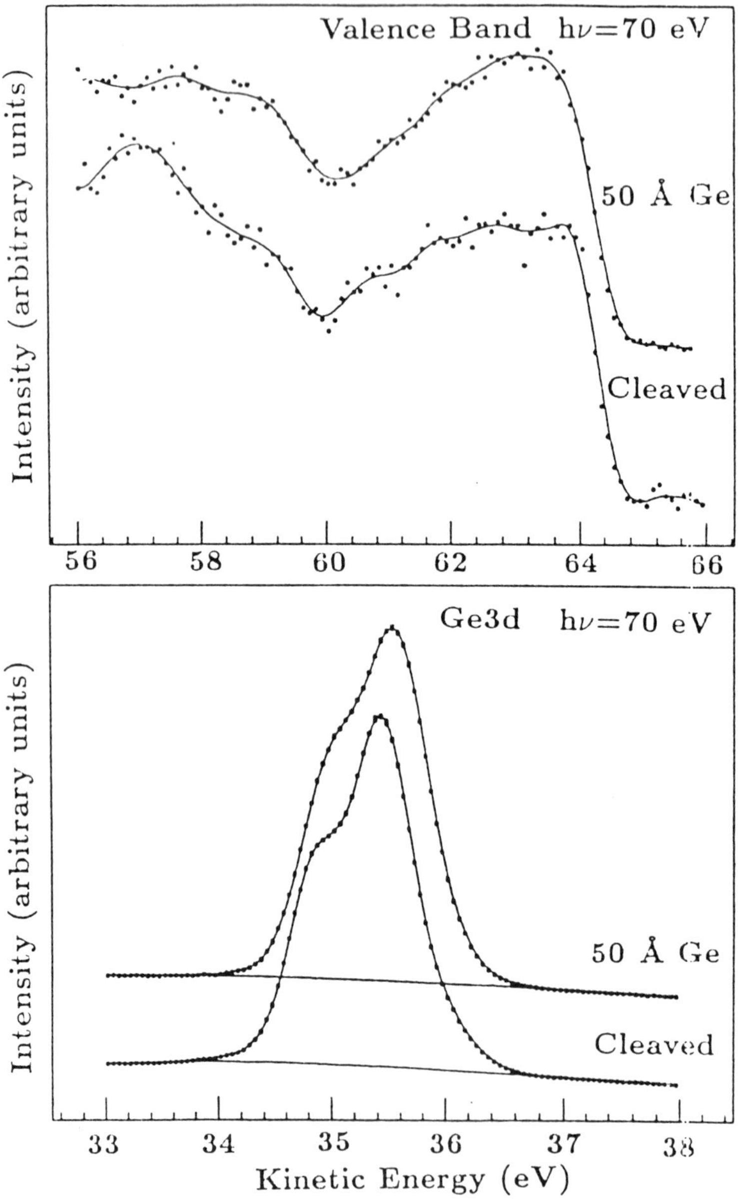


**Figure 20.** Fifty angstroms a-Ge film grown on c-Ge(111).

## 4.5 Creation of Homojunction Band Discontinuities: Some Preliminary Conclusions and Future Directions

The most important conclusion of these tests is again the most elementary: evidence has been obtained for dipole intralayer induced valence band discontinuities in Ge homojunctions. We have also found our results to be in reasonable agreement with a theoretical alchemy model. The behaviors dependent on growth sequence are explained by morphological arguments. Amorphous Ge homojunctions were examined finding consistent results.

The Ga-As in Ge system is considered prototypical, but this technique for controlling band lineups could be extended to other systems. There are several systems of particular interest. LRT predicts that the Al-As in Ge dipole will be the same as the Ga-As in Ge dipole. The former system may have certain advantages experimentally: because of its reactivity, Al forms smooth, flat, abrupt layers—potentially avoiding the complex interface morphology that occurs for the Ga-As in Ge system for the Ga-first case. Furthermore, like GaAs, AlAs is well lattice-matched with Ge. Because of the preeminent position of Si in semiconductor technology, the Ga-P and Al-P in Si systems may be of considerable importance. Again, these materials are well latticed-matched. We have already conducted preliminary studies of Cs-O dipole intralayers in an attempt to explore the effects of using elements of extremely varied electronegativity; this will be reported elsewhere.

Farther into the future, more sophisticated experiments can be imagined. Since order can be restored on thin Ge films with crystalline lattice-matched substrates by applying as little as 300°C for 5 min,[42] the effect of crystallinity on the dipole could be ascertained easily. This would also give information concerning the long-term stability of the dipole. More ionic combinations such as Zn-Se in Ge would enable us to determine the dependence on ionicity. LRT predicts a dipole twice as large as for the Ga-As in Ge system. The Phillips ionicity of ZnSe is also almost exactly twice that of GaAs.[41] Ferroelectric intralayer materials such as $BaTiO_3$ would seem to be a natural direction of interest, too; a ferroelectric intralayer would not have to be a double monolayer but could be of arbitrary thickness and thus arbitrary magnitude. Another very simple but important question is whether the effects of multiple dipole layers are additive. With all of this said, it is clear that many questions remain to be answered. We fully expect the subject of dipole intralayers to be a germinant new research field.

## 5.0 DETAILED DISCUSSION OF THEORETICAL MODELS FOR DIPOLE INTRALAYERS

Though the concept of double layer dipoles at homojunctions is still very young, it has already attracted extensive theoretical investigation, indicating its potential importance.[34][35][36] In using the term *double layer,* we are distinguishing between the ideal case of two adjacent monolayers and the more general idea of a dipole intralayer. This second term is more descriptive of experimental systems in which the intralayer may be of arbitrary thickness and may not even be abrupt. As a caveat, we mention two major weaknesses inherent in all the models to be presented:

1. A single ideal crystal lattice is assumed to span the interface, e.g., the Ga-As in Ge system is assumed to have the same lattice as pure Ge.
2. An ideal double layer with no intermixing of atoms from different layers is assumed.

The first assumption amounts to discounting interfacial strain. This is not a severe approximation since we have used well lattice-matched materials. Furthermore, we can estimate the size of any strain-related effects that do exist because they will be proportional to the differences in planar spacings between the real and ideal systems. By summing the covalent radii, we find that Ga-As bonds are about the same length as the ideal crystal's Ge-Ge bonds, whereas Ga-Ge bonds are $\approx$ 2% longer and Ge-As bonds are $\approx$ 2% shorter; any strain-related effects must be correspondingly small. Our expansive knowledge of heterojunctions also indicates that strain may be neglected; a recent review by Cardona and Christensen[43] indicates that strain corrections are required only for badly lattice-mismatched systems.

Other evidence for the minimal importance of strain in our system comes from studies of single monolayers of As deposited on cleaved Ge(111) 2x1 ; the ideal 1x1 reconstruction resulted, and the Ge-As bond lengths were near those of ideal Ge-Ge bonds.[40] Thus, a monolayer As cap actually *lessens* the surface strain. The reason for this is fairly straightforward. Cleaved Ge(111) has unpaired electrons at its surface, the so called *dangling bonds.* These bonds "heal" themselves by forming surface dimers. The resulting surface stress brings about the 2x1 reconstruction. If we cap the Ge with one monolayer of pentavalent As (or for that matter, trivalent Ga), all electrons are paired. No surface dimers are needed, and the ideal 1x1 reconstruction occurs.

The second assumption used in all the theories to be presented is that there is no intermixing at the interface. This is not as sound an assumption as the neglect of strain: the experimental system evidenced some signs of diffusion. Furthermore, the electric field associated with a ≈ 0.4 eV dipole occurring over 1 - 10 Å is enormous, giving strong incentive for interdiffusion. If a defect due to intermixing is responsible for the remaining discrepancy between experiment and theory, it is interesting that the experimentally determined dipole magnitude was unaffected by the dramatically differing degrees of diffusion present in the As-first and Ga-first growth sequences.

As we shall see, the theoretical predictions are very sensitive to the microscopic details of the interface. Interdiffusion effects can be modeled by using an increased number of transition planes to describe the interface, as done by Harrison for semiconductor heterojunctions.[20] Harrison found new atomic arrangements that lessen or entirely remove the interface dipole. Several papers discuss the energies of atomic substitutions and exchange reactions that might produce energetically favorable multitransition plane geometries at Ge-GaAs interfaces, but such a discussion is beyond the scope of this paper.[22][44] One conclusion that we can draw is that the ideal double layer geometry will always err to the side of overestimating the dipole, since the other configurations tend to lead to smaller dipoles.

## 5.1 Harrison's Theoretical Alchemy Model

A surprisingly simple and elegant model proposed by Harrison predicts both the sign and approximate magnitude of the intralayer-induced dipole.[36] In Harrison's "Theoretical Alchemy" (HTA) model, we calculate the effect that would occur if protons could be transferred between some of the nuclei in an infinite solid. For example, starting with pure Ge, we could imagine transferring a single proton per atom between two adjacent (111) planes. The plane that gained protons would "become" As (the atomic number of As is one greater than Ge) and the plane that lost protons would "become" Ga (the atomic number of Ga is one less than Ge). As a first approximation, we neglect the relaxation of the electron distribution that would occur after the protonic dipole forms. By simply applying Gauss' law, we find that the side of the junction that the protons were transferred towards steps up in electric potential by:

$$\Delta\phi = \frac{n d \sigma |e|}{\varepsilon \varepsilon_o} \text{ (MKS)}$$

$$\Delta\phi = \frac{4\pi n d \sigma |e|}{\varepsilon} \text{ (cgs)}$$

$$\sigma = \frac{4}{\sqrt{3}a_o^2}$$

$$d_{far} = \frac{\sqrt{3}a_o}{4}$$

$$d_{near} = \frac{\sqrt{3}a_o}{12}$$

where n is the number of protons/atom transferred, d is the double layer planar spacing, $\sigma$ is the (111) atomic areal density, $a_0$ is the lattice constant, and e is the interface dielectric constant. If we use 15.6 and 5.65 Å for $\varepsilon_{Ge}$ and $a_0$, respectively, then HTA predicts that the As side of the (111)-near Ga-As in Ge system will be raised 0.69 volts above the Ga side.

Several features of HTA should be pointed out. The scheme is not limited to double layer problems but can also be applied to other systems; Harrison originally applied it to heterojunctions.[36] A second point is that only interfaces of materials within an isoelectronic sequence can be formed (e.g., Ge crystals can be transformed into GaAs, ZnSe or CuBr crystals but not into AlAs). Because isoelectronic sequences tend to have nearly constant bond lengths,[41] transforming one side of an interface into, say, GaAs and the other side into Ge does not affect the lattice structure—the ideal lattice approximation is accurate. Lastly, the use of the Ge dielectric constant rather than some kind of "interface" dielectric constant, though a reasonable approximation, could be supplanted. There are experimental methods for measuring interface dielectric constants. For example, Brillson used interface specific plasmon modes in an Electron Energy Loss study of the CdS-Al interface to derive its interface dielectric constant.[5]

## 5.2 Self-Consistent Dipole Theory

In a recent paper, Muñoz et al. carried out ab initio SCD calculations using the Harrison theoretical alchemy framework on a Ge-Ge double layer in GaAs.[36] Muñoz picked up where HTA had left off by using a self-consistent approach to determine the electronic relaxation that occurs after the protonic dipole is formed. He found the (111)-near interface to have the smallest interface dipole with the As-terminated side of the interface at 0.74 eV higher binding energy. This is a 0.24 eV reduction from the HTA value of 0.98 eV—apparently an estimate of the electronic counter dipole.

Though this calculation does not directly address our experimental work, the Ge-Ge in GaAs system is related to it and worth mentioning in lieu of SCD calculations specifically on Ga-As in Ge. The Ge-Ge in GaAs system can be formed by moving one proton per atom from an As-(111) plane to an adjacent Ga-(111) plane in pure GaAs. This mirrors the operation used to form the Ga-As in Ge system from pure Ge. The only difference between the electrostatics of the two systems, as pictured within theoretical alchemy, is the dielectric constant.

If we scale with the dielectric constant, we can use the SCD result for Ge-Ge in GaAs to estimate a Ga-As in Ge dipole of 0.44 eV to 0.52 eV. The smaller and larger numbers correspond to the dielectric constants of amorphous and crystalline Ge, respectively. This is in close agreement with the experimental result of 0.35 - 0.45eV in which a crystalline Ge substrate and an amorphous Ge overlayer were used. The dielectric constants of GaAs, a-Ge, and c-Ge are 10.9, 18.5, and 15.6, respectively.

## 5.3 Linear Response Theory

We expect that many people will be skeptical about the imaginary proton transfers used by HTA and the work by Muñoz. Linear Response Theory (LRT), puts theoretical alchemy on a firmer theoretical footing by forming the Ga-As double layer in Ge by *perturbing* a pure Ge crystal. Describing a "real" system as the perturbation of a "virtual" crystal[34][35] is a time honored practice. Heterojunction band lineups, in particular, have been extensively considered. The result has been a set of surprisingly simple and accurate (≈ 0.01 eV) analytic expressions based on readily available bulk parameters. We shall first discuss LRT's application to heterojunctions, then move on to the double layer at a homojunction problem.

Baroni et al.[34][35] argue that charge transfers and screening dipoles, concepts commonly used in theories of band lineups, are too vaguely defined. The principal difficulty that they cite is the ill-defined nature of the absolute electrostatic potential in an infinite periodic system. For example, the conventional use of removal energies to determine the absolute average potential in an infinite solid is an ill-defined operation—there is no "outside" to which the particle can be removed.

Baroni et al. surmounted this problem by noticing that it is unnecessary to know the absolute average potential of the virtual crystal. Instead, the difference between the average potential of the virtual crystal and the average potential of the perturbed crystal is all that is necessary to determine the lineup. Poisson's equation leads to the following expression for the dipole moment of the interface charge density profile:

$$\Delta V = (e^2 \varepsilon \varepsilon_o) \int z\,(n_{real} - n_{virtual})\,dz$$

Here, $n_{real}$ is the macroscopic averaged electron density of the real crystal, $n_{virtual}$ may be taken as eight electrons per unit cell (for zincblende semiconductors), and the z-direction is perpendicular to the interface. In LRT, $n_{real}$ - $n_{virtual}$ is equated to the first order perturbation of $n_{virtual}$

There are many possible ways of defining the virtual crystal, but the optimal choice is one that makes the virtual crystal as similar as possible to the real system. We can illustrate how the optimal virtual lattice is chosen through the use of an example. Let us say that we wished to study the GaAs-AlAs interface. If we choose a virtual cation lattice that has the arithmetic average properties of pure Ga and pure Al sublattices, then perturbations of the same size but opposite direction transform a virtual cation into Ga or into Al. If we choose the virtual crystal to be GaAs, a much larger perturbation is required to form an Al cation. Since perturbation theory performs best when the perturbations are kept small, the more symmetric choice is mandated.

Since the virtual crystal is an infinite periodic system, its microscopic charge distribution can be determined by conventional means. The real system's microscopic charge density is then determined by low order perturbation theory. For lattice-matched heterojunctions, the macroscopic charge density is the average over a unit cell. The valence band offset (VBO) calculation takes place in two stages. Distinct bulk calculations are carried out for both materials. This determines the position of the valence band maximum, $E_v$, relative to the average electrostatic potential. When the materials are joined at an interface, charge transfer misaligns the average electrostatic potentials. This potential difference, $\Delta V$, is related to the macroscopic charge density profile by Poisson's equation. The VBO is now given by (see Fig. 21):

$$VBO = \Delta E_v + \Delta V$$

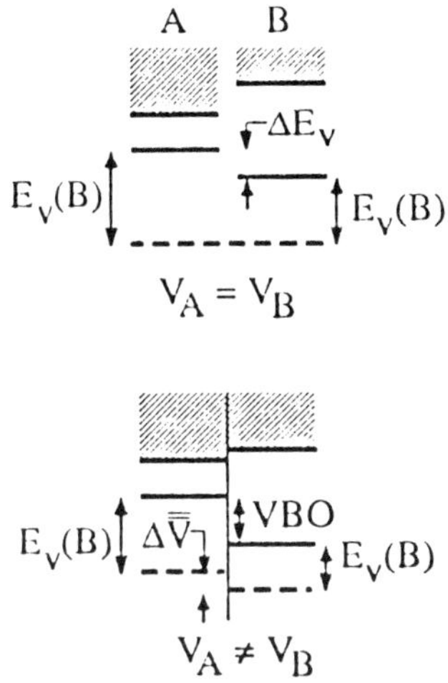


**Figure 21.** Schematic of LRT. LRT determines the VBO in two distinct stages, as discussed in the text. This schematic was taken from Ref. 34.

Baroni et al. divide heterojunctions into broad classes according to whether a particular interface is lattice-matched or strained, isovalent or heterovalent, polar or nonpolar. Their conclusion is that, for lattice-matched cases, isovalent and non-polar heterovalent interfaces are independent of the details of the interface itself, while polar heterovalent interfaces can depend on the interface geometry.

We will start with isovalent lattice-matched systems which are the least difficult to handle. In these, the $\Delta V$ part of the VBO depends only on the properties of the bulk materials. Thus, it is independent of crystallographic orientation of the interface and its abruptness. Furthermore, it exhibits transitivity—that is:

$$\Delta V(A,C) = \Delta V(A,B) + \Delta V(B,C)$$

Transitivity of the VBO is observed experimentally for a wide number of different materials,[10][45] and is considered a basic test of all linear models.[46]

A heterovalent heterojunction is one in which the cations (anions) on either side of the interface come from different columns of the periodic table. This category has an extra complication with respect to the isovalent case. In the virtual crystal, each unit cell has eight electrons divided between the cation and anion. When it is perturbed to form isovalent and nonpolar heterovalent interfaces, this remains true. However, in polar heterovalent interfaces, some unit cells gain electrons while others lose them. This is most easily seen by way of example. Consider a GaAs-Ge interface. With a polar interface, one of the unit cells will contain a Ga and a Ge atom (seven electrons) while a neighboring cell across the interface contains a Ge and an As atom (nine electrons). Since the charge excesses are across the interface, they contribute to the dipole. (Note: Charge excesses also occur in nonpolar heterovalent interfaces but the resulting dipoles are all parallel to the interface so that no contribution to the interface dipole results. In isovalent cases, of course, there are no charge excesses.) Thus, $\Delta V$ is broken up into two pieces. One part is like the isovalent $\Delta V$. The second describes the effects of the excess electrons by treating them as classical point charges centered on the appropriate lattice sites; we shall call this the heteropolar part.

We now turn to the Ga-As in Ge system which falls into the lattice-matched polar heterovalent category. The best choice for the virtual crystal is pure Ge. Since this is a homojunction, the isovalent-like part of $\Delta V$, which gives the difference in average potential between the two identical semi-infinite Ge bulks, is zero. Similarly, the position of the valence band maximum with respect to the average electrostatic potential will be the same on both sides of the interface, and thus, $\Delta E_v$ will be zero. All that remains is the heteropolar part of $\Delta V$.

The heteropolar part of $\Delta V$ is simply the step potential due to two consecutive (111) planes, one with -e charge per lattice site and the other with +e charge per lattice site. Since this is the same charge configuration as the one used in HTA, *the LRT and HTA results agree exactly:*

$$\Delta V = - d\sigma e^2/\varepsilon\varepsilon_o$$

where the direction of the step is from the Ga to the As side of the interface. The parameters are defined in the same way as in HTA. The Ga-As in Ge dipole lowers the As side of the junction by 0.69 eV as compared to 0.35 - 0.45 eV experimentally. The source of the discrepancy is believed to be due to the microscopic details of the interface. If a configuration other than the ideal double layer was assumed, LRT would make a different prediction. As mentioned before, alternate geometries can be found that dramatically change the interface dipole.[20] Interfacial strain cannot explain the discrepancy, because a ≈ 30% variation in the Ga-As bond length would be required which is not physically realistic.

In cases of heterovalent heterojunctions (including GaAs-Ge) where LRT has been checked against self-consistent supercell calculations, agreement is typically within 0.01 eV. There are several reasons to believe that the theoretical situation worsens for the double layer system and that the accuracy limit is more like 0.1 eV.[47] LRT predicts a dipole of 0.98 eV for the Ge-Ge in GaAs system, which compares to 0.74 eV for the full self-consistent calculation[36]—a difference of 0.24 eV. The source of additional error probably lies in two places.[47] Foremost, is the fact that the self-consistent results use the Local-Density Approximation (LDA) which is unable to predict semiconductor dielectric constants accurately. Use of the LDA dielectric constant in LRT improves the agreement. The second major reason that LRT performs more poorly is that a less than optimal virtual crystal is used in the double layer case. In the heterojunction case, the virtual crystal is an average of Ge and GaAs; a virtual cation has a valence charge of 3.5 electrons, requiring a change of only 0.5 electron to convert it into either Ga or Ge. In the double layer system the virtual crystal is Ge, a virtual ion has a valence of 4, requiring a larger perturbation of 1 electron to convert it into Ga or As. Obviously, the larger the perturbation, the more important higher order terms neglected by LRT become.

By the nature of the heteropolar part of ΔV, it is independent of the choice of virtual crystal. For example, a more symmetric virtual crystal would be $(Ga_{0.5}Ge_{0.5})(Ge_{0.5}As_{0.5})$. This choice puts alternating ± e/2 charges on successive Ge planes, -e/2 on the Ga plane, and +e/2 on the As plane, making a somewhat physically-unrealistic oscillatory potential in the bulk Ge. However, the total dipole is unchanged.

Before moving on to the next model of the Ga-As in Ge system, we point out that LRT can be applied to the more general case of a binary host with a binary intralayer material (e.g., Ga-As in ZnSe). The virtual crystal is taken to be the host material (ZnSe). One virtual cation plane is replaced by intralayer cations, and one neighboring virtual anion plane is replaced by intralayer anions. The electric potential step that occurs as we traverse the interface from the host anion-terminated side to the host cation-terminated side is:

$$\Delta\phi_{total}^{111\text{-near}} = -\frac{|e|}{6\varepsilon\varepsilon_o a_o}\left[(Z_c^I - Z_c^H) - (Z_a^I - Z_a^H)\right]$$

where the superscripts I and H refer to the intralayer and host, respectively. We can also replace virtual anions with intralayer cations and virtual cations with intralayer anions. This "anion-anion-cation-cation" sequence gives a dipole of:

$$\Delta\phi_{total}^{111\text{-near}} = -\frac{|e|}{6\varepsilon\varepsilon_o a_o}\left[(Z_a^I - Z_c^H) - (Z_c^I - Z_a^H)\right]$$

The (111)-far cases are the same except that the 6 is replaced by a 2.

Several comments can be made about the generalized results. Unlike HTA, which applies only to materials in horizontal sequences, i.e., isoelectronic sequences), LRT also works with non-horizontal sequences (e.g., Ga-P in Si). It is also interesting to observe how reversing the orientation of the intralayer components effects the dipole. If the host material is elemental (e.g., Ga-As in Ge), then reversing the orientation (As-Ga in Ge) produces a dipole with the same magnitude but opposite sign. However, if the host material is a II-VI and the intralayer material a III-V (e.g., Ga-As in ZnSe), then reversing the orientation produces a dipole with the *same* direction and three times the magnitude. Table 7 gives the results of LRT for a large number of potentially interesting lattice-matched systems.

**Table 7.** $\Delta\phi_{(111)\text{-near}}$ by LRT and IBP

| Host Material (band gap (eV) dielectric constant; lattice constant ( Å)) | Intralayer Material | LRT | IBP (bond polarity term alone) |
|---|---|---|---|
| Ge (0.67; 15.6; 5.65) | Ga-As | +0.69 | +0.73 (+0.043) |
| | Al-As | +0.69 | +0.74 (+0.054) |
| | Zn-Se | +1.37 | +1.61 (+0.239) |
| | Cu-Br | +2.06 | +2.60 (+0.543) |
| Si (1.11; 12.0; 5.45) | Al-P | +0.92 | +1.03 (+0.103) |
| | Ga-P | +0.92 | +1.01 (+0.083) |
| GaAs (1.35; 10.9; 5.65) | Ge-Ge | -0.98 | -0.92 (+0.062) |
| | Al-As | 0 | -0.006 (-0.006) |
| | Zn-Se | +0.98 | +1.07 (+0.088) |
| | Cu-Br | +1.96 | +2.27 (+0.312) |
| GaP (2.24; 9.1; 5.45) | Si-Si | -1.22 | -1.11 (+0.110) |
| | Al-P | 0 | -0.004 (-0.004) |
| ZnSe (2.58; 5.9; 5.65) | Ge-Ge | -3.62 | -2.99 (+0.631) |
| | Ga-As | -1.81 | -1.65 (+0.163) |
| CdTe (1.44; 7.2; 6.48) | Sn-Sn (α-Sn) | -2.59 | -2.17 (+0.420) |
| | In-Sb | -1.30 | -1.19 (+0.109) |
| | Hg-Te | 0 | -0.004 (-0.004) |

## 5.4 Interface Bond Polarity Model

In the past simple capacitor models using electron charge transfer schemes have proven surprisingly successful at predicting semiconductor heterojunction band lineups.[3][22][26][29] In this spirit, we developed a simple capacitor model applicable to the present problem. In our model, the excess charge accumulation on each (111) plane is determined by applying Harrison's bond polarity arguments to the individual bonds near the interface;[48] the amount of charge in units of |e| transferred to an atom from its four nearest neighbors is:

$$N = z - 4 \pm \sum_{i=1}^{4} \alpha_i$$

where Z is the valence and $\alpha_i$ are the bond polarities. Bond polarities are always written positive, and the direction of charge transfer is indicated explicitly by the sign on $\alpha$, positive if the atom is a cation and negative if it is an anion.

In Harrison's original work, the bond polarities were derived from bulk bond-orbital parameters. A more recent bond polarity scale by Lambrecht and Segall[28] determined the polarities of interfacial bonds at heterojunctions using self-consistent interface calculations. These interface bond polarities are probably more descriptive of the bonds in our double layer system than are Harrison's original bulk-derived bond polarities, hence the model presented here is referred to as the Interface Bond Polarity (IBP) model.

According to IBP, all the charge transfers contributing to the interface dipole occur in the double layer and the nearest adjoining layers of the host material (see Fig. 22). Atoms from atomic layers farther from the interface will have the same environments as any other equivalent bulk atom, so that bond polarity arguments predict them to have bulk charge levels, and they do not contribute to the interface dipole. As evidence that this is an acceptable approximation, we point to the final self-consistent solutions of Lambrecht and Segall for semiconductor heterojunctions: except for planes immediately bordering an interface, near bulk-like charge levels exist.

In order to determine the charge accumulation per atom on each of these four planes, we use the bond polarity equation and the fact that there are three bonds per atom between closely spaced planes and one per atom between widely spaced planes. If we designate layers 1, 2, 3, and 4 as in Fig. 22, then:

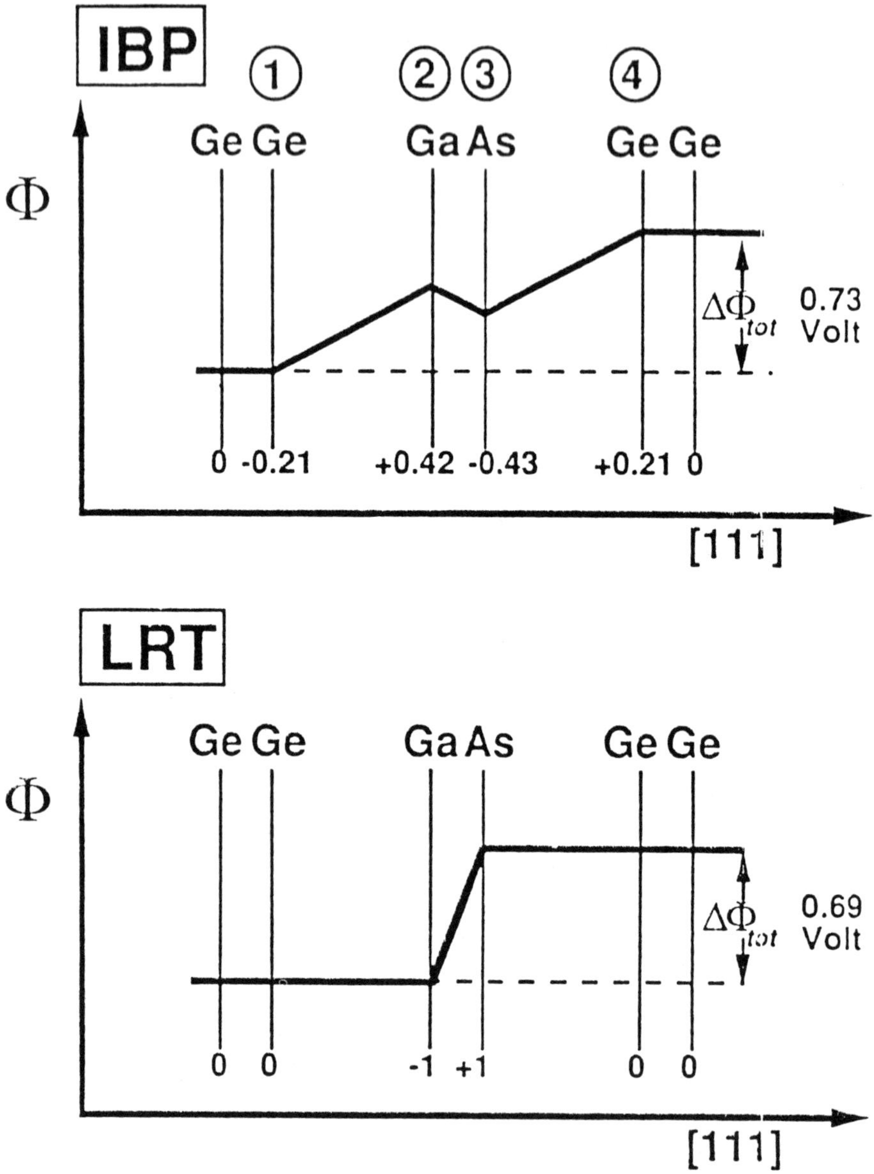


**Figure 22.** IBP model: (111)-near Ga-As in Ge. The charge/atom (units of |e|) and the electrostatic potential profile predicted by LRT and IBP are dramatically different. LRT predicts a +0.69 volt step occurring over 0.82 Å, while IBP predicts a +0.72 volt step occurring over 5.7 Å so that the maximum electric field is ≈ 5 times less in the IBP case.

$$\text{Ge:}\ N_1 = 4 - 4 - 3\alpha_{GeGe} - 1\alpha_{GaGe} = -0.212$$
$$\text{Ga:}\ N_2 = 3 - 4 + 1\alpha_{GaGe} - 1\alpha_{GaAs} = +0.424$$
$$\text{As:}\ N_3 = 5 - 4 - 3\alpha_{GaAs} - 1\alpha_{GeAs} = -0.425$$
$$\text{Ge:}\ N_4 = 4 - 4 + 1\alpha_{GeAs} + 3\alpha_{GeGe} = +0.213$$

where $N_i$ is the charge per atom on plane i in units of $|e|$. The values of $\alpha$ are taken from Ref. 28:

$$\alpha_{GeGe} = 0;\ \ \alpha_{GaGe} = 0.212;\ \ \alpha_{GaAs} = 0.404;\ \ \alpha_{GeAs} = 0.213$$

The interface charge profile that we have just derived is dramatically different from the one used by LRT, in which the atoms of planes 1 through 4 have charges of 0, -1, +1, and 0, respectively. Consequently, the predicted electrostatic potential profiles are also different (see Fig. 22). Curiously, the total dipoles are almost identical.

Now that we know the interface charge profile, it is trivial to calculate the electric potential. To highlight the difference between IBP and LRT, we are going to track the potential at each atomic plane as the interface is crossed. Simple electrostatics relates the electric field on the right and left of a sheet of charge:

$$E_{i+} = E_{i-} + \sigma_i/\varepsilon\varepsilon_o$$

In the present case, the areal charge densities are given by:

$$\sigma_i = \frac{4Ni|e|}{\sqrt{3}a_o^2}$$

The potential step that occurs between sheet i and sheet i+1 to its right at distance $d_i$ is:

$$\Delta\phi_{i,i+1} = -d_i E_{i+}$$

We simply iterate the process to determine the potential step between layers i+1 and i+2:

$$E_{i+} = E_{(i+1)-}$$

$$E_{(i+1)+} = E_{(i+1)-} + \sigma_{i+1}/\varepsilon\varepsilon_o$$

$$\Delta\phi_{i+1,i+2} = -d_{i+1}E_{(i+1)+}$$

and so on. For the system shown in Fig. 22, the areal charge densities on the four planes sum to zero which means that the electric field is zero outside the four layer region. Thus:

$$E_{i-} = 0$$

$$\Delta_{1,2} = -d_{far}[0 + (\sigma_1/\varepsilon\varepsilon_o)] = +0.436 \text{ volt}$$

$$\Delta_{2,3} = -d_{near}(\sigma_1 + \sigma_2)/\varepsilon\varepsilon_o = -0.145 \text{ volt}$$

$$\Delta_{3,4} = -d_{far}(\sigma_1 + \sigma_2 + \sigma_3)/\varepsilon\varepsilon_o = +0.438 \text{ volt}$$

$$\Delta_{total} = -\ (|e|\,/\varepsilon\varepsilon_o a_o)\ [(1/3) + a_{GaGe} + a_{GeAs} + a_{GaAs}]$$

$$= +0.728 \text{ volt}$$

The total dipole differs only slightly from the 0.69 volts estimated by LRT. The IBP dipole, however, is predominantly supported across the widely spaced planes while all of the LRT dipole is across the Ga-As double layer (see Fig. 22). It is also interesting that the functional form of the dipole is identical to that of LRT except for the addition of a small term linearly dependent on the bond polarities which amounts to about 6% of dipole.

IBP can be applied to the more general case of a binary host with a binary intralayer material (e.g., Ga-As in ZnSe). First, the layer indices are assigned as in Fig. 23, to form an anion-cation-anion-cation sequence. The derivation of the total dipole is slightly more difficult than for the Ga-As in Ge system. In binary materials, charge is transferred between consecutive cation and anion (111) planes in the bulk materials; thus, the electric potential rises at cation planes and falls at anion planes by an amount:

$$\Delta\phi_o = \frac{-|e|}{\varepsilon\varepsilon_o a_o}\left(\alpha + \frac{Z_4 - Z_1}{8}\right)$$

This fluctuation can be sizable: in GaAs, it is ≈ 0.45 eV. The total dipole is the difference between the average potentials on opposite sides of the interface; we must include not only the dipole from layer 1 to layer 4 but also a term due to the fluctuation in the bulk. This is shown in the figure.

Incidentally, this is equivalent to taking the dipole from layer 1 to layer 5 (i.e., from a host anion on one side of the interface to a host anion on the other side of the interface). We will not go through the derivation but will simply state that the total dipole is given by:

$$\Delta\phi_{\text{total}}^{111\text{-near}} = -\frac{|e|}{6\varepsilon\varepsilon_o a_o}[(Z_1 + Z_2 - Z_3 + Z_4) + (\alpha_{12} + \alpha_{34} - \alpha_{23} - \alpha_{14})]$$

The (111)-far dipole is the same except that the 6 becomes a 2.

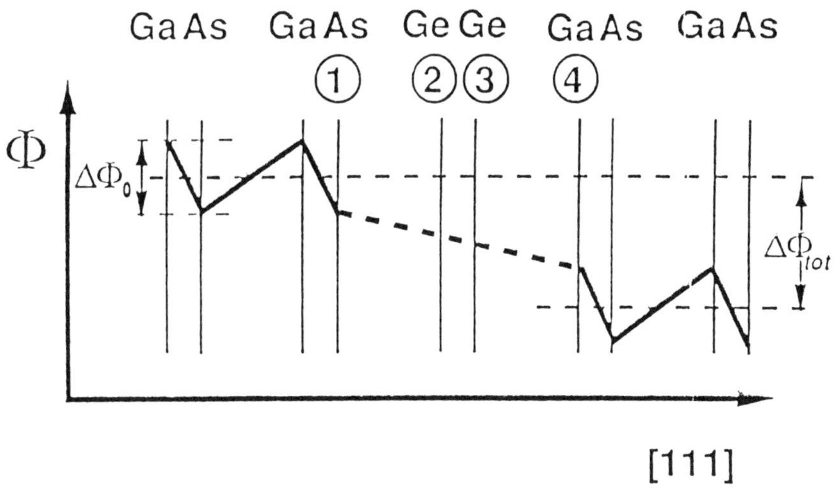


**Figure 23.** IBP model: (111)-near interface with binary host. The total dipole is the difference between the *average* potentials across the interface, given by the horizontal dashed lines. The total dipole includes a term due to the dipole from layer 1 to layer 4 and a second term due to the fluctuation of the potential in the bulk host material. In order to highlight the difference between the elemental and binary host cases, we have not shown the modulation occurring as we cross layers 1 through 4.

The most salient feature of the generalized IBP result is that its zero order term is identical to the generalized LRT result. In IBP, however, only bond counting arguments have been used. The second feature to be noticed is the existence of a term linearly dependent on the bond polarities; this term is typically quite small. The bond polarity term is the same for both the (111)-near and (111)-far cases. We are tempted to give it the physical interpretation of being the "electronic-counter" dipole neglected in HTA. When the host material is more ionic than the intralayer material (e.g., Ge-Ge in GaAs), it opposes the LRT-like term. Unfortunately, our simple interpretation does not hold up when the intralayer material is more ionic than host (e.g., Ga-As in Ge); in that case, it aligns with the LRT-like term. A final observation is that the IBP result simplifies if either the host or intralayer material is elemental since this increases the reflection symmetry. If the inner material is elemental, $\alpha_{23}$ and $Z_2$ - $Z_3$ become zero; if the outer material is elemental, $\alpha_{14}$ and $Z_1$ - $Z_4$ become zero. We have applied the generalized formula to a large number of potentially interesting lattice-matched double layer systems and given the results in Table 7.

## 5.5 Dipole Magnitudes for Other Double Layer Systems

The IBP model can be checked numerically against LRT and SCD. For the (111)-near Ga-As in Ge system, IBP predicts 0.73 eV while LRT predicts 0.69 eV. IBP and LRT predict that the (111)-far dipoles will be ≈ 2.9 and 3 times as large as the corresponding (111)-near dipoles, respectively. IBP can also be checked against fully self-consistent calculations for the Ge-Ge in GaAs system: SCD predicts 0.74 eV, IBP predicts 0.92 eV, and LRT predicts 0.98 eV.

Another interesting comparison is between Al-P and Ga-P in Si. Non-horizontal sequences (GaP and Si) do not usually have as good a lattice-match as isoelectronic sequences (AlP and Si), but in this case, both have a mismatch of 0.4%.[25] This is useful because it prevents strain-related effects from complicating the comparison. LRT predicts identical dipoles of 0.92 eV for both systems, while IBP predicts 1.01 eV and 1.03 eV for Al-P and Ga-P, respectively. Unfortunately, the difference predicted by IBP is far below the accuracy limit of present measurement techniques, ± 0.1 eV.

In order to see the trends, we calculated the LRT and IBP predictions for $\Delta\phi_{(111)\text{-near}}$ for a wide variety of potential double layer systems (see Table 7). The sign on the dipole assumes that the interface is traversed from the host anion to the host cation side of the junction (i.e., an anion-cation-anion-cation sequence). All the materials are well lattice-matched, and most are semiconductors with technological importance. Except for the Ge dielectric constant taken from Ref. 37, all dielectric constants and lattice constants are from Ref. 28; all semiconductor forbidden energy gaps are from Ref. 25. From Table 7, it is clear that the dipole magnitudes are most strongly dependent upon the difference in ionicity of the host and intralayer materials. Combinations of I-VII and IV-IV materials form enormous dipoles. We also noticed that these dipoles generally scaled with the experimental valence band discontinuity of the corresponding semiconductor heterojunctions (see Fig. 24).

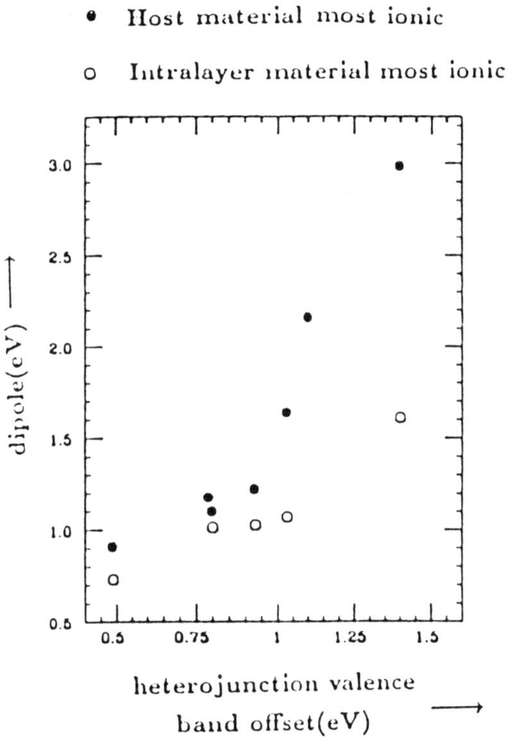


**Figure 24.** Double layer dipole vs. heterojunction $\Delta E_V$. Since the heterojunction valence band discontinuity is a fundamental interface parameter, it would be interesting to see if it correlates with the predicted double layer dipoles. Changes in the slope of the lines correlate to the difference in ionicity of the host and intralayer materials used in the IBP model. Values of $\Delta E_V$ are from Ref. 1, and are dominantly experimental.

For some of the systems presented, the estimates of the dipole exceed the host material's forbidden energy gap. Potentially, a rapid "discontinuity-like" potential step larger than the host material's band gap is of technological importance. If the potential step occurs over a distance shorter than an electron scattering distance and the electron falls through a step larger than the band gap, it can ionize other Ge atoms producing more conduction electrons. Impact-ionization devices of this kind are described in Ref. 23 and Ref. 49. One application is a staircase electron multiplier.

Another effect associated with "broken-gap" band lineups is spontaneous carrier production. Since the valence band maximum of one material is higher than the conduction band minimum of the other, electrons from the valence band of the higher material tunnel into the conduction band of the lower. Eventually, of course, a sufficient space charge barrier builds up to prevent further charge transfer. This effect presumably tends to prevent the total dipole from exceeding the band gap. Since a space charge dipole has the dimensions of a Debye length, 100 - 1000 Å, it will not interfere with the previously mentioned impact-ionization effect.

If device quality Ga-As in Ge double layers can be made, it will be interesting to see how they affect the use of the common p-n junction in very high speed electronics. Presently, the rectifying action is due to a ≈ 0.6 eV dipole spread broadly over a few Debye lengths. If a Ga-As intralayer was oriented so that it aligned with the direction of the normal space charge dipole, ≈ 0.4 eV of the dipole would be supported at the interface. The normal barrier mechanism would make up the difference. Rectification, however, would be occurring over a much shorter distance.

## 5.6 Dipole Intralayer Theory: Conclusions

Several approaches based on theoretical alchemy as well as one based on bond polarity arguments were found to predict the sign and approximate magnitude of the experimentally determined dipole for the Ga-As in Ge system. LRT in particular gives theoretical alchemy a firm theoretical basis. IBP gives a simple physical reason for the dipole: over and under filled bonds. Remarkably, the theoretical alchemy and bond polarity based models predict very different interface charge profiles, while arriving at very similar results for the total dipole.

All of the theories overestimate the dipole that was measured experimentally. Apparently, an unknown mechanism still exists which reduces the experimental value from the expected result for the ideal interface. It is likely that this mechanism is a modified interface atomic configuration brought about by microdiffusion processes; configurations other than the ideal double layer system are expected to reduce the dipole. With this in mind, the next theoretical step must be to explore alternate interface geometries. Experimentally, the next step will be the use of device quality growth methods to produce truly epitaxial Ga-As layers.

## REFERENCES

1. F. Capasso and G. Margaritondo, *Heterojunction Band Discontinuities: Physics and Device Applications,* North-Holland, Amsterdam (1987)
2. F. Flores and C. Tejedor, *J. Phys.,* C20:145 (1987)
3. D. W. Niles, M. Tang, J. McKinley, R. Zanoni, and G. Margaritondo, *Phys. Rev.,* B38:10949 (1988); L. Sorba, G. Bratina, G. Ceccone, A. Antonini, J. F. Walker, M. Micovic and A. Franciosi, *Phys. Rev.* (in press)
4. D. W. Niles, E. Colavita, G. Margaritondo, P. Perfetti, C. Quaresima, and M. Capozi, *J. Vac. Sci. Technol.,* A4:962 (1985)
5. L. J. Brillson, *Phys. Rev. Lett.,* 31:245 (1977)
6. J. C. Phillips, *J. Vac. Sci. Technol.,* 6:947 (1974)
7. L. J. Brillson, *Thin Solid Films,* 89:461 (1982); L. J. Brillson, *Phys. Rev. Lett.,* 40:260 (1978)
8. R. T. Sanderson, *Chemical Bond and Bond Energy,* Reinhold, New York (1967)
9. P. Perfetti, F. Patella, F. Sette, C. Quaresima, C. Capasso, A. Savoia, and G. Margaritondo, *Phys. Rev.,* B30:4533 (1984)
10. A. D. Katnani and G. Margaritondo, *Phys. Rev.,* B28:1944 (1983)
11. S. P. Kowalczyk, E. A. Kraut, J. R. Waldrop, and R. W. Grant, *J. Vac. Sci. Technol.,* 21:482 (1982)
12. D. W. Niles, PhD Thesis, University of Wisconsin-Madison (1988)
13. P. Chiaradia, L. J. Brillson, M. Slade, R. E. Viturro, D. Kilday, N. Tache, M. Kelly, and G. Margaritondo, *J. Vac. Sci. Technol.,* B5:1075 (1987)
14. J. Tersoff, *Phys. Rev. Lett.,* 52:465 (1984); J. Tersoff, *Phys. Rev.,* B30:4874 (1984)
15. R. E. Viturro, S. Chang, J. L. Shaw, L. J. Brillson, A. Terrasi, Y. Hwu, G. Margaritondo, P. D. Kirchner, and J. M. Woodall, *J. Vac. Sci.,* B7:1007 (1989); S. Chang, J. L. Shaw, R. E. Viturro, L. J. Brillson, P. D. Kirchner, and J. M. Woodall, *J. Vac. Sci. Technol.,* A8:3803 (1990)
16. R. L. Anderson, *Solid State Electron,* 5:341 (1962)
17. W. E. Spicer, P. W. Chye, P. R. Skeath, and I. Lindau, *J. Vac. Sci. Technol.,* 16:1422 (1979); H. Wieder, *J. Vac. Sci. Technol.,* 15:1498 (1978)
18. B. I. Boltaks, *Diffusion in Semiconductors,* Academic Press, New York (1963)
19. J. C. Durán, A. Muñoz, and F. Flores, *Phys. Rev.,* B35:7721 (1987)

20. W. A. Harrison, E. A. Kraut, J. R. Waldrop, and R. W. Grant, *Phys. Rev.,* B18:4402 (1978)

21. E. A. Kraut and W. A. Harrison, *J. Vac. Sci. Technol.,* B3:1267 (1985)

22. D. W. Niles, M. Tang, J. McKinley, R. Zanoni, and G. Margaritondo, *Appl. Surface Sci.,* 41/42:139 (1989)

23. J. Tersoff, private communication

24. W. Mönch, *Phys. Rev. Lett.,* 58:1260 (1987)

25. *CRC Handbook of Chemistry and Physics,* 61st Edition, CRC Press, Boca Raton (1980 - 1981)

26. P. Perfetti, *Surf. Sci.,* 189/190:362 (1987)

27. W. R. Frensley and H. Kroemer, *Phys. Rev.,* B16:2642 (1977); *J. Vac. Sci. Technol.,* 13:810 (1976)

28. W. R. L. Lambrecht and B. Segall, *Phys. Rev.,* B41:2832 (1990).

29. P. Perfetti, C. Quaresima, C. Coluzza, C. Fortunato, and G. Margaritondo, *Phys. Rev. Lett.,* 57:2065 (1986)

30. C. A. Mead and W. G. Spitzer, *Phys. Rev.,* 134:A713 (1967)

31. G. Margaritondo, *Surf. Sci.,* 168:438 (1986)

32. E. H. Rhoderick and R. H. Williams, *Metal-Semiconductor Contacts,* Oxford University, New York (1988)

33. J. T. McKinley, Y. Hwu, D. Rioux, A. Terrasi, F. Zanini, G. Margaritondo, U. Debska, and J. K. Furdyna, *J. Vac. Sci. Technol.,* A8:1917 (1990)

34. S. Baroni, R. Resta, A. Baldereschi, and M. Peressi, *Spectroscopy of Semiconductor Heterojunctions,* (G. Gasol, A. Fasolino, and P. Lugli, eds.), NATO Asi Ser. B, 206:251, Plenum, New York (1989)

35. S. Baroni, R. Resta, and A. Baldereschi, *Proc. of the Nineteenth International Conf. of the Physics of Semiconductors,* (W. Zawadzki, ed.), Institute of Physics, Polish Academy of Sciences, Wroclaw (1988)

36. A. Muñoz, N. Chetty, and R. M. Martin, *Phys. Rev.,* B41:2976 (1990)

37. F. Patella, F. Sette, P. Perfetti, C. Quaresima, C. Capasso, M. Capozi, A. Savoia, and F. Evangelisti, *Sol. State Comm.,* 49:749 (1984)

38. A. D. Katnani, P. Chiaradia, H. W. Sang, Jr., P. Zurcher, and R. S. Bauer, *Phys. Rev.,* B31:2146 (1985)

39. R. Z. Bachrach, R. D. Bringans, and M. A. Olmstead, *Current Trends in the Physics of Materials,* Italian Physical Society Proceedings, International School of Physics Enrico Fermi - Course CVI, Lerici, La Spezia, Italy (1988)

40. R. D. Bringans, R. I. G. Uhrberg, R. Z. Bachrach, and J. E. Northrup, *Phys. Rev. Lett.*, 55:533 (1985)

41. J. C. Phillips, *Bonds and Bands in Semiconductors,* Academic Press, New York (1973)

42. G. Margaritondo, C. Capasso, F. Patella, P. Perfetti, C. Quaresima, A. Savoia, and F. Sette, *J. Vac. Sci. Technol.,* A2:508 (1984)

43. M. Cardona and N. E. Christensen, *Phys. Rev.,* B35:6182 (1987)

44. R. W. Grant, J. R. Waldrop, S. P. Kowalczyk, and E. A. Kraut, *J. Vac. Sci. Technol.,* B3:1295 (1985)

45. A. D. Katnani and G. Margaritondo, *J. Appl. Phys.,* 54:2522 (1983)

46. H. Kroemer, *Proc. of the Nato Advanced Study Institute on Molecular Beam Epitaxy (MBE) and Heterostructures,* Erice, Italy, 1983, (L. L. Lang and K. Ploog, eds.), Martinus Nijhof, Dordrecht (1985)

47. S. Baroni and R. Resta, private communication

48. W. A. Harrison, *Electronic Structure and Properties of Solids,* Freeman, San Francisco (1980)

49. D. D. Coon, *J. Vac. Sci. Technol.,* A8:2950 (1990)

# Index

## A

**D**

T



U